Applied Thermodynamics

26

For Engineering Technologists

…its

…ition

T. D. EASTOP

B.Sc., Ph.D., C.Eng.,
F.I.Mech.E., F.C.I.B.S.E.,
M.I.Mar.E.

Head of the School of Engineering at Wolverhampton Polytechnic

A. McCONKEY

B.Sc., Ph.D., C.Eng.,
F.I.Mech.E.

Head of the Department of Mechanical and Industrial Engineering at Dundee College of Technology

Longman
London and New York

Longman Group Limited
Longman House, Burnt Mill, Harlow
Essex CM20 2JE, England
Associated companies throughout the world

Published in the United States of America
by Longman Inc., New York

First published 1963
Second Edition 1969
Third Edition 1978
Second impression 1979
Third impression (with corrections) 1981
Fourth impression 1982
Fifth impression 1983
Sixth impression 1984
Seventh impression 1985
Fourth Edition 1986

British Library Cataloguing in Publication Data

Eastop, T.D.
 Applied thermodynamics for engineering
 technologists: S.I. units.——4th ed.
 1. Thermodynamics
 I. Title   II. McConkey, A.
 536′.7′0246   QC311

ISBN 0-582-30535-7

Library of Congress Cataloging in Publication Data

Eastop T. D. (Thomas D.)
 Applied thermodynamics for engineering technologists.

 Bibliography: p.
 Includes index.
 1. Thermodynamics.   I. McConkey, A. (Allen),
1927–   . II. Title.
TJ265.E23   1986   621.402′1   85-16632
ISBN 0-582-30535-7

Produced by Longman Group (FE) Ltd
Printed in Hong Kong

Contents

v

Acknowledgements

We are grateful to Basil Blackwell & Mott Ltd. for permission to include extracts from the Mayhew and Rogers *Thermodynamic Tables*. Fig. 14.4 is reproduced by permission of the Chartered Institution of Building Services Engineers; copies of the chart (size A3) for record purposes may be obtained from CIBSE, 222 Balham High Road, London SW12 9BS.

The following sources have been drawn on for information: Section 7.12(c) and Table 7.3 are adapted from data contained in *Nuclear Power Reactors* published by the Information Services Branch, United Kingdom Atomic Energy Authority. Section 8.1 and Table 8.1 use data from *Design and Management for Energy Conservation* by P. S. O'Callaghan, published by Pergamon Press. Section 8.2 contains material adapted from an article by J. Roberts in Rolls Royce Magazine No. 21, June 1984. Section 8.3 draws on information contained in the *Fuel Efficiency Booklet 7: Degree Days* published by HMSO for the Department of Energy, and 'Energy: a major factor in vehicle production and competitiveness' by Marcus Jacobson published in *Automotive Engineer*, December 1981. Figs. 18.22 and 18.23 are adapted from *The Internal Combustion Engine in Theory and Practice* by C. F. Taylor, MIT Press. Section 18.13 includes material adapted from *Exhaust Emissions Handbook* published by Cussons Ltd. The data for fig. 9.28 was provided by J. S. Milne, Senior Lecturer in the Department of Mechanical and Industrial Engineering, Dundee College of Technology, from an original test carried out by him. The problem used for Worked Example 8.2 was originally set as an examination question at Wolverhampton Polytechnic by R. A. Briggs, Senior Lecturer in the School of Engineering.

Preface to the Fourth Edition

We acknowledge the generous support and acceptance our book has received over the three editions since it was first published in 1963, despite the several changes in the philosophy of engineering education for the degree, diploma and certificate courses in the United Kingdom over that long period. Now, in the post-Finniston (i.e. since 1981) period, and with the development of the 'applications oriented' philosophy for the future, we have been encouraged by the generous comments we have received on the book to believe that in its modified form it will continue to be accepted as a suitable text to support the subject of applied thermodynamics in those courses designed for the formation of professional engineers and engineering technicians in the United Kingdom and abroad.

Prior to the preparation of the fourth edition the publishers, Longman Group Limited, arranged for a comprehensive review of the third edition by several qualified users in the Polytechnics who are aware of the requirements of the new philosophy. We are appreciative of the thoroughness of the reviews and the many helpful comments and suggestions made by the reviewers, most of which have been incorporated in the new edition. We hope that the changes made will reaffirm the usefulness of the book to students and lecturers alike.

The objective is to provide a text which deals with the essential principles which emerge from the extensive and fundamental science of the subject, leading to an understanding of the practical problems which will be met in design, development and research work in engineering by professional engineers and technicians. The text enables some selection of topics to be studied as required by different courses and individual students and yet provides sufficient reading to enable the student to appreciate the broader perspective of the subject.

Particular reference must be made to the far reaching influence of the new technologies on thermodynamics which is no less than it has been for any other engineering subject. The move in engineering education towards the application of the theory to the solution of real engineering problems leads to the use of assignments as a better

ix

method of learning; these assignments will become more and more computer oriented as time passes. The decision has been taken in revising the book to maintain the basic coverage of principles and practice, to introduce numerical methods where these are now an integral part of the topic (e.g. conduction heat transfer), but not to attempt to include actual computer programs or assignment briefs. We believe that students will be required, by necessity, to apply their computing knowledge and skills to their understanding and application of the basic principles dealt with in this book, and that lecturers will wish to set their own assignments to achieve this. When we consider the impact of modern methods of instrumentation, measurement, data collection and handling, and the many control applications at the heart (or brain) of which is the inevitable microprocessor it is clear that it is not possible to do justice to this range of topics in a book on Thermodynamics. We hope that the references made in the text will encourage readers to pursue these extremely interesting subjects in their own right to the benefit of their understanding of thermodynamics.

The main changes from the third edition are a new Chapter 8 on 'The sources, use and management of energy'; an updating and extension of Chapter 14 on Psychrometry; an updating of Chapter 17, 'Heat Transfer', to include modern methods of heat exchanger analysis and an extensive treatment of the applications of heat transfer by conduction; Chapter 18 has been updated and the work on turbo-charging extended. Other, smaller, changes have been made in the re-arrangement of material to complete what is hoped is a logical revision.

T.D.E.
A.McC.
1985

1. Introduction

All living things depend on energy for survival, and modern civilizations can continue to thrive only if existing sources of energy can be developed to meet the growing demands. Energy exists in many forms, from the energy locked up in the atoms of matter itself to the intense radiant heat emitted by the sun, and between these limits energy sources are available such as the chemical energy of fuels and the potential energy of large water masses evaporated by the sun. The picture presented is one of Man tapping into Nature and taking off energy at every level he can, to put it to his purpose. Many sources of energy exist; many are known, some perhaps unknown; but when an energy source exists means must first be found to transform the energy into a form convenient to Man's purpose. The potential energy of large masses of water is converted into electrical energy as it passes through water turbines on its way from the mountains to the sea; the energy of combustion of coal is used to produce steam which passes through turbines to generate electrical energy; the energy of combustion of petroleum fuels is used to heat air which expands and pushes a piston in an internal-combustion engine to develop mechanical work; uranium atoms are bombarded asunder and the nuclear energy released is used as heat to produce steam and generate electricity. The machinery for such energy transformations has been developed over the last two centuries, mainly by practical engineering, followed closely, but sometimes more distantly, by theoretical analysis and research. Refinements made over the years to fundamental practical ideas have transformed, for example, the steam plant of Watt into the highly complex modern power station.

APPLIED THERMODYNAMICS is the science of the relationship between heat, work, and the properties of systems. It is concerned with the means necessary to convert heat energy from available sources such as chemical fuels or nuclear piles into mechanical work. A HEAT ENGINE is the name given to a system which by operating in a cyclic manner produces net work from a supply of heat. The laws of Thermodynamics are natural hypotheses based on observations of the world in which we live. It is observed that heat and work are two mutually convertible forms of energy, and this is the basis of the First Law of Thermodynamics. It is also observed that heat never flows unaided from an object at a low temperature to one at a high temperature, in the same way that a river never flows unaided uphill. This observation is the basis of the Second Law of Thermodynamics, which can be used to show that a heat engine cannot convert all the heat supplied to it into mechanical work but must always reject some heat at a lower temperature. These ideas will be discussed and developed in due course, but first some fundamental definitions must be made.

1.1 Heat, work, and the system

In order to deal with the subject of Applied Thermodynamics rigorously it is necessary to define the concepts used.

Heat is a form of energy which is transferred from one body to another body at a lower temperature, by virtue of the temperature difference between the bodies.

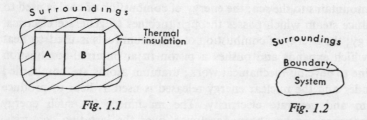

Fig. 1.1   *Fig. 1.2*

For example, when a body A at a certain temperature, say 20°C, is brought into contact with a body B at a higher temperature, say 21°C, then there will be a transfer of heat from B to A until the temperatures of A and B are equal (fig. 1.1). When the temperature of A is the same as the temperature of B no heat transfer takes place between the bodies, and they are said to be in *thermal equilibrium*. Heat is apparent during the process only and is therefore transitory energy. Since heat energy flows from B to A there is a reduction in the intrinsic energy possessed by B and an increase in the intrinsic energy possessed by A. This intrinsic energy of a body, which is a function of temperature at least, must not be confused with heat. *Heat can never be contained in a body or possessed by a body.*

A *system* may be defined as a collection of matter within prescribed and identifiable boundaries (fig. 1.2). The boundaries are not necessarily inflexible; for instance the fluid in the cylinder of a reciprocating engine during the expansion stroke may be defined as a system whose boundaries are the cylinder walls and the piston crown. As the piston moves so do the boundaries move (fig. 1.3).

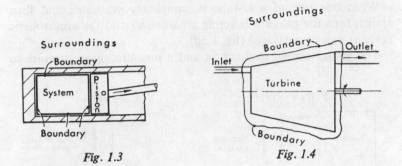

Fig. 1.3   Fig. 1.4

This type of system is known as a *closed* system. An *open* system is one in which there is a transfer of mass across the boundaries; for instance, the fluid in a turbine at any instant may be defined as an open system whose boundaries are as shown in fig. 1.4. At present only closed systems will be dealt with; more will be said of open systems in Section 2.3.

The *pressure* of a system is the force exerted by the system on unit

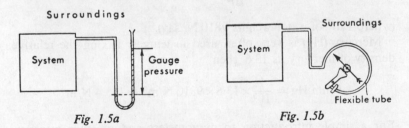

Fig. 1.5a   Fig. 1.5b

area of its boundaries. Units of pressure are, for example, N/m^2 or bar; the symbol p will be used for pressure. Pressure as defined here is called *absolute pressure*. A gauge for measuring pressure (e.g. as shown in fig. 1.5a and b, records the pressure above atmospheric. This is called *gauge pressure*, i.e. absolute pressure equals gauge pressure plus atmospheric pressure.

The gauge shown in fig. 1.5b is called a Bourdon gauge. The absolute pressure of the system in a closed elliptical section tube forces the tube out of position against the pressure of the atmosphere. The tube's displacement is recorded by a pointer on a circular scale, which can be calibrated directly in bars.

When the pressure of a system is below atmospheric, it is called *vacuum pressure* (fig. 1.5c).

When one side of a U-tube is completely evacuated and then sealed, then the gauge will act as a *barometer* and the atmospheric pressure can be measured (fig. 1.5d).

The gauges shown in fig. 1.5a and c measure gauge pressure in

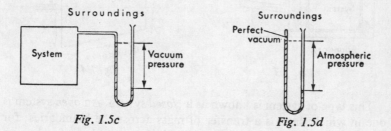

Fig. 1.5c   *Fig. 1.5d*

mm of a liquid of known relative density, and are called manometers.

For example, when water is the liquid, then

$$1 \text{ mm of water} = \frac{1}{10^3} \times 9806 \cdot 65 \text{ N/m}^2 = 9 \cdot 81 \text{ N/m}^2$$

(where 1 m^3 of water weighs 9810 N, say).

Mercury (Hg) is very often used in gauges. Taking the relative density of mercury as 13·6, then

$$1 \text{ mm Hg} = \frac{1}{10^3} \times 13 \cdot 6 \times 9810 \text{ N/m}^2 = 133 \cdot 4 \text{ N/m}^2$$

For a simple introduction to manometers and pressure measurement see reference 1.1.

The *specific volume* of a system is the volume occupied by unit mass of the system. The symbol used is v and the units are, for example, m^3/kg. The symbol V will be used for volume. (Note that the specific volume is the reciprocal of density.)

Work is defined as the product of a force and the distance moved in the direction of the force. When a boundary of a closed system moves in the direction of the force acting on it, then the system does work on its surroundings. When the boundary is moved inwards the work is done on the system by its surroundings. The units of work are, for example, N m. If work is done on unit mass of a fluid, then the work done per kg of fluid has units of N m/kg.

Work is observed to be energy in transition. *It is never contained in a body or possessed by a body.*

Heat and work are both transitory energies and must not be confused with the intrinsic energy possessed by a system. For example, when a gas contained in a well-lagged cylinder (fig. 1.6a) is compressed by moving the piston to the left, the pressure and temperature of the gas are observed to increase, and hence the intrinsic energy of the gas increases. Since the cylinder is well lagged, no heat can flow into or out of the gas. The increase in intrinsic energy of the gas

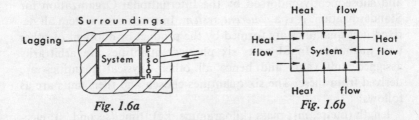

Fig. 1.6a   *Fig. 1.6b*

has therefore been caused by the work done by the piston on the gas.

As another example, consider a gas contained in a rigid container and heated (fig. 1.6b). Since the boundaries of the system are rigidly fixed then no work is done on or by the system. The pressure and temperature of the gas are observed to increase, and hence the intrinsic energy of the gas will increase. The increase in intrinsic energy has been caused by the heat added to the system.

In the example of fig. 1.6a the work done on the system is energy which is apparent only during the actual process of compression. There is an intrinsic energy of the system initially and an intrinsic energy finally, but the work done appears only in transition from the initial to the final condition. Similarly in the example of fig. 1.6b, the heat supplied appears only in transition from one state of the gas to another.

Another way in which work may be transferred to a system is

illustrated in fig. 1.7. The paddle wheel imparts a change of momentum to the fluid and a work input is required to turn the shaft. The kinetic energy attained by the fluid is dissipated by internal fluid friction, and friction between the fluid and the container. When the container is well lagged, all the work input goes to increasing the intrinsic energy of the system.

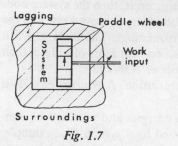

Fig. 1.7

1.2 Units

SI units will be used throughout this book. The International System of Units (Système International d'Unités, abbreviation SI), was adopted by the General Conference of Weights and Measures in 1960 and subsequently endorsed by the International Organization for Standardization. It is a *coherent system*. In a coherent system all derived unit quantities are formed by the product or quotient of other unit quantities. In SI units six physical quantities are arbitrarily assigned unit value and hence all other physical quantities are derived from these. The six quantities chosen and their units are as follows:

length (metre, m); mass (kilogramme, kg); time (second, s); electric current (ampere, A); thermodynamic temperature (degree Kelvin, K); luminous intensity (candela, cd).

Then, for example, velocity = length/time, has units of m/s; acceleration = velocity/time has units of m/s²; volume = length × length × length, has units of m³; specific volume = volume/mass has units of m³/kg.

Force, Energy, and Power

Newton's second law may be written as force ∝ mass × acceleration, for a body of constant mass.

i.e.

$$F = kma$$

(where m is the mass of a body accelerated with an acceleration a, by a force F; k is a constant).

In a coherent system of units such as SI, $k = 1$, hence:

$$F = ma$$

The SI unit of force is therefore kg m/s². This composite unit is called the *Newton*, N. i.e. 1 N is the force required to give a mass of 1 kg an acceleration of 1 m/s².

It follows that the SI unit of work (=force × distance) is the Newton metre, N m. As stated earlier heat and work are both forms of energy, and hence both can have the units of kg m²/s² or N m. A general unit for energy is introduced by giving the Newton metre the name *Joule*, J.

i.e.   1 Joule, J = 1 Newton × 1 metre

or,   1 J = 1 N m

The use of additional names for composite units is extended further by introducing the *Watt*, W as the unit of power.

i.e.   1 Watt, W = 1 J/s = 1 N m/s

Note also that:

$$1 \text{ Watt, W} = 1 \text{ ampere, A} \times 1 \text{ volt, V}$$

Pressure

The unit of pressure (force per unit area), is N/m² and this unit is sometimes called the *Pascal*, Pa. For most cases occurring in Thermodynamics the pressure expressed in Pascals would be a very small number; a new unit is defined as follows:

$$1 \text{ bar} = 10^5 \text{ N/m}^2 = 10^5 \text{ Pa}$$

The advantage of using a unit such as the bar is that it is approximately equal to atmospheric pressure. In fact the standard atmospheric pressure is exactly 1·013 25 bar.

As indicated in Section 1.1 it is often convenient to express a pressure as a head of a liquid. We have:

Standard atmospheric pressure = 1·013 25 bar = 0·76 m Hg

Temperature

The variation of an easily measurable property of a substance with temperature can be used to provide a temperature-measuring instrument. For example, the length of a column of mercury will vary with temperature due to the expansion and contraction of the mercury. The instrument can be calibrated by marking the length of the column when it is brought into thermal equilibrium with the vapour of boiling water at atmospheric pressure and again when it is in thermal equilibrium with ice at atmospheric pressure. On the Celsius (or Centigrade) scale 100 divisions are made between the two fixed points and the zero is taken at the ice point.

The change in volume at constant pressure, or the change in pressure at constant volume, of a fixed mass of gas which is not easily liquified (e.g. oxygen, nitrogen, helium, etc.) can be used as a measure of temperature. Such an instrument is called a *Gas Thermometer*. It is found that for all gases used in such thermometers that if the graph of temperature against volume in the constant pressure gas thermometer is extrapolated beyond the ice point to the point at which the volume of the gas would become zero, then the temperature at this

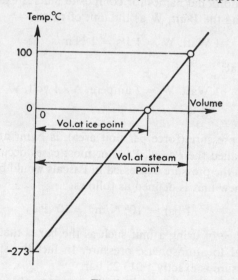

Fig. 1.8

point is $-273°C$ approximately (fig 1.8). Similarly if the graph of temperature against pressure in the constant volume gas thermometer is extrapolated to zero pressure, then the same zero of temperature is found. An absolute zero of temperature has therefore been fixed, and an absolute scale of temperature can be defined. Temperature on the absolute Celsius scale can be obtained by adding 273 to all temperatures on the Celsius scale; this scale is called the *Kelvin* scale. The unit of temperature is the degree Kelvin and is given the symbol 'K', but since the Celsius scale which is used in practice has a different zero the temperature in degrees Celsius is given the symbol ('C' (e.g. $20°C) = 293$ K approximately; also, $30°C - 20°C = 10$ K). It is customary to use capital T for absolute temperature and small t for other temperatures.

In Chapter 6 an absolute scale of temperature will be introduced as a direct consequence of the Second Law of Thermodynamics. It is found that the gas thermometer absolute scales approach the ideal scale as a limit. Also, with regard to the practical absolute temperature scale, there is an internationally agreed working scale which gives temperatures in terms of more practicable, and more accurate, instruments than the gas thermometer (see references 1.2 and 1.3).

Multiples and sub-multiples

Multiples and sub-multiples of the basic units are formed by means of prefixes, and the ones most commonly used are shown in the following table:

Multiplying factor	Prefix	Symbol
one million, 10^6	mega	M
one thousand, 10^3	kilo	k
one thousandth, 10^{-3}	milli	m
one millionth, 10^{-6}	micro	μ

A complete list of multiples can be found in reference 1.4. For most purposes the multiplying factors shown in the above table are sufficient but in some cases it may be more convenient to use some other multiple. For example, power can usually be satisfactorily expressed in either megawatts, MW, or kilowatts, kW, or watts, W. In the measurement of length the millimetre, mm, and the metre, m, are often adequate and in fact the British Standards Institution have advised against the use of centimetre, cm, which is one hundredth (10^{-2}), of a metre. For areas, the difference in size between the square millimetre, mm^2 and the square metre, m^2 is large (a factor of 10^6), and an intermediate size is useful; the square centimetre, cm^2, is recommended for limited use only. For volumes, the difference between the cubic millimetre, mm^3, and the cubic metre, m^3, is much too great (a factor of 10^9), and the most commonly used intermediate unit is the cubic decimetre, dm^3, which is equal to one thousandth of a cubic metre (i.e. $1 \ dm^3 = 10^{-3} \ m^3$). The cubic decimetre can also be called the *litre*, 1,

i.e. $1 \ \text{litre}, 1 = 1 \ dm^3 = 10^{-3} \ m^3$

(Note, for very precise measurements, 1 litre $= 1 \cdot 000\ 028$ dm³.)

Certain exceptions to the general rule of multiplying factors are inevitable. The most obvious example is in the case of the unit of time. Instead of using the centisecond, kilosecond, or megasecond, for instance, the minute, hour, day, etc. will continue to be used. Similarly, a mass flow rate may be expressed in kilogrammes per hour, kg/h, if this gives a more convenient number than when expressed in kilogrammes per second, kg/s. Also the speed of road vehicles is expressed in kilometres per hour, km/h, since this is more convenient than the normal unit of velocity which is metres per second, m/s.

1.3 The state of the working fluid

In all problems in Applied Thermodynamics we are concerned with energy transfers to or from a system. In practice the matter contained within the boundaries of the system can be liquid, vapour, or gas, and is known as the *working fluid*. At any instant the *state* of the working fluid may be defined by certain characteristics called its *properties*. Many properties have no significance in Thermodynamics

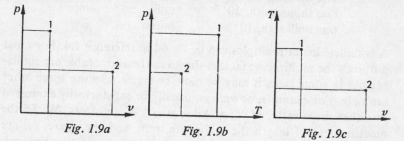

Fig. 1.9a   *Fig. 1.9b*   *Fig. 1.9c*

(e.g. electrical resistance), and will not be considered. The thermodynamic properties introduced in this book are pressure, temperature, specific volume, specific internal energy, specific enthalpy, and specific entropy. It has been found that, for any pure working fluid, only two independent properties are necessary to define completely the state of the fluid. Since any two independent properties suffice to define the state of a system, it is possible to represent the state of a system by a point situated on a diagram of properties. For example, a cylinder containing a certain fluid at pressure p_1 and specific volume v_1 is at state 1, defined by point 1 on a diagram of p against v (fig. 1.9a). Since the state is defined, then the temperature of the fluid, T, is fixed and the state point can be located on a diagram of p against T and T against v (fig. 1.9b and c). At any other instant the piston may be moved in the cylinder such that the pressure and specific volume are changed to p_2 and v_2. State 2 can then be marked on the diagrams. Diagrams of properties are used continually in Applied Thermodynamics to plot state changes. The most important are the pressure-volume and temperature-entropy diagrams, but enthalpy-entropy and pressure-enthalpy diagrams are also used frequently.

1.4 Reversibility

In Section 1.3 it was shown that the state of a fluid can be represented by a point located on a diagram using two properties as coordinates. When a system changes state in such a way that at any instant during the process the state point can be located on the diagram, then the process is said to be *reversible*. The fluid undergoing the process passes through a continuous series of equilibrium states. A reversible process between two states can therefore be drawn as a line on any diagram of properties (fig. 1.10). In practice, the fluid undergoing

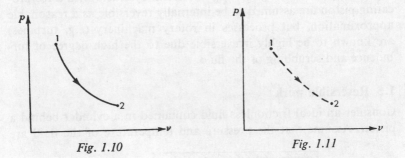

Fig. 1.10   Fig. 1.11

a process cannot be kept in equilibrium in its intermediate states and a continuous path cannot be traced on a diagram of properties. Such real processes are called *irreversible processes*. An irreversible process is usually represented by a dotted line joining the end states to indicate that the intermediate states are indeterminate (fig. 1.11).

A more rigorous definition of reversibility is as follows: *When a fluid undergoes a reversible process, both the fluid and its surroundings can always be restored to their original state.* The criteria of reversibility are as follows:

(a) *The process must be frictionless*

The fluid itself must have no internal friction and there must be no mechanical friction (e.g. between cylinder and piston).

(b) *The difference in pressure between the fluid and its surroundings during the process must be infinitely small*

This means that the process must take place infinitely slowly, since the force to accelerate the boundaries of the system is infinitely small.

(c) *The difference in temperature between the fluid and its surroundings during the process must be infinitely small*

This means that the heat supplied or rejected to or from the fluid must be transferred infinitely slowly.

It is obvious from the above criteria that no process in practice is truly reversible. However, in many practical processes a very close approximation to an *internal reversibility* may be obtained. In an internally reversible process, although the *surroundings* can never be restored to their original state, the *fluid* itself is at all times in an equilibrium state and the path of the process can be exactly retraced to the initial state. In general, processes in cylinders with a reciprocating piston are assumed to be internally reversible as a reasonable approximation, but processes in rotary machinery (e.g. turbines) are known to be highly irreversible due to the high degree of turbulence and scrubbing of the fluid.

1.5 Reversible work

Consider an ideal frictionless fluid contained in a cylinder behind a piston. Assume that the pressure and temperature of the fluid are

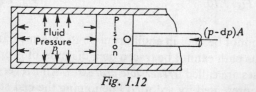

Fig. 1.12

uniform and that there is no friction between the piston and the cylinder walls. Let the cross-sectional area of the piston be A, and let the pressure of the fluid at any instant be p (fig. 1.12). Let the restraining force exerted by the surroundings on the piston be $(p - dp)A$. Let the piston move under the action of the force exerted by the fluid a distance dl to the right. Then work done by the fluid on the piston is given by force times the distance moved,

i.e.   Work done by fluid $= (pA) \times \mathrm{dl} = p\,\mathrm{d}V$

(where $\mathrm{d}V$ is a small increase in volume).

Or considering unit mass

$$\text{Work done} = p\,\mathrm{d}v$$

(where v is the specific volume).

This is only true when criteria (a) and (b) hold as stated in Section 1.4. Hence when a reversible process takes place between state 1 and state 2 we have,

$$\text{Work done by unit mass of fluid} = \int_1^2 p\,\mathrm{d}v \qquad 1.2$$

When a fluid undergoes a reversible process a series of state points can be joined up to form a line on a diagram of properties. The work done by the fluid during any reversible process is therefore given by the area under the line of the process plotted on a p-v diagram (fig. 1.13),

i.e.   Work done $=$ shaded area on fig. 1.13 $= \int_1^2 p\,\mathrm{d}v$

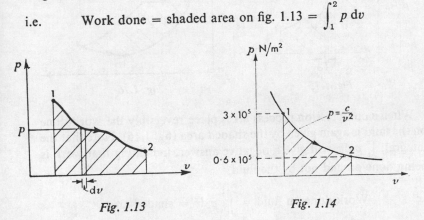

Fig. 1.13   Fig. 1.14

When p can be expressed in terms of v then the integral, $\int_1^2 p\,\mathrm{d}v$, can be evaluated.

Example 1.1

A fluid at a pressure of 3 bar, and with specific volume of 0·18 m³/kg, contained in a cylinder behind a piston expands reversibly to a pressure of 0·6 bar according to a law, $p = c/v^2$ where c is a constant. Calculate the work done by the fluid on the piston.

Referring to fig. 1.14,

$$\text{Work done} = \text{shaded area} = \int_1^2 p \, dv$$

i.e.
$$\text{Work done} = c \int_{v_1}^{v_2} \frac{dv}{v^2} = c \left[-\frac{1}{v} \right]_{v_1}^{v_2}$$

also
$$c = pv^2 = 3 \times 0·18^2 = 0·0972 \text{ bar } (m^3/kg)^2$$

and
$$v_2 = \sqrt{\frac{c}{p_2}} = \sqrt{\frac{0·0972}{0·6}} = 0·402 \text{ m}^3/\text{kg}$$

$$\therefore \text{ Work done} = 0·0972 \times 10^5 \left(\frac{1}{0·18} - \frac{1}{0·402} \right) \text{ N m/kg}$$

$$= 29\,840 \text{ N m/kg}$$

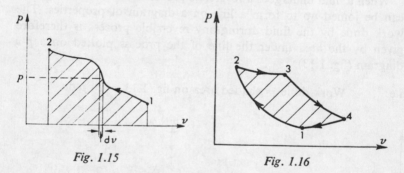

Fig. 1.15   Fig. 1.16

When a compression process takes place reversibly the work done on the fluid is again given by the shaded area (fig. 1.15). Note that the integral, $\int_1^2 p \, dv$, will give a negative answer, indicating that work is being done on and not by the fluid,

i.e.
$$\text{Work done on fluid} = \int_2^1 p \, dv = \text{shaded area}$$

The rule is that a process from left to right on the p-v diagram is one in which the fluid does work on the surroundings (i.e. W is positive). Conversely a process from right to left is one in which the fluid has work done on it by the surroundings (i.e. W is negative).

When a fluid undergoes a series of processes and finally returns to its initial state, then it is said to have undergone a thermodynamic *cycle*. A cycle which consists only of reversible processes is a reversible cycle. A cycle plotted on a diagram of properties forms a closed figure, and a reversible cycle plotted on a *p-v* diagram forms a closed figure the area of which represents the net work of the cycle. For example, a reversible cycle consisting of four reversible processes 1 to 2, 2 to 3, 3 to 4, and 4 to 1 is shown in fig. 1.16. The net work out is equal to the shaded area. If the cycle were described in the reverse direction (i.e. 1 to 4, 4 to 3, 3 to 2, and 2 to 1), then the shaded area would represent net work put into the system. The rule is that the enclosed area of a reversible cycle represents work out (i.e. work done by the system) when the cycle is described in a clockwise manner, and the enclosed area represents work in (i.e. work done on the system) when the cycle is described in an anticlockwise manner.

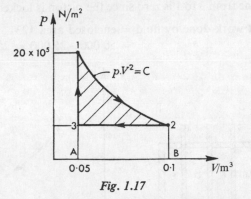

Fig. 1.17

Example 1.2

1 kg of a certain fluid is contained in a cylinder at an initial pressure of 20 bar. The fluid is allowed to expand reversibly behind a piston according to a law $pV^2 =$ constant until the volume is doubled. The fluid is then cooled reversibly at constant pressure until the piston regains its original position; heat is then supplied reversibly with the piston firmly locked in position until the pressure rises to the original value of 20 bar. Calculate the net work done by the fluid, for an initial volume of 0·05 m³.

Referring to fig. 1.17,

$$p_1 V_1^2 = p_2 V_2^2$$

$$\therefore \; p_2 = p_1 \left(\frac{V_1}{V_2}\right)^2 = \frac{20}{2^2} = 5 \text{ bar}$$

Work done by fluid from 1 to 2 = area 12BA1

$$= \int_1^2 p \, dV \text{ from equation 1.2}$$

i.e. $W_{1-2} = \int_{v_1}^{v_2} \frac{c}{V^2} \, dV$ where $c = p_1 V_1^2 = 20 \times 0.05^2$ bar m^6

$$\therefore W_{1-2} = 10^5 \times 20 \times 0.0025 \left[-\frac{1}{V} \right]_{0.05}^{0.1} =$$

$$10^5 \times 20 \times 0.0025 \left(\frac{1}{0.05} - \frac{1}{0.1} \right) = 50\,000 \text{ N m}$$

Work done on fluid from 2 to 3 = area 32BA3 = $p_2(V_2 - V_3)$
$$= 10^5 \times 5 \times (0.1 - 0.05)$$
$$= 25\,000 \text{ N m}$$

Work done from 3 to 1 is zero since the piston is locked in position.

$\therefore$ Net work done by fluid = enclosed area 1231
$$= 50\,000 - 25\,000 = 25\,000 \text{ N m}$$

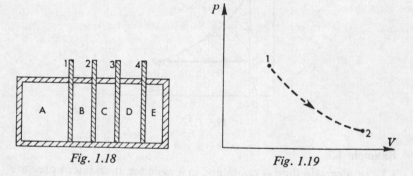

Fig. 1.18   *Fig. 1.19*

It has been stated above that work done is given by $\int p \, dv$ for a reversible process only. It can easily be shown that $\int p \, dv$ is not equal to the work done if a process is irreversible. For example, consider a cylinder divided into a number of compartments by sliding partitions (fig. 1.18). Initially, compartment A is filled with a mass of fluid at pressure p_1. When the sliding partition 1 is removed quickly, then the fluid expands to fill compartments A and B. When the system settles down to a new equilibrium state the pressure and volume are fixed and the state can be marked on the p-V diagram (fig. 1.19). Sliding partition 2 is now removed and the fluid expands to occupy compartments A, B, and C. Again the equilibrium state can be marked on the diagram. The same procedure can be adopted with partitions 3 and 4 until finally the fluid is at p_2 and occupies a volume V_2 when filling compartments A, B, C, D, and E. The area under the curve 1–2 on fig. 1.19 is given by $\int_1^2 p \, dV$, but no work has been done by the fluid. No piston has been moved, no turbine wheel has been revolved; in other words, no external force has been moved through a distance. This is the extreme case of an irreversible process in which $\int p \, dV$ has a value and yet the work done is zero. When a fluid expands without a restraining force being exerted by the surroundings, as in the example above, the process is known as *free expansion*. Free expansion is highly irreversible by criterion (b) Section 1.4. In many practical expansion processes some work is done by the fluid which is less than $\int p \, dv$ and in many practical compression processes work is done on the fluid which is greater than $\int p \, dv$. For example, the paddle wheel work input shown in fig. 1.7 is irreversible. It is important to represent all irreversible processes by dotted lines on diagrams of properties.

PROBLEMS

1.1 A certain fluid at 10 bar is contained in a cylinder behind a piston, the initial volume being 0.05 m^3. Calculate the work done by the fluid when it expands reversibly,

(a) At constant pressure to a final volume of 0.2 m^3.

(b) According to a linear law to a final volume of 0.2 m^3 and a final pressure of 2 bar.

(c) According to a law $pV = $ constant to a final volume of 0.1 m^3.

(d) According to a law $pV^3 = $ constant to a final volume of 0.06 m^3.

(e) According to a law $p = (A/V^2) - (B/V)$ to a final volume of 0.1 m^3 and a final pressure of 1 bar. A and B are constants.

Sketch all processes on the p-V diagram.
(150 000; 90 000; 34 700; 7640; 19 200 N m)

1.2 1 kg of a fluid is compressed reversibly according to a law $pv = 0.25$ where p is in bar and v is in m^3/kg. The final volume is $\frac{1}{4}$ of the initial volume. Calculate the work done on the fluid and sketch the process on a p-v diagram.
(34 660 N m)

1.3 0.05 m^3 of a gas at 6·9 bar expand reversibly in a cylinder behind a piston according to the law $pv^{1·2} = \text{constant}$ until the volume is 0.08 m^3. Calculate the work done by the gas and sketch the process on a p-v diagram.   (15 300 N m)

1.4 1 kg of a fluid expands reversibly according to a linear law from 4·2 bar to 1·4 bar. The initial and final volumes are 0.004 m^3 and 0.02 m^3 respectively. The fluid is then cooled reversibly at constant pressure and finally compressed reversibly according to a law $pv = \text{constant}$ back to the initial conditions of 4·2 bar and 0.004 m^3. Calculate the work done in each process stating whether it is done on or by the fluid and calculate the net work of the cycle. Sketch the cycle on a p-v diagram.   (4480; − 1120; − 1845; 1515 N m)

1.5 0.09 m^3 of a fluid at 0·7 bar are compressed reversibly to a pressure of 3·5 bar according to a law $pv^n = \text{constant}$. The fluid is then heated reversibly at a constant volume until the pressure is 4 bar; the specific volume is then $0.5 \text{ m}^3/\text{kg}$. A reversible expansion according to a law $pv^2 = \text{constant}$ restores the fluid to its initial state. Calculate the mass of fluid present, the value of n in the first process, and the net work done on or by the fluid in the cycle. Sketch the cycle on a p-v diagram.   (0·0753 kg; 1·85; 676 N m)

1.6 A fluid is heated reversibly at a constant pressure of 1·05 bar until it has a specific volume of $0.1 \text{ m}^3/\text{kg}$. It is then compressed reversibly according to a law $pv = \text{constant}$ to a pressure of 4·2 bar, then allowed to expand reversibly according to a law $pv^{1·3} = \text{constant}$, and is finally heated at constant volume back to the initial conditions. The work done in the constant pressure process is 515 N m and the mass of fluid present is 0·2 kg. Calculate the net work done on or by the fluid in the cycle and sketch the cycle on a p-v diagram.

                                                              (− 422 N m)

REFERENCES

1.1 DOUGLAS, J. F., GASIOREK, J. M. and SWAFFIELD, J. A., *Fluid Mechanics* (Pitman, 1985).

1.2 *Code for Temperature Measurement*, B.S. 1041:1943.

1.3 *Units and Standards of Measurement*, IV—*Temperature*, National Physical Laboratory (H.M.S.O., 1966).

1.4 *The International System (SI) Units*, B.S. 3763:1970.

1.5 *Changing to the Metric System*, National Physical Laboratory (H.M.S.O., 1967).

1.6 *International Practical Temperature Scale* (H.M.S.O., 1969).

2. The First Law of Thermodynamics

2.1 Conservation of energy

The concept of energy and the hypothesis that it can be neither created nor destroyed were developed by scientists in the early part of the nineteenth century, and became known as the *Principle of the Conservation of Energy*. The First Law of Thermodynamics is merely one statement of this general principle with particular reference to heat energy* and mechanical energy (i.e. work).

It was shown in Section 1.5 that when a system is made to undergo a complete cycle then net work is done on or by the system. Consider a cycle in which net work is done by the system. Since energy cannot be created, this mechanical energy must have been supplied from some source of energy. Now the system has been returned to its initial state, therefore its intrinsic energy is unchanged, and hence the mechanical energy has not been provided by the system itself. The only other energy involved in the cycle is the heat which was supplied and rejected in the various processes. Hence, by the principle of the conservation of energy, the net work done by the system is equal to the net heat supplied to the system. The First Law of Thermodynamics can therefore be stated as follows:

When a system undergoes a thermodynamic cycle then the net heat supplied to the system from its surroundings is equal to the net work done by the system on its surroundings.

In symbols,

$$\sum dQ = \sum dW \qquad\qquad 2.1$$

where $\sum$ represents the sum for a complete cycle.

* Heat was considered to be a fluid (known as caloric) by scientists in the eighteenth century, and it was not conclusively shown to be a form of energy until Joule, about 1840, proved by a series of experiments that heat and work are mutually convertible forms of the same thing. The concept of heat as a fluid may seem strange now, but the theory did appear to explain many phenomena, and it was widely believed for many years prior to the experiments of Joule (see reference 2.1).

Example 2.1

In a certain steam plant the turbine develops 1000 kW. The heat supplied to the steam in the boiler is 2800 kJ/kg, the heat rejected by the system to cooling water in the condenser is 2100 kJ/kg and the feed pump work required to pump the condensate back into the boiler is 5 kW. Calculate the steam flow round the cycle in kg/s. The cycle is shown diagrammatically in fig. 2.1. A boundary is shown which encompasses the entire plant. Strictly, this boundary should be thought of as encompassing the working fluid only.

$$\sum dQ = 2800 - 2100 = 700 \text{ kJ/kg}$$

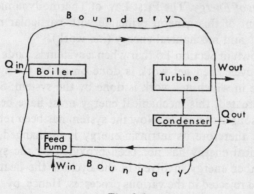

Fig. 2.1

Let the steam flow be $\dot{m}$ kg/s

$$\therefore \sum dQ = 700 \, \dot{m} \text{ kJ/s}$$

$$\sum dW = 1000 - 5 = 995 \text{ kW} = 995 \text{ kJ/s}$$

Then in equation 2.1

$$\sum dQ = \sum dW$$

i.e. $\qquad 700 \times \dot{m} = 995 \qquad \therefore \dot{m} = \dfrac{995}{700} = 1 \cdot 421 \text{ kg/s}$

i.e. $\qquad\qquad$ Steam flow required $= 1 \cdot 421$ kg/s

2.2 The non-flow equation

In the previous section it was stated that when a system possessing a certain intrinsic energy is made to undergo a cycle by heat and work transfer, then the net heat supplied is equal to the net work output.

This is true for a complete cycle when the final intrinsic energy of the system is equal to its initial value. Consider now a process in which the intrinsic energy of the system is finally greater than the initial intrinsic energy. The difference between the net heat supplied and the net work output has increased the intrinsic energy of the system,

i.e.   Gain in intrinsic energy

= Net heat supplied − net work output

When the net effect is to transfer energy from the system, then there will be a loss in the intrinsic energy of the system.

When a fluid is not in motion then its intrinsic energy per unit mass is known as the *specific internal energy* of the fluid and is given the symbol u. The internal energy of a fluid depends on its pressure and temperature, and is itself a property. The simple proof that internal energy is a property is given in reference 2.2. The internal energy of mass, m, of a fluid is written as U, i.e. $mu = U$. The units of internal energy, U, are usually written as kJ.

Since internal energy is a property then gain in internal energy in changing from state 1 to state 2 can be written $U_2 - U_1$.

Also, gain in internal energy = net heat supplied − net work output,

i.e.
$$U_2 - U_1 = \sum_1^2 \mathrm{d}Q - \sum_1^2 \mathrm{d}W$$

This equation is true for a process or series of processes between state 1 and state 2 provided there is no flow of fluid into or out of the system. In any one non-flow process there will be either heat supplied or heat rejected, but not both; similarly there will be either work output or work input, but not both. Hence, denoting heat which is supplied to the system as positive and work which is done by the system (i.e. work output) as positive, we have,

$$U_2 - U_1 = Q - W \quad \text{for a non-flow process}$$

i.e.
$$Q = (U_2 - U_1) + W$$

or for 1 kg,
$$Q = (u_2 - u_1) + W \qquad 2.2$$

This equation is known as the *non-flow energy equation*. Equation 2.2 is very often written in differential form. For a small amount of heat supplied dQ, a small amount of work done by the fluid dW, and a small gain in specific internal energy du, then,

$$dQ = du + dW \qquad 2.3$$

Example 2.2

In the compression stroke of an internal-combustion engine the heat rejected to the cooling water is 45 kJ/kg and the work input is 90 kJ/kg. Calculate the change in specific internal energy of the working fluid stating whether it is a gain or a loss.

$$Q = -45 \text{ kJ/kg}$$

($-$ve sign since heat is rejected).

$$W = -90 \text{ kJ/kg}$$

($-$ve sign since work is a work input to the system).
Using equation 2.2,

$$Q = (u_2 - u_1) + W$$

$$\therefore \quad -45 = (u_2 - u_1) - 90$$

$$\therefore \quad u_2 - u_1 = +90 - 45 = 45 \text{ kJ/kg}$$

$$\therefore \quad \text{Gain in internal energy} = 45 \text{ kJ/kg}$$

Example 2.3

In the cylinder of an air motor the compressed air has a specific internal energy of 420 kJ/kg at the beginning of the expansion and a specific internal energy of 200 kJ/kg after expansion. Calculate the heat flow to or from the cylinder when the work done by the air during the expansion is 100 kJ/kg.

From equation 2.2

$$Q = (u_2 - u_1) + W$$

$$\therefore \ Q = (200 - 420) + 100 \ = \ -220 + 100 \ = \ -120 \text{ kJ/kg}$$

i.e.   Heat rejected by the air = 120 kJ/kg

It is important to note that equations 2.1, 2.2, and 2.3 are true whether the process is reversible or not. These are energy equations. For a reversible non-flow process we have from equation 1.2,

$$W = \int_1^2 p \, dv$$

or for small quantities, $dW = p \, dv$

Hence for any reversible non-flow process, substituting in equation 2.3,

$$dQ = du + p \, dv \qquad\qquad 2.4$$

or substituting in equation 2.2,

$$Q = (u_2 - u_1) + \int_1^2 p \, dv \qquad\qquad 2.5$$

Equations 2.4 and 2.5 can only be used for ideal reversible non-flow processes.

2.3 The steady flow equation

In Section 2.2, the internal energy of a fluid was said to be the intrinsic energy of the fluid due to its thermodynamic properties. When 1 kg of a fluid with specific internal energy, u is moving with velocity C and is a height Z above a datum level, then it posseses a total energy of $u + (C^2/2) + Zg$, where $C^2/2$ is the kinetic energy of 1 kg of the fluid and Zg is the potential energy of 1 kg of the fluid.

In most practical problems the rate at which the fluid flows through a machine or piece of apparatus is constant. This type of flow is called *steady flow*.

Consider 1 kg of a fluid flowing in steady flow through a piece of apparatus (fig. 2.2). This constitutes an open system as defined in

Section 1.2. The boundary is shown cutting the inlet pipe at section 1 and the outlet pipe at section 2. This boundary is sometimes called a *control surface*, and the system encompassed, a *control volume*.

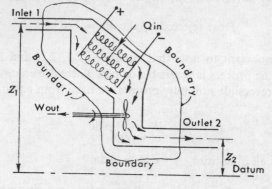

Fig. 2.2

Let it be assumed that a steady flow of heat Q units per kg of fluid is supplied, and that each kg of fluid does W units of work as it passes

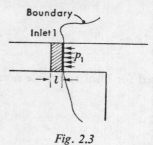

Fig. 2.3

through the apparatus. Now in order to introduce 1 kg of fluid across the boundary an expenditure of energy is required; similarly in order to push 1 kg of the fluid across the boundary at exit, an expenditure of energy is required. The inlet section is shown enlarged in fig. 2.3. Consider an element of fluid, length l, and let the cross-sectional area of the inlet pipe be A_1. Then we have,

Energy required to push element across boundary $= (p_1 A_1) \times l$

$$= p_1 \times (\text{volume of fluid element})$$

$\therefore$ Energy required for 1 kg of fluid $= p_1 v_1$

(where v_1 is the specific volume of the fluid at section 1).

Similarly it can be shown that,

Energy required at exit to push 1 kg of fluid
$$\text{across the boundary} = p_2 v_2$$

Consider now the energy entering and leaving the system. The energy entering the system consists of the energy of the flowing fluid at inlet $\left(u_1 + \dfrac{C_1^2}{2} + Z_1 g\right)$, the energy term $p_1 v_1$, and the heat supplied Q. The energy leaving the system consists of the energy of the flowing fluid at the outlet section $\left(u_2 + \dfrac{C_2^2}{2} + Z_2 g\right)$, the energy term $p_2 v_2$, and the work done by the fluid W. Since there is steady flow of fluid into and out of the system, and there are steady flows of heat and work, then the energy entering must exactly equal the energy leaving.

$$u_1 + \frac{C_1^2}{2} + Z_1 g + p_1 v_1 + Q = u_2 + \frac{C_2^2}{2} + Z_2 g + p_2 v_2 + W \qquad 2.6$$

In nearly all problems in Applied Thermodynamics, changes in height are negligible and the potential energy terms can be omitted from the equation. The terms in u and pv occur on both sides of the equation and always will do so in a flow process, since a fluid always possesses a certain internal energy, and the term pv always occurs at inlet and outlet as seen in the above proof. The sum of specific internal energy and the pv term is given the symbol h, and is called *specific enthalpy*,

i.e. *Specific enthalpy,* $h = u + pv$ $\qquad$ 2.7

The enthalpy of a fluid is a property of the fluid, since it consists of the sum of a property and the product of two properties. Since enthalpy is a property like internal energy, pressure, specific volume. and temperature, it can be introduced into any problem whether the process is a flow process or a non-flow process. The enthalpy of mass, m, of a fluid can be written as H (i.e. $mh = H$). The units of h are the same as those of internal energy.

Substituting equation 2.7 in equation 2.6,

$$h_1 + \frac{C_1^2}{2} + Q = h_2 + \frac{C_2^2}{2} + W \qquad 2.8$$

Equation 2.8 is known as the *steady-flow energy equation*. In steady flow the rate of mass flow of fluid at any section is the same as at any other section. Consider any section of cross-sectional area A, where the fluid velocity is C, then the rate of volume flow past the section is CA. Also, since mass flow is volume flow divided by specific volume,

$$\text{Mass flow rate, } \dot{m} = \frac{CA}{v} = \rho C A \qquad 2.9$$

(where v = specific volume at the section; ρ = density at the section). This equation is known as the *continuity of mass equation*.

With reference to fig. 2.2,

$$\dot{m} = \frac{C_1 A_1}{v_1} = \frac{C_2 A_2}{v_2}$$

Example 2.4

In the turbine of a gas turbine unit the gases flow through the

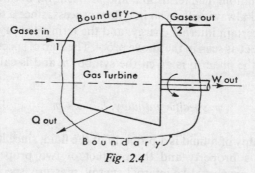

Fig. 2.4

turbine at 17 kg/s and the power developed by the turbine is 14 000 kW. The enthalpies of the gases at inlet and outlet are 1200 kJ/kg and 360 kJ/kg respectively, and the velocities of the gases at inlet and outlet are 60 m/s and 150 m/s respectively. Calculate the rate at which heat is rejected from the turbine. Find also the area of the inlet pipe given that the specific volume of the gases at inlet is 0·5 m³/kg.

A diagrammatic representation of the turbine is shown in fig. 2.4. From equation 2.8,

$$h_1 + \frac{C_1^2}{2} + Q = h_2 + \frac{C_2^2}{2} + W$$

$$\text{Kinetic energy at inlet} = \frac{C_1^2}{2} = \frac{60^2}{2}\ \text{m}^2/\text{s}^2 = \frac{60^2}{2}\ \frac{\text{kg m}^2}{\text{s}^2\ \text{kg}}$$

$$= 1800\ \text{N m/kg} = 1\cdot 8\ \text{kJ/kg}$$

$$\text{Kinetic energy at outlet} = \frac{C_2^2}{2}$$

$$= 2\cdot 5^2 \times (\text{kinetic energy at inlet})$$

$$= 11\cdot 25\ \text{kJ/kg} \ (\text{since } C_2 = 2\cdot 5C_1)$$

$$W = \frac{14\ 000}{17}\ \text{kJ/kg} = 823\cdot 5\ \text{kJ/kg}$$

Substituting in equation 2.8,

$$1200 + 1\cdot 8 + Q = 360 + 11\cdot 25 + 823\cdot 5$$

$$\therefore\ Q = -7\cdot 02\ \text{kJ/kg}$$

i.e.   Heat rejected $= +7\cdot 02\ \text{kJ/kg} = 7\cdot 02 \times 17\ \text{kJ/s} = 119\cdot 3\ \text{kW}$

To find the inlet area, use equation 2.9,

i.e.
$$\dot{m} = \frac{CA}{v} \qquad \therefore\ A = \frac{v\dot{m}}{C}$$

$$\therefore\ \text{Inlet area, } A_1 = \frac{17 \times 0\cdot 5}{60} = 0\cdot 142\text{m}^2$$

Example 2.5

Air flows steadily at the rate of 0·4 kg/s through an air compressor, entering at 6 m/s with a pressure of 1 bar and a specific volume of 0·85 m³/kg, and leaving at 4·5 m/s with a pressure of 6·9 bar and a specific volume of 0·16 m³/kg. The specific internal energy of the air leaving is 88 kJ/kg greater than that of the air entering. Cooling water in a jacket surrounding the cylinder absorbs heat from the air at the rate of 59 kJ/s. Calculate the power required to drive the compressor and the inlet and outlet pipe cross-sectional areas.

In this problem it is more convenient to write the flow equation as in equation 2.6, omitting the Z terms

i.e.
$$u_1 + \frac{C_1^2}{2} + p_1 v_1 + Q = u_2 + \frac{C_2^2}{2} + p_2 v_2 + W$$

A diagrammatic representation of the compressor is shown in

fig. 2.5. Note that the heat rejected across the boundary is equivalent to the heat removed by the cooling water from the compressor.

$$\frac{C_1^2}{2} = \frac{6 \times 6}{2} \text{ J/kg} = 18 \text{ J/kg}$$

$$\frac{C_2^2}{2} = \frac{4 \cdot 5 \times 4 \cdot 5}{2} \text{ J/kg} = 10 \cdot 1 \text{ J/kg}$$

$$p_1 v_1 = 1 \times 10^5 \times 0 \cdot 85 = 85\,000 \text{ J/kg}$$

$$p_2 v_2 = 6 \cdot 9 \times 10^5 \times 0 \cdot 16 = 110\,400 \text{ J/kg}$$

$$u_2 - u_1 = 88 \text{ kJ/kg}$$

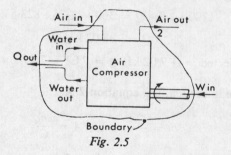

Fig. 2.5

$$\text{Heat rejected} = 59 \text{ kJ/s} = \frac{59}{0 \cdot 4} = 147 \cdot 5 \text{ kJ/kg}$$

Now
$$W = (u_1 - u_2) + (p_1 v_1 - p_2 v_2) + \left(\frac{C_1^2}{2} - \frac{C_2^2}{2} \right) + Q$$

i.e.

$$W = -88 + 85 - 110 \cdot 4 + 0 \cdot 018 - 0 \cdot 0101 - 147 \cdot 5 = -260 \cdot 9 \text{ kJ/kg}$$

(Note that the change in kinetic energy is negligibly small in comparison with the other terms.)

i.e.
$$\text{Work input required} = 260 \cdot 9 \text{ kJ/kg}$$
$$= 260 \cdot 9 \times 0 \cdot 4 \text{ kJ/s}$$

i.e.
$$\text{Work input required} = 260 \cdot 9 \times 0 \cdot 4 = 104 \cdot 4 \text{ kW}$$

From equation 2.9,

$$\dot{m} = \frac{CA}{v}$$

i.e. $$A_1 = \frac{0\cdot4 \times 0\cdot85}{6} \text{ m}^2 = 0\cdot057 \text{ m}^2$$

i.e. Inlet pipe cross-sectional area $= 0\cdot057 \text{ m}^2$

Similarly, $$A_2 = \frac{0\cdot4 \times 0\cdot16}{4\cdot5} = 0\cdot014 \text{ m}^2$$

i.e. Outlet pipe cross-sectional area $= 0\cdot014 \text{ m}^2$

In example 2.5 the steady flow energy equation has been used, despite the fact that the compression consists of: suction of air; compression in a closed cylinder; and discharge of air. The steady flow equation can be used because the cycle of processes takes place many times in a minute, and therefore the average effect is steady flow of air through the machine.

PROBLEMS

2.1 In an air compressor the compression takes place at constant internal energy and 50 kJ of heat are rejected to the cooling water for every kg of air. Find the work required for the compression stroke per kg of air. (50 kJ/kg)

2.2 In the compression stroke of a gas engine the work done on the gas by the piston is 70 kJ/kg and the heat rejected to the cooling water is 42 kJ/kg. Find the change of internal energy, stating whether it is a gain or a loss. (28 kJ/kg Gain.)

2.3 A mass of gas with an internal energy of 1500 kJ is contained in a cylinder which has perfect thermal insulation. The gas is allowed to expand behind a piston until its internal energy is 1400 kJ. Calculate the work done by the gas. If the expansion follows a law $pv^2 =$ constant, and the initial pressure and volume of the gas are 28 bar and 0·06 m³ respectively, calculate the final pressure and volume.
(100 kJ; 4·59 bar; 0·148 m³)

2.4 The gases in the cylinder of an internal-combustion engine have an internal energy of 800 kJ/kg and a specific volume of 0·06 m³/kg at the beginning of expansion. The expansion of the gases may be assumed to take place according to a reversible law $pv^{1\cdot5} =$ constant, from 55 bar to 1·4 bar. The internal energy after expansion is 230 kJ/kg. Calculate the heat rejected to the cylinder cooling water per kg of gases during the expansion stroke. (104 kJ/kg)

2.5 A steam turbine receives a steam flow of 1·35 kg/s and delivers 500 kW. The heat loss from the casing is negligible.

(a) Find the change of specific enthalpy across the turbine when the velocities at entrance and exit and the difference in elevation at entrance and exit are negligible.

(b) Find the change of specific enthalpy across the turbine when the velocity at entrance is 60 m/s, the velocity at exit is 360 m/s, and the inlet pipe is 3 m above the exhaust pipe.   (370 kJ/kg; 433 kJ/kg)

2.6 A steady flow of steam enters a condenser with an enthalpy of 2300 kJ/kg and a velocity of 350 m/s. The condensate leaves the condenser with an enthalpy of 160 kJ/kg and a velocity of 70 m/s. Find the heat transfer to the cooling fluid per kg of steam condensed.

(−2199 kJ/kg)

2.7 A turbine operating under steady flow conditions receives steam at the following state: pressure 13·8 bar; specific volume 0·143 m³/kg; internal energy 2590 kJ/kg; velocity 30 m/s. The state of the steam leaving the turbine is: pressure 0·35 bar, specific volume 4·37 m³/kg, internal energy 2360 kJ/kg, velocity 90 m/s. Heat is lost to the surroundings at the rate of 0·25 kJ/s. If the rate of steam flow is 0·38 kg/s, what is the power developed by the turbine?

(102·8 kW)

2.8 A nozzle is a device for increasing the velocity of a steadily flowing stream of fluid. At the inlet to a certain nozzle the enthalpy of the fluid is 3025 kJ/kg and the velocity is 60 m/s. At the exit from the nozzle the enthalpy is 2790 kJ/kg. The nozzle is horizontal and there is negligible heat loss from it.

(a) Find the velocity at the nozzle exit.

(b) If the inlet area is 0·1 m² and the specific volume at inlet is 0·19 m³/kg, find the rate of flow of fluid.

(c) If the specific volume at the nozzle exit is 0·5 m³/kg, find the exit area of the nozzle.   (688 m/s; 31·6 kg/s; 0·0229 m²)

REFERENCES

2.1 JEANS, SIR JAMES, *The Growth of Physical Science* (Century Bookbindery, 1982).

2.2 ROGERS, G. F. C., and MAYHEW, Y. R., *Engineering Thermodynamics, Work andn Heat Transfer,* S I Units (Longmans, 1980).

3. The Working Fluid

In Section 1.3 the matter contained within the boundaries of a system is defined as the working fluid, and it is stated that when two independent properties of the fluid are known then the thermodynamic state of the fluid is defined. In thermodynamic systems the working fluid can be in the liquid, vapour, or gaseous phase. All substances can exist in any one of these phases, but we tend to identify all substances with the phase in which they are in equilibrium at atmospheric pressure and temperature. For instance, substances such as oxygen and nitrogen are thought of as gases; H_2O is thought of as liquid or vapour (i.e. water or steam); mercury is thought of as a liquid. All these substances can exist in different phases: oxygen and nitrogen can be liquefied; H_2O can become a gas at very high temperatures; mercury can be vaporized and will act as a gas if the temperature is raised high enough.

3.1 Liquid, vapour, and gas

Consider a p-v diagram for any substance. The solid phase is not important in engineering thermodynamics, being more the province of the metallurgist or physicist. When a liquid is heated at any one constant pressure there is one fixed temperature at which bubbles of vapour form in the liquid; this phenomenon is known as boiling. The higher the pressure of the liquid then the higher the temperature at which boiling occurs. It is also found that the volume occupied by 1 kg of a boiling liquid at a higher pressure is slightly larger than the volume occupied by 1 kg of the same liquid when it is boiling at a low pressure. A series of boiling points plotted on a p-v diagram will appear as a sloping line, as shown in fig. 3.1. The points P, Q, and R represent the boiling points of a liquid at pressures p_P, p_Q, and p_R respectively.

When a liquid at boiling point is heated further at constant pressure the additional heat supplied changes the phase of the substance from liquid to vapour; during this change of phase the pressure and temperature remain constant. The heat supplied is called the *latent heat of vaporization*. It is found that the higher the pressure then the smaller is the amount of latent heat required. There is a definite value of specific volume of the vapour at any one pressure, at the point at which vaporization is complete, hence a

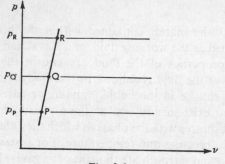

Fig. 3.1

series of points such as P′, Q′, and R′ can be plotted and joined to form a line as shown in fig. 3.2

When the two curves already drawn are extended to higher pressures they form a continuous curve, thus forming a loop (see fig. 3.3). The pressure at which the turning point occurs is called the *critical pressure* and the turning point itself is called the *critical point*

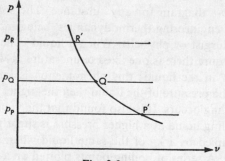

Fig. 3.2

(point C on fig. 3.3). It can be seen that at the critical point the latent heat is zero. The substance existing at a state point inside the loop consists of a mixture of liquid and dry vapour and is known as a *wet vapour*. *A saturation state* is defined as a state at which a change of phase may occur without change of pressure or temperature. Hence the boiling points P, Q, and R are saturation states, and a series of such boiling points joined up is called the *saturated liquid line*. Similarly the points P′, Q′, and R′, at which the liquid is completely changed into vapour, are saturation states, and a series of such

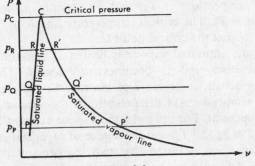

Fig. 3.3

points joined up is called the *saturated vapour line*. The word saturation as used here refers to energy saturation. For example, a slight addition of heat to a boiling liquid changes some of it into a vapour, and it is no longer a liquid but is now a wet vapour. Similarly when a substance just on the saturated vapour line is cooled slightly, droplets

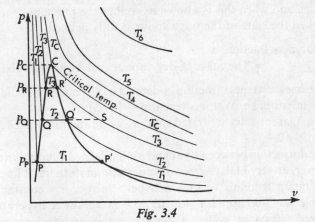

Fig. 3.4

of liquid will begin to form, and the saturated vapour becomes a wet vapour. A saturated vapour is usually called *dry saturated* to emphasize the fact that no liquid is present in the vapour in this state.

Lines of constant temperature, called isothermals, can be plotted on a *p-v* diagram as shown in fig. 3.4. The temperature lines become horizontal between the saturated liquid line and the saturated vapour line (e.g. between P and P', Q and Q', R and R'). Thus there is a corresponding *saturation temperature* for each *saturation pressure*. At pressure p_P the saturation temperature is T_1, at pressure p_Q the saturation temperature is T_2, and at pressure p_R the saturation temperature is T_3. The critical temperature line T_C just touches the top of the loop at the critical point C.

When a dry saturated vapour is heated at constant pressure its temperature rises and it becomes *superheated*. The difference between the actual temperature of the superheated vapour and the saturation temperature at the pressure of the vapour is called the *degree of superheat*. For example, the vapour at point S (fig. 3.4) is superheated at p_Q and T_3, and the degree of superheat is $T_3 - T_2$.

In Section 1.5 it is stated that two independent properties are sufficient to define the state of a substance. Now between P and P', Q and Q', R and R' the temperature and pressure are not independent since they remain constant for a range of values of v. For example, a substance at p_Q and T_2 (fig. 3.4) could be a saturated liquid, a wet vapour, or a dry saturated vapour. The state cannot be defined until one other property (e.g. specific volume) is given. The condition or quality of a wet vapour is most frequently defined by its *dryness fraction*, and when this is known as well as the pressure or temperature then the state of the wet vapour is fully defined.

Dryness fraction, x
= The mass of dry vapour in 1 kg of the mixture

(Sometimes a wetness fraction is defined as the mass of liquid in 1 kg of the mixture, i.e. Wetness fraction $= 1 - x$.)

Note that for a dry saturated vapour, $x = 1$; and that for a saturated liquid, $x = 0$.

The distinction between a gas and a superheated vapour is not rigid. However, at very high degrees of superheat an isothermal line on the p-v diagram tends to become a hyperbola (i.e. $pv =$ constant). For example the isothermal T_6 on fig. 3.4 is almost a hyperbola. An idealized substance called a *perfect gas* is assumed to have an equation of state $pv/T =$ constant. It can be seen that when a line of constant temperature follows a hyperbolic law then the equation $pv/T =$ constant is satisfied. All substances tend towards a perfect gas at very high degrees of superheat. Substances which are thought of as gases (e.g. oxygen, nitrogen, hydrogen, etc.) are highly superheated at normal atmospheric conditions. For example, the critical temperatures of oxygen, nitrogen, and hydrogen are approximately $-119°C$, $-147°C$, and $-240°C$ respectively. Substances normally existing as vapours must be raised to high temperatures before they begin to act as a perfect gas. For example, the critical temperatures of ammonia, sulphur dioxide, and water vapour are $130°C$, $157°C$, and $374.15°C$ respectively.

The working fluid in practical engineering problems is either a substance which is approximately a perfect gas (e.g. air), or a substance which exists mainly as liquid and vapour, such as steam and the refrigerant vapours (e.g. ammonia, freon, methyl chloride, etc.). For the substances which approximate to perfect gases certain laws relating the properties can be assumed. For the substances in the liquid and vapour phases the properties are not related by definite laws, and values of the properties are determined empirically and tabulated in a convenient form.

3.2 The use of vapour tables

Tables are available for a wide variety of substances which normally exist in the vapour phase (e.g. steam, ammonia, freon, etc.). The tables which will be used in this book are those arranged by Mayhew and Rogers (reference 3.1), which are suitable for student use. For more comprehensive tables for steam, reference 3.2 should be consulted. The tables of Mayhew and Rogers are mainly concerned with steam, but some properties of ammonia and freon-12 are also given.

Saturation state properties

The saturation pressures and corresponding saturation temperatures of steam are tabulated in parallel columns in the first table, for pressures ranging from 0.006112 bar to the critical pressure of 221.2 bar. The specific volume, internal energy, enthalpy, and entropy are also tabulated for the dry saturated vapour at each pressure and corresponding saturation temperature. The suffix g is used to denote the dry saturated state. A specimen row from the tables is shown in fig. 3.5. For example at 0.34 bar the saturation tempera-

p	t_s	v_g	u_f	u_g	h_f	h_{fg}	hg	s_f	s_{fg}	s_g
bar	°C	m³/kg	kJ/kg		kJ/kg			kJ/kg K		
0.34	72.0	4.649	302	2472	302	2328	2630	0.980	6.745	7.725

Fig. 3.5

ture is 72°C, the specific volume of dry saturated vapour, v_g, at this pressure is 4·649 m³/kg, the internal energy of dry saturated vapour, u_g, is 2472 kJ/kg, and the enthalpy of dry saturated vapour, h_g, is 2630 kJ/kg. The steam is in the state represented by point A on fig. 3.6. At point B dry saturated steam at a pressure of 100 bar and saturation temperature 311°C has a specific volume, v_g, of 0·01802 m³/kg, internal energy, u_g, of 2545 kJ/kg and enthalpy, h_g, of 2725 kJ/kg.

The specific internal energy, specific enthalpy, and specific entropy of saturated liquid are also tabulated, the suffix f being used for this state. For example at 4 bar and the corresponding saturation

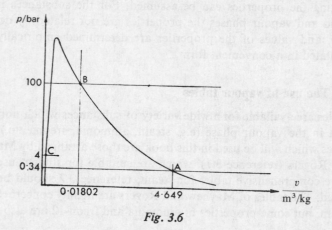

Fig. 3.6

temperature 143·6°C, saturated water has an internal energy, u_f, of 605 kJ/kg, and an enthalpy, h_f, of 605 kJ/kg. This state corresponds to point C on fig. 3.6. The specific volume of saturated water, v_f, is tabulated in a separate table, but it is usually negligibly small in comparison with the specific volume of the dry saturated vapour, and its variation with temperature is very small; the saturated liquid line on a p-v diagram is very nearly coincident with the pressure axis in comparison with the width of the wet loop (see fig. 3.6). As seen from the table, values of v_f vary from about 0.001 m³/kg at 0·01°C to about 0·0011 m³/kg at 160°C; as the pressure approaches the critical value, the increase of v_f is more marked, and at the critical temperature of 374·15°C the value of v_f is 0·00317 m³/kg.

The change in specific enthalpy from h_f to h_g is given the symbol

h_{fg}. When saturated water is changed to dry saturated vapour, from equation 2.2,

$$Q = (u_2 - u_1) + W = (u_g - u_f) + W$$

Also W is represented by the area under the horizontal line on the p-v diagram,

i.e. $$W = (v_g - v_f)p$$

$$\therefore \ Q = (u_g - u_f) + p(v_g - v_f)$$

$$= (u_g + pv_g) - (u_f + pv_f)$$

From equation 2.7

$$h = u + pv$$

$$\therefore \ Q = h_g - h_f = h_{fg}$$

The heat required to change a saturated liquid to a dry saturated vapour is called the latent heat. Hence, latent heat is given in the tables as h_{fg}.

In the case of steam tables, the internal energy of saturated liquid is taken to be zero at the Triple point (i.e. at $0.01°C$ and 0.006112 bar). Then since, from equation 2.7, $h = u + pv$, we have,

$$h \text{ at } 0.01°C \text{ and } 0.006112 \text{ bar} = 0 + \frac{0.006112 \times 10^5 \times 0.0010002}{10^3}$$

(where v_f at $0.01°C$ is 0.0010002 m³/kg)

i.e. $$h = 6.112 \times 10^{-4} \text{ kJ/kg}$$

This is negligibly small and hence the zero for enthalpy may be taken as $0.01°C$.

Note that at the other end of the pressure range tabulated in the first table the pressure of 221.2 bar is the critical pressure, $374.15°C$ is the critical temperature, and the latent heat, h_{fg}, is zero.

Properties of wet vapour

For a wet vapour the total volume of the mixture is given by the volume of liquid present plus the volume of dry vapour present.

Therefore the specific volume is given by,

$$v = \frac{\text{volume of liquid} + \text{volume of dry vapour}}{\text{total mass of wet vapour}}$$

Now for 1 kg of wet vapour there are x kg of dry vapour and $(1-x)$ kg of liquid, where x is the dryness fraction as defined earlier. Hence,

$$v = v_f(1-x) + v_g x$$

The volume of the liquid is usually negligibly small compared to the volume of dry saturated vapour, hence for most practical problems,

$$v = xv_g \qquad 3.1$$

The enthalpy of a wet vapour is given by the sum of the enthalpy of the liquid plus the enthalpy of the dry vapour,

i.e. $$h = (1-x)h_f + xh_g$$
$$\therefore \ h = h_f + x(h_g - h_f)$$

i.e. $$h = h_f + xh_{fg} \qquad 3.2$$

Similarly, the internal energy of a wet vapour is given by the internal energy of the liquid plus the internal energy of the dry vapour,

i.e. $$u = (1-x)u_f + xu_g \qquad 3.3$$

or $$u = u_f + x(u_g - u_f) \qquad 3.4$$

Equation 3.4 can be expressed in a form similar to equation 3.2, but equations 3.3 and 3.4 are more convenient since u_g and u_f are tabulated and the difference, $u_g - u_f$, is not tabulated.

Example 3.1

Find the specific volume, specific enthalpy, and specific internal energy of wet steam at 18 bar, dryness fraction 0·9.

From equation 3.1 $\qquad v = xv_g$

$$\therefore \ v = 0 \cdot 9 \times 0 \cdot 1104 = 0 \cdot 0994 \ \text{m}^3/\text{kg}$$

From equation 3.2 $\qquad h = h_f + xh_{fg}$

$$\therefore \ h = 885 + 0 \cdot 9 \times 1912 = 2605 \cdot 8 \ \text{kJ/kg}$$

From equation 3.3 $\quad u = (1-x)u_f + xu_g$

$$\therefore\ u = (1-0{\cdot}9)883 + 0{\cdot}9 \times 2598 = 2426{\cdot}5 \text{ kJ/kg}$$

Example 3.2

Find the dryness fraction, specific volume and specific internal energy of steam at 7 bar and enthalpy 2600 kJ/kg.

At 7 bar, $h_g = 2764$ kJ/kg, hence since the actual enthalpy is given as 2600 kJ/kg, the steam must be in the wet vapour state. From equation 3.2, $h = h_f + xh_{fg}$

i.e. $\qquad\qquad\qquad 2600 = 697 + x2067$

$$\therefore\ x = \frac{2600 - 697}{2067} = 0{\cdot}921$$

Then from equation 3.1

$$v = xv_g = 0{\cdot}921 \times 0{\cdot}2728 = 0{\cdot}2515 \text{ m}^3/\text{kg}$$

From equation 3.3 $\quad u = (1-x)u_f + xu_g$

i.e. $\qquad u = (1-0{\cdot}921)696 + 0{\cdot}921 \times 2573 = 55 + 2365$

i.e. $\qquad\qquad\qquad u = 2420 \text{ kJ/kg}$

Properties of superheated vapour

For steam in the superheat region temperature and pressure are independent properties. When the temperature and pressure are given for superheated steam then the state is defined and all the other properties can be found. For example, steam at 2 bar and 200°C is superheated since the saturation temperature at 2 bar is 120·2°C, which is less than the actual temperature. The steam in this state has a degree of superheat of $200 - 120{\cdot}2 = 79{\cdot}8$ K. The tables of properties of superheated steam range in pressure from 0·006112 bar to the critical pressure of 221·2 bar, and there is an additional table of supercritical pressures up to 1000 bar. At each pressure there is a range of temperatures up to high degrees of superheat, and the values of specific volume, internal energy, enthalpy, and entropy are tabulated at each pressure and temperature for pressures up to and including 70 bar; above this pressure the internal energy is not tabulated. For reference the saturation temperature is inserted in brackets under each pressure in the superheat tables and values of v_g, u_g, h_g and s_g are also given. A specimen row of values is shown in

p/bar (t_s)	$\dfrac{t}{\degree C}$	250	300	350	400	450	500	600
20 (212·4)	v	0·1115	0·1255	0·1386	0·1511	0·1634	0·1756	0·1995
	u	2681	2774	2861	2946	3030	3116	3291
	h	2904	3025	3138	3248	3357	3467	3690
	s	6·547	6·768	6·957	7·126	7·283	7·431	7·701

Fig. 3.7

fig. 3.7. For example, from superheat tables at 20 bar and 400°C the specific volume is 0·1511 m³/kg and the enthalpy is 3248 kJ/kg.

For pressures above 70 bar the internal energy can be found when required using equation 2.7. For example, steam at 80 bar, 400°C has an enthalpy, h, of 3139 kJ/kg and a specific volume, v, of $3·428 \times 10^{-2}$ m³/kg, therefore,

$$u = h - pv = 3139 - \frac{80 \times 10^5 \times 0.03428}{10^3}$$

i.e. $$u = 3139 - 274·2 = 2864·8 \text{ kJ/kg}$$

Example 3.3

Steam at 110 bar has a specific volume of 0·0196 m³/kg, find the temperature, the specific enthalpy and the specific internal energy.

First it is necessary to decide whether the steam is wet, dry saturated, or superheated. At 110 bar, $v_g = 0·01598$ m³/kg, which is less than the actual specific volume of 0·0196 m³/kg, and hence the steam is superheated. The state of the steam is shown as point A of fig. 3.8.

From the superheat tables at 110 bar, the specific volume is 0·0196

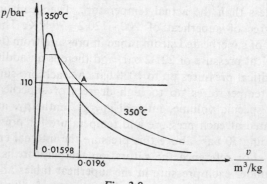

Fig. 3.8

m³/kg at a temperature of 350°C. Hence this is the isothermal which passes through point A as shown. The degree of superheat in this case is $350 - 318 = 32$ K. From tables the enthalpy, h, is 2889 kJ/kg. Then using equation 2.7, we have,

$$u = h - pv = 2889 - \frac{110 \times 10^5 \times 0.0196}{10^3}$$

i.e. $\qquad u = 2889 - 215.6 = 2673.4 \text{ kJ/kg}$

Example 3.4

Steam at 150 bar has a specific enthalpy of 3309 kJ/kg, find the temperature, the specific volume and the specific internal energy.

At 150 bar, $h_g = 2611$ kJ/kg, which is less than the actual enthalpy of 3309 kJ/kg, and hence the steam is superheated. From superheat tables at 150 bar, $h = 3309$ kJ/kg at a temperature of 500°C. The specific volume is $v = 0.02078$ m³/kg. Using equation 2.7,

$$u = h - pv = 3309 - \frac{150 \times 10^5 \times 0.02078}{10^3} = 2997.3 \text{ kJ/kg}$$

Interpolation

For properties which are not tabulated exactly in the tables it is necessary to interpolate between the values tabulated. For example, to find the temperature, specific volume, internal energy, and enthalpy of dry saturated steam at 9·8 bar, it is necessary to interpolate between the values given in the tables.

At 9·8 bar, the saturation temperature, t, is equal to the saturation temperature at 9 bar, plus $\left(\frac{9.8 - 9}{10 - 9}\right) \times$ (saturation temperature at 10 bar − saturation temperature at 9 bar). Note that this assumes a linear variation between the two values (see fig. 3.9),

i.e. $\qquad t = 175.4 + \left(\frac{9.8 - 9}{10 - 9}\right) \times (179.9 - 175.4)$

$\therefore \ t = 175.4 + 0.8 \times 4.5 = 179°C$

Similarly,

$\qquad h_g$ at 9·8 bar $= h_g$ at 9 bar $+ 0.8$
$\qquad\qquad\qquad \times (h_g$ at 10 bar $- h_g$ at 9 bar)

i.e. $\qquad h_g$ at 9·8 bar $= 2774 + 0.8 \times (2778 - 2774)$
$\qquad\qquad\qquad\qquad = 2774 + 0.8 \times 4 = 2777.2 \text{ kJ/kg}$

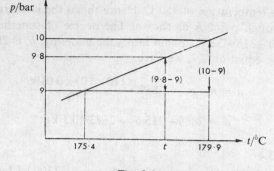

Fig. 3.9

Also, u_g at 9·8 bar $= 2581 + 0·8(2584 - 2581)$
$$= 2581 + 0·8 \times 3 = 2583·4 \text{ kJ/kg}$$

As another example consider steam at 5 bar and 320°C. The steam is superheated since the saturation temperature at 5 bar is 151·8°C, but to find the specific volume and enthalpy an interpolation is necessary,

i.e. $v = (v$ at 5 bar and 300°C)

$\qquad + \dfrac{20}{50} (v$ at 5 bar and 350°C $- v$ at 5 bar and 300°C)

$\therefore v = 0·5226 + 0·4(0·5701 - 0·5226)$
$$= 0·5226 + 0·019 = 0·5416 \text{ m}^3/\text{kg}$$

Similarly,

$\qquad h = 3065 + 0·4(3168 - 3065) = 3065 + 41·2$
i.e. $h = 3106·2 \text{ kJ/kg}$

In some cases a double interpolation is necessary. For example, to

p	t	400	432	450
15	h	3256	?	3364
18·5	h		?	
20	h	3248	?	3357

Fig. 3.10

find the enthalpy of superheated steam at 18·5 bar and 432°C an interpolation between 15 bar and 20 bar is necessary, and an interpolation between 400°C and 450°C is also necessary. A tabular presentation is usually better in such cases (fig. 3.10). First find the enthalpy at 15 bar and 432°C.

i.e.

$$h = 3256 + \frac{32}{50}(3364 - 3256) = 3256 + 0.64 \times 108$$

i.e.

$$h = 3325 \cdot 1 \text{ kJ/kg}$$

Now find the enthalpy at 20 bar, 432°C,

i.e.

$$h = 3248 + 0.64(3357 - 3248) = 3248 + 0.64 \times 109$$

i.e.

$$h = 3317 \cdot 8 \text{ kJ/kg}$$

Now interpolate between h at 15 bar, 432°C, and h at 20 bar, 432°C in order to find h at 18·5 bar, 432°C.

i.e.
$$h = 3325 \cdot 1 - \frac{3 \cdot 5}{5}(3325 \cdot 1 - 3317 \cdot 8)$$

(Note the negative sign in this case since h at 15 bar, 432°C is larger than h at 20 bar, 432°C)

i.e.

$$h \text{ at } 18 \cdot 5 \text{ bar, } 432°C = 3325 \cdot 1 - 0 \cdot 7 \times 7 \cdot 3 = 3320 \text{ kJ/kg}$$

Example 3.5

Sketch a pressure-volume diagram for steam and mark on it the following points, labelling clearly the pressure, specific volume and temperature of each point.

(a) $p = 20$ bar, $t = 250°C$
(b) $t = 212 \cdot 4°C$, $v = 0 \cdot 09957 \text{ m}^3/\text{kg}$
(c) $p = 10$ bar, $h = 2650 \text{ kJ/kg}$
(d) $p = 6$ bar, $h = 3166 \text{ kJ/kg}$

Point A:

At 20 bar the saturation temperature is 212·4°C, hence the steam is superheated at 250°C. Then from tables, $v = 0 \cdot 1115 \text{ m}^3/\text{kg}$.

Point B:

At 212·4°C the saturation pressure is 20 bar and v_g is 0·09957 m³/kg. Therefore the steam is just dry saturated since $v = v_g$.

Point C:

At 10 bar, h_g is 2778 kJ/kg, therefore the steam is wet since $h = 2650$ kJ/kg. Since the steam is wet, the temperature is the saturation temperature at 10 bar, i.e. $t = 179·9°C$. The dryness fraction can be found from equation 3.2,

$$h = h_f + xh_{fg}$$

$$\therefore\ x = \frac{2650 - 763}{2015} = \frac{1887}{2015} = 0·937$$

Then from equation 3.1

$$v = xv_g$$
$$v = 0·937 \times 0·1944 = 0·182 \text{ m}^3/\text{kg}$$

Point D:

At 6 bar, h_g is 2757 kJ/kg, therefore the steam is superheated, since it is given that $h = 3166$ kJ/kg. Hence from tables at 6 bar and $h = 3166$ kJ/kg the temperature is 350°C, and the specific volume is 0·4743 m³/kg.

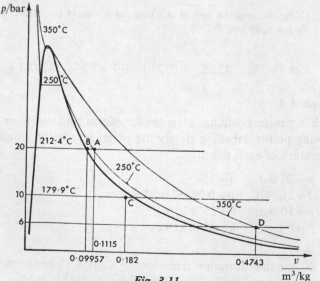

Fig. 3.11

The points A, B, C, and D can now be marked on a p-v diagram as shown in fig. 3.11.

Example 3.6

Calculate the internal energy for each of the four states given in example 3.5.

(a) The steam is superheated at 20 bar, 250°C,

i.e. $$u = 2681 \text{ kJ/kg}$$

(b) The steam is dry saturated at 20 bar,

i.e. $$u = u_g = 2600 \text{ kJ/kg}$$

(c) The steam is wet at 10 bar with $x = 0.937$.

$$\therefore u = (1-x)u_f + xu_g \quad \text{from equation 3.3}$$

i.e. $$u = (1-0.937)762 + 0.937 \times 2584 = 2470 \text{ kJ/kg}$$

(d) The steam is superheated at 6 bar, 350°C.

i.e. $$u = 2881 \text{ kJ/kg}$$

3.3 The perfect gas

The characteristic equation of state

At temperatures that are considerably in excess of the critical temperature of a fluid, and also at very low pressures, the vapour of the fluid tends to obey the equation

$$\frac{pv}{T} = \text{constant} = R$$

No gases in practice obey this law rigidly, but many gases tend towards it. An imaginary ideal gas which obeys the law is called a *perfect gas*, and the equation, $pv/T = R$, is called the characteristic equation of state of a perfect gas. The constant, R, is called the *specific gas constant*. The units of R are N m/kg K or kJ/kg K. Each perfect gas has a different specific gas constant.

The characteristic equation is usually written

$$pv = RT \qquad 3.5$$

or for a mass, m, occupying a volume, V,

$$pV = mRT \qquad 3.6$$

Another form of the characteristic equation can be derived using the *mole* as a unit. The mole was defined by the 1971 General Conference of Weights and Measures (CGPM) as follows:

'The mole is the amount of substance of a system which contains as many elementary entities as there are atoms in 0.012 kg of Carbon-12. When the mole is used, the elementary entities must be specified and may be atoms, molecules, ions, electrons, or other particles, or specified groups of such particles.'

The unit symbol used for the mole is 'mol'. In SI it is convenient to use the kilomole (unit symbol 'kmol').

The mass per kilomole of any substance is known as the *molar mass*, M, i.e.

$$m = nM \qquad 3.7$$

(where m is in kg; n is the number of kilomoles; and M is in kg/kmol).

Relative masses of the various elements are commonly used, and physicists and chemists agreed in 1960 to give the value of 12 to the isotope 12 of Carbon (this led to the definition of the mole as above). A scale is thus obtained of *relative atomic mass* or *relative molecular mass* (e.g. the relative atomic mass of the element Oxygen is approximately 16; the relative molecular mass of Oxygen gas, O_2, is approximately 32).

The relative molecular mass is numerically equal to the molar mass, M, but is dimensionless.

Substituting for m from equation 3.7 in equation 3.6 gives

$$pV = nMRT \quad \text{or} \quad MR = \frac{pV}{nT}$$

Now *Avogadro's hypothesis* states that the volume of 1 mole of any gas is the same as the volume of 1 mole of any other gas, when the gases are at the same temperature and pressure. Therefore V/n is the same for all gases at the same value of p and T. That is the quantity pV/nT is a constant for all gases. This constant is called the *molar gas constant*, or *universal gas constant*, and is given the symbol, R_0,

i.e.
$$MR = R_0 = \frac{pV}{nT} \quad \text{or} \quad pV = nR_0T \qquad 3.8$$

or since $MR = R_0$ then,
$$R = \frac{R_0}{M} \qquad 3.9$$

The value of R_0 has been shown to be 8314·4 N m/kmol K.

From equation 3.9 the specific gas constant for any gas can be found when the molar mass is known, e.g. for oxygen of molar mass 32 kg/kmol, the specific gas constant,

$$R = \frac{R_0}{M} = \frac{8314·4}{32} = 259·8 \text{ N m/kg K}$$

Example 3.7

A vessel of volume 0·2 m³ contains nitrogen at 1·013 bar and 15°C. If 0·2 kg of nitrogen is now pumped into the vessel, calculate the new pressure when the vessel has returned to its initial temperature. The molar mass of nitrogen is 28 kg/kmol, and it may be assumed to be a perfect gas.

From equation 3.9

$$\text{Specific gas constant, } R = \frac{R_0}{M} = \frac{8314·4}{28} = 296·9 \text{ N m/kg K}$$

From equation 3.6, for the initial conditions,

$$p_1 V_1 = m_1 R T_1$$

$$\therefore m_1 = \frac{p_1 V_1}{R T_1} = \frac{1·013 \times 10^5 \times 0·2}{296·9 \times 288} = 0·237 \text{ kg}$$

(where $T_1 = 15 + 273 = 288$ K).

0·2 kg of nitrogen are added, hence $m_2 = 0·2 + 0·237 = 0·437$ kg. Then from equation 3.6, for the final conditions,

$$p_2 V_2 = m_2 R T_2$$

but $V_2 = V_1$ and $T_2 = T_1$,

$$\therefore p_2 = \frac{m_2 R T_2}{V_2} = \frac{0·437 \times 296·9 \times 288}{0·2}$$

i.e.

$$p_2 = \frac{0·437 \times 296·9 \times 288}{10^5 \times 0·2} = 1·87 \text{ bar}$$

Example 3.8

0·01 kg of a certain perfect gas occupies a volume of 0·003 m³ at a pressure of 7 bar and a temperature of 131°C. Calculate the molar mass of the gas. When the gas is allowed to expand until the pressure is 1 bar the final volume is 0·02 m³. Calculate the final temperature.

From equation 3.6

$$p_1 V_1 = mRT_1$$

$$\therefore \ R = \frac{p_1 V_1}{m T_1} = \frac{7 \times 10^5 \times 0 \cdot 003}{0 \cdot 01 \times 404} = 520 \ \text{N m/kg K}$$

(where $T_1 = 131 + 273 = 404$ K).

Then from equation 3.9

$$R = \frac{R_0}{M}$$

$$\therefore \ M = \frac{R_0}{R} = \frac{8314}{520} = 16 \ \text{kg/kmol}$$

i.e.   molar mass $= 16$ kg/kmol

From equation 3·6

$$p_2 V_2 = mRT_2$$

$$\therefore \ T_2 = \frac{p_2 V_2}{mR} = \frac{1 \times 10^5 \times 0 \cdot 02}{0 \cdot 01 \times 520} = 384 \cdot 5 \ \text{K}$$

i.e.   Final temperature $= 384 \cdot 5 - 273 = 111 \cdot 5°\text{C}$.

Specific heats

The specific heat of a solid or liquid is usually defined as the heat required to raise unit mass through one degree temperature rise. For small quantities we have $dQ = mc \, dT$, where m is the mass, dT is the increase in temperature, and c is the specific heat. For a gas there are an infinite number of ways in which heat may be added between any two temperatures, and hence a gas could have an infinite number of specific heats. However, only two specific heats for gases are defined; the specific heat at constant volume, c_v, and the specific heat at constant pressure, c_p.

Note that in the equation defining specific heat (i.e. $dQ = mc \, dT$), the temperature rise, dT, may be partly due to a work input. The definition must be restricted to reversible non-flow processes, since irreversibilities can cause temperature changes which are indistinguishable from those due to reversible heat and work quantities. Specific heat can be introduced more rigorously as a property of a fluid; the reader is recommended to reference 3.3 for a more complete treatment.

We have

$$dQ = mc_p \, dT \quad \text{for a reversible non-flow process} \atop \text{at constant pressure} \qquad 3.10$$

and
$$dQ = mc_v \, dT \quad \text{for a reversible non-flow process} \atop \text{at constant volume} \qquad 3.11$$

For a perfect gas the values of c_p and c_v are constant for any one gas at all pressures and temperatures. Hence integrating equations 3.10 and 3.11 we have,

Heat flow in a reversible constant pressure process
$$= mc_p(T_2 - T_1) \qquad 3.12$$

Heat flow in a reversible constant volume process
$$= mc_v(T_2 - T_1) \qquad 3.13$$

For real gases, c_p and c_v vary with temperature, but for most practical purposes a suitable average value may be used.

Joule's Law

Joule's law states that the internal energy of a perfect gas is a function of the absolute temperature only, i.e. $u = f(T)$. To evaluate this function let 1 kg of a perfect gas be heated at constant volume. From the non-flow energy equation, 2.3,

$$dQ = du + dW$$

Since the volume remains constant then no work is done, i.e. $dW = 0$,

$$\therefore \ dQ = du$$

At constant volume for a perfect gas, from equation 3.11, for 1 kg,

$$dQ = c_v \, dT$$

Therefore, $dQ = du = c_v \, dT$, and integrating,

$$u = c_v T + K$$

(where K is a constant).

Joule's law states that $u = f(T)$, hence it follows that the internal energy varies linearly with absolute temperature. Internal energy can be made zero at any arbitrary reference temperature. For a perfect gas it can be assumed that $u=0$ when $T=0$, hence the constant K is zero,

i.e.   Specific internal energy, $u=c_vT$ for a perfect gas   3.14

or for mass, m, of a perfect gas,

$$\text{Internal energy, } U = mc_vT \qquad 3.15$$

In any process for a perfect gas, between states 1 and 2, we have from equation 3.15,

$$\text{Gain in internal energy, } U_2 - U_1 = mc_v(T_2 - T_1) \qquad 3.16$$

The gain of internal energy for a perfect gas between two states is always given by equation 3.16, *for any process, reversible or irreversible.*

Relationship between the specific heats

Let a perfect gas be heated at constant pressure from T_1 to T_2. From the non-flow equation 2.2, $Q=(U_2-U_1)+W$. Also, for a perfect gas, from equation 3.16, $U_2-U_1=mc_v(T_2-T_1)$. Hence,

$$Q = mc_v(T_2-T_1)+W$$

In a constant pressure process the work done by the fluid is given by the pressure times the change in volume, i.e. $W=p(V_2-V_1)$. Then using equation 3.6, $pV_2=mRT_2$ and $pV_1=mRT_1$, we have

$$W=mR(T_2-T_1).$$

Therefore substituting,

$$Q = mc_v(T_2-T_1)+mR(T_2-T_1) = m(c_v+R)(T_2-T_1)$$

But for a constant pressure process from equation 3.12,

$$Q = mc_p(T_2-T_1)$$

Hence by equating the two expressions for the heat flow, Q, we have

$$m(c_v+R)(T_2-T_1) = mc_p(T_2-T_1)$$

$$\therefore\ c_v+R = c_p$$

or $c_p-c_v = R$ 3.17

Enthalpy of a perfect gas

From equation 2.7, enthalpy, $h = u + pv$.

For a perfect gas, from equation 3.5, $pv = RT$. Also for a perfect gas, from Joule's law, equation 3.14, $u = c_v T$. Hence, substituting,

$$h = c_v T + RT = (c_v + R)T$$

But from equation 3.17

$$c_p - c_v = R \quad \text{or} \quad c_v + R = c_p$$

Therefore, specific enthalpy, h, for a perfect gas is given by

$$h = c_p T \qquad\qquad 3.18$$

For mass, m, of a perfect gas

$$H = m c_p T \qquad\qquad 3.19$$

(Note that, since it has been assumed that $u = 0$ at $T = 0$, then $h = 0$ at $T = 0$.)

Ratio of specific heats

The ratio of the specific heat at constant pressure to the specific heat at constant volume is given the symbol γ (gamma)

i.e. $$\gamma = \frac{c_p}{c_v} \qquad\qquad 3.20$$

Note that since $c_p - c_v = R$, from equation 3.17, it is clear that c_p must be greater than c_v for any perfect gas. It follows therefore that the ratio, $c_p/c_v = \gamma$, is always greater than unity. In general, γ is about 1·4 for diatomic gases such as carbon monoxide (CO), hydrogen (H_2), nitrogen (N_2), and oxygen (O_2). For monatomic gases such as argon (A), and helium (He), γ is about 1·6, and for triatomic gases such as carbon dioxide (CO_2), and sulphur dioxide (SO_2), γ is about 1·3. For some hydro-carbons the value of γ is quite low (e.g. for ethane (C_2H_6), $\gamma = 1·22$, and for iso-butane (C_4H_{10}), $\gamma = 1·11$).

Some useful relationships between c_p, c_v, R, and γ can be derived. From equation 3.17

$$c_p - c_v = R$$

Dividing through by c_v

$$\frac{c_p}{c_v} - 1 = \frac{R}{c_v}$$

Therefore using equation 3.17, $\gamma = c_p/c_v$, then,

$$\gamma - 1 = \frac{R}{c_v}$$

$$\therefore \quad c_v = \frac{R}{(\gamma - 1)} \qquad 3.21$$

Also from equation 3.20, $c_p = \gamma c_v$, hence substituting in equation 3.21,

$$c_p = \gamma c_v = \frac{\gamma R}{(\gamma - 1)}$$

i.e.
$$c_p = \frac{\gamma R}{(\gamma - 1)} \qquad 3.22$$

Example 3.9

A certain perfect gas has specific heats as follows

$$c_p = 0.846 \text{ kJ/kg K} \quad \text{and} \quad c_v = 0.657 \text{ kJ/kg K}$$

Find the gas constant and the molar mass of the gas.
From equation 3.17

$$c_p - c_v = R \quad \therefore \quad R = c_p - c_v$$

i.e. $\qquad R = 0.846 - 0.657 = 0.189 \text{ kJ/kg K}$

or $\qquad\qquad R = 189 \text{ N m/kg K}$

From equation 3.9

$$R = \frac{R_0}{M} \quad \therefore \quad M = \frac{R_0}{R}$$

i.e. $\qquad M = \frac{8314}{189} = 44 \text{ kg/kmol}$

Example 3.10

A perfect gas has a molar mass of 26 kg/kmol and a value of $\gamma = 1.26$. Calculate the heat rejected per kg of gas:

(a) when the gas is contained in a rigid vessel at 3 bar and 315°C, and is then cooled until the pressure falls to 1.5 bar

(b) when the gas enters a pipeline at 280°C, and flows steadily to the end of the pipe where the temperature is 20°C. Neglect changes in velocity of the gas in the pipeline.

From equation 3.9

$$R = \frac{R_0}{M} = \frac{8314}{26} = 319 \cdot 8 \text{ N m/kg K}$$

From equation 3.21

$$c_v = \frac{R}{(\gamma - 1)} = \frac{319 \cdot 8}{10^3 (1 \cdot 26 - 1)} = 1 \cdot 229 \text{ kJ/kg K}$$

Also from equation 3.20

$$\frac{c_p}{c_v} = \gamma$$

$$\therefore \; c_p = \gamma c_v = 1 \cdot 26 \times 1 \cdot 229 = 1 \cdot 548 \text{ kJ/kg K}$$

(a) The volume remains constant for the mass of gas present, and hence the specific volume remains constant. From equation 3.5,

$$p_1 v_1 = RT_1 \quad \text{and} \quad p_2 v_2 = RT_2$$

Therefore since $v_1 = v_2$ we have

$$T_2 = T_1 \frac{p_2}{p_1} = 588 \times \frac{1 \cdot 5}{3} = 294 \text{ K}$$

(where $T_1 = 315 + 273 = 588$ K).

Then from equation 3.13

Heat rejected per kg of gas $= c_v(T_2 - T_1) = 1 \cdot 229(588 - 294)$
$$= 1 \cdot 229 \times 294 = 361 \text{ kJ/kg}$$

(b) From the steady flow energy equation, 2.8,

$$h_1 + \frac{C_1^2}{2} + Q = h_2 + \frac{C_2^2}{2} + W$$

In this case we are told that changes in velocity are negligible; also there is no work done on, or by, the gas.

Therefore we have

$$h_1 + Q = h_2 \quad \text{or} \quad Q = (h_2 - h_1)$$

For a perfect gas, from equation 3.18,

$$h = c_p T$$

$$\therefore \quad Q = c_p(T_2 - T_1)$$

or Heat rejected per kg $= c_p(T_1 - T_2) = 1 \cdot 548(280 - 20)$

i.e. Heat rejected $= 1 \cdot 548 \times 260 = 403$ kJ/kg

Note that it is not necessary to convert $t_1 = 280°C$ and $t_2 = 20°C$ into degrees Kelvin, since the temperature difference $(t_1 - t_2)$ in degrees Celsius, is numerically the same as the temperature difference $(T_1 - T_2)$ K.

PROBLEMS

[Note: the answers to these problems have been evaluated using the tables of Rogers and Mayhew (reference 3.1). The values of R, c_p, c_v, and γ for air may be assumed to be as given on page 24 of the afore-mentioned tables (i.e. $R = 0 \cdot 287$ kJ/kg K; $c_p = 1 \cdot 005$ kJ/kg K; $c_v = 0 \cdot 718$ kJ/kg K; and $\gamma = 1 \cdot 4$). For any other perfect gas the values of R, c_p, c_v, and γ, if required, must be calculated from the information given in the problem.]

3.1 Complete the following table using steam tables. Insert a dash for irrelevant items, and interpolate where necessary.

p bar	t °C	v m³/kg	x	degree of superheat	h kJ/kg	u kJ/kg
	90	2·364				
20					2799	
5		0·3565				
	188					2400
34			0·9			
	81·3		0·85			
3	200					
15		0·152				
130					3335	
	250	1·601				
38·2			0·8			
	297		0·95			
2·3	300					
44	420					

(The completed table is given on page 56.)

3.2 A vessel of volume 0.03 m^3 contains dry saturated steam at 17 bar. Calculate the mass of steam in the vessel and the enthalpy of this mass. (0·257 kg; 718 kJ)

3.3 Steam at 7 bar and 250°C enters a pipeline and flows along it at constant pressure. If the steam rejects heat steadily to the surroundings, at what temperature will droplets of water begin to form in the vapour? Using the steady flow energy equation, and neglecting changes in velocity of the steam, calculate the heat rejected per kg of steam flowing. (165°C; 191 kJ/kg)

3.4 0.05 kg of steam at 15 bar is contained in a rigid vessel of volume 0.0076 m^3. What is the temperature of the steam? If the vessel is cooled, at what temperature will the steam be just dry saturated? Cooling is continued until the pressure in the vessel is 11 bar, calculate the final dryness fraction of the steam, and the heat rejected between the initial and the final states.

(250°C; 191·4°C; 0·857; 18·5 kJ)

3.5 The relative molecular mass of carbon dioxide, CO_2, is 44. In an experiment the value of γ for CO_2 was found to be 1.3. Assuming that CO_2 is a perfect gas, calculate the specific gas constant, R, and the specific heats at constant pressure and constant volume, c_p and c_v.

(0·189 kJ/kg K; 0·63 kJ/kg K; 0·819 kJ/kg K)

3.6 Calculate the internal energy and enthalpy of 1 kg of air occupying 0.05 m^3 at 20 bar. If the internal energy is increased by 120 kJ/kg as the air is compressed to 50 bar, calculate the new volume occupied by 1 kg of the air. (250·1 kJ/kg; 350·1 kJ/kg; 0·0296 m³)

3.7 Oxygen, O_2, at 200 bar is to be stored in a steel vessel at 20°C. The capacity of the vessel is 0.04 m^3. Assuming that O_2 is a perfect gas, calulate the mass of oxygen that can be stored in the vessel. The vessel is protected against excessive pressure by a fusible plug which will melt if the temperature rises too high. At what temperature must the plug melt to limit the pressure in the vessel to 240 bar? The molar mass of oxygen is 32 kg/kmol. (10·5 kg; 78·6°C)

3.8 When a certain perfect gas is heated at constant pressure from 15°C to 95°C, the heat required is 1136 kJ/kg. When the same gas is heated at constant volume between the same temperatures the heat required is 808 kJ/kg. Calculate c_p, c_v, γ, R and the relative molecular mass of the gas.

(14·2 kJ/kg K; 10·1 kJ/kg K; 1·405; 4·1 kJ/kg K; 2·028)

3.9 In an air compressor the pressures at inlet and outlet are 1 bar and 5 bar respectively. The temperature of the air at inlet is 15°C and the volume at the beginning of compression is three times that at the end of compression. Calculate the temperature of the air at outlet and the increase of internal energy per kg of air.

(207°C; 138 kJ/kg)

3.10 A quantity of a certain perfect gas is compressed from an initial state of 0·085 m³, 1 bar to a final state of 0·034 m³, 3·9 bar. The specific heat at constant volume is 0·724 kJ/kg K, and the specific heat at constant pressure is 1·02 kJ/kg K. The observed temperature rise is 146 K. Calculate the specific gas constant, R, the mass of gas present, and the increase of internal energy of the gas.

(0·296 kJ/kg K; 0·11 kg; 11·63 kJ)

Solution to Problem 3.1 on page 54

p bar	t °C	v m³/kg	x	degree of superheat	h kJ/kg	u kJ/kg
0·70	90	2·364	1	0	2660	2494
20	212·4	0·09957	1	0	2799	2600
5	151·8	0·3565	0·951	—	2646	2471
12	188	0·1461	0·895	—	2576	2400
34	240·9	0·0529	0·9	—	2627	2447
0·5	81·3	2·75	0·85	—	2300	2165
3	200	0·7166	—	66·5	2866	2651
15	250	0·152	—	51·7	2925	2697
130	500	0·02447	—	169·2	3335	3017
1·5	250	1·601	—	138·6	2973	2733
38·2	247·6	0·0417	0·8	—	2456·5	2296
82·38	297	0·0216	0·95	—	2683	2505
2·3	300	1·184	—	175·8	3071	2808
44	420	0·0696	—	164·3	3254	2952

REFERENCES

3.1 ROGERS, G. F. C. and MAYHEW, Y. R., *Thermodynamic and Transport Properties of Fluids*, S I Units (Basil Blackwell, 1980).

3.2 National Engineering Laboratory, *Steam Tables 1964* (H.M.S.O., 1964).

3.3 ROGERS, G. F. C. and MAYHEW, Y. R., *Engineering Thermodynamics, Work and Heat Transfer*, S I Units (Longmans, 1980).

4. Reversible and Irreversible Processes

In the previous three chapters the energy equations for non-flow and flow processes are derived, the concepts of reversibility and irreversibility introduced, and the properties of vapours and perfect gases discussed. It is the purpose of this chapter to consider processes which are approximated to in practice, and to combine this with the work of the previous three chapters.

4.1 Reversible non-flow processes

Constant volume process

In a constant volume process the working substance is contained in a rigid vessel, hence the boundaries of the system are immovable and no work can be done on or by the system, other than paddle-wheel work input. It will be assumed that 'constant volume' implies zero work unless stated otherwise.

From the non-flow energy equation, 2.2,

$$Q = (u_2 - u_1) + W$$

Since no work is done, we therefore have

$$Q = u_2 - u_1 \qquad 4.1$$

or for mass, m, of the working substance

$$Q = U_2 - U_1 \qquad 4.2$$

All the heat supplied in a constant volume process goes to increasing the internal energy.

A constant volume process for a vapour is shown on a p-v diagram in fig. 4.1a. The initial and final states have been chosen to be in the wet region and superheat region respectively. In fig. 4.1b a constant volume process is shown on a p-v diagram for a perfect gas. For a perfect gas we have from equation 3.13,

$$Q = mc_v(T_2 - T_1)$$

Constant pressure process

It can be seen from figs. 4.1a and 4.1b that when the boundary of the system is inflexible as in a constant volume process, then the pressure rises when heat is supplied. Hence for a constant pressure process the boundary must move against an external resistance as heat is supplied; for instance a fluid in a cylinder behind a piston can be made to undergo a constant pressure process. Since the piston is pushed through a certain distance by the force exerted by the fluid, then work is done by the fluid on its surroundings.

From equation 1.2

$$W = \int_{v_1}^{v_2} p \, dv \quad \text{for any reversible process}$$

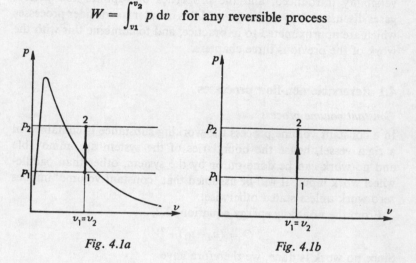

Fig. 4.1a   Fig. 4.1b

Therefore, since p is constant,

$$W = p \int_{v_1}^{v_2} dv = p(v_2 - v_1)$$

From the non-flow energy equation, 2.2,

$$Q = (u_2 - u_1) + W$$

Hence for a reversible constant pressure process

$$Q = (u_2 - u_1) + p(v_2 - v_1) = (u_2 + pv_2) - (u_1 + pv_1)$$

Now from equation 2.7, enthalpy, $h = u + pv$, hence,

$$Q = h_2 - h_1 \qquad 4.3$$

or for mass, m, of a fluid,

$$Q = H_2 - H_1 \qquad 4.4$$

A constant pressure process for a vapour is shown on a p-v diagram in fig. 4.2a. The initial and final states have been chosen to be in the wet region and the superheat region respectively. In fig. 4.2b a constant pressure process for a perfect gas is shown on a p-v diagram. For a perfect gas we have from equation 3.12,

$$Q = mc_p(T_2 - T_1)$$

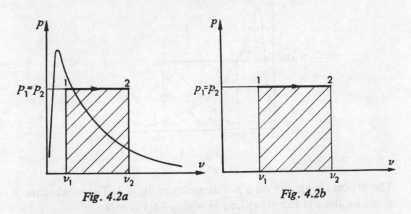

Fig. 4.2a   Fig. 4.2b

Note that in figs. 4.2a and 4.2b the shaded areas represent the work done by the fluid, $p(v_2 - v_1)$.

Example 4.1

0·05 kg of a certain fluid is heated at a constant pressure of 2 bar until the volume occupied is 0·0658 m³. Calculate the heat supplied and the work done,

    (a) When the fluid is steam, initially dry saturated.

    (b) When the fluid is air, initially at 130°C.

  (a) Initially the steam is dry saturated at 2 bar, hence,

$$h_1 = h_g \text{ at } 2 \text{ bar} = 2707 \text{ kJ/kg}$$

Finally the steam is at 2 bar and the specific volume is given by

$$v_2 = \frac{0\cdot0658}{0\cdot05} = 1\cdot316 \text{ m}^3/\text{kg}$$

Hence the steam is superheated finally. From superheat tables at 2 bar and $1\cdot316$ m³/kg the temperature of the steam is 300°C, and the enthalpy is $h_2 = 3072$ kJ/kg.

Then from equation 4.4

$$Q = H_2 - H_1 = m(h_2 - h_1) = 0\cdot05(3072 - 2707)$$

i.e.   Heat supplied $= 0\cdot05 \times 365 = 18\cdot25$ kJ

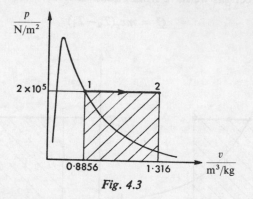

Fig. 4.3

The process is shown on a p-v diagram in fig. 4.3. The work done is given by the shaded area; i.e., $W = p(v_2 - v_1)$ N m/kg.

Now $v_1 = v_g$ at 2 bar $= 0\cdot8856$ m³/kg, and $v_2 = 1\cdot316$ m³/kg.

$$\therefore\ W = 2 \times 10^5(1\cdot316 - 0\cdot8856) = 2 \times 10^5 \times 0\cdot4304 \text{ N m/kg}$$

i.e.

Work done by the total mass present $= 0\cdot05 \times 2 \times 10^5 \times 0\cdot4304 \times 10^{-3}$
$$= 4\cdot304 \text{ kJ}$$

(b) Using equation 3.6,

$$T_2 = \frac{p_2 V_2}{mR} = \frac{2 \times 10^5 \times 0\cdot0658}{0\cdot05 \times 0\cdot287 \times 10^3} = 917 \text{ K}$$

For a perfect gas undergoing a constant pressure process we have, from equation 3.12,

$$Q = mc_p(T_2 - T_1)$$

i.e.   Heat supplied $= 0.05 \times 1.005(917 - 403)$

(where $T_1 = 130 + 273 = 403$ K),

i.e.   Heat supplied $= 0.05 \times 1.005 \times 514 = 25.83$ kJ

The process is shown on a p-v diagram in fig. 4.4. The work done is given by the shaded area, i.e., $W = p(v_2 - v_1)$ N m/kg. From equation 3.5, $pv = RT$

$$\therefore \text{ Work done} = R(T_2 - T_1) = 0.287(917 - 403) \text{ kJ/kg}$$

i.e. Work done by the mass of gas present $= 0.05 \times 0.287 \times 514$
$$= 7.38 \text{ kJ}$$

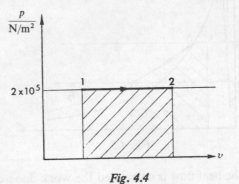

Fig. 4.4

Constant temperature or isothermal process

A process at constant temperature is called an isothermal process. When a fluid in a cylinder behind a piston expands from a high pressure to a low pressure there is a tendency for the temperature to fall. In an isothermal expansion heat must be added continuously in order to keep the temperature at the initial value. Similarly in an isothermal compression heat must be removed from the fluid continuously during the process. An isothermal process for a vapour is shown on a p-v diagram in fig. 4.5. The initial and final states have been chosen in the wet region and superheat region respectively. From state 1 to state A the pressure remains at p_1, since in the wet region the pressure and temperature are the corresponding saturation values. It can be seen therefore that an isothermal process for wet steam is also at constant pressure and equations 4.3 and 4.4 can be used (e.g. heat supplied from state 1 to state A per kg of steam $= h_A - h_1$). In the superheat region the pressure falls to p_2 as shown in fig. 4.5, and the procedure is not so simple. When states 1 and 2 are fixed then the internal energies u_1 and u_2 may be obtained from tables. The work done is given by the shaded area in fig. 4.5. This can only be evaluated by plotting the process and measuring the area graphically. However, when the property entropy, s, is introduced in Chapter 5, a convenient way of evaluating the heat supplied will be

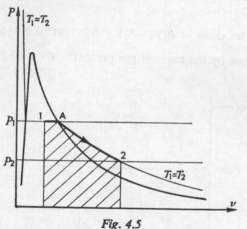

Fig. 4.5

shown. When the heat flow is calculated the work done can then be obtained using the non-flow energy equation 2.2,

$$Q = (u_2 - u_1) + W$$

Example 4.2

Steam at 7 bar and dryness fraction 0·9 expands in a cylinder behind a piston isothermally and reversibly to a pressure of 1·5 bar. Calculate the change of internal energy and the change of enthalpy per kg of steam. The heat supplied during the process is found to be 547 kJ/kg, by the method of Chapter 5. Calculate the work done per kg of steam.

The process is shown in fig. 4.6. The saturation temperature corresponding to 7 bar is 165°C. Therefore the steam is superheated at state 2. The internal energy at state 1 is found by using equation 3.3,

i.e. $u_1 = (1-x)u_f + xu_g = (1-0.9) \times 696 + (0.9 \times 2573)$

∴ $u_1 = 69.6 + 2315.7 = 2385.3$ kJ/kg

Interpolating from superheat tables at 1·5 bar and 165°C, we have,

$$u_2 = 2580 + \frac{15}{50}(2656 - 2580) = 2580 + 22·8$$

i.e. $u_2 = 2602·8 \text{ kJ/kg}$

Therefore,

Gain in internal energy $= u_2 - u_1 = 2602·8 - 2385·3$
$$= 217·5 \text{ kJ/kg}$$

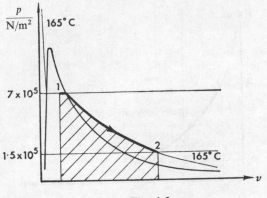

Fig. 4.6

$$h_1 = h_f + x h_{fg} = 697 + 0·9 \times 2067$$

$$\therefore \; h_1 = 697 + 1860·3 = 2557·3 \text{ kJ/kg}$$

Interpolating from superheat tables at 1·5 bar and 165°C, we have,

$$h_2 = 2773 + \frac{15}{50}(2873 - 2773) = 2773 + 30$$

i.e. $h_2 = 2803 \text{ kJ/kg}$

i.e. $h_2 - h_1 = 2803 - 2557·3 = 245·7 \text{ kJ/kg}$

From the non-flow energy equation, 2.2,

$$Q = (u_2 - u_1) + W$$

$$\therefore \quad W = Q - (u_2 - u_1) = 547 - 217{\cdot}5 = 329{\cdot}5 \text{ kJ/kg}$$

i.e.

Work done by the steam $= 329{\cdot}5$ kJ/kg

(The work done is also given by the area on fig. 4.6 ($\int_{v_1}^{v_2} p \, dv$); this can only be evaluated graphically.)

An isothermal process for a perfect gas is more easily dealt with than an isothermal process for a vapour, since there are definite laws for a perfect gas relating p, v, and T, and the internal energy u. We have, from equation 3.5,

$$pv = RT$$

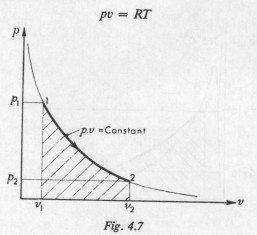

Fig. 4.7

Now when the temperature is constant as in an isothermal process then we have

$$pv = RT = \text{constant}$$

Therefore for an isothermal process for a perfect gas,

$$pv = \text{constant} \qquad\qquad 4.5$$

i.e.

$$p_1 v_1 = p_2 v_2 = p_3 v_3 \quad \text{etc.}$$

In fig. 4.7 an isothermal process for a perfect gas is shown on a p-v diagram. The equation of the process is $pv = $ constant, which is the equation of a hyperbola. It must be stressed that an isothermal process is only of the form $pv = $ constant *for a perfect gas*, because it is only for a perfect gas that an equation of state, $pv = RT$, can be applied.

The work done by a perfect gas in expanding from state 1 to state 2 isothermally and reversibly is given by the shaded area on fig. 4.7.

From equation 1.2 we have

$$W = \int_1^2 p \, dv$$

In this case, $pv = $ constant, or $p = c/v$ (where $c = $ constant).

$$\therefore \quad W = \int_{v_1}^{v_2} c \, \frac{dv}{v} = c[\log_e v]_{v_1}^{v_2} = c \log_e \frac{v_2}{v_1}$$

The constant c can either be written as $p_1 v_1$ or as $p_2 v_2$, since $p_1 v_1 = p_2 v_2 = $ constant, c

i.e. $\qquad W = p_1 v_1 \log_e \frac{v_2}{v_1}$ per unit mass of gas $\qquad$ 4.6

or $\qquad W = p_2 v_2 \log_e \frac{v_2}{v_1}$ per unit mass of gas

For mass, m, of the gas

$$W = p_1 V_1 \log_e \frac{v_2}{v_1} \qquad\qquad 4.7$$

Also, since $p_1 v_1 = p_2 v_2$, then

$$\frac{v_2}{v_1} = \frac{p_1}{p_2}$$

Hence, substituting in equation 4.6,

$$W = p_1 v_1 \log_e \frac{p_1}{p_2} \quad \text{per unit mass of gas} \qquad 4.8$$

or for mass, m, of the gas

$$W = p_1 V_1 \log_e \frac{p_1}{p_2} \qquad\qquad 4.9$$

Using equation 3.5

$$p_1 v_1 = RT$$

Hence, substituting in equation 4.8,

$$W = RT \log_e \frac{p_1}{p_2} \quad \text{per unit mass of gas} \qquad 4.10$$

or for mass, m, of the gas

$$W = mRT \log_e \frac{p_1}{p_2} \qquad 4.11$$

There are clearly a large number of equations for the work done, and no attempt should be made to memorize these since they can all be derived very simply from first principles.

For a perfect gas from Joule's law, equation 3.16, we have,

$$U_2 - U_1 = mc_v(T_2 - T_1)$$

Hence for an isothermal process for a perfect gas, since $T_2 = T_1$, then

$$U_2 - U_1 = 0$$

i.e. the internal energy remains constant in an isothermal process for a perfect gas.

From the non-flow energy equation, 2.2,

$$Q = (u_2 - u_1) + W$$

Therefore, since $u_2 = u_1$, then

$$Q = W \qquad 4.12$$

for an isothermal process for a perfect gas.

Note that the heat flow is equivalent to the work done in an isothermal process *for a perfect gas only*. From example 4.2 for steam it is seen that, although the process is isothermal, the change in internal energy is 217·5 kJ/kg, and the heat supplied is *not* equivalent to the work done.

Example 4.3

1 kg of nitrogen (molar mass 28 kg/kmol) is compressed reversibly and isothermally from 1·01 bar, 20°C to 4·2 bar. Calculate the work done and the heat flow during the process. Assume nitrogen to be a perfect gas.

From equation 3.9, for nitrogen,

$$R = \frac{R_0}{M} = \frac{8\cdot314}{28} = 0\cdot297 \text{ kJ/kg K}$$

The process is shown on a p-v diagram in fig. 4.8. In Section 1.6 it was pointed out that when a process takes place from right to left on a p-v diagram then the work done by the fluid is negative. That is, work is done on the fluid.

From equation 4.10

$$W = RT \log_e \frac{p_1}{p_2} = 0.297 \times 293 \times \log_e \frac{1.01}{4.2}$$

i.e. $W = -0.297 \times 293 \times \log_e \frac{4.2}{1.01} = -0.297 \times 293 \times 1.425$

(where $T = 20 + 273 = 293$ K).

$$\therefore \text{ Work input} = +0.297 \times 293 \times 1.425 = 124 \text{ kJ/kg}$$

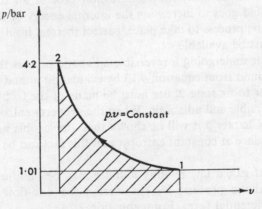

Fig. 4.8

From equation 4·12, for an isothermal process for a perfect gas,

$$Q = W = -124 \text{ kJ/kg}$$

i.e.   Heat rejected $= +124$ kJ/kg

4.2 Reversible adiabatic non-flow process

An *adiabatic* process is one in which no heat is transferred to or from the fluid during the process. Such a process can be reversible or irreversible. The reversible adiabatic non-flow process will be considered in this section.

From the non-flow equation, 2.2,

$$Q = (u_2 - u_1) + W$$

and for an adiabatic process

$$Q = 0$$

Therefore we have

$$W = u_1 - u_2 \quad \text{for any adiabatic process} \qquad 4.13$$

Equation 4.13 is true for an adiabatic process whether the process is reversible or not. In an adiabatic expansion, the work done by the fluid is at the expense of a reduction in the internal energy of the fluid. Similarly in an adiabatic compression process all the work done on the fluid goes to increasing the internal energy of the fluid. For an adiabatic process to take place, perfect thermal insulation for the system must be available.

For a vapour undergoing a reversible adiabatic process the work done can be found from equation 4.13 by evaluating u_1 and u_2 from tables. In order to fix state 2, use must be made of the fact that the process is reversible and adiabatic. When the property entropy, s, is introduced in Chapter 5 it will be shown that a reversible adiabatic process takes place at constant entropy, and this fact can be used to fix state 2.

For a perfect gas, a law relating p and v may be obtained for a reversible adiabatic process, by considering the non-flow energy equation in differential form. From equation 2.2

$$dQ = du + dW$$

Also for a reversible process $dW = p \, dv$, hence for an adiabatic process

$$d Q = du + p \, dv = 0 \qquad 4.14$$

Since $h = u + pv$

then, $dh = du + p \, dv + v \, dp$

i.e. $du + p \, dv = dh - v \, dp$

and hence,

$$dQ = dh - v \, dp = 0 \qquad 4.15$$

Hence,

$$du + \frac{RT\,dv}{v} = 0$$

From equation 3.14

$$u = c_v T \quad \text{or} \quad du = c_v\,dT$$

$$\therefore \quad c_v\,dT + \frac{RT\,dv}{v} = 0$$

Dividing through by T to give a form that can be integrated,

i.e.

$$c_v\frac{dT}{T} + \frac{R\,dv}{v} = 0$$

Integrating

$$c_v\log_e T + R\log_e v = \text{constant}$$

Using equation 3.5 we have $T=(pv)/R$, therefore substituting,

$$c_v\log_e \frac{pv}{R} + R\log_e v = \text{constant}$$

Dividing through by c_v

$$\log_e \frac{pv}{R} + \frac{R}{c_v}\log_e v = \text{constant}$$

Also, from equation 3.21,

$$c_v = \frac{R}{(\gamma - 1)} \quad \text{or} \quad \frac{R}{c_v} = \gamma - 1$$

Hence substituting

$$\log_e \frac{pv}{R} + (\gamma - 1)\log_e v = \text{constant}$$

or

$$\log_e \frac{pv}{R} + \log_e v^{\gamma - 1} = \text{constant}$$

$$\therefore \log_e \frac{pvv^{\gamma - 1}}{R} = \text{constant}$$

i.e.

$$\log_e \frac{pv^\gamma}{R} = \text{constant}$$

i.e.

$$\frac{pv^\gamma}{R} = e^{(\text{constant})} = \text{constant}$$

or

$$pv^\gamma = \text{constant} \qquad 4.16$$

We therefore have a simple relationship between p and v for any perfect gas undergoing a reversible adiabatic process, each perfect gas having its own value of γ.

Using equation 3.5, $pv = RT$, relationships between T and v, and T and p, may be derived,

i.e.
$$pv = RT$$

$$\therefore p = \frac{RT}{v}$$

Substituting in equation 4.16

$$\frac{RT}{v} v^\gamma = \text{constant}$$

i.e.
$$Tv^{\gamma - 1} = \text{constant} \qquad 4.17$$

Also, $v = (RT)/p$; hence substituting in equation 4.16

$$p\left(\frac{RT}{p}\right)^\gamma = \text{constant}$$

$$\therefore \frac{T^\gamma}{p^{\gamma - 1}} = \text{constant}$$

or
$$\frac{T}{p^{(\gamma - 1)/\gamma}} = \text{constant} \qquad 4.18$$

Therefore for a reversible adiabatic process for a perfect gas between states 1 and 2 we can write:

from equation 4.16

$$p_1 v_1^\gamma = p_2 v_2^\gamma \quad \text{or} \quad \frac{p_1}{p_2} = \left(\frac{v_2}{v_1}\right)^\gamma \qquad 4.19$$

from equation 4.17

$$T_1 v_1^{\gamma - 1} = T_2 v_2^{\gamma - 1} \quad \text{or} \quad \frac{T_1}{T_2} = \left(\frac{v_2}{v_1}\right)^{\gamma - 1} \qquad 4.20$$

from equation 4.18

$$\frac{T_1}{p_1^{(\gamma - 1)/\gamma}} = \frac{T_2}{p_2^{(\gamma - 1)/\gamma}} \quad \text{or} \quad \frac{T_1}{T_2} = \left(\frac{p_1}{p_2}\right)^{(\gamma - 1)/\gamma} \qquad 4.21$$

From equation 4.13 the work done in an adiabatic process per kg of gas is given by $W = (u_1 - u_2)$. The gain in internal energy of a perfect gas is given by equation 3.16,

i.e.   for 1 kg $u_2 - u_1 = c_v(T_2 - T_1)$

$$\therefore\ W = c_v(T_1 - T_2)$$

Also, from equation 3.21,

$$c_v = \frac{R}{(\gamma - 1)}$$

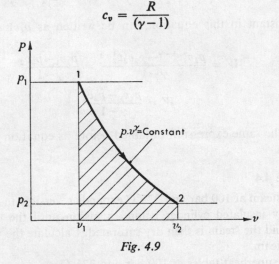

Fig. 4.9

Hence substituting

$$W = \frac{R(T_1 - T_2)}{(\gamma - 1)} \qquad 4.22$$

Using equation 3.5, $pv = RT$,

$$W = \frac{p_1 v_1 - p_2 v_2}{\gamma - 1} \qquad 4.23$$

A reversible adiabatic process for a perfect gas is shown on a p-v diagram in fig. 4.9. The work done is given by the shaded area, and this area can be evaluated by integration,

i.e. $$W = \int_{v_1}^{v_2} p\ dv$$

Therefore, since $pv^\gamma = \text{constant}, c,$ then

$$W = \int_{v_1}^{v_2} \frac{c\,\mathrm{d}v}{v^\gamma}$$

i.e. $$W = c\int_{v_1}^{v_2} \frac{\mathrm{d}v}{v^\gamma} = c\left[\frac{v^{-\gamma+1}}{-\gamma+1}\right]_{v_1}^{v_2}$$

$$= c\left(\frac{v_2^{-\gamma+1}-v_1^{-\gamma+1}}{1-\gamma}\right) = c\left(\frac{v_1^{-\gamma+1}-v_2^{-\gamma+1}}{\gamma-1}\right)$$

The constant in this equation can be written as $p_1v_1^\gamma$ or as $p_2v_2^\gamma$. Hence,

$$W = \frac{p_1v_1^\gamma v_1^{1-\gamma}-p_2v_2^\gamma v_2^{1-\gamma}}{\gamma-1} = \frac{p_1v_1-p_2v_2}{\gamma-1}$$

i.e. $$W = \frac{p_1v_1-p_2v_2}{\gamma-1}$$

This is the same expression obtained before as equation 4.23.

Example 4.4

1 kg of steam at 100 bar and 375°C expands reversibly in a perfectly thermally insulated cylinder behind a piston until the pressure is 38 bar and the steam is then dry saturated. Calculate the work done by the steam.

From superheat tables at 100 bar and 375°C,

$$h_1 = 3017 \text{ kJ/kg} \quad \text{and} \quad v_1 = 0{\cdot}02453 \text{ m}^3/\text{kg}$$

Using equation 2.7

$$u = h-pv$$

$$\therefore\ u_1 = 3017 - \frac{100\times10^5\times0{\cdot}02453}{10^3} = 2771{\cdot}7 \text{ kJ/kg}$$

Also, $$u_2 = u_g \text{ at 38 bar} = 2602 \text{ kJ/kg}$$

Since the cylinder is perfectly thermally insulated then no heat flows to or from the steam during the expansion; the process is therefore adiabatic. Using equation 4.13,

$$W = u_1-u_2 = 2771{\cdot}7-2602$$

$$\therefore\ W = 169{\cdot}7 \text{ kJ/kg}$$

The process is shown on a p-v diagram in fig. 4.10, the shaded area representing the work done.

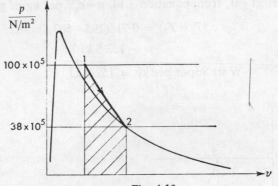

Fig. 4.10

Example 4.5

Air at 1·02 bar, 22°C, initially occupying a cylinder volume of 0·015 m³, is compressed reversibly and adiabatically by a piston to a pressure of 6·8 bar. Calculate the final temperature, the final volume, and the work done on the mass of air in the cylinder.

From equation 4.21

$$\frac{T_1}{T_2} = \left(\frac{p_1}{p_2}\right)^{(\gamma-1)/\gamma} \quad \text{or} \quad T_2 = T_1 \times \left(\frac{p_2}{p_1}\right)^{(\gamma-1)/\gamma}$$

i.e.

$$T_2 = 295 \times \left(\frac{6\cdot8}{1\cdot02}\right)^{(1\cdot4-1)/1\cdot4} = 295 \times 6\cdot67^{0\cdot286} = 295 \times 1\cdot72 = 507\cdot5 \text{ K}$$

(where $T_1 = 22 + 273 = 295$ K; γ for air = 1·4),

i.e.   Final temperature = $507\cdot5 - 273 = 234\cdot5$°C

From equation 4.19

$$\frac{p_1}{p_2} = \left(\frac{V_2}{V_1}\right)^{\gamma} \quad \text{or} \quad \frac{V_1}{V_2} = \left(\frac{p_2}{p_1}\right)^{1/\gamma}$$

$$\therefore \frac{0\cdot015}{V_2} = \left(\frac{6\cdot8}{1\cdot02}\right)^{1/1\cdot4} = 6\cdot67^{0\cdot714} = 3\cdot87$$

$$\therefore V_2 = \frac{0\cdot015}{3\cdot87} = 0\cdot00388 \text{ m}^3$$

i.e.   Final volume = 0·00388 m³

From equation 4.13, for an adiabatic process,

$$W = u_1 - u_2$$

and for a perfect gas, from equation 3.14, $u = c_v T$ per kg of gas,

$$\therefore \ W = c_v(T_1 - T_2) = 0.718(295 - 507.5)$$
$$= -152.8 \text{ kJ/kg}$$

i.e.   Work input per kg $= 152.8$ kJ

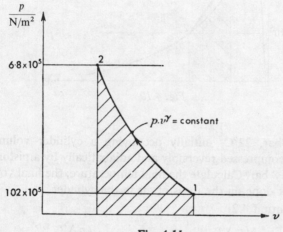

Fig. 4.11

The mass of air can be found using equation 3.6, $pV = mRT$.

$$\therefore \ m = \frac{p_1 v_1}{RT_1} = \frac{1.02 \times 10^5 \times 0.015}{0.287 \times 10^3 \times 295} = 0.0181 \text{ kg}$$

i.e.   Total work done $= 0.0181 \times 152.8 = 2.76$ kJ

The process is shown on a p-v diagram in fig. 4.11, the shaded area representing the work done per kg of air.

4.3 Polytropic processes

It is found that many processes in practice approximate to a reversible law of the form $pv^n = $ constant, where n is a constant. Both vapours and perfect gases obey this type of law closely in many non-flow processes. Such processes are internally reversible.

From equation 1.2 for any reversible process,

$$W = \int p \, dv$$

For a process in which $pv^n = $ constant, we have $p = c/v^n$, where c is a constant.

$$\therefore \; W = c \int_{v_1}^{v_2} \frac{dv}{v^n} = c\left[\frac{v^{-n+1}}{-n+1}\right] = c\left(\frac{v_2^{-n+1} - v_1^{-n+1}}{-n+1}\right)$$

i.e.
$$W = c\left(\frac{v_1^{1-n} - v_2^{1-n}}{n-1}\right) = \frac{p_1 v_1^n v_1^{1-n} - p_2 v_2^n v_2^{1-n}}{n-1}$$

(since the constant, c, can be written as $p_1 v_1^n$ or as $p_2 v_2^n$),

i.e.
$$\text{Work done} = \frac{p_1 v_1 - p_2 v_2}{n-1} \qquad\qquad 4.24$$

Equation 4.24 is true for any working substance undergoing a reversible polytropic process. It follows also that for any polytropic process we can write

$$\frac{p_1}{p_2} = \left(\frac{v_2}{v_1}\right)^n \qquad\qquad 4.25$$

Example 4.6

In a steam engine the steam at the beginning of the expansion process is at 7 bar, dryness fraction 0·95, and the expansion follows the law $pv^{1·1} = $ constant, down to a pressure of 0·34 bar. Calculate the work done per kg of steam during the expansion, and the heat flow per kg of steam to or from the cylinder walls during the expansion.

At 7 bar, $\qquad\qquad v_g = 0\cdot2728 \text{ m}^3/\text{kg}$

Therefore, using equation 3.1,

$$v_1 = xv_g = 0\cdot95 \times 0\cdot2728 = 0\cdot259 \text{ m}^3/\text{kg}$$

Then from equation 4.25

$$\frac{p_1}{p_2} = \left(\frac{v_2}{v_1}\right)^n \quad \text{or} \quad v_2 = v_1 \left(\frac{p_1}{p_2}\right)^{1/n}$$

$$\therefore \; v_2 = 0\cdot259 \left(\frac{7}{0\cdot34}\right)^{1/1\cdot1} = 20\cdot59^{0\cdot909} \times 0\cdot259$$

$$= 15\cdot64 \times 0\cdot259 = 4\cdot05 \text{ m}^3/\text{kg}$$

From equation 4.24

$$W = \frac{p_1 v_1 - p_2 v_2}{n-1} = \frac{7 \times 10^5 \times 0 \cdot 259 - 0 \cdot 34 \times 10^5 \times 4 \cdot 05}{1 \cdot 1 - 1}$$

i.e. $W = \dfrac{10^5}{0 \cdot 1}(1 \cdot 813 - 1 \cdot 377) = \dfrac{10^5 \times 0 \cdot 436}{0 \cdot 1}$ N m/kg

i.e.   Work done = 436 kJ/kg

At 0·34 bar, $v_g = 4 \cdot 649$ m³/kg, therefore the steam is wet at state 2, and using equation 3.1, we have,

$$x_2 = \frac{4 \cdot 05}{4 \cdot 649} = 0 \cdot 873$$

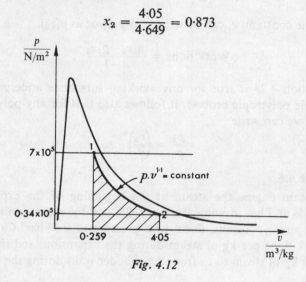

Fig. 4.12

The expansion is shown on a p-v diagram in fig. 4.12, the area under 1–2 representing the work done per kg of steam.

From equation 3.3

$$u_1 = (1-x_1)u_f + x_1 u_g = (1-0 \cdot 95)696 + 0 \cdot 95 \times 2573$$

i.e. $u_1 = 34 \cdot 8 + 2442 = 2476 \cdot 8$ kJ/kg

and $u_2 = (1-x_2)u_f + x_2 u_g = (1-0 \cdot 873)302 + 0 \cdot 873 \times 2472$

i.e. $u_2 = 38 \cdot 35 + 2158 = 2196 \cdot 4$ kJ/kg

From the non-flow energy equation, 2.2,

$$Q = (u_2 - u_1) + W = (2196 \cdot 4 - 2476 \cdot 8) + 436$$

i.e. $\qquad\qquad Q = -280 \cdot 4 + 436 = 155 \cdot 6 \text{ kJ/kg}$

i.e. $\qquad\qquad$ Heat supplied $= 155 \cdot 6 \text{ kJ/kg}$

Consider now the polytropic process for a perfect gas. From equation 3.5

$$pv = RT \quad \text{or} \quad p = \frac{RT}{v}$$

Hence, substituting in the equation $pv^n = \text{constant}$, we have

$$\frac{RT}{v} v^n = \text{constant} \quad \text{or} \quad Tv^{n-1} = \text{constant} \qquad 4.26$$

Also, writing $v = (RT)/p$, we have

$$p \left(\frac{RT}{p} \right)^n = \text{constant} \quad \text{or} \quad \frac{T}{p^{(n-1)/n}} = \text{constant} \qquad 4.27$$

It can be seen that these equations are exactly similar to the equations 4.17 and 4.18 for a reversible adiabatic process for a perfect gas. In fact the reversible adiabatic process for a perfect gas is a particular case of a polytropic process with the index, n, equal to γ.

Equations 4.26 and 4.27 can be written as

$$\frac{T_1}{T_2} = \left(\frac{v_2}{v_1} \right)^{n-1} \qquad 4.28$$

and $\qquad\qquad \dfrac{T_1}{T_2} = \left(\dfrac{p_1}{p_2} \right)^{(n-1)/n} \qquad 4.29$

Note that equations 4.26, 4.27, 4.28, and 4.29 do *not* apply to a vapour undergoing a polytropic process, since the characteristic equation of state, $pv = RT$, which was used in the derivation of the equations, applies only to a perfect gas.

For a perfect gas expanding polytropically it is sometimes more convenient to express the work done in terms of the temperatures at the end states. From equation 4.24, $W = (p_1 v_1 - p_2 v_2)/(n-1)$, then, from equation 3.5, $p_1 v_1 = RT_1$ and $p_2 v_2 = RT_2$. Hence,

$$W = \frac{R(T_1 - T_2)}{n-1} \qquad 4.30$$

or for mass, m, $\qquad\qquad W = \dfrac{mR(T_1 - T_2)}{n-1} \qquad 4.31$

Using the non-flow energy equation, 2.2, the heat flow during the process can be found,

i.e. $$Q = (u_2 - u_1) + W = c_v(T_2 - T_1) + \frac{R(T_1 - T_2)}{(n-1)}$$

i.e. $$Q = \frac{R(T_1 - T_2)}{(n-1)} - c_v(T_1 - T_2)$$

From equation 3.21

$$c_v = \frac{R}{(\gamma - 1)}$$

Hence substituting,

$$Q = \frac{R}{(n-1)}(T_1 - T_2) - \frac{R}{(\gamma - 1)}(T_1 - T_2)$$

i.e. $$Q = R(T_1 - T_2)\left(\frac{1}{n-1} - \frac{1}{\gamma - 1}\right) = \frac{R(T_1 - T_2)(\gamma - 1 - n + 1)}{(\gamma - 1)(n-1)}$$

$$\therefore Q = \left(\frac{\gamma - n}{\gamma - 1}\right)\frac{R(T_1 - T_2)}{(n-1)}$$

Now from equation 4.30, $W = R(T_1 - T_2)/(n-1)$ per unit mass of gas, therefore

$$Q = \left(\frac{\gamma - n}{\gamma - 1}\right)W \qquad\qquad 4.32$$

Equation 4.32 is a convenient and concise expression relating the heat supplied and the work done in a polytropic process. In an expansion, work is done by the gas, and hence the term W is positive. Therefore it can be seen from equation 4.32 that when the polytropic index n is less than γ, in an expansion, then the right-hand side of the equation is positive (i.e. heat is supplied during the process). Conversely, when n is greater than γ in an expansion, then heat is rejected by the gas. Similarly, the work done in a compression process is negative, therefore when n is less than γ, in compression, heat is rejected; and when n is greater than γ, in compression, heat must be supplied to the gas during the process. It was shown in Section 3.3 that γ for all perfect gases has a value greater than unity.

Example 4.7

1 kg of a perfect gas is compressed from 1·1 bar, 27°C according to a law $pv^{1·3} = $ constant, until the pressure is 6·6 bar. Calculate the heat flow to or from the cylinder walls,

(a) When the gas is ethane (molar mass 30 kg/kmol), which has $c_p = 2·10$ kJ/kg K.

(b) When the gas is argon (molecular mass 40 kg/kmol), which has $c_p = 0·520$ kJ/kg K.

From equation 4.29, for both ethane and argon,

$$\frac{T_1}{T_2} = \left(\frac{p_1}{p_2}\right)^{(n-1)/n} \quad \text{or} \quad T_2 = T_1 \left(\frac{p_2}{p_1}\right)^{(n-1)/n}$$

i.e.

$$T_2 = 300 \left(\frac{6·6}{1·1}\right)^{1·3-1/1·3} = 300 \times 6^{0·231} = 300 \times 1·512 = 453·6 \text{ K}$$

(where $T_1 = 27 + 273 = 300$ K).

(a) From equation 3.9, $R = R_0/M$, therefore, for ethane

$$R = \frac{8·314}{30} = 0·277 \text{ kJ/kg K}$$

Then from equation 3.17, $c_p - c_v = R$, therefore

$$c_v = 2·10 - 0·277 = 1·823 \text{ kJ/kg K}$$

(where $c_p = 1·75$ kJ/kg K for ethane).

Then from equation 3.20

$$\gamma = \frac{c_p}{c_v} = \frac{2·10}{1.823} = 1·152$$

From equation 4.30

$$W = \frac{R(T_1 - T_2)}{n-1} = \frac{0·277 \times (300 - 453·6)}{1·3 - 1} = -141·8 \text{ kJ/kg}$$

Then from equation 4.32

$$Q = \left(\frac{\gamma - n}{\gamma - 1}\right) W = \left(\frac{1·152 - 1·3}{1·152 - 1}\right) \times -141·8 = -\frac{0·148}{0·152} \times -141·8$$

$$\therefore Q = +\frac{0·148 \times 141·8}{0.152} = 138·1 \text{ kJ/kg}$$

i.e.   Heat supplied $= 138·1$ kJ/kg

(b) Using the same method for Argon we have,

$$R = \frac{8\cdot314}{40} = 0\cdot208 \text{ kJ/kg K}$$

Also, $c_v = 0\cdot520 - 0\cdot208 = 0\cdot312 \text{ kJ/kg degC}$

$$\therefore \ \gamma = \frac{0\cdot520}{0\cdot312} = 1\cdot667$$

Then the work done is given by

$$W = \frac{R(T_1 - T_2)}{n-1} = \frac{0\cdot208 \times (300 - 453\cdot6)}{1\cdot3 - 1} = -106\cdot5 \text{ kJ/kg}$$

Then,

$$Q = \left(\frac{\gamma - n}{\gamma - 1}\right) W = \left(\frac{1\cdot667 - 1\cdot3}{1\cdot667 - 1}\right) \times -106\cdot5 = -\frac{0\cdot367 \times 106\cdot5}{0\cdot667}$$

$$\therefore \ Q = -58\cdot6 \text{ kJ/kg}$$

i.e.   Heat rejected $= 58\cdot6 \text{ kJ/kg}$

In a polytropic process the index n depends only on the heat and work quantities during the process. The various processes considered in Sections 4.1 and 4.2 are special cases of the polytropic process for a perfect gas. For example,

* when $n = 0$ pv^0 = constant, i.e. p = constant

  when $n = \infty$ pv^∞ = constant

              or $p^{1/\infty}v$ = constant, i.e. v = constant

  when $n = 1$ pv = constant, i.e. T = constant

        (since $(pv)/T$ = constant for a perfect gas)

  when $n = \gamma$ pv^γ = constant, i.e. reversible adiabatic

This is illustrated on a p-v diagram in fig. 4.13. Thus,

    state 1 to state A is constant pressure cooling $(n=0)$;

    state 1 to state B is isothermal compression $(n=1)$;

    state 1 to state C is reversible adiabatic compression $(n=\gamma)$;

    state 1 to state D is constant volume heating $(n=\infty)$.

Similarly, 1 to A' is constant pressure heating; 1 to B' is isothermal expansion; 1 to C′ is reversible adiabatic expansion; 1 to D′ is constant volume cooling. Note that, since γ is always greater than unity, then process 1 to C must lie between processes 1 to B and 1 to D; similarly, process 1 to C′ must lie between processes 1 to B′ and 1 to D′.

For a vapour a generalization such as the above is not possible.

One important process for a vapour should be mentioned here. A vapour may undergo a process according to a law $pv =$ constant. In this case, since the characteristic equation of state, $pv = RT$, does not apply to a vapour, then the process is *not* isothermal. Tables must be used to find the properties at the end states, making use of the fact that $p_1 v_1 = p_2 v_2$.

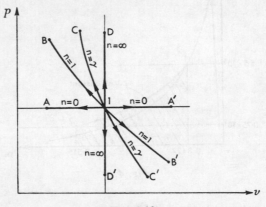

Fig. 4.13

Example 4.8

In the cylinder of a steam engine the steam expands from 5·5 bar to 0·75 bar according to a hyperbolic law $pv =$ constant. If the steam is initially dry saturated, calculate the work done per kg of steam, and the heat flow to or from the cylinder walls.

At 5·5 bar,

$$v_1 = v_g = 0{\cdot}3427 \text{ m}^3/\text{kg}$$

Then,

$$p_1 v_1 = p_2 v_2$$

$$\therefore v_2 = \frac{p_1 v_1}{p_2} = \frac{5{\cdot}5 \times 0{\cdot}3427}{0{\cdot}75} = 2{\cdot}515 \text{ m}^3/\text{kg}$$

At 0·75 bar, $v_g = 2·217$ m^3/kg, hence the steam is superheated at state 2. Interpolating from superheat tables at 0·75 bar we have

$$u_2 = 2510 + \left(\frac{2·515 - 2·271}{2·588 - 2·271}\right)(2585 - 2510)$$

i.e.
$$u_2 = 2510 + 57·7$$

$$= 2567·7 \text{ kJ/kg}$$

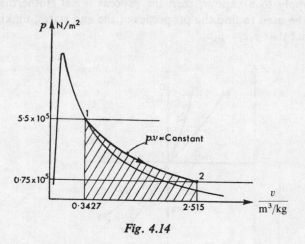

Fig. 4.14

For dry saturated steam at 5·5 bar,

$$u_1 = u_g = 2565 \text{ kJ/kg}$$

Hence,

$$\text{Gain in internal energy} = 2567·7 - 2565$$

$$= 2·7 \text{ kJ/kg}$$

The process is shown on a *p-v* diagram in fig. 4.14, the shaded area representing the work done. From equation 1.2

$$W = \int_{v_1}^{v_2} p \, dv = \int_{v_1}^{v_2} \left(\frac{\text{constant}}{v}\right) dv$$

$$= (\text{constant})[\log_e v]_{v_1}^{v_2}$$

The constant is either p_1v_1 or p_2v_2
i.e.

$$W = 5 \cdot 5 \times 10^5 \times 0 \cdot 3427 \times \log_e \frac{v_2}{v_1} = 5 \cdot 5 \times 10^5 \times 0 \cdot 3427 \times \log_e \frac{p_1}{p_2}$$

$$\therefore \quad W = 5 \cdot 5 \times 10^5 \times 0 \cdot 3427 \times \log_e \frac{5 \cdot 5}{0 \cdot 75} = 375\,500 \text{ N m/kg}$$

From the non-flow energy equation, 2.2,

$$Q = (u_2 - u_1) + W = 2 \cdot 7 + \frac{375\,500}{10^3} = 2 \cdot 7 + 375 \cdot 5 = 378 \cdot 2$$

i.e.   Heat supplied $= 378 \cdot 2$ kJ/kg

4.4 Irreversible processes

The criteria of reversibility were stated in Section 1.4. The equations of Sections 4.1, 4.2, and 4.3 can only be used when the process obeys the criteria of reversibility to a close approximation. In processes in which a fluid is enclosed in a cylinder behind a piston, friction effects can be assumed to be negligible. However, in order to satisfy criterion (c) heat must never be transferred to or from the system through a finite temperature difference. Only in an isothermal process is this conceivable, since in all other processes the temperature of the system is continually changing during the process; in order to satisfy criterion (c) the temperature of the cooling or heating medium external to the system would be required to change correspondingly. Ideally a way of achieving reversibility can be imagined, but in practice it cannot even be approached as an approximation. Nevertheless, if we accept inevitable irreversibilities in the surroundings, we can still have processes which are internally reversible. That is, the system undergoes a process which can be reversed, but the surroundings undergo an irreversible change. Most processes occurring in a cylinder behind a piston can be assumed to be internally reversible to a close approximation, and the equations of sections 4.1, 4.2, and 4.3 can be used where applicable. Certain processes cannot be assumed to be internally reversible, and the important cases will now be briefly discussed.

Unresisted, or free, expansion

This process was mentioned in Section 1.5 in order to show that in an irreversible process the work done is *not* given by $\int p \, dv$. Consider two vessels A and B, interconnected by a short pipe with a valve X, and perfectly thermally insulated (see fig. 4.15). Initially let the vessel A be filled with a fluid at a certain pressure, and let B be completely evacuated. When the valve X is opened the fluid in A will expand rapidly to fill both vessels A and B. The pressure finally will be lower than the initial pressure in vessel A. This is known as an unresisted expansion or a free expansion. The process is not reversible, since external work would have to be done to restore the fluid to its initial condition. The non-flow energy equation, 2.2, can be applied between the initial and final states,

i.e. $$Q = (u_2 - u_1) + W$$

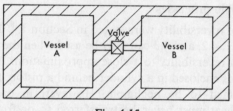

Fig. 4.15

Now in this process no work is done on or by the fluid, since the boundary of the system does not move. No heat flows to or from the fluid since the system is well lagged The process is therefore adiabatic, but irreversible.

i.e. $$u_2 - u_1 = 0 \quad \text{or} \quad u_2 = u_1$$

In a free expansion therefore the internal energy initially equals the internal energy finally.

For a perfect gas, we have, from equation 3.14,

$$u = c_v T$$

Therefore for a free expansion of a perfect gas

$$c_v T_1 = c_v T_2$$

i.e. $$T_1 = T_2$$

That is, for a perfect gas undergoing a free expansion, the initial temperature is equal to the final temperature.

Example 4.9

Air at 20 bar is initially contained in vessel A of fig. 4.15, the volume of which can be assumed to be 1 m^3. The valve X is opened and the air expands to fill vessels A and B. Assuming that the vessels are of equal volume, calculate the final pressure of the air.

For a perfect gas for a free expansion, $T_1 = T_2$. Also from equation 3.6, $pV = mRT$, hence $p_1 V_1 = p_2 V_2$.

Now V_2 is the combined volumes of vessels A and B,

i.e. $V_2 = V_A + V_B = 1 + 1 = 2 \text{ m}^3$ and $V_1 = 1 \text{ m}^3$

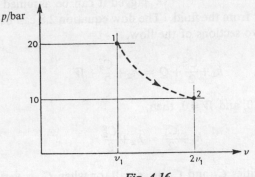

Fig. 4.16

Therefore we have

$$p_2 = p_1 \times \frac{V_1}{V_2} = 20 \times \frac{1}{2} = 10 \text{ bar}$$

i.e.   Final pressure = 10 bar

The process is shown on a p-v diagram in fig. 4.16. State 1 is fixed at 20 bar and 1 m^3 when the mass of gas is known; state 2 is fixed at 10 bar and 2 m^3 for the same mass of gas. The process between these states is irreversible and must be drawn dotted. The points 1 and 2 lie on an isothermal line, but the process between 1 and 2 cannot be called isothermal, since the intermediate temperatures are not the same throughout the process. There is no work done during the process, and the area under the dotted line does *not* represent work done.

Throttling

A flow of fluid is said to be throttled when there is some restriction to the flow, when the velocities before and after the restriction are either equal or negligibly small, and when there is a negligible heat loss to the surroundings. The restriction to flow can be a partly open valve, an orifice, or any other sudden reduction in the cross-section of the flow.

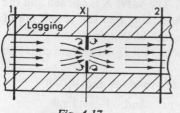

Fig. 4.17

An example of throttling is shown in fig. 4.17. The fluid, flowing steadily along a well-lagged pipe, passes through an orifice at section X. Since the pipe is well lagged it can be assumed that no heat flows to or from the fluid. The flow equation 2.8 can be applied between any two sections of the flow,

i.e.
$$h_1 + \frac{C_1^2}{2} + Q = h_2 + \frac{C_2^2}{2} + W$$

Now since $Q = 0$, and $W = 0$, then,

$$h_1 + \frac{C_1^2}{2} = h_2 + \frac{C_2^2}{2}$$

When the velocities C_1 and C_2 are small, or when C_1 is very nearly equal to C_2, then the kinetic energy terms may be neglected. (Note that sections 1 and 2 can be chosen well upstream and well downstream of the disturbance to the flow, so that this latter assumption is justified.)

Then,
$$h_1 = h_2$$

Therefore for a throttling process, the enthalpy initially is equal to the enthalpy finally.

The process is adiabatic, but is highly irreversible because of the eddying of the fluid round the orifice at X. Between sections 1 and X the enthalpy drops and the kinetic energy increases as the fluid accelerates through the orifice. Between sections X and 2 the enthalpy increases as the kinetic energy is destroyed by fluid eddies.

For a perfect gas, from equation 3.18, $h = c_p T$, therefore,

$$c_p T_1 = c_p T_2 \quad \text{or} \quad T_1 = T_2$$

For throttling of a perfect gas, therefore, the temperature initially equals the temperature finally.

Example 4.10

Steam at 19 bar is throttled to 1 bar and the temperature after throttling is found to be 150°C. Calculate the initial dryness fraction of the steam.

From superheat tables at 1 bar and 150°C we have $h_2 = 2777$ kJ/kg. Then for throttling, $h_1 = h_2 = 2777$ kJ/kg.

Using equation 3.2,

$$h_1 = h_f + x_1 h_{fg}$$

i.e.
$$2777 = 897 + x_1 \times 1901$$

$$\therefore \; x_1 = \frac{1880}{1901} = 0.989$$

i.e.
$$\text{Initial dryness fraction} = 0.989$$

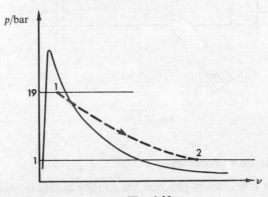

Fig. 4.18

The process is shown on a p-v digram in fig. 4.18. States 1 and 2 are fixed, but the intermediate states are indeterminate; the process must be drawn dotted, as shown. No work is done during the process, and the area under the line 1–2 is *not* equal to work done.

For a vapour, throttling can be used as a means of finding the dryness fraction of wet steam, as in example 4.10. This will be dealt with more fully in section 7.4.

Adiabatic mixing

The mixing of two streams of fluid is quite common in engineering practice, and can usually be assumed to occur adiabatically. Consider two streams of a fluid mixing as shown in fig. 4.19. Let the streams have mass flow rates $\dot{m}_1$ and $\dot{m}_2$, and temperatures T_1 and T_2. Let the resulting mixed stream have a temperature T_3. There is no heat flow to or from the fluid, and no work is done, hence from the flow equation, we have, neglecting changes in kinetic energy,

$$H_1 + H_2 = H_3 \quad \text{or} \quad \dot{m}_1 h_1 + \dot{m}_2 h_2 = (\dot{m}_1 + \dot{m}_2) h_3 \qquad 4.33$$

For a perfect gas, from equation 3.18, $h = c_p T$, hence,

$$\dot{m}_1 c_p T_1 + \dot{m}_2 c_p T_2 = (\dot{m}_1 + \dot{m}_2) c_p T_3$$

i.e.
$$\dot{m}_1 T_1 + \dot{m}_2 T_2 = (\dot{m}_1 + \dot{m}_2) T_3 \qquad 4.34$$

The mixing process is highly irreversible due to the large amount of eddying and churning of the fluid that takes place.

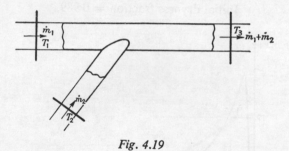

Fig. 4.19

4.5 Reversible flow processes

Although flow processes in practice are usually highly irreversible it is sometimes convenient to assume that a flow process is reversible in order to provide an ideal comparison. An observer travelling with the flowing fluid would appear to see a change in thermodynamic properties as in a non-flow process. For example in a reversible adiabatic process for a perfect gas, an observer travelling with the gas would appear to see a process $pv^\gamma = \text{constant}$ taking place, but the work done by the gas would *not* be given by $\int p \, dv$, or by the change in internal energy as given by equation 4.13. Some work is done on or by the gas by virtue of the forces acting between the moving gas and its surroundings. For example, for a reversible adiabatic flow process for a perfect gas, from the flow equation, 2.8,

$$h_1 + \frac{C_1^2}{2} + Q = h_2 + \frac{C_2^2}{2} + W$$

Then, since $Q = 0$,

$$W = (h_1 - h_2) + \left(\frac{C_1^2 - C_2^2}{2}\right)$$

Also, since the process is assumed to be reversible, then for a perfect gas, $pv^\gamma = $ constant. This equation can be used to fix the end states. Note that, even if the kinetic energies terms are negligibly small, the work done in a reversible adiabatic flow process between two states is *not* equal to the work done in a reversible adiabatic non-flow process between the same states (i.e. $W = (u_1 - u_2)$ as in equation 4.13).

Example 4.11

A gas turbine receives gases from the combustion chamber at 7 bar and 650°C, with a velocity of 9 m/s. The gases leave the turbine at 1 bar with a velocity of 45 m/s. Assuming that the expansion is adiabatic and reversible in the ideal case, calculate the work done per kg of gas. For the gases take $\gamma = 1 \cdot 333$ and $c_p = 1 \cdot 11$ kJ/kg K.

Using the flow equation, for an adiabatic process,

$$W = (h_1 - h_2) + \left(\frac{C_1^2 - C_2^2}{2}\right)$$

For a perfect gas from equation 3.18, $h = c_p T$, therefore,

$$W = c_p(T_1 - T_2) + \left(\frac{C_1^2 - C_2^2}{2}\right)$$

To find T_2 use equation 4.21,

$$\frac{T_1}{T_2} = \left(\frac{p_1}{p_2}\right)^{(\gamma-1)/\gamma}$$

i.e. $\dfrac{T_1}{T_2} = \left(\dfrac{7}{1}\right)^{(1\cdot333-1)/1\cdot333} = 7^{0\cdot25} = 1\cdot627$

$$\therefore T_2 = \frac{T_1}{1\cdot627} = \frac{923}{1\cdot627} = 567 \text{ K}$$

(where $T_1 = 650 + 273 = 923$ K).

Hence substituting,

$$W = 1\cdot11(923 - 567) + \left(\frac{9^2 - 45^2}{2 \times 10^3}\right)$$

i.e. $W = 395\cdot2 - 0\cdot97 = 394\cdot2$ kJ/kg

Note that the kinetic energy change is small compared with the enthalpy change. This is often the case in problems on flow processes, and the change in kinetic energy can sometimes be taken to be negligible.

4.6 Nonsteady-flow processes

There are many cases in practice when the rate of mass flow crossing the boundary of a system at inlet is not the same as the rate of mass flow crossing the boundary of the system at outlet. Also, the rate at which work is done on or by the fluid, and the rate at which heat is transferred to or from the system is not necessarily constant with time. In a case of this kind the total energy of the system within the boundary is no longer constant, as it is in a steady flow process, but varies with time.

Let the total energy of the system within the boundary at any instant be E. During a small time interval let the mass entering the system be δm_1, and let the mass leaving the system be δm_2; let the heat transferred and the work done during the same time be δQ and δW respectively. Consider a similar system to the one shown in fig. 2.2. Now, as shown in section 2.3 (page 24), work is done at inlet and outlet in introducing and expelling mass across the system boundaries.

i.e.　　at inlet,　　energy required $= \delta m_1 p_1 v_1$

　　and at outlet,　　energy required $= \delta m_2 p_2 v_2$

Also, as before, the energy of unit mass of the flowing fluid is given by $(u_1 + C_1^2/2 + Z_1 g)$ at inlet, and by $(u_2 + C_2^2/2 + Z_2 g)$ at outlet. Hence,

Energy entering system $= \delta Q + \delta m_1(u_1 + C_1^2/2 + Z_1 g) + \delta m_1 p_1 v_1$

and

Energy leaving system $= \delta W + \delta m_2(u_2 + C_2^2/2 + Z_2 g) + \delta m_2 p_2 v_2$

Then applying the first law:

Energy entering $-$ Energy leaving $=$ Increase of energy of the system, δE

$$\therefore\ \delta Q + \delta m_1(u_1 + C_1^2/2 + Z_1 g + p_1 v_1) - \delta W - \delta m_2(u_2 + C_2^2/2 + Z_2 g + p_2 v_2) = \delta E$$

During a finite time the total heat transferred is given by $\sum \delta Q = Q$, and the total work done is given by $\sum \delta W = W$.

Let the initial mass within the system boundaries be m', and the initial internal energy be u'; let the final mass within the boundaries at the end of the time interval be m'', and the final internal energy be u''.

$$\therefore \sum \delta E = m''u'' - m'u'$$

Therefore we have:

$$Q + \sum \delta m_1(u_1 + p_1v_1 + C_1^2/2 + Z_1g)$$
$$= W + \sum \delta m_2(u_2 + p_2v_2 + C_2^2/2 + Z_2g) + (m''u'' - m'u') \quad 4.35$$

or,

$$Q + \sum \delta m_1(h_1 + C_1^2/2 + Z_1g) = W + \sum \delta m_2(h_2 + C_2^2/2 + Z_2g)$$
$$+ (m''u'' - m'u') \quad 4.36$$

Also, from continuity of mass:

Mass entering $-$ Mass leaving
$\qquad\qquad$ = Increase of mass within system boundary

i.e. $\qquad\qquad \sum \delta m_1 - \sum \delta m_2 = m'' - m' \qquad\qquad 4.37$

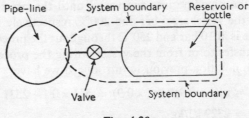

Fig. 4.20

One of the most commonly occurring problems involving the non-steady-flow equation is the filling of a bottle or reservoir from a source which is large in comparison with the bottle or reservoir. Figure 4.20 shows a typical example. It is assumed that the condition of the fluid in the pipe-line is unchanged during the filling process. In this case there is no work done on the system boundary; also, no mass leaves the system during the process, hence $\delta m_2 = 0$.

Applying equation 4.36, making the additional assumption that changes in potential energy are zero, and that the kinetic energy, $C_1^2/2$, is small compared with the enthalpy, h_1, we have:

$$Q + \sum \delta m_1 h_1 = m''u'' - m'u'$$

Or, since h_1 is constant during the process:

$$Q + h_1 \sum \delta m_1 = m''u'' - m'u'$$

In this case equation 4.37 becomes:

$$\sum \delta m_1 = m'' - m'$$

Hence substituting:

$$Q + h_1(m'' - m') = m''u'' - m'u' \qquad 4.38$$

It is often possible to assume that the process is adiabatic, and in that case we have,

$$h_1(m'' - m') = m''u'' - m'u'$$

Or, in words:

Enthalpy of mass which enters the bottle = Increase of internal energy of the system.

Example 4.12

A rigid vessel of volume 10 m³ containing steam at 2·1 bar and dryness fraction 0·9, is connected to a pipe-line and steam is allowed to flow from the pipe-line into the vessel until the pressure and temperature in the vessel are 6 bar and 200°C respectively. The steam in the pipe-line is at 10 bar and 250°C throughout the process. Calculate the heat transfer to or from the vessel during the process.

Using the notation previously introduced we have:

$$u' = u_f'(1 - 0·9) + (u_g' \times 0·9) = 511 \times 0·1 + 2531 \times 0·9$$

i.e. $\qquad u' = 2329 \text{ kJ/kg}$

Also,

$$m' = V/v' = 10/0·9 \, v_g = 10/0·9 \times 0·8461 = 13·13 \text{ kg}$$

The steam is superheated finally at 6 bar and 200°C, therefore:

$$u'' = 2640 \text{ kJ/kg}$$

and $\qquad v'' = 0·3522 \text{ m}^3/\text{kg}$

i.e. $\qquad m'' = V/v'' = 10/0·3522 = 28·4 \text{ kg}$

The steam in the pipe-line is superheated at 10 bar and 250°C, hence:

$$h_1 = 2944 \text{ kJ/kg}$$

Then using equation 4.38:

$$Q + 2944(28·4 - 13·13) = (28·4 \times 2640) - (13·13 \times 2329)$$

$$\therefore \quad Q = 74\,980 - 30\,590 - 44\,940 = -550 \text{ kJ}$$

i.e. $\qquad$ Heat rejected from vessel = 550 kJ

Another commonly occurring example of the nonsteady-flow process is the case in which a vessel is opened to a large space and fluid is allowed to escape (fig. 4.21). There is no work done and in this case $\delta m_1 = 0$ since no mass enters the system. Neglecting changes in potential energy and applying equation 4.36:

$$Q = \sum \delta m_2 (h_2 + C_2^2/2) + (m''u'' - m'u')$$

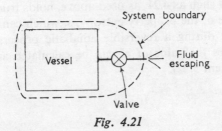

Fig. 4.21

The difficulty arising in this analysis is that the state 2 of the mass leaving the vessel is continually changing, and hence it is impossible to evaluate the term $\sum \delta m_2 (h_2 + C_2^2/2)$. An approximation can be made in order to find the mass of fluid which leaves the vessel as the pressure drops to a given value. It can be assumed that the fluid remaining in the vessel undergoes a reversible adiabatic expansion. This is a good approximation if the vessel is well-lagged, or if the duration of the process is short. Using this assumption the end state of the fluid in the vessel can be found, and hence the mass remaining in the vessel, m'', can be calculated.

Example 4.13

An air receiver of volume 6 m³ contains air at 15 bar and 40·5°C. A valve is opened and some air is allowed to blow out to atmosphere. The pressure of the air in the receiver drops rapidly to 12 bar when the valve is then closed. Calculate the mass of air which has left the receiver.

Initially: $\quad m' = p'V/RT' = \dfrac{15 \times 10^5 \times 6}{0\cdot287 \times 10^3 \times 313\cdot5} = 100 \text{ kg}$

Assuming that the mass in the receiver undergoes a reversible adiabatic process, then using equation 4.21:

$$\frac{T'}{T''} = \left(\frac{p'}{p''}\right)^{\gamma - 1/\gamma} = \left(\frac{15}{12}\right)^{0\cdot4/1\cdot4} = 1\cdot25^{0\cdot286} = 1\cdot066$$

$$\therefore \quad T'' = 313\cdot5/1\cdot066 = 294\cdot2 \text{ K}$$

Hence, $m'' = p''V/RT'' = \dfrac{12 \times 10^5 \times 6}{0 \cdot 287 \times 10^3 \times 294 \cdot 2} = 85 \cdot 3$ kg

Therefore,

Mass of air which left receiver $= 100 - 85 \cdot 3 = 14 \cdot 7$ kg

In the case of a vapour undergoing a reversible adiabatic expansion no equation such as 4.21, as used above, holds true. It is necessary to make use of the property entropy, s, which can be shown to remain constant during a reversible adiabatic process, i.e. $s' = s''$. Then using tables the value of v'' can be calculated and hence m'' found (see Problem 5.22).

Example 4.14

At the beginning of the induction stroke of a petrol engine of compression ratio 8/1, the clearance volume is occupied by residual gas at a temperature of 840°C and pressure 1·034 bar. The volume of mixture induced during the stroke, measured at atmospheric conditions of 1·013 bar and 15°C, is 0·75 of the cylinder swept volume. The mean pressure and temperature in the induction manifold during induction is 0·965 bar and 27°C respectively, and the mean pressure in the cylinder during the induction stroke is 0·828 bar. Calculate the temperature of the mixture at the end of the induction stroke assuming the process to be adiabatic. Calculate also the final pressure in the cylinder. For the induced mixture and final mixture take $c_v = 0 \cdot 718$ kJ/kg K and $R = 0 \cdot 2871$ kJ/kg K; for the residual gas take $c_v = 0 \cdot 84$ kJ/kg K and $R = 0 \cdot 296$ kJ/kg K.

Let swept volume be V_s and clearance volume be V_c. Then,

$$\text{Compression ratio} = \frac{V_s + V_c}{V_c} = 8 \quad \text{(see page 161)}$$

i.e. $V_s = 7V_c$

Initially the residual gas occupies the volume, $V_c = V_s/7$

$$\therefore \; m' = \frac{p'V_c}{RT'} = \frac{1 \cdot 034 \times 10^5 \times V_s}{0 \cdot 296 \times 1113 \times 7 \times 10^3} = 0 \cdot 0448 V_s \text{ kg}$$

(where $T' = 840 + 273 = 1113$ K).

Also using equation 4.37,

$$m'' - m' = \sum \delta m_1 - \sum \delta m_2$$

and noting that in this example, $\sum \delta m_2 = 0$, we have:

$$m'' - m' = m_1 = \frac{1 \cdot 013 \times 10^5 \times 0 \cdot 75 V_s}{0 \cdot 2871 \times 288 \times 10^3} = 0 \cdot 919 V_s \text{ kg}$$

$$\therefore \ m'' = 0 \cdot 919 V_s + 0 \cdot 0448 V_s = 0 \cdot 9638 V_s \text{ kg}$$

Changes in kinetic and potential energy can be neglected, and the process is adiabatic (i.e. $Q = 0$), hence applying equation 4.36 we have:

$$m_1 h_1 = W + m'' u'' - m' u'$$

Also, the temperature of the mixture in the induction manifold is constant throughout the stroke, i.e. $h_1 = c_p T_1 = $ constant.

i.e. $$m_1 c_p T_1 = W + m'' c_v T'' - m' c_v T'$$

The work done is given by,

$W = $ mean pressure in cylinder during induction $\times$ swept volume
$= 0 \cdot 828 \times 10^5 \times V_s = 82\ 800 V_s \text{ N m} = 82 \cdot 8 V_s \text{ kJ}$

i.e.

$$V_s \times 1 \cdot 0051 \times 300 = 82 \cdot 8 V_s + 0 \cdot 9638 V_s \times 0 \cdot 718 \times T''$$
$$- 0 \cdot 0448 V_s \times 0 \cdot 84 \times 1113$$

(where for the mixture induced, $c_p = c_v + R = 0 \cdot 718 + 0 \cdot 2871 = 1 \cdot 0051$ kJ/kg K)

$$\therefore \ T'' = \frac{236 \cdot 1}{0 \cdot 692} = 341 \text{ K} = 68°C$$

i.e. Final temperature $= 68°C$

Then,

$$p'' = \frac{m'' R T''}{V_s \times V_c} = \frac{0 \cdot 9638 V_s \times 0 \cdot 2871 \times 341 \times 10^3}{8 V_s / 7} = 82\ 700 \text{ N/m}^2$$

i.e. Final pressure $= 0 \cdot 827$ bar

PROBLEMS

4.1 1 kg of air enclosed in a rigid container is initially at 4·8 bar and 150°C. The container is heated until the temperature is 200°C. Calculate the pressure of the air finally and the heat supplied during the process. (5·37 bar; 35·9 kJ/kg)

4.2 A rigid vessel of volume 1 m³ contains steam at 20 bar and 400°C. The vessel is cooled until the steam is just dry saturated. Calculate the mass of steam in the vessel, the final pressure of the steam, and the heat removed during the process.

(6·62 kg; 13·01 bar; 2355 kJ)

4.3 Oxygen (molar mass 32 kg/kmol) expands reversibly in a cylinder behind a piston at a constant pressure of 3 bar. The volume initially is $0·01 \, m^3$ and finally is $0·03 \, m^3$; the initial temperature is 17°C. Calculate the work done by the oxygen and the heat flow to or from the cylinder walls during the expansion. Assume oxygen to be a perfect gas and take $c_p = 0·917$ kJ/kg K.   (6 kJ; 21·16 kJ)

4.4 Steam at 7 bar, dryness fraction 0·9, expands reversibly at constant pressure until the temperature is 200°C. Calculate the work done and heat supplied per kg of steam during the process.

(38·2 kJ/kg; 288·7 kJ/kg)

4.5 0·05 m³ of a perfect gas at 6·3 bar undergoes a reversible isothermal process to a pressure of 1·05 bar. Calculate the heat flow to or from the gas.   (56·4 kJ)

4.6 Dry saturated steam at 7 bar expands reversibly in a cylinder behind a piston until the pressure is 0·1 bar. If heat is supplied continuously during the process in order to keep the temperature constant, calculate the change of internal energy per kg of steam.

(37·2 kJ/kg)

4.7 1 kg of air is compressed isothermally and reversibly from 1 bar and 30°C to 5 bar. Calculate the work done on the air and the heat flow to or from the air.   (140 kJ/kg; −140 kJ/kg)

4.8 1 kg of air at 1 bar, 15°C is compressed reversibly and adiabatically to a pressure of 4 bar. Calculate the final temperature and the work done on the air.   (155°C; 100·5 kJ/kg)

4.9 Nitrogen (molar mass 28 kg/kmol) expands reversibly in a perfectly thermally insulated cylinder from 3·5 bar, 200°C to a volume of 0·09 m³. If the initial volume occupied was 0·03 m³, calculate the work done during the expansion. Assume nitrogen to be a perfect gas and take $c_v = 0·741$ kJ/kg K.   (9·31 kJ)

4.10 A certain perfect gas is compressed reversibly from 1 bar, 17°C to a pressure of 5 bar in a perfectly thermally insulated cylinder, the final temperature being 77°C. The work done on the gas during the compression is 45 kJ/kg. Calculate γ, c_v, R, and the relative molecular mass of the gas. (1·132; 0·75 kJ/kg K; 0·099 kJ/kg K; 84)

4.11 1 kg of steam in a cylinder expands reversibly behind a piston according to a law $pv=$constant, from 7 bar to 0·75 bar. If the steam is initially dry saturated, find the temperature finally, the work done by the steam, and the heat flow to or from the cylinder walls.
(144°C; 427 kJ/kg; 430 kJ/kg)

4.12 1 kg of air at 1·02 bar, 20°C is compressed reversibly according to a law $pv^{1 \cdot 3}=$constant, to a pressure of 5·5 bar. Calculate the work done on the air and the heat flow to or from the cylinder walls during the compression. (133·5 kJ/kg; −33·38 kJ/kg)

4.13 Oxygen (molar mass 32 kg/kmol) is compressed reversibly and polytropically in a cylinder from 1·05 bar, 15°C to 4·2 bar in such a way that one-third of the work input is rejected as heat to the cylinder walls. Calculate the final temperature of the oxygen. Assume oxygen to be a perfect gas and take $c_v=0·649$ kJ/kg K. (113°C)

4.14 0·05 kg of carbon dioxide (molar mass 44 kg/kmol), occupying a volume of 0·03 m³ at 1·025 bar, is compressed reversibly until the pressure is 6·15 bar. Calculate the final temperature, the work done on the CO_2, and the heat flow to or from the cylinder walls,

  (a) When the process is according to a law $pv^{1 \cdot 4}=$constant,
  (b) When the process is isothermal,
  (c) When the process takes place in a perfectly thermally insulated cylinder.

Assume carbon dioxide to be a perfect gas, and take $\gamma=1·3$.
(270°C; 5·138 kJ; 1·713 kJ; 52·6°C; 5·51 kJ; −5·51 kJ; 219°C; 5·25 kJ; 0 kJ)

4.15 In a steam jacketed cylinder, steam expands from 5 bar to 1·2 bar according to a law $pv^{1 \cdot 05}=$constant. Assuming that the initial dryness fraction is 0·9, calculate the work done and the heat supplied per kg of steam during the expansion.
(221·8 kJ/kg; 197·5 kJ/kg)

4.16 Steam at 17 bar, dryness fraction 0·95, expands slowly in a cylinder behind a piston until the pressure is 4 bar. Calculate,

  (a) The final specific volume and the final temperature of the steam when the expansion follows the law $pv=$constant,

(b) The final specific volume and the final temperature when the working substance is air expanding according to the law $pv = $ constant between the same pressures as in part (a) and from the same initial temperature.   (0·471 m³/kg; 150°C; 0·343 m³/kg; 204·3°C)

4.17 The pressure in a steam main is 12 bar. A sample of steam is drawn off and passed through a throttling calorimeter, the pressure and temperature at exit from the calorimeter being 1 bar and 140°C respectively. Calculate the dryness fraction of the steam in the main, stating any assumptions made in the throttling process.   (0·986)

4.18 Air at 6·9 bar, 260°C is throttled to 5·5 bar before expanding through a nozzle to a pressure of 1·1 bar. Assuming that the air flows reversibly in steady flow through the nozzle, and that no heat is rejected, calculate the velocity of the air at exit from the nozzle when the inlet velocity is 100 m/s.   (637 m/s)

4.19 225 kg/h of air at 40°C enter a mixing chamber where it mixes with 540 kg/h of air at 15°C. Calculate the temperature of the air leaving the chamber, assuming steady flow conditions. Assume that the heat loss is negligible.   (22·4°C)

4.20 Steam from a superheater at 7 bar, 300°C is mixed in steady adiabatic flow with wet steam at 7 bar, dryness fraction 0·9. Calculate the mass of wet steam required per kg of superheated steam to produce steam at 7 bar, dry saturated.   (1·43 kg)

4.21 A rigid cylinder contains helium (molar mass 4 kg/kmol) at a pressure of 5 bar and a temperature of 15°C. The cylinder is now connected to a large source of helium at 10 bar and 15°C, and the valve connecting the cylinder is closed when the cylinder pressure has risen to 8 bar. Calculate the final temperature of the helium in the cylinder assuming that the heat transfer during the process is negligibly small. Take c_v for helium as 3·12 kJ/kg K.   (61·5°C)

4.22 A well-lagged vessel of volume 1 m³, containing 1·25 kg of steam at a pressure of 2·2 bar, is connected via a valve to a large source of steam at 20 bar. The valve is opened and the pressure in the vessel is allowed to rise until the steam in the vessel is just dry saturated at 4 bar and the valve is then closed. Calculate the dryness fraction of the steam supplied.   (0·905)

4.23 An air receiver contains 10 kg of air at 7 bar. A blow-off valve is opened in error and closed again within seconds, but the pressure is observed to drop to 6 bar. Calculate the mass of air which has escaped from the receiver stating clearly any assumptions made.

Calculate also the pressure of the air in the receiver some time after the valve has been closed such that the air temperature has attained its original value. (1·04 kg; 6·27 bar)

4.24 A vertical cylinder of cross-sectional area 6450 mm² is open to the atmosphere at one end and connected to a large storage vessel at the other end by means of a pipe-line and valve. A frictionless piston, of weight 100 N, is fitted into the cylinder and the initial cylinder volume is zero. The valve is then opened and air is slowly admitted from the large storage vessel into the cylinder until the piston has moved *very slowly* a distance of 0·6 m, when the valve is shut. If the temperature of the air in the cylinder is 30°C at the end of the operation and the temperature of the air in the large storage vessel is constant at 90°C, calculate:

(a) the pressure of the air in the cylinder during the process;
(b) the work done by the air in the cylinder;
(c) the work done on the piston;
(d) the heat transfer to or from the air in the cylinder during the process.

Take the atmospheric pressure as 1·013 bar.

(1·168 bar; 452 N m; 60 N m; −0·31 kJ)

5. The Second Law

In Chapter 2 it is stated that, according to the First Law of Thermodynamics, when a system undergoes a complete cycle then the net heat supplied is equal to the net work done. This is based on the conservation of energy principle, which follows from observation of natural events. The Second Law of Thermodynamics, which is also a natural law, indicates that, although the net heat supplied in a cycle is equal to the net work done, the *gross* heat supplied must be greater than the net work done; some heat must always be rejected by the system. To enable the second law to be considered more fully the heat engine must be discussed.

5.1 The heat engine

A heat engine is a system operating in a complete cycle and developing net work from a supply of heat. The second law implies that a source of heat supply and a sink for the rejection of heat are both necessary, since some heat must always be rejected by the system. A diagrammatic representation of a heat engine is shown in fig. 5.1. The heat supplied from the source is Q_1, the work done is W, and the heat rejected is Q_2. By the first law, in a complete cycle,

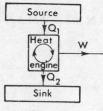

Fig. 5.1

Net heat supplied = Net work done

Then from equation 2.1, $\sum dQ = \sum dW$, we have, referring to fig. 5.1,

$$Q_1 - Q_2 = W \qquad\qquad 5.1$$

By the second law, the gross heat supplied must be greater than the net work done,

i.e. $$Q_1 > W$$

The *thermal efficiency* of a heat engine is defined as the ratio of the net work done in the cycle to the gross heat supplied in the cycle. It is usually expressed as a percentage.

Referring to fig. 5.1,

$$\text{Thermal efficiency, } \eta = \frac{W}{Q_1} \qquad 5.2$$

Substituting from equation 5.1,

$$\eta = \frac{Q_1 - Q_2}{Q_1} = 1 - \frac{Q_2}{Q_1} \qquad 5.3$$

It can be seen that the second law implies that the thermal efficiency of a heat engine must always be less than 100%.

From the definition of heat given in Section 1.1, a temperature

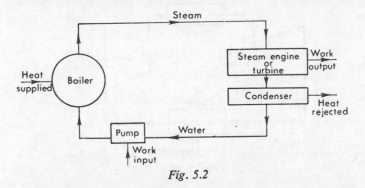

Fig. 5.2

difference is necessary for heat to flow. It follows therefore that the source of heat in fig. 5.1 must be at a higher temperature than the sink. The source can be thought of as a hot reservoir and the sink as a cold reservoir. The second law shows that a temperature difference, no matter how small, is necessary before net work can be produced in a cycle. This leads to a statement of the second law as follows:

It is impossible for a heat engine to produce net work in a complete cycle if it exchanges heat only with bodies at a single fixed temperature.

The restriction imposed by the second law is made more obvious if an attempt is made to think of a system which is not bound by the law. For instance, there is nothing in the first law to indicate that the internal energy of the sea could not be converted into mechanical work in a continuous manner. The sea represents a huge amount of energy with many millions of tonnes of water at a temperature well above absolute zero. However, no ship can be devised whose engines will run by tapping the energy of the sea. From the second law as stated above, we see that a second reservoir of energy at a lower temperature is essential before work can be developed.

One good example in practice of the heat engine as defined at the beginning of this section, is the simple steam cycle. This cycle has already been used to illustrate the first law in example 2.1. Referring to fig. 5.2, heat is supplied in the boiler, work is developed in a steam engine or turbine, heat is rejected in a condenser, and a small amount of work input is required for the pump. The hot reservoir is the furnace of the boiler, the cold reservoir is the cooling water circulating in the condenser, and the system itself is the steam.

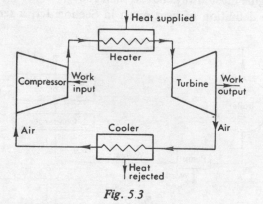

Fig. 5.3

Another example of a heat engine is the *closed cycle* gas turbine plant as shown in fig. 5.3. The system in this case is air. Heat is supplied to the air by hot gases in a heat exchanger, work is developed by the turbine, heat is rejected to cooling water in a cooler, and work is done on the air in a compressor. The hot reservoir is the hot gas circulating round the air in the heat exchanger; the cold reservoir is the cooling water circulating in the cooler.

In an *open cycle* gas turbine plant the energy is supplied by spraying fuel into the air stream in a combustion chamber; the resulting gases expand in the turbine and are then exhausted to atmosphere, (see fig. 5.4). This cycle is not a heat-engine cycle according to the definition given, since the system is not restored to its original state, and in fact undergoes a chemical change by combustion. Similarly in an internal-combustion reciprocating engine the air is mixed with fuel and burned in the cylinder, and the resulting gases after expansion are exhausted to the atmosphere. However, the open cycle gas turbine plant and the internal-combustion engine are important power producers in engineering and they are usually called heat

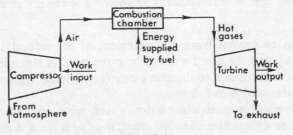

Fig. 5.4

engines. It is usually possible to neglect the mass of fuel in comparison with the mass of air, and the heat rejected may be taken as the energy of the exhaust gas less the energy of the air at inlet (i.e. the heat rejected if the exhaust were cooled to inlet conditions and then re-circulated).

The first and second laws apply equally well to cycles working in

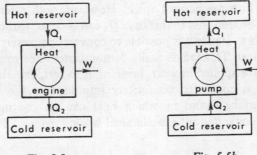

Fig. 5.5a   *Fig. 5.5b*

the reverse direction to those of the heat engine. In the case of a reversed cycle, net work is done on the system which is equal to the net heat rejected by the system. Such cycles occur in heat pumps and refrigerators. The equivalent diagrams of the heat engine and the heat pump (or refrigerator) are shown in fig. 5.5a and fig. 5.5b. In the heat pump (or refrigerator) cycle an amount of heat, Q_2, is supplied from the cold reservoir, and an amount of heat, Q_1, is rejected to the hot reservoir. By the first law we have

$$Q_1 = Q_2 + W \qquad \qquad 5.4$$

By the second law we can say that the work input is essential in order that heat be transferred from the cold to the hot reservoir, i.e. $$W > 0$$

This can be proved from the statement of the second law given previously, but the proof will not be given here. A statement of the second law in relation to the heat pump (or refrigerator) is attributed to Clausius, and is as follows:

It is impossible to construct a device that, operating in a cycle, will produce no effect other than the transfer of heat from a cooler to a hotter body.

This statement is easily verified by experience of natural processes: heat is never observed to flow from a cold body to a hot body; a refrigerator requires an input of energy in order to abstract heat from the cold chamber and reject it at a higher temperature.

When the two statements of the second law are considered, an interesting fact emerges. By reference to fig. 5.5a and the first statement of the second law it is clear that Q_2 cannot be zero. In other words, it is impossible to convert continuously a supply of heat completely into mechanical work. However, with reference to fig. 5.5b, it can be seen that in this case Q_2 can be zero, without violating the second law. Hence it is possible to convert completely mechanical work into heat. This fact is easily demonstrated. For example, when the brakes are applied in a car, bringing it to rest, the kinetic energy of the car is converted completely into heat at the wheels. No example can be found in which heat can be continuously and completely converted into mechanical work.

5.2 Entropy

In Section 2.2, an important property, internal energy, was found to arise as a consequence of the First Law of Thermodynamics. Another important property, *entropy*, follows from the second law.

Consider a reversible adiabatic process for any system on a *p-v* diagram. This is represented by line AB on fig. 5.6. Let us suppose that it is possible for the system to undergo a reversible isothermal

process at temperature T_1 from B to C and then be restored to its
original state by a second reversible adiabatic process from C to A.
Now by definition an adiabatic process is one in which no heat flows
to or from the system. Hence the only heat transferred is from B to C
during the isothermal process. The work done by the system is given
by the enclosed area (see Section 1.6). We therefore have a system
undergoing a cycle and developing net work while drawing heat
from a reservoir at one fixed temperature. This is impossible because
it violates the second law, as stated in Section 5.1. Therefore the
original supposition is wrong, and it is not possible to have two
reversible adiabatic processes passing through the same state A.

Now one of the characteristics of a property of a system is that
there is one unique line which represents a value of the property on a

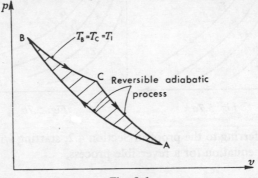

Fig. 5.6

diagram of properties. (For example, the line BC on fig. 5.6 represents
the isothermal at T_1.) Hence there must be a property represented
by a reversible adiabatic process. This property is called entropy, s.

It follows that there is no change of entropy in a reversible
adiabatic process. Each reversible adiabatic process represents a
unique value of entropy. On a p-v diagram a series of reversible
adiabatic processes appear as shown in fig. 5.7a, each line representing one value of entropy. This is similar to fig. 5.7b in which a series
of isothermals is drawn, each representing one value of temperature.

In order to be able to define entropy in terms of the other thermodynamic properties a rigorous approach is necessary. It is not the
intention of this book to deal with the basic theory in a truly rigorous
manner, and a much simplified approach has been adopted. For a
more rigorous approach references 5.1 or 5.2 should be consulted.

In Section 4.2 a reversible adiabatic process for a perfect gas was shown to follow a law $pv^\gamma = $ constant. Now the law $pv^\gamma = $ constant is a unique line on a p-v diagram, so that the proof given in Section 4.2 for a perfect gas is a similar proof to that given above (i.e. the proof that a reversible adiabatic process occupies a unique line on a diagram of properties). The proof given above depends on the second law and has been used to introduce entropy as a property. It follows therefore that the proof of $pv^\gamma = $ constant in Section 4.2 must imply the fact that the entropy does not change during a reversible adiabatic

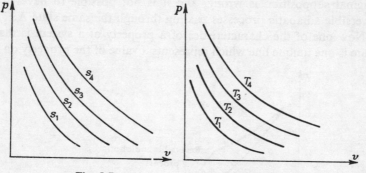

Fig. 5.7a   *Fig. 5.7b*

process. Referring to the proof in Section 4.2, starting with the non-flow energy equation for a reversible process,

$$dQ = du + p\,dv$$

and for a perfect gas,

$$dQ = c_v\,dT + RT\frac{dv}{v}$$

This equation can be integrated after dividing through by T,

i.e.
$$\frac{dQ}{T} = \frac{c_v\,dT}{T} + \frac{R\,dv}{v}$$

Also for an adiabatic process, $dQ = 0$,

i.e.
$$\frac{dQ}{T} = \frac{c_v\,dT}{T} + \frac{R\,dv}{v} = 0 \qquad 5.5$$

Now apart from mathematical manipulation and the introduction of the relationship between R, c_p, c_v, and γ, there are no other major steps in the proof. This must mean that dividing through by T is the one step which implies the restriction of the second law, and the important fact that the change of entropy is zero. We can say, therefore, $dQ/T = 0$ for a reversible adiabatic process. For any other reversible process $dQ/T \neq 0$.

This result can be shown to apply to all working substances,

i.e. $$ds = \frac{dQ}{T} \quad \text{for all working substances*} \qquad 5.6$$

(where s is entropy).

Note that since equation 5.5 is for a reversible process, then dQ in equation 5.6 is the heat added reversibly.

The change of entropy is more important than its absolute value, and the zero of entropy can be chosen quite arbitrarily. For example, in steam tables the entropy is put equal to zero at $0 \cdot 01°C$; in tables of refrigerants the entropy is put equal to zero at $-40°C$.

Integrating equation 5.6 gives

$$s_2 - s_1 = \int_1^2 \frac{dQ}{T} \qquad 5.7$$

Considering 1 kg of fluid, the units of entropy are given by kJ/kg divided by K. That is the units of specific entropy, s, are kJ/kg K. The symbol S will be used for the entropy of mass, m, of a fluid,

i.e. $$S = ms$$

Re-writing equation 5.6 we have

$$dQ = T\,ds$$

or for any reversible process

$$Q = \int_1^2 T\,ds \qquad 5.8$$

This equation is analogous to equation 1.2,

$$W = \int_1^2 p\,dv \quad \text{for any reversible process}$$

Thus, as there is a diagram on which areas represent work done in a

* The argument in this section does not constitute a proof of $ds = dQ/T$. For such a proof the reader is recommended to reference 5.1.

reversible process, there is also a diagram on which areas represent heat flow in a reversible process. These diagrams are the p-v and the T-s diagrams respectively, as shown in figs. 5.8a and 5.8b. For a reversible process 1-2 in fig. 5.8a, the shaded area $\int_1^2 p \, dv$, represents work done; for a reversible process 1-2 in fig. 5.8b, the shaded area $\int_1^2 T \, ds$, represents heat flow. Therefore one great use of the property entropy is that it enables a diagram to be drawn on which areas

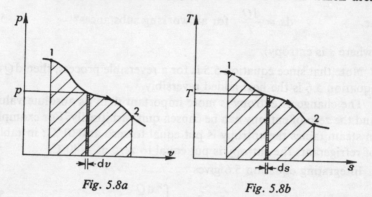

Fig. 5.8a   Fig. 5.8b

represent heat flow in a reversible process. In the next section the T-s diagram will be considered for a vapour and for a perfect gas.

5.3 The T-s diagram

(a) *For a vapour*

As mentioned earlier, the zero for entropy is taken as 0·01°C for steam and as −40°C for refrigerants. The T-s diagram for steam only will be considered here; the diagram for a refrigerant is exactly similar with the important exception of the zero of entropy. The T-s diagram for steam is shown in fig. 5.9. Three lines of constant pressure $(p_1, p_2,$ and $p_3)$ are shown (i.e. lines ABCD, EFGH, and JKLM). The pressure lines in the liquid region are practically coincident with the saturated liquid line (i.e. portions AB, EF, and JK), and the difference is usually neglected. The pressure remains constant with temperature when the latent heat is added, hence the pressure lines are horizontal in the wet region (i.e. portions BC, FG, and KL). The pressure lines curve upwards in the superheat region as shown (i.e. portions CD, GH, and LM). Thus the temperature

rises as heating continues at constant pressure. One constant volume
line (shown chain-dotted) is drawn in fig. 5.9. Lines of constant
volume are concave down in the wet region and slope up more steeply than pressure lines in the superheat region.

In steam tables the entropy of the saturated liquid and the dry
saturated vapour are represented by s_f and s_g respectively. The
difference, $s_g - s_f = s_{fg}$, is also tabulated. The entropy of wet
steam is given by the entropy of the water in the mixture plus the

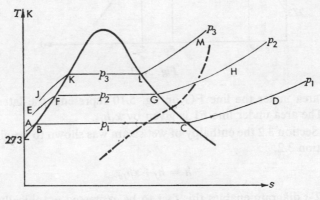

Fig. 5.9

entropy of the dry steam in the mixture. For wet steam with dryness
fraction, x, we have

$$s = (1-x)s_f + xs_g \qquad 5.9$$

or

$$s = s_f + x(s_g - s_f)$$

i.e.

$$s = s_f + xs_{fg} \qquad 5.10$$

Then the dryness fraction is given by

$$x = \frac{s - s_f}{s_{fg}} \qquad 5.11$$

It can be seen from equation 5.11 that the dryness fraction is proportional to the distance of the state point from the liquid line on a
T-s diagram. For example, for state 1 on fig. 5.10 the dryness
fraction,

$$x_1 = \frac{\text{distance F1}}{\text{distance FG}} = \frac{s_1 - s_{f_1}}{s_{fg_1}}$$

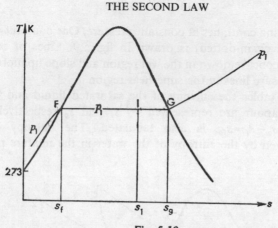

Fig. 5.10

The area under the line FG on fig. 5.10 represents the latent heat h_{fg}. The area under line F1 is given by $x_1 h_{fg}$.

In Section 3.2 the enthalpy of wet steam was shown to be given by equation 3.2,

$$h = h_f + x h_{fg}$$

The *T-s* diagram enables this fact to be expressed graphically, since areas on the diagram represent heat flow. Assuming that the pressure line in the liquid region is coincident with the saturated liquid line, then the enthalpy can be represented on the diagram. Referring to fig. 5.11, when water at any pressure *p*, at 0·01°C, is heated at constant pressure it follows the line AB approximately; the point B is at

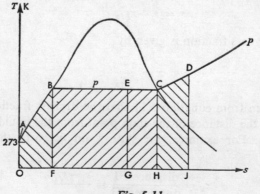

Fig. 5.11

the saturation temperature T at which the water boils at the pressure p. From equation 4.3, at constant pressure,

$$Q = h_B - h_A = h_B$$

(since h_A at $0.01°C$ is approximately zero).

Therefore we have

$$\text{area ABFOA} = h_B = h_f \quad \text{at pressure } p$$

At point B, if heating continues, the water changes gradually into steam until at C the steam is just dry saturated. Thus we have

$$\text{area BCHFB} = \text{latent heat} = h_{fg} \text{ at pressure } p = h_C - h_B$$

Then at point C the enthalpy is given by

$$h_C = \text{area ABFOA} + \text{area BCHFB} = h_g \quad \text{at pressure } p$$

For wet steam at point E,

$$h_E = h_B + x_E h_{fg}$$

i.e. $$h_E = \text{area ABEGOA}$$

When dry saturated steam is further heated it becomes superheated.

The heat added from C to D at constant pressure, p, is given by

$$Q = h_D - h_C = \text{area CDJHC}$$

Then the enthalpy at D is

$$h_D = h_C + \text{area CDJHC} = \text{area ABCDJOA}$$

Example 5.1

1 kg of steam at 7 bar, entropy 6.5 kJ/kg K, is heated reversibly at constant pressure until the temperature is $250°C$. Calculate the heat supplied, and show on a *T-s* diagram the area which represents the heat flow.

At 7 bar, $s_g = 6.709$ kJ/kg K, hence the steam is wet, since the actual entropy, s, is less than s_g.

From equation 5.11

$$x_1 = \frac{s_1 - s_{f_1}}{s_{fg_1}} = \frac{6.5 - 1.992}{4.717} = 0.955$$

Then from equation 3.2

$$h_1 = h_{f_1} + x_1 h_{fg_1} = 697 + 0.955 \times 2067$$

i.e. $$h_1 = 697 + 1975 = 2672 \text{ kJ/kg}$$

At state 2 the steam is at 250°C at 7 bar, and is therefore super-heated. From superheat tables, $h_2 = 2955$ kJ/kg.

At constant pressure from equation 4.3,

$$Q = h_2 - h_1 = 2955 - 2672 = 283 \text{ kJ/kg}$$

i.e.   Heat supplied $= 283$ kJ/kg

The *T-s* diagram showing the process is given in fig. 5.12, the shaded area representing the heat flow.

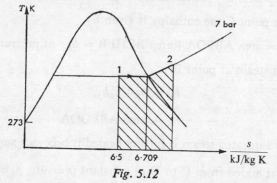

Fig. 5.12

Example 5.2

A rigid cylinder of volume 0.025 m³ contains steam at 80 bar and 350°C. The cylinder is cooled until the pressure is 50 bar. Calculate the state of the steam after cooling and the amount of heat rejected by the steam. Sketch the process on a *T-s* diagram indicating the area which represents the heat flow.

Steam at 80 bar and 350°C is superheated, and the specific volume from tables is 0.02994 m³/kg. Hence the mass of steam in the cylinder is given by

$$m = \frac{0.025}{0.02994} = 0.835 \text{ kg}$$

For superheated steam above 80 bar the internal energy is found from equation 2.7,

$$u_1 = h_1 - p_1 v_1 = 2990 - \frac{80 \times 10^5 \times 0.02994}{10^3} = 2990 - 239.5$$

i.e. $$u_1 = 2750.5 \text{ kJ/kg}$$

At state 2, $p_2 = 50$ bar and $v_2 = 0.02994$ m³/kg, therefore the steam is wet, and the dryness fraction is given by equation 3.1,

$$x_2 = \frac{v_2}{v_{g_2}} = \frac{0.02994}{0.03944} = 0.758$$

From equation 3.3

$$u_2 = (1-x_2)u_{f_2} + x_2 u_{g_2} = 0.242 \times 1149 + 0.758 \times 2597$$

i.e. $u_2 = 278 + 1969 = 2247$ kJ/kg

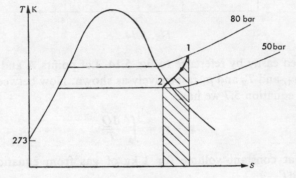

Fig. 5.13

At constant volume from equation 4.2,

$$Q = U_2 - U_1 = m(u_2 - u_1) = 0.835(2247 - 2750.5)$$

i.e. $Q = -0.835 \times 503.5 = -420$ kJ

i.e.   Heat rejected $= 420$ kJ

Fig. 5.13 shows the process drawn on a *T-s* diagram, the shaded area representing the heat rejected by the steam.

(b) *For a perfect gas*

It is useful to plot lines of constant pressure and constant volume on a *T-s* diagram for a perfect gas. Since changes of entropy are of more direct application than the absolute value, the zero of entropy can be chosen at any arbitrary reference temperature and pressure. In fig. 5.14 the pressure line p_1 and the volume line v_1 have been drawn passing through the state point 1. Note that a line of constant pressure slopes less steeply than a line of constant volume. This can

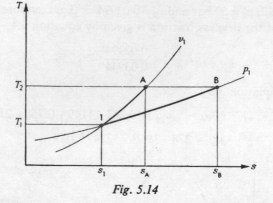

Fig. 5.14

be proved easily by reference to fig. 5.14. Let points A and B be at T_2 and v_1, and T_2 and p_1 respectively as shown. Now between 1 and A from equation 5.7 we have

$$s_A - s_1 = \int_1^A \frac{dQ}{T}$$

Also at constant volume for 1 kg of gas from equation 3.11, $dQ = c_v\,dT$.

$$\therefore\ s_A - s_1 = \int_1^A \frac{c_v\,dT}{T} = c_v\,\log_e \frac{T_A}{T_1} = c_v\,\log_e \frac{T_2}{T_1}$$

Similarly, at constant pressure for 1 kg of gas, $dQ = c_p\,dT$. Hence,

$$s_B - s_1 = \int_1^B \frac{c_p\,dT}{T} = c_p\,\log_e \frac{T_B}{T_1} = c_p\,\log_e \frac{T_2}{T_1}$$

Now since c_p is greater than c_v for any perfect gas, then $s_B - s_1$ is greater than $s_A - s_1$. Point A must therefore lie to the left of point B

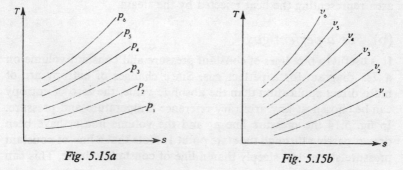

Fig. 5.15a   *Fig. 5.15b*

on the diagram, and hence a line of constant pressure slopes less steeply than a line of constant volume. Fig. 5.15a shows a series of constant pressure lines on a *T-s* diagram, and fig. 5.15b shows a series of constant volume lines on a *T-s* diagram. Note that in fig. 5.15a, $p_6 > p_5 > p_4 > p_3$, etc.; and in fig. 5.15b, $v_1 > v_2 > v_3$, etc. As the pressure rises, the temperature rises and the volume decreases; conversely as the pressure and temperature fall, the volume increases.

Example 5.3

Air at 15°C and 1·05 bar occupies 0·02 m³. The air is heated at constant volume until the pressure is 4·2 bar, and then cooled at constant pressure back to the original temperature. Calculate the net

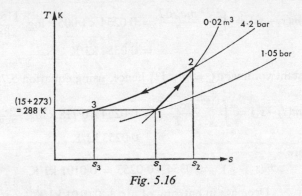

Fig. 5.16

heat flow to or from the air and the net entropy change. Sketch the process on a *T-s* diagram.

The processes are shown on a *T-s* diagram in fig. 5.16.

From equation 3.6, for a perfect gas,

$$m = \frac{pV}{RT} = \frac{1·05 \times 10^5 \times 0·02}{0·287 \times 10^3 \times 288} = 0·0254 \text{ kg}$$

(where $T_1 = 15 + 273 = 288$ K).

For a perfect gas at constant volume, $p_1/T_1 = p_2/T_2$, hence,

$$T_2 = \frac{4·2 \times 288}{1·05} = 1152 \text{ K}$$

From equation 3.13, at constant volume

$$Q = mc_v(T_2 - T_1) = 0·0254 \times 0·718(1152 - 288)$$

i.e. $Q_{1-2} = 15·75$ kJ

From equation 3.12, at constant pressure

$$Q = mc_p(T_3 - T_2) = 0.0254 \times 1.005(288 - 1152)$$

i.e. $$Q_{2-3} = -22.05 \text{ kJ}$$

$\therefore$ Net heat flow $= Q_{1-2} + Q_{2-3} = 15.75 - 22.05 = -6.3 \text{ kJ}$

i.e.   Heat rejected $= 6.3 \text{ kJ}$

Referring to fig. 5.16,

Net decrease in entropy $= s_1 - s_3 = (s_2 - s_3) - (s_2 - s_1)$

At constant pressure, $dQ = mc_p \, dT$, hence, using equation 5.7,

$$m(s_2 - s_3) = \int_{288}^{1152} \frac{mc_p \, dT}{T} = 0.0254 \times 1.005 \times \log_e \frac{1152}{288}$$

$$= 0.0354 \text{ kJ/K}$$

At constant volume, $dQ = mc_v \, dT$, hence, using equation 5.7,

$$m(s_2 - s_1) = \int_{288}^{1152} \frac{mc_v \, dT}{T} = 0.0254 \times 0.718 \times \log_e \frac{1152}{288}$$

$$= 0.0253 \text{ kJ/K}$$

Therefore,

$$m(s_1 - s_3) = 0.0354 - 0.0253 = 0.0101 \text{ kJ/K}$$

i.e.   Decrease in entropy of air $= 0.0101 \text{ kJ/K}$

Note that since entropy is a property, the decrease in entropy in example 5.3, given by $s_1 - s_3$, is independent of the processes undergone between states 1 and 3. The change $s_1 - s_3$ can also be found by imagining a reversible isothermal process taking place between 1 and 3. The isothermal process on the $T\text{-}s$ diagram will be considered in the next section.

5.4 Reversible processes on the $T\text{-}s$ diagram

The various reversible processes dealt with in Chapter 4 will now be considered in relation to the $T\text{-}s$ diagram. The constant volume and constant pressure processes have been represented on the $T\text{-}s$ diagram in Section 5.3, and will therefore not be discussed again in this section.

Reversible isothermal process

A reversible isothermal process will appear as a straight line on a *T-s* diagram, and the area under the line must represent the heat flow during the process. For example, fig. 5.17 shows a reversible isothermal expansion of wet steam into the superheat region. The shaded area represents the heat supplied during the process,

i.e. Heat supplied $= T(s_2 - s_1)$

Note that the absolute temperature must be used. The temperature tabulated in steam tables is $t°C$, and care must be taken to convert this into TK.

When the isothermal process for a vapour was considered in Section 4.1, no method was then available for the evaluation of the

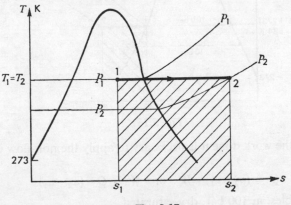

Fig. 5.17

heat flow. The introduction of the *T-s* diagram enables the heat flow to be found, as shown in the following example.

Example 5.5

Dry saturated steam at 100 bar expands isothermally and reversibly to a pressure of 10 bar. Calculate the heat supplied and the work done per kg of steam during the process.

The process is shown in fig. 5.18, the shaded area representing the heat supplied.

From tables at 100 bar, dry saturated

$$s_1 = s_g = 5 \cdot 615 \text{ kJ/kg K} \quad \text{and} \quad t_1 = 311°C$$

At 10 bar and 311°C the steam is superheated, hence interpolating

$$s_2 = 7\cdot124 + \left(\frac{311-300}{350-300}\right)(7\cdot301 - 7\cdot124)$$

i.e.
$$s_2 = 7\cdot124 + 0\cdot039 = 7\cdot163 \text{ kJ/kg K}$$

Then we have,

$$\text{Heat supplied} = \text{shaded area} = T(s_2 - s_1)$$
$$= 584(7\cdot163 - 5\cdot615) = 584 \times 1\cdot548$$

(where $T = 311 + 273 = 584$ K).

i.e.
$$\text{Heat supplied} = 584 \times 1\cdot548 = 904 \text{ kJ/kg}$$

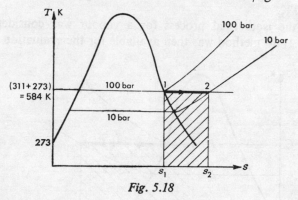

Fig. 5.18

To find the work done it is necessary to apply the non-flow energy equation,

i.e.
$$Q = (u_2 - u_1) + W \quad \text{or} \quad W = Q - (u_2 - u_1)$$

From tables, at 100 bar, dry saturated,

$$u_1 = u_g = 2545 \text{ kJ/kg}$$

At 10 bar and 311°C, interpolating,

$$u_2 = 2794 + \left(\frac{311-300}{350-300}\right)(2875 - 2794)$$
$$= 2794 + 17\cdot8$$

i.e.
$$u_2 = 2811\cdot8 \text{ kJ/kg}$$

Then,

$$W = Q - (u_2 - u_1)$$
$$= 904 - (2811\cdot8 - 2545)$$
$$= 904 - 266\cdot8$$

i.e.
$$W = 637\cdot2 \text{ kJ/kg}$$

i.e.
$$\text{Work done by the steam} = 637\cdot2 \text{ kJ/kg}$$

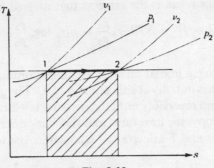

Fig. 5.19

A reversible isothermal process for a perfect gas is shown on a T-s diagram in fig. 5.19. The shaded area represents the heat supplied during the process,

i.e. $$Q = T(s_2 - s_1)$$

For a perfect gas undergoing an isothermal process it is possible to evaluate $s_2 - s_1$. From the non-flow equation, 2.4, we have, for a reversible process,

$$dQ = du + p\,dv$$

Also for a perfect gas from Joule's law $du = c_v\,dT$,

i.e. $$dQ = c_v\,dT + p\,dv$$

For an isothermal process, $dT = 0$, hence,

$$dQ = p\,dv$$

Then, since $pv = RT$, we have

$$dQ = RT\frac{dv}{v}$$

Now from equation 5.7

$$s_2 - s_1 = \int_1^2 \frac{dQ}{T} = \int_{v_1}^{v_2} \frac{RT\,dv}{Tv} = R\int_{v_1}^{v_2} \frac{dv}{v}$$

i.e. $$s_2 - s_1 = R\log_e\frac{v_2}{v_1} = R\log_e\frac{p_1}{p_2} \qquad 5.12$$

Therefore the heat supplied is given by

$$Q = T(s_2 - s_1) = RT\log_e\frac{v_2}{v_1} = RT\log_e\frac{p_1}{p_2}$$

Note that this result is the same as that derived in section 4.1,

i.e. $Q = W = RT \log_e \dfrac{p_1}{p_2} = p_1 v_1 \log_e \dfrac{p_1}{p_2}$,   etc.

Example 5.6

0.03 m³ of nitrogen (molar mass 28 kg/kmol) contained in a cylinder behind a piston is initially at 1.05 bar and 15°C. The gas is compressed isothermally and reversibly until the pressure is 4.2 bar. Calculate the change of entropy, the heat flow, and the work done, and sketch the process on a p-v and T-s diagram. Assume nitrogen to act as a perfect gas.

The process is shown on a p-v and a T-s diagram in figs. 5.20a and

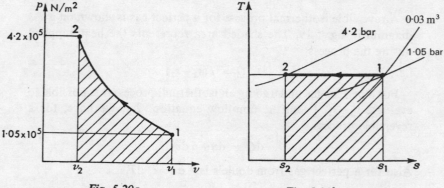

Fig. 5.20a   Fig. 5.20b

5.20b, respectively. The shaded area on fig. 5.20a represents work input, and the shaded area on fig. 5.20b represents heat rejected.

From equation 3.9

$$R = \frac{R_0}{M} = \frac{8314}{28} = 297 \text{ N m/kg K}$$

Then, since $pV = mRT$, we have

$$m = \frac{pV}{RT} = \frac{1.05 \times 10^5 \times 0.03}{297 \times 288} = 0.0368 \text{ kg}$$

(where $T = 15 + 273 = 288$ K).

Then from equation 5.12, for m kg,

$$S_2 - S_1 = mR \log_e \frac{p_1}{p_2} = \frac{0.0368 \times 297}{10^3} \log_e \frac{1.05}{4.2}$$

i.e. $S_2 - S_1 = -\dfrac{0.0368 \times 297}{10^3} \log_e \dfrac{4.2}{1.05} = -0.01516 \text{ kJ/K}$

∴ Decrease in entropy, $S_1 - S_2 = 0.01516$ kJ/K

Heat rejected = shaded area on fig. 5.20b = $T(S_1 - S_2)$
$$= 288 \times 0.01516 = 4.37 \text{ kJ}$$

Then for an isothermal process for a perfect gas, from equation 4.12,
$$W = Q = 4.37 \text{ kJ}$$
i.e.   Work input = 4.37 kJ

Reversible adiabatic process (or isentropic process)

For a reversible adiabatic process the entropy remains constant, and hence the process is called an *isentropic process*. Note that for a process to be isentropic it need not be either adiabatic or reversible, but the process will always appear as a vertical line on a *T-s* diagram. Cases in which an isentropic process is not both adiabatic and reversible occur infrequently and will be ignored throughout this book.

An isentropic process for superheated steam expanding into the wet region is shown in fig. 5.21. When the reversible adiabatic process was considered in Section 4.1, it was stated that no simple method was available for fixing the end states. Now, using the fact that the entropy remains constant, the end states can be found easily from tables. This is illustrated in the following example.

Example 5.7

Steam at 100 bar, 375°C expands isentropically in a cylinder behind a piston to a pressure of 10 bar. Calculate the work done per kg of steam.

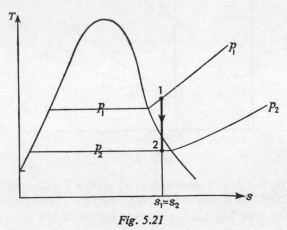

Fig. 5.21

From superheat tables, at 100 bar, 375°C, we have

$$s_1 = s_2 = 6.091 \text{ kJ/kg K}$$

At 10 bar and $s_2 = 6.091$, the steam is wet, since s_2 is less than s_{g_2}. Then from equation 5.11

$$x_2 = \frac{s_2 - s_{f_2}}{s_{fg_2}} = \frac{6.091 - 2.138}{4.448} = 0.889$$

Then from equation 3.3

$$u_2 = (1 - x_2)u_{f_2} + x_2 u_{g_2} = (0.111 \times 762) + (0.889 \times 2584)$$

i.e. $$u_2 = 84.6 + 2297 = 2381.6 \text{ kJ/kg}$$

At 100 bar, 375°C, we have from tables, $h_1 = 3017$ kJ/kg, and $v_1 = 0.02453$ m³/kg. Then using equation 2.7

$$u_1 = h_1 - p_1 v_1 = 3017 - \frac{100 \times 10^5 \times 0.02453}{10^3} = 3017 - 245.3$$

i.e. $$u_1 = 2771.7 \text{ kJ/kg}$$

For an adiabatic process from equation 4.13

i.e. $$W = u_1 - u_2$$

$$\therefore \text{ Work done by the steam} = 2771.7 - 2381.6$$
$$= 390.1 \text{ kJ/kg}$$

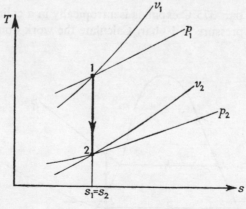

Fig. 5.22

For a perfect gas an isentropic process on a T-s diagram is shown in fig. 5.22. It was shown in Section 4.1 that for a reversible adiabatic process for a perfect gas the process follows a law $pv^\gamma = $ constant.

Since a reversible adiabatic process occurs at constant entropy, and is known as an isentropic process, the index γ, is known as the *isentropic index* of the gas.

Polytropic process

To find the change of entropy in a polytropic process for a vapour when the end states have been fixed using $p_1v_1^n = p_2v_2^n$, the entropy values at the end states can be read straight from tables.

Example 5.8

In a steam engine the steam at the beginning of the expansion process is at 7 bar, dryness fraction 0·95, and the expansion follows the law

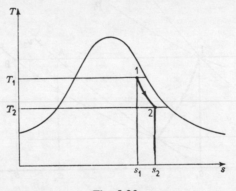

Fig. 5.23

$pv^{1.1} = $ constant, down to a pressure of 0·34 bar. Calculate the change of entropy per kg of steam during the process.

(Note that this is the data of example 4.6.)

At 7 bar, $v_g = 0.2728 \text{ m}^3/\text{kg}$, then from equation 3.1

$$v_1 = x_1 v_{g_1} = 0.95 \times 0.2728 = 0.26 \text{ m}^3/\text{kg}$$

Then from equation 4.25

$$\frac{p_1}{p_2} = \left(\frac{v_2}{v_1}\right)^{1.1} \quad \text{or} \quad \frac{v_2}{v_1} = \left(\frac{p_1}{p_2}\right)^{1/1.1}$$

$$\therefore \ v_2 = 0.26 \times \left(\frac{7}{0.34}\right)^{0.909} = 0.26 \times 20.59^{0.909} = 4.06 \text{ m}^3/\text{kg}$$

At 0·34 bar, and $v_2 = 4.06 \text{ m}^3/\text{kg}$, the steam is wet, since $v_g = 4.649$.

From equation 3.1

$$x_2 = \frac{v_2}{v_{g_2}} = \frac{4 \cdot 06}{4 \cdot 649} = 0 \cdot 876$$

Then from equation 5.10

$$s_1 = s_{f_1} + x_1 s_{fg_1} = 1 \cdot 992 + 0 \cdot 95 \times 4 \cdot 717 = 6 \cdot 472 \text{ kJ/kg K}$$

and

$$s_2 = s_{f_2} + x_2 s_{fg_2} = 0 \cdot 98 + 0 \cdot 876 \times 6 \cdot 745 = 6 \cdot 889 \text{ kJ/kg K}$$

$$\therefore \text{ Increase in entropy } (s_2 - s_1) = 6 \cdot 889 - 6 \cdot 472$$
$$= 0 \cdot 417 \text{ kJ/kg K}$$

The process is shown on a T-s diagram in fig. 5.23.

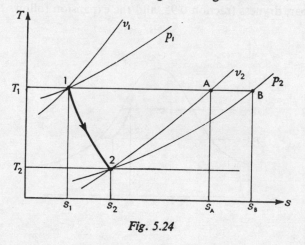

Fig. 5.24

It was shown in Section 4.1 that the polytropic process is the general case for a perfect gas. To find the entropy change for a perfect gas in the general case, consider the non-flow energy equation for a reversible process, equation 2.4,

$$dQ = du + p \, dv$$

Also for unit mass of a perfect gas from Joule's law $du = c_v \, dT$, and from equation 3.5, $pv = RT$.

$$\therefore \quad dQ = c_v \, dT + \frac{RT \, dv}{v}$$

Then from equation 5.6

$$ds = \frac{dQ}{T} = \frac{c_v \, dT}{T} + \frac{R \, dv}{v}$$

Hence between any two states 1 and 2

$$s_2 - s_1 = c_v \int_{T_1}^{T_2} \frac{dT}{T} + R \int_{v_1}^{v_2} \frac{dv}{v} = c_v \log_e \frac{T_2}{T_1} + R \log_e \frac{v_2}{v_1} \qquad 5.13$$

This can be illustrated on a T-s diagram as in fig. 5.24. Since in the process in fig. 5.24, $T_2 < T_1$, then it is more convenient to write

$$s_2 - s_1 = R \log_e \frac{v_2}{v_1} - c_v \log_e \frac{T_1}{T_2} \qquad 5.14$$

The first part of the expression for $s_2 - s_1$ in equation 5.14 is the change of entropy in an isothermal process from v_1 to v_2,

i.e. from equation 5.12

$$s_A - s_1 = R \log_e \frac{v_2}{v_1} \quad \text{(see fig. 5.24)}$$

Also the second part of the expression for $s_2 - s_1$ in equation 5.14 is the change of entropy in a constant volume process from T_1 to T_2,

i.e. referring to fig. 5.24,

$$s_A - s_2 = c_v \log_e \frac{T_1}{T_2}$$

It can be seen therefore that in calculating the entropy change in a polytropic process from state 1 to state 2 we have in effect replaced the process by two simpler processes; from 1 to A and then from A to 2. It is clear from fig. 5.24 that

$$s_2 - s_1 = (s_A - s_1) - (s_A - s_2)$$

Any two processes can be chosen to replace a polytropic process in order to find the entropy change. For example, going from 1 to B and then from B to 2 as in fig. 5.24, we have

$$s_2 - s_1 = (s_B - s_1) - (s_B - s_2)$$

At constant temperature between p_1 and p_2, using equation 5.12,

$$s_B - s_1 = R \log_e \frac{p_1}{p_2}$$

and at constant pressure between T_1 and T_2 we have

$$s_B - s_2 = c_p \log_e \frac{T_1}{T_2}$$

Hence,

$$s_2 - s_1 = R \log_e \frac{p_1}{p_2} - c_p \log_e \frac{T_1}{T_2}$$

or $$s_2 - s_1 = c_p \log_e \frac{T_2}{T_1} + R \log_e \frac{p_1}{p_2} \qquad 5.15$$

Equation 5.15 can also be derived easily from equation 5.13. There are obviously a large number of possible equations for the change of entropy in a polytropic process, and it is stressed that no attempt should be made to memorize all such expressions. Each problem can be dealt with by sketching the T-s diagram and replacing the process by two other simpler reversible processes, as in fig. 5.24.

Example 5.9

Calculate the change of entropy of 1 kg of air expanding polytropically in a cylinder behind a piston from 6·3 bar and 550°C to 1·05 bar. The index of expansion is 1·3.

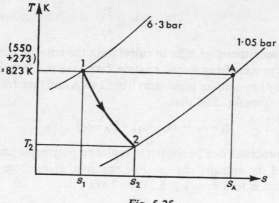

Fig. 5.25

The process is shown on a T-s diagram in fig. 5.25. From equation 4.29,

$$\frac{T_1}{T_2} = \left(\frac{p_1}{p_2}\right)^{(n-1)/n} = \left(\frac{6\cdot3}{1\cdot05}\right)^{(1\cdot2-1)/1\cdot3} = 6^{0\cdot231} = 1\cdot512$$

$$\therefore \ T_2 = \frac{823}{1\cdot512} = 544 \text{ K}$$

(where $T_1 = 550 + 273 = 823$ K).

Now replace the process 1 to 2 by two processes, 1 to A and A to 2. Then at constant temperature from 1 to A, from equation 5.12,

$$s_A - s_1 = R \log_e \frac{p_1}{p_2} = 0.287 \log_e \frac{6.3}{1.05}$$

$$= 0.287 \times 1.792 = 0.515 \text{ kJ/kg K}$$

At constant pressure from A to 2

$$s_A - s_2 = c_p \log_e \frac{T_1}{T_2} = 1.005 \log_e \frac{823}{544}$$

$$= 1.005 \times 0.413 = 0.415$$

Then $s_2 - s_1 = 0.515 - 0.415 = 0.1 \text{ kJ/kg K}$

i.e.   Increase in entropy $= 0.1 \text{ kJ/kg K}$

Note that if in this problem $s_A - s_2$ happened to be greater than

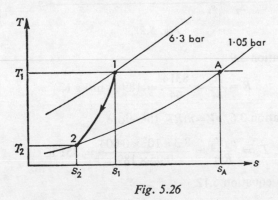

Fig. 5.26

$s_A - s_1$, this would mean that s_1 was greater than s_2, and the process should appear as in fig. 5.26.

Example 5.10

0·05 kg of carbon dioxide (molar mass 44 kg/kmol) is compressed from 1 bar, 15°C, until the pressure is 8·3 bar, and the volume is then 0.004 m³. Calculate the change of entropy. Take c_p for carbon dioxide as 0·88 kJ/kg K, and assume carbon dioxide to be a perfect gas.

The two end states are marked on a T-s diagram in fig. 5.27. The process is not specified in the example and no information about it is necessary. States 1 and 2 are fixed and hence $s_2 - s_1$ is fixed. The process between 1 and 2 could be reversible or irreversible; the change of entropy is the same between the end states given. With reference to fig. 5.27, to find $s_1 - s_2$ we can first find $s_A - s_2$ and then subtract $s_A - s_1$. First of all it is necessary to find R and then T_2.

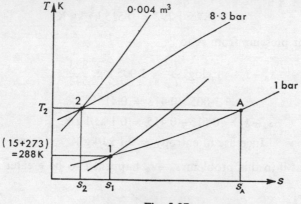

Fig. 5.27

From equation 3.9

$$R = \frac{R_0}{M} = \frac{8314}{44} = 189 \text{ N m/kg K}$$

From equation 3.6, $pV = mRT$, therefore,

$$T_2 = \frac{p_2 V_2}{Rm} = \frac{8\cdot3 \times 10^5 \times 0\cdot004}{0\cdot05 \times 189} = 351 \text{ K}$$

Then from equation 5.12

$$s_A - s_2 = R \log_e \frac{p_2}{p_A} = 0\cdot189 \log_e \frac{8\cdot3}{1} = 0\cdot4 \text{ kJ/kg K}$$

Also at constant pressure from 1 to A

$$s_A - s_1 = c_p \log_e \frac{T_2}{T_1} = 0\cdot88 \log_e \frac{351}{288} = 0\cdot174 \text{ kJ/kg K}$$

(where $T_1 = 15 + 273 = 288$ K).

Then

$$s_1 - s_2 = 0\cdot4 - 0\cdot174 = 0\cdot226 \text{ kJ/kg K}$$

Hence for 0·05 kg of carbon dioxide,

Decrease in entropy $= 0\cdot05 \times 0\cdot226 = 0\cdot0113 \text{ kJ/K}$

5.5 Entropy and irreversibility

In the previous section it was pointed out that, since entropy is a property, the change of entropy depends only on the end states and not on the process between the end states. Therefore, provided an irreversible process gives enough information to fix the end states then the change of entropy can be found. This can best be illustrated by some examples.

Example 5.11

Steam at 7 bar, dryness fraction 0·96, is throttled down to 3·5 bar. Calculate the change of entropy per kg of steam.

At 7 bar, dryness fraction 0·96, using equation 5.10 we have

$$s_1 = s_{f_1} + x_1 s_{fg_1} = 1\cdot992 + 0\cdot96 \times 4\cdot717$$

i.e.
$$s_1 = 6\cdot522 \text{ kJ/kg K}$$

In Section 4.4 it was shown that for a throttling process, $h_1 = h_2$. From equation 3.2

$$h_2 = h_1 = h_{f_1} + x_1 h_{fg_1} = 697 + 0\cdot96 \times 2067 = 2682 \text{ kJ/kg}$$

At 3·5 bar and $h_2 = 2682$ kJ/kg the steam is still wet, since $h_{g_2} > h_2$. From equation 3.2, $h_2 = h_{f_2} + x_2 . h_{fg_2}$, therefore,

$$x_2 = \frac{h_2 - h_{f_2}}{h_{fg_2}} = \frac{2682 - 584}{2148} = 0\cdot977$$

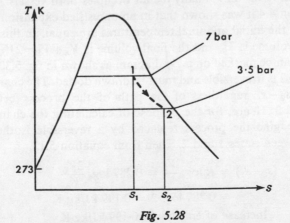

Fig. 5.28

Then from equation 5.10

$$s_2 = s_{f_2} + x_2 s_{fg_2} = 1 \cdot 727 + 0 \cdot 977 \times 5 \cdot 214 = 6 \cdot 817 \text{ kJ/kg K}$$

Therefore,

$$\text{Increase of entropy} = 6 \cdot 817 - 6 \cdot 522 = 0 \cdot 295 \text{ kJ/kg K}$$

The process is shown on a T-s diagram in fig. 5.28. Note that the process is shown dotted, and the area under the line does *not* represent heat flow; a throttling process assumes no heat flow, but there is a change in entropy because the process is irreversible.

Example 5.12

Two vessels of equal volume are connected by a short length of pipe containing a valve; both vessels are well lagged. One vessel contains air and the other is completely evacuated. Calculate the change of entropy per kg of air in the system when the valve is opened and the air is allowed to fill both vessels.

Initially the vessel A contains air and the vessel B is completely

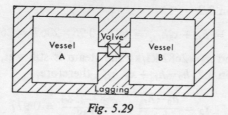

Fig. 5.29

evacuated, as in fig. 5.29; finally the air occupies both vessels A and B. In Section 4.4 it was shown that in an unresisted expansion for a perfect gas, the initial and final temperatures are equal. In this case, the initial volume is V_A and the final volume is $V_A + V_B = 2V_A$. The end states can be marked on a T-s diagram as shown in fig. 5.30. The process 1 to 2 is irreversible and must be drawn dotted. The change of entropy is $s_2 - s_1$, regardless of the path of the process between states 1 and 2. Hence, for the purpose of calculating the change of entropy, imagine the process replaced by a reversible isothermal process between states 1 and 2. Then from equation 5.12,

$$(s_2 - s_1) = R \log_e \frac{V_2}{V_1} = 0 \cdot 287 \log_e \frac{2V_A}{V_A}$$
$$= 0 \cdot 287 \log_e 2 = 0 \cdot 199 \text{ kJ/kg K}$$

i.e.
$$\text{Increase of entropy} = 0 \cdot 199 \text{ kJ/kg K}$$

Note that the process is drawn dotted in fig. 5.30, and the area under the line has no significance; the process is adiabatic and there is a change in entropy since the process is irreversible.

It is important to remember that the equation 5.6, $ds = dQ/T$, is true only for reversible processes. In the same way, the equation $dW = p \, dv$, or $dv = dW/p$, is true only for reversible processes. In example 5.12 the volume of the air increased from V_A to $2V_A$, and yet no work was done by the air during the process,

i.e. $$dW = 0 \quad \text{yet} \quad v_2 + v_1 = 2V_A - V_A = V_A$$

Hence in the irreversible process of example 5.12, $dv \neq dW/p$. Similarly, the entropy in example 5.12 increased by 0.199 kJ/kg K and yet the heat flow was zero, i.e. $ds \neq dQ/T$. No confusion should

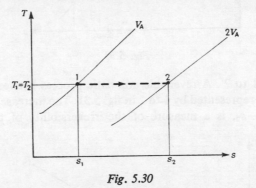

Fig. 5.30

be caused if the T-s and/or the p-v diagram is drawn for each problem and the state points marked in their correct positions. Then, when a process between two states is reversible, the lines representing the process can be drawn in as full lines, and the area under the line represents heat flow on the T-s diagram and represents work done on the p-v diagram. When the process between the states is irreversible, the line must be drawn dotted, and the area under the line has no significance on either diagram.

It can be shown from the second law that the entropy of a thermally isolated system must either increase or remain the same. For instance, a system undergoing an adiabatic process is thermally isolated from its surroundings since no heat flows to or from the system. We have seen that in a reversible adiabatic process the entropy remains the same. In an irreversible adiabatic process the entropy must always increase, and the gain of entropy is a measure of the irreversibility of the process. The processes in examples 5.11 and 5.12 illustrate this fact. As another example, consider an irreversible adiabatic expansion in a steam turbine as shown in fig. 5.31

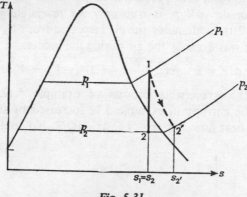

Fig. 5.31

as process 1 to 2′. A reversible adiabatic process between the same pressures is represented by 1 to 2 in fig. 5.31. The increase of entropy, $s_{2'} - s_1 = s_{2'} - s_2$, is a measure of the irreversibility of the process.

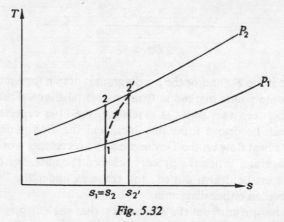

Fig. 5.32

Similarly, in fig. 5.32, an irreversible adiabatic compression in a rotary compressor is shown as process 1 to 2′. A reversible adiabatic process between the same pressures is represented by 1 to 2. As before, the increase of entropy shows the irreversibility of the process.

Example 5.13

In an air turbine the air expands from 6·8 bar and 430°C to 1·013 bar and 150°C. The heat loss from the turbine can be assumed to be negligible. Show that the process is irreversible, and calculate the change of entropy per kg of air.

Since the heat loss is negligible, the process is adiabatic. For a reversible adiabatic process for a perfect gas, using equation 4.21,

$$\frac{T_1}{T_2} = \left(\frac{p_1}{p_2}\right)^{(\gamma-1)/\gamma}$$

i.e. $$\frac{703}{T_2} = \left(\frac{6·8}{1·013}\right)^{(1·4-1)/1·4}$$

(where $T_1 = 430 + 273 = 703$ K)

i.e. $T_2 = \dfrac{703}{6·71^{0·286}} = \dfrac{703}{1·724} = 408$ K $= 408 - 273 = 135$°C

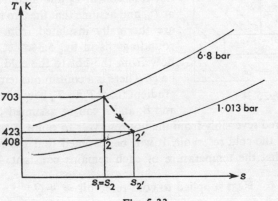

Fig. 5.33

But the actual temperature is 150°C at the pressure of 1·013 bar, hence the process is irreversible. The process is shown as 1 to 2′ in fig. 5.33; the ideal isentropic process 1 to 2 is also shown. It is not possible for process 1 to 2′ to be reversible, because in that case the area under line 1–2′ would represent heat flow and yet the process is adiabatic.

The change of entropy, $s_{2'} - s_1$, can be found by considering a reversible constant pressure process between 2 and 2′. Then from equation 5.6, $ds = dQ/T$, and at constant pressure for 1 kg of a perfect gas we have, $dQ = c_p \, dT$, therefore,

$$s_{2'} - s_2 = \int_2^{2'} \frac{c_p \, dT}{T} = c_p \log_e \frac{T_{2'}}{T_2}$$

$$= 1 \cdot 005 \log_e \frac{423}{408} = 0 \cdot 0355 \text{ kJ/kg K}$$

i.e.   Increase of entropy, $s_{2'} - s_1 = 0 \cdot 0355$ kJ/kg K

Consider now the case when a system is not thermally isolated from its surroundings. The entropy of such a system can increase, decrease, or remain the same, depending on the heat crossing the boundary. However, if the boundary is extended to include the source or sink of heat with which the system is in communication, then the entropy of this new system must either increase or remain the same. To illustrate this, consider a hot reservoir at T_1 and a cold reservoir at T_2, and assume that the two reservoirs are thermally insulated from the surroundings as in fig. 5.34. Let the heat flow from the hot to the cold reservoir be Q. There is a continuous temperature gradient from T_1 to T_2 between points A and B, and it can be assumed that heat is transferred reversibly from the hot reservoir to point A, and from point B to the cold reservoir. It will be assumed that the reservoirs are such that the temperature of each remains constant. Then we have

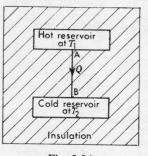

Fig. 5.34

Heat supplied to cold reservoir $= + Q$

Hence from equation 5.7,

Increase of entropy of cold reservoir $= + \dfrac{Q}{T_2}$

Also,

Heat supplied to hot reservoir $= - Q$

$\therefore$ Increase of entropy of hot reservoir $= - \dfrac{Q}{T_1}$

i.e.   Net increase of entropy of system, $\Delta s = \left(\dfrac{Q}{T_2} - \dfrac{Q}{T_1} \right)$

Since $T_1 > T_2$ it can be seen that Δs is positive, and hence the entropy of the system must increase. In the limit when the difference in temperature is infinitely small, then $\Delta s = 0$. This confirms the principle that the entropy of an isolated system must either increase or remain the same. In Section 1.4, criterion (c) for reversibility was stated as follows:

The difference in temperature between a system and its surroundings must be infinitely small during a reversible process.

In the above example, when $T_1 > T_2$, then the heat flow between the reservoirs is irreversible by the above criterion. Thus the entropy of the system increases when the heat flow process is irreversible but remains the same when the process is reversible. The increase of

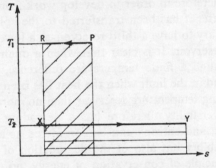

Fig. 5.35

entropy is a measure of the irreversibility. The processes occurring in the above example can be drawn on a T-s diagram as in fig. 5.35. The two processes have been superimposed on the same diagram. Process P–R represents the transfer of Q units of heat from the hot reservoir, and the area under P–R is equal to Q. Process X–Y represents the transfer of Q units of heat to the cold reservoir, and the area under X–Y is equal to Q. The area under P–R is equal to the area under X–Y, and hence it can be seen from the diagram that the entropy of the cold reservoir must always increase more than the entropy of the hot reservoir decreases. Thus the entropy of the combined system must increase. Note that, since in this example both processes P–R and X–Y are reversible, then the irreversibility occurs between A and B on fig. 5.34. That is, the irreversibility is caused by the heat transfer process between A and B. Whenever heat is transferred through a finite temperature difference the process is irreversible and there is an increase of entropy of the system and its surroundings.

In certain processes the irreversibility may occur in the surroundings, then the process is internally reversible, and areas on the p-v and T-s diagrams approximate closely to the work done and heat flow respectively. Internal reversibility was mentioned earlier, in Section 1.5. In most problems when a process is assumed to be reversible it is internal reversibility which is implied. Conversely, most processes in practice which are said to be irreversible, are internally irreversible due to eddying and churning of the working fluid, as in example 5.13.

Referring to fig. 5.34, if a heat engine were interposed between the hot and cold reservoirs, some work could be developed. The second law states that heat can never flow unaided from a cold reservoir to a hot reservoir, therefore in order to develop work from the quantity of energy, Q, after it has been transferred to the cold reservoir, it would be necessary to have a third reservoir at a lower temperature than the cold reservoir. It is clear that when a quantity of heat is transferred through a finite temperature difference, its usefulness becomes less, and in the limit when the heat has been transferred to the lowest existing temperature reservoir then no more work can be developed. Irreversibility therefore has a degrading effect on the energy available, and entropy can be considered as a measure not only of irreversibility but also of the degradation of energy. Note that, by the principle of conservation of energy, no energy can be destroyed; by the second law of thermodynamics, energy can only become less useful and never more useful. Systems tend naturally to states of lower grade energy; any system moving to a state of higher grade energy without an external supply of energy would be violating the second law. The second law can be seen to imply a direction or a gradient of usefulness of energy. Work is more useful than heat; the higher the temperature of a reservoir of energy, the more useful is the amount of energy available. Applying this latter conclusion to a heat engine it can be deduced that, for a given cold reservoir (e.g. the atmosphere), then the higher the temperature of the hot reservoir, the higher will be the thermal efficiency of the heat engine. This will be discussed more fully in the next chapter.

5.6 Availability

The theoretical maximum amount of work which can be obtained from a system at any state p_1 and T_1 when operating with a reservoir at the constant pressure and temperature p_0 and T_0 is called the *availability*.

(a) *Non-flow systems*

Consider a system consisting of a fluid in a cylinder behind a piston, the fluid expanding reversibly from initial conditions of p_1 and T_1 to final atmospheric conditions of p_0 and T_0. Imagine also that the system works in conjunction with a reversible heat engine which receives heat reversibly from the fluid in the cylinder such that the working substance of the heat engine follows the cycle 01A0 as shown in figures 5.36a and 5.36b, where $s_1 = s_A$ and $T_0 = T_A$*. The work done by this engine is given by:

$$W_{\text{Engine}} = \text{Heat supplied} - \text{Heat rejected}$$

$$= Q - T_0(s_1 - s_0)$$

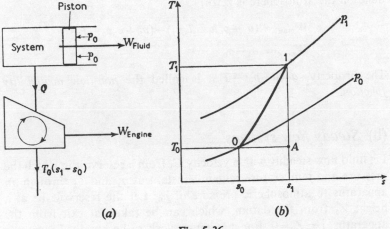

(a)   *(b)*

Fig. 5.36

The heat supplied to the engine is equal to the heat rejected by the

---

* The only possible way in which this could occur would be if an infinite number of reversible heat engines were arranged in parallel, each operating on a Carnot cycle (see Chapter 6), each one receiving heat at a different constant temperature and each one rejecting heat at T_0.

fluid in the cylinder. Therefore for the fluid in the cylinder under-going the process 1 to 0, we have,

$$-Q = (u_0 - u_1) + W_{\text{Fluid}}$$

i.e.
$$W_{\text{Fluid}} = (u_1 - u_0) - Q$$

Therefore adding the two equations,

$$W_{\text{Fluid}} + W_{\text{Engine}} = (u_1 - u_0) - T_0(s_1 - s_0)$$

The work done by the fluid on the piston is less than the total work done by the fluid, since there is work done on the atmosphere which is at the constant pressure p_0 (see Problem 4.24).

i.e.   Work done on atmosphere $= p_0(v_0 - v_1)$

Hence,

Maximum work available $= (u_1 - u_0) - T_0(s_1 - s_0) - p_0(v_0 - v_1)$

(Note: when a fluid undergoes a complete cycle then the net work done on the atmosphere is zero.)

$$W_{\text{max}} = (u_1 + p_0 v_1 - T_0 s_1) - (u_0 + p_0 v_0 - T_0 s_0)$$
$$\therefore\ W_{\text{max}} = a_1 - a_0$$

The property, $a = u + p_0 v - T_0 s$, is called the *non-flow availability function*.

(b) *Steady flow systems*

Let fluid flow steadily with a velocity C_1 from a reservoir in which the pressure and temperature remain constant at p_1 and T_1 through an apparatus to atmospheric pressure of p_0. Let the reservoir be at a height Z_1 from the datum, which can be taken at exit from the apparatus, i.e. $Z_0 = 0$. For maximum work to be obtained from the apparatus the exit velocity, C_0, must be zero. It can be shown as for (a) above that a reversible heat engine working between the limits would reject $T_0(s_1 - s_0)$ units of heat, where T_0 is the atmospheric temperature.

Therefore we have:

$$W_{\text{max}} = (h_1 + C_1^2/2 + Z_1 g) - h_0 - T_0(s_1 - s_0)$$

In many thermodynamic systems the kinetic and potential energy terms are negligible.

i.e. $W_{max} = (h_1 - T_0 s_1) - (h_0 - T_0 s_0) = b_1 - b_0$

The property, $b = h - T_0 s$, is called the *steady-flow availability function*.

Effectiveness

Instead of comparing a process to some imaginary ideal process, as is done in the case of isentropic efficiency for instance (see page 176), it is a better measure of the usefulness of the process to compare the useful output of the process with the loss of availability of the system. The useful output of a system is given by the increase of availability of the surroundings.

i.e.

$$\text{Effectiveness, } \varepsilon = \frac{\text{Increase of availability of surroundings}}{\text{Loss of availability of the system}} \qquad 5.16$$

For a compression or heating process the effectiveness becomes,

$$\varepsilon = \frac{\text{Increases of availability of the system}}{\text{Loss of availability of the surroundings}}$$

Example 5.14

Steam expands adiabatically in a turbine from 20 bar, 400°C to 4 bar, 250°C. Calculate:

(a) the isentropic efficiency of the process;
(b) the loss of availability of the system assuming an atmospheric temperature of 15°C;
(c) the effectiveness of the process.

Neglect changes in kinetic and potential energy.

(a) Initially the steam is superheated at 20 bar and 400°C, hence from tables,

$$h_1 = 3248 \text{ kJ/kg} \quad \text{and} \quad s_1 = 7 \cdot 126 \text{ kJ/kg K}$$

Finally the steam is superheated at 4 bar and 250°C, hence from tables,

$$h_{2'} = 2965 \text{ kJ/kg} \quad \text{and} \quad s_{2'} = 7 \cdot 379 \text{ kJ/kg K}$$

The process is shown as 1 to 2' in fig. 5.37.

$$s_1 = s_2 = 7 \cdot 126 \text{ kJ/kg K}$$

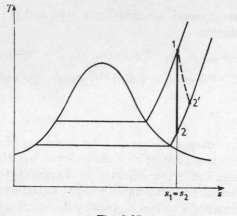

Fig. 5.37

Hence interpolating,

$$h_2 = 2753 + \left(\frac{7\cdot126 - 6\cdot929}{7\cdot172 - 6\cdot929}\right)(2862 - 2753)$$

$$= 2841\cdot4 \text{ kJ/kg}$$

$$\text{Isentropic efficiency} = \frac{\text{Actual work output}}{\text{Isentropic work}}$$

$$= \frac{h_1 - h_{2'}}{h_1 - h_2} = \frac{3248 - 2965}{3248 - 2841\cdot4} = \frac{283}{406\cdot6} = 69\cdot6\%$$

(b)   Loss of availability $= b_1 - b_{2'} = h_1 - h_{2'} + T_0(s_{2'} - s_1)$

$$= 283 + 288(7\cdot379 - 7\cdot126)$$

$$= 355\cdot9 \text{ kJ/kg}$$

(c)   Effectiveness, $\varepsilon = \dfrac{W}{b_1 - b_{2'}} = \dfrac{h_1 - h_{2'}}{b_1 - b_{2'}}$

i.e.
$$\varepsilon = \frac{283}{355\cdot9} = 79\cdot6\%$$

Example 5.15

Air at 15°C is to be heated to 40°C by mixing it in steady flow with a quantity of air at 90°C. Assuming that the mixing process is adiabatic and neglecting changes in kinetic and potential energy, calculate the ratio of the mass flow of air initially at 90°C to that initially at 15°C. Calculate also the effectiveness of the heating process, if the atmospheric temperature is 15°C.

Let the ratio of mass flows required be y; let the air at 15°C be stream 1, the air at 90°C be stream 2, and the mixed air stream at 40°C be stream 3.

Then,
$$c_p T_1 + y c_p T_2 = (1+y) c_p T_3$$

or,
$$y c_p (T_2 - T_3) = c_p (T_3 - T_1)$$

i.e.
$$(90 - 40) = 40 - 15$$

$$\therefore y = \frac{25}{50} = 0{\cdot}5$$

Let the system considered be a stream of air of unit mass, heated from 15°C to 40°C.

Increase of availability of system $= b_3 - b_1 = (h_3 - h_1) - T_0(s_3 - s_1)$
$$= 1{\cdot}005(40 - 15) - 288(s_3 - s_1)$$

Also, $s_3 - s_1 = c_p \log_e \dfrac{T_3}{T_1} = 1{\cdot}005 \log_e \dfrac{313}{288} = 0{\cdot}0831$ kJ/kg K

$\therefore$ Increase of availability of system $= 1{\cdot}005 \times 25 - 288 \times 0{\cdot}0831$
$$= 1{\cdot}195 \text{ kJ/kg}$$

The system, which is the air being heated, is 'surrounded' by the air stream being cooled. Therefore, the loss of availability of the surroundings is given by, $y(b_2 - b_3)$.
i.e.

Loss of availability of surroundings
$$= 0{\cdot}5\{(h_2 - h_3) - T_0(s_2 - s_3)\}$$
$$= 0{\cdot}5\left(1{\cdot}005(90 - 40) - 288 \times 1{\cdot}005 \log_e \frac{363}{313}\right)$$
$$= 3{\cdot}65 \text{ kJ/kg}$$

Therefore,   Effectiveness $= \dfrac{1{\cdot}195}{3{\cdot}65} = 0{\cdot}327$ or 32·7%

The low figure for the effectiveness is an indication of the highly irreversible nature of the mixing process.

Example 5.16

A liquid of specific heat 6·3 kJ/kg K is heated at approximately constant pressure from 15°C to 70°C by passing it through tubes which are immersed in a furnace. The furnace temperature is constant at 1400°C. Calculate the effectiveness of the heating process when the atmospheric temperature is 10°C.

Increase of availability of the liquid

$$= b_2 - b_1 = (h_2 - h_1) - T_0(s_2 - s_1)$$

i.e. $$b_2 - b_1 = 6·3(70 - 15) - 283 \times 6·3 \log_e \frac{343}{288}$$

$$= 34·7 \text{ kJ/kg}$$

Now the heat rejected by the furnace is equal to the heat supplied to the liquid, $(h_2 - h_1)$. If this quantity of heat were supplied to a heat engine operating on the Carnot cycle its thermal efficiency would be $\left(1 - \dfrac{T_0}{1400 + 273}\right)$ (for the thermal efficiency of a Carnot cycle see page 150). Therefore, work which could be obtained from a heat engine is given by the product of the thermal efficiency and the heat supplied.

i.e.   Possible work of a heat engine $= (h_2 - h_1)\left(1 - \dfrac{283}{1673}\right)$

The possible work from a heat engine is a measure of the loss of availability of the furnace.
i.e.

Loss of availability of surroundings $= 6·3(70 - 15)\left(1 - \dfrac{283}{1673}\right)$

$$= 288 \text{ kJ/kg}$$

Then,   Effectiveness $= \dfrac{34·7}{288} = 0·121$ or $12·1\%$

The very low value of effectiveness reflects the irreversibility of the transfer of heat through a large temperature difference. If the furnace temperature were much lower the process would be much more effective, although the heat transferred to the liquid would remain the same.

For further reading on this topic see references 5.2 and 5.3.

PROBLEMS

5.1 1 kg of steam at 20 bar, dryness fraction 0·9, is heated reversibly at constant pressure to a temperature of 300°C. Calculate the heat supplied, and the change of entropy, and show the process on a T-s diagram, indicating the area which represents the heat flow.

(415 kJ/kg; 0·8173 kJ/kg K)

5.2 Steam at 0·05 bar, 100°C is to be condensed completely by a reversible constant pressure process. Calculate the heat to be removed per kg of steam, and the change of entropy. Sketch the process on a T-s diagram and shade in the area which represents the heat flow. (2550 kJ/kg; 8·292 kJ/kg K)

5.3 0·05 kg of steam at 10 bar, dryness fraction 0·84, is heated reversibly in a rigid vessel until the pressure is 20 bar. Calculate the change of entropy and the heat supplied. Show the area which represents the heat supplied on a T-s diagram.

(0·0704 kJ/kg K; 36·85 kJ)

5.4 A rigid cylinder containing 0·006 m³ of nitrogen (molar mass 28 kg/kmol) at 1·04 bar, 15°C, is heated reversibly until the temperature is 90°C. Calculate the change of entropy and the heat supplied. Sketch the process on a T-s diagram. Take the isentropic index, γ, for nitrogen as 1·4, and assume that nitrogen is a perfect gas.

(0·001 25 kJ/K; 0·407 kJ)

5.5 1 m³ of air is heated reversibly at constant pressure from 15°C to 300°C, and is then cooled reversibly at constant volume back to the initial temperature. The initial pressure is 1·03 bar. Calculate the net heat flow and the overall change of entropy, and sketch the processes on a T-s diagram. (101·5 kJ; 0·246 kJ/K)

5.6 1 kg of steam undergoes a reversible isothermal process from 20 bar and 250°C to a pressure of 30 bar. Calculate the heat flow, stating whether it is supplied or rejected, and sketch the process on a T-s diagram. (−135 kJ/kg)

5.7 1 kg of air is allowed to expand reversibly in a cylinder behind a piston in such a way that the temperature remains constant at 260°C while the volume is doubled. The piston is then moved in, and heat is rejected by the air reversibly at constant pressure until the volume is the same as it was initially. Calculate the net heat flow and the overall change of entropy. Sketch the processes on a T-s diagram.

(−161·9 kJ/kg; −0·497 kJ/kg K)

5.8 Steam at 5 bar, 250°C, expands isentropically to a pressure of 0·7 bar. Calculate the final condition of the steam.   (0·967)

5.9 Steam expands reversibly in a cylinder behind a piston from 6 bar dry saturated, to a pressure of 0·65 bar. Assuming that the cylinder is perfectly thermally insulated, calculate the work done during the expansion per kg of steam. Sketch the process on a T-s diagram.

(323·8 kJ/kg)

5.10 1 kg of a fluid at 30 bar, 300°C, expands reversibly and isothermally to a pressure of 0·75 bar. Calculate the heat flow, and the work done

(a) when the fluid is air;
(b) when the fluid is steam.

Sketch each process on a T-s diagram.

(607 kJ/kg; 607 kJ/kg; 1035 kJ/kg; 975 kJ/kg)

5.11 1 kg of a fluid at 30 bar, 300°C, expands according to a law pv = constant to a pressure of 0·75 bar. Calculate the heat flow and the work done,

(a) when the fluid is air;
(b) when the fluid is steam.

Sketch each process on a T-s diagram.

(607 kJ/kg; 607 kJ/kg; 891·2 kJ/kg; 899 kJ/kg)

5.12 1 kg of air at 1·013 bar, 17°C, is compressed according to a law $pv^{1·3}$ = constant, until the pressure is 5 bar. Calculate the change of entropy and sketch the process on a T-s diagram, indicating the area which represents the heat flow.   (−0·0885 kJ/kg K)

5.13 0·06 m³ of ethane (molar mass 30 kg/kmol), at 6·9 bar and 260°C, is allowed to expand isentropically in a cylinder behind a piston to a pressure of 1·05 bar and a temperature of 107°C. Calculate γ, R, c_p, c_v, for ethane, and calculate the work done during the expansion. Assume ethane to be a perfect gas.

The same mass of ethane at 1·05 bar, 107°C, is compressed to 6·9 bar according to a law $pv^{1·4}$ = constant. Calculate the final temperature of the ethane and the heat flow to or from the cylinder walls during the compression. Calculate also the change of entropy during the compression, and sketch both processes on a p-v and a T-s diagram.

(1·219; 0·277 kJ/kg K; 1·542 kJ/kg K; 1·265 kJ/kg K; 54·2 kJ; 378°C; 43·4 kJ; 0·0867 kJ/K)

5.14 A steam engine receives steam at 4 bar, dryness fraction 0·8, and expands it according to a law $pv^{1·05}$ = constant to a condenser pressure of 1 bar. Calculate the change of entropy per kg of steam during the expansion, and sketch the process on a T-s diagram.

(0·381 kJ/kg K)

5.15 A certain perfect gas for which $\gamma = 1·26$ and the molar mass is 26 kg/kmol, expands reversibly from 727°C, 0·003 m^3 to 2°C, 0·6 m^3, according to a linear law on the T-s diagram. Calculate the work done per kg of gas and sketch the process on a T-s diagram.

(959·3 kJ/kg)

5.16 1 kg of air at 1·02 bar, 20°C, undergoes a process in which the pressure is raised to 6·12 bar, and the volume becomes 0·25 m^3. Calculate the change of entropy and mark the initial and final states on a T-s diagram. (0·087 kJ/kg K)

5.17 Steam at 15 bar is throttled to 1 bar and a temperature of 150°C. Calculate the initial dryness fraction and the change of entropy. Sketch the process on a T-s diagram and state the assumptions made in the throttling process. (0·992; 1·202 kJ/kg K)

5.18 Two vessels, one exactly twice the volume of the other, are connected by a valve and immersed in a constant temperature bath of water. The smaller vessel contains hydrogen (molar mass 2 kg/kmol), and the other is completely evacuated. Calculate the change of entropy per kg of gas when the valve is opened and conditions are allowed to settle. Sketch the process on a T-s diagram. Assume hydrogen to be a perfect gas. (4·57 kJ/kg K)

5.19 A turbine is supplied with steam at 40 bar, 400°C, which expands through the turbine in steady flow to an exit pressure of 0·2 bar, and a dryness fraction of 0·93. The inlet velocity is negligible, but the steam leaves at high velocity through a duct of 0·14 m^2 cross-sectional area. If the mass flow is 3 kg/s, and the mechanical efficiency is 90%, calculate the horse power output of the turbine. Show that the process is irreversible and calculate the change of entropy. Heat losses from the turbine are negligible.   (2048 kW; 0·643 kJ/kg K)

5.20 In a centrifugal compressor the air is compressed through a pressure ratio of 4 to 1, and the temperature of the air increases by a factor of 1·65. Show that the process is irreversible and calculate the change of entropy per kg of air. Assume that the process is adiabatic. Sketch the process on a T-s diagram.   (0·105 kJ/kg K)

5.21 In a gas turbine unit the gases enter the turbine at 550°C and 5 bar and leave at 1 bar. The process is approximately adiabatic, but the entropy changes by 0·174 kJ/kg K. Calculate the exit temperature of the gases. Assume the gases to act as a perfect gas, and take $\gamma = 1·333$ and $c_p = 1·11$ kJ/kg K. Sketch the process on a T-s diagram.   (370°C)

5.22 A rigid, well-lagged vessel of 0·3 m³ capacity contains 0·762 kg of steam at 6 bar. A valve is opened and the pressure falls to 1·4 bar before the valve is shut again. Calculate the condition of the steam remaining in the vessel, and the mass of steam which has escaped.   (0·99; 0·517 kg)

5.23 A rigid vessel contains 0·5 kg of a perfect gas of specific heat at constant volume 1·1 kJ/kg K. A stirring paddle is inserted into the vessel and 11 kJ of work are done on the paddle by the stirrer motor. Assuming that the vessel is well-lagged and that the gas is initially at the temperature of the surroundings which are at 17°C, calculate the effectiveness of the process.   (3%)

5.24 The identical vessel of Problem 5.23 is heated through the same temperature difference by immersing it in a furnace of constant temperature 100°C. Calculate the effectiveness of the process.
          (13·5%)

5.25 Steam enters a turbine at 70 bar, 500°C and leaves at 2 bar in a dry saturated state. Calculate the isentropic efficiency and the effectiveness of the process. Neglect changes of kinetic and potential energy and assume that the process is adiabatic. The atmospheric temperature is 17°C.   (84·4%; 88%)

5.26 In an open type feed heater (see page 198), steam enters at 15 bar, 200°C. The feed water enters the heater at 130°C and the feed water leaves the heater at the saturation temperature corresponding to the pressure in the heater of 15 bar. Calculate the mass of steam entering per unit mass of feed water entering the heater.

Calculate also the loss of availability of the steam per unit mass and the effectiveness of the heater. Assume that there is no heat loss from the heater and that the atmospheric temperature is 20°C. State any other assumptions made.   (0·1533 kg; 738 kJ/kg; 87·9%)

REFERENCES

5.1 ROGERS, G. F. C., and MAYHEW, Y. R., *Engineering Thermodynamics, Work and Heat Transfer*, S I Units (Longmans, 1980).

5.2 KEENAN, J. H., *Thermodynamics* (MIT Press, 1970).

5.3 BRUGES, E. A., *Available Energy and the Second Law Analysis* (Academic Press, 1959).

6. The Heat Engine Cycle

In this chapter the heat engine cycle is discussed more fully and gas power cycles are considered. It can be shown that there is an ideal theoretical cycle which is the most efficient conceivable; this cycle is called the Carnot cycle. The highest thermal efficiency possible for a heat engine in practice is only about half that of the ideal theoretical Carnot cycle, between the same temperature limits. This is due to irreversibilities in the actual cycle, and to deviations from the ideal cycle, which are made for various practical reasons. The choice of a power plant in practice is a compromise between thermal efficiency and various factors such as the size of the plant for a given power requirement, mechanical complexity, operating cost, and capital cost.

6.1 The Carnot cycle

It can be shown from the Second Law of Thermodynamics that no heat engine can be more efficient than a reversible heat engine working between the same temperature limits (see reference 6.1). Carnot, a French engineer, showed in a paper written in 1824* that the most efficient possible cycle is one in which all the heat supplied is supplied at one fixed temperature, and all the heat rejected is rejected at a lower fixed temperature. The cycle therefore consists of two isothermal processes joined by two adiabatic processes. Since all processes are reversible, then the adiabatic processes in the cycle are also isentropic. The cycle is most conveniently represented on a T-s diagram as shown in fig. 6.1.

Process 1 to 2 is isentropic expansion from T_1 to T_2.
Process 2 to 3 is isothermal heat rejection.
Process 3 to 4 is isentropic compression from T_2 to T_1.
Process 4 to 1 is isothermal heat supply.

* This paper, called 'Reflections on the Motive Power of Heat' (see reference 6.2) was written by Carnot before the enunciation of the First and Second Laws of Thermodynamics. It is a remarkable piece of original thinking, and it laid the foundations for the work of Kelvin, Clausius, and others on the second law and its corollaries.

The cycle is completely independent of the working substance used.

The thermal efficiency of a heat engine, defined in Section 5.1, was shown to be given by equation 5.3,

$$\eta = 1 - \frac{Q_2}{Q_1}$$

In the Carnot cycle, with reference to fig. 6.1, it can be seen that the heat supplied, Q_1, is given by the area 41BA4,

i.e. $Q_1 = \text{area } 41BA4 = T_1(s_B - s_A)$

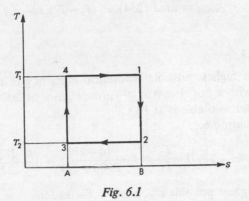

Fig. 6.1

Similarly the heat rejected, Q_2, is given by the area 23AB2,

i.e. $Q_2 = \text{area } 23AB2 = T_2(s_B - s_A)$

Hence we have

Thermal efficiency of Carnot cycle, $\eta_{\text{Carnot}} = 1 - \dfrac{T_2(s_B - s_A)}{T_1(s_B - s_A)}$

i.e. $\eta_{\text{Carnot}} = 1 - \dfrac{T_2}{T_1}$ 6.1

If a sink for heat rejection is available at a fixed temperature T_2 (e.g. a large supply of cooling water), then the ratio T_2/T_1 will decrease as the temperature of the source T_1 is increased. From equation 6.1 it can be seen that as T_2/T_1 decreases, then the thermal efficiency increases. Hence for a fixed lower temperature for heat rejection, the upper temperature at which heat is supplied must be made as high as possible. The maximum possible thermal efficiency between any two temperatures is that of the Carnot cycle.

The work output of the Carnot cycle can be found very simply from the T-s diagram. From the first law,

$$\sum Q = \sum W$$

therefore, the work output of the cycle is given by

$$W = Q_1 - Q_2$$

Hence for the Carnot cycle, referring to fig. 6.1,

$$W_{\text{Carnot}} = \text{area } 12341 = (T_1 - T_2)(s_{\text{B}} - s_{\text{A}})$$

Example 6.1

What is the highest possible theoretical efficiency of a heat engine operating with a hot reservoir of furnace gases at 2000°C when the cooling water available is at 10°C?

From equation 6.1,

$$\eta_{\text{Carnot}} = 1 - \frac{T_2}{T_1} = 1 - \frac{10 + 273}{2000 + 273} = 1 - \frac{283}{2273}$$

i.e.   Highest possible efficiency $= 1 - 0.1246$

$$= 0.8754 \quad \text{or} \quad 87.54\%$$

It should be noted that a system in practice operating between similar temperatures (e.g. a steam generating plant) would have a thermal efficiency of about 30%. The discrepancy is due to losses due

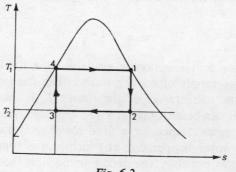

Fig. 6.2

to irreversibility in the actual plant, and also because of deviations from the ideal Carnot cycle made for various practical reasons.

It is difficult in practice to devise a system which can receive and reject heat at constant temperature. A wet vapour is the only working substance which can do this conveniently, since for a wet vapour the pressure and temperature remain constant as the latent heat is supplied or rejected. A Carnot cycle for a wet vapour is as shown in fig. 6.2. Although this cycle is the most efficient possible vapour cycle, it is not used in steam plant. The theoretical cycle on which steam cycles are based is known as the Rankine cycle. This will be discussed in detail in Chapter 7, and the reasons for using it in preference to the Carnot cycle will be given.

6.2 Absolute temperature scale

In the preceding chapters a temperature scale based on the perfect gas thermometer has been assumed. Using the Second Law of Thermodynamics it is possible to establish a temperature scale which is independent of the working substance.

We have, for any heat engine from equation 5.3,

$$\eta = 1 - \frac{Q_2}{Q_1} \qquad 6.2$$

Also the efficiency of an engine operating on the Carnot cycle depends only on the temperatures of the hot and cold reservoirs. Denoting temperature on any arbitrary scale by X, we have

$$\eta = \phi(X_1, X_2) \qquad 6.3$$

(where ϕ is a function, and X_1 and X_2 are the temperatures of the hot and cold reservoirs).

Combining equations 6.2 and 6.3 we have

$$\frac{Q_2}{Q_1} = F(X_1, X_2)$$

(where F is a new function).

There are a large number of possible temperature scales which are all independent of the working substance. Any working scale can be chosen by suitably selecting the value of the function F. The function can be chosen such that

$$\frac{Q_2}{Q_1} = \frac{X_2}{X_1} \qquad 6.4$$

Also from equation 6.1, we have

$$\eta = 1 - \frac{T_2}{T_1}$$

Hence using equation 6.2,

$$\eta = 1 - \frac{Q_2}{Q_1} = 1 - \frac{T_2}{T_1}$$

or

$$\frac{Q_2}{Q_1} = \frac{T_2}{T_1} \qquad\qquad 6.5$$

Comparing equations 6.4 and 6.5 it can be seen that the temperature X is equivalent to the temperature T. Thus by suitably choosing the function F, the ideal temperature scale is made equivalent to the scale based on the perfect gas thermometer.

6.3 The Carnot cycle for a perfect gas

A Carnot cycle for a perfect gas is shown on a T-s diagram in fig. 6.3. Note that the pressure of the gas changes continuously from p_4 to p_1

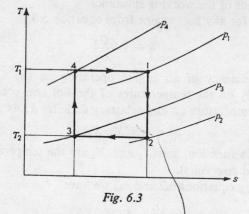

Fig. 6.3

during the isothermal heat supply, and from p_2 to p_3 during the isothermal heat rejection. In practice it is much more convenient to heat a gas at approximately constant pressure or at constant volume, hence it is difficult to attempt to operate an actual heat engine on the Carnot cycle using a gas as working substance. Another important reason for not attempting to use the Carnot cycle in practice is illustrated by drawing the cycle on a p-v diagram, as in fig. 6.4. The net work of the cycle is given by the area 12341. This is a small quantity compared with the gross work of the expansion processes of the cycle, given by area 412BA4. The'work of the compression processes (i.e. work done on the gas) is given by the area 234AB2. The ratio of the net work output to the gross work output of the system is called the *work ratio*. The Carnot cycle, despite its high thermal efficiency, has a low work ratio.

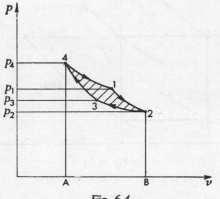

Fig. 6.4

Example 6.2

A hot reservoir at 800°C and a cold reservoir at 15°C are available. Calculate the thermal efficiency and the work ratio of a Carnot cycle using air as the working fluid, if the maximum and minimum pressures in the cycle are 210 bar and 1 bar.

The cycle is shown on a *T-s* and *p-v* diagram in figs. 6.5a and 6.5b respectively.

Using equation 6.1,

$$\eta_{\text{Carnot}} = 1 - \frac{T_2}{T_1} = 1 - \frac{15 + 273}{800 + 273} = 1 - 0.268$$

i.e. $\eta_{\text{Carnot}} = 0.732$ or 73.2%

In order to find the work output and the work ratio it is necessary to find the entropy change, $(s_1 - s_4)$.

For an isothermal process from 4 to A, using equation 5.12,

$$s_A - s_4 = R \log_e \frac{p_4}{p_2} = 0.287 \log_e \frac{210}{1} = 1.535 \text{ kJ/kg K}$$

At constant pressure from A to 2, we have

$$s_A - s_2 = c_p \log_e \frac{T_1}{T_2} = 1.005 \log_e \frac{1073}{288} = 1.321 \text{ kJ/kg K}$$

$$\therefore \quad s_1 - s_4 = 1.535 - 1.321 = 0.214 \text{ kJ/kg K}$$

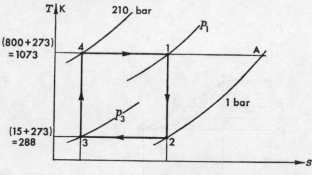

Fig. 6.5a

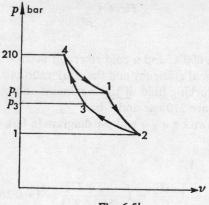

Fig. 6.5b

Then,

$$\text{Net work output} = (T_1 - T_2)(s_1 - s_4) = \text{area } 12341$$
$$= (1073 - 288) \times 0.214 = 168 \text{ kJ/kg}$$

Gross work of expansion

$$= \text{work done 4 to 1} + \text{work done 1 to 2}$$

From equation 4.12, for an isothermal process, $Q = W$

i.e. $\quad W_{4-1} = Q_{4-1} =$ area under line 4–1 on fig. 6.5a

$$= (s_1 - s_4) \times T_1 = 0.214 \times 1073$$

$$= 229.6 \text{ kJ/kg}$$

For an isentropic process from 1 to 2, from equation 4.13, $W = (u_1 - u_2)$, therefore for a perfect gas

$$W_{1-2} = c_v(T_1 - T_2)$$

$$= 0.718(1073 - 288) = 563.6 \text{ kJ/kg}$$

$\therefore$ Gross work $= 229.6 + 563.6 = 793.2 \text{ kJ/kg}$

i.e. $\quad$ Work ratio $= \dfrac{\text{Net work}}{\text{Gross work}} = \dfrac{168}{793.2} = 0.212$

6.4 The constant pressure cycle

In this cycle the heat supply and heat rejection processes occur reversibly at constant pressure. The expansion and compression processes are isentropic. The cycle is shown on a T-s diagram and a

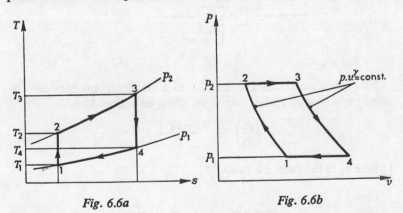

Fig. 6.6a $\qquad$ *Fig. 6.6b*

p-v diagram in figs. 6.6a and 6.6b. This cycle was at one time used as the ideal basis for a hot-air reciprocating engine, and the cycle was known as the Joule or Brayton cycle. Nowadays the cycle is the ideal for the closed cycle gas turbine unit. A simple line diagram of the plant is shown in fig. 6.7, with the numbers corresponding to those of figs. 6.6a and 6.6b. The working substance is air which flows in steady flow round the cycle, hence, neglecting velocity changes, and applying the steady flow energy equation to each part of the cycle, we have

Work input to compressor $= (h_2 - h_1) = c_p(T_2 - T_1)$
Work output from turbine $= (h_3 - h_4) = c_p(T_3 - T_4)$
Heat supplied in heater, $Q_1 = (h_3 - h_2) = c_p(T_3 - T_2)$
Heat rejected in cooler, $Q_2 = (h_4 - h_1) = c_p(T_4 - T_1)$

Then from equation 5.3

$$\eta = 1 - \frac{Q_2}{Q_1} = 1 - \frac{c_p(T_4 - T_1)}{c_p(T_3 - T_2)} = 1 - \frac{T_4 - T_1}{T_3 - T_2}$$

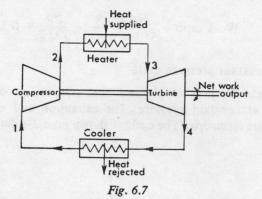

Fig. 6.7

Now since processes 1 to 2 and 3 to 4 are isentropic between the same pressures p_2 and p_1, we have, using equation 4.21,

$$\frac{T_2}{T_1} = \left(\frac{p_2}{p_1}\right)^{(\gamma-1)/\gamma} = \frac{T_3}{T_4} = r_p^{(\gamma-1)/\gamma}$$

$\left(\text{where } r_p \text{ is the pressure ratio, } \dfrac{p_2}{p_1}\right).$

Then,

$$T_3 = T_4 r_p^{(\gamma-1)/\gamma} \quad \text{and} \quad T_2 = T_1 r_p^{(\gamma-1)/\gamma}$$

i.e. $$T_3 - T_2 = r_p^{(\gamma-1)/\gamma}(T_4 - T_1)$$

Hence substituting in the expression for the efficiency,

$$\eta = 1 - \frac{T_4 - T_1}{(T_4 - T_1)r_p^{(\gamma-1)/\gamma}} = 1 - \frac{1}{r_p^{(\gamma-1)/\gamma}} \qquad 6.6$$

Thus for the constant pressure cycle, the thermal efficiency depends only on the pressure ratio. In the ideal case the value of γ for air is constant and equal to 1·4. In practice, due to the eddying of the air as it flows through the compressor and turbine which are both rotary machines, the actual thermal efficiency is greatly reduced compared to that given by equation 6.6.

The work ratio of the constant pressure cycle may be found as follows:

$$\text{Work ratio} = \frac{\text{Net work}}{\text{Gross work}} = \frac{c_p(T_3-T_4)-c_p(T_2-T_1)}{c_p(T_3-T_4)}$$

$$= 1 - \frac{T_2-T_1}{T_3-T_4}$$

Now as before,

$$\frac{T_2}{T_1} = r_p^{(\gamma-1)/\gamma} = \frac{T_3}{T_4}$$

$$\therefore \ T_2 = T_1 r_p^{(\gamma-1)/\gamma} \quad \text{and} \quad T_4 = \frac{T_3}{r_p^{(\gamma-1)/\gamma}}$$

Hence substituting,

$$\text{Work ratio} = 1 - \frac{T_1(r_p^{(\gamma-1)/\gamma}-1)}{T_3[1-(1/r_p^{(\gamma-1)/\gamma})]} = 1 - \frac{T_1}{T_3}\left(\frac{r_p^{(\gamma-1)/\gamma}-1}{r_p^{(\gamma-1)/\gamma}-1}\right)r_p^{(\gamma-1)/\gamma}$$

i.e. $$\text{Work ratio} = 1 - \frac{T_1}{T_3} r_p^{(\gamma-1)/\gamma} \qquad 6.7$$

It can be seen from equation 6.7 that the work ratio depends, not only on the pressure ratio but also on the ratio of the minimum and maximum temperatures. For a given inlet temperature, T_1, the maximum temperature, T_3, must be made as high as possible for a high work ratio.

For an open cycle gas turbine unit the actual cycle is not such a good approximation to the ideal constant pressure cycle, since fuel is burned with the air, and a fresh charge is continuously induced into the compressor. The ideal cycle provides nevertheless a good basis for comparison, and in many calculations for an ideal open cycle gas turbine the effects of the mass of fuel and the change in the working fluid are neglected.

Example 6.3

In a gas turbine unit air is drawn in at 1·02 bar and 15°C, and is compressed to 6·12 bar. Calculate the thermal efficiency and the work ratio of the ideal constant pressure cycle, when the maximum cycle temperature is limited to 800°C.

The ideal cycle is shown on a T-s diagram in fig. 6.8. From equation 6.6

$$\text{Thermal efficiency, } \eta = 1 - \frac{1}{r_p^{(\gamma-1)/\gamma}}$$

i.e. $\eta = 1 - \left(\frac{1\cdot02}{6\cdot12}\right)^{(\gamma-1)/\gamma} = 1 - \frac{1}{6^{0\cdot286}} = 1 - 0\cdot598$

$$\therefore \text{ Thermal efficiency} = 0\cdot402 \quad \text{or} \quad 40\cdot2\%$$

The net work of the cycle is given by the work done by the turbine minus the work done on the air in the compressor,

i.e. $$\text{Net work} = c_p(T_3 - T_4) - c_p(T_2 - T_1)$$

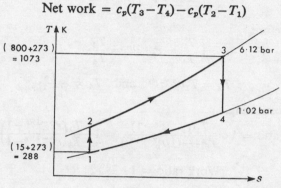

Fig. 6.8

From equation 4.21

$$\frac{T_2}{T_1} = \left(\frac{p_2}{p_1}\right)^{(\gamma-1)/\gamma} = \frac{T_3}{T_4} = \left(\frac{6\cdot12}{1\cdot02}\right)^{(\gamma-1)/\gamma} = 6^{0\cdot286} = 1\cdot67$$

$$\therefore \ T_2 = 1\cdot67 \times T_1 = 1\cdot67 \times 288 = 481 \text{ K}$$

(where $T_2 = 15 + 273 = 288$ K) and

$$T_4 = \frac{T_3}{1\cdot67} = \frac{1073}{1\cdot67} = 643 \text{ K}$$

(where $T_3 = 800 + 273 = 1073$ K).

$$\therefore \ \text{Net work} = 1\cdot005(1073 - 643) - 1\cdot005(481 - 288)$$
$$= (1\cdot005 \times 430 - 1\cdot005 \times 193) = 238 \text{ kJ/kg}$$

$$\text{Gross work} = \text{Work of the turbine} = c_p(T_3 - T_4)$$
$$= 1\cdot005(1073 - 643) = 432 \text{ kJ/kg}$$

Then, $$\text{Work ratio} = \frac{\text{Net work}}{\text{Gross work}} = \frac{238}{432} = 0\cdot55$$

6.5 The air standard cycle

It was pointed out in Section 5.1 that cycles in which the fuel is burned directly in the working fluid are not heat engines in the true meaning of the term. In practice such cycles are used frequently and are called internal-combustion cycles. The fuel is burned directly in the working fluid, which is normally air. The main advantage of such power units is that high temperatures of the fluid can be attained, since heat is not transferred through metal walls to the fluid. It was seen from equation 6.1 $\eta = 1 - (T_2/T_1)$, that for a given sink for the rejection of heat at T_2, the temperature of the source, T_1 must be as high as possible. This applies to all heat engines. By supplying fuel inside the cylinder as in the internal-combustion engine, higher temperatures for the working fluid can be attained. The maximum temperature of all cycles is limited by the metallurgical limit of the materials used (e.g. in gas turbines the limiting temperature is about 800°C). The fluid in an internal-combustion engine may reach a temperature of as high as 2750°C. This is made possible by externally cooling the cylinder by water or air cooling; also, due to the intermittent nature of the cycle, the working fluid reaches its maximum temperature for only an instant during each cycle.

Examples of internal-combustion cycles are the open cycle gas turbine unit, the petrol engine, the diesel engine or oil engine, and the gas engine. The open cycle gas turbine unit, although an internal combustion cycle, is nevertheless in a different category to the other internal-combustion engines. The cycle was mentioned in Section 5.1 and a diagram of the plant shown in fig. 5.4. It can be seen that the cycle is a steady flow cycle in which the working fluid flows from one component to another round the cycle. It will be assumed therefore that the gas turbine unit, whether operating on the open or the closed cycle, can be satisfactorily compared with the ideal constant pressure cycle, dealt with in Section 6.4.

In the petrol engine a mixture of air and petrol is drawn into the cylinder, compressed by the piston, then ignited by an electric spark. The hot gases expand, pushing the piston back, and are then swept out to exhaust, and the cycle recommences with the induction of a fresh charge of petrol and air. In the diesel or oil engine the oil is sprayed under pressure into the compressed air at the end of the compression stroke, and combustion is spontaneous due to the high temperature of the air after compression. In a gas engine a mixture of gas and air is induced into the cylinder, compressed, and then ignited as in the petrol engine, by an electric spark. To give a basis of comparison for the actual internal-combustion engine the air standard cycle is defined.

In an *air standard cycle* the working substance is assumed to be air throughout, all processes are assumed to be reversible, and the source of heat supply and the sink for heat rejection are assumed to be external to the air. The cycle can be represented on any diagram of properties, and is usually drawn on the *p-v* diagram, since this allows a more direct comparison to be made with the actual engine machine cycle which can be obtained from an indicator diagram. It must be stressed that an air standard cycle on a *p-v* diagram is a true thermodynamic cycle, whereas an indicator diagram taken from an actual engine is a record of pressure variations in the cylinder against piston displacement. Indicator diagrams and their use and significance are considered in more detail in Chapter 18.

6.6 The Otto cycle

The Otto cycle is the ideal air standard cycle for the petrol engine,

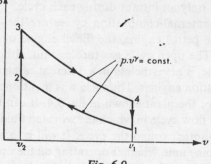

Fig. 6.9

the gas engine, and the high-speed oil engine. The cycle is shown on a *p-v* diagram in fig. 6.9.

    Process 1 to 2 is isentropic compression.
    Process 2 to 3 is reversible constant volume heating.
    Process 3 to 4 is isentropic expansion.
    Process 4 to 1 is reversible constant volume cooling.

To give a direct comparison with an actual engine the ratio of the specific volumes, v_1/v_2, is taken to be the same as the compression ratio of the actual engine, i.e.

$$\text{Compression ratio, } r_v = \frac{v_1}{v_2}$$

$$= \frac{\text{swept volume} + \text{clearance volume}}{\text{clearance volume}} \qquad 6.8$$

The thermal efficiency of the Otto cycle can be found using equation 5.3,

$$\eta = 1 - \frac{Q_2}{Q_1}$$

The heat supplied, Q_1, at constant volume between T_2 and T_3 is given by equation 3.13, per kg of air,

$$Q_1 = c_v(T_3 - T_2)$$

Similarly the heat rejected, Q_2, at constant volume between T_4 and T_1 is given by equation 3.13, per kg of air,

$$Q_2 = c_v(T_4 - T_1)$$

The processes 1 to 2 and 3 to 4 are isentropic and therefore there is no heat flow during these processes.

$$\therefore \ \eta = 1 - \frac{Q_2}{Q_1} = 1 - \frac{c_v(T_4 - T_1)}{c_v(T_3 - T_2)} = 1 - \left(\frac{T_4 - T_1}{T_3 - T_2}\right)$$

Now since processes 1 to 2 and 3 to 4 are isentropic, then using equation 4.21,

$$\frac{T_2}{T_1} = \left(\frac{v_1}{v_2}\right)^{\gamma-1} = \left(\frac{v_4}{v_3}\right)^{\gamma-1} = \frac{T_3}{T_4} = r_v^{\gamma-1}$$

(where r_v is the compression ratio from equation 6.8).

Then, $$T_3 = T_4 r_v^{\gamma-1} \quad \text{and} \quad T_2 = T_1 r_v^{\gamma-1}$$

Hence substituting

$$\eta = 1 - \frac{T_4 - T_1}{(T_4 - T_1) r_v^{\gamma-1}} = 1 - \frac{1}{r_v^{\gamma-1}} \qquad 6.9$$

It can be seen from equation 6.9 that the thermal efficiency of the Otto cycle depends only on the compression ratio, r_v.

Example 6.4

Calculate the ideal air standard thermal efficiency based on the Otto cycle for a petrol engine with a cylinder bore of 50 mm and a stroke of 75 mm, and a clearance volume of 21·3 cm³.

$$\text{Swept volume} = \frac{\pi}{4} \times 50^2 \times 75 = 147\,200 \text{ mm}^3 = 147\cdot2 \text{ cm}^3$$

$$\therefore \text{ Total cylinder volume} = 147\cdot2 + 21\cdot3 = 168\cdot5 \text{ cm}^3$$

i.e. $$\text{Compression ratio, } r_v = \frac{168\cdot5}{21\cdot3} = 7\cdot92/1$$

Then using equation 6.9

$$\eta = 1 - \frac{1}{r_v^{\gamma-1}} = 1 - \frac{1}{7\cdot92^{0\cdot4}} = 1 - 0\cdot437 = 0\cdot563 \quad \text{or} \quad 56\cdot3\%$$

6.7 The diesel cycle

The engines in use today which are called diesel engines are far removed from the original engine invented by Diesel in 1892. Diesel worked on the idea of spontaneous ignition of powdered coal, which was blasted into the cylinder by compressed air (see reference 6.3). Oil became the accepted fuel used in compression-ignition engines, and the oil was originally blasted into the cylinder in the same way that Diesel had intended to inject the powdered coal. This gave a cycle of operation which has as its ideal counterpart the ideal air standard diesel cycle shown in fig. 6.10.

As before, the compression ratio, r_v, is given by the ratio v_1/v_2.

Process 1 to 2 is isentropic compression.
Process 2 to 3 is reversible constant pressure heating.
Process 3 to 4 is isentropic expansion.
Process 4 to 1 is reversible constant volume cooling.

From equation 5.3

$$\eta = 1 - \frac{Q_2}{Q_1}$$

At constant pressure from equation 3.12, per kg of air,

$$Q_1 = c_p(T_3 - T_2)$$

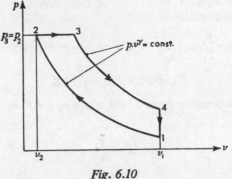

Fig. 6.10

Also at constant volume from equation 3.13, per kg of air,

$$Q_2 = c_v(T_4 - T_1)$$

There is no heat flow in processes 1 to 2 and 3 to 4 since these processes are isentropic. Hence by substituting for Q_1 and Q_2 in the expression for thermal efficiency the following equation may be derived:

$$\eta = 1 - \frac{\beta^\gamma - 1}{(\beta - 1)r_v^{\gamma - 1}\gamma} \qquad 6.10$$

(where $\beta = v_3/v_2 =$ cut-off ratio).

Equation 6.10 shows that the thermal efficiency depends not only on the compression ratio but also on the heat supplied between 2 and 3, which fixes the ratio, v_3/v_2. The equation 6.10 is derived by expressing each temperature in terms of T_1 and r_v or β. The derivation is not given here, because it is believed that the best method of working out the thermal efficiency is to calculate each temperature individually round the cycle, and then apply equation 5.3, $\eta = 1 - (Q_2/Q_1)$. This is illustrated in the following example.

Example 6.5
A diesel engine has an inlet temperature and pressure of 15°C and 1 bar respectively. The compression ratio is 12/1 and the maximum cycle temperature is 1100°C. Calculate the air standard thermal efficiency based on the diesel cycle.

Referring to fig. 6.11, $T_1 = 15 + 273 = 288$ K and $T_3 = 1100 + 273 = 1373$ K. From equation 4.20

$$\frac{T_2}{T_1} = \left(\frac{v_1}{v_2}\right)^{\gamma-1} = r_v^{\gamma-1} = 12^{0 \cdot 4} = 2 \cdot 7$$

i.e. $$T_2 = 2 \cdot 7 \times 288 = 778 \text{ K}$$

At constant pressure from 2 to 3, since $pv = RT$, for a perfect gas, then

$$\frac{T_3}{T_2} = \frac{v_3}{v_2}$$

i.e. $$\frac{v_3}{v_2} = \frac{1373}{778} = 1 \cdot 765$$

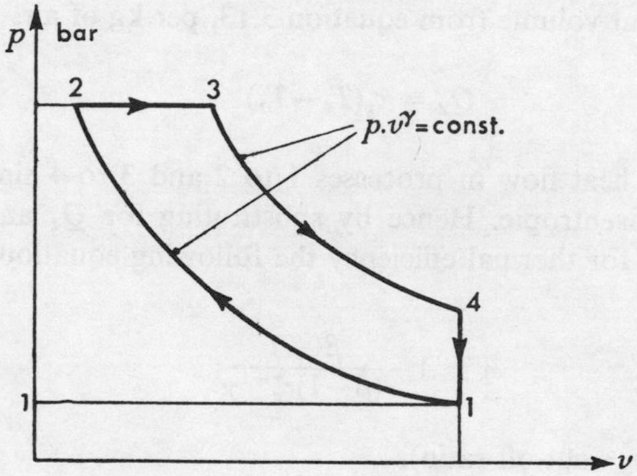

Fig. 6.11

Therefore,

$$\frac{v_4}{v_3} = \frac{v_4}{v_2}\frac{v_2}{v_3} = \frac{v_1}{v_2}\frac{v_2}{v_3} = 12 \times \frac{1}{1 \cdot 765} = 6 \cdot 8$$

Then using equation 4.20

$$\frac{T_3}{T_4} = \left(\frac{v_4}{v_3}\right)^{\gamma-1} = 6 \cdot 8^{0 \cdot 4} = 2 \cdot 153$$

i.e. $$T_4 = \frac{1373}{2 \cdot 153} = 638 \text{ K}$$

Then from equation 3.12, per kg of air

$$Q_1 = c_p(T_3 - T_2) = 1 \cdot 005(1373 - 778) = 598 \text{ kJ/kg}$$

Also, from equation 3.13, per kg of air

$$Q_2 = c_v(T_4 - T_1) = 0{\cdot}718(638 - 288) = 251 \text{ kJ/kg}$$

Therefore from equation 5.3

$$\eta = 1 - \frac{Q_2}{Q_1} = 1 - \frac{251}{598} = 0{\cdot}58 \quad \text{or} \quad 58\%$$

6.8 The dual combustion cycle

Modern oil engines although still called diesel engines are more closely derived from an engine invented by Ackroyd-Stuart in 1888 (see reference 6.3). All oil engines today use solid injection of the fuel; the fuel is injected by a spring-loaded injector the fuel pump being operated by a cam driven from the engine crankshaft (see Section

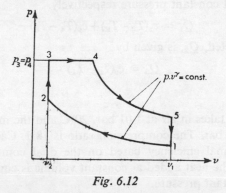

Fig. 6.12

18.10). The ideal cycle used as a basis for comparison is called the dual combustion cycle or the mixed cycle, and is shown on a *p-v* diagram in fig. 6.12.

Process 1 to 2 is isentropic compression.
Process 2 to 3 is reversible constant volume heating.
Process 3 to 4 is reversible constant pressure heating.
Process 4 to 5 is isentropic expansion.
Process 5 to 1 is reversible constant volume cooling.

The heat is supplied in two parts, the first part at constant volume and the remainder at constant pressure, hence the name 'dual combustion'. In order to fix the thermal efficiency completely three factors are necessary. These are: the compression ratio, $r_v = v_1/v_2$; the ratio of pressures, $k = p_3/p_2$; and the ratio of volumes, $\beta = v_4/v_3$.

Then it can be shown that

$$\eta = 1 - \frac{k\beta^\gamma - 1}{[(k-1) + \gamma k(\beta - 1)]r_v^{\gamma - 1}} \qquad 6.11$$

Note that when $k = 1$ (i.e. $p_3 = p_2$), then the equation 6.11 reduces to the thermal efficiency of the diesel cycle given by equation 6.10. The efficiency of the dual combustion cycle depends not only on the compression ratio but also on the relative amounts of heat supplied at constant volume and at constant pressure. The equation 6.11 is much too cumbersome to use, and the best method of calculating thermal efficiency is to evaluate each temperature round the cycle and then use equation 5.3, $\eta = 1 - (Q_2/Q_1)$. The heat supplied, Q_1, is found by using equation 3.13 and 3.12 for the heat added at constant volume and at constant pressure respectively.

i.e. $$Q_1 = c_v(T_3 - T_2) + c_p(T_4 - T_3)$$

The heat rejected, Q_2, is given by

$$Q_2 = c_v(T_5 - T_1)$$

Example 6.6

An oil engine takes in air at 1·01 bar, 20°C and the maximum cycle pressure is 69 bar. The compression ratio is 18/1. Calculate the air standard thermal efficiency based on the dual combustion cycle. Assume that the heat added at constant volume is equal to the heat added at constant pressure.

The cycle is shown on a p-v diagram in fig. 6.13. Using equation 4.20,

$$\frac{T_2}{T_1} = \left(\frac{v_1}{v_2}\right)^{\gamma - 1} = 18^{0\cdot 4} = 3\cdot 18$$

i.e. $$T_2 = 3\cdot 18 \times T_1$$

i.e. $$T_2 = 3\cdot 18 \times 293 = 931 \text{ K}$$

(where $T_1 = 20 + 273 = 293$ K).

From 2 to 3 the process is at constant volume, hence,

$$\frac{p_3}{p_2} = \frac{T_3}{T_2} \quad \left(\text{since } \frac{p_3 v_3}{T_3} = \frac{p_2 v_2}{T_2}, \text{ and } v_3 = v_2\right)$$

i.e. $$T_3 = \frac{p_3}{p_2} \times T_2 = \frac{69 \times 931}{p_2}$$

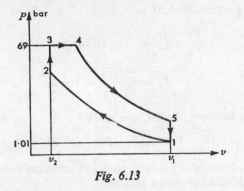

Fig. 6.13

To find p_2, use equation 4.19,

i.e.
$$\frac{p_2}{p_1} = \left(\frac{v_1}{v_2}\right)^{\gamma} = 18^{1\cdot4} = 57\cdot2$$

i.e.
$$p_2 = 57\cdot2 \times 1\cdot01 = 57\cdot8 \text{ bar}$$

Then substituting,

$$T_3 = \frac{69 \times 931}{57\cdot8} = 1112 \text{ K}$$

Now the heat added at constant volume is equal to the heat added at constant pressure in this example, therefore

$$c_v(T_3 - T_2) = c_p(T_4 - T_3)$$

i.e.
$$0\cdot718(1112 - 931) = 1\cdot005(T_4 - 1112)$$

$$\therefore T_4 = \frac{0\cdot718 \times 181}{1\cdot005} + 1112$$

i.e.
$$T_4 = 1241\cdot4 \text{ K}$$

To find T_5 it is necessary to know the value of the volume ratio, v_5/v_4. At constant pressure from 3 to 4,

$$\frac{v_4}{v_3} = \frac{T_4}{T_3} = \frac{1241\cdot4}{1112} = 1\cdot116$$

Therefore

$$\frac{v_5}{v_4} = \frac{v_1}{v_4} = \frac{v_1}{v_2}\frac{v_3}{v_4} = 18 \times \frac{1}{1\cdot116} = 16\cdot14$$

Then using equation 4.20

$$\frac{T_4}{T_5} = \left(\frac{v_5}{v_4}\right)^{\gamma-1} = 16\cdot14^{0\cdot4} = 3\cdot04$$

i.e.

$$T_5 = \frac{1241\cdot4}{3\cdot04} = 408 \text{ K}$$

Now the heat supplied, Q_1, is given by

$$Q_1 = c_v(T_3 - T_2) + c_p(T_4 - T_3) \quad \text{or} \quad Q_1 = 2c_v(T_3 - T_2)$$

(since in this example the heat added at constant volume is equal to the heat added at constant pressure).

$$\therefore \quad Q_1 = 2 \times 0\cdot718 \times (1112 - 931) = 260 \text{ kJ/kg}$$

The heat rejected, Q_2, is given by

$$Q_2 = c_v(T_5 - T_1) = 0\cdot718(408 - 293) = 82\cdot6 \text{ kJ/kg}$$

Then from equation 5.3

$$\eta = 1 - \frac{Q_2}{Q_1} = 1 - \frac{82\cdot6}{260} = 1 - 0\cdot318 = 0\cdot682 \quad \text{or} \quad 68\cdot2\%$$

It should be mentioned here that the modern high-speed oil engine operates on a cycle for which the Otto cycle is a better basis of comparison. Also, since the Otto cycle calculation for thermal efficiency is much simpler than that of the dual combustion cycle, then this is another reason for using the Otto cycle as a standard of comparison.

6.9 Mean effective pressure

The term work ratio was defined in Section 6.3, and this was shown to be a useful criterion for practical power plants. For internal-combustion engines the term work ratio is not such a useful concept, since the work done on and by the working fluid takes place inside one cylinder. In order to compare reciprocating engines another term is defined called the *mean effective pressure*. The mean effective pressure is defined as the height of a rectangle having the same length and area as the cycle plotted on a *p-v* diagram. This is illustrated for an Otto cycle in fig. 6.14. The rectangle ABCDA is the same length as the cycle 12341, and area ABCDA is equal to area 12341. Then the

$$V_1 - V_2 = (r_v - 1)V$$

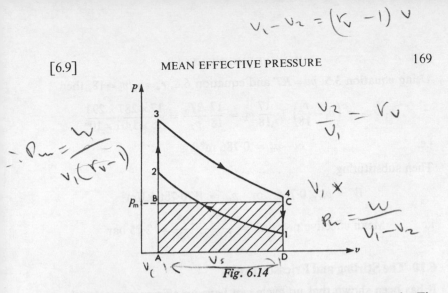

$$\frac{V_2}{V_1} = r_v$$

$$P_m = \frac{W}{V_1(r_v - 1)}$$

$$V_1 \times$$

$$P_m = \frac{W}{V_1 - V_2}$$

Fig. 6.14

mean effective pressure, p_m, is the height AB of the rectangle. The work done per kg of air can therefore be written as

$$W = \text{area ABCDA} = p_m(v_1 - v_2) \qquad 6.12$$

The term $(v_1 - v_2)$ is proportional to the swept volume of the cylinder, hence it can be seen from equation 6.12 that the mean effective pressure gives a measure of the work output per swept volume. It can therefore be used to compare similar engines of different size. The mean effective pressure discussed in this section is for the air standard cycle. It will be shown in Chapter 18 that the indicated mean effective pressure of an actual engine can be measured from an indicator diagram and used to evaluate the indicated work done by the engine.

Example 6.7

Calculate the mean effective pressure for the cycle of example 6.6.

In example 6.6 the heat supplied, Q_1, and the thermal efficiency were found to be 260 kJ/kg and 68·2% respectively. From equation 5.2

$$\eta = \frac{W}{Q_1}$$

therefore,

$$W = \eta Q_1 = 0 \cdot 682 \times 260 = 177 \text{ kJ/kg}$$

Now from the definition of mean effective pressure, and equation 6.12, we have

$$W = p_m(v_1 - v_2)$$

Using equation 3.5, $pv = RT$ and equation 6.8, $r_v = v_1/v_2 = 18$, then

$$v_1 - v_2 = \left(v_1 - \frac{v_1}{18}\right) = \frac{17}{18} v_1 = \frac{17}{18} \frac{RT_1}{p_1} = \frac{17 \times 287 \times 293}{18 \times 1 \cdot 01 \times 10^5}$$

i.e.
$$v_1 - v_2 = 0 \cdot 786 \text{ m}^3/\text{kg}$$

Then substituting,

$$W = p_m \times 0 \cdot 786 \quad \text{or} \quad p_m = W/0 \cdot 786 \text{ kJ/m}^3$$

i.e.   Mean effective pressure $= \dfrac{177 \times 10^3}{10^5 \times 0 \cdot 786} = 2 \cdot 25$ bar

6.10 The Stirling and Ericsson cycles

It has been shown that no cycle can have an efficiency greater than that of the Carnot cycle working between given temperature limits T_1 and T_2. Cycles which have an efficiency *equal to* that of the Carnot

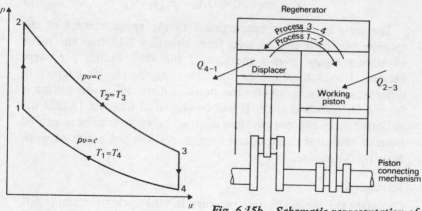

Fig. 6.15a   The Stirling cycle.

Fig. 6.15b   Schematic representation of the Stirling engine.

cycle have been defined and are known as the Stirling* and Ericsson† cycles and they are superior to the Carnot cycle in that they have higher work ratios.

The Stirling cycle is shown in the *p-v* diagram in fig. 6.15a and is represented diagrammatically in fig. 6.15b: it must be emphasized

---

* Reverend Robert Stirling (1790–1878), a Scottish minister of the parish of Galston, Ayrshire, who invented a practical engine in 1817.

† John Ericsson (1803–89), a Swede who worked in England and later in the United States of America.

that this is not a physical description of a Stirling engine but one which may help to give an understanding of the way the processes which make up the cycle are related.

Heat is supplied to the working fluid, usually hydrogen or helium, from an external source, process 2–3, as the gas expands isothermally $(T_2 = T_3)$, and heat is rejected to an external sink, process 4–1, as the gas is compressed isothermally $(T_1 = T_4)$. The two isothermals are connected by the reversible constant volume processes 1–2 and 3–4 during which the temperature changes are equal to $(T_2 - T_1)$. The heat rejected during process 3–4, $Q_{34} = c_v (T_2 - T_1)$, is used to heat the gas during process 1–2, i.e. $Q_{12} = c_v (T_2 - T_1) = Q_{34}$, and this is assumed to take place ideally and reversibly in a *regenerator*. The regenerator requires a matrix of material which separates the heating and cooling gases but allows the temperatures to change progressively by infinitesimal and corresponding amounts during the processes. This regenerative process takes place at constant volume and is *internal* to the cycle.

The Ericsson cycle is similar to the Stirling cycle except that the two isothermals are connected by constant pressure processes, as shown in fig. 6.16

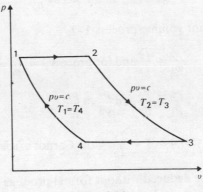

Fig. 6.16

The efficiency of the Stirling cycle is obtained by considering the heat transfers between the system and the bodies external to it, i.e. a high temperature heat supply and a low temperature sink to which heat is rejected.

Heat supplied from the hot source, using equations 4.11 and 4.12,

$$Q_{2-3} = W_{2-3} = RT_2 \log_e \frac{p_2}{p_3}, \text{ per unit mass of gas.}$$

Similarly, the heat rejected to the cold sink,

$$Q_{4-1} = W_{4-1} = RT_1 \log_e \frac{p_1}{p_4}$$

and for the complete system,

$$\text{Net work done} = \text{Net heat supplied}$$
$$W = Q_{2-3} - Q_{4-1}$$

and as the cycle efficiency, $\eta = \dfrac{W}{Q_{2-3}}$

$$\therefore \quad \eta = \frac{Q_{2-3} - Q_{4-1}}{Q_{2-3}}$$

$$= 1 - \frac{Q_{4-1}}{Q_{2-3}}$$

$$= 1 - \frac{RT_1 \log_e \dfrac{p_2}{p_3}}{RT_2 \log_e \dfrac{p_1}{p_4}}$$

For the constant volume process 1–2,

$$\frac{p_2}{p_1} = \frac{T_2}{T_1} \text{ and for process 3–4, } \frac{p_3}{p_4} = \frac{T_3}{T_4} = \frac{T_2}{T_1}$$

$$\therefore \frac{p_2}{p_1} = \frac{p_3}{p_4} \text{ and } \frac{p_2}{p_3} = \frac{p_1}{p_4}$$

$$\therefore \eta = 1 - \frac{T_1}{T_2} = \text{the Carnot efficiency.}$$

(This result can be deduced without formal proof as the heat supply and rejection processes take place at constant temperatures.)

$$\text{The Work ratio} = \frac{W_{2-3} - W_{4-1}}{W_{2-3}}$$

$$= 1 - \frac{W_{4-1}}{W_{2-3}} = 1 - \frac{Q_{4-1}}{Q_{2-3}} = 1 - \frac{T_1}{T_2}$$

and is equal in value to the cycle efficiency.

The practical interpretation of the ideal cycle will not be described in detail here and the reader is advised to consult the specialist literature for the mechanical arrangements employed and the performance assessments (see reference 6.4). Figure 6.15b gives a simplified representation of the engine and shows the necessity for two pistons, a working piston and a displacing piston, which in fact work in different parts of the same cylinder and not as represented. It is necessary to the ideal cycle for the pistons to move discontinuously and this is only approximated to by the mechanisms employed. The result is that the processes of the ideal cycle are not achieved and there is a considerable 'rounding off' of the ideal p-v diagram as the heating and cooling processes merge to depart considerably from the constant volume heating concept.

The early attempts to build a practical 'Stirling' engine were not successful and the achievements of the internal combustion engine rendered it obsolete. Since 1938 when Philips of Eindhoven started to develop the cycle, interest in the practical possibilities of the Stirling engine has increased. The attractions are that the engine can utilize any form of heat from conventional or indigenous fuel, solar or nuclear sources, provided the temperature created is high enough. The engines are quiet, with an efficiency equal to or better than the best internal combustion engines and with little vibration due to the nature of the drive needed to give the differential movements between the working and displacing pistons. The possible range of application of the Stirling engine is wide and includes marine use. electricity generation for peak loads and as a stand-by unit, automotive purposes particularly in comparison with the diesel engine, and for situations when unconventional fuels or heat sources can, or must, be used. The Stirling engine has been considered for use in space using energy from the sun, for non-nuclear submarines and for torpedoes. The most important applications up to now have been as air engines and as a refrigerator; with the Stirling cycle reversed, it is capable of reaching the low temperatures of the cryogenic regions. Machines have been built and used for the liquefaction of gases, and since 1958 the General Motors Corporation of America has built and tested Stirling engines for automotive purposes and a considerable assessment experience has been obtained.

PROBLEMS

6.1 What is the highest thermal efficiency possible for a heat engine operating between 800°C and 15°C? (73·2%)

6.2 Two reversible heat engines operate in series between a source at 527°C and a sink at 17°C. If the engines have equal efficiencies and the first rejects 400 kJ to the second, calculate:

(a) The temperature at which heat is supplied to the second engine,
(b) The heat taken from the source,
(c) The work done by each engine.

Assume that each engine operates on the Carnot cycle.
(209°C; 664 kJ; 264 kJ; 159·2 kJ)

6.3 In a Carnot cycle operating between 307°C and 17°C the maximum and minimum pressures are 62·4 bar and 1·04 bar. Calculate the thermal efficiency and the work ratio. Assume air to be the working fluid. (50%; 0·287)

6.4 A closed cycle gas turbine unit operating with maximum and minimum temperatures of 760°C and 20°C has a pressure ratio of 7/1. Calculate the ideal thermal efficiency and the work ratio.
(42·7%; 0·503)

6.5 In an air standard Otto cycle the maximum and minimum temperatures are 1400°C and 15°C. The heat supplied per kg of air is 800 kJ. Calculate the compression ratio and the thermal efficiency. Calculate also the ratio of maximum to minimum pressures in the cycle. (5·26/1; 48·6%; 30·5/1)

6.6 A four-cylinder petrol engine has a swept volume of 2000 cm³, and the clearance volume in each cylinder is 60 cm³. Calculate the air standard thermal efficiency. If the induction conditions are 1 bar and 24°C, and the maximum cycle temperature is 1400°C, calculate the mean effective pressure based on the air standard cycle.
(59%; 5·27 bar)

6.7 Calculate the thermal efficiency and mean effective pressure of an air standard diesel cycle with a compression ratio of 15/1, and maximum and minimum cycle temperatures of 1650°C and 15°C respectively. The maximum cycle pressure is 45 bar.
(59·1%; 8·39 bar)

6.8 In a dual combustion cycle the maximum temperature is 2000°C and the maximum pressure is 70 bar. Calculate the thermal efficiency and the mean effective pressure when the pressure and temperature at the start of compression are 1 bar and 17°C respectively. The compression ratio is 18/1. (63·6%; 10·5 bar)

6.9 An air standard dual combustion cycle has a mean effective pressure of 10 bar. The minimum pressure and temperature are 1 bar and 17°C respectively, and the compression ratio is 16/1. Calculate the maximum cycle temperature when the thermal efficiency is 60%. The maximum cycle pressure is 60 bar. (1959°C)

REFERENCES

6.1 ROGERS, G. F. C., and MAYHEW, Y. R., *Engineering Thermodynamics, Work and Heat Transfer*, S I Units (Longmans, 1980).

6.2 MENDOZA, E. (translator), *Reflections on the Motive Power of Fire; and Other Papers on the Second Law by Clapeyron and Clausius* (Dover, 1960).

6.3 WILLIAMS, T. T. (editor), *A History of Technology v. VII, 20: The Internal Combustion Engine* (Clarendon Press, 1978).

6.4 WALKER, G., *Stirling Engines* (Oxford Univ. Press, 1980).

6.5 I. Mech. E. Conference: *Stirling engines—progress towards reality* (MEP, 1982).

7. Steam Plant

The heat engine cycle was discussed in Chapter 6, and it was shown that the most efficient cycle is the Carnot cycle for given temperatures of source and sink. This applies to both gases and vapours, and the cycle for a wet vapour is shown in fig. 7.1. A brief summary of the essential features is as follows:

4 to 1: heat is supplied at constant temperature and pressure.

1 to 2: the vapour expands isentropically from the high pressure and temperature to the low pressure. In doing so it does work, which is the purpose of the cycle.

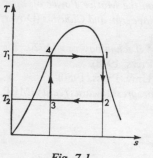

Fig. 7.1

2 to 3: the vapour, which is wet at 2, has to be cooled to state point 3 such that isentropic compression from 3 will return the vapour to its original state at 4. From 4 the cycle is repeated.

The cycle described shows the different types of processes involved in the complete cycle and the changes in the thermodynamic properties of the vapour as it passes through the cycle. The four processes are physically very different from each other and thus they each require particular equipment. The heat supply, 4–1, can be made in a boiler, generator, or nuclear reactor. The work output, 1–2, can be obtained by expanding the vapour through an engine or turbine. The vapour is condensed, 2–3, in a condenser, and to raise the pressure of the wet vapour, 3–4, requires a pump or compressor.

Thus the components of the plant are as shown in fig. 7.3, but before this is discussed further, the deficiencies of the Carnot cycle as the ideal cycle for a vapour must be considered.

7.1 The Rankine cycle

It is stated in Section 6.3 that, although the Carnot cycle is the most efficient cycle, its work ratio is low. Further, there are practical difficulties in following it. Considering the Carnot cycle for steam as shown in fig. 7.1, at state 3 the steam is wet at T_2. It is difficult to stop condensation at the point 3 and then compress it just to state 4. It is more convenient to allow the condensation process to proceed to completion, as in fig. 7.2. The working fluid is water at the new state point 3 in fig. 7.2, and this can be conveniently pumped to boiler pressure as shown at state point 4. The pump has much smaller dimensions than it would have if it had to pump a wet vapour, the compression process is carried out more efficiently, and the equipment required is simpler and less expensive. One of the features of the Carnot cycle has thus been departed from by the modification to the condensation process. At state 4 the water is not at the saturation temperature corresponding to the boiler pressure. Thus heat must be supplied to change the state from water at 4 to saturated water at 5; this is a constant pressure process, but is not at constant temperature. Hence the efficiency of this modified cycle is not as high as that of the Carnot cycle. This ideal cycle, which is more suitable as a criterion for actual steam cycles than the Carnot cycle, is called the *Rankine cycle*.

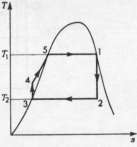

Fig. 7.2

The plant required for the Rankine cycle is shown in fig. 7.3, and the numbers refer to the state points of fig. 7.2. The steam at inlet to the turbine may be wet, dry saturated, or superheated, but only the dry saturated condition is shown in fig. 7.2. The steam flows round the cycle and each process may be analysed using the steady flow energy equation; changes in kinetic energy and potential energy may be neglected.

i.e.
$$h_1 + Q = h_2 + W$$

In this statement of the equation the subscripts 1 and 2 refer to the initial and final state points of the process; each process in the cycle can be considered in turn as follows:

Boiler:
$$h_4 + Q_{451} = h_1 + W$$

Therefore, since $W = 0$, $Q_{451} = h_1 - h_4$

7.1

Turbine: The expansion is adiabatic (i.e. $Q=0$), and isentropic (i.e. $s_1=s_2$), and h_2 can be calculated using this latter fact. Then

$$h_1 + Q_{12} = h_2 + W_{12}$$
$$\therefore W_{12} = (h_1 - h_2) \qquad 7.2$$

Condenser:

$$h_2 + Q_{23} = h_3 + W$$

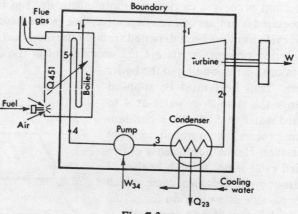

Fig. 7.3

Therefore, since $W=0$

$$Q_{23} = h_3 - h_2 \qquad \text{i.e.} \quad Q_{23} = -(h_2 - h_3)$$
$$\therefore \text{Heat rejected in condenser} = h_2 - h_3 \qquad 7.3$$

Pump:

$$h_3 + Q_{34} = h_4 + W_{34}$$

The compression is isentropic (i.e. $s_3=s_4$), and adiabatic (i.e. $Q=0$).

$$\therefore W_{34} = (h_3 - h_4) = -(h_4 - h_3)$$

i.e. $$\text{Work input to pump} = (h_4 - h_3) \qquad 7.4$$

This is the feed pump term, and as it is a small quantity in comparison with the turbine work, W_{12}, it is usually neglected, especially when boiler pressures are low.

$$\text{Net work done in the cycle, } W = W_{12} + W_{34}$$

i.e. $$W = (h_1 - h_2) - (h_4 - h_3) \qquad 7.5$$

Or, if the feed pump work is neglected,

$$W = (h_1 - h_2) \qquad 7.6$$

The heat supplied in the boiler, $Q_{451} = h_1 - h_4$. Then we have

$$\text{Rankine efficiency, } \eta_R = \frac{\text{Net work output}}{\text{Heat supplied in the boiler}} \qquad 7.7$$

i.e.

$$\eta_R = \frac{(h_1 - h_2) - (h_4 - h_3)}{h_1 - h_4}$$

or

$$\eta_R = \frac{(h_1 - h_2) - (h_4 - h_3)}{(h_1 - h_3) - (h_4 - h_3)} \qquad 7.8$$

If the feed pump term, $h_4 - h_3$, is neglected equation 7.8 becomes

$$\eta_R = \frac{h_1 - h_2}{h_1 - h_3} \qquad 7.9$$

When the feed pump term is to be included it is necessary to evaluate the quantity, W_{34}.

From equation 7.4

$$\text{Pump work} = -W_{34} = (h_4 - h_3)$$

It can be shown that for a liquid, which is assumed to be incompressible (i.e. $v = $ constant), the increase in enthalpy for isentropic compression is given by

$$(h_4 - h_3) = v(p_4 - p_3)$$

The proof is as follows:

For a reversible adiabatic process, from equation 4.15,

$$dQ = dh - v\, dp = 0$$

$$\therefore\ dh = v\, dp$$

i.e.

$$\int_3^4 dh = \int_3^4 v\, dp$$

For a liquid, since v is approximately constant, we have

$$h_4 - h_3 = v \int_3^4 \mathrm{d}p = v(p_4 - p_3)$$

i.e. $(h_4 - h_3) = v(p_4 - p_3)$

$\therefore$ Pump work input $= (h_4 - h_3) = v(p_4 - p_3)$ 7.10

(where v can be taken from tables for water at the pressure p_3).

The *efficiency ratio* of a cycle is the ratio of the actual efficiency to the ideal efficiency. In vapour cycles the efficiency ratio compares the actual cycle efficiency to the Rankine cycle efficiency,

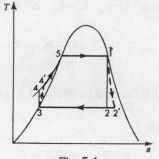

Fig. 7.4

i.e.   Efficiency ratio $= \dfrac{\text{cycle efficiency}}{\text{Rankine efficiency}}$ 7.11

The actual expansion process is irreversible, as shown by line 1–2' in fig. 7.4. Similarly the actual compression of the water is irreversible, as indicated by line 3–4'. The *isentropic efficiency* of a process is defined by

Isentropic efficiency

$= \dfrac{\text{actual work}}{\text{isentropic work}}$, for an expansion process

and

Isentropic efficiency

$= \dfrac{\text{isentropic work input}}{\text{actual work input}}$, for a compression process.

Hence,

$$\text{Turbine isentropic efficiency} = \frac{W_{12'}}{W_{12}} = \frac{h_1 - h_{2'}}{h_1 - h_2} \qquad 7.12$$

It has been stated that the efficiency of the Carnot cycle is the maximum possible, but that the cycle has a low work ratio. Both efficiency and work ratio are criteria of performance. By the definition of work ratio in Section 6.3,

$$\text{Work ratio} = \frac{\text{net work}}{\text{gross work}} \qquad 7.13$$

Another criterion of performance in steam plant is the *specific steam consumption*. It relates the power output to the steam flow necessary to produce it. The steam flow indicates the size of plant and its component parts, and the specific steam consumption is a means whereby the relative sizes of different plants can be compared.

The specific steam consumption is the steam flow in kg/h required to develop 1 kW,

i.e. $W \times$ (specific steam consumption, s.s.c.) $= 1 \times 3600$ kJ/h

(where W is in kJ/kg),

i.e. $$\text{s.s.c.} = \frac{3600}{W} \text{ kg/kW h} \qquad 7.14$$

Neglecting the feed pump work, from equation 7.6 we have $W = (h_1 - h_2)$, therefore

$$\text{s.s.c.} = \frac{3600}{h_1 - h_2} \text{ kg/kW h} \qquad 7.15$$

(where h_1 and h_2 are in kJ/kg).

Example 7.1

A steam power plant operates between a boiler pressure of 42 bar and a condenser pressure of 0·035 bar. Calculate for these limits the cycle efficiency, the work ratio, and the specific steam consumption:

(a) for a Carnot cycle using wet steam;
(b) for a Rankine cycle with dry saturated steam at entry to the turbine; and

(c) for the Rankine cycle of (b), when the expansion process has an isentropic efficiency of 80%.

(a) A Carnot cycle is shown in fig. 7.5.

$$T_1 \text{ saturation temperature at 42 bar}$$
$$= 253 \cdot 2 + 273 = 526 \cdot 2 \text{ K}$$
$$T_2 = \text{saturation temperature at } 0 \cdot 035 \text{ bar}$$
$$= 26 \cdot 7 + 273 = 299 \cdot 7 \text{ K}$$

Then from equation 6.1

$$\eta_{\text{Carnot}} = \frac{T_1 - T_2}{T_1} = \frac{526 \cdot 2 - 299 \cdot 7}{526 \cdot 2} = 0 \cdot 432 \quad \text{or} \quad 43 \cdot 2\%$$

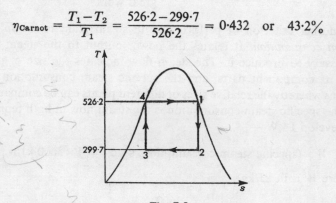

Fig. 7.5

Also,

$$\text{Heat supplied} = h_1 - h_4 = h_{fg} \text{ at 42 bar} = 1698 \text{ kJ/kg}$$

Then, $$\eta_{\text{Carnot}} = \frac{W}{Q} = 0 \cdot 432 \qquad \therefore \quad W = 0 \cdot 432 \times 1698$$

i.e. $$W = 734 \text{ kJ/kg}$$

To find the gross work of the expansion process it is necessary to calculate h_2, using the fact that $s_1 = s_2$.

From tables,

$$h_1 = 2800 \text{ kJ/kg} \quad \text{and} \quad s_1 = s_2 = 6 \cdot 049 \text{ kJ/kg K}$$

Using equation 5.10

$$s_2 = 6 \cdot 049 = s_{f_2} + x_2 s_{fg_2} = 0 \cdot 391 + x_2 8 \cdot 13$$

$$\therefore \quad x_2 = \frac{6 \cdot 049 - 0 \cdot 391}{8 \cdot 13} = 0 \cdot 696$$

Then using equation 3.2

$$h_2 = h_{f_2} + x_2 h_{fg_2} = 112 + 0.696 \times 2438 = 1808 \text{ kJ/kg}$$

Hence, from equation 7.2

$$W_{12} = (h_1 - h_2) = (2800 - 1808) = 992 \text{ kJ/kg}$$

Therefore we have, using equation 7.13,

$$\text{Work ratio} = \frac{\text{net work}}{\text{gross work}} = \frac{734}{992} = 0.739$$

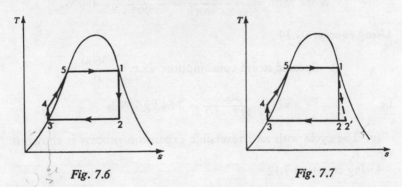

Fig. 7.6   Fig. 7.7

Using equation 7.14

$$\text{Specific steam consumption, s.s.c.} = \frac{3600}{W} = \frac{3600}{734}$$

i.e.   s.s.c. $= 4.9 \text{ kg/kW h}$

(b) The Rankine cycle is shown in fig. 7.6.

As in part (a)

$$h_1 = 2800 \text{ kJ/kg} \quad \text{and} \quad h_2 = 1808 \text{ kJ/kg}$$

Also, $h_3 = h_f$ at 0.035 bar $= 112 \text{ kJ/kg}$

Using equation 7.10, with $v = v_f$ at 0.035 bar

$$\text{Pump work} = v_f(p_4 - p_3) = 0.001 \times (42 - 0.035) \times \frac{10^5}{10^3}$$
$$= 4.2 \text{ kJ/kg}$$

Using equation 7.2

$$W_{12} = h_1 - h_2 = 2800 - 1808 = 992 \text{ kJ/kg}$$

Then using equation 7.8

$$\eta_R = \frac{(h_1 - h_2) - (h_4 - h_3)}{(h_1 - h_3) - (h_4 - h_3)} = \frac{992 - 4 \cdot 2}{(2800 - 112) - 4 \cdot 2} = 0 \cdot 368$$

i.e. $\qquad \eta_R = 36 \cdot 8\%$

Using equation 7.13

$$\text{Work ratio} = \frac{\text{net work}}{\text{gross work}} = \frac{992 - 4 \cdot 2}{992} = 0 \cdot 996$$

Using equation 7.14

$$\text{Specific steam consumption, s.s.c.} = \frac{3600}{W}$$

i.e. $\qquad \text{s.s.c.} = \frac{3600}{992 - 4 \cdot 2} = 3 \cdot 64 \text{ kg/kW h}$

(c) The cycle with an irreversible expansion process is shown in fig. 7.7.

Using equation 7.12

$$\text{Isentropic efficiency} = \frac{h_1 - h_{2'}}{h_1 - h_2} = \frac{W_{12'}}{W_{12}}$$

$$\therefore \quad 0 \cdot 8 = \frac{W_{12'}}{992}$$

i.e. $\qquad W_{12'} = 0 \cdot 8 \times 992 = 793 \cdot 6 \text{ kJ/kg}$

Then the cycle efficiency is given by

$$\text{Cycle efficiency} = \frac{(h_1 - h_{2'}) - (h_4 - h_3)}{\text{heat supplied}}$$

$$= \frac{793 \cdot 6 - 4 \cdot 2}{(2800 - 112) - 4 \cdot 2} = 0 \cdot 294$$

i.e. $\qquad \text{Cycle efficiency} = 29 \cdot 4\%$

$$\text{Work ratio} = \frac{W_{12'} - \text{pump work}}{W_{12'}} = \frac{793 \cdot 6 - 4 \cdot 2}{793 \cdot 6} = 0 \cdot 995$$

Also,

$$\text{s.s.c.} = \frac{3600}{793 \cdot 6 - 4 \cdot 2} = \frac{3600}{789 \cdot 4} = 4 \cdot 56 \text{ kg/kW h}$$

The feed pump term has been included in the above calculations, but an inspection of the comparative values shows that it could have been neglected without having a noticeable effect on the results.

It is instructive to carry out these calculations for different boiler pressures and to represent the results graphically against boiler pressure, as in fig. 7.8. As the boiler pressure increases the latent heat decreases, thus less heat is transferred at the maximum cycle temperature. Although the efficiency increases with boiler pressure over

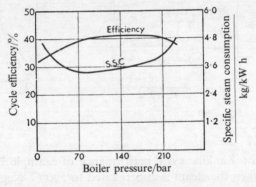

Fig. 7.8

the first part of the range, due to the maximum cycle temperature being raised, it is affected by the lowering of the mean temperature at which heat is transferred. The graph for the efficiency rises therefore, reaches a maximum, and then falls.

7.2 Rankine cycle with superheat

The average temperature at which heat is supplied in the boiler can be increased by superheating the steam. Usually the dry saturated steam from the boiler drum is passed through a second bank of smaller bore tubes within the boiler. This bank is situated such that it is heated by the hot gases from the furnace until the steam reaches the required temperature.

The Rankine cycle with superheat is shown in fig. 7.9a and fig.

7.9b. Fig. 7.9a includes a *steam receiver* which can receive steam from other boilers. In modern plant a receiver is used with one boiler and is placed between the boiler and the turbine. Since the quantity of feed water varies with the different demands on the boiler, it is necessary to provide a storage of condensate between the condensate and boiler feed pumps. This storage may be either a *surge tank* or *hot well*. A hot well is shown dotted in fig. 7.9a.

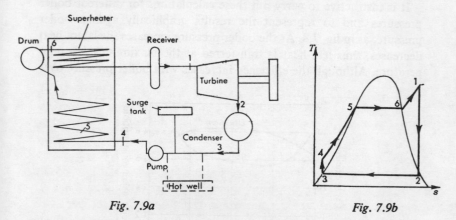

Fig. 7.9a   Fig. 7.9b

Example 7.2

Compare the Rankine cycle performance of example 7.1 with that obtained when the steam is superheated to 500°C. Neglect the feed pump work.

From tables, by interpolation, at 42 bar:

$$h_1 = 3442 \cdot 6 \text{ kJ/kg} \quad \text{and} \quad s_1 = s_2 = 7 \cdot 066 \text{ kJ/kg K}$$

Using equation 5.10

$$s_2 = s_{f_2} + x_2 s_{fg_2} \qquad \therefore \quad 0 \cdot 391 + x_2 8 \cdot 13 = 7 \cdot 066$$

i.e. $x_2 = 0 \cdot 821$

Using equation 3.2

$$h_2 = h_{f_2} + x_2 h_{fg_2} = 112 + (0 \cdot 821 \times 2438) = 2113 \text{ kJ/kg}$$

From tables $h_3 = 112 \text{ kJ/kg}$

Then, using equation 7.2

$$W_{12} = h_1 - h_2 = 3442 \cdot 6 - 2113 = 1329 \cdot 6 \text{ kJ/kg}$$

Neglecting the feed pump term, we have

Heat supplied $= h_1 - h_3 = 3442 \cdot 6 - 112 = 3330 \cdot 6$ kJ/kg

Using equation 7.9

Thermal efficiency $= \dfrac{h_1 - h_2}{h_1 - h_3} = \dfrac{1329 \cdot 6}{3330 \cdot 6} = 0 \cdot 399$ or $39 \cdot 9\%$

Also, using equation 7.14

$$\text{s.s.c.} = \frac{3600}{W_{12}} = \frac{3600}{1329 \cdot 6} = 2 \cdot 71 \text{ kg/kW h}$$

The thermal efficiency has increased due to superheating and the improvement in specific steam consumption is even more marked. This indicates that for a given power output the plant using superheated steam will be of smaller proportions than that using dry saturated steam.

The condenser heat loads for different plants can be compared by calculating the rate of heat removal in the condenser, per unit power output. This is given by the product, s.s.c. $\times (h_2 - h_3)$, where $h_2 - h_3$ is the heat removed in the condenser by the cooling water, per kg of steam. Comparing the condenser heat loads for the Rankine cycles of examples 7.1 and 7.2, we have

With dry saturated steam at entry to turbine:

Condenser heat load $= 3 \cdot 64(1808 - 112) = 6175$ (kJ/h)/kW

With superheated steam at entry to the turbine:

Condenser heat load $= 2 \cdot 71(2113 - 112) = 5420$ (kJ/h)/kW

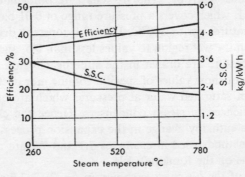

Fig. 7.10

For given boiler and condenser pressures, as the superheat temperature increases, the Rankine cycle thermal efficiency increases, and the specific steam consumption decreases, as shown in fig. 7.10.

There is also a practical advantage in using superheated steam. For the data of the Rankine cycle of examples 7.1 and 7.2, the steam leaves the turbine with dryness fractions of 0·696 and 0·821, respectively. The presence of water during the expansion is undesirable, since the droplets are denser than the remainder of the working fluid and therefore have different flow characteristics. The result is the physical erosion of the turbine blades, and a reduction in isentropic efficiency.

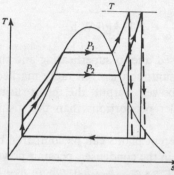

Fig. 7.11

The modern tendency is to use higher boiler pressures, and a comparison of cycles on the *T-s* diagram shows that for a given steam temperature at turbine inlet, the higher pressure plant will have the wetter steam at turbine exhaust (see fig. 7.11 in which $p_1 > p_2$). It is usual to design for a dryness fraction of not less than 0·9 at the turbine exhaust.

7.3 The enthalpy-entropy chart (*h-s* chart)

In this chapter, and in later ones, we are concerned with changes in enthalpy. It is convenient to have a chart on which enthalpy is plotted against entropy. The *h-s* chart recommended is that of reference 7.1, which covers a pressure range of 0·01 bar to 1000 bar, and temperatures up to 800°C. Lines of constant dryness fraction are drawn in the wet region to values less than 0·5, and lines of constant temperature are drawn in the superheat region. *h-s* charts in general do not show values of specific volume, nor do they show the enthalpies of saturated water at pressures which are of the order of those experienced in steam condensers. Hence the chart is useful only for the enthalpy change in the expansion process of the steam cycle; the methods used in examples 7.1 and 7.2 are recommended for problems on the Rankine cycle.

A sketch of the *h-s* chart is shown in fig. 7.12. Lines of constant

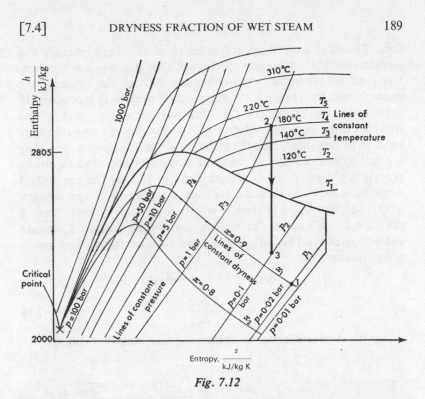

Fig. 7.12

pressure are indicated by p_1, p_2, etc.; lines of constant temperature by T_1, T_2, etc. Any two independent properties which appear on the chart are sufficient to define the state (e.g. p_1 and x_1 define state 1, and h_1 can be read off the vertical axis). In the superheat region, pressure and temperature can define the state (e.g. p_3 and T_4 define the state 2, and h_2 can be read off). A line of constant entropy between two state points 2 and 3 defines the properties at all points during an isentropic process between the two states.

7.4 Dryness fraction of wet steam

In Section 3.1 the definition of the dryness fraction of a vapour is given. It is often necessary in steam plant work to determine the condition or state of the steam, and for wet steam this entails finding the dryness fraction. In Section 1.5 it is stated that to define the state of a pure substance, two independent properties are required. Since the directly observable properties are pressure, temperature, and specific volume, two of these are required and must be independent. The most convenient properties to measure are pressure and temperature. If the steam is superheated, then pressure and temperature readings are sufficient to define the state of the steam; if the steam is wet, the pressure and temperature are not independent. In Section 4.4 the process of throttling was described, and it was seen that if the velocities before and after the process are sufficiently small to be negligible, then the enthalpy of the fluid before throttling is equal to the enthalpy of the fluid after throttling. The throttling process is shown on an h-s diagram in fig. 7.13 by the line 1–2. If steam initially wet is throttled through a sufficiently large pressure drop, then the steam at state 2 will become superheated. State 2 can then be defined, as described above, by the measured pressure and temperature. The enthalpy, h_2, can then be found, and hence, using equation 3.2,

$$h_2 = h_1 = h_{f_1} + x_1 h_{fg_1}$$

or
$$x_1 = \frac{h_2 - h_{f_1}}{h_{fg_1}}$$
7.16

Hence the dryness fraction is determined and state 1 is defined.

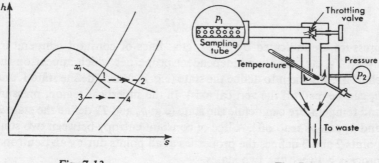

Fig. 7.13   *Fig. 7.14*

This determination can be carried out using a throttling calorimeter which is illustrated diagrammatically in fig. 7.14. The steam to be sampled is taken from the pipe by means of a suitably positioned and dimensioned sampling tube. It passes into an insulated container and is throttled through an orifice to atmospheric pressure. Here the temperature is taken and the steam ideally should have about 5·5 K of superheat.

If the steam is very wet, then throttling to atmospheric pressure may not be sufficient to ensure superheated steam at exit. In this case it is necessary to dry the steam partially, before throttling. This is done by passing the steam sample from the main through a separating calorimeter as in fig. 7.15. The steam is made to change direction

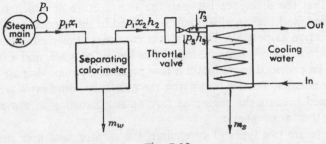

Fig. 7.15

suddenly, and the water, being denser than the dry vapour, is separated out. The quantity of water which is separated out is measured at the separator, and the steam remaining, which now has a higher dryness fraction, is passed through the throttling calorimeter. With the combined separating and throttling calorimeter it is necessary to condense the steam after throttling and measure the amount of condensate. If a throttling calorimeter only is sufficient, there is no need to measure condensate, the pressure and temperature measurements at exit being sufficient.

Dryness fraction at 2 is x_2, therefore the mass of dry vapour leaving the separating calorimeter is equal to $x_2 m_s$, and this must be the mass of dry vapour in the sample drawn from the main at state 1.

Hence,

$$\text{Dryness fraction in main, } x_1 = \frac{\text{Mass of dry vapour}}{\text{Total mass}}$$

$$= \frac{x_2 m_s}{m_w + m_s}$$

The dryness fraction, x_2, can be determined using equation 3.2,

$$h_3 = h_2 = h_{f_2} + x_2 h_{fg_2} \qquad \therefore \ x_2 = \frac{h_3 - h_{f_2}}{h_{fg_2}}$$

The values h_{f_2} and h_{fg_2} are read from tables at pressure p_2. The pressure loss in the separator is small so that p_1 is approximately equal to p_2.

7.5 Steam condensers

The function of the condenser is to receive the exhaust steam from the turbine or engine, condense it, and deliver the condensate to the feed pump. In order to increase the efficiency of the plant it is necessary that the difference between the temperature at which heat is supplied, and that at which heat is rejected in the condenser, should be as large as possible. For a steam condenser this means that the pressure in the condenser should be as low as possible, and a means must be provided to maintain a low pressure by removing air from the condenser. Besides improving the efficiency some extra work is obtained because the engine or turbine is exhausting to a pressure lower than atmospheric.

There are two types of condenser, the *surface condenser* and the *jet condenser*, but there are a number of variations within the general classification. A shell and tube type surface condenser is shown in fig. 7.16. Cooling water, from a cooling tower (see Section 14.5), or

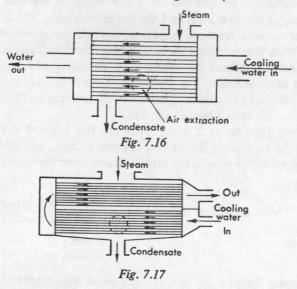

Fig. 7.16

Fig. 7.17

river, passes through the condenser tubes, and the steam condenses on the outside of the tubes. The average temperature difference between the condensing steam and the cooling water is about 12 K.

The water may make two passes through the condenser as shown in fig. 7.17. The two pass arrangement gives a more compact condenser and is more usual if the cooling water is in short supply.

The heat exchanging process is more efficient in the two pass system, but against this must be set the extra pumping power required to pass the water through the tubes. The selection of a condenser for a particular plant is thus a compromise.

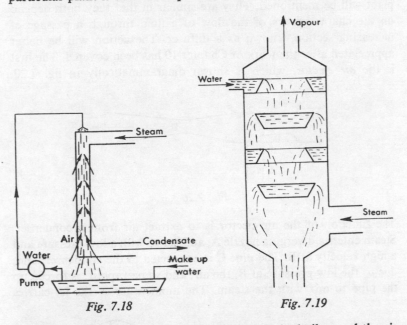

Fig. 7.18   Fig. 7.19

The condensate has to be pumped away to the boiler, and the air may be removed from the condenser by the same pump, or by a separate extraction pump. In the former case, since it is mainly liquid which is being pumped, the power expenditure is small. The air may be removed by an air pump or an air ejector. The condenser contains a mixture of water, steam, and air, and an analysis of the condenser operating characteristics will be left until the properties of mixtures have been considered (see Section 13.7).

There is another type of surface condenser called the *evaporative condenser* (see fig. 7.18). The steam passes through a series of tubes over which cooling water falls, while a stream of air passes over the tubes.

With jet condensers the cooling water is mixed with the steam, and the two streams may be arranged to meet at the top of the condenser, or the water may enter at the top and the steam enter at the bottom (see fig. 7.19). The counter current type is the only one of importance.

The mixing of vapour and cooling water may be a disadvantage if the water is untreated and is not of sufficiently good quality for boiler feed water.

Two pieces of equipment which sometimes form part of a vapour plant will be mentioned. They are similar in that they both depend on the characteristics of the flow of a fluid through a passage of increasing section, known as a diffuser. The action will be better appreciated after the work of Chapter 10 has been covered. The first is the *air ejector*, which is shown diagrammatically in fig. 7.20.

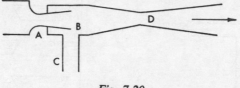

Fig. 7.20

The function of the air ejector is to extract air from a condenser. Steam enters a diverging nozzle A, and leaves with a low pressure and a high velocity at B. The pipe C is connected to the condenser, and, due to the low pressure at B, the air, and some vapour, is forced up the pipe to mix with the steam. The momentum of the jet carries

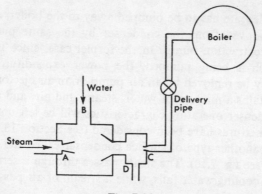

Fig. 7.21

the mixture through the diffuser D. Here the velocity decreases and the pressure increases to the outlet value. The ejection process can be carried out in one or more stages (up to four). It is usual to obtain a vacuum of about 0·90 bar with a single stage ejector, and absolute pressures of the order of 0·003 bar can be obtained in the condenser using a four-stage ejector. The second piece of equipment is the *injector* (see fig. 7.21), which is used to feed water into the boiler, and operates usually from boiler steam. The steam expands through the nozzle A at entry and makes contact with the water from the feed tank B. The steam is condensed in the converging pipe and the resulting hot water enters the delivery pipe. At the inlet end of the delivery pipe the cross-sectional area is a minimum and the velocity is a maximum. In the diffusing section of the pipe kinetic energy is lost and the pressure increases. The pressure reaches a value which is greater than the boiler pressure, and so the water is admitted to the boiler. The vessel D is to allow for the overflow of water on starting up the injector.

7.6 The reheat cycle

It is desirable to increase the average temperature at which heat is supplied to the steam, and also to keep the steam as dry as possible in the lower pressure stages of the turbine. The wetness at exhaust should be no greater than 10%. The considerations of Section 7.2 show that high boiler pressures are required for high efficiency, but that expansion in one stage can result in exhaust steam which is wet. This is a condition which is improved by superheating the steam. The exhaust steam condition can be improved most effectively by *reheating* the steam, the expansion being carried out in two stages. Referring to fig. 7.22, 1–2 represents isentropic expansion in the high-pressure turbine, and 6–7 represents isentropic expansion in the low-pressure turbine. The steam is reheated at constant pressure in process 2–6. The reheat can be carried out by returning the steam to the boiler, and passing it through a special bank of tubes, the reheat bank of tubes being situated in the proximity of the superheat tubes. Alternatively, the reheat may take place in a separate reheater situated near the turbine; this arrangement reduces the

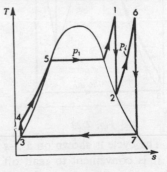

Fig. 7.22

amount of pipe work required. The use of reheat cycles has encouraged the development of higher pressure, forced circulation boilers, since the specific steam consumption is improved, and the dryness fraction of the exhaust steam is increased.

The T-s diagram for the reheat cycle is shown in fig. 7.22, and the analysis is as follows:

$$\text{Heat supplied} = Q_{451} + Q_{26}$$

Neglecting the feed pump work,

$$Q_{451} = h_1 - h_3$$

Also, for the reheat process,

$$Q_{26} = h_6 - h_2$$

$$\text{Work output} = W_{12} + W_{67}$$

And,

$$W_{12} = h_1 - h_2 \quad \text{and} \quad W_{67} = h_6 - h_7$$

i.e.

$$\text{Cycle efficiency} = \frac{W_{12} + W_{67}}{Q_{451} + Q_{26}}$$

$$= \frac{(h_1 - h_2) + (h_6 - h_7)}{(h_1 - h_3) + (h_6 - h_2)}$$

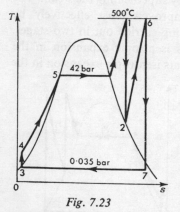

Fig. 7.23

Example 7.3

Calculate the new cycle efficiency and specific steam consumption if reheat is included in the plant of example 7.2. The steam conditions at inlet to the turbine are 42 bar and 500°C, and the condenser pressure is 0·035 bar as before. Assume that the steam is just dry saturated on leaving the first turbine, and is reheated to its initial temperature. Neglect the feed pump term.

The cycle is shown on a T-s diagram in fig. 7.23.

It is convenient to read off the values of enthalpy from the h-s chart, i.e. $h_1 = 3442 \cdot 6$ kJ/kg; $h_2 = 2713$ kJ/kg (at 2·3 bar); $h_6 = 3487$ kJ/kg (at 2·3 bar and 500°C); $h_7 = 2535$ kJ/kg.

From tables,

$$h_3 = 112 \text{ kJ/kg}$$

Then,

$$\text{Turbine work} = (h_1 - h_2) + (h_6 - h_7)$$
$$= (3443 - 2713) + (3487 - 2535)$$

i.e.   Turbine work = 1682 kJ/kg

$$\text{Heat supplied} = (h_1 - h_3) + (h_6 - h_2)$$
$$= (3443 - 112) + (3487 - 2713)$$

i.e.   Heat supplied = 4105 kJ/kg

$$\therefore \text{ Cycle efficiency} = \frac{1682}{4105} = 0 \cdot 41 \quad \text{or} \quad 41\%$$

Also,

$$\text{Specific steam consumption, s.s.c.} = \frac{3600}{W} = \frac{3600}{1682}$$

i.e.   s.s.c. = 2·14 kg/kW h

Comparing these answers with the results of example 7.2 it can be seen that the specific steam consumption has been improved considerably by reheating (i.e. reduced from 2·71 kg/kW h to 2·14 kg/ kW h). The effect on the efficiency is very small (i.e. increased from 39·9% to 41%). If reheating takes place at a low pressure, then the thermal efficiency will be reduced by reheating, since the average temperature during heating will then be low.

7.7 The regenerative cycle

In order to achieve the Carnot efficiency it is necessary to supply and reject heat at single fixed temperatures. One method of doing this, and at the same time having a work ratio comparable to the Rankine cycle, is by raising the feed water to the saturation temperature corresponding to the boiler pressure before it enters the boiler. This method is not a practical proposition, but is of academic interest. The feed water is passed from the pump through the turbine in counter-flow to the steam, as shown in fig. 7.24a. The feed water enters the turbine at t_3 and is heated to the steam temperature at inlet to the turbine. If at all points the temperature difference

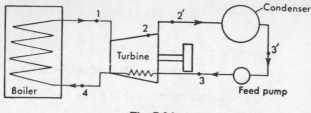

Fig. 7.24a

between the steam and the water is negligibly small, then the heat transfer takes place in an ideal reversible manner. Assuming dry saturated steam at turbine inlet, the expansion process is represented by line 1–2–2' in fig. 7.24b. The heat rejected by the steam, area

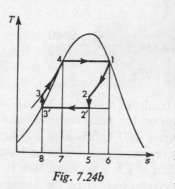

Fig. 7.24b

12561, is equal to the heat taken up by the water, area 34783. The heat supplied in the boiler is given by area 41674, and the heat rejected in the condenser is given by area 3'2'583'. This *regenerative* cycle has an efficiency equal to the Carnot cycle, since the heat supplied and rejected externally is done at constant temperature.

This cycle is clearly not a practical proposition, and in addition it can be seen that the turbine operates with wet steam which is to be avoided if possible. However the Rankine efficiency can be improved upon in practice by bleeding off some of the steam at an intermediate pressure during the expansion, and mixing this steam with feed water which has been pumped to the same pressure. The mixing process is carried out in a *feed heater*, and the arrangement is represented in figs. 7.25a and 7.25b. Only one feed heater is shown but several could be used.

The steam expands from condition 1 through the turbine. At the pressure corresponding to point 6, a quantity of steam, say y kg per kg of steam supplied from the boiler, is bled off for feed heating purposes. The rest of the steam $(1-y)$ kg, completes the expansion and is exhausted at state 2. This amount of steam is then condensed and pumped to the same pressure as the bleed steam. The bleed steam and the feed water are mixed in the feed heater, and the quantity of bleed steam, y kg, is such that after mixing and being pumped in a second

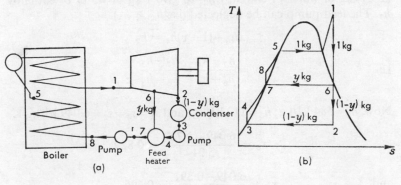

Fig. 7.25

feed pump, the condition is as defined by state 8. The heat to be supplied in the boiler is then given by $(h_1 - h_8)$ kJ/kg of steam; this heat is supplied between the temperatures T_8 and T_1.

If this procedure could be repeated an infinite number of times, then the ideal regenerative cycle would be approached.

It is necessary to determine the bleed pressure when one or more feed heaters are used, and this can be based on the assumption that the bleed temperature to obtain maximum efficiency for such a cycle is approximately the mean of the temperatures at 5 and 2 (see fig. 7.25b),

i.e. $$t_{\text{bleed}} = \frac{t_5 + t_2}{2} \qquad 7.17$$

Example 7.4

If the Rankine cycle of example 7.1 is modified to include one feed heater, calculate the cycle efficiency and the specific steam consumption.

The steam enters the turbine at 42 bar, dry saturated, and the condenser pressure is 0·035 bar.

At 42 bar, $t_1 = 253 \cdot 2°C$; and at 0·035 bar, $t_2 = 26 \cdot 7°C$.

Therefore by equation 7.17,

$$t_6 = \frac{253 \cdot 2 + 26 \cdot 7}{2} = 140°C$$

Selecting the nearest saturation pressure from the tables gives the bleed pressure as 3·5 bar (i.e. $t_6 = 138.9°C$).

To determine the fraction y, consider the adiabatic mixing process at the feed heater, in which y kg of steam of enthalpy h_6, mix with $(1-y)$ kg of water of enthalpy h_3, to give 1 kg of water of enthalpy h_7. The feed pump can be neglected (i.e. $h_4 = h_3$).

$$\therefore \; yh_6 + (1-y)h_4 = h_7$$

i.e.

$$y = \frac{h_7 - h_4}{h_6 - h_4} = \frac{h_7 - h_3}{h_6 - h_3}$$

Now, $h_7 = 584$ kJ/kg; $h_3 = 112$ kJ/kg; and $s_1 = s_6 = s_2 = 6.049$ kJ/kg K.

$$\therefore \; x_6 = \frac{6.049 - 1.727}{5.214} = 0.829$$

and

$$x_2 = \frac{6.049 - 0.391}{8.130} = 0.696$$

Hence,

$$h_6 = h_{f_6} + x_6 h_{fg_6} = 584 + (0.829 \times 2148) = 2364 \text{ kJ/kg}$$

and

$$h_2 = h_{f_2} + x_2 h_{fg_2} = 112 + (0.696 \times 2438) = 1808 \text{ kJ/kg}$$

$$\therefore \; y = \frac{584 - 112}{2364 - 112} = 0.21 \text{ kg}$$

Neglecting the second feed pump term (i.e. $h_7 = h_8$), we have,

$$\text{Heat supplied in boiler} = (h_1 - h_7) = 2800 - 584$$
$$= 2216 \text{ kJ/kg}$$

Total work output, $W = W_{16} + W_{62} = (h_1 - h_6) + (1-y)(h_6 - h_2)$

i.e.

$$\text{Work output} = (2800 - 2364)$$
$$+ (1 - 0.21)(2364 - 1808)$$
$$= 876 \text{ kJ per kg of steam}$$
$$\text{delivered by the boiler}$$

Therefore,

$$\text{Cycle efficiency} = \frac{W}{Q} = \frac{876}{2216} = 0.396 \quad \text{or} \quad 39.6\%$$

and

$$\text{s.s.c.} = \frac{3600}{W} = \frac{3600}{876} = 4.12 \text{ kg/kW h}$$

Comparing these results with those of example 7.1, it can be seen that the addition of one feed heater has increased the thermal efficiency from 36·8% to 39·6%, but the specific steam consumption has increased from 3·64 kg/kW h to 4·12 kg/kW h. The thermal efficiency continues to be increased with the addition of further heaters, but the capital expenditure is also increased considerably since a feed pump is required with each feed heater. Because of the number of feed pumps required, the heating of the feed water by mixing is dispensed with, and *closed heaters* are used. The method is indicated in fig. 7.26 for two feed heaters, but the number used could be as high as eight. (An efficient modern plant with reheating and five regenerative heaters has a cycle efficiency of 35%.) Referring to fig. 7.26, the feed water is passed at boiler pressure through the feed heaters 2 and

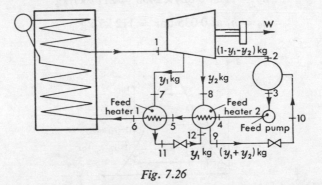

Fig. 7.26

1 in series. An amount of bleed steam, y_1, is passed to feed heater 1, and the feed water receives heat from it by the transfer of heat through the separating tubes. The condensed steam is then throttled to the next feed heater which is also supplied with a second quantity of bleed steam, y_2, and a lower temperature heating of the feed water is carried out. When the final feed heating has been accomplished the condensed steam is then fed to the condenser. The temperature differences between successive heaters are constant, and the heating process at each is considered to be complete (i.e. the feed water leaves the feed heater at the temperature of the bleed steam supplied to it).

Example 7.5

In a regenerative steam cycle employing two closed feed heaters the steam is supplied to the turbine at 40 bar and 500°C and is exhausted to the condenser at 0·035 bar. The intermediate bleed pressures are obtained such that the *saturation* temperature intervals are approximately equal, giving pressures of 10 and 1·1 bar.

Calculate the amount of steam bled at each stage, the work output of the plant in kJ/kg of boiler steam and the thermal efficiency of the plant. Assume ideal processes where required.

Referring to fig. 7.26 and the *T-s* diagram of fig. 7.27, from tables:

$$h_1 = 3445 \text{ kJ/kg} \quad \text{and} \quad s_1 = 7·089 \text{ kJ/kg K} = s_2$$

At state 2, $\qquad 0·391 + x_2 \times 8·13 = 7·089$

$$\therefore \; x_2 = \frac{6·698}{8·13} = 0·824$$

i.e. $\qquad h_2 = 112 + 0·824 \times 2438 = 2117 \text{ kJ/kg}$

Also, $\qquad h_3 = h_f \text{ at } 0·035 \text{ bar} = 112 \text{ kJ/kg}$

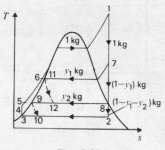

Fig. 7.27

For the first stage of expansion, 1–7, $s_7 = s_1 = 7·089$ kJ/kg K, and from tables at 10 bar $s_g < 7·089$, hence the steam is superheated at state 7. By interpolation between 250°C and 300°C at 10 bar we have:

$$h_7 = 2944 + \left(\frac{7·089 - 6·926}{7·124 - 6·926} \right)(3052 - 2944) = 2944 + \frac{0·163}{0·198} \times 108$$

i.e.

$$h_7 = 3032·9 \text{ kJ/kg}$$

For the throttling process, 11–12, we have:

$$h_6 = h_{11} = h_{12} = 763 \text{ kJ/kg}$$

For the second stage of expansion, 7–8, $s_7 = s_8 = s_1 = 7.089$ kJ/kg K, and from tables at 1·1 bar $s_g > 7.089$ kJ/kg K, hence the steam is wet at state 8. Therefore,

$$1.333 + (x_8 \times 5.994) = 7.089$$

$$\therefore \ x_8 = 0.961$$

i.e. $\qquad h_8 = 429 + (0.961 \times 2251) = 2591$ kJ/kg

For the throttling process, 9–10:

$$h_5 = h_9 = h_{10} = 429 \text{ kJ/kg}$$

Applying an energy balance to the first feed heater, remembering that there is no work or heat transfer:

$$y_1 h_7 + h_5 = y_1 h_{11} + h_6$$

i.e. $\qquad y_1 = \dfrac{h_6 - h_5}{h_7 - h_{11}} = \dfrac{763 - 429}{3032.9 - 763} = 0.147$

Similarly for the second heater, taking $h_4 = h_3$:

$$y_2 h_8 + y_1 h_{12} + h_4 = h_5 + (y_1 + y_2) h_9$$

i.e. $\qquad y_2(h_8 - h_9) + y_1 h_{12} + h_4 = h_5 + y_1 h_9$

$$y_2(2591 - 429) + (0.147 \times 763) + 112 = 429 + (0.147 \times 429)$$

$$\therefore \ y_2 = \frac{267.8}{2162} = 0.124$$

The heat supplied to the boiler, Q_1, per kg of boiler steam is given by:

$$Q_1 = h_1 - h_6 = 3445 - 763 = 2682 \text{ kJ/kg}$$

The work output, neglecting pump work, is given by:

$$\begin{aligned}
W &= (h_1 - h_7) + (1 - y_1)(h_7 - h_8) + (1 - y_1 - y_2)(h_8 - h_2) \\
&= (3445 - 3032.9) + (1 - 0.147)(3032.9 - 2591) \\
&\qquad\qquad\qquad\qquad + (1 - 0.147 - 0.124)(2591 - 2117) \\
&= 412.1 + 376.9 + 345.5 = 1134.5 \text{ kJ/kg}
\end{aligned}$$

Then,

$$\text{Thermal efficiency} = \frac{W}{Q_1} = \frac{1134.5}{2682} = 0.423 \quad \text{or} \quad 42.3\%$$

7.8 Further considerations of plant efficiency

Up to now the considerations of efficiency have been based on the heat which is actually supplied to the steam, and not the heat which has been produced by the combustion of fuel in the boiler. The heat is transferred to the steam from gases which are at a higher temperature than the steam, and the exhaust gases pass to the atmosphere at a high temperature.

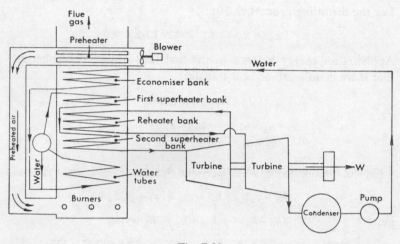

Fig. 7.28

To utilize some of the energy in the flue gas an *economizer* can be fitted. This consists of a coil situated in the flue gas stream. The cold feed water enters at the top of the coil, and as it descends it is heated, and continues to meet higher temperature gas. For the Carnot, the ideal regenerative, and complete feed heating cycles, no use can be made of an economizer since the feed water enters the boiler at the saturation temperature corresponding to the boiler pressure.

To cool the flue gas even further and improve the plant efficiency, the air which is required for the combustion of the fuel can be preheated. For a given temperature of combustion gases, the higher the initial temperature of the air then the less will be the energy input required, and hence less fuel will be used.

Plants which have both economizer and pre-heater coils in the boiler usually require a forced draught for the flue gas, and the power input to the fan, which is a comparatively small quantity, must be taken into account in the energy balance for the plant. Fig. 7.28 represents diagrammatically a plant with economizer, pre-heater, and a re-heater.

The boiler efficiency is the heat supplied to the steam in the boiler expressed as a percentage of the chemical energy of the fuel which is available on combustion;

i.e.   Boiler efficiency $= \dfrac{h_1 - (\text{enthalpy of the feed water})}{m_f \times (\text{H.C.V. or L.C.V.})}$ 7.18

(where h_1 is the enthalpy of the steam entering the turbine; m_f is the mass of fuel burned per kg of steam delivered from the boiler).

The H.C.V. and L.C.V. are the higher and lower calorific values of the fuel, and the determination of these quantities is considered in Section 15.11. The calorific value of a fuel is the energy released by unit quantity of the fuel on complete combustion as obtained by a practical test under defined conditions. The H.C.V. is that value obtained when the vapour of combustion has been condensed, and the L.C.V. is the H.C.V. minus an allowance for the latent heat of vapourization of the vapour. In actual plant the vapour formed on combustion is not condensed before the gases are exhausted to the flue and hence the heat released is nearer to the L.C.V. than the H.C.V. Both H.C.V. and L.C.V. are used in the definition of equation 7.18, and it is necessary to state which one has been chosen in any particular calculation. See Section 15.11 for further work.

The size of the boiler, or its *capacity*, is quoted as the rate in kg/h at which the steam is generated. A comparison is sometimes made by an *equivalent evaporation*, which is defined as the quantity of steam produced per unit quantity of fuel burned when the evaporation process takes place from and at 100°C.

7.9 The binary vapour cycle

It has been stated that the efficiency of the Rankine cycle increases as the maximum temperature increases. The present limit on temperature is that imposed by the materials of which the boiler and turbine are constructed. The maximum value at present is about 600°C. The critical temperature of steam is 374·15°C, which is well within this limit; the corresponding saturation pressure is high at 221·2 bar.

These considerations show some of the limitations of steam as the working substance. It would be better to have a working substance which has a critical temperature well above 600°C, and a low corresponding saturation pressure. Mercury is such a substance; from tables, at $p = 24$ bar, $t = 604·6$°C, and $v_g = 0·01518$ m³/kg. At the normal temperature of condensation (i.e. about 30°C) the vapour pressure is very low and v_g is very high; at 109·2°C, $p = 0·0006$ bar and $v_g = 259·6$ m³/kg.

It is evident then that mercury is satisfactory at the upper temperature limit but at the lower temperatures the properties of steam are preferable. This indicates the possibility of combining the desirable properties of mercury and steam in one plant. This is done in the *binary vapour plant*, and the *T-s* diagram for such a system is shown in fig. 7.29. The upper cycle is a Rankine cycle for mercury,

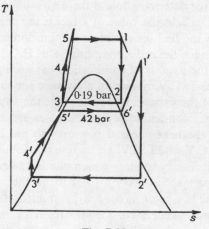

Fig. 7.29

and the lower is a Rankine cycle for steam. The two cycles have different scales, and the connecting link between the two is that the heat required to evaporate the steam is obtained by condensing the mercury, the temperature difference, $(T_3 - T_{5'}) = (T_2 - T_{6'})$, being that required for the heat transfer process. The superheat to the steam is provided by the hot gases in the mercury boiler, as is the preheating of the feed water.

If the steam pressure is 42 bar, say, the temperature of steam generation is 253·2°C. Allowing 22 K for the heat transfer process gives a temperature of 275·2°C for the mercury condensation, with a corresponding saturation pressure of 0·19 bar approximately.

Several binary vapour plants using mercury and steam have been built, but there are several disadvantages in such plants. Mercury is very expensive, heavy, and poisonous, so that the capital cost of plant is high. Mercury in its normal state does not wet steel, and this impairs the transfer of heat to the mercury; this can be improved by adding magnesium and titanium to the mercury.

The increases in plant efficiency in recent years made by increasing the maximum pressures and temperatures of the normal steam cycle have rendered the binary vapour plant unnecessary.

7.10 Modern boiler plant

Progress over the years has seen the demand for higher power outputs from boiler and associated plant. Most of the power stations in this country are coal fired, and the higher power demands have been made against a background of rising fuel cost with fuel of decreasing calorific value. Also, official regulations governing the operation of plant with regard to atmospheric pollution have to be complied with. These factors make necessary an increase in boiler and turbine plant efficiency.

The operating temperatures and pressures of boilers have risen, and this has been possible because of the developments of materials, especially alloy steels. Methods have been discussed in previous sections of improving plant efficiency, but the application of the principle produces practical problems which must be solved. The efficiency of a plant is increased by using lower temperatures in the condenser, and this has been possible only because the means of producing and maintaining a high vacuum have become available. As an improvement in plant design the modern lay-out is of a single boiler-turbine combination rather than several boilers operating with multiple piping systems. The trend is to higher pressure, higher temperature, larger capacity boilers, designed to meet particular specifications of power generation or process work, or both. For given steam conditions and boiler size there is not much variation in efficiency between different types, and the widest scope left to the designer is in increasing plant economy by making use of the high temperature flue gases. The gases should be cooled to 140°C to 200°C, the usual value at outlet from the air pre-heater being about 170°C.

The modern design of boilers for power generation are for steam capacities of 27 000 to 600 000 kg/h and above, with pressures up to

170 bar and maximum steam temperatures of about 600°C. There are exceptions to this range, and one re-heat boiler with natural water circulation has a maximum pressure of 190 bar. The larger boilers are of the *radiant type*, and have a moderately rated, water-

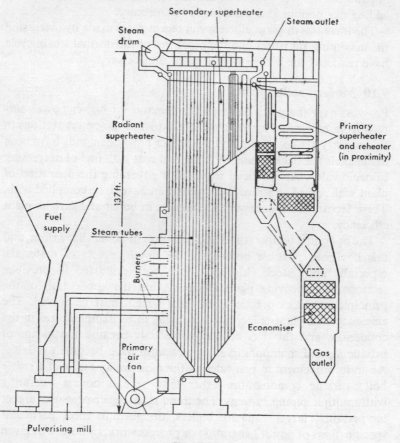

Fig. 7.30

cooled combustion chamber in which the heat is transferred to the evaporating surfaces by direct radiation (see fig. 7.30). These boilers may burn pulverized coal or have *cyclone* furnace (see fig. 7.31), or oil fuel burning equipment. Some coal fired boilers have oil burners for starting and for low load demands. Some radiant boilers are stoker fired. On all but the smaller boilers the stokers are mechanical

and consist of an endless articulated grate, driven forward at a
uniform rate. Coal is fed on to one end of the grate to a regulated
thickness, and combustion air passes through the spaces between the
grate links. The coal burns as the stoker moves, so that at the end of
its travel the burnt out fuel is passed over ash dumping bars into a
hopper.

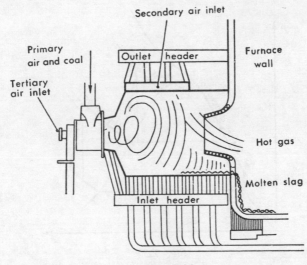

Fig. 7.31

The achieving of higher efficiencies is dependent upon efficient
control, which involves the preparation of the fuel and air, the combustion process, and the efficient transfer of heat to the steam.
Equipment is required to obtain the combustion of low grade fuel
with a minimum of excess air (this can be as little as 10% with a
cyclone furnace), in order that flue gas temperatures will not be
reduced. Pulverizing mills for pulverized fuel plant are required and
also the means of keeping heat transfer surfaces clean by means of
soot blowers and shot blasters. It is usual to handle boiler ash and
dust by hydraulic or pneumatic methods.

It is impossible here to deal with the various types of boilers within
the general classification, and not all of these will be given. Boilers
have been built in the U.S.A. which are in the super-critical range;
that is they have pressures in excess of the critical pressure for steam
of 221·2 bar. The design of the boiler is fundamentally different from
the sub-critical types due to the different properties of steam in the

critical region. Lower pressure boilers have a means of separating water from steam but in the super-critical boiler this is not required since the phenomenon of boiling does not occur. When the water is heated at pressures above the critical at constant pressure, the temperature is never constant, and no distinction between the gaseous and liquid states can be observed (see the *T-s* diagram in fig. 7.32).

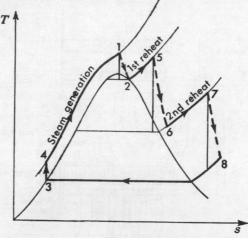

Fig. 7.32

The liquid is assumed to become a gas when it passes the critical temperature. The boiler thus becomes a continuous tube, which is heated along its length, with water going in at one end and superheated steam leaving at the other. About 65% of the heat transferred is used in superheating and reheating the steam, and to limit gas temperature a radiant type of superheater surface is used. With this type of boiler the preparation of feed water is more important than in the conventional type. At the point of transition from water to steam the deposition of solids and the oxidation of surfaces is likely. This boiler is one which necessitates the application of close automatic control. Modern plants working on the Rankine cycle, or modifications to it, have efficiencies of over 30%; the efficiency of super-critical plants is of the order of 40%. To increase the efficiency beyond this, consideration must be given to gas turbine or mixed steam turbine and gas turbine cycles.

Marine steam boilers and auxiliaries have a wide application, and several types are necessary to satisfy all requirements. The *header* boiler (see fig. 7.33), which has straight tubes and makes a single gas pass, is used for the main propulsion steam where means of correct water treatment and skilled maintenance may not be available. Where improved feed water treatment is available, and frequent boiler tube replacement is not necessary, *bent tube* boilers are fitted.

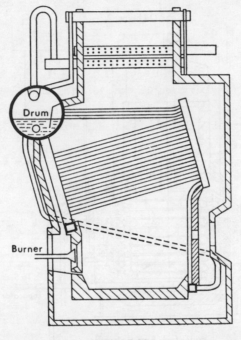

Fig. 7.33

These are more compact than the header type and have a higher output, and therefore are used for large tankers and passenger vessels. The bent tube boiler is produced in different designs to meet the different aspects of modern high performance requirements.

A range of steam generators is available for the rapid raising of steam for applications in industry, for marine purposes and in diesel locomotives for heating purposes. These have a forced circulation of water through a single, coiled water tube, and having no drum, the only water in the generator is that in the coil. The generators are automatically controlled to deliver steam at better than 99% dry, and to meet fluctuating demands down to 25% of the full demand by continuous operation, and below this by cycling on and off. The generators are oil fired with electric ignition. One generator with a rating of 2050 kg/h at 10·4 bar from feed water at 16°C has an efficiency of 78 to 82%. The steam can be superheated by passing it

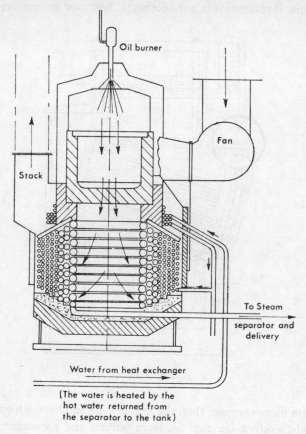

Fig. 7.34

through a separate superheater which may be fired by oil, gas or electricity. Steam is raised at the required condition within two or three minutes from starting. Fig. 7.34 shows the arrangement of such a generator.

Electrode boilers are suitable for the rapid raising of steam at moderate pressures and moderate rates of flow. The usual range for this type of boiler is 35 to 2150 kW, and pressure ranges up to 84 bar may be produced for special purposes. The steam capacities range from 45 to 4300 kg/h. The principle of the electrode boiler is that heat is generated within the water and is not transferred to it through an external wall. The electrodes are situated in the water which forms part of the circuit (see fig. 7.35). Should the water level fall so

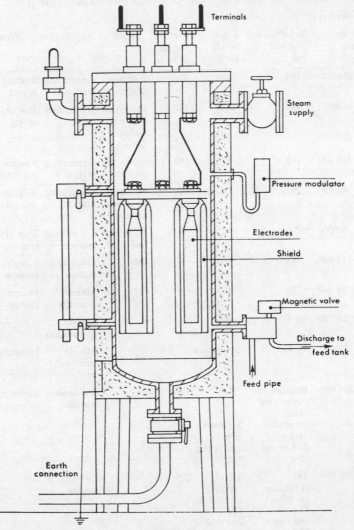

Fig. 7.35

that it is no longer in contact with the electrodes, then the circuit is broken and the boiler shuts down. This gives a valuable safety control. The steam output is controlled by raising or lowering the water level. The boiler is automatic, adjusting itself to a fluctuating demand and controlling the steam pressure.

Table 7.1 gives a selection of boiler performance figures. See ref. 7.2 for a comprehensive treatment of steam power generation.

Table 7.1

Steam rate kg/h	Superheater pressure bar	Final steam temp. °C	Turbine output MW	Fuel	Application	Type
600 000	169	570	200	Pulverised coal	Generating station	Radiant with reheat
380 000	110	540	120	Pulverised coal with cyclone furnace	Generating station	Radiant with reheat
380 000	110	575	100	Pulverised coal	Generating station	Radiant without reheat
85 500	43	460	—	Stoked by travelling grate	Generating station	Radiant without reheat
250 000	65·5	495	60	Pulverised fuel	Generating station	Two-drum type
143 000	43	460	—	Oil	Generating station	Integral furnace
43 200	59	460	—	Oil	Marine (tanker)	Integral furnace
26 300	31·5	480	—	—	Marine (passenger)	—
11 400	15·2	Saturation			Marine (motor driven)	Header type
306 000	Pressure at the turbine 310 1st reheat 79·5 2nd reheat 11·4	620 1st reheat 570 2nd reheat 540	120	—	Generating station	Super critical
700 000	345 at turbine 1st reheat 72·5 2nd reheat 17·3	650 at turbine 1st reheat 570 2nd reheat 570	275	—	Generating station	Super critical

7.11 Steam for heating and process work

A modern power plant working continuously on full load may have an overall efficiency of 30%. This is low due to the large percentage of energy which is rejected to the condenser. The power plant efficiency can be considerably increased if it is part of a plant which requires a low temperature supply of heat for heating or process work. Factories generating their own power may do so at an efficiency much less than 30%, but if heating and/or process steam is required then the overall efficiency may be as high as 85%.

Heating by steam is efficient, and the only losses incurred are those due to leakage and to radiation. The energy distribution in such plant can be illustrated by means of a Sankey diagram, and two of these are given in fig. 7.36. Fig. 7.36a shows the distribution for a condensing

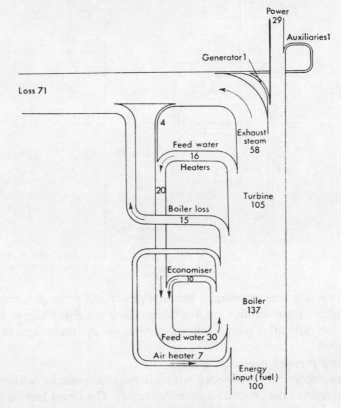

Fig. 7.36a

power plant, and fig. 7.36b shows the distribution for a combined power and process plant.

If the sole purpose of the steam plant is heating, then the function of the boiler is to transfer the energy of the high temperature combustion to the steam. The steam is then passed to a condenser or *calorifier* in which the steam is condensed by the transfer of heat to the water of the central heating system, or to the process liquid. The condensate is then returned to the boiler. In a combined plant the process heat exchanger takes the place of the power plant condenser.

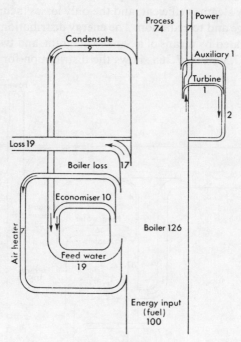

Fig. 7.36b

In reaching the compromise between power and process demands a number of possibilities are available. These are listed below, but the units are often combined to obtain an optimum operating condition.

(a) *Back pressure engine or turbine*
The engine or turbine works with an exhaust pressure which is appropriate to the process steam requirement. The steam leaving the turbine is not condensed but is passed to process work.

(b) *Pass-out engine or turbine* (also *Bleeder or Extractor*)
Steam is bled from the turbine at some point or points between
inlet and exhaust, and is passed to process work.

(c) *Mixed pressure turbine*
This is a turbine which receives steam at more than one pressure,
the supplies being fed into the turbine at the appropriate points.

(d) *Exhaust turbine*
A turbine which takes its supply from the exhaust of another
turbine or engine.

(e) *Vacuum turbine*
A turbine which takes steam from a process at below atmospheric
pressure.

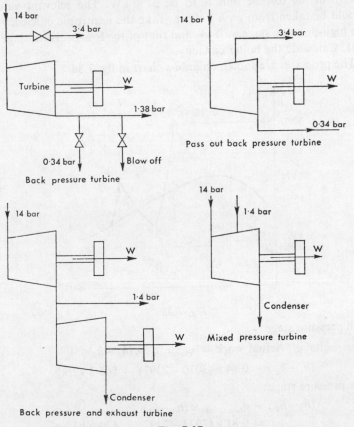

Fig. 7.37

To enable process steam to be taken from high pressure lines it is necessary to employ reducing valves which drop the pressure to that required. It may also be necessary to blow off some of the steam to waste.

Some of the above arrangements or combinations of them are shown in fig. 7.37.

Example 7.6

A pass-out two stage turbine receives steam at 50 bar and 350°C. At 1·5 bar the high pressure stage exhausts and 12 000 kg of steam per hour are taken at this stage for process purposes. The remainder is reheated at 1·5 bar to 250°C and then expanded through the low pressure turbine to a condenser pressure of 0·05 bar. The power output from the turbine unit is to be 3750 kW. The relevant values should be taken from an *h-s* chart. Take the isentropic efficiency of the high-pressure stage as 0·84, and that of the low-pressure stage as 0·81. Calculate the boiler capacity.

The processes are shown on an *h-s* chart in fig. 7.38.

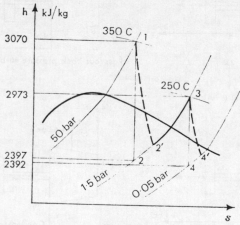

Fig. 7.38

High pressure stage:

$$\text{Actual work} = \eta_{\text{isentropic}} \times (h_1 - h_2)$$

i.e. $(h_1 - h_{2'}) = 0.84 \times (3070 - 2397) = 565.3$ kJ/kg

Low pressure stage:

$$(h_3 - h_{4'}) = \eta_{\text{isentropic}} \times (h_3 - h_4)$$
$$= 0.81 \times (2973 - 2392) = 470.6 \text{ kJ/kg}$$

$$\text{Process steam flow} = \frac{12\,000}{3600} = 3\cdot33 \text{ kg/s}$$

Steam flow through the boiler $= \dot{m}$ kg/s

Steam flow through L.P. stage $= (\dot{m}-3\cdot33)$ kg/s

$$\text{Turbine power output} = 3750 \text{ kW}$$
$$= 3750 \text{ kJ/s}$$

$$\therefore \dot{m}(h_1-h_{2'})+(\dot{m}-3\cdot33)(h_3-h_{4'}) = 3750$$

i.e. $\qquad \dot{m} \times 565\cdot3+(\dot{m}-3\cdot33) \times 470\cdot6 = 3750$

$$\therefore \dot{m} = 5\cdot14 \text{ kg/s}$$

i.e. $\qquad$ Boiler capacity $= 18\,500$ kg of steam per hour

7.12 Nuclear power plant

The years since the 1939–1945 war have seen intensive developments in the application of nuclear reaction to power production. The reactor and heat exchanger take the place of the conventional boiler, and the steam generated is then expanded through a conventional turbine. With the nuclear power station there are problems of reactor physics, thermodynamics and engineering materials, which are inter-related, and a sound compromise must be made in the design if the ultimate and all important economic result is to be acceptable.

The improvement in the efficiency of conventional power plant is one of diminishing return, and this, together with the rising cost of a fuel which has a diminishing supply, makes additional sources of power desirable. Also the by-products of a nuclear reactor are useful for military and other purposes where fissionable materials are required. In some circumstances the power production may be the by-product and the main purpose the production of plutonium.

The science of nuclear physics has made great strides and is the province of the specialist, but an elementary picture of the basic reaction will be given here. The heat supplied in a nuclear reaction is exactly similar in effect to that from the combustion of a fuel. It is possible to consider the principles and accept the results without considering or even knowing all the intricacies of the reaction.

(a) Nuclear fission

Some difficulty arises in appreciating the nuclear reaction because the basic unit of the reaction, the atom, is of such small dimensions (e.g. one million billion atoms would only cover a pin-head). Each atom is itself like a miniature universe. At the centre of the system is a *nucleus* (see fig. 7.39). Around the nucleus in orbits or shells, travel negatively charged particles called *electrons*, at high velocity. The nucleus, at which most of the mass of the atom is concentrated,

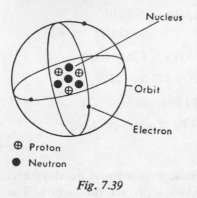

⊕ Proton
● Neutron

Fig. 7.39

consists of positively charged particles called *protons*, and *neutrons* which have no charge at all. The electrons are of very small mass relative to the proton (about 1/1800). The number of electrons is equal to the number of protons and the atom is electrically neutral. It should be realized that most of the volume enclosed by the outer orbital shell is space, with the mass concentrated at the nucleus.

The nuclear reaction is concerned with the nucleus. The nucleus has a high cohesive force binding the protons and neutrons, and this force is much stronger than that holding the electrons in orbit. The nuclear reaction involves splitting the nucleus and is accompanied by a large release of energy. The action of splitting the atom is called a *fission*. The only substance occurring naturally which is fissionable, is *uranium*. Only a part of natural uranium will take part in fission. There are three isotopes of uranium in the natural metal, uranium-234, uranium-235, and uranium-238, and only uranium-235 is fissionable. The essential differences are as shown in Table 7.2.

Table 7.2

Isotope	% in natural uranium	Proton	Neutron	Mass number
U-234	Trace	92	142	234
U-235	0·7	92	143	235
U-238	99·3	92	146	238

The uranium metal must be extracted from the ore, purified, and prepared for the reaction. An atom bomb requires pure U-235, but the fuel for a reactor is natural uranium enriched with U-235. The isotope U-238 is a fertile material, which means that it can be converted to a fissionable material by absorbing a neutron. The U-238 then becomes the fissionable *plutonium* (Pu-239).

The nucleus can only be approached by a mass which has no charge (i.e. a neutron), and if a neutron strikes a nucleus there is a physical 'probability' that the nucleus will be split and new atoms of fission products will be formed. Other neutrons will be released (about 2·5 per atom of U-235), and some will continue the reaction with other atoms. Some will be captured by U-238 (to give later Pu-239), and the rest will escape. The process is illustrated in a simplified form in fig. 7.40.

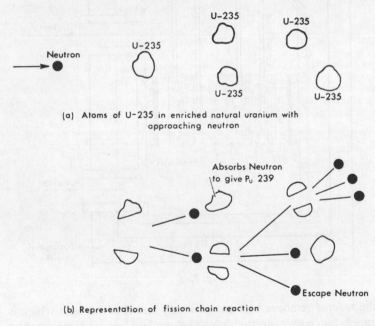

(a) Atoms of U-235 in enriched natural uranium with approaching neutron

(b) Representation of fission chain reaction

Fig. 7.40

A simplified sketch of a reactor is given in fig. 7.41. The core of the reactor consists of a moderating material (e.g. carbon), which slows down the neutrons to speeds which will give controlled fission. The holes in the core carry the fuel elements in suitable casings (e.g.

magnox casings). Through some of the holes are passed control rods which are made of a material which will absorb neutrons and so control the rate at which the reaction takes place. The function of the reflector is to bounce back escaping neutrons into the core, and the shielding is to prevent the transmission of harmful particle radiation, such as α and β particles, and γ radiation. The whole of the reactor vessel is surrounded by shielding of steel and concrete. The coolant

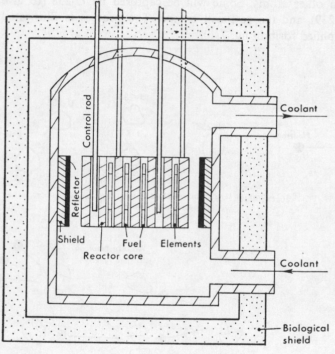

Fig. 7.41

to the reactor removes heat continuously and it is this heat which is required for power production. The coolant may be air, carbon dioxide, hydrogen, helium, water, or liquid metals. The coolant follows a closed circuit, part of which is through the heat exchanger.

(b) *Nuclear cycles*

Fig. 7.42a shows the diagrammatic arrangement of a single pressure cycle and in fig. 7.42b is shown the variation in coolant gas temperature and water/steam temperature as they pass through the boiler (or heat exchanger).

The heat transfer in the boiler is not made at the highest temperature available, and to make use of this, a second pressure level is necessary. This increases the efficiency of the plant at the expense of a

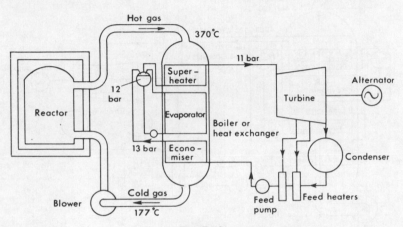

Fig. 7.42a

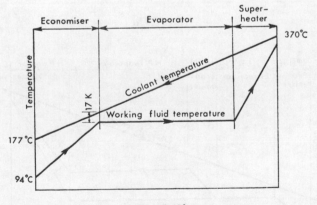

Fig. 7.42b

more complicated boiler design. There are really two separate boiler systems within the heat exchanger and the steam produced is fed to the turbine at the appropriate stage. The number of variables which have to be considered is increased, which complicates the selection of optimum conditions. The *dual pressure* system is shown in fig. 7.43a, and the temperature plot in fig. 7.43b.

The improvement in station overall efficiency of the dual pressure

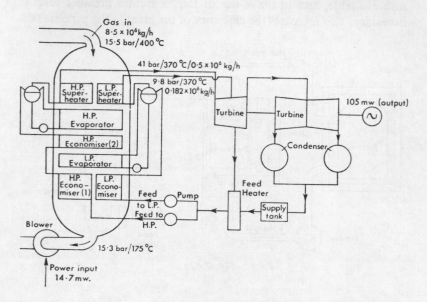

Fig. 7.43a

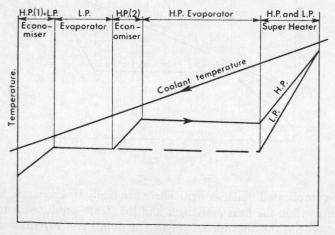

Fig. 7.43b

cycle over the single pressure cycle is from 2% to 2·5% for a plant with high pressure steam conditions of 41·5 bar and maximum temperature of 340°C.

(c) Power generation by nuclear reactors

Fig. 7.41 is a simplified sketch of the first type of thermal reactor to be installed in the U.K. This was commissioned in 1956 at Calder Hall in Cumbria and is of the Magnox type of 200 MW output. In all there are eleven* twin reactor magnox stations in Britain ranging in output from 200 to 840 MW. They are fuelled by uranium metal clad in a magnesium alloy (Magnox) case which has a low neutron absorption. The core is made of graphite and the coolant is carbon dioxide. The coolant pressure is 20 bar, its outlet temperature is 400°C and the steam cycle efficiency is about 30%.

The second generation of thermal reactors in the UK was the advanced gas cooled reactor, the AGR, and the total programme includes seven twin reactors,† six of 1320 MW output and one of 1200 MW. The fuel is uranium dioxide contained in a stainless steel cladding which allows higher temperatures and pressures (650°C and 40 bar) and gives higher heat outputs than the Magnox type resulting in a smaller-size reactor and improved steam-cycle efficiencies up to 42%.

The most widely-used reactor outside the U.K. (U.S.A. and Russia) is the pressurized water reactor (PWR) fuelled with uranium dioxide clad in Zircaloy, an alloy of zirconium. Water is the moderator and coolant at a pressure of 154 bar and an outlet temperature of 317°C. The steam-cycle efficiency is 32% for an output of 700 MW.

The spent fuel from thermal reactors contains a high proportion of uranium and some plutonium created during the operation. This is a fuel and can be reprocessed and used in a fast reactor giving a very efficient use of the original uranium. In Britain fast reactors have been under development, the first being at Dounreay in Caithness which was closed in 1977 and succeeded by a prototype fast reactor (PRF)

* The stations are: Calder Hall (200 MW); Chapelcross (200 MW); Berkeley (276 MW); Bradwell (250 MW); Dungeness (410 MW); Hinkley Point 'A' (430 MW); Hunterston 'A' (300 MW); Oldbury on Severn (416 MW); Sizewell 'A' (420 MW); Transfynydd (390 MW); Wylfa (840 MW).

† The stations are: Hinkley Point 'B'; Hunterston 'B'; Hartlepool; Heysham 'A'; Heysham 'B'; and Torness – all of 1320 MW output – and Dungeness 'B' with 1200 MW.

STEAM PLANT

Table 7.3

Reactor type	Fuel	Moderator	Coolant	Power (MW)	Coolant pressure (bar) (temp °C)	Steam cycle efficiency (%)	Country of origin
Boiling water	Uranium dioxide in Zircaloy	Water	Water	600	72·4 (286)	32	Several
Candu	Uranium dioxide in Zircaloy	Heavy water (D₂O)	Heavy water	600	88·6 (305)	30	Canada
Steam generating heavy water (SGHW)	Uranium dioxide in Zircaloy	Heavy water (D₂O)	Water	600	62 (272)	32	Japan Italy U.K.
High temperature reactor (experimental)	Uranium dioxide in graphite	Graphite	Helium gas	1300	49 (720)	39	U.K. U.S.A.
Leningrad	Uranium dioxide	Graphite	Water	1000	70 (284)	31	Germany Russia

with an output of 250 MW, and commercial reactors of 1250 MW output are anticipated. The fuel is a mixture of plutonium and uranium dioxide in stainless steel. There is no moderator and the coolant is sodium which then passes through a heat exchanger to heat sodium in a secondary circuit which includes the steam generator. The coolant pressure is 0·34 bar, its outlet temperature 620°C and the steam-cycle efficiency 44%.

Table 7.3 gives a brief summary of other types of reactors which are under construction, at the experimental stage or used outside the U.K.

The difference in action between the thermal and fast reactors lies in the nature of the fuel and the means of increasing the probability of fission. Natural uranium is mainly U-238 with a small amount of isotope U-235 and there is a higher probability of fission of U-235 by slow neutrons than there is of U-238 by fast neutrons, hence the neutrons are slowed down to give fission of U-235. This is the function of the moderator, graphite or water. In enriched uranium the U-235 content is raised to 2·3%.

In fast reactors the fuel is plutonium which is a by-product of thermal reactors and is sufficiently concentrated for the reaction to proceed with fast neutrons and no moderator. This leads to a compact reactor requiring an efficient cooling medium and sodium is used. If the core is surrounded by a uranium blanket uranium 'breeds' so fast reactors can be breeder reactors.

The study of nuclear power can be extended by the use of references 7.3 to 7.8 inclusive.

PROBLEMS

7.1 (i) Steam is supplied, dry saturated at 40 bar to a turbine and the condenser pressure is 0·035 bar. If the plant operates on the Rankine cycle, calculate, per kg of steam:

(a) the work output neglecting the feed pump work;
(b) the work required for the feed pump;
(c) the heat transferred to the condenser cooling water, and the amount of cooling water required through the condenser if the temperature rise of the water is assumed to be 5·5 K;
(d) the heat supplied;

(e) the Rankine efficiency;

(f) the specific steam consumption.

(ii) For the same steam conditions calculate the efficiency and the specific steam consumption for a Carnot cycle operating with wet steam.

(982·4 kJ; 4 kJ; 1706·6 kJ; 74·1 kg; 2685 kJ; 36·6%; 3·66 kg/kW h; 43%; 4·88 kg/kW h)

7.2 Repeat problem 7.1(i) for a steam supply condition of 40 bar and 350°C and the same condenser pressure of 0·035 bar.

(1125 kJ; 4 kJ; 1857 kJ; 80·5 kg; 2978 kJ; 37·8%; 3·2 kg/kW h)

7.3 Steam is supplied to a two-stage turbine at 40 bar and 350°C. It expands in the first turbine until it is just dry saturated, then it is reheated to 350°C and expanded through the second-stage turbine. The condenser pressure is 0·035 bar. Calculate the work output and the heat supplied per kg of steam for the plant, assuming ideal processes and neglecting the feed pump term. Calculate also the specific steam consumption and the cycle efficiency.

(1290 kJ; 3362 kJ; 2·79 kg/kW h; 38·4%)

7.4 If the expansion processes in the turbines of problem 7.3 have isentropic efficiencies of 84% and 78%, respectively, in the first and second stages, calculate the work output and the heat supplied per kg of steam, the thermal efficiency, and the specific steam consumption.

Compare the efficiencies and specific steam consumptions obtained from problems 7.1, 7.2, 7.3, and 7.4. Compare also the wetness of the steam leaving the turbines in each case.

(1026 kJ; 3311 kJ; 31·1%; 3·51 kg/kW h)
(Dryness fractions at condenser in each case: 0·7, 0·76, 0·85, and 0·94.)

7.5 A generating station is to give a power output of 200 MW. The superheat outlet pressure of the boiler is to be 170 bar and the temperature 600°C. After expansion through the first stage turbine to a pressure of 40 bar, 15% of the steam is extracted for feed heating. The remainder is reheated to 600°C and is then expanded through the second turbine stage to a condenser pressure of 0·035 bar. For preliminary calculations it is assumed that the actual cycle will have an efficiency ratio of 70% and that the generator mechanical and electrical efficiency is 95%. Calculate the maximum continuous rating of the boiler in kg/h.

(633 000 kg/h)

7.6 A steam turbine is to operate on a simple regenerative cycle. Steam is supplied dry saturated at 40 bar, and is exhausted to a condenser at 0·07 bar. The condensate is pumped to a pressure of 3·5 bar at which it is mixed with bleed steam from the turbine at 3·5 bar. The resulting water which is at saturation temperature is then pumped to the boiler. For the ideal cycle calculate, neglecting feed pump work,

(a) the amount of bleed steam required per kg of supply steam;
(b) the thermal efficiency of the plant;
(c) the specific steam consumption.

(0·1906; 37%; 4·39 kg/kW h)

7.7 Steam is supplied to a two-stage turbine at 40 bar and 500°C. In the first stage the steam expands isentropically to 3·0 bar at which pressure 2500 kg/h of steam is extracted for process work. The remainder is reheated to 500°C and then expanded isentropically to 0·06 bar. The by-product power from the plant is required to be 6000 kW. Calculate the amount of steam required from the boiler, and the heat supplied in kW. Neglect feed pump terms, and assume that the process condensate returns at the saturation temperature to mix adiabatically with the condensate from the condenser.

(15 000 kg/h; 15 620 kW)

7.8 For the plant of problem 7.7 it is required to improve the efficiency by employing regenerative feed heating by taking off the necessary bleed steam at the same point as the process steam. The process steam is not returned to the boiler but make-up water at 15°C is supplied. The bleed steam is mixed with the condensate and make-up water at 3·0 bar such that the resultant water is at the saturation temperature corresponding to 3·0 bar. Calculate the steam supply necessary to meet the same power and process requirements, and the amount of bleed steam. Neglect feed pump terms.

Calculate also the heat supplied in kW.

(16 500 kg/h; 2670 kg/h; 15 500 kW)

7.9 A combined separating and throttling calorimeter is used to determine the dryness fraction of steam in a main. The pressure of the steam in the main and the separator is 6·9 bar. After throttling to 1·5 bar the temperature is 127°C. During a ten-minute test 0·09 kg of water is collected at the separator and 1·53 kg of condensate is collected after throttling. Calculate the dryness fraction of the steam in the main.

(0·927)

7.10 In a regenerative steam cycle employing three closed feed heaters the steam is supplied to the turbine at 42 bar and 500°C and is exhausted to the condenser at 0·035 bar. The bleed steam for feed heating is taken at pressures of 15, 4, and 0·5 bar.

Calculate the amount of steam bled at each stage, the work output of the plant in kJ/kg of boiler steam, and the thermal efficiency. Assume ideal processes.

(0·105 kg; 0·108 kg; 0·081 kg; 1120 kJ; 43·1%)

7.11 The heat exchangers of a nuclear power plant deliver steam at two pressures, 50 bar and 20 bar, and the steam temperature is 400°C at both pressures. The high pressure steam flow rate is 1×10^6 kg/h and the low pressure steam rate is 0.5×10^6 kg/h. The high pressure steam expands through the turbine with an isentropic efficiency of 0·75 and is then mixed, adiabatically, with the low pressure steam at 20 bar. The combined steam flow then expands through the low pressure turbine with an isentropic efficiency of 0·78 to a condenser pressure of 0·035 bar, after which the condensate is returned, through pumps, to the heat exchangers. Show diagrammatically the arrangement required and calculate the power generated and the thermodynamic efficiency of the cycle neglecting the feed pump terms.

(386 MW, 29·9%)

7.12 A pressurized water reactor power plant is shown in fig. 7.44 and the operating pressures and temperatures are included.

The steam leaving the heat exchanger at 50 bar is dry saturated and the condensate is saturated liquid. The feed water is preheated to the saturation temperature of the bleed steam in an open feed heater. Assume all processes to be ideal; pressure, heat losses and pump work to be negligible, and calculate:

(a) the ratio of the mass flow rates of the working fluids in the two circuits;

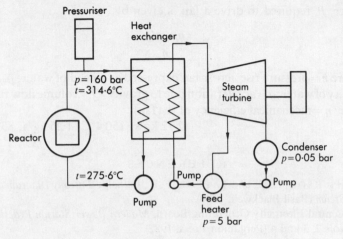

Fig. 7.44

(b) the fraction of steam supply which is bled from the turbine to the feed heater;

(c) the cycle efficiency.   (water to steam 10/1; 0·22; 39%)

7.13 A boiler plant, see fig. 7.28, incorporates an economizer and an air preheater, and generates steam at 40 bar and 300°C with fuel of calorific value 33 000 kJ/kg burned at a rate of 400 kg/h. The temperature of the feed water is raised from 40°C to 125°C in the economizer, and the flue gases are cooled at the same time from 395°C to 225°C. The flue gases then enter the air preheater in which the temperature of the combustion air is raised by 75 K. A forced-draught fan delivers the air to the preheater at a pressure of 1·02 bar and a temperature of 16°C with a pressure rise across the fan of 180 mm of water. The power input to the fan is 5 kW and it has a mechanical efficiency of 78%.

Calculate

(a) the mass of air used per unit time;
(b) the temperature of the flue gases leaving the plant;
(c) the mass flow rate of the steam;
(d) the efficiency of the boiler plant.

Neglect heat losses and take $c_p = 1·01$ kJ/kg K for the flue gases. The power, P, required to drive a fan is given by,

$$P = \frac{h\rho_w gQ}{\eta}$$

(where h = pressure rise across fan expressed as a head of water; ρ_w = density of water; g = acceleration due to gravity; Q = volume flow rate of air; η = mechanical efficiency of the fan).

(2·72 kg/s; 150·4°C; 1·25 kg/s; 83%)

REFERENCES

7.1 HICKSON, D. C., and TAYLOR, F. R., *Enthalpy-Entropy Diagram for Steam* (Basil Blackwell, 1980).

7.2 Central Electricity Generating Board, *Modern Power Station Practice*, vols 2, 3 and 4 (Pergamon Press, 1982).

7.3 United Kingdom Atomic Energy Authority, *Publications on Atomic Energy* (1980).

7.4 ANTHONY, V., and NERO, J. R., *A Guidebook to Nuclear Reactors* (University of California Press, 1979).

7.5 WINTERTON, R. H. S., *Thermal Design of Nuclear Reactors* (Pergamon Press, 1981).

7.6 MURRAY, R. L., *Nuclear Energy* (Pergamon Press, 1983).

7.7 BURN, D., *Nuclear Power and the Energy Crisis* (Macmillan Press, 1980).

7.8 BENNET, D. J., *Elements of Nuclear Power* (Longman, 1981).

8. The Sources, Use and Management of Energy

The most common source of heat energy is the chemical energy of substances called fuels which is released upon the combustion of the fuel in air as described in Chapter 15. Fuels such as coal, oil and natural gas are created by natural processes deep down in the earth after many thousands of years and as such are dscribed as non-renewable fossil fuels. However large the world's resources of fossil fuels may be they are being consumed at a high rate and one day, however far away, the fuel resources will become so depleted that the normal existence of energy-dependent countries will be seriously disrupted unless other energy sources have become available on the scale necessary to meet world demand. The fossil fuels are also sources of chemical substances to be used, other than as fuels, in the manufacture of goods such as plastics or chemicals for agriculture and animal foodstuffs. A comprehensive understanding of the supply problem includes all natural products including fuels. Many warnings have been given over the years about the world rate of fuel consumption and the prospects of a fuel shortage, but each time the crisis has passed and the world has continued on its way, hopefully more aware of the importance of its energy sources, the need for the efficient use of fuels and materials and for long-term planning for the future.

In the early 1970s the oil industry warned users that the exponential growth in oil consumption, which was doubling every seven years, could not be sustained. In 1979 the oil suppliers doubled the price of oil and shocked those countries, including Britain and the U.S.A., which had based their industrial economy on the cheap oil available from the Middle Eastern countries. In equivalent costs in dollars per barrel of oil the relative costs of fuels in 1978 was: Middle East oil 1·3; North Sea oil 5·3; imported coal (N.W. Europe) 8·3; indigenous coal (N.W. Europe) 11·3; nuclear 7·6; imported natural gas 15; natural gas from indigenous coal 24·3; biomass (crops grown for fuel) 43; solar hot water 40.

The increase in the cost of oil in the late 1970s resulted in a fall in demand and a world-wide recession in the manufacturing industries leading to a considerable conservation not only of fuels but of other basic manufacturing materials such as iron, tin, zinc, aluminium, silver, gold, lead and copper which are also in finite supply.

It was estimated in 1973 that by the mid 1980s the world oil consumption would be three times as much as it actually became and the electricity consumption would be twice as much. Instead there was a surplus of oil tankers and bulk carriers, the demands on oil refineries and power stations decreased and so did the demand for steel, cement and building materials leading to a further reduction in the demand for energy. During this period Britain became an oil and natural-gas producing nation due to the exploitation of finds under the North Sea.

A greater awareness of the value of natural resources grew in the mid 1970s, and probably an even greater one for the influence of the cost of fuels on manufactured goods, for it was realized that an efficient use of fuels could affect industry and society substantially in relation to basic costs.

A cheap supply of energy does not encourage an economical attitude to energy use and methods of energy conservation, but the balance between demand, supply and cost is one which can change rapidly and must be kept under constant review by nations and their industries. If the balance of supply from various sources is to be changed, the technology required to make the change must be available and a new technology can take a long time and need high investments. Since the 1970s a greater interest in alternative sources of energy has grown, many developments have been made and considerable research has taken place, but by the mid 1980s no alternative solutions to existing sources had appeared which made any real impact on the energy scene—other than those for nuclear energy which began in Britain in the 1950s. The development work done since the 1970s on alternative forms of energy may show benefit in the future.

The following sections will cover some aspects of energy supply, demand, use, conservation and management, etc., but such a wide-ranging subject cannot be give a comprehensive treatment in this book and readers are advised to pursue the subject by reading some of the many specialist texts and the current reports of government departments, such as the Department of Energy (reference 8.1), and research institutions, such as the Research Councils, published by the Department of Education and Science.

8.1 Sources of energy supply and energy demands

The primary source of energy is the sun from which all of the earth's energy requirement is finally obtained. Only a negligible fraction of the daily need is obtained directly, even including the secondary supplies from power generation by winds, waves and rivers. The vast majority is obtained from the combustion of fossil fuels including coal, oil, natural gas, peat, wood, natural waste, etc. and a lesser amount from nuclear reactions. Additional known sources of energy are the tides and the natural geothermal gradient of the earth. It has been estimated that the amount of solar energy falling on the earth in three days is equal to the known fossil-fuel reserves of the world. The sun is an ample provider of energy for the earth; the problem is how to collect and store solar energy so that it can be released in the right form to meet the world demand for heat and power.

It is highly desirable at any time to be able to say what the world's fuel resources are, and the corresponding demand and the rate at which it is changing. The determination of this information is a complex exercise but it is done from time to time and the most authoritative source available should be consulted. Any estimate is likely to be out of date when it is published. New finds of coal, gas and oil are being discovered and old sources abandoned; the extent of deposits in known sources is unknown in many places. Not many years ago there was no indication of gas and oil under the North Sea; now it is possible that known reserves are only a fraction of the total. It is believed that the experience of the North Sea will be repeated elsewhere in the world. On the mainland of Britain large deposits of coal have been found and there is an increasing possibility of oil and gas being found in commercially viable quantities.

The energy demands are more easily measured than the possible supplies, as all of the fuel has to be procured and is sold at a price to the consumer who requires that the quantities be very carefully measured. Estimates for the future based upon previous years are made but they can receive severe disruption as by the oil producers' embargo in 1974 and the industrial recession which followed.

In 1974 the estimates of the total world resources of coal, oil, natural gas, tars, shale oil and other fossil fuels (in units of 10^{15} MJ) ranged from 47·5 to 169, the extractable amounts from 9 to 70 and the annual rate of consumption was 0·287. Estimates on the life of resources based on (a) the 1974 figures and (b) on a 5% growth, by different authorities gave ranges of 32 to 586 and 20 to 65 years respectively. The energy available from tidal power, water, wind, solar, geothermal and nuclear fuels (other than uranium) were not included in the estimates.

The energy requirements of the U.K., the U.S.A. and the U.S.S.R. in 1974 are summarized in Table 8.1 and are shown related to the total world requirements and the populations of the countries. The figures are taken from ref. 8.2.

The breakdown of the U.K. energy distribution gives: industry 42% ; transport 24%; domestic and business 33%. Industry covers all products for home use and export, electrical power, agriculture, iron, steel and materials manufacture, food, engineering, chemicals, textiles, paper, bricks, etc. Transport covers road, rail, water and air. Domestic and business covers heating, lighting, cooking, private vehicles, entertainment. Thirty-three per cent of the total is used in the form of electrical energy which has to be generated.

The dependence on energy can be moved from one source to another as the supply and demand changes, e.g.: 55% of Europe's energy supply comes from oil and of this amount 22% cannot be replaced by substitute fuels as it is used for transport, lubricants, bitumen and chemical feedstocks; 33% could be replaced by coal, nuclear fuel, solar and possibly wave and wind power sources.

The energy calculations may be speculative but the essential truth is that known energy sources will be consumed one day; energy should be used efficiently; other energy sources should be sought; alternative methods of power generation are desirable. There is widespread acceptance of these facts and many countries, including Britain, are implementing energy policies and encouraging research and development work to contribute to an energy efficient future.

The role of the engineer is to be seen clearly in the energy problem and its solution from the assessment of supply and demand; thermodynamic analysis; new technologies; power supply and control, etc. There are, however, additional factors to take into account which make the problem even more complex including:

1. The world-wide distribution of fuel resources is very different from the pattern of demand for energy.

Table 8.1 Energy Requirements of the U.K., the U.S.A. and the U.S.S.R. in units of 10^{12} MJ

Country	U.K. (pop. 56×10^6)			U.S.A. (pop. 203×10^6)			U.S.S.R. (pop. 257×10^6)		
Fuel	$\times 10^{12}$ MJ	U.K. %	World %	$\times 10^{12}$ MJ	U.S.A. %	World %	$\times 10^{12}$ MJ	U.S.S.R. %	World %
Coal	3·2	36	3·7	10·8	16·8	12·5	13·1	28	15·2
Oil	4·3	48	3·3	31·6	47·3	24·5	24·6	53	19
Natural gas	1·06	12	1·9	17·6	26·3	32	4·9	12·8	10·7
Others*	0·25	2·8	1·4	7·0	10·5	41	2·2	4·7	13
Totals	8·81	100	3·1	67	100	23	46	100	16

* Includes nuclear and water power in the ratio of 5 to 1. pop – population

2. The accessibility of fuels and the cost of exploration.
3. Different countries have different energy situations, e.g. Switzerland is low in natural fuel supplies whilst Britain is sitting upon large reserves of coal, oil and gas. One country may therefore have to develop new technologies as an urgency but another may rely on cheap fuel supplies and neglect its technology.
4. The effect of the demands for energy by the developing countries.
5. Supplies from different parts of the world are subject to political attitudes and the political stability of supplying nations.
6. Distant suppliers are vulnerable to enemy action and sabotage.
7. Storage capacity is limited.
8. Demand is not really controlled and is subject to variations over the day and over the year.
9. Government decisions can affect the demand for and provision of basic fuels.
10. Changes in technology are not easily regulated to need.
11. The growing unpopularity of fuels which may pollute the environment, e.g. nuclear fuels.
12. The price of fuel is an important feature in its selection but this is subject very much to the world market forces.

Although 'alternative methods' of power generation and heat supply are attractive in principle, and it is essential to know what they can offer as part of energy policy development, there are good reasons for supporting existing methods of energy supply based on fossil and nuclear fuels, such as

(a) The resources are known to a reliable extent.
(b) The technology is established and the economics are understood, i.e. the cost-demand balance and labour needs are known.
(c) There is a reasonable variety in form (gas, solid and liquid), use and main features.
(d) Fundamental to all considerations is the high potential energy per unit mass of the fossil and nuclear fuels.

A high proportion of fuel is used in generating electricity (33% in the U.K.) and it was shown in Chapter 7 that about 40% of the fuel energy is converted into electricity, i.e. 60% goes to waste! If the overall losses could be reduced to 40% the useful output would be increased by 50%. The invitation to improve known methods of power generation is evident but there are fundamental limitations to gains to be obtained from conventional cycles as described in Chapter 7.

Figs. 7.8 and 7.10 show the variations in efficiency with boiler pressure and steam temperature respectively. It is seen that increases in efficiency by raising steam pressure and temperature is likely to be very expensive in terms of capital investment and the amount of improvement in efficiency small.

Prospects for energy saving and the efficient use of fuels are not necessarily to be found only in areas of completely new technology. A great deal of research work has been done over the years to re-establish coal as a primary source of fuel particularly if it is an alternative to the more expensive fuel oil. One of the developments has been the firing of coal in a fluid bed as an alternative to the normal combustion in a steam boiler which creates soot, ash and noxious gases like oxides of sulphur (SO_x) and oxides of nitrogen (NO_x).

In fluid bed firing the coal is mixed with fine particles of sand, limestone and ash and is burned in a suspended state in one or more fluid compartments. The combustion air is supplied through a nozzle causing the air to swirl and hence the coal mixture also. Water tubes are immersed in the fluid bed which absorb heat directly and control the temperature of the bed. The heat transfer rate is good due to the swirling movement of the charge. The fluid bed is convective in action also as in conventional combustion.

At the combustion temperature, 800–900°C, the sulphur combines with the limestone to give a dry waste product which is disposed of with the ash removing 80–90% of the sulphur content. The NO_x emission level is low because of the low combustion temperature. Dust particles are removed by cyclone separators, electrostatic precipitators and filters.

The attraction of the fluid bed coalfired boiler is the relatively low price of coal, particularly the lower quality grades which it burns very well. The capital outlay for such a boiler, the ash and coal handling equipment, control and maintenance, etc., is higher than that of an oil or gas fired boiler.

The *fluidized bed boiler* is an economical venture in the 10–20 MW range which is particularly useful for district heating systems and industrial energy plants.

8.2* Combined cycles and the total energy concept

The most obvious line of development is improving and using established methods and techniques. Apart from the objective of using present fuels more wisely the methods used will utilize known, and therefore cheap, engineering, applied readily and to calculable capital costs. In considering power cycles the starting point must be the attainable efficiency of conventional units which is, at best, about 40% or, alternatively, *with 60% energy wastage*. If the losses could be reduced to 40% an increase of 50% in efficiency would be obtained and this, in itself, would have a remarkable influence on the energy situation, industrial economics and the life of energy sources such as coal, gas and oil.

The complete use of the energy available to a system is called the 'total energy' approach and has the objective of using all of the heat energy in a power system at the different temperature levels at which it becomes available, to produce work, or steam, or the heating of air or water, thereby rejecting a minimum of waste energy. The most direct approach is the use of 'combined cycles' and these will now be considered but it must be emphasized that different kinds of industrial plant have widely different requirements and must be considered individually; the requirements of a plant producing electricity only are quite different from plants, like paper mills, which have combined power and process requirements which, in themselves, vary with the type of product and the production rates required at different times.

A particular type of combined cycle, the binary vapour cycle, was described in section 7.9 and although this could be considered to be the forerunner of combined cycles it did not become commercially significant. The cycle, as described in section 7.9, employed two Rankine cycles in series but modern considerations are based on the constant pressure (or Joule or Brayton) cycle in combination with the Rankine cycle. This brings together the gas and steam turbines as basic combinations. The gas turbine is the higher temperature unit and the gas leaving the turbine is at a sufficiently high temperature to be used as a source of heat for the production of steam at a suitable pressure and temperature. The simplest form of energy transfer from one cycle to the other is by means of a recuperative heat exchanger, as

---

* For this and other sections in this chapter readers will need to have an understanding of other chapters as the separate references will indicate.

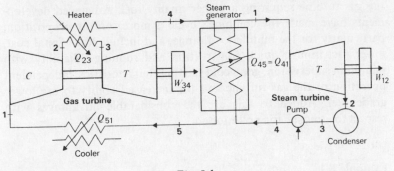

Fig. 8.1

shown in fig. 8.1 in which a closed cycle gas turbine and a steam turbine are combined. Another characteristic of the gas turbine, its high air-to-fuel ratio, can be exploited if, with an open cycle gas turbine, the exhaust gas, with its high oxygen content, is used as the inlet gas to the steam generator where the combustion of additional fuel takes place. This is shown in fig. 8.2 and this combination

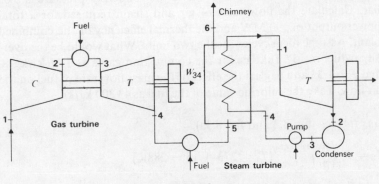

Fig. 8.2

allows nearer equality between the power outputs of the two units than is obtained with the simple recuperative heat exchanger. For a given total power output the energy input is reduced and the installed cost of a gas turbine per unit of power output is about one quarter of that of the steam turbine. The disadvantages of the combined cycles up to now has been greater complexity, the possible loss of flexibility and reliability and the different fuel requirements between the two, the gas turbine requiring a higher quality fuel. Advanced developments have made the combined cycle a more serious proposition, particularly for the middle load range, and include low capital cost, lower fuel inputs, quick construction and running reliability with minimum supervision, good cold starting characteristics, independent operation of the gas turbine and the increased ability to use lower grade fuel. Added to this are the considerable advantages with respect to atmospheric pollution.

Example 8.1

A combined power plant consists of a gas turbine unit and a steam turbine unit, the exhaust gas from the turbine being the supply gas to the steam generator at which a further supply of fuel is burned in the gas.

The pressure ratio for the gas turbine is 8 to 1, the inlet air temperature is 15°C and the maximum cycle temperature is 800°C.

Combustion at the steam generator raises the gas temperature to 800°C and the gas leaves the generator at 100°C. The steam pressure at supply is at 60 bar and 600°C and the condenser pressure is 0·05 bar. Calculate the flow rates of air and steam required for a total power output of 190 MW and the thermal efficiency of the combined plant. Assume ideal cycles for the two units. What would be the overall air/fuel ratio? Take c_p for the combustion gases as 1·11 kJ/kg K and $\gamma = 1·33$ and neglect the effect of the mass flow rate of fuel on the air flow. Take the calorific value of the fuel as 43 300 kJ/kg.

Gas turbine (fig. 8.2 and fig. 8.3a)

$$T_2 = T_1 \left(\frac{p_2}{p_1}\right)^{\gamma-1/\gamma} \quad (T_1 = 273 + 15 = 288\text{K})$$

$$= 288 \times 8^{\,1·4-1/1·4}$$

$$= 288 \times 8^{\,0·286} = 288 \times 1·813 = 522 \text{ K } (249°\text{C})$$

$$T_4 = \frac{T_3}{\left(\frac{p_2}{p_1}\right)^{\gamma-1/\gamma}} \quad (T_3 = 273 + 800 = 1073 \text{ K})$$

$$= \frac{1073}{8^{0·333/1·333}} = \frac{1073}{1·682} = 638 \text{ K } (365°\text{C})$$

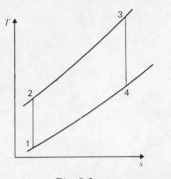

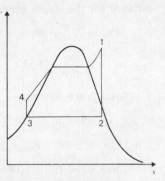

Fig. 8.3a   *Fig. 8.3b*

$$W = c_{p_g}(T_3 - T_4) - c_{p_a}(T_2 - T_1)$$

$$= 1\cdot11\,(1073 - 638) - 1\cdot005\,(522 - 288)$$

$$= 249 \text{ kJ/kg}$$

Energy supplied to the gas, $Q_{23} = c_{p_g}(T_3 - T_2)$

$$= 1\cdot11\,(1073 - 522) = 612 \text{ kJ/kg}$$

Energy supplied to the gas, $Q_{45} = c_{p_g}(T_5 - T_4) = c_{p_g}(t_5 - t_4)$

$$= 1\cdot11\,(800 - 365) = 483 \text{ kJ/kg}$$

Steam turbine (fig. 8.2 and fig. 8.3b)

From tables or $h - s$ chart

$h_1 = 3657 \text{ kJ/kg}$ $h_2 = 2183 \text{ kJ/kg}$

$h_3 = 138 \text{ kJ/kg} = h_4$, neglecting the feed pump term.

$W = h_1 - h_2 = 3657 - 2183 = 1472 \text{ kJ/kg}$.

Change in enthalpy of combustion gas
$$= 1\cdot11\,(800 - 100) = 777 \text{ kJ/kg}$$

Increase in enthalpy of steam $= h_1 - h_3$

$$= 3657 - 138$$

$$= 3519 \text{ kJ/kg}$$

If $\dot{m}_a$ and $\dot{m}_s$ are the mass flow rates of air and steam then an energy balance for the steam generator gives

$$\dot{m}_a \times 777 = \dot{m}_s \times 3519$$

$$\therefore \frac{\dot{m}_a}{\dot{m}_s} = 4\cdot53$$

The total power output $= 190$ MW

$$\therefore \dot{m}_a \times 249 + \dot{m}_s \times 1472 = 190 \times 10^6$$

$$4\cdot53 \, \dot{m}_s \times 249 + \dot{m}_s \times 1472 = 190 \times 10^6$$

$$2560 \, \dot{m}_s = 190 \times 10^6$$

$$\dot{m}_s = 7\cdot42 \times 10^4 \text{ kg/s}$$

$$\therefore \dot{m}_a = 4\cdot53 \times 7\cdot42 \times 10^4 = 33\cdot6 \times 10^4 \text{ kg/s}$$

Fuel energy input $= 33\cdot6 \times 10^4 \times 612 + 33\cdot6 \times 10^4 \times 483$
$$= 33\cdot6 \times 10^4 \times 1095 = 368 \times 10^6 = 368 \text{ MW}$$

$$\therefore \text{ thermal efficiency} = \frac{\text{total work output}}{\text{fuel energy input}}$$

$$= \frac{190}{368} \times 100 = 51\cdot6\%$$

If the flow rate of the fuel is $\dot{m}_f$, then

$$\dot{m}_f \times 43300 = \dot{m}_a \times 1095$$

$$\therefore \frac{\dot{m}_a}{\dot{m}_f} = \frac{43\,300}{1095} = 39\cdot5$$

i.e. Overall air/fuel ratio $= 39\cdot5/1$

The combined power units can be extended to use the 'waste' heat, which would normally be rejected to the atmosphere, to produce hot water for industrial process work or district heating for domestic and commercial buildings. Two arrangements of combined heat and power units, CHPs, are shown in fig. 8.4 and 8.5 based on nuclear reactors as the source of heat. The first, in fig. 8.4, uses the reactor to generate steam which then expands through turbines with steam extraction at a suitable pressure to provide heat for the hot water which is circulated to the district at about 120°C. The second, in fig. 8.5, uses helium (He) as the working fluid which is expanded through

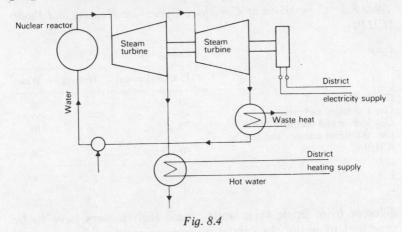

Fig. 8.4

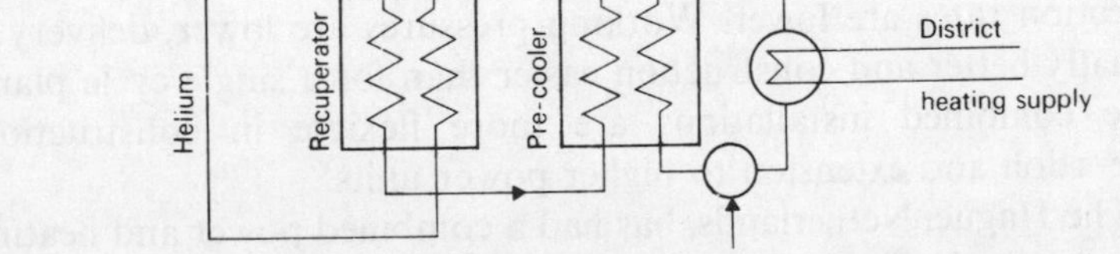

Fig. 8.5

the turbine and then returned, through the compressor, to the reactor. Other possibilities range from the utilization of established power stations to the incorporation of large incinerators used to dispose of refuse collected from a wide area.

A number of notable plants exist over the world which are based on the gas turbine used in series with a steam turbine. A comparison of the simple power generating cycle with the combined power cycle and the combined power generating and heating (cogeneration) or CHP cycles is summarized in Table 8.2.

Combined cycles have efficiencies of 44 to 50% in comparison with 40% for a good steam unit. In addition, the plant economics are quite

Table 8.2   Comparison of Combined Cycles for Heating and Power (CHP)

Cycle	Energy distribution (%)		
	Electrical power	Heating	Waste
Gas turbines	30	—	70
Gas and steam turbines	45	—	55
Gas turbine and heating (CHP)	30	60	10
Gas and steam turbines and heating (CHP)	40	40	20

different from single cycle units where high powers have to be produced to justify the capital investment required for economic operation, i.e. 50 MW units can have efficiencies normally associated with 500 MW units. The use of exhaust gas for space heating and process work can raise the efficiency to 85%. The cost of a combined plant, per kW of power, is between that of a gas turbine and steam turbine unit and arises because the conventional high temperature generator is replaced by a lower duty heat exchanger. The condenser and cooling water system capacities are also less because the heat rejection rates are lower. Working pressures are lower, delivery is usually better and construction easier than for a single cycle plant. The combined installations are more flexible in construction, operation and extension to higher power units.

The Hague, Netherlands, has had a combined power and heating plant in service since 1982. It is powered by two Rolls-Royce Olympus SK30 gas turbine sets each rated at 25·5 MW with two exhaust recovery boilers and a Delaval Stork 26 MW steam turbine. The maximum heat output available is 20 MW for heating buildings and the distribution between steam generated power and heating capacity is proportioned to meet the demands, a flexibility which is improved by using two turbines as indicated by the performance figures below although it has operated at 86% efficiency. A diagrammatic representation is shown in fig. 8.6.

Electrical power	Heating power	Thermal efficiency
77·3 MW	—	44%
77·3 MW	19·4 MW	55%
69·6 MW	60 MW	74%

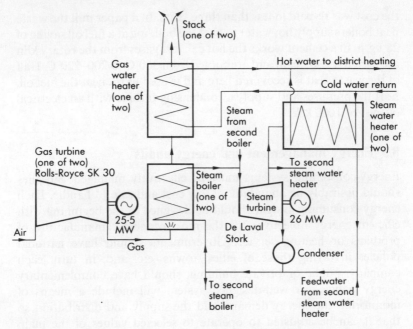

*Fig. 8.6 Diagrammatic representation of a CHP plant at The Hague,
Netherlands*

Experience with combined cycles has been obtained in Europe and
the U.S.A. although the practical and economic advantages do not
seem to have been applied on a large scale.

Waste heat boilers have to be designed to meet the particular
conditions of the plant in which they are to be installed and their
utilization depends very much on the temperature of the supply gas
and the acceptable temperature of the low pressure steam or hot
water which will be generated by the process. Low output, i.e.
<0·5 MW units, may be economically acceptable with a pay back
period of three to five years, and stack temperatures as low as 130°C,
from natural gas, can be utilized.

An example of the use of waste heat boilers in refuse incineration is
quoted in Section 8.5 but others can be quoted for different kinds of
plant. A chemical plant using natural gas has waste heat boilers with
steam outputs of from 45 to 80 tonnes/h and the stack temperature is
lowered from 130°C to 48°C as the plant make up water is heated
from 25 to 73°C. This saved 800 to 1200 tonnes/annum of fuel oil and

the cost was re-paid in less than three years. In a paper mill the waste heat boilers supply hot water to heat the building at a fuel oil saving of 28 kg/h. In a cement works the hot exhaust gases from the rotary kiln are used to raise steam and are cooled from 360°C to 200–220°C. Half of the waste heat is recovered here and is used to preheat the fuel oil, heat the premises and supply a condensing turbine with an electrical output of 1000 kW.

8.3 Energy management and energy audits

Energy-consuming systems must be effectively managed. Supplies should desirably be abundant, of low cost and safe to handle. Each energy-consuming system should be designed in the beginning with efficient energy objectives for the most economic manufacture of products or heating service. Governments should have national policies involving those of cities, towns, etc. and, in turn, each company, public or private building, should have complementary energy policies. A well-designed system will include a means of monitoring the energy demand and the supply and distribution so that it can be adjusted to operate to selected values of the main parameters which will usually be the temperatures at the main points in the system. This requires the services of technologists in the design and operation of energy systems who will require a knowledge of thermodynamics, heat and mass transfer, fuels, fluid flow, systems analysis and control. Some of these topics are dealt with in other chapters in this book and the remainder will be touched upon in this and later sections.

Most energy-using systems and buildings could be made more efficient immediately by an application of common sense methods of preventing heat loss to the surroundings and lowering temperature levels where possible as heat loss is proportionate to temperature difference (Chapter 17).

A basis of organization of an energy policy is the energy audit which is a formal account of the energy consumption and costs of a building or company over a period of a year or shorter periods if necessary. The account or audit is broken down into different sections. Table 8.3 shows the form of an energy consumption and cost distribution account.

Numerical values are substituted for A, B–F for the actual quantities used and the current prices/unit for a, b–f. The totals

Table 8.3 Energy Consumption and Cost

Energy	Quantity	Heating value	Price/unit	Cost £	Energy MJ	Cost per MJ
Coal	A tonnes	27 500 MJ/tonne	a £/tonne	A × a	27 500 × A	a/27 500
Heavy oil	B tonnes	43 200 MJ/tonne	b £/tonne	B × b	43 200 × B	b/43 200
Medium oil	C tonnes	43 600 MJ/tonne	c £/tonne	C × c	43 600 × c	c/43 600
Gas oil	D tonnes	45 480 MJ/tonne	d £/tonne	D × d	45 480 × D	d/45 480
Gas	E m^3	38·5 MJ/m^3	e £/m^3	E × e	38·5 × E	e/38·5
Electricity	F kW h	3·6 MJ/kW h	f £/kW H	F × f	3·6 F	f/3·6
			Totals			

evaluated on this basis are readily computed but perhaps what is important is the cost for each form of energy used as this suggests which may be the best on which to economise or perhaps consider changing to a different, cheaper form of energy.

The next part of the audit is a breakdown of the different forms of energy and where it is consumed. An example for a manufacturing company is shown in Table 8.4.

Table 8.4   Energy Distribution Through Company A

Energy	Workshop or department				
	Workshop 1	Workshop 2	Boiler house	Stores	Office complex
Electricity kWh					
Machines					
Lighting					
Compressed air					
Gas (MJ)					
Oil					
Coal					

The above is a simplified example of a way in which the energy distribution can be examined and an actual audit could be more detailed. Another audit would be made for the transport used by the company as follows:

Table 8.5   Fuels and Lubricating Oils Used for the Transport for Company A Including Garaging

Energy (quantity)	Internal	External delivery, staff	Cost (£)
Petrol (l)			
Diesel (l)			
Lubricant (l)			
Electricity (kW h)			
Vehicle miles			
Load (tonnes)			

Tables 8.4 and 8.5 should be completed monthly and comparisons can then be made between the months and across the years. Changes

should be explained and any corrective action taken to reduce fuel consumption. Such action may involve improving the maintenance standards, checking control equipment and level settings, modifying plant or operation control, installing new plant, changing process methods, installing more meters etc.

Table 8.6 gives some energy values for different fuels and useful conversion factors for the kinds of calculations covered in this chapter. Energy values vary with the quality of the particular fuel and the values quoted may be different from those given elsewhere (see Table 15.8). The values quoted here are taken from reference 8.1.

Table 8.6   Energy Values and Conversion Factors

Energy source	Energy content
Coal	
Anthracite	32 000 MJ/tonne
Good bituminous	30 000 MJ/tonne
Average industrial	28 000 MJ/tonne
Poor industrial	21 000 MJ/tonne
Oil (relative density)	
Gas (0·835)	45 600 MJ/tonne
Light (0·935)	43 500 MJ/tonne
Medium (0·95)	43 000 MJ/tonne
Heavy (0·97)	42 600 MJ/tonne
Gas	
North Sea	38·5 MJ/m^3

Conversion factors

1 Btu = 1·055 kJ	1 Therm = 100 000 Btu
1 Btu/ft^3 = 37·259 kJ/m^3	1 Therm = 105·506 MJ
1 kW = 3·6 MJ	1 tonne = 1000 kg
1 lb = 0·4536 kg	1 ton = 1·016 tonne
1 imperial gallon = 4·546 l	1 Therm/ton = 103·839 kJ/kg

Degree days

A comparison of energy audits for different periods may have limited significance and may even be misleading as the conditions under which they were obtained may have been quite different.

In the UK it is assumed that heating in buildings is necessary only when the outside temperature falls below 15·5°C (60°F). The normal inside temperature is taken as 18·3°C (65°F), and the heat from internal sources such as occupants, lighting, machinery, plant, etc. is assumed to be sufficient to maintain the inside at 18·3°C when the outside temperature is at 15·5°C.

When the outside temperature falls below 15·5°C heating is necessary and the fuel used is proportional to the temperature difference between 15·5°C and the actual outside temperature. This is so since the building heat losses are proportional to the temperature difference between the inside and outside temperatures. If the outside temperature averaged over 24 hours was 14·5°C, say, then this would represent 1 degree day. If maintained for a week the cumulative heat loss would be proportional to seven degree days and so on. If degree days and the fuel consumed are measured over the same period the quantity of litres of fuel/degree day can be obtained and is a criterion of the energy consumption corrected for climatic conditions. Outside temperatures do not remain constant for 24-hour periods and averages are taken by one of three formulae recommended by the Meteorological Office, see reference 8.1.

Degree day measurements are taken at seventeen meteorological stations spread over England, Wales, Scotland and Northern Ireland and the reports for the regions are available from the Meteorological Office. Table 8.7 shows the average values of degree tays taken over a 20-year period for a selection of the seventeen stations. The complete table is given in reference 8.1.

The reference temperature of 15·5°C is usually used but for some buildings higher inside temperatures than 18·3°C are necessary, e.g. in industrial buildings or hospitals. The degree days tabulated may be referred to another base temperature as described in reference 8.1.

The fuel used per degree day is a useful guide to the fuel consumption figures for buildings with similar use and size. Degree days should not be used for short period tests and at least monthly figures are necessary. If combined heating and hot water systems are installed the two parts of the heating load as expressed in degree days cannot be separated due to the fluctuations in demand. If the incidental heating is not equivalent to 2·8 K the degree day method will be less reliable. Heat losses and hence fuel consumptions are affected by other influences such as prevailing winds, humidity, solar radiation, cloud, fluctuating demands, thermal capacity, etc. By their nature degree days need to be used with caution in calculating fuel consumptions and cannot be used to forecast weather conditions.

Table 8.7 Degree Days Averaged over a 20-year Period

Region	Jan.	Feb.	March	April	May	June	July	Aug.	Sept.	Oct.	Nov.	Dec.	Total
1 Thames Valley	354	307	287	202	117	51	25	28	57	129	252	332	2141
6 Midlands	382	340	322	239	158	81	48	53	93	171	286	360	2533
14 East Scotland	387	350	330	261	195	104	72	74	109	186	312	371	2751
16 Wales	335	307	301	236	165	87	51	46	77	137	239	305	2286
17 Northern Ireland	367	329	319	243	171	89	62	65	102	168	289	346	2550
Average for 17 regions	368	329	313	236	160	81	51	53	88	159	279	347	2462

Example 8.2

A central heating boiler-house provides steam for both the process and space-heating requirements of a factory complex. The boilers are fired with gas oil costing £3/GJ and run at an average efficiency of 71%. The monthly fuel consumptions and corresponding degree days, D (reference 15·5°C), for a previous period are as follows:

Month	Consumption (GJ)	D days	20-year average D days
S	12 040	102	94
O	13 100	156	171
N	NA*	300	286
D	17 460	370	360
J	NA	370	379
F	17 000	350	343
M	NA	312	320
A	NA	215	238
M	12 600	132	156
J	NA	60	79
J	NA	28	48
A	10 600	30	53
			2527

* NA denotes Not Available.

Recent modifications at the factory have included the installation of a new process consuming 1000 GJ steam/month and the installation of an economiser to increase the boiler efficiency to 75%.

Using the degree days as being applicable to all space heating loads, state any assumptions made and determine:

(i) the expected annual fuel consumption of the modified factory;
(ii) the maximum monthly fuel consumption of the modified factory;
(iii) the payback period in years for the economiser if the capital cost was £40 000.

A graph of the fuel consumption in GJ against the recorded degree days is drawn as shown in fig. 8.7 and gives a base load at zero degree days of 10 000 GJ/month which is a measure of the process energy required. The slope of the graph gives a space heat factor of 20 GJ/D

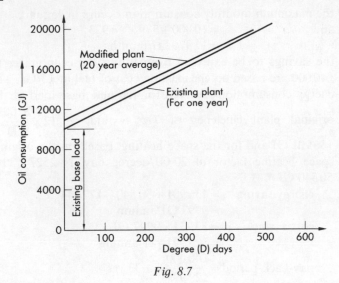

Fig. 8.7

day. The actual energy requirements taking into account the boiler efficiency of 71% are:

space heat factor $= 0.71 \times 20 = 14.2$ GJ/D
base load $= 0.71 \times 10\,000 = 7100$ GJ/month
base load for the new process $= 7100 + 1000 = 8100$ GJ/month

and for a new boiler efficiency of 75% the new fuel consumption will be:

for the base load $= \dfrac{8100}{0.75} = 10\,800$ GJ/month

for space heating $= \dfrac{14.2}{0.75} = 18.933$ GJ/D

The projected annual figures are based on the 20-year average degree day figure quoted:

(i) the annual consumption $=$ 12 months $\times$ 10 800 GJ/month
(base load + space $+ 2527$ total average D's
heating) $\times$ 18.933 GJ/D
          $= 129\,600 + 47\,844$
          $= 177\,444$ GJ/annum

(ii)   the maximum monthly consumption occurs in January

and $= 10\,800 + 379 \times 18{\cdot}933$

$= 17\,976$ GJ/month

(iii) the savings to be expected from fitting the economizer at £40 000 are based on the existing cost of fuel at £3/GJ. The energy consumption required for the new base load at the original plant efficiency of 71% would be $12 \times \dfrac{8100}{0{\cdot}71} =$ 136 901 GJ and for the space heating, based on the original space heating factor of 20 GJ/degree day, is $2527 \times 20 = 50\,540$ GJ

$\therefore$ energy saving $= 136\,901 + 50\,540 - 177\,444$

$= 9997$ GJ/annum

$\therefore$ cost saving $= £3/GJ \times 9997$ GJ

$= £29\,991$ per annum

$\therefore$ pay-back period $= \dfrac{£40\,000}{£29\,991} = 1{\cdot}33$ yrs

Energy management

Efficient energy management must cover the whole industrial, domestic or service system used and a consideration here of the requirements will be based on an industrial unit as a comprehensive example. Energy considerations should be taken into account at the start of a new industrial project and may even be influential in deciding on the geographical location in the first place. The problem of improving the energy audit of an existing plant is quite different and has inherent constraints. Even for a new enterprise energy conservation may not be able to be taken to its limit because of the investment capital involved even though the long-term saving may be attractive, so from the outset a compromise solution may have to be accepted. The energy analysis includes the energy required for the manufacture of the product, the energy used in the manufacturing process, the energy needed to support the manufacturing environment and the effect of the industrial unit on its external surroundings including the economic disposal of waste. With existing plant the opportunities for improving the energy consumption may be more restricted and relatively more expensive if constructional changes, the replacement of plant and the changing of fixed attitudes and practices are concerned. Sankey diagrams, as shown in Section 7.11, illustrate energy use.

The industrial recession in western countries in the mid 1970s created an active interest in the relative industrial performances of the Western nations and Japan, which emerged as a leading industrial nation producing an increasing range of high quality manufactured goods at low prices. The manufacture of motor vehicles can be taken as an example of particular significance to the U.K. and the U.S.A. and this example has been subject to some analysis. One of the features of the study was the energy consumption, although it is not the main factor in relation to others like the productivity of more modern plants, labour costs and employee attitudes and practices.

Japan had to be energy conscious as 90% of its energy was imported – about 75% being oil – and so its industry implemented conservation techniques more rigorously than did industry in the West. Waste heat recovery systems, the burning of waste paint solvents, the purification and recycling of heated air contributed to the energy costs of Japanese products being 20% less than those of many of their competitors. Between 1973 and 1978 the energy consumption per vehicle dropped from 13·6 to 10·8 GJ. For a similar vehicle built in the U.S.A. in 1978 the energy consumption was 31·44 GJ/vehicle, an improvement on 40·32 GJ/vehicle for 1972, and even taking into account the mass ratio of 1·8 for the U.S. to Japanese vehicles the energy consumption for the Japanese is better by 1·9 to 1. For one U.K. manufacturer the figure in the same year was 22 GJ/car and for another, working at less than full production capacity, the energy consumed was 58 GJ/car. The relative cost advantages to the Japanese car were estimated at £45 to £85 depending on the capacity working. This difference is significant but is swamped by the much lower labour cost savings of £1200–£1600 of a high production unit employing efficient manufacturing methods.

Some of the factors contributing to better energy use by the Japanese manufacturers are instructive.

(a) The holding of buffer stocks of parts and materials is held to a very low period of 3–4 hours, where U.K. manufacturers hold several weeks supply of the main items. Cash flow is reduced and savings are made in buildings, heating, lighting and employees.

(b) The factories are compact and do not have the long lines connecting some U.K. companies where car bodies are made in one place and then transported long distances for assembly.

(c) The factories are sited in a temperate climate giving a 20% saving on similar size plants in the U.S.A. and 15% on the U.K. plants. They have better low grade heat recovery, better insulation and more use of recirculated air.

(d) The working periods are continuous – two shifts instead of one – therefore there are shorter shut-down periods, giving about 30% reduction in the energy for the working environment.

(e) The energy cost of the basic materials such as steel is less for the Japanese industry due to a high investment in modern energy-saving plant which give higher yields than their western competitors of the high quality steels used in vehicle manufacture. The relative energy costs for a tonne of sheet steel was U.K. 40 GJ, U.S. 38 GJ and Japan 28·5 GJ.

It is hoped that the case just described suggests to the reader the economic importance of energy to a manufacturing nation which has to be competitive in world markets. It is not possible to give a complete treatment of the subject but the following points should be given:

1. Energy production itself is the greatest user of fuel energy.
2. The production of materials is the second largest consumer of energy.
3. Every product has an energy cost, e.g. 8 MJ for a milk bottle, 25 000 MJ for a colour TV. The energy savings of double glazing are enthusiastically quoted by salesmen but an overall analysis would take into account that it costs 6000 MJ/m^2 to manufacture double glazing units.
4. The energy cost of manufactured items should be constantly reviewed as the opportunities for waste and hence saving are many, complex and inter-related. Manufactured waste should be a minimum and different methods of forming should be investigated, e.g. casting and machining, extrusion, drawing or continuous casting.
5. The reduction of losses by insulation and using 'waste' heat for useful applications.
6. Space heating takes about 30% of the total energy used and an estimate is that 50% of this could be saved by using lower room temperatures, preventing draughts, insulation, controlled ventilation rates, efficient air and water distribution, and planned systematic maintenance.

7. The need for an energy-conscious transport system – a massive study in itself.

8. The possibilities for re-use, recycling and reclamation of discarded commodities.

9. The pollution of the environment by thermal output, solid waste disposal, oil, smoke and gases such as sulphur dioxide, nitrogen oxide and carbon monoxide. Radioactive wastes are a particular problem, being gaseous, liquid and solid.

8.4 The technology of energy saving

Some of the aspects of energy saving involving the expertise of the engineer include:

1. An economic use of steam in process heating between the requirements of higher temperature (and hence pressure) to provide a good heat transfer rate and lower temperature (and pressure) to increase the latent heat of vaporisation available.

2. Operating heating plant at full load for shorter periods and shutting down when no processing is required. Moisture-removal processes before drying by heating should be as complete as possible.

3. The use of simple thermostatic control systems to limit the temperatures reached in process work and space heating.

4. Keeping the proper balance between fresh air and recirculated air between summer and winter conditions.

5. The recovery of flash steam or vapour and the use of condensate for heating purposes or return as boiler feed.

6. The prevention of heat losses from steam leakage at valves and joints, e.g. steam at 7 bar leaking through a hole 0·8 mm diameter gives a waste equal to 1500 l of oil per year. Effective lagging of pipes and flanges is essential.

7. The separating out of steam and air in steam heating processes by the use of automatic air vents or steam traps. Trapped air interferes with the heating process and effectively lowers the steam temperature.

8. The separation of water from steam in the condensate as water forms an insulating layer on heating surfaces. This requires efficient steam trapping with thermostatic control and keeping the steam clean of pipe scale and dirt.

9. The proper use of compressed air, which uses 10% of the industrial electrical power, by creating optimum conditions for

efficient compression (see Chapter 9), preventing leakage, seeing the air-driven equipment is efficient, using low pressures and recovering the heat of the compression process which is transferred to the jacket-cooling water. This requires good initial selection and installation of plant and continuous maintenance and is described fully in reference 8.1. The heat of compression can be used for space heating or as a supply to air dryers.

10. The prevention of heat loss by insulation has been mentioned before and the extent of such losses can be calculated using the principles of heat transfer dealt with in Chapter 17. Insulation is costly and the individual cases must be considered from an economic viewpoint. This is shown in fig. 8.8 as a graph of cost against the thickness of insulation.

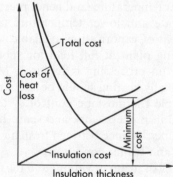

Fig. 8.8   Economic thickness of insulation

11. The application of automatic control and metering systems, which in principle are simple but can be complex in application when the processes or environments to be controlled are themselves complicated. In the simplest version, such as maintaining a room at a given temperature, the temperature is controlled by a 'thermostat' which measures the room temperature and compares it with a 'set' value. If the room temperature is higher than the reference value the thermostat switches off the supply of heat and vice versa.

12. The selection of good lighting to suit the illumination required for different purposes and described in standard codes of practice (see reference 8.1).

13. The recovery of waste heat using heat exchangers of different types, e.g. shell and tube, recuperators and regenerators (see Chapter 17); run-round coil systems, plate type, heat (thermal) wheels, heat pipes and thermosyphons (see reference 8.1); and heat pumps (see Chapter 16).

14. The economic use of oil, gas or coal fired boiler plant. The proper functioning of every boiler or generator according to its specification and application is fundamental to the economic operation of the whole plant. Installation and service criteria must be met and continuous performance monitoring and preventive maintenance carried out.

15. Just as plant should be economically insulated, so should the whole industrial building. The extent to which this is done depends on the annual savings in the cost of energy, the initial cost of insulation and the repayment period in relation to the savings made. See references 8.1, 8.2 and 8.9.

8.5 Alternative energy supplies

The first consideration must be to see if there are other ways in which better use can be made of the energy which is generated by known methods and applied to known devices. As described previously, there is great potential in using the waste heat from electricity generating stations for district heating purposes. In some countries (e.g. Switzerland) this has been done and is economically justified; in Britain there is not yet sufficient cause to put in the very high capital investment district heating systems would require.

The energy demand over the day, week, month and year is very variable. If society was organized on an 'energy cycle' for the best use of energy our way of life may be considerably different, e.g. traffic would be organized to move freely without jams at peak periods, but the industrial plus domestic demand for energy would be smoothed out to give an optimum use of energy. There is no doubt great savings could be made in this way if the motivation was sufficient to bring it about.

One of the outcomes of a developing energy programme over recent years has been the renewed interest in using municipal and industrial waste as a source of heat energy. Whilst exploitation of this resource in Great Britain has been slow there are some notable examples of successful refuse incineration heat recovery plants at

Edmonton, Sheffield and Coventry and experience in Europe and Eastern countries is more extensive. Solid waste has to be disposed of and collection and disposal is expensive so the financial return possible from burning the waste to give a useful supply of heat is worth considering. The nature of the waste material brings its own problems when it is to be used as a fuel which is highly heterogeneous and produces, on combustion, dusty and corrosive hot gases. The removal of these from the hot flue gases is a critical factor in the economic supply of energy by this means. The nature of the fuel creates parameters for the design of boilers and ancillary equipment which are quite different from those for plant using conventional fuels.

Refuse incineration plants discharge gases at temperatures between 270° and 300°C and these can be lowered to about 200°C without corrosion of the boiler heating surfaces. In one such plant in Switzerland three low-pressure boilers give an extra thermal output of 3·7 MW for 6000 hours per annum which is equivalent to 2200 tonnes of fuel oil. An overall figure indicates that burning 5 tonnes of refuse is equivalent to 1 tonne of fuel oil.

Consideration will be given to some of the alternative forms of energy supply which are very different in form and have quite different attractions and technological problems. It is difficult to compare such widely different systems but one of the main attributes of power generation from fossil or nuclear fuels is the high potential energy per unit mass of fuel. In a hydroelectric scheme the equivalent mass of water to be handled for an equal energy output requires a very large-scale structure to store and handle the water. Collection, storage and distribution are other factors which have to be taken into account. (See references 8.2, 8.5, 8.7 and 8.8.)

Heat pumps

The principles and the theory of the 'heat pump' are described in Chapter 16 and the heat pump for many years has offered considerable prospects for the useful application of energy. The heat pump was patented in 1852 by Sir William Thomson and first appeared as an air machine.

In Example 16.7 it is shown, for the cycle specified, that for every kW of power input to drive the compressor 6 kW of heat are supplied to the application at a temperature in excess of the 30°C required for the air. The ammonia is delivered at 103·5°C, condenses at 38°C and is undercooled to 26°C, the air being received at 5°C. This means that $6 - 1 = 5$ kW are obtained from a low grade cheap supply of heat such as a pond, river, ground, the atmosphere, sewage farm or an industrial source of 'waste' heat. The investment in heat pump plant would be expected to be justified on theoretical figures like this.

However, the appearance of heat pumps over the years has been erratic on both the domestic and industrial scales. When oil was cheap heat pump installations were not economic but escalating oil prices have revived interest in the heat pump. Some public buildings are heated in part by heat pumps and include Churchill College, Cambridge, and the Royal Festival Hall, London. Generally heat pump installations have been oversized by being designed to meet the peak load. If the heat pump is designed for 60% of the maximum load a lower investment cost is incurred and a better coefficient of performance is obtained. The peak demands are best met by a conventional boiler. Close attention must also be paid to the proportioning of condensers and evaporators to meet the temperature variations caused by the environmental conditions and to the control requirements on both sides of the plant. An interesting and successful plant is described in reference 8.4 and includes a representation of the annual pattern of operation.

Solar energy

The earth receives energy directly from the sun. It is silent, inexhaustible and non-polluting. The means of collecting and distributing solar energy are known but the cost is about twice that of conventional electricity generation. The sun's rays fall on 'collectors' which are mirrors and reflect the rays to a central receiver. It is necessary for the collectors to be able to 'track' the sun to ensure a continuous maximum reception. Solar energy can be used on small-scale units, and even in Britain solar heating panels are being fitted into the roofs of houses and other buildings to contribute about 50% of the water heating load. In other, sunnier, countries the use of solar heating is more widespread. The application to large power plants is likely to be slow. An ambitious study has been made of using a solar satellite which is continuously in direct sunlight to collect the energy, convert it to electricity and direct a microwave beam to a receiver on earth where it would be reconverted to electricity. The cost of such a scheme is likely to prohibit its realization.

Three experimental solar power stations have been erected in Almeria, Spain, one of 1000 kW and two of 500 kW rating. The irradiation density is about 1 kW/m² and the locality has 3000 hours of sunshine per annum. The 500 kW station has 93 collectors or heliostats which are computer controlled, each of them having twelve mirrors directed to reflect sunlight over the year into the receiver which is mounted on a tower 43 m high. The collector consists of a bank of tubes behind which is a ceramic wall that absorbs the radiation passing the tubes, amounting to about 5%, and serves as a heat store which is insulated to prevent loss. Liquid sodium is circulated through the receiver tubes where the temperature is raised to 530°C and then through the steam generator to a cold storage tank before returning to the receiver at 270°C for heating. The steam drives a steam engine which drives the power generator. At the receiver the thermal output is 2·7 MW, the mass flow rate of sodium is 7.34 kg/s. Sodium is used because of its good heat transfer properties and its ability to be stored. Thus the plant handling the sodium is compact.

It is anticipated that in some areas the solar power station should compete with oil but perhaps not with large coal or nuclear stations. However, there are areas in the world, as in the developing countries, where conventional power stations are less likely to be chosen and the solar station could be very attractive. In the U.K. sunshine hours are about 1500 per annum in comparison with 3000 h/annum elsewhere. (See references 8.5 and 8.6.)

In some parts of the world the sea temperatures are such that at the surface readings of 24° to 32°C are measured and at depths of 300 to 400 m temperatures of 4° to 7°C exist. This temperature gradient is a source of power. A boiler plant using propane or ammonia under high pressure is required, the hot water heating the boiler and the condenser situated in the cold water. The potential power by this means is very high, about 300 times the world's present power consumption, but the associated engineering problems have to be overcome which includes transmitting the power developed to the land.

Wind power

Winds possess high kinetic energy and windmills have been used for many years to drive mill mechanisms. The search for alternative power sources has led to the rediscovery of wind power and many wind-driven power stations, large and small, have been built, mainly as prototypes, and are generating power. The modern windmills are much more technically sound than their historical counterparts and have benefited from established knowledge of aerodynamic blade design. They include automatic control of the rotor position to suit changing wind directions and for adjustment to the blade pitch. The behaviour of the unit can be controlled and monitored by computers.

A joint project by German companies built Growian near Marne in 1982 on the North Sea coast which has a 3 MW output. The tower is 100 m high with two blades 50 m long, 5·2 m at the root and 1·3 m at the top. It generates 6·3 kV at 50 Hz for a speed of 1500 rev/min with wind speed from 6·3 to 24 m/s (nominal speed 12 m/s). In Denmark two 630 kW wind power plants have been constructed near Nibe and were commissioned in 1979 and 1980. The reasons for wind power developments vary between different countries and localities within the countries and quite different generators have been evolved to meet the demands. There is a particular market for small systems for use in remote areas where diesel generators and thermo-electric generators have been used traditionally but both need fuel to be transported to them. The wind generator is taking its place with the other two in an integrated system. For this application the cost per kW h is quoted as £1·30 for solar generators, £0·35 for wind, £1·25 for thermoelectric and £0·80 for diesel generators showing the wind generators to be competitive.

In the U.K. a Wind Energy Group of Taylor Woodrow Construction Ltd, Greenford have designed and built a 20 m diameter wind turbine generator (250 kW) which was erected on Orkney at Burgar Hill and produced power in August 1983. This is the forerunning of a 3 MW unit with a 60 m diameter blade. Both units are extensively monitored by a microprocessor-based system to give blade loads, power levels, gearbox, nacelle and tower accelerations, blade pitch angles, etc. (See reference 8.3 for complete details.) Machines of 20 and 60 kW are being built by other companies for small power users. (See references 8.3 and 8.5.)

Water power

The movement of large quantities of water naturally or by design affords the opportunity to generate electrical power. Water flowing from reservoirs or in rivers passes through hydro-electric generators to produce electricity. In the U.S.S.R. stations of 4000–6000 MW output have been constructed. In the U.K. most potential sites have been developed but the hydroelectric power is a small proportion of the total used.

In the U.K. pumped-storage systems have been constructed, e.g. at Dmorwic, Snowdonia and on Loch Ness in Scotland. They are not net producers of power but are used to smooth out the load requirements on conventional plant as they pump water to the high storage reservoir when the external load is low and return it through the hydro-generators when the external load is high.

Tidal power has been traditionally harnessed, along with windmills and watermills, for small power units and several are still in use in England and Wales. The fluctuations in tidal behaviour are not compatible with the continuous operation mode of the mill. Modern tidal power systems are large-scale constructions and there are two, one in the Rance estuary in France (544×10^6 kW h/annum from twenty four 10 MW units) and at Kislaya near Murmansk, Russia (400 kW). There are a number of countries which have suitable sites for barrage schemes including Morecambe Bay, the Solway Firth and the Wash in Britain but these could also be freshwater storage sites.

The site requires a natural coastal basin formed by a short dam to separate it from the sea which should have a mean tidal range > 7 m. The water-driven generators are sited in the dam and are driven by the water as it passes in each direction between high and low tide. The potential power is very high but so is the capital investment and probably the prospects of nuclear power are more attractive.

Wave generators received a great deal of attention in Britain in the 1970s although devices to extract power from wave movement have been proposed for over a hundred years. In some parts of the world, e.g. the Orkneys (77 kW/m), the estimates of possible power outputs are high, particularly in winter when the energy demands are greatest, but the variation in output would mean that other systems would be required as well. The purpose of the wave generators is to accept the movement of the waves in some way and use this to generate electricity by a mechanical means or by displacing air from a bag through an air turbine into another bag. There are many different designs built on different principles such as Cockerell's rafts, Masuda's ring buoy and, perhaps the most interesting wave energy machine, the Salter duck. The Salter duck is mechanically sophisticated with hydraulic pumps, motors and generators and is used in banks of units about 500 m long with power take-off links to reach the shore. Perhaps the greatest weakness of the wave generators is their lack of robustness to stand the severe conditions which are required for power generation. Some devices have been designed to operate below the surface where there is still a substantial disturbance. (See reference 8.5.)

PROBLEMS

(Before attempting these problems the reader is advised to read the later chapters, particularly Chapter 16.)

8.1 A combined power plant consisting of a closed cycle gas turbine unit, using air as the working fluid, and a steam turbine is to be designed such that the heat rejected at the gas turbine unit is to be utilized to produce steam at the generator for the steam turbine. The air leaving the generator is at 200°C and it is cooled to the compressor inlet temperature by a second cooler which rejects the heat to waste.

Draw a line diagram for the plant and calculate, on the basis of ideal cycles, and neglecting the feed pump term of the Rankine cycle, the mass flow rate of steam per kg/s of air flow, the total power output per kg/s of air flow and the thermal efficiency of the cycle. Take c_p for air as 1·005 kJ/kg K.

The particulars of each cycle are as follows:

		Gas turbine cycle	Steam turbine cycle
pressures	minimum	1	0·07
(bar)	maximum	5	30
temperatures	minimum	20	–
°C	maximum	830	300

(0·079 kg/s, 313 kW, 48·8%)

8.2 A combined power plant consists of a gas turbine unit and a steam turbine unit. The exhaust gas from the open cycle gas turbine is the supply gas to the steam generator of the steam turbine cycle at which additional fuel is burned in the gas. Draw a line diagram of the unit.

The pressure ratio for the gas turbine unit is 7·5 to 1, the air inlet temperature is 15°C and the maximum cycle temperature is 750°C. Combustion in the steam generator raises the gas temperature to 750°C and the gas leaves the generator to atmosphere at 100°C. The steam is supplied to the turbine at 50 bar and 600°C and the condenser pressure is 0·1 bar.

Calculate the flow rates of air and steam required for a total power output of 200 MW, the power output of each turbine and the thermal efficiency of the plant.

The calculation should be based on ideal processes in the cycles. For the combustion gases take $c_p = 1·11$ kJ/kg K and $\gamma = 1·33$ and neglect the effect of the mass flow rate of the fuel on that of the air.

$(39·3 \times 10^4$ kg/s; $8·19 \times 10^4$ kg/s; $88·7$ MW; $112·1$ MW; $47·8\%)$

8.3 The fuel and electricity consumptions of a company in 1985 quoted in the normal commercial units were:

Gas   170 348 ft³
Coal (Grade 1)   250 tons (note: not tonnes)
Oil   350 000 l
Electricity   351 223 kW h

10% of this was consumed at night (23.30 to 07.30)
Monthly consumption equal between summer and winter (November to March inclusive).
The fuel and electricity tariffs were:

Gas Up to 25 000 therms/annum   37p/therm
      over 25 000 therms/annum   32·4p/therm
      Calorific value 38·5 MJ/m³ (1 therm = 105·58 MJ)
Coal Two grades of suitable coal are quoted:

Grade	Pit-head cost	Transport cost	Calorific value
1	£60·80/ton	£3·50/ton	27·9 MJ/kg
2	£65·60/ton	£3·50/ton	32·55 MJ/kg

Oil   For a full load of 12 500 l or more 21·35p/l. Calorific value 45 480 MJ/tonne; relative density of oil = 0·854.

*Electri*city
Normal tariff – no maximum demand

Period	Cost/kW h
Summer day	2·75p
Winter day	3·62p
Night 23.30–07.30	1·93p

Calculate

(a) The total cost of fuel and electricity for the year;
(b) The relative cost of each supply per MJ;
(c) The saving in changing to Grade 2 coal.

What is meant by a maximum demand tariff?
The costs are for 1985. What would the costs be now?
Gas £650·83; coal £16 075; oil £74 725; electricity £10 042·34. Total £101 493·17: respectively 0·35, 0·22, 0·54, 0·79 p/MJ; saving £1268.)

8.4 An office building is space heated between October and April (inclusive). The degree days and the corresponding fuel consumptions in units of 1000 l for two consecutive years are shown in the table. In between the two seasons the heating plant was modernized and insulation work was carried out which cost £27 000. The cost of fuel was 22p/l when the costing was done. Plot the fuel consumption/D day characteristics for the two seasons and calculate:

(a) the average fuel consumptions for the two seasons in *l*/degree day;
(b) the saving to be expected per annum based on the average degree day figures for the district tabulated, for the same months, over a twenty-year period;
(c) the period of repayment for the cost of the modifications.

(82·8, 50 l/D day; £13 525; 2 years)

| Month | 1st year | | 2nd year | | Average |
	D days	Fuel × 1000 l	D days	Fuel × 1000 l	D days over 20 years
Oct.	175	14·5	240	12	129
Nov.	290	24	240	12	252
Dec.	320	26·5	220	11	332
Jan.	280	23·25	255	12·75	354
Feb.	275	23	290	14·5	307
Mar.	300	25	315	15·75	287
Apr.	200	16·5	215	10·75	202
Totals	1840	152·75	1775	88·75	1863

8.5 Fig. 8.9 shows an arrangement whereby a supply of process water at 55°C is obtained from river water initially at 10°C. The steam turbine drives a heat pump which uses Refrigerant 12.

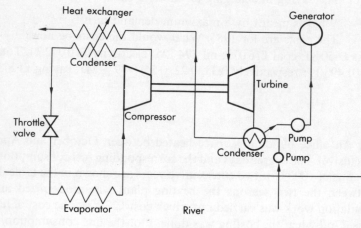

Fig. 8.9

The heat pump uses the river as the source of heat and operates between evaporator and condenser *saturation* temperatures of 0°C and 65°C. The refrigerant is dry saturated at the compressor intake and the liquid is just saturated at inlet to the expansion valve.

The steam supply is at 60 bar and 450°C, the condenser pressure is 0·07 bar and the condensate is saturated liquid at inlet to the feed pump, the power input to which can be neglected. The process water, drawn from the river, is pumped through the steam condenser as cooling water before passing to the heat pump as the condensing medium. Neglect the pump work required for the process water.

Compare the performance of the plant with direct heating of the water assuming that all processes are ideal. You are advised to work to a basis of 1 kg of boiler steam.

(Heat pump to direct heating advantage 1·87)

8.6 A sports centre includes an ice rink and swimming pools and it is proposed to use a mechanical heat pump to provide the combined effect of maintaining the ice in the rink and some of the heating requirement for the water of the swimming pools and for the buildings.

The power input to each of the compressors is 150 kW and the mechanical efficiency of the compressor is 96%. The refrigerant is ammonia, the evaporation temperature for which is $-7°C$ and the condensation temperature is 48°C. The ammonia can be assumed to be dry saturated at entry to the compressor and the liquid ammonia is cooled to 46°C on leaving the condenser. The refrigeration rate in the ice rink for this power input is 460 kW. Calculate:

(a) the mass flow rate of refrigerant in kg/min;
(b) the isentropic efficiency of the compression process;
(c) the heat available for heating purposes.

It is necessary to compare this method to providing the heating by the direct heating of water in a boiler and this should be done for 1 kW of work input using the following data:

Electrical power is generated at a power station overall efficiency of 40% and the transmission efficiency to the sports centre is 83%. The water boiler has an overall efficiency of 90%.

What other factors would be taken into account before reaching a decision on the proposed method?

(26·64 kg/min; 79%; 610 kW; 4·07 kW by heat pump, 2·65 kW for direct heating)

8.7 For process purposes a factory requires a supply of air at normal atmospheric pressure and at 35°C at a rate of 10 m³/s for 16 hours per day, 5 days per week and 48 weeks per year. The average air temperature at supply is 11°C.

Two heating systems are to be considered:

(a) Direct heating by a gas-fired heater which is 80% efficient and uses gas costing 1·25p/kW h.
(b) A vapour compression heat pump using Refrigerant 12 driven by an electric motor. The heat is taken from an outside source which has an average temperature over the year of 8°C. Electricity costs 3·0p/kW h.

Allow a 10 K temperature difference for heat transfer at the evaporator and condenser and assume a real coefficient of performance of 0·6 of that based on a simple heat pump cycle.

It is estimated that the cost of equipment and installation will be £250 per kW of heating load more for the heat pump than it would for the gas heating.

Calculate:

(a) the average rate of heating required;
(b) the cost/annum for gas and electricity;
(c) The payback time for the heat pump installation based on fuel cost savings alone.

What other factors would you take into account before making a recommendation on the choice of installation?

(277 kW; £16 625, £9390; 9·6 years)

8.8 A manufacturer quoted the following energy balance for a diesel engine at full power:

Fuel energy supplied kW	Power kW	Coolant kW	Oil kW	Exhaust kW	Radiation kW
190·6	64·7	30·6	5·1	77·3	12·9

The atmospheric pressure during the test was 1013 mbar and the temperature was 20°C. The specific fuel consumption was 236 g/kW h for gas oil of relative density 0·854. The air flow to the engine was 4·6 m³/min and the exhaust gas temperature was 640°C. cp for the exhaust gas can be taken as 1·175 kJ/kg K.

A scheme to supply hot water to a heating system is to be considered using the engine to drive the compressor of a heat pump which takes heat from a large pond at 10°C. The heat pump uses

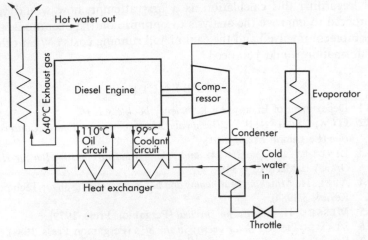

Fig. 8.10

Refrigerant 12 operating between saturation temperatures of $-5°C$ and $45°C$. Assume that the actual coefficient of performance is 0.7 of that for the idealized vapour compression cycle with dry vapour entering the compressor and saturated liquid leaving the condenser. The scheme is shown in fig. 8.10 with water at $10°C$ entering the condenser as a coolant and then passing through two heat exchangers where it receives additional heating from the engine coolant and lubrication circuits before receiving heat from the engine exhaust gas. If the exhaust gas is to be cooled to $150°C$ and the water is required at $90°C$ calculate:

(a) The flow of water that can be heated in kg/s;
(b) The heating available for the water in kW;
(c) The ratio of the heat available to the fuel energy supplied;
(d) The cost advantage to the engine/heat pump proposal over direct heating in an oil fired water boiler of 80% efficiency;
(e) The relative costs per hour of the engine/heat pump and a gas-fired water boiler of 80% efficiency if the fuel tariffs quoted in Problem 8.3 are to be used. It is assumed that the plant will run on average for eight hours per day, five days per week, forty eight weeks per year. Take any additional data required from Problem 8.3.

(0.97 kg/s, 325.2 kW, 1.7, 2.13, £3.77 and £4.49 per hour)

Regarding this calculation as a first attempt, how would you proceed to improve the analysis to optimize on the operations, the equipment involved and the capital and running costs? What other information would you need?

REFERENCES

8.1 Department of Energy, *Fuel efficiency booklets 1–16.*

8.2 O'CALLAGHAN, P. W., *Design and Management for Energy Conservation* (Pergamon Press, 1981).

8.3 LINDLEY, D., *The 250 kW and 3 MW wind turbines on Burgar Hill, Orkney* (I Mech E, 1984).

8.4 ABEL, H., *Matching heat pumps and heating systems* (Sultzer Technical Review 3/1984).

8.5 MESSEL, H., *Energy for survival* (Pergamon Press, 1979).

8.6 STAMBOLIS, C., *Solar energy in the 80's* (Pergamon Press, 1981).

8.7 Energy policy. *A consultative document* (H.M.S.O., 1978).

8.8 HOWDON, D., *The energy crisis – ten years after* (Croom Helm, 1984).

8.9 MURPHY, W. R., and McKAY, G., *Energy management* (Butterworth, 1982).

8.10 HARLOCH, J. H., and OWEN, R. G., *Thermodynamics and economics of district heating using combined heat and power plant* (British Association Paper No. A21, 1978).

9. Positive Displacement Machines

The function of a compressor is to take a definite quantity of fluid (usually a gas, and most often air) and deliver it at a required pressure. The most efficient machine is one which will accomplish this with the minimum input of mechanical work. Both reciprocating and rotary positive displacement machines are used for a variety of purposes. On the basis of performance a general distinction can be made between the two types by defining the reciprocating type as having the characteristics of a low mass rate of flow and high-pressure ratios, and the rotary type as having a high mass rate of flow and low-pressure ratios. The pressure range of atmospheric to about 9 bar is common to both types.

Some rotary machines are only suitable for low-pressure ratio work, and are applied to the scavenging and supercharging of engines, and the various applications of exhausting and vacuum pumping. For pressures above 9 bar the vane-type rotary machine can be used to supply boost pressures, but for sustained high-pressure work up to 485 bar and above, for special purposes, the reciprocating type is used.

Both basic types exist in many different forms each having its own characteristics. They may be single or multi-stage, and have either air or water cooling. The reciprocating machine is pulsating in action which limits the rate at which fluid can be delivered, but the rotary machine is continuous in action and does not have this disadvantage. The rotary machines are smaller in size for a given flow, lighter in weight and mechanically simpler than their reciprocating counterparts. Figs. 9.1 and 9.22 (page 306), show diagrammatically the various types to be discussed. The treatment and scope of the following sections is fundamental and is not exhaustive. Many compressors are designed to overcome the deficiencies of the basic machines and to satisfy special requirements. For descriptions of these machines the excellent literature supplied by the manufacturers concerned, should be consulted.

The compression process has already been dealt with as an example of the steady flow energy equation in example 2.2. For a compressor which operates in a cyclic or pulsating manner, such as a reciprocating compressor, the properties at inlet and outlet are the average values taken over the cycle. Alternatively the boundary of the control volume is chosen such that states 1 and 2 are constant with

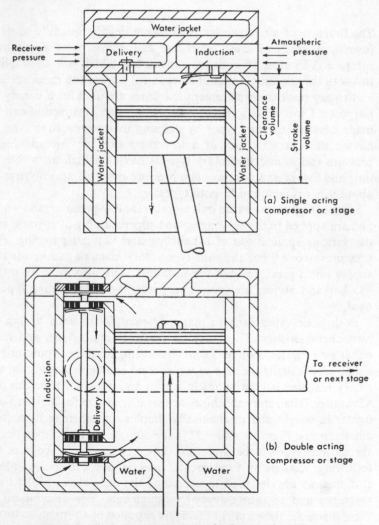

(a) Single acting compressor or stage

(b) Double acting compressor or stage

Fig. 9.1   Reciprocating positive displacement compressors

time, the positions selected being remote from the pulsating disturbance. The type analyses which follow consider the cycle in detail.

9.1 Reciprocating machines

The mechanism involved is the basic piston, connecting-rod, crank, and cylinder arrangement. Initially the clearance volume in the cylinder will be considered negligible. Also the working fluid will be assumed to be a perfect gas. The cycle takes one revolution of the crankshaft for completion and the basic indicator diagram is shown in fig. 9.2.

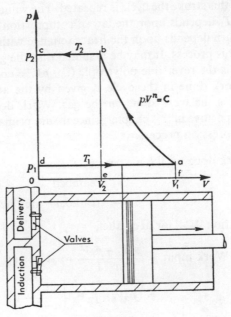

Fig. 9.2

The valves employed in most air compressors are designed to give automatic action. They are of the spring-loaded type operated by a small difference in pressure across them, the light spring pressure giving a rapid closing action. The lift of the valve to give the required air flow should be as small as possible and should operate without shock.

In fig. 9.2 the line d–a represents the induction stroke. The mass in the cylinder increases from zero at d to that required to fill the cylinder at a. The temperature is constant at T_1 for this process and there is no heat exchange with the surroundings in the ideal case. Induction commences when the pressure difference across the valve is sufficient to open it. Line a–b–c represents the compression and delivery stroke. As the piston begins its return stroke the pressure in the cylinder rises and closes the inlet valve. The pressure rise continues with the returning piston as shown by line a–b until the pressure is reached at which the delivery valve opens (a value decided by the valve and the pressure in the receiver). The delivery takes place as shown by the line b–c, which is a process at constant temperature T_2, constant pressure p_2, zero heat exchange and decreasing mass. At the end of this stroke the cycle is repeated. The value of the delivery temperature T_2 depends upon the law of compression between a and b, which in turn depends upon the heat exchange with the surroundings during this process. It may be assumed that the general form of compression is the reversible polytropic (i.e. $pV^n = $ constant).

The net work done in the cycle is given by the area of the p-V diagram and is the work done *on* the gas. Work done on the gas will be called positive in this chapter, since we are primarily concerned with the compression process.

Indicated work done on the air per cycle

$$= \text{area abcd}$$

$$= \text{area abef} + \text{area bcoc} - \text{area adof}$$

Using equation 4.24, for area abef,

i.e.

$$\text{Work input} = \frac{p_2 V_b - p_1 V_a}{n-1} + p_2 V_b - p_1 V_a$$

$$= (p_2 V_b - p_1 V_a)\left(\frac{1}{n-1} + 1\right)$$

i.e.

$$\text{Work input} = (p_2 V_b - p_1 V_a)\frac{1 + n - 1}{n-1}$$

$$= \frac{n}{n-1}(p_2 V_b - p_1 V_a) \qquad 9.1$$

From equation 3.6 we can write

$$p_1 V_a = \dot{m} R T_1 \quad \text{and} \quad p_2 V_b = \dot{m} R T_2$$

(where $\dot{m}$ is the mass induced and delivered per cycle).

Then,   Work input per cycle $= \dfrac{n}{n-1}\,\dot{m}R(T_2-T_1)$ **9.2**

Work done on the air per unit time is equal to the work done per cycle times the number of cycles per unit time. The rate of mass flow is more often used than the mass per cycle; and if the rate of mass flow is also given the symbol m, then equation 9.2 gives the rate at which work is done on the air, or the indicated power.

The working fluid changes state between a and b in fig. 9.2, from p_1 and T_1 to p_2 and T_2, the change being shown in fig. 9.3, which is a diagram of properties (i.e. p against v).

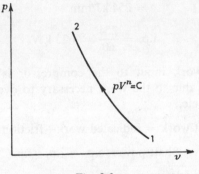

Fig. 9.3

The delivery temperature is given by the equation 4.27,

i.e. $$T_2 = T_1\left(\frac{p_2}{p_1}\right)^{(n-1)/n}$$

Example 9.1

A single-stage reciprocating compressor takes 1 m³ of air per minute at 1·013 bar and 15°C and delivers it at 7 bar. Assuming that the law of compression is $pV^{1\cdot35}=$ constant, and that clearance is negligible, calculate the indicated power.

Mass delivered per min, $\dot{m} = \dfrac{p_1V_1}{RT_1}$

$$= \frac{1\cdot013 \times 1 \times 10^5}{287 \times 288} = 1\cdot226 \text{ kg/min}$$

(where $T_1 = 15 + 273 = 288$ K).

Delivery temp., $T_2 = T_1\left(\dfrac{p_2}{p_1}\right)^{(n-1)/n} = 288\left(\dfrac{7}{1\cdot013}\right)^{(1\cdot35-1)/1\cdot35}$

$$= 288 \times 6\cdot91^{0\cdot259} = 288 \times 1\cdot65 = 475\cdot2 \text{ K}$$

From equation 9.2,

$$\text{Indicated work} = \frac{n}{n-1}\,\dot{m}R(T_2 - T_1) \text{ kJ/min}$$

(where $\dot{m}$ is the mass flow in kg/min).

i.e.   Indicated work $= \dfrac{1\cdot35 \times 1\cdot226 \times 287 \times (475\cdot2 - 288)}{10^3 \times (1\cdot35 - 1)}$

$$= 254 \text{ kJ/min}$$

i.e. $\text{i.p.} = \dfrac{254}{60} = 4\cdot23 \text{ kW}$

The actual work input to the compressor is larger than the indicated work, due to the work necessary to overcome the losses due to friction, etc.,

i.e.   Shaft work = indicated work + friction work

or   Shaft power, s.p. = i.p. + f.p.   9.3

(where f.p. is the friction power).

The mechanical efficiency of the machine is given by

Compressor mechanical efficiency

$$= \frac{\text{indicated work or i.p.}}{\text{shaft work or s.p.}} \qquad 9.4$$

To determine the power input required, the efficiency of the driving motor must be taken into account in addition to the mechanical efficiency.

Then,

$$\text{Input power} = \frac{\text{s.p.}}{\text{efficiency of motor and drive}} \qquad 9.5$$

Example 9.2

If the compressor of example 9.1 is to be driven at 300 rev/min and is a single-acting, single-cylinder machine, calculate the cylinder bore required, assuming a stroke to bore ratio of 1·5/1. Calculate the power of the motor required to drive the compressor if the mechanical efficiency of the compressor is 85% and that of the motor transmission is 90%.

Volume dealt with per minute at inlet = 1 m³/min

$$\therefore \text{ Volume drawn in per cycle} = \frac{1}{300} = 0.00333 \text{ m}^3/\text{cycle}$$

i.e.   Cylinder volume = 0.00333 m³

$$\therefore \frac{\pi}{4} d^2 L = 0.00333$$

(where d = bore; L = stroke),

i.e. $$\frac{\pi}{4} d^2 (1.5 \times d) = 0.00333$$

$$\therefore d^3 = 0.00283 \text{ m}^3$$

i.e.   Cylinder bore = 141·5 mm

$$\text{Power input to the compressor} = \frac{4.23}{0.85} = 4.98 \text{ kW}$$

$$\therefore \text{ Motor power} = \frac{4.98}{0.9} = 5.53 \text{ kW}$$

Proceeding from equation 9.2, other expressions for the indicated work can be derived.

$$\text{Indicated power} = \frac{n}{n-1} \dot{m} R (T_2 - T_1) = \frac{n}{n-1} \dot{m} R T_1 \left(\frac{T_2}{T_1} - 1 \right)$$

Also, from equation 4.29,

$$\frac{T_2}{T_1} = \left(\frac{p_2}{p_1} \right)^{(n-1)/n}$$

Therefore,

$$\text{Indicated power} = \frac{n}{n-1} \dot{m} R T_1 \left\{ \left(\frac{p_2}{p_1} \right)^{(n-1)/n} - 1 \right\} \qquad 9.6$$

or $$\text{Indicated power} = \frac{n}{n-1} p_1 \dot{V} \left\{ \left(\frac{p_2}{p_1} \right)^{(n-1)/n} - 1 \right\} \qquad 9.7$$

(where V is the volume induced per unit time).

9.2 The condition for minimum work

The work done on the air is given by the area of the indicator diagram, and the work done will be a minimum when the area of the diagram is a minimum. The height of the diagram is fixed by the required pressure ratio (when p_1 is fixed), and the length of the line da is fixed by the cylinder volume, which is itself fixed by the required induction of gas. The only process which can influence the area of the diagram is the line ab. The position taken by this line is decided by the value of the index n; fig. 9.4 shows the limits of the possible processes.

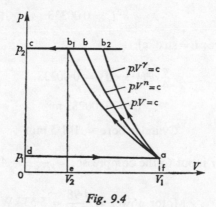

Fig. 9.4

line ab_1 is according to the law pV = const. (i.e. isothermal)
line ab_2 „   „   „ „ „ pV^γ = const. (i.e. isentropic)

Both processes are reversible.

Isothermal compression is the most desirable process between a and b, giving the minimum work to be done on the air. This means that in an actual compressor the gas temperature must be kept as close as possible to its initial value, and a means of cooling the gas is always provided, either by air or by water.

The indicated work done when the gas is compressed isothermally is given by the area ab_1cd.

$$\text{area } ab_1cd = \text{area } ab_1ef + \text{area } b_1coe - \text{area } adof$$

$$\text{area } ab_1ef = p_2 V_{b_1} \log_e \frac{p_2}{p_1} \quad \text{(from equation 4.9)}$$

i.e.   indicated work per cycle $= p_2 V_{b_1} \log_e \dfrac{p_2}{p_1} + p_2 V_{b_1} - p_1 V_a$

also $p_1 V_a = p_2 V_{b_1}$, since the process ab_1 is isothermal.

$$\therefore \text{ indicated work per cycle } = p_2 V_{b_1} \log_e \frac{p_2}{p_1} \qquad 9.8$$

$$= p_1 V_a \log_e \frac{p_2}{p_1} \qquad 9.9$$

$$= \dot{m} RT \log_e \frac{p_2}{p_1} \qquad 9.10$$

When $\dot{m}$ and V_a in equations 9.9 and 9.10 are the mass and volume induced, respectively, per unit time, then these equations give the isothermal power.

9.3 Isothermal efficiency

By definition, based on the indicator diagram,

$$\text{Isothermal efficiency} = \frac{\text{isothermal work}}{\text{indicated work}} \qquad 9.11$$

Example 9.3

Using the data of example 9.1 calculate the isothermal efficiency of the compressor.

From equation 9.10,

$$\text{Isothermal power} = \dot{m} RT \log_e \frac{p_2}{p_1} = 1\cdot226 \times 0\cdot287 \times 288 \times \log_e \frac{7}{1\cdot013}$$

$$= 196 \text{ kJ/min}$$

From example 9.1,

$$\text{Indicated work} = 254 \text{ kJ/min}$$

Therefore using equation 9.11 above,

$$\text{Isothermal efficiency} = \frac{196}{254} = 0\cdot772 \quad \text{or} \quad 77\cdot2\%$$

The least desirable form of compression in reciprocating compressors is that given by the isentropic process (see fig. 9.4). The actual form of compression will usually be one between these two limits. The three processes are shown represented on a T-s diagram in fig. 9.5.

POSITIVE DISPLACEMENT MACHINES

1–2′ represents isothermal compression

1–2″ represents isentropic compression

1–2 represents compression according to a law $pv^n = $ constant

n is usually between 1·2 and 1·3 for a reciprocating air compressor. The main method used for cooling the air is by surrounding the cylinder by a water jacket and designing for the best ratio of surface area to volume of the cylinder.

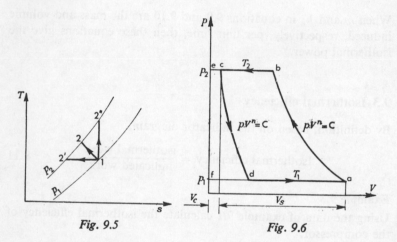

Fig. 9.5

Fig. 9.6

9.4 Reciprocating compressors including clearance

Clearance is necessary in a compressor to give mechanical freedom to the working parts and allow the necessary space for valve operations.

Fig. 9.6 shows the ideal indicator diagram with the clearance volume included. For good-quality machines the clearance volume is about 6% of the swept volume, and with a sleeve valve machine it can be as low as 2%, but machines with clearances of 30–35% are also common.

When the delivery stroke bc is completed the clearance volume V_c is full of gas at pressure p_2 and temperature T_2. As the piston proceeds on the next induction stroke the air expands behind it until the pressure p_1 is reached. Ideally as soon as the pressure reaches p_1, the induction of fresh gas will begin and continue to the end of this stroke at a. The gas is then compressed according to the law $pV^n = C$ (in general), and delivery begins at b as controlled by the valves. The effect of clearance is to reduce the induced volume at p_1 and T_1 from V_s to $(V_a - V_d)$. The masses of gas at the four principal points are such that, $\dot{m}_a = \dot{m}_b$ and $\dot{m}_c = \dot{m}_d$. The mass delivered per cycle is given by $(\dot{m}_b - \dot{m}_c)$, which is equal to that induced, given by $(\dot{m}_a - \dot{m}_d)$. The properties of the working fluid change in processes a–b and c–d as shown in fig. 9.7.

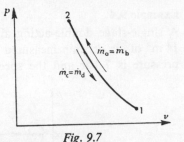

Fig. 9.7

Referring to fig. 9.6 the indicated work done is given by the area of the p–V diagram.

$$\text{Indicated work} = \text{area abcd}$$
$$= \text{area abef} - \text{area cefd}$$

Then, using equation 9.2,

$$\text{Indicated power} = \frac{n}{n-1} \dot{m}_a R(T_2 - T_1) - \frac{n}{n-1} \dot{m}_d R(T_2 - T_1)$$

i.e.

$$\text{Indicated power} = \frac{n}{n-1} R(\dot{m}_a - \dot{m}_d)(T_2 - T_1)$$

$$= \frac{n}{n-1} R\dot{m}(T_2 - T_1) \qquad 9.12$$

(where $\dot{m} = $ the mass induced per unit time $= (\dot{m}_a - \dot{m}_d)$).

A comparison of equations 9.12 and 9.2 shows that they are identical. The work done on compressing the mass of gas $\dot{m}_c$ (or $\dot{m}_d$) on compression, a–b, is returned when the gas expands from c to d. Hence the work done per unit mass of air delivered is unaffected by the size of the clearance volume.

Other expressions can be derived as before. From equation 9.7.

$$\text{Indicated power} = \frac{n}{n-1} p_1 \dot{V} \left\{ \left(\frac{p_2}{p_1}\right)^{(n-1)/n} - 1 \right\}$$

$$\therefore \quad \text{Indicated power} = \frac{n}{n-1} p_1 (V_a - V_d) \left\{ \left(\frac{p_2}{p_1}\right)^{(n-1)/n} - 1 \right\} \qquad 9.13$$

The mass delivered per unit time can be increased by designing the machine to be double acting, i.e. gas is dealt with on both sides of the piston, the induction stroke for one side being the compression stroke for the other (see fig. 9.1).

Example 9.4

A single-stage double-acting air compressor is required to deliver 14 m³ of air per min measured at 1·013 bar and 15°C. The delivery pressure is 7 bar and the speed 300 rev/min. Take the clearance

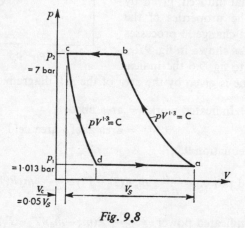

Fig. 9.8

volume as 5% of the swept volume with a compression index of $n = 1\cdot3$. Calculate the swept volume of the cylinder, the delivery temperature and the indicated power.

Referring to fig. 9.8,

$$\text{Swept volume} = (V_a - V_c) = V_s$$

and

$$\text{Clearance volume, } V_c = 0\cdot05V_s$$

$$V_a = 1\cdot05V_s$$

$$\text{Volume induced per cycle} = (V_a - V_d)$$

and

$$(V_a - V_d) = \frac{14}{300 \times 2} = 0\cdot0233 \text{ m}^3$$

(Cycles per min = Rev per min × cycles per rev).

Now,

$$V_a = 1\cdot05V_s \quad \text{and} \quad V_d = V_c\left(\frac{p_2}{p_1}\right)^{1/n} = 0\cdot05V_s\left(\frac{7}{1\cdot013}\right)^{1/1\cdot3}$$

i.e.

$$V_d = 0\cdot221V_s$$

$$\therefore \; (V_a - V_d) = 1 \cdot 05 V_s - 0 \cdot 221 V_s = 0 \cdot 0233 \; \text{m}^3$$

$$\therefore \; V_s = \frac{0 \cdot 0233}{0 \cdot 829} = 0 \cdot 0281 \; \text{m}^3$$

Delivery temp. $T_2 = T_1 \left(\dfrac{p_2}{p_1}\right)^{(n-1)/n}$ from equation 4.27

and $\qquad\qquad\qquad\qquad T_1 = 15 + 273 = 288 \; \text{K}$

i.e. $\qquad\quad T_2 = 288 \left(\dfrac{7}{1 \cdot 013}\right)^{(1 \cdot 3 - 1)/1 \cdot 3} = 288 \times 6 \cdot 91^{0 \cdot 231}$

$$= 288 \times 1 \cdot 563 = 450 \; \text{K}$$

$$\therefore \; \text{Delivery temp.} = 177°\text{C}$$

Using equation 9.13,

Indicated power:

$$= \frac{n}{n-1} p_1 (V_a - V_d) \left\{ \left(\frac{p_2}{p_1}\right)^{(n-1)/n} - 1 \right\}$$

$$= \frac{1 \cdot 3}{0 \cdot 3} \times \frac{1 \cdot 013 \times 10^5 \times 14}{10^3 \times 60} \left\{ \left(\frac{7}{1 \cdot 013}\right)^{(1 \cdot 3 - 1)/1 \cdot 3} - 1 \right\} \; \text{kW}$$

i.e. $\qquad\qquad\qquad$ Indicated power $= 57 \cdot 65 \; \text{kW}$

The approach used for a particular problem depends on how the data is stated and the quantities evaluated during the solution. In some problems it is better to evaluate $\dot{m}$ and T_2 and then use equation 9.12 for the indicated power; e.g. in example 9.4 above, T_2 has been calculated, and the mas induced is given by

$$\dot{m} = \frac{1 \cdot 013 \times 14 \times 10^5}{0 \cdot 287 \times 288 \times 10^3} = 17 \cdot 16 \; \text{kg/min}$$

Then, using equation 9.12,

$$\text{Indicated power} = \frac{n}{n-1} \dot{m} R (T_2 - T_1)$$

$$= \frac{1 \cdot 3}{0 \cdot 3} \times 17 \cdot 16 \times 0 \cdot 287 (450 - 288)$$

$$= 3459 \; \text{kJ/min}$$

i.e. $\quad$ Indicated power $= \dfrac{3459}{60} \; \text{kW} = 57 \cdot 65 \; \text{kW}$ (as before)

9.5 Volumetric efficiency, η_V

It has been shown that one of the effects of clearance is to reduce the induced volume to a value less than that of the swept volume. This means that for a required induction the cylinder size must be increased over that calculated on the assumption of zero clearance. The volumetric efficiency is defined as:

η_V = The mass of air delivered, divided by the mass of air which would fill the swept volume at the free air conditions of pressure and temperature   9.14

Or,

η_V = The volume of air delivered measured at the free air pressure and temperature, divided by the swept volume of the cylinder   9.15

The volume of air dealt with by an air compressor is quoted as the free air delivery (F.A.D.), and is the volume delivered, measured at the pressure and temperature of the atmosphere in which the machine is situated.

[Equations 9.14 and 9.15 can be shown to be identical,
i.e. if the F.A.D. is V, at p and T, then the mass delivered is

$$\dot{m} = \frac{pV}{RT}$$

The mass required to fill the swept volume, V_s, at p and T is given by

$$\dot{m}_s = \frac{pV_s}{RT}$$

Therefore by equation 9.14, $\eta_V = \dfrac{\dot{m}}{\dot{m}_s} = \dfrac{pV}{RT} \times \dfrac{RT}{pV_s} = \dfrac{V}{V_s}$]

The volumetric efficiency can be obtained from the indicator diagram. Referring to fig. 9.9,

$$\text{Volume induced} = V_a - V_d = V_s + V_c - V_d$$

and using equation 4.25,

$$\frac{V_d}{V_c} = \left(\frac{p_2}{p_1}\right)^{1/n} \quad \text{i.e.} \quad V_d = V_c\left(\frac{p_2}{p_1}\right)^{1/n}$$

$$\therefore \text{ Volume induced } = V_s + V_c - V_c \left(\frac{p_2}{p_1}\right)^{1/n}$$

$$= V_s - V_c \left\{\left(\frac{p_2}{p_1}\right)^{1/n} - 1\right\}$$

Hence using equation 9.15,

$$\eta_v = \frac{V_a - V_d}{V_s} = \frac{V_s - V_c\{(p_2/p_1)^{1/n} - 1\}}{V_s}$$

i.e. $$\eta_v = 1 - \frac{V_c}{V_s}\left\{\left(\frac{p_2}{p_1}\right)^{1/n} - 1\right\}.$$ 9.16

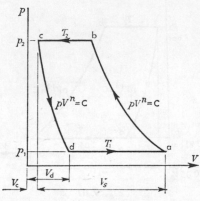

Fig. 9.9

It is important to note that this definition of volumetric efficiency is only consistent with that of 9.14 and 9.15 if the conditions of pressure and temperature in the cylinder during the induction stroke are identical with those of the free air. In fact the gas will be heated by the cylinder walls, and there will be a reduction in pressure due to the pressure drop required to induce the gas into the cylinder against the inevitable resistance to flow. These modifications to the ideal case require a more careful application of the formulae previously derived.

For example, when as before the F.A.D. per cycle is denoted by V at p and T, then,

$$\dot{m} = \frac{pV}{RT} = \frac{p_1(V_a - V_d)}{RT_1}$$

i.e.   F.A.D./cycle, $V = (V_a - V_d)\frac{T}{T_1} \cdot \frac{p_1}{p}$ 9.17

(where p_1 and T_1 are the suction conditions).

Example 9.5

A single-stage, double-acting air compressor has a F.A.D. of 14 m³/ min measured at 1·013 bar and 15°C. The pressure and temperature in the cylinder during induction are 0·95 bar and 32°C. The delivery pressure is 7 bar and the index of compression and expansion, $n = 1·3$. Calculate the indicated power required and the volumetric efficiency. The clearance volume is 5% of the swept volume.

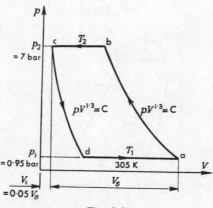

Fig. 9.10

The p-V diagram is shown in fig. 9.10.

$$\text{Mass delivered per minute, } \dot{m} = \frac{pV}{RT}$$

(where the F.A.D. per min. is V at p and T),

i.e. $$\dot{m} = \frac{1·013 \times 14 \times 10^5}{0·287 \times 288 \times 10^3} = 17·16 \text{ kg/min}$$

(where $T = 15 + 273 = 288$ K).

$$T_2 = T_1\left(\frac{p_2}{p_1}\right)^{(n-1)/n} \quad \text{from equation 4.29}$$

i.e. $$T_2 = 305 \times \left(\frac{7}{0·95}\right)^{(1·3-1)/1·3} = 305 \times 1·586 = 483·7 \text{ K}$$

(where $T_1 = 32 + 273 = 305$ K).

From equation 9.12,

$$\text{Indicated power} = \frac{n}{n-1}\, \dot{m} R(T_2 - T_1)$$

$$= \frac{1 \cdot 3}{0 \cdot 3} \times 17 \cdot 16 \times 0 \cdot 287(483 \cdot 7 - 305)$$

$$= 3813 \text{ kJ/min}$$

$$\therefore \text{ i.p.} = \frac{3813}{60} = 63 \cdot 55 \text{ kW}$$

As before,

$$V_d = V_c \left(\frac{p_2}{p_1}\right)^{1/n}$$

i.e.
$$V_d = 0 \cdot 05 V_s \left(\frac{7}{0 \cdot 95}\right)^{1/1 \cdot 3} = 0 \cdot 05 V_s \times 7 \cdot 369^{0 \cdot 769}$$

$$= 0 \cdot 05 V_s \times 4 \cdot 65 = 0 \cdot 233 V_s$$

$$\therefore\ V_a - V_d = V_a - 0 \cdot 233 V_s = 1 \cdot 05 V_s - 0 \cdot 233 V_s = 0 \cdot 817 V_s$$

Using equation 9.17,

$$\text{F.A.D./cycle} = (V_a - V_d)\frac{T}{T_1} \cdot \frac{p_1}{p}$$

i.e.
$$\text{F.A.D./cycle} = 0 \cdot 817 V_s \times \frac{288}{305} \times \frac{0 \cdot 95}{1 \cdot 013} = 0 \cdot 724 V_s$$

Then from equation 9.15,

$$\eta_v = \frac{V}{V_s} = \frac{0 \cdot 724 V_s}{V_s} = 0 \cdot 724 \quad \text{or} \quad 72 \cdot 4\%$$

Note that if the volumetric efficiency in the above example is evaluated using equation 9.16 then,

$$\eta_v = 1 - \frac{V_c}{V_s}\left\{\left(\frac{p_2}{p_1}\right)^{1/n} - 1\right\} = 1 - \frac{0 \cdot 05 V_s}{V_s}\left\{\left(\frac{7}{0 \cdot 95}\right)^{1/1 \cdot 3} - 1\right\}$$

i.e.

$$\eta_v = 1 - 0 \cdot 05(4 \cdot 65 - 1) = 1 - 0 \cdot 183 = 0 \cdot 817 \quad \text{or} \quad 81 \cdot 7\%$$

There is a considerable difference between the two values, since the latter answer ignores the difference in temperature and pressure between the free air conditions and the suction conditions. One of the advantages of the volumetric efficiency as defined by equation 9.16 is that it can be determined conveniently by scaling V_s and $(V_a - V_d)$ from an indicator diagram.

9.6 Actual indicator diagram

The diagrams previously shown have been ideal diagrams. An actual indicator diagram is similar to the ideal except for the induction and delivery processes which are modified by valve action. This is shown

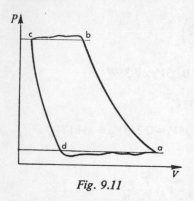

Fig. 9.11

in fig. 9.11. The waviness of the lines da and bc is due to valve movement because of bounce, or variation in the back pressures. Automatic valves are less definite in their action than mechanically operated valves and will give more throttling of the gas than valves which follow a definite cam displacement. The induction stroke da is a mixing process, the induced air mixing with that in the cylinder.

The indicated power can be determined from the area of the indicator diagram if the pressure and volume scales are known. The most convenient way is to find the average height of the diagram and multiply this value by the pressure scale, the quantity obtained being the mean effective pressure.

i.e.

$$\text{Average height of diagram} = \frac{\text{area of diagram}}{\text{length of diagram}}$$

and

Mean effective pressure, p_m

$$= (\text{average height}) \times (\text{pressure scale})$$

Then, $\quad\quad\quad\quad\quad\quad$ i.p. $= p_m A L f$ $\quad\quad\quad\quad$ 9.18

(where $p_m =$ mean effective pressure; A = cylinder cross-sectional area; $L =$ stroke; $f =$ cycles per unit time).

For a single-acting machine the number of cycles per unit time is equal to the rotational speed.

Mechanical indicators are most often used to obtain the diagram, and the way in which the pressure scale is obtained varies somewhat with different makes. Usually an indicator fitted with a spring of No. 100, say, gives a pressure scale on the diagram of 100 bar per m of height (i.e. 0·1 bar per mm deflection of the stylus). This takes into account the indicator piston size, the spring rating, and the magnification factor of the mechanical linkage between the spring and the stylus. If the instrument can be sleeved and a smaller piston fitted, the pressure scale obtained with the same spring will be changed. Fig. 9.12 shows the two cases, p_1 is the pressure exerted by the spring on the

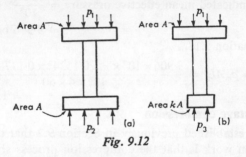

Fig. 9.12

upper side of the indicator piston. The load due to this pressure is in equilibrium with the gas load on the cylinder side of the indicator piston. Referring to fig. 9.12a, $p_1 A = p_2 A$, and referring to fig. 9.12b, $p_1 A = p_3 kA$. Hence for a given spring rate, the cylinder pressures to give a certain deflection of the stylus are related by the equation,

$$p_2 A = p_3 kA \quad \text{or} \quad p_2 = kp_3 \qquad 9.19$$

Usually the alternative piston has an area one-half that of the standard (i.e. $k = \frac{1}{2}$), and is called a 'half piston'. With the half-piston and a No. 100 spring the pressure scale will be 100 divided by $\frac{1}{2}$, i.e. pressure scale = 200 bar/m.

Example 9.6

A mechanical indicator was calibrated with a piston of area A and a spring No. of 325 and gave a pressure scale of 325 bar/m. For another purpose this piston was removed and a sleeve carrying a piston of area A/2 was fitted instead. Calculate the new pressure scale with the same spring fitted.

The indicator diagram obtained with this instrument had an area of 11·6 cm² and a length of 76·2 mm. The compressor, of bore 153·4 mm and stroke 177·8 mm was single acting and ran at 250 rev/min. Calculate the indicated power.

$$\text{The new pressure scale} = \frac{325 \text{ bar/m}}{k}$$

where $k = (A/2)/A = \frac{1}{2}$.

Therefore, the new pressure scale $= 325/\frac{1}{2} = 650$ bar/m.

$$\text{Mean height of diagram} = \frac{11·6 \times 10^2}{76·2} = 15·24 \text{ mm}$$

Therefore, indicated mean effective pressure $= \dfrac{15·24}{10^3} \times 650$

$$= 9·906 \text{ bar.}$$

From equation 9.18,

$$\text{i.p.} = p_m A L f = \frac{9·906 \times 10^5 \times \pi \times 0·1534^2 \times 0·1778 \times 250}{10^3 \times 4 \times 60}$$

$$= 13·56 \text{ kW}$$

9.7 Multi-stage compression

It has been established previously in Section 9.2 that the condition for minimum work is that the compression process should be isothermal. In general the temperature after compression is given by equation 4.27, $T_2 = T_1(p_2/p_1)^{(n-1)/n}$. The delivery temperature increases with the pressure ratio. Further, from equation 9.16,

$$\eta_v = 1 - \frac{V_c^*}{V_s}\left\{\left(\frac{p_2}{p_1}\right)^{1/n} - 1\right\}$$

it can be seen that as the pressure ratio increases the volumetric efficiency decreases. This is illustrated in fig. 9.13.

For compression from p_1 to p_2 the cycle is abcd and the F.A.D. per cycle is $V_a - V_d$; for compression from p_1 to p_3 the cycle is ab'c'd' and the F.A.D. per cycle

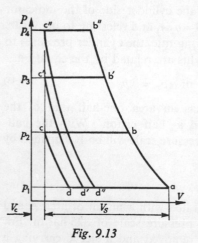

Fig. 9.13

is $V_a - V_{d'}$; for compression from p_1 to p_4 the cycle is ab″c″d″ and the F.A.D. per cycle is $V_a - V_{d''}$. Therefore for a required F.A.D. the cylinder size would have to increase as the pressure ratio increases.

The volumetric efficiency can be improved by carrying out the compression in two stages. After the first stage of compression the fluid is passed into a smaller cylinder in which the gas is compressed to the required final pressure. If the machine has two stages, the gas will be delivered at the end of this stage, but it could be delivered to a third cylinder for higher pressure ratios. The cylinders of the successive stages are proportioned to take the volume of gas delivered from the previous stage.

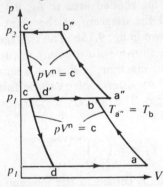

Fig. 9.14

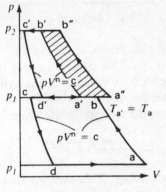

Fig. 9.15a

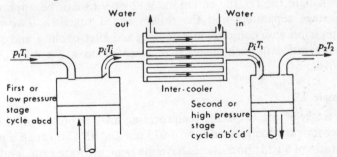

Fig. 9.15b

The indicator diagram for a two-stage machine is shown in fig. 9.14. In this diagram it is assumed that the delivery process from the first or low-pressure stage and the induction process of the second or high-pressure stage, are at the same pressure.

The ideal isothermal compression can only be obtained if ideal cooling is continuous. This is difficult to obtain during normal compression. With multi-stage compression the opportunity presents

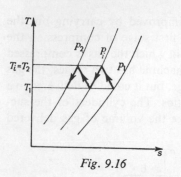

Fig. 9.16

itself for the gas to be cooled as it is being transferred from one cylinder to the next, by passing it through an intercooler. If inter-cooling is complete, the gas will enter the second stage at the same temperature at which it entered the first stage. The saving in work obtained by intercooling is shown by the shaded area in fig. 9.15a, and the diagram of the plant is shown in fig. 9.15b. The two indi-cator diagrams abcd and a'b'c'd' are shown with a common pres-sure, p_i. This does not occur in a real machine as there is a small pressure drop between the cylinders. An after-cooler can be fitted after the delivery process to cool the gas. The delivery temperatures from the two stages are given by

$$T_i = T_1\left(\frac{p_i}{p_1}\right)^{(n-1)/n} \quad \text{and} \quad T_2 = T_1\left(\frac{p_2}{p_i}\right)^{(n-1)/n}$$

respectively. This assumes that the gas is cooled in the intercooler back to the inlet temperature, and is called complete intercooling. To calculate the i.p. the equations 9.12 or 9.13 can be applied to each stage separately and the results added together. Two-stage compression with complete intercooling and after-cooling, and equal pressure ratios in each stage, is represented on a *T-s* diagram in fig. 9.16.

Example 9.7

In a single-acting two-stage reciprocating air compressor 4·5 kg of air per min are compressed from 1·013 bar and 15°C through a pressure ratio of 9 to 1. Both stages have the same pressure ratio, and the law of compression and expansion in both stages is $pV^{1·3} = $ constant. If intercooling is complete, calculate the indicated power and the cylinder swept volumes required. Assume that the clearance volumes of both·stages are 5% of their respective swept volumes and that the compressor runs at 300 rev/min.

The two indicator diagrams are shown superimposed in fig. 9.17. The low-pressure stage cycle is abcd and the high-pressure cycle is a'b'c'd'.

Now $p_2/p_1 = 9$,

$$\therefore \quad p_2 = 9p_1$$

also $p_i/p_1 = p_2/p_i$,

$$\therefore \quad p_i^2 = p_1p_2 = p_1 \cdot 9p_1$$

$$\therefore \quad p_i^2 = 9 \cdot p_1^2 \qquad \therefore \quad p_i/p_1 = \sqrt{9} = 3$$

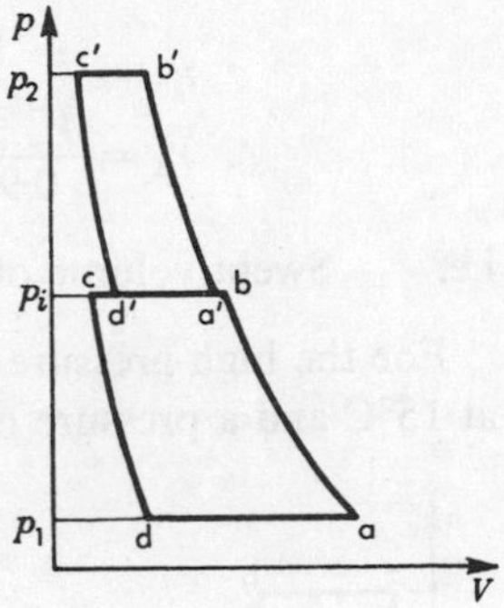

Using equation 4.29,

$$\frac{T_i}{T_1} = \left(\frac{p_i}{p_1}\right)^{(n-1)/n} \qquad \therefore \quad \frac{T_i}{288} = 3^{(1 \cdot 3 - 1)/1 \cdot 3}$$

Fig. 9.17

(where $T_1 = 15 + 273 = 288$ K, and T_i is the temperature of the air entering the intercooler)

i.e. $$T_i = 288 \times 1 \cdot 289 = 371 \text{ K}$$

Now as n, $\dot{m}$, and the temperature difference are the same for both stages, then the work done in each stage is the same.

i.e., using equation 9.12,

$$\text{Total work required per min} = 2 \times \frac{n}{n-1} \dot{m}R(T_i - T_1)$$

$$= 2 \times \frac{1 \cdot 3}{1 \cdot 3 - 1} \times 4 \cdot 5 \times 0 \cdot 287(371 - 288)$$

$$= 2 \times 4 \cdot 33 \times 4 \cdot 5 \times 0 \cdot 287 \times 83$$

$$= 930 \text{ kJ/min}$$

$$\therefore \quad \text{Indicated power} = \frac{930}{60} = 15 \cdot 5 \text{ kW}$$

The mass induced per cycle is,

$$\dot{m} = \frac{4 \cdot 5}{300} = 0 \cdot 015 \text{ kg/cycle}$$

This mass is passed through each stage in turn.
For the low-pressure cylinder, referring to fig. 9.18,

$$V_a - V_d = \frac{\dot{m}RT_1}{p_1} = \frac{0 \cdot 015 \times 287 \times 288}{1 \cdot 013 \times 10^5} = 0 \cdot 0122 \text{ m}^3/\text{cycle}$$

Using equation 9.16,

$$\eta_V = \frac{V_a - V_d}{V_s} = 1 - \frac{V_c}{V_s}\left\{\left(\frac{p_i}{p_1}\right)^{1/n} - 1\right\} = 1 - 0 \cdot 05(3^{0 \cdot 769} - 1)$$

$$\therefore \eta_V = 1 - 0 \cdot 066 = 0 \cdot 934$$

$$\therefore V_s = \frac{V_a - V_d}{0 \cdot 934} = \frac{0 \cdot 0122}{0 \cdot 934} = 0 \cdot 0131 \text{ m}^3/\text{cycle}$$

i.e.   Swept volume of low-pressure cylinder = $0 \cdot 0131$ m³

For the high-pressure stage, a mass of $0 \cdot 015$ kg/cycle is drawn in at 15°C and a pressure of $p_i = 3 \times 1 \cdot 013 = 3 \cdot 039$ bar.

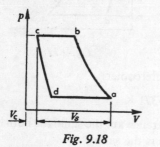

Fig. 9.18

i.e.

Volume drawn in =

$$\frac{0 \cdot 015 \times 287 \times 288}{3 \cdot 039 \times 10^5}$$

$$= 0 \cdot 00406 \text{ m}^3/\text{cycle}$$

Using equation 9.16 for the high-pressure stage,

$$\eta_V = 1 - \frac{V_c}{V_s}\left\{\left(\frac{p_2}{p_i}\right)^{1/n} - 1\right\}$$

and since V_c/V_s is the same as for the low-pressure stage and also $p_2/p_i = p_i/p_1$ then η_V is $0 \cdot 934$ as above.

$$\therefore \text{ Swept volume of high-pressure stage} = \frac{0 \cdot 00406}{0 \cdot 934} = 0 \cdot 00436 \text{ m}^3$$

Note that the clearance ratio is the same in each cylinder, and the suction temperatures are the same since intercooling is complete, therefore the swept volumes are in the ratio of the suction pressures.

i.e. $$V_H = \frac{V_L}{3} = \frac{0 \cdot 0131}{3} = 0 \cdot 00436 \text{ m}^3$$

9.8 The ideal intermediate pressure

The value chosen for the intermediate pressure p_i, influences the work to be done on the gas and its distribution between the stages. The condition for the work done to be a minimum will be proved for two-stage compression but can be extended to any number of stages.

Total work = low-pressure stage work + high pressure stage work, i.e., using equation 9.6,

Total work =

$$\frac{n}{n-1} \dot{m}RT_1 \left\{ \left(\frac{p_i}{p_1}\right)^{(n-1)/n} - 1 \right\} + \frac{n}{n-1} \dot{m}RT_1 \left\{ \left(\frac{p_2}{p_i}\right)^{(n-1)/n} - 1 \right\}$$

It is assumed that intercooling is complete and therefore the temperature at the start of each stage is T_1.

$$\therefore \text{ Total work} = \frac{n}{n-1} \dot{m}RT_1 \left\{ \left(\frac{p_i}{p_1}\right)^{(n-1)/n} - 1 + \left(\frac{p_2}{p_i}\right)^{(n-1)/n} - 1 \right\}$$

If p_1, T_1, and p_2 are fixed, then the optimum value of p_i which makes the work a minimum can be obtained by equating d (Work)/(dp_i) to zero,
i.e. optimum value of p_i when

$$\frac{d}{dp_i} \left\{ \left(\frac{p_i}{p_1}\right)^{(n-1)/n} + \left(\frac{p_2}{p_i}\right)^{(n-1)/n} - 2 \right\} = 0$$

i.e. when,

$$\frac{d}{dp_i} \left\{ \left(\frac{1}{p_1}\right)^{(n-1)/n} p_i^{(n-1)/n} + p_2^{(n-1)/n} \left(\frac{1}{p_i}\right)^{(n-1)/n} - 2 \right\} = 0$$

$$\therefore p_1^{-(n-1)/n} \left(\frac{n-1}{n}\right) p_i^{(n-1)/n-1} + p_2^{(n-1)/n} \left(\frac{1-n}{n}\right) p_i^{(1-n)/n-1} = 0$$

$$\therefore p_1^{-(n-1)/n} \left(\frac{n-1}{n}\right) p_i^{-1/n} = p_2^{(n-1)/n} \left(\frac{n-1}{n}\right) p_i^{(1-2n)/n}$$

$$\therefore p_i^{\{2(n-1)\}/n} = (p_1 p_2)^{(n-1)/n}$$

$$\therefore p_i^2 = p_1 p_2 \qquad 9.20$$

or

$$\frac{p_i}{p_1} = \frac{p_2}{p_i} \qquad 9.21$$

i.e. the pressure ratio is the same for each stage.

Hence total minimum work = 2 × (work required for one stage)

$$= 2 \times \frac{n \dot{m}RT_1}{n-1} \left\{ \left(\frac{p_i}{p_1}\right)^{(n-1)/n} - 1 \right\}$$

Or in terms of the overall pressure ratio p_2/p_1, we have, using equation 9.20,

$$\frac{p_i}{p_1} = \frac{\sqrt{p_1 p_2}}{p_1} = \sqrt{\frac{p_2}{p_1}}$$

$$\therefore \text{ total minimum work} = 2 \times \frac{n\dot{m}RT_1}{n-1}\left\{\left(\frac{p_2}{p_1}\right)^{(n-1)/2n} - 1\right\}$$

This can be shown to extend to z stages giving in general,

$$\text{Total minimum work} = z\frac{n}{n-1}\dot{m}RT_1\left\{\left(\frac{p_2}{p_1}\right)^{(n-1)/zn} - 1\right\} \quad 9.22$$

Also,

$$\text{Pressure ratio for each stage} = \left(\frac{p_2}{p_1}\right)^{1/z} \quad 9.23$$

Hence the condition for minimum work is that the pressure ratio in each stage is the same and that intercooling is complete.

(Note that in example 9.7 the information given implies minimum work.)

Example 9.8

A three-stage single-acting air compressor running in an atmosphere at 1·013 bar and 15°C has a free air delivery of 2·83 m³/min. The suction pressure and temperature are 0·98 bar and 32°C respectively. Calculate the indicated power required, assuming complete intercooling, $n = 1·3$, and that the machine is designed for minimum work. The delivery pressure is to be 70 bar.

$$\text{Mass of air delivered} = \frac{pV}{RT} = \frac{1·013 \times 10^5 \times 2·83}{287 \times 288} = 3·47 \text{ kg/min}$$

(where $T = 15 + 273 = 288$ K).

Then using equation 9.22,

Total indicated work:

$$= z\frac{n}{n-1}\dot{m}RT_1\left\{\left(\frac{p_2}{p_1}\right)^{(n-1)/zn} - 1\right\}$$

$$= 3 \times \frac{1·3}{0·3} \times 3·47 \times 0·287 \times 288\left\{\left(\frac{70}{0·98}\right)^{(1·3-1)/(3 \times 1·3)} - 1\right\}$$

$$= 3·728 \times 10^3(1·389 - 1) \text{ kJ/min}$$

$$\therefore \text{ Indicated power} = \frac{3·728 \times 10^3 \times 0·389}{60} = 24·2 \text{ kW}$$

Besides the benefits of multi-stage compression already dealt with there are also mechanical advantages. The higher pressures are confined to the smaller cylinders and a multi-cylinder machine has less variation in rotational speed and requires a smaller flywheel.

9.9 Energy balance for a two stage machine with intercooler

Referring to fig. 9.19, the steady-flow energy equation 2.8 can be applied to the low-pressure stage, the intercooler, and the high-

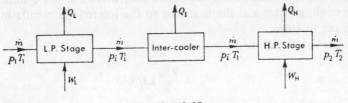

Fig. 9.19

pressure stage, in turn. Changes in kinetic energy can be neglected, i.e. from equation 2.8,

$$h_1 + \frac{C_1^2}{2} + Q = h_2 + \frac{C_2^2}{2} + W$$

for the low-pressure stage, for unit mass,

$$h_1 - Q_L = h_i - W_L$$

or for mass flow $\dot{m}$ $\dot{m}c_p T_1 - Q_L = \dot{m}c_p T_i - W_L$

$$\therefore\ Q_L = W_L - \dot{m}c_p(T_i - T_1) \qquad\qquad 9.24$$

for the intercooler, for unit mass,

$$h_i - Q_I = h_1$$

or for mass flow $\dot{m}$ $\dot{m}c_p T_i - Q_I = \dot{m}c_p T_1$

$$\therefore\ Q_I = \dot{m}c_p(T_i - T_1) \qquad\qquad 9.25$$

for the high-pressure stage, for unit mass,

$$h_1 - Q_H = h_2 - W_H$$

or for mass flow $\dot{m}$ $\dot{m}c_pT_1 - Q_H = \dot{m}c_pT_2 - W_H$

$$\therefore\ Q_H = W_H - \dot{m}c_p(T_2 - T_1) \qquad\qquad 9.26$$

With complete intercooling, as assumed in fig. 9.19, and the compressor designed for minimum work, then, from equation 9.12,

$$W_L = W_H = \frac{n}{n-1}\,\dot{m}R(T_2 - T_1)$$

Example 9.9

Using the data of example 9.7 determine the heat loss to the cylinder jacket cooling water and the heat loss to the intercooler circulating water, per minute.

From example 9.7 we have,

$$W_L = W_H = \frac{930}{2}\,\text{kJ/min}$$

and $\qquad\qquad T_2 = T_i = 371\ \text{K}$

Then, from equation 9.24,

$$Q_L = W_L - \dot{m}c_p(T_2 - T_1)$$

$$\therefore\ Q_L = \frac{930}{2} - 4\cdot5 \times 1\cdot005(371 - 288)\ \text{kJ/min}$$

i.e. $\qquad\qquad Q_L = 465 - 375 = 90\ \text{kJ/min}$

From equation 9.26,

$$Q_H = W_H - \dot{m}c_p(T_2 - T_1)$$

But $\qquad\qquad W_H = W_L \quad\text{and}\quad T_2 = T_i$

$$\therefore\ Q_H = Q_L = 90\ \text{kJ/min}$$

i.e. Heat loss from the cylinder in each stage = 90 kJ/min

From equation 9.25,

$$Q_I = \dot{m}c_p(T_i - T_1) = 4\cdot5 \times 1\cdot005 \times (371 - 288)$$

i.e. Heat to intercooler circulating water = 375 kJ/min

The quantities W_L and W_H, as defined by fig. 9.19, are the amounts of work done on the air as determined from the indicator diagram. The actual work inputs exceed this by the amounts necessary to overcome frictional resistance to the moving parts of the machine. It can be assumed that about 50% of the friction power goes to increasing the energy transferred to the cooling water in addition to the heat transferred to the cooling water from the air in the cylinder.

9.10 The steady flow analysis of the compression process

In section 9.4 an expression was obtained, equation 9.12, for the indicated power required to take a mass flow rate of gas, $\dot{m}$, in state 1

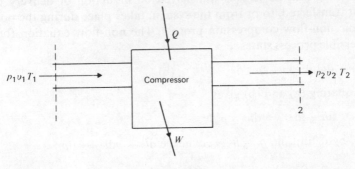

Fig. 9.20

and deliver it at a higher pressure in state 2. This was done by analysing the internal processes of the machine. Another approach is to consider the compression process as one of steady flow, as shown in fig. 9.20, with the change of state from 1 to 2 being achieved by a

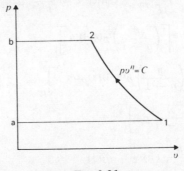

Fig. 9.21

non-flow process of polytropic compression, as indicated in the property diagram of fig. 9.21.

The steady flow energy equation for the system shown in fig. 9.20, neglecting changes in potential and kinetic energy and for unit mass flow rate, is

$$h_1 + Q = W + h_2$$

$$\therefore \qquad Q = h_2 - h_1 + W$$

or for an elemental process

$$dQ = dh + dW \qquad (a)$$

Assuming that no heat is transferred on induction or delivery the heat transferred, to or from the system, takes place during the polytropic non-flow compression process. The non-flow equation for a reversible process states

$$dQ = du + p \cdot dv \qquad (b)$$

Equating (a) and (b) gives

$$dh + dW = du + p\,dv$$

and, by definition, $h = u + pv$, hence $dh + p\,dv + v\,dp$

$$\therefore \quad du + p\,dv + v\,dp + dW = du + p\,dv$$

$$\therefore \quad dW = -v\,dp$$

$$\therefore \quad W = -\int_1^2 v \cdot dp = \text{area 12 b a 1 in fig. 9.21,}$$

i.e. $\qquad W = -C^{1/n} \int_1^2 \dfrac{dp}{p^{1/n}} \quad$ (since $v = \dfrac{C^{1/n}}{p^{1/n}}$ if $pv^n = C$)

$$= -C^{1/n} \left[\left(\frac{n}{n-1} \right) p^{(n-1)/n} \right]_1^2$$

$$= -\left[\left(\frac{n}{n-1} \right) p^{(n-1)/n}\, p^{1/n}\, v \right]_1^2$$

$$= -\left[\frac{n}{n-1}\, pv \right]_1^2$$

$$= -\frac{n}{n-1} \left(p_2 v_2 - p_1 v_1 \right)$$

i.e.   Work input, $W_t = \dfrac{n}{n-1}(p_2v_2 - p_1v_1)$

and, as $p_1v_1 = RT_1$ and $p_2v_2 = RT_2$ then,

$$W_t = \frac{nR}{n-1}(T_2 - T_1)$$

9.11 Air control

Reciprocating compressors require a means of controlling the flow rate to meet the demand. A balanced valve can be fitted to the air inlet; this is governed by an air governor or pilot unloading valve. When a pre-set delivery pressure in the receiver is reached the valve closes automatically, and is opened again when the receiver pressure falls by a predetermined amount. The air is delivered to meet the demand and the settings of the governor can be made adjustable. The result is automatic control and an economic use of the power input. A hand control can be fitted for use in starting up the machine, to relieve the driving motor of the full load on starting. In portable installations the air governor is connected with the engine driving the compressor and the engine is governed to the air load.

9.12 Rotary machines

Because of the continuous rotary action, the rotary positive displacement machine is smaller for a given flow than its reciprocating counterpart. The machines in this category are generally uncooled and as the compression is carried out at a high rate the conditions are approximately adiabatic. Examples of this type are:

(i) The Roots blower.   (ii) Vane type.

See fig. 9.22.

(i) *Roots blower*

The two-lobe type is shown in fig. 9.22a, but three- and four-lobe versions are in use for higher pressure ratios. One of the rotors is connected to the drive and the second rotor is gear driven from the first. In this way the rotors rotate in phase and the profile of the lobes is of cycloidal or involute form giving correct mating of the lobes to seal the delivery side from the inlet side. This sealing continues

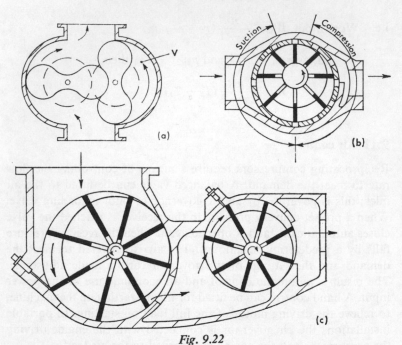

Fig. 9.22

(a) *Roots blower, two lobe rotors.*
(b) *Rotary sliding vane type with a floating drum.*
(c) *Rotary sliding vane two stage machine, the vanes being in contact with the cylinder walls.*

until delivery commences. There must be some clearance between the lobes, and between the casing and the lobes, to reduce wear; this clearance forms a leakage path which has an increasingly adverse effect on efficiency as the pressure ratio increases.

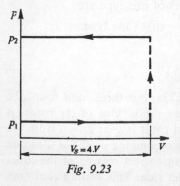

Fig. 9.23

As each side of each lobe faces its side of the casing a volume of gas V, at pressure p_1, is displaced towards the delivery side at constant pressure. A further rotation of the rotor opens this volume to the receiver, and the gas flows back from the receiver, since this gas is at a higher pressure. The gas induced is compressed irreversibly by that from the receiver, to the pressure p_2, and then delivery begins. This process is carried out four times per revolution of the driving shaft.

The p-V diagram for this machine is shown in fig. 9.23, in which the pressure rise from p_1 to p_2 is shown as an irreversible process at constant volume.

$$\text{Work done per cycle} = (p_2 - p_1)V$$

$$\therefore \text{ Work done per rev} = 4(p_2 - p_1)V \qquad 9.27$$

If V_s is the volume dealt with per minute at p_1 and T_1, then

$$\text{Work done per minute} = (p_2 - p_1)V_s \qquad 9.28$$

The ideal compression process from p_1 to p_2 is a reversible adiabatic (i.e. isentropic) process. The work done per minute ideally is thus given by equation 9.7 with $n = \gamma$,

i.e. $\qquad$ Work done per min $= \dfrac{\gamma}{\gamma - 1} p_1 V_s \left\{ \left(\dfrac{p_2}{p_1}\right)^{(\gamma - 1)/\gamma} - 1 \right\}$

Then a comparison may be made on the basis of a Roots efficiency,

i.e. $\qquad$ Roots efficiency $= \dfrac{\text{work done isentropically}}{\text{actual work done}}$

i.e. $\qquad$ Roots efficiency $= \dfrac{\dfrac{\gamma}{\gamma - 1} p_1 V_s \left\{ \left(\dfrac{p_2}{p_1}\right)^{(\gamma - 1)/\gamma} - 1 \right\}}{V_s(p_2 - p_1)}$

$$= \dfrac{\dfrac{\gamma}{\gamma - 1} p_1 V_s \{ r^{(\gamma - 1)/\gamma} - 1 \}}{p_1 V_s(r - 1)}$$

(where $r = \dfrac{p_2}{p_1} =$ pressure ratio).

From equation 3.22, we can write

$$\frac{\gamma}{\gamma - 1} = \frac{c_p}{R}$$

$$\therefore \text{ Roots efficiency} = \frac{c_p}{R} \left\{ \frac{r^{(\gamma - 1)/\gamma} - 1}{(r - 1)} \right\} \qquad 9.29$$

For a Roots air blower values of pressure ratio, r, of 1·2, 1·6, and 2 give values for the Roots efficiency of 0·945, 0·84, and 0·765 respectively. These values show that the efficiency decreases as the pressure ratio increases.

The actual compression process is not quite as simple as that described. When the displacement volume V is opened to the delivery space a pressure wave enters which increases with the opening and moves at the velocity of sound. This wave is reflected from the approaching lobe to the delivery space. The pressure oscillations set up unsteady conditions in the delivery space which vary considerably from one design to another. The actual torque and loading on the rotors are higher than is suggested by the p-V diagram, and fluctuate with high frequency. This fluctuation is transmitted to the drive and creates difficulties due to vibrations. This machine has a number of imperfections but is well suited to such tasks as the scavenging and supercharging of I.C. engines.

Roots blowers are built for capacities of from 0·14 m³/min to 1400 m³/min, and pressure ratios of the order of 2 to 1 for a single-stage machine and 3 to 1 for a two-stage machine. Other designs have been produced to improve on the Roots blower, one of these being the Bicera compressor, designed by the British Internal Combustion Engineering Research Association (B.I.C.E.R.A.).

(ii) *Vane type*

The simple vane type is shown in fig. 9.22c and consists of a rotor mounted eccentrically in the body, and supported by ball and roller bearings in the end covers of the body. The rotor is slotted to take the blades which are of a non-metallic material, usually fibre or carbon. As each blade moves past the inlet passage, compression begins due to decreasing volume between the rotor and casing. Delivery begins with the arrival of each blade at the delivery passage. This type of compression differs from that of the Roots blower in that some or all of the compression is obtained before the trapped volume is opened to delivery. Further compression can be obtained by the back-flow of air from the receiver which occurs in an irreversible manner.

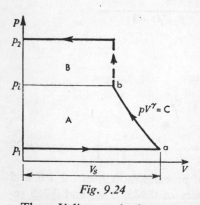

Fig. 9.24

The p-V diagram is shown in fig. 9.24. V_s is the induced volume at pressure p_1 and temperature T_1. Compression occurs to the pressure

p_i, the ideal form for an uncooled machine being isentropic. At this pressure the displaced gas is opened to the receiver and gas flowing back from the receiver raises the pressure irreversibly to p_2. The work done per minute is given by the sum of the areas A and B, referring to fig. 9.24. Comparing the areas of fig. 9.23 and 9.24 it can be seen that for a given air flow and given pressure ratio the vane type requires less work input than the Roots blower.

Example 9.10

Compare the work inputs required for a Roots blower and a Vane type compressor having the same induced volume of 0·03 m³/rev, the inlet pressure being 1·013 bar and the pressure ratio 1·5 to 1. For the Vane type assume that internal compression takes place through half the pressure range.

$$p_1 = 1.013 \text{ bar}$$

$$\therefore \ p_2 = 1.013 \times 1.5 = 1.52 \text{ bar}$$

For the Roots blower, referring to fig. 9.23,

$$\text{Work done per rev} = (p_2 - p_1)V_s$$

$$= (1.52 - 1.013) \times \frac{10^5 \times 0.03}{10^3} = 1.52 \text{ kJ}$$

For the vane type,

$$p_i = \frac{1.52 + 1.013}{2} = 1.266 \text{ bar}$$

Referring to fig. 9.24,

$$\text{Work required} = (\text{area A} + \text{area B})$$

Now using equation 9.7 with $n = \gamma$,

$$\text{area A} = \frac{\gamma}{\gamma - 1} p_1 V_s \left\{ \left(\frac{p_i}{p_1} \right)^{(\gamma-1)/\gamma} - 1 \right\}$$

$$= \frac{1.4}{0.4} \times \frac{1.013 \times 10^5 \times 0.03}{10^3} \left\{ \left(\frac{1.266}{1.013} \right)^{0.4/1.4} - 1 \right\} \text{kJ/rev}$$

$$= 3.5 \times 1.013 \times 100 \times 0.03 \times 0.066 = 0.703 \text{ kJ/rev}$$

$$\text{area B} = (p_2 - p_i)V_b$$

(where V_b is given by equation 4.19),

i.e.

$$V_b = V_s \left(\frac{p_1}{p_i} \right)^{1/\gamma} = 0.03 \times \left(\frac{1.013}{1.266} \right)^{1/1.4}$$

$$= \frac{0.03}{1.249^{0.714}} = \frac{0.03}{0.173} = 0.0256 \text{ m}^3$$

i.e. area B $= (1 \cdot 52 - 1 \cdot 266) \times 10^2 \times 0 \cdot 0256$ kJ/rev

$\qquad = 0 \cdot 65$ kJ/rev

$\therefore$ Work required $= 0 \cdot 703 + 0 \cdot 65 = 1 \cdot 353$ kJ/rev

Rotary sliding vane compressors are used with free air deliveries of up to 150 m³/min and pressure ratios up to 8·5 to 1. For special applications and boosting, pressure ratios of the order of 20 to 1 have been obtained from this type. The larger machines are usually water cooled.

Lubrication is important with vane-type machines and is accomplished by injecting oil to the vane tips in contact with the casing. Some machines, having carbon vanes, require no lubrication. Another version is designed to reduce the friction between vane and casing. This employs a floating drum which rotates between the rotor and casing, and does not allow the vanes to make contact with the

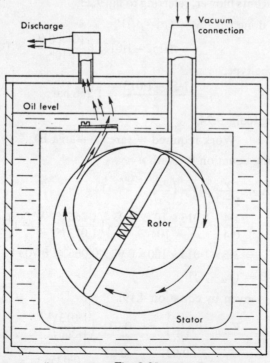

Fig. 9.25

casing. The only movement of the blades relative to the floating drum is along the slots. See fig. 9.22b.

9.13 Vacuum pumps

Rotary positive displacement pumps are used to produce a vacuum or to scavenge a vessel. An example of this type of pump is shown in fig. 9.25. The rotor is eccentrically mounted in the stator and carries two blades which sweep the space between the rotor and stator. The gas being exhausted enters through the vacuum connection and is compressed before delivery through the discharge valve. The efficiency of such pumps is impaired by the presence of condensable vapours, and means must be provided to deal with these if necessary. The vapours tend to condense before delivery through the discharge valve and mix with the sealing oil. The liquid eventually evaporates into the vacuum system and lowers the vacuum obtainable, as well as impairing the sealing and lubricating properties of the oil.

9.14 Air motors

Compressed air is used in a wide variety of applications in industry. For some purposes air-operated motors are the most suitable forms of power, especially where there are safety requirements to be met as in mining applications.

Pneumatic breakers, picks, spades, rammers, vibrators, riveters, etc., form a range of hand tools which have wide applications in constructional work. They are light in construction and suitable for operation in remote situations for which other forms of power tools may not be suitable. The action required of such tools, with the associated simplicity and robustness of construction, is obtained with air-operated design.

Basically the cycle in the reciprocating expander is the reverse of that in the reciprocating compressor. Air is supplied to the air motor from an air receiver in which the air is at approximately ambient temperature. There is a pressure drop in the air line between the receiver and the motor. The air expands in the motor cylinder to atmospheric pressure in a manner which is polytropic (i.e. the expansion is internally reversible and the 'law of expansion is pv^n = constant, where $n < \gamma$, and is usually about 1·3). If the air is initially at ambient temperature, then this form of expansion will bring about a reduction in the air temperature as lower pressures are reached. The temperatures reached may be sufficiently low to be below the dew point of the moisture in the air (see Section 14.2); this may be condensed, and the water formed may even be cooled to its freezing point. This may lead to the formation of ice in the cylinder with the consequence of blocked valves. To prevent this condition it may be necessary to pre-heat the air to an initial temperature which is high enough to prevent the formation of ice. This heating of the air causes an increase in volume at the supply pressure and reduces the demand from the compressor. Further, the temperature at which the heat transfer is required is low, and a low grade supply of heat or 'waste heat' may be utilized for the purpose.

A hypothetical indicator diagram for an air motor is shown in fig. 9.26. In this case the air expands from 1 to the pressure p_2 at the end of the stroke. There is then a *blow-down* of air from 2 to 3. Air is exhausted from 3 to 4, and at 4 compression of the trapped or *cushion air* begins. Air at the supply pressure, p_6, is admitted to the cylinder at the point 5 where it mixes irreversibly with the cushion air. The pressure in the cylinder is rapidly brought up to the inlet value, p_6. The further supply of air is made at constant pressure behind the moving piston to the point of cut-off at 1. The cut-off ratio is given by

$$\text{Cut-off ratio} = \frac{V_1 - V_6}{V_3 - V_6}$$

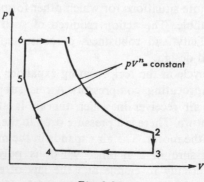

Fig. 9.26

The effect of the cushion air is to give a smoother running motor. The position of the point 5 depends on the point of initial compression 4, and on the law of compression $pV^n = $ constant. The conditions may be such that the points 5 and 6 coincide.

The analysis of such a diagram is best carried out from basic principles, as illustrated in the following example.

Example 9.11

The cylinder of an air motor has a bore of 63·5 mm and a stroke of 114 mm. The supply pressure is 6·3 bar, the supply temperature 24°C, and the exhaust pressure is 1·013 bar. The clearance volume is 5% of the swept volume and the cut-off ratio is 0·5. The air is compressed by the returning piston after it has travelled through 0·95 of its stroke. The law of compression and expansion is $pV^{1·3} = $ constant. Calculate the temperature at the end of expansion and the indicated power of the motor which runs at 300 rev/min. Calculate also the air supplied per minute.

$$\text{Swept volume} = \frac{\pi \times 63·5^2 \times 114}{4 \times 10^9} = 0·361 \times 10^{-3} \text{ m}^3$$

Referring to the cycle of fig. 9.26,

$$\text{Clearance volume} = V_6 = V_5 = 0·05 \times 0·361 \times 10^{-3}$$
$$= 0·018 \times 10^{-3} \text{ m}^3$$
$$\therefore V_1 = \frac{0·361 \times 10^{-3}}{2} + 0·018 \times 10^{-3} = 0·198 \times 10^{-3} \text{ m}^3$$
$$V_2 = 0·361 \times 10^{-3} + 0·018 \times 10^{-3} = 0·379 \times 10^{-3} \text{ m}^3$$
$$V_4 - V_5 = 0·05 \times 0·361 \times 10^{-3} = 0·018 \times 10^{-3} \text{ m}^3$$
$$\therefore V_4 = 0·018 \times 10^{-3} + 0·018 \times 10^{-3} = 0·036 \times 10^{-3} \text{ m}^3$$
$$p_1 V_1^n = p_2 V^n$$
$$\therefore p_2 = p_1 \left(\frac{V_1}{V_2}\right)^n = 6·3 \left(\frac{0·198}{0·379}\right)^{1·3} = 2·71 \text{ bar}$$

also, $$T_2 = T_1 \left(\frac{V_1}{V_2}\right)^{n-1} = 297 \left(\frac{0·198}{0·379}\right)^{0·3} = 244·4 \text{ K}$$

i.e.   Temperature after expansion $= 244·4 - 273 = -28·6°\text{C}$

Then,

$$p_5 = p_4\left(\frac{V_4}{V_5}\right)^n = 1 \cdot 013\left(\frac{0 \cdot 036}{0 \cdot 018}\right)^{1 \cdot 3} = 2 \cdot 494 \text{ bar}$$

Work done per cycle = area 1234561

i.e.   Work done $= p_1(V_1 - V_6) + \left(\dfrac{p_1 V_1 - p_2 V_2}{n-1}\right)$

$$-p_3(V_3 - V_4) - \left(\frac{p_5 V_5 - p_4 V_4}{n-1}\right)$$

$\therefore$ Work done per cycle $= 10^2\Bigg\{6 \cdot 3(0 \cdot 198 \times 10^{-3} - 0 \cdot 018 \times 10^{-3})$

$$+\left(\frac{6 \cdot 3 \times 0 \cdot 198 \times 10^{-3} - 2 \cdot 71 \times 0 \cdot 379 \times 10^{-3}}{0 \cdot 3}\right)$$

$$-1 \cdot 013(0 \cdot 379 \times 10^{-3} - 0 \cdot 036 \times 10^{-3})$$

$$-\left(\frac{2 \cdot 494 \times 0 \cdot 018 \times 10^{-3} - 1 \cdot 013 \times 0 \cdot 036 \times 10^{-3}}{0 \cdot 3}\right)\Bigg\}$$

i.e. Work done per cycle $= 10^{-1}(1 \cdot 134 + 0 \cdot 733 - 0 \cdot 347 - 0 \cdot 028)$

$$= 0 \cdot 149 \text{ kJ/cycle}$$

$$\therefore \text{ Power developed} = \frac{0 \cdot 149 \times 300}{60} = 0 \cdot 745 \text{ kW}$$

The mass induced per cycle is given by $(\dot{m}_1 - \dot{m}_4)$. It is necessary to determine the temperature of the air at 4, which can be taken as equal to that at 3. It is assumed that the air in the cylinder at the point 2 expands isentropically to the exhaust pressure.

$$\therefore T_3 = T_2\left(\frac{p_3}{p_2}\right)^{(\gamma-1)/\gamma} = 244 \cdot 4\left(\frac{1 \cdot 013}{2 \cdot 71}\right)^{0 \cdot 4/1 \cdot 4} = 184 \cdot 4 \text{ K}$$

i.e. $\dot{m}_4 = \dfrac{p_4 V_4}{RT_4} = \dfrac{1 \cdot 013 \times 10^5 \times 0 \cdot 0361}{287 \times 184 \cdot 4} = 0 \cdot 069 \text{ g}$

Also, $\dot{m}_1 = \dfrac{p_1 V_1}{RT_1} = \dfrac{6 \cdot 3 \times 10^5 \times 0 \cdot 198}{287 \times 297} = 1 \cdot 463 \text{ g}$

$\therefore$ Induced mass per cycle $= 1 \cdot 463 - 0 \cdot 069 = 1 \cdot 394 \text{ g}$

i.e. Mass of air supplied $= 1 \cdot 394 \times 10^{-3} \times 300 = 0 \cdot 418 \text{ kg/min}$

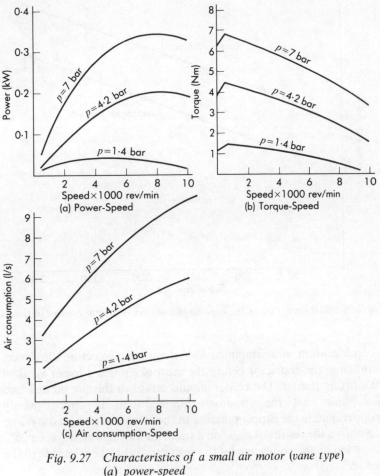

Fig. 9.27   *Characteristics of a small air motor (vane type)*
(a) *power-speed*
(b) *torque-speed*
(c) *air consumption-speed*

Air motors can be rotary in action and are similar in form to their compressor counterparts, see Section 9.12. Fig. 9.27(a), (b) and (c) show the forms of the performance characteristics of a small, 0.3 kW, vane type air motor in terms of power/speed, torque/speed and air consumption/speed. An air motor which receives air from a constant pressure supply can be controlled to meet the load requirements by fitting a restrictor either before or after the motor. It can be shown by

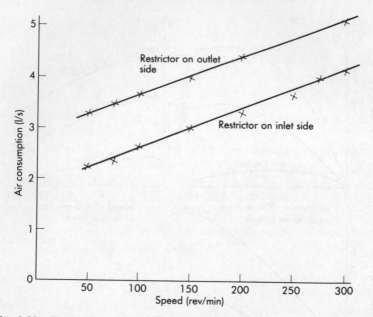

Fig. 9.28   Test results on a vane-type air motor with restrictor control (no load)

a consideration of a simplified P-V diagram, neglecting clearance, that fitting the restrictor before the motor requires a lower air flow than fitting it after. The reader should establish this for himself and also show that the air flow rate required is approximately proportional to the supply pressure to the motor for a given duty. Fig. 9.28 shows the results of a test on a small air motor which gives a 25% reduction in air requirement if the restrictor is on the inlet side to the motor.

PROBLEMS

9.1 Air is to be compressed in a single-stage reciprocating compressor from 1·013 bar and 15°C to 7 bar. Calculate the indicated power required for a free air delivery of 0·3 m³/min, when the compression process is:

(a) Isentropic;
(b) Reversible isothermal;
(c) Polytropic, with $n = 1.25$.

What will be the delivery temperature in each case?

(1·31 kW; 0·98 kW; 1·19 kW; 228°C; 15°C; 151°C)

9.2 The compressor of problem 9.1 is to run at 1000 rev/min. If the compressor is single-acting and has a stroke/bore ratio of 1·2/1, calculate the bore size required. (68·3 mm)

9.3 A single-stage, single-acting air compressor running at 1000 rev/min delivers air at 25 bar. For this purpose the induction and free air conditions can be taken as 1·013 bar and 15°C, and the free air delivery as 0·25 m³/min. The clearance volume is 3% of the swept volume and the stroke/bore ratio is 1·2/1. Calculate the bore and stroke and the volumetric efficiency of this machine. Take the index of compression and expansion as 1·3. Calculate also the indicated power and the isothermal efficiency.

(73·2 mm; 87·84 mm; 67·6%; 2 kW; 67·5%)

9.4 The compressor of problem 9.3 has actual induction conditions of 1 bar and 40°C, and the delivery pressure is 25 bar. Taking the bore and stroke as calculated in problem 9.3, calculate the free-air delivery referred to 1·013 bar and 15°C and the indicated power required. What value for the volumetric efficiency would be measured from the indicator diagram?

(0·225 m³/min; 1·97 kW; 60·9%; 67%)

9.5 A single-acting compressor is required to deliver air at 70 bar from an induction pressure of 1 bar, at the rate of 2·4 m³/min measured at free-air conditions of 1·013 bar and 15°C. The temperature at the end of the induction stroke is 32°C. Calculate the indicated power required if the compression is carried out in two stages with an ideal intermediate pressure and complete inter-cooling. The index of compression and expansion for both stages is 1·25. What is the saving in power over single-stage compression?

If the clearance volume is 3% of the swept volume in each cylinder, calculate the swept volumes of the cylinders. The speed of the compressor is 750 rev/min.

If the mechanical efficiency of the compressor is 85%, calculate the power output in kilowatts of the motor required.

(22·7 kW; 6 kW; 0·00396 m³; 0·000474 m³; 26·75 kW)

9.6 For the compressor of problem 9.5 calculate the heat rejected per minute to the jacket cooling water of each stage, and the heat rejected per minute to the intercooler. Assume that 50% of the friction power in each stage is transferred to the jacket cooling water.

(325 kJ/min; 476 kJ/min;)

9.7 A single-cylinder, single-acting air compressor of 200 mm bore by 250 mm stroke is constructed so that its clearance can be altered by moving the cylinder head, the stroke being unaffected.

(a) Calculate the free-air delivery at 300 rev/min when the clearance volume is set at 700 cm^3 and the delivery pressure is 5 bar. Assume that $n = 1 \cdot 25$ and the suction pressure and temperature are 1 bar and 32°C, respectively. Find also the power required assuming a mechanical efficiency of 80%. The free-air conditions are 1·013 bar and 15°C.

(b) To what minimum value may the clearance volume be reduced when the delivery pressure is 4·2 bar, assuming that the same driving power is available and that the suction pressure, speed, value of n, and the mechanical efficiency remain unaltered?

(1·68 m^3/min; 7·2 kW; 453 cm^3)

9.8 A single-cylinder, single-acting air compressor running at 300 rev/min is driven by a 23 kW electric motor. The mechanical efficiency of the drive between motor and compressor is 87%. The air inlet conditions are 1·013 bar and 15°C and the delivery pressure is 8 bar. Calculate the free-air delivery in m^3/min, the volumetric efficiency, and the bore and stroke of the compressor. Assume that the index of compression and expansion is $n = 1 \cdot 3$, that the clearance volume is 7% of the swept volume and that the bore is equal to the stroke.

(4·47 m^3/min; 73%; 296 mm)

9.9 A two-stage air compressor consists of three cylinders having the same bore and stroke. The delivery pressure is 7 bar and the free air delivery is 4·2 m^3/min. Air is drawn in at 1·013 bar, 15°C and an intercooler cools the air to 38°C. The index of compression is 1·3 for all three cylinders. Neglecting clearance calculate:

(a) The intermediate pressure.
(b) The power required to drive the compressor.
(c) The isothermal efficiency.

(2·19 bar; 16·3 kW; 84·5%)

9.10 A four-stage compressor works between limits of 1 bar and 112 bar. The index of compression in each stage is 1·28, the temperature at the start of compression in each stage is 32°C, and the intermediate pressures are so chosen that the work is divided equally among the stages. Neglecting clearance find:

(a) The volume of free air delivered per kW h at 1·013 bar and 15°C.

(b) The temperature at delivery from each stage.

(c) The isothermal efficiency. (6·24 m³/kW h; 122°C; 87·8%)

9.11 The indicator fitted to a reciprocating compressor has a rating of 0·275 bar/mm when fitted with a piston of 650 mm² area. The instrument when sleeved has a piston area of 325 mm². Calculate the indicator rating when operated with the smaller piston and the same spring.

The indicator cards taken from a double-acting single-stage compressor gave areas of 1300 mm² for the cover end and 1250 mm² for the crank end of the cylinder. Both diagrams had a length of 76 mm and were taken with the 'half' piston fitted. The machine has a bore of 152 mm and a stroke of 177 mm with a piston rod of 25 mm diameter. Calculate the mean effective pressure for each end and the total indicated power for a speed of 300 rev/min. Assuming an overall mechanical efficiency of the compressor and motor drive of 85%, calculate the output power of the driving motor.

(0·55 bar/mm; 9·41 bar; 9·05 bar; 29·2 kW; 34·4 kW)

9.12 Air at 1·013 bar and 15°C is to be compressed at the rate of 5·6 m³/min to 1·75 bar. Two machines are considered: (a) the Roots blower; and (b) a sliding vane rotary compressor. Compare the powers required, assuming for the vane type that internal compression takes place through 75% of the pressure rise before delivery takes place, and that the compressor is an ideal uncooled machine. (6·88 kW; 5·75 kW)

9.13 Air is compressed in a two-stage vane type compressor from 1·013 bar to 8·75 bar. Assuming equal pressure ratios in each stage, calculate the power required. Assume that in each stage compression is complete and that intercooling between stages is 75% complete. Calculate also the capacity of the high-pressure stage in cubic metres per minute for a free air delivery of 42 m³/min measured at 1·013 bar and 15°C. The machine is uncooled except for the intercooler and operates in an ideal manner. (187 kW; 15·6 m³/min)

9.14 The following particulars refer to a single-acting air motor: cylinder diameter 380 mm; stroke 610 mm; speed 200 rev/min; supply pressure and temperature, 6·2 bar and 150°C; back pressure 1·03 bar; index of expansion and compression, 1·35; cut-off ratio,

0·46; clearance volume, 20% of swept volume; mechanical efficiency 95%.

Assuming that the temperature and pressure of the air in the clearance space at the beginning of admission are 6·2 bar and 150°C, calculate:

(a) the air consumption;
(b) the shaft power developed;
(c) the air temperature after blow-down;
(d) the fraction of stroke travelled by the piston before re-compression begins.   (0·54 kg/s; 71·8 kW; −14·3°C; 0·444)

REFERENCES

9.1 BS 1571, *Testing of positive displacement compressors and exhausters:* Part I (1975), Part II (1984).

10. Nozzles

A nozzle is a duct of smoothly varying cross-sectional area in which a steadily flowing fluid can be made to accelerate by a pressure drop along the duct. There are many applications in practice which require a high-velocity stream of fluid, and the nozzle is the best means of obtaining this. For example, nozzles are used in steam and gas turbines, in jet engines, in rocket motors, in flow measurement, and in many other applications. When a fluid is decelerated in a duct, causing a rise in pressure along the stream, then the duct is called a *diffuser*; two applications in practice in which a diffuser is used are the centrifugal compressor and the ram jet.

The analysis presented in this chapter will be restricted to *one-dimensional flow*. In one-dimensional flow it is assumed that the fluid velocity, and the fluid properties, change only in the direction of the flow. This means that the fluid velocity is assumed to remain constant at a mean value across the cross-section of the duct. The effects of friction will not be analysed fundamentally, suitable efficiencies or coefficients being adopted to allow for the departure from the ideal frictionless case. The analysis of fluid flow involving friction has become of increasing importance due to the development of the turbo-jet, the ram jet, and the rocket, and the introduction of high-speed flight. For a fundamental approach to the topic a study of fluid dynamics is required, and the reader is recommended to books on *gas dynamics*, such as reference 10.1.

10.1 Nozzle shape

Consider a stream of fluid at pressure p_1, enthalpy h_1, and with a low velocity C_1. It is required to find the shape of duct which will cause the fluid to accelerate to a high velocity as the pressure falls along the duct. It can be assumed that the heat loss from the duct is negligibly small (i.e. adiabatic flow, $Q=0$), and it is clear that no work is done on or by the fluid (i.e. $W=0$). Applying the steady-flow energy equation, 2.8, between section 1 and any other section

X–X where the pressure is p, the enthalpy is h, and the velocity is C, we have,

$$h_1 + \frac{C_1^2}{2} = h + \frac{C^2}{2}$$

i.e. $$C^2 = 2(h_1 - h) + C_1^2$$

or $$C = \sqrt{\{2(h_1 - h) + C_1^2\}} \qquad 10.1$$

If the area at the section X–X is A, and the specific volume is v, then, using equation 2.9,

$$\text{Mass flow, } \dot{m} = \frac{CA}{v}$$

or $$\text{Area per unit mass flow, } \frac{A}{\dot{m}} = \frac{v}{C} \qquad 10.2$$

Then substituting for the velocity C, from equation 10.1,

$$\text{Area per unit mass flow} = \frac{v}{\sqrt{\{2(h_1 - h) + C_1^2\}}} \qquad 10.3$$

It can be seen from equation 10.3 that in order to find the way in which the area of the duct varies it is necessary to be able to evaluate the specific volume, v, and the enthalpy, h, at any section X–X. In order to do this, some information about the process undergone between section 1 and section X–X must be known. For the ideal frictionless case, since the flow is adiabatic and reversible, the process undergone is an isentropic process, and hence,

$$s_1 = (\text{entropy at any section X–X}) = s, \text{ say}$$

Now using equation 10.2, and the fact that $s_1 = s$, it is possible to plot the variation of the cross-sectional area of the duct against the pressure along the duct. For a vapour this can be done using tables; for a perfect gas the procedure is simpler, since we have $pv^\gamma = $ constant, for an isentropic process. In either case, choosing fixed inlet conditions, then the variation in the area, A, the specific volume, v, and the velocity, C, can be plotted against the pressure along the duct. Typical curves are shown in fig. 10.1. It can be seen that the area decreases initially, reaches a minimum, and then increases again. This can also be seen from equation 10.2,

i.e. $$\text{Area per unit mass flow} = \frac{v}{C}$$

When v increases less rapidly than C, then the area decreases.
When v increases more rapidly than C, then the area increases.

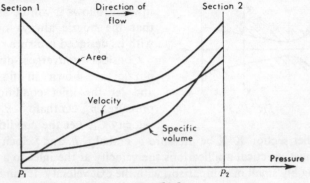

Fig. 10.1

A nozzle, the area of which varies as in fig. 10.1, is called a *convergent-divergent* nozzle. A cross-section of a typical convergent-divergent nozzle is shown in fig. 10.2. The section of minimum area is called the *throat* of the nozzle.

It will be shown later in Section 10.2 that the velocity at the throat of a nozzle operating at its designed pressure ratio is the velocity of sound at the throat conditions. The flow up to the throat is *sub-sonic*; the flow after the throat is *supersonic*. It should be noted that a sonic or a super-

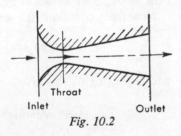

Fig. 10.2

sonic flow requires a diverging duct to accelerate it.

The specific volume of a liquid is constant over a wide pressure range, and therefore nozzles for liquids are always convergent, even at very high exit velocities (e.g. a fire-hose uses a convergent nozzle).

10.2 Critical pressure ratio

It has been stated in Section 10.1 that the velocity at the throat of a correctly designed nozzle is the velocity of sound. In the same way, for a nozzle that is convergent only, then the fluid will attain sonic velocity at exit if the pressure drop across the nozzle is large enough. The ratio of the pressure at the section where sonic velocity is attained

to the inlet pressure of a nozzle is called the *critical pressure ratio*.
Cases in which nozzles operate off the design conditions of pressure

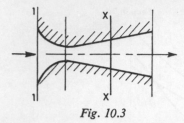

will be considered in Section 10.4;
in what follows it will be assumed
that the nozzle always operates
with its designed pressure ratio.

Consider a convergent-divergent
nozzle as shown in fig. 10.3,
and let the inlet conditions be
pressure p_1, enthalpy h_1, and
velocity C_1. Let the conditions at

Fig. 10.3

any other section X–X be pressure p, enthalpy h, and velocity C.

In most practical applications the velocity at the inlet to a nozzle
is negligibly small in comparison with the exit velocity. It can be seen
from equation 10.2, $A/\dot{m} = v/C$, that a negligibly small velocity
implies a very large area, and most nozzles are
in fact shaped at inlet in such a way that the
nozzle converges rapidly over the first fraction
of its length; this is illustrated in the diagram
of a nozzle inlet shown in fig. 10.4.

Now from equation 10.1 we have,

$$C = \sqrt{\{2(h_1 - h) + C_1^2\}}$$

and neglecting C_1 this gives,

$$C = \sqrt{\{2(h_1 - h)\}} \qquad 10.4$$

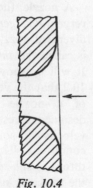

Fig. 10.4

Since enthalpy is usually expressed in kJ/kg,
then an additional constant of 10^3 will appear within the root sign
if C is to be expressed in m/s,

i.e. $\qquad\qquad\qquad \sqrt{(2 \times 10^3)} = 44 \cdot 72$

(where 1 kJ $= 10^3$ N m).

Hence,

$$C = \text{const.} \sqrt{(h_1 - h)}$$

(where const. $= 44 \cdot 72 \dfrac{m}{s} \left(\dfrac{kg}{kJ}\right)^{1/2}$.)

Then, substituting from equation 10.4 in equation 10.2, we have,

$$\text{Area per unit mass flow, } \frac{A}{\dot{m}} = \frac{v}{C} = \frac{v}{\sqrt{\{2(h_1 - h)\}}} \qquad 10.5$$

As stated in Section 10.1 the area can be evaluated at any section where the pressure is p, by assuming that the process is isentropic (i.e. $s_1 = s$). When this is done for a series of pressures, the area can be plotted against pressure along the duct, or against pressure ratio, and the critical pressure can thus be found graphically. This method can be used for vapours and an example of the procedure is given in Section 10.7. For a perfect gas it is possible to simplify equation 10.5 by making use of the perfect gas laws.

From equation 3.18, $h = c_p T$ for a perfect gas, therefore,

$$\text{Area per unit mass flow rate} = \frac{v}{\sqrt{\{2c_p(T_1 - T)\}}}$$

$$= \frac{v}{\sqrt{\left\{2c_p T_1 \left(1 - \frac{T}{T_1}\right)\right\}}}$$

From equation 3.5, $v = RT/p$, therefore,

$$\text{Area per unit mass flow rate} = \frac{RT}{\sqrt{\left\{2c_p T_1 \left(1 - \frac{T}{T_1}\right)\right\}}}$$

Let the pressure ratio, p/p_1, be x. Then using equation 4.21, for a perfect gas,

$$\frac{T}{T_1} = \left(\frac{p}{p_1}\right)^{(\gamma-1)/\gamma} = x^{(\gamma-1)/\gamma}$$

for an isentropic process.

Substituting for $p = xp_1$, for $T = T_1 x^{(\gamma-1)/\gamma}$, and for $T/T_1 = x^{(\gamma-1)/\gamma}$, we have,

$$\text{Area per unit mass flow rate} = \frac{RT_1 x^{(\gamma-1)/\gamma}}{p_1 x \sqrt{\{2c_p T_1(1 - x^{(\gamma-1)/\gamma})\}}}$$

For fixed inlet conditions (i.e. p_1 and T_1 fixed), we have,

$$\text{Area per unit mass flow rate} = \text{constant} \times \frac{x^{(\gamma-1)/\gamma}}{x\sqrt{(1-x^{(\gamma-1)/\gamma})}}$$

$$= \text{constant} \times \frac{1}{x^{1/\gamma}\sqrt{(1-x^{(\gamma-1)/\gamma})}}$$

i.e.
$$= \frac{\text{constant}}{\sqrt{(x^{2/\gamma} - x^{2/\gamma}x^{(\gamma-1)/\gamma})}}$$

$$\text{Area per unit mass flow rate} = \frac{\text{constant}}{\sqrt{(x^{2/\gamma} - x^{(\gamma+1)/\gamma})}} \qquad 10.6$$

To find the value of the pressure ratio, x, at which the area is a minimum it is necessary to differentiate equation 10.6 with respect to x and equate the result to zero,
i.e. for minimum area

$$\frac{\mathrm{d}}{\mathrm{d}x}\left\{\frac{1}{(x^{2/\gamma} - x^{(\gamma+1)/\gamma})^{1/2}}\right\} = 0$$

i.e.
$$-\frac{\left\{\dfrac{2}{\gamma}x^{(2/\gamma)-1} - \left(\dfrac{\gamma+1}{\gamma}\right)x^{((\gamma+1)/\gamma)-1}\right\}}{2(x^{2/\gamma} - x^{(\gamma+1)/\gamma})^{3/2}} = 0$$

Hence the area is a minimum when,

$$\frac{2}{\gamma}x^{(2/\gamma)-1} = \left(\frac{\gamma+1}{\gamma}\right)x^{((\gamma+1)/\gamma)-1}$$

i.e.
$$x^{((\gamma+1)/\gamma)-1-(2/\gamma)+1} = \frac{2}{\gamma+1}$$

$$\therefore\ x = \left(\frac{2}{\gamma+1}\right)^{\gamma/(\gamma-1)}$$

i.e. Critical pressure ratio, $\dfrac{p_c}{p_1} = \left(\dfrac{2}{\gamma+1}\right)^{\gamma/(\gamma-1)}$ \qquad 10.7

It can be seen from equation 10.7 that for a perfect gas the pressure ratio required to attain sonic velocity in a nozzle depends only on the value of γ for the gas. For example, for air $\gamma = 1\cdot4$, therefore,

$$\frac{p_c}{p_1} = \left(\frac{2}{1\cdot4+1}\right)^{1\cdot4/0\cdot4} = 0\cdot528$$

Hence for air at 10 bar, say, a convergent nozzle requires a back pressure of 5·28 bar, in order that the flow should be sonic at exit and for a correctly designed convergent-divergent nozzle with inlet pressure 10 bar, the pressure at the throat is 5·28 bar. For carbon dioxide, $\gamma = 1\cdot3$, therefore,

$$\frac{p_c}{p_1} = \left(\frac{2}{1\cdot3+1}\right)^{1\cdot3/0\cdot3} = 0\cdot546$$

Hence, for carbon dioxide at 10 bar, a convergent nozzle requires a back pressure of 5·46 bar for sonic flow at exit, and the pressure at the throat of a convergent-divergent nozzle with inlet pressure 10 bar, is 5·46 bar.

The ratio of the temperature at the section of the nozzle where the velocity is sonic to the inlet temperature is called the *critical temperature ratio*,

i.e. Critical temperature ratio, $\dfrac{T_c}{T_1} = \left(\dfrac{p_c}{p_1}\right)^{(\gamma-1)/\gamma} = \dfrac{2}{\gamma+1}$

i.e. $$\frac{T_c}{T_1} = \frac{2}{\gamma+1} \qquad\qquad 10.8$$

Equations 10.7 and 10.8 apply to perfect gases only, and not to vapours. However, it is found that a sufficiently close approximation is obtained for a steam nozzle if it is assumed that the expansion follows a law $pv^k = \text{constant}$. The process is assumed to be isentropic, and therefore the index k is an approximate isentropic index for steam. When the steam is initially dry saturated then $k = 1\cdot135$; when the steam is initially superheated then $k = 1\cdot3$. Note that equation 10.8 cannot be used for a wet vapour, since no simple relationship between p and T is known for a wet vapour undergoing an isentropic process. More will be said of this in Sections 10.7 and 10.8.

The critical velocity at the throat of a nozzle can be found for a perfect gas by substituting in equation 10.1,

i.e. $$C_c = \sqrt{\{2(h_1-h)+C_1^2\}}$$

Putting $C_1 = 0$, as before, and using equation 3.18 for a perfect gas, $h = c_p T$, we have,

$$C_c = \sqrt{\{2c_p(T_1-T_c)\}} = \sqrt{\left\{2c_pT_c\left(\frac{T_1}{T_c}-1\right)\right\}}$$

From equation 10.8, $\dfrac{T_c}{T_1} = \dfrac{2}{\gamma+1}$, hence,

$$C_c = \sqrt{\left[2c_pT_c\left\{\left(\frac{\gamma+1}{2}\right)-1\right\}\right]} = \sqrt{\{c_pT_c(\gamma-1)\}}$$

Also, from equation 3.22,

$$c_p = \frac{\gamma R}{(\gamma - 1)} \quad \text{or} \quad c_p(\gamma - 1) = \gamma R$$

Hence substituting,

$$C_c = \sqrt{(\gamma R T_c)}$$

i.e.   Critical velocity, $C_c = \sqrt{(\gamma R T_c)}$ 10.9

The critical velocity given by equation 10.9, is the velocity at the throat of a correctly designed convergent-divergent nozzle, or the velocity at the exit of a convergent nozzle when the pressure ratio across the nozzle is the critical pressure ratio.

It can be shown that the critical velocity is the velocity of sound at the critical conditions.

The velocity of sound, a, is defined by the equation,

$$a^2 = \frac{dp}{d\rho} \quad \text{at constant entropy}$$

(where p is pressure and ρ is the density).

A proof of this expression can be found in reference 10.2.
Now $\rho = 1/v$, where v is the specific volume.

$$\therefore \quad d\rho = d(1/v) = -\frac{1}{v^2} \, dv$$

Hence, $$a^2 = -\frac{dp}{dv} v^2$$

For a perfect gas undergoing an isentropic process, $pv^\gamma = \text{constant}$

i.e. $$p = \frac{K}{v^\gamma} \quad \text{where } K \text{ is a constant}$$

$$\therefore \quad \frac{dp}{dv} = -\frac{\gamma K}{v^{\gamma + 1}}$$

Therefore substituting,

$$a^2 = \frac{v^2 \gamma K}{v^{\gamma + 1}}$$

Also, $K = pv^\gamma$, hence,

$$a^2 = \frac{\gamma p v^\gamma v^2}{v^{\gamma + 1}} = \gamma p v$$

i.e.   Velocity of sound, $a = \sqrt{(\gamma p v)} = \sqrt{(\gamma R T)}$ 10.10

It can be seen that the critical velocity of a perfect gas in a nozzle, as given by equation 10.9, is the velocity of sound in the gas at the critical temperature.

Equations 10.9 and 10.10 cannot be applied to a vapour; however, if an approximate isentropic law, $pv^k =$ constant, is assumed for a vapour, then the critical velocity can be taken as, $C_c = \sqrt{(kpv)}$. This expression is derived in Section 10.7. (Note that for a vapour the critical velocity cannot be expressed in terms of the temperature.) It is usually more convenient to evaluate the critical velocity of a vapour using equation 10.4, $C_c = \sqrt{\{2(h_1 - h_c)\}}$, where $(h_1 - h_c)$ is the enthalpy drop from the inlet to the throat, which can be evaluated from tables or by using an h-s chart.

Example 10.1

Air at 8·6 bar and 190°C expands at the rate of 4·5 kg/s through a convergent-divergent nozzle into a space at 1·03 bar. Assuming that the inlet velocity is negligible, calculate the throat and the exit cross-sectional areas of the nozzle.

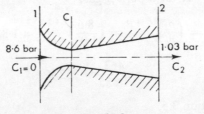

Fig. 10.5

The nozzle is shown diagrammatically in fig. 10.5. From equation 10.7, the critical pressure ratio is given by,

$$\frac{p_c}{p_1} = \left(\frac{2}{\gamma + 1}\right)^{\gamma/(\gamma - 1)} = \left(\frac{2}{2\cdot4}\right)^{1\cdot4/0\cdot4} = 0\cdot528$$

i.e. $p_c = 0\cdot528 \times 8\cdot6 = 4\cdot54$ bar

Also, from equation 10.8,

$$\frac{T_c}{T_1} = \frac{2}{\gamma + 1} = \frac{1}{1\cdot2}$$

i.e. $T_c = \frac{190 + 273}{1\cdot2} = 385\cdot8$ K

From equation 3.5,

$$v_c = \frac{RT_c}{p_c} = \frac{287 \times 385 \cdot 8}{10^5 \times 4 \cdot 54} = 0 \cdot 244 \text{ m}^3/\text{kg}$$

Also, from equation 10.9,

$$C_c = \sqrt{(\gamma R T_c)} = \sqrt{(1 \cdot 4 \times 287 \times 385 \cdot 8)} = 394 \text{ m/s}$$

[or, from equation 10.4,

$$C_c = \sqrt{\{2(h_1 - h_c)\}} = \sqrt{\{2c_p(T_1 - T_c)\}}$$

i.e. $$C_c = 44 \cdot 72\sqrt{\{1 \cdot 005(463 - 385 \cdot 8)\}} = 394 \text{ m/s}]$$

To find the area of the throat, using equation 2.9, we have

$$A_c = \frac{\dot{m}v_c}{C_c} = \frac{4 \cdot 5 \times 0 \cdot 244}{394} = 0 \cdot 00279 \text{ m}^2$$

i.e.   Area of throat $= 0 \cdot 00279 \times 10^6 = 2790 \text{ mm}^2$

Using equation 4.21 for a perfect gas,

$$\frac{T_1}{T_2} = \left(\frac{p_1}{p_2}\right)^{(\gamma-1)/\gamma} = \left(\frac{8 \cdot 6}{1 \cdot 03}\right)^{0 \cdot 4/1 \cdot 4} = 1 \cdot 835$$

i.e. $$T_2 = \frac{463}{1 \cdot 835} = 252 \text{ K}$$

Then from equation 3.5,

$$v_2 = \frac{RT_2}{p_2} = \frac{287 \times 252}{10^5 \times 1 \cdot 03} = 0 \cdot 702 \text{ m}^3/\text{kg}$$

Also, from equation 10.4,

$$C_2 = \sqrt{\{2(h_1 - h_2)\}} = \sqrt{\{2c_p(T_1 - T_2)\}}$$

i.e. $$C_2 = 44 \cdot 72\sqrt{\{1 \cdot 005(463 - 252)\}} = 651 \text{ m/s}$$

Then to find the exit area, using equation 2.9,

$$A_2 = \frac{\dot{m}v_2}{C_2} = \frac{4 \cdot 5 \times 0 \cdot 702}{651} = 0 \cdot 00485 \text{ m}^2$$

i.e.   Exit area $= 0 \cdot 00485 \times 10^6 = 4850 \text{ mm}^2$

10.3 Maximum mass flow or choked flow

Consider a convergent nozzle expanding into a space, the pressure of which can be varied, while the inlet pressure remains fixed. The nozzle is shown diagrammatically in fig. 10.6. When the back pressure, p_b, is equal to p_1, then no fluid can flow through the nozzle. As p_b is reduced the mass flow through the nozzle increases, since the enthalpy drop, and hence the velocity, increases. However, when the back pressure reaches the critical value, it is found that no

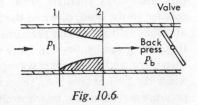

Fig. 10.6.

further reduction in back pressure can affect the mass flow. When the back pressure is exactly equal to the critical pressure, p_c, then the velocity at exit is sonic and the mass flow through the nozzle is at a maximum. If the back pressure is reduced below the critical value then the mass flow remains at the maximum value, the exit pressure remains at p_c, and the fluid expands violently outside the nozzle down to the back pressure. It can be seen that the maximum mass flow through a convergent nozzle is obtained when the pressure ratio across the nozzle is the critical pressure ratio. Also, for a convergent-divergent nozzle, with sonic velocity at the throat, the cross-sectional area of the throat fixes the mass flow through the nozzle for fixed inlet conditions.

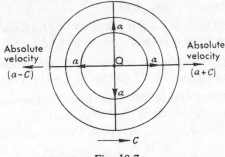

Fig. 10.7

When a nozzle operates with the maximum mass flow it is said to be *choked*. A correctly designed convergent-divergent nozzle is always choked.

An attempt can be made to explain the phenomenon of choking, by considering the velocity of any small disturbance in the stream. Any small disturbance in the flow is propagated as small pressure waves travelling at the velocity of sound in the fluid in all directions from the centre of the disturbance. This is illustrated in fig. 10.7; the pressure waves emanate from point Q at the velocity of sound relative to the fluid, a, while the fluid moves with a velocity, C. The absolute velocity of the pressure waves travelling back upstream is therefore given by, $(a - C)$. Now when the fluid velocity is sub-sonic, then $C < a$, and the pressure waves can move back upstream; however, when the flow is sonic, or supersonic (i.e. $C = a$ or $C > a$), then the pressure waves cannot be transmitted back upstream. It follows from this reasoning that in a nozzle in which sonic velocity has been attained no alteration in the back pressure can be transmitted back upstream. For example, when air at 10 bar expands in a nozzle, the critical pressure can be shown to be 5·28 bar. When the back pressure of the nozzle is 4 bar, say, then the nozzle is choked and is passing the maximum mass flow. If the back pressure is reduced to 1 bar, say, the mass flow through the nozzle remains unchanged. Even if the air were allowed to expand into an evacuated space, the mass flow would be no greater than that through the nozzle when the back pressure is 5·28 bar.

Example 10.2

A fluid at 6·9 bar and 93°C enters a convergent nozzle with negligible velocity, and expands isentropically into a space at 3·6 bar. Calculate the mass flow per m² of exit area.

(a) When the fluid is helium   ($c_p = 5\cdot19$ kJ/kg K).
(b) When the fluid is ethane   ($c_p = 1\cdot88$ kJ/kg K).

Assume that both helium and ethane are perfect gases, and take the respective molar masses as 4 kg/kmol and 30 kg/kmol.

(a) It is necessary first to calculate the critical pressure in order to discover whether the nozzle is choked or not.

From equation 3.9, $R = R_0/M$, therefore for helium,

$$R = \frac{8314}{4} = 2079 \text{ N m/kg K}$$

Then from equation 3.22,

$$c_p = \frac{\gamma R}{(\gamma - 1)}$$

i.e.
$$\frac{\gamma - 1}{\gamma} = \frac{R}{c_p} = \frac{2079}{10^3 \times 5\cdot19} = 0\cdot4$$

$$\therefore \ \gamma = \frac{1}{1 - 0\cdot4} = 1\cdot667$$

Then using equation 10.7,

$$\frac{p_c}{p_1} = \left(\frac{2}{\gamma + 1}\right)^{\gamma/(\gamma - 1)} = \left(\frac{2}{2\cdot667}\right)^{1\cdot667/0\cdot667} = 0\cdot487$$

i.e. $p_c = 0\cdot487 \times 6\cdot9$ bar

i.e.   Critical pressure $p_c = 3\cdot36$ bar

The actual back pressure is 3·6 bar, hence in this case the fluid does not reach the critical conditions and the nozzle is not choked.

The nozzle is shown diagrammatically in fig. 10.8.

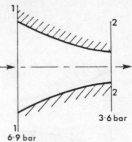

Using equation 4.21,

$$\frac{T_1}{T_2} = \left(\frac{p_1}{p_2}\right)^{(\gamma - 1)/\gamma} = \left(\frac{6\cdot9}{3\cdot6}\right)^{0\cdot4} = 1\cdot297$$

Fig. 10.8

i.e. $T_2 = \frac{93 + 273}{1\cdot297} = 282\cdot2$ K

Then from equation 10.4,

$$C_2 = \sqrt{\{2(h_1 - h_2)\}} = \sqrt{2c_p(T_1 - T_2)}$$

i.e. $C_2 = 44\cdot72\sqrt{\{5\cdot19(366 - 282\cdot2)\}} = 933$ m/s

Also, from equation 3.5,

$$v_2 = \frac{RT_2}{p_2} = \frac{2079 \times 383\cdot2}{10^5 \times 3\cdot6} = 1\cdot63 \ \text{m}^3/\text{kg}$$

Hence from equation 2.9,

$$\dot{m} = \frac{A_2 C_2}{v_2} = \frac{1 \times 933}{1\cdot63} = 572 \ \text{kg/s}$$

i.e.   Mass flow per m² exit area = 572 kg/s

(b) Using the same procedure for ethane, we have,

$$R = \frac{R_0}{M} = \frac{8314}{30} = 277 \cdot 1 \text{ N m/kg K}$$

and

$$\frac{\gamma - 1}{\gamma} = \frac{R}{c_p} = \frac{277 \cdot 1}{10^3 \times 1 \cdot 88} = 0 \cdot 147$$

i.e.

$$\gamma = \frac{1}{1 - 0 \cdot 147} = 1 \cdot 172$$

Then,

$$\frac{p_c}{p_1} = \left(\frac{2}{\gamma + 1}\right)^{\gamma/(\gamma - 1)} = \left(\frac{2}{2 \cdot 172}\right)^{1 \cdot 172/0 \cdot 172} = 0 \cdot 57$$

$$\therefore \; p_c = 0 \cdot 57 \times 6 \cdot 9 \text{ bar}$$

i.e.

Critical pressure, $p_c = 3 \cdot 93$ bar

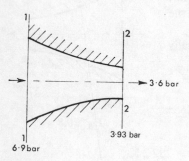

Fig. 10.9

The actual back pressure is 3·6 bar, hence in this case the fluid reaches critical conditions at exit and the nozzle is choked. The expansion from the exit pressure of 3·93 bar down to the back pressure of 3·6 bar must take place outside the nozzle.

The nozzle is shown diagrammatically in fig. 10.9.

Since the nozzle is choked, from equation 10.8, we have

$$\frac{T_c}{T_1} = \frac{2}{\gamma + 1} = \frac{2}{2 \cdot 172}$$

i.e.

$$T_2 = T_c = \frac{2 \times 366}{2 \cdot 172} = 337 \text{ K}$$

Also, from equation 10.9,

$$C_2 = C_c = \sqrt{(\gamma R T_c)} = \sqrt{(1 \cdot 172 \times 277 \cdot 1 \times 337)} = 331 \text{ m/s}$$

From equation 3.5,

$$v_2 = \frac{R T_2}{p_2} = \frac{277 \cdot 1 \times 337}{10^5 \times 3 \cdot 93} = 0 \cdot 238 \text{ m}^3/\text{kg}$$

Then using equation 2.9,

$$\dot{m} = \frac{A_2 C_2}{v_2} = \frac{1 \times 331}{0\cdot238} = 1391 \text{ kg/s}$$

i.e.   Mass flow per m^2 exit area $= 1391$ kg/s

10.4 Nozzles off the design pressure ratio

When the back pressure of a nozzle is below the design value the nozzle is said to *underexpand*. In underexpansion the fluid expands to the design pressure in the nozzle and then expands violently and irreversibly down to the back pressure on leaving the nozzle (e.g. the nozzle in example 10.2b shown in fig. 10.9 is underexpanding).

When the back pressure of a nozzle is above the design value the nozzle is said to *overexpand*. In overexpansion in a convergent nozzle the exit pressure is greater than the critical pressure and the effect is to reduce the mass flow through the nozzle. In overexpansion in a convergent-divergent nozzle there is always an expansion followed by a re-compression. The two types of nozzle can be considered separately.

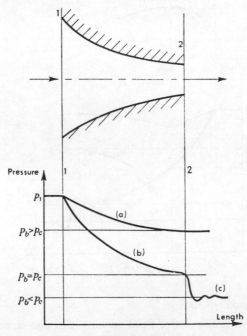

Fig. 10.10

(i) *Convergent nozzle*

The pressure variations of a fluid flowing through a convergent nozzle are shown in fig. 10.10. Assuming that the design back pressure is the critical pressure, p_c, then when the back pressure is above this value the nozzle is overexpanding as shown by line (a), and the mass flow is some value below the maximum. When the back pressure is equal to the critical pressure the expansion follows the line (b), the nozzle is choked, and the mass flow is a maximum. When the back pressure is below the critical pressure the expansion in the nozzle still follows the line (b), but there is an additional expansion from p_c down to the back-pressure, p_b, outside the nozzle. It can be seen from fig. 10.10 that in the expansion outside the nozzle the pressure oscillates violently and in fact a shock wave is formed. In this latter case the nozzle is underexpanding.

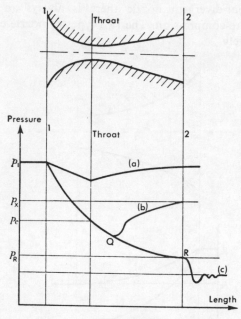

Fig. 10.11

(ii) *Convergent-divergent nozzle*

The pressure variations of a fluid flowing through a convergent-divergent nozzle are shown in fig. 10.11. When the mass flow through the nozzle is very low, the pressure at the throat of the nozzle is well above the critical pressure and therefore the divergent portion acts as a diffuser, as shown by line (a) in fig. 10.11. The nozzle is then acting as a venturimeter (see Section 10.11).

When the back pressure is above the design value at some value, $p_\mathbf{x}$, as shown in fig. 10.11, then the fluid expands from p_1 down to point Q and is then recompressed along line (b). Whenever a supersonic stream is decelerated a shock wave results, and hence the recompression process of line (b) is an irreversible compression through a shock wave. Both (a) and (b) are cases of overexpansion.

When the back pressure is below the design value, $p_\mathbf{R}$, then there is an expansion outside the nozzle as shown by line (c). The nozzle is then underexpanding and the expansion outside the nozzle consists of a series of irreversible compressions through shock waves, alternated with irreversible expansions, until the back pressure is reached.

10.5 Nozzle efficiency

Due to friction between the fluid and the walls of the nozzle, and to friction within the fluid itself, the expansion process is irreversible, although still approximately adiabatic. In nozzle design it is usual to base all calculations on isentropic flow and then to make an allowance for friction by using a coefficient or an efficiency. Typical

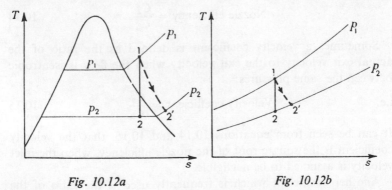

Fig. 10.12a   *Fig. 10.12b*

expansions between p_1 and p_2 in a nozzle are shown on a T-s diagram in figs. 10.12a and 10.12b for a vapour and for a perfect gas, respectively. The line 1–2 on each diagram represents the ideal isentropic expansion, and the line 1–2′ represents the actual irreversible adiabatic expansion.

The *nozzle efficiency* is defined by the ratio of the actual enthalpy drop to the isentropic enthalpy drop between the same pressures,

i.e.   Nozzle efficiency $= \dfrac{h_1 - h_{2'}}{h_1 - h_2}$ 10.11

For a perfect gas this equation reduces to

$$\text{Nozzle efficiency} = \frac{c_p(T_1 - T_{2'})}{c_p(T_1 - T_2)} = \frac{T_1 - T_{2'}}{T_1 - T_2} \qquad 10.12$$

If the actual velocity at exit from the nozzle is $C_{2'}$, and the velocity at exit when the flow is isentropic is C_2, then using the steady flow energy equation, in each case we have

$$h_1 + \frac{C_1^2}{2} = h_2 + \frac{C_2^2}{2} \quad \text{or} \quad h_1 - h_2 = \frac{C_2^2 - C_1^2}{2}$$

and $$h_1 + \frac{C_1^2}{2} = h_{2'} + \frac{C_{2'}^2}{2} \quad \text{or} \quad h_1 - h_{2'} = \frac{C_{2'}^2 - C_1^2}{2}$$

Therefore substituting in equation 10.12,

$$\text{Nozzle efficiency} = \frac{C_{2'}^2 - C_1^2}{C_2^2 - C_1^2} \qquad 10.13$$

When the inlet velocity, C_1, is negligibly small then,

$$\text{Nozzle efficiency} = \frac{C_{2'}^2}{C_2^2} \qquad 10.14$$

Sometimes a velocity coefficient is defined as the ratio of the actual exit velocity to the exit velocity when the flow is isentropic between the same pressures,

i.e.   Velocity coefficient $= \dfrac{C_{2'}}{C_2}$ 10.15

It can be seen from equations 10.14 and 10.15, that the velocity coefficient is the square root of the nozzle efficiency, when the inlet velocity is assumed to be negligible.

Another coefficient which is frequently used is the ratio of the actual mass flow through the nozzle, $\dot{m}'$, to the mass flow which would be passed if the flow were isentropic, $\dot{m}$; this is called the *coefficient of discharge*,

i.e.   Coefficient of discharge $= \dfrac{\dot{m}'}{\dot{m}}$ 10.16

If the angle of divergence of a convergent-divergent nozzle is made too large, then break-away of the fluid from the duct walls is liable to occur, with consequent increased friction losses. The included angle of a divergent duct is usually kept below about 20°. It follows that for a given pressure ratio across a convergent-divergent nozzle the divergent portion must be long compared to the convergent portion. Now because the divergent portion of the nozzle is comparatively long, and since a diverging flow is more susceptible to losses, and the velocities in this portion are higher, it follows that the bulk of the friction losses occur in the divergent portion. In fact it is sometimes assumed that all the friction losses occur after the throat of the nozzle. This latter assumption implies that the coefficient of discharge is unity, since any friction after the throat cannot affect the mass flow through a nozzle which is choked.

Nozzles in practice are used with a variety of shapes and cross-sections. The cross-section can be either circular or rectangular, and the axis of the nozzle can be straight or curved. A typical circular section, straight axis nozzle is shown in fig. 10.13a, and a series of typical plate-type, curved-axis steam nozzles is shown in fig. 10.13b.

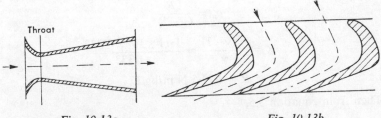

Fig. 10.13a   Fig. 10.13b

Example 10.3

Gases expand in a propulsion nozzle from 3·5 bar and 425°C down to a back pressure of 0·97 bar, at the rate of 18 kg/s. Taking a co-efficient of discharge of 0·99 and a nozzle efficiency of 0·94, calculate the required throat and exit areas of the nozzle. For the gases take $\gamma = 1·333$ and $c_p = 1·11$ kJ/kg K. Assume that the inlet velocity is negligible.

The critical pressure is given by equation 10.7,

$$\frac{p_c}{p_1} = \left(\frac{2}{\gamma+1}\right)^{\gamma/(\gamma-1)} = \left(\frac{2}{2·333}\right)^{1·333/0·333} = 0·541$$

i.e. $p_c = 0.541 \times 3.5$ bar

i.e.   Critical pressure $= 1.894$ bar

The nozzle is therefore choking and a convergent-divergent nozzle is required. The mass flow is determined by the throat of the nozzle. Using equation 10.8,

$$\frac{T_c}{T_1} = \frac{2}{\gamma+1} = \frac{1}{1.1665}$$

i.e. $T_c = \frac{425+273}{1.1665} = 598$ K

Then from equation 10.4,

$$C_c = \sqrt{\{2(h_1 - h_c)\}} = \sqrt{2\{c_p(T_1 - T_c)\}}$$

i.e. $C_c = 44.72\sqrt{\{1.11(698 - 598)\}} = 471$ m/s

(Note that C_c can also be found using equation 10.9,

$$C_c = \sqrt{(\gamma R T_c)}$$

The specific gas constant, R, for the gases can be found from equation 3.22,

$$c_p = \frac{\gamma R}{(\gamma - 1)}$$

$$\therefore \ R = \frac{c_p(\gamma - 1)}{\gamma} = \frac{1.11 \times 10^3 \times 0.333}{1.333}$$

i.e. $R = 277.5$ N m/kg K

Then from equation 3.5,

$$v_c = \frac{RT_c}{p_c} = \frac{277.5 \times 598}{10^5 \times 1.894} = 0.875 \text{ m}^3/\text{kg}$$

Now the mass flow for isentropic flow, $\dot{m}$, is given from equation 10.17 as,

$$0.99 = \frac{18}{\dot{m}} \quad \text{i.e.} \quad \dot{m} = \frac{18}{0.99} = 18.18 \text{ kg/s}$$

Then using equation 2.9,

$$A_c = \frac{\dot{m}v_c}{C_c} = \frac{18.18 \times 0.875}{471} = 0.0338 \text{ m}^2$$

i.e.   Throat area $= 0.0338$ m²

For an isentropic expansion from the inlet conditions down to the back pressure, the temperature at exit is T_2, given by equation 4.21,

$$\frac{T_1}{T_2} = \left(\frac{p_1}{p_2}\right)^{(\gamma-1)/\gamma} \quad \text{i.e.} \quad \frac{698}{T_2} = \left(\frac{3\cdot5}{0\cdot97}\right)^{0\cdot333/1\cdot333} = 1\cdot379$$

$$\therefore \quad T_2 = \frac{698}{1\cdot379} = 506 \text{ K}$$

The expansion is shown on a T-s diagram in fig. 10.14, line 1–c–2 representing the isentropic expansion, and line 1–2′ representing the actual expansion.

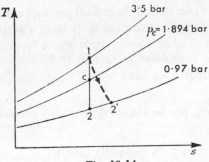

Fig. 10.14

From equation 10.13,

$$\text{Nozzle efficiency} = 0\cdot94 = \frac{T_1 - T_{2'}}{T_1 - T_2} = \frac{698 - T_{2'}}{698 - 506}$$

i.e. $T_{2'} = 698 - 0\cdot94(698 - 506) = 698 - 180\cdot5 = 517\cdot5$ K

Then from equation 3.5,

$$v_{2'} = \frac{RT_{2'}}{p_2} = \frac{277\cdot5 \times 517\cdot5}{10^5 \times 0\cdot97} = 1\cdot48 \text{ m}^3/\text{kg}$$

Also from equation 10.4,

$$C_{2'} = \sqrt{\{2(h_1 - h_{2'})\}} = \sqrt{\{2c_p(T_1 - T_{2'})\}}$$
$$= 44\cdot72\sqrt{\{1\cdot11(698 - 517\cdot5)\}} = 633 \text{ m/s}$$

Then substituting in equation 2.9,

$$A_2 = \frac{\dot{m}'v_{2'}}{C_{2'}} = \frac{18 \times 1\cdot48}{633} = 0\cdot0422 \text{ m}^2$$

It should be noted that in example 10.3, the critical pressure, $p_c = 1 \cdot 894$ bar, is the pressure at the throat of the nozzle when the flow is isentropic. The actual conditions at the nozzle throat are unknown, since no information is given about the proportion of the friction losses which occurs in the convergent portion. In order to allow for the friction present in the convergent portion the coefficient of discharge has been used. If the flow to the throat is isentropic the coefficient of discharge is unity.

10.6 The steam nozzle

The properties of steam can be obtained from tables or from the h-s chart, but in order to find the critical pressure ratio, and hence the critical velocity and the maximum mass flow, a graphical method is necessary for an exact solution. Approximate formulae can be used, and these will be considered in Section 10.7.

From equation 10.5 we have

$$\text{Area per unit mass flow} = \frac{v}{\sqrt{\{2(h_1 - h)\}}}$$

(when the inlet velocity is negligibly small).

In order to find the throat area for a given mass flow it is necessary to find the specific volume, v, and the enthalpy, h, for a series of pressures at various cross-sections of the nozzle, thus finding the variation of area through the nozzle. This is best illustrated by an example.

Example 10.4

Estimate graphically the critical pressure and the throat area per kg/s mass flow of a convergent-divergent nozzle expanding steam from 10 bar, dry saturated, down to atmospheric pressure of 1 bar. Assume that the inlet velocity is negligible and that the expansion is isentropic.

The procedure is to choose a series of pressures, say 9 bar, 8 bar, 7 bar, 6 bar, 5 bar, 4 bar, and 3 bar, and to calculate the cross-sectional area, A, at each of the pressures chosen.

From tables at 10 bar, dry saturated, we have

$$h_1 = h_g = 2778 \text{ kJ/kg} \quad \text{and} \quad s_1 = s_g = 6 \cdot 586 \text{ kJ/kg K}$$

Table 10.1

$p/[\text{bar}]$	9	8	7	6	5	4	3
x	0·992	0·983	0·974	0·964	0·952	0·939	0·924
$h/[\text{kJ/kg}]$	2758	2736	2711	2682	2649	2609	2560
$v/[\text{m}^3/\text{kg}]$	0·213	0·236	0·266	0·304	0·357	0·434	0·560
$(h_1-h)/[\text{kJ/kg}]$	20	42	67	96	129	169	218
$\sqrt{(h_1-h)}$	4·47	6·48	8·19	9·80	11·36	13·00	14·77
$A/\dot{m}/[\text{mm}^2\text{ s/kg}]$ $= \dfrac{v \times 10^6}{44·72\sqrt{(h_1-h)}}$	1065	802·5	726·3	694·2	704	745	847·2

The calculations are tabulated in table 10.1, and a specimen calculation for a section where the pressure is 7 bar is as follows:

At 7 bar and $s = 6·586$ kJ/kg K, the dryness fraction can be found from equation 5.11,

$$x = \frac{s-s_f}{s_{fg}} = \frac{6·586-1·992}{4·717} = 0·974$$

Then the specific volume, v, is given by

$$v = xv_g = 0·974 \times 0·2728 = 0·266 \text{ m}^3/\text{kg}$$

Also the enthalpy, h, is given by equation 3.2,

$$h = h_f + xh_{fg} = 697 + 0·974 \times 2067 = 2711 \text{ kJ/kg}$$

Hence from equation 10.5,

$$A/\dot{m} = \frac{v}{\sqrt{\{2(h_1-h)\}}} = \frac{0·266 \times 10^6}{44·72\sqrt{(2778-2711)}} = 726·3$$

i.e.   Area per kg/s mass flow $= 726·3$ mm^2

When the areas at the various sections chosen have been found in the same way as the above, and tabulated in table 10.1, the variation of area with pressure through the nozzle can then be plotted as shown in fig. 10.15.

From the graph, fig. 10.15, the critical pressure is 5·8 bar and the area per kg/s mass flow is 689 mm^2,

i.e.   Throat area per kg/s mass flow $= 689$ mm^2

For greater accuracy it would be necessary to choose additional points between 7 bar and 5 bar. The procedure can be speeded up, at the expense of some accuracy, by taking the enthalpy drops and the dryness fraction from the h-s chart.

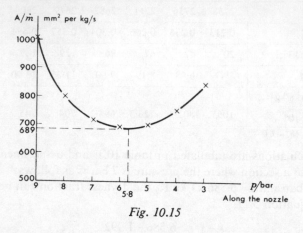

Fig. 10.15

10.7 Approximations for the steam nozzle

The method used in example 10.4 is rather long, and a shorter way of solving the problem can be found by making suitable approximations. It is found that it is a good approximation to assume that the steam follows an isentropic law $pv^k = $ constant, where k is an isentropic index for steam. (Although k is an isentropic index it is *not* a ratio of specific heats as is γ.) For steam initially dry saturated, $k = 1\cdot135$; for steam initially superheated, $k = 1\cdot3$.

Then equation 10.7 can be written as,

$$\frac{p_c}{p_1} = \left(\frac{2}{k+1}\right)^{k/(k-1)} \qquad 10.17$$

Therefore when the steam entering a nozzle is dry saturated,

$$\frac{p_c}{p_1} = \left(\frac{2}{1\cdot135+1}\right)^{1\cdot135/0\cdot135} = 0\cdot577$$

and when the steam entering a nozzle is superheated,

$$\frac{p_c}{p_1} = \left(\frac{2}{1\cdot3+1}\right)^{1\cdot3/0\cdot1} = 0\cdot546$$

As mentioned earlier, it is not possible to approximate a relationship between temperatures, especially for wet steam, since there is no simple relationship between p and T for a vapour. The temperature at the throat (i.e. the critical temperature) can be found from tables at p_c, and $s_c = s_1$. The critical velocity can be found from equation 10.4, as before,

$$C_c = \sqrt{\{2(h_1 - h_c)\}}$$

(where h_c is read from tables or the h-s chart, at p_c and $s_c = s_1$).

The case of steam initially superheated expanding into the wet region requires special treatment, and will be considered in Section 10.8.

Example 10.5

Re-calculate example 10.4 using the approximation $pv^k = $ constant, taking $k = 1.135$ for wet steam.

Using equation 10.18,

$$\frac{p_c}{p_1} = \left(\frac{2}{k+1}\right)^{k/(k-1)} = \left(\frac{2}{1.135+1}\right)^{1.135/0.135} = 0.577$$

i.e.   Critical pressure $= 0.577 \times 10 = 5.77$ bar

From the h-s chart at 5.77 bar and on a vertical line below point 1 at 10 bar, dry saturated, we have

$$h_c = 2675 \text{ kJ/kg} \quad \text{and} \quad x_c = 0.962$$

$$\therefore \ v_c = 0.962 \times (v_g \text{ at } 5.77 \text{ bar})$$

The specific volume must be found from tables, since no values of specific volume are given on the chart.

Interpolating,

$$v_g \text{ at } 5.77 \text{ bar} = 0.3427 - \left(\frac{5.77 - 5.5}{6.0 - 5.5}\right)(0.3427 - 0.3156)$$

i.e. $v_g = 0.3427 - 0.0146 = 0.3281 \text{ m}^3/\text{kg}$

$$\therefore \ v = 0.962 \times 0.3281 = 0.316 \text{ m}^3/\text{kg}$$

From equation 10.4,

$$C_c = \sqrt{\{2(h_1 - h_c)\}} = 44.72\sqrt{(2778 - 2675)} = 454 \text{ m/s}$$

Then using equation 2.9,

$$A/\dot{m} = \frac{v}{C} = \frac{0\cdot316\times10^6}{454} = 696 \text{ mm}^2$$

i.e.   Throat area per kg/s mass flow $= 696 \text{ mm}^2$

It is possible to express the enthalpy drop and the velocity at any section in terms of the index k, when the approximation $pv^k = \text{constant}$ is made. The expressions can be derived as follows:

For any small length of the duct, where the pressure changes by $\mathrm{d}p$, and the velocity changes by $\mathrm{d}C$, then, neglecting friction,

$$\text{Force to accelerate the fluid} = -A\,\mathrm{d}p$$

Also this force is equal to the rate of change of momentum of the fluid.

$$\text{Rate of change of momentum} = (\text{mass flow} \times \text{change of velocity})$$
$$= \dot{m}\,\mathrm{d}C$$

Therefore we have

$$-A\,\mathrm{d}p = \dot{m}\,\mathrm{d}C$$

Using equation 2.9, $\dot{m} = AC/v$, then,

$$-A\,\mathrm{d}p = \frac{AC\,\mathrm{d}C}{v} \quad \text{or} \quad -v\,\mathrm{d}p = C\,\mathrm{d}C$$

Assuming that

$$pv^k = \text{constant} \quad \text{or} \quad v = \frac{\text{constant}}{p^{1/k}}$$

then, $-(\text{constant})\times\dfrac{\mathrm{d}p}{p^{1/k}} = C\,\mathrm{d}C$

Integrating,

$$-(\text{constant})\times\int_{p_1}^{p_c}\frac{\mathrm{d}p}{p^{1/k}} = \int_{C_1}^{C_c} C\,\mathrm{d}C$$

i.e. $-(\text{constant})\times\left(\dfrac{p_c^{(k-1)/k}-p_1^{(k-1)/k}}{(k-1)/k}\right) = \dfrac{C_c^2-C_1^2}{2}$

Now the constant in the above is equal to $p_1^{1/k}\times v_1$ or $p_c^{1/k}\times v_c$, hence,

$$\frac{-k}{k-1}\left(p_c^{1/k}\times v_c\times p_c^{(k-1)/k}-p_1^{1/k}\times v_1\times p_1^{(k-1)/k}\right) = \frac{C_c^2-C_1^2}{2}$$

or $$\frac{k}{k-1}(p_1v_1 - p_cv_c) = \frac{C_c^2 - C_1^2}{2}$$

The term on the right-hand side of the equation is the change of kinetic energy, and from the steady flow energy equation we have

$$h_1 - h_c = \frac{C_c^2 - C_1^2}{2}$$

i.e. $$h_1 - h_c = \frac{C_c^2 - C_1^2}{2} = \frac{k}{k-1}(p_1v_1 - p_cv_c) \qquad 10.18$$

(Equation 10.18 can be applied between the inlet and any other section of the nozzle.)

To find the critical velocity in terms of the critical pressure, assuming that the inlet velocity is negligibly small, we have,

$$\frac{C_c^2}{2} = \frac{k}{k-1}(p_1v_1 - p_cv_c) = \frac{kp_cv_c}{k-1}\left(\frac{p_1v_1}{p_cv_c} - 1\right)$$

Since,

$$p_1^{1/k} \times v_1 = p_c^{1/k} \times v_c \quad \text{then} \quad \frac{v_1}{v_c} = \left(\frac{p_c}{p_1}\right)^{1/k}$$

i.e. $$\frac{C_c^2}{2} = \frac{kp_cv_c}{k-1}\left\{\frac{p_1}{p_c}\left(\frac{p_c}{p_1}\right)^{1/k} - 1\right\} = \frac{kp_cv_c}{k-1}\left\{\left(\frac{p_1}{p_c}\right)^{1-1/k} - 1\right\}$$

Also, from equation 10.17,

$$\frac{p_c}{p_1} = \left(\frac{2}{k+1}\right)^{k/(k-1)} \quad \text{or} \quad \left(\frac{p_c}{p_1}\right)^{(k-1)/k} = \frac{2}{k+1}$$

Substituting we have,

$$\frac{C_c^2}{2} = \frac{kp_cv_c}{k-1}\left\{\left(\frac{k+1}{2}\right) - 1\right\} = \frac{kp_cv_c(k-1)}{(k-1)\times 2} = \frac{kp_cv_c}{2}$$

i.e $$C_c^2 = kp_cv_c$$

i.e. Critical velocity, $$C_c = \sqrt{(kp_cv_c)} \qquad 10.19$$

(Note that equation 10.19 is similar to the equation for the velocity of sound in a perfect gas, given by equation 10.10, $a = \sqrt{(\gamma pv)}$.)

Example 10.6

Steam at 30 bar and specific volume 0·0993 m³/kg enters a convergent-divergent nozzle with negligible inlet velocity, and expands into a space at 4 bar. Calculate the throat and exit areas for a mass flow of 0·2 kg/s. Assume that the steam expands isentropically according to a law $pv^{1·3} = $ constant, and assume that there is no underexpansion or overexpansion. Do not use tables or the h-s chart for this example.

From equation 10.17,

$$\frac{p_c}{p_1} = \left(\frac{2}{k+1}\right)^{k/(k-1)} = \left(\frac{2}{2·3}\right)^{1·3/0·3} = 0·546$$

i.e.
$$p_c = 0·546 \times 30 = 16·38 \text{ bar}$$

Also,
$$\frac{v_c}{v_1} = \left(\frac{p_1}{p_c}\right)^{1/k} = \left(\frac{1}{0·546}\right)^{1/1·3} = 1·591$$

i.e.
$$v_c = 1·591 \times 0·0993 = 0·158 \text{ m}^3/\text{kg}$$

Then using equation 10.19,

$$C_c = \sqrt{(kp_cv_c)} = \sqrt{(1·3 \times 10^5 \times 16·38 \times 0·158)}$$
$$= 580 \text{ m/s}$$

From equation 2.9,

$$A_c = \frac{\dot{m}v_c}{C_c} = \frac{0·2 \times 0·158 \times 10^6}{580} = 54·5 \text{ mm}^2$$

i.e.
$$\text{Throat area} = 54·5 \text{ mm}^2$$

To find the specific volume at exit from the nozzle, we have,

$$\frac{v_2}{v_1} = \left(\frac{p_2}{p_1}\right)^{1/k} = \left(\frac{30}{4}\right)^{1/1·3} = 4·71$$

$$\therefore v_2 = 4·71 \times 0·158 = 0·744 \text{ m}^3/\text{kg}$$

Then using equation 10.18,

$$\frac{C_2^2}{2} = \frac{k}{k-1}(p_1v_1 - p_2v_2) = \frac{1·3 \times 10^5}{0·3}(30 \times 0·158 - 4 \times 0·745)$$

i.e.
$$\frac{C_2^2}{2} = \frac{1·3 \times 10^5 \times 1·768}{0·3}$$

$$\therefore C_2 = \sqrt{\left(\frac{2 \times 1·3 \times 10^5 \times 1·768}{0·3}\right)} = 1236 \text{ m/s}$$

Therefore from equation 2.9,

$$A_2 = \frac{\dot{m}v_2}{C_2} = \frac{0 \cdot 2 \times 0 \cdot 744 \times 10^6}{1236} = 120 \text{ mm}^2$$

i.e.   Exit area = 120 mm²

Example 10.7

Re-calculate example 10.6 using the h-s chart, and taking a coefficient of discharge of 0·98 and a coefficient of velocity of 0·9. Assume that the critical pressure ratio is 0·546 as before.

At state 1 the pressure is $p_1 = 30$ bar and the specific volume is $v_1 = 0 \cdot 0993$ m³/kg. In order to locate state 1 on the chart it is necessary to find the temperature t_1; this can be read from tables.

From superheat tables at 30 bar, and 0·0993 m³/kg, we have $t_1 = 400°C$.

The expansion is shown as a dotted line, 1–2′, on a sketch of the h-s chart in fig. 10.16; the full vertical line, 1–c–2, represents an

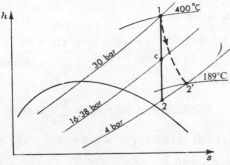

Fig. 10.16

isentropic process between the same pressures. The enthalpy at inlet can be read from the chart, i.e. $h_1 = 3231$ kJ/kg. The specific volume at c, v_c, can be found by interpolating from superheat tables at 16·38 bar and 313°C approximately. However, it is just as accurate in this case to use the relationship,

$$p_1 v_1^k = p_c v_c^k \quad \text{or} \quad \frac{v_c}{v_1} = \left(\frac{p_1}{p_c}\right)^{1/k}$$

i.e. as before,

$$v_c = 0 \cdot 0993 \times \left(\frac{1}{0 \cdot 546}\right)^{1/1 \cdot 3} = 0 \cdot 158 \text{ m}^3/\text{kg}$$

Then using equation 10.5,

Area required per unit mass isentropic flow $= \dfrac{v_c}{\sqrt{\{2(h_1 - h_c)\}}}$

$$= \frac{0 \cdot 158 \times 10^6}{44 \cdot 72 \sqrt{(3231 - 3065)}}$$

$$= 274 \text{ mm}^2 \text{ per kg/s}$$

Now, from equation 10.16,

$$\text{Coefficient of discharge} = \frac{\dot{m}'}{\dot{m}} = 0 \cdot 98$$

i.e.   Isentropic mass flow, $\dot{m} = \dfrac{0 \cdot 2}{0 \cdot 98} = 0 \cdot 204$ kg/s

Hence,   Throat area $= 274 \times 0 \cdot 204 = 55 \cdot 9$ mm^2

From the chart, $h_2 = 2745$ kJ/kg.

i.e. $(h_1 - h_2) = (3231 - 2745) = 486$ kJ/kg

Then, using equation 10.4,

Exit velocity for isentropic flow, $C_2 = 44 \cdot 72 \sqrt{486}$

$$= 986 \text{ m/s}$$

From equation 10.15,

$$\text{Coefficient of velocity} = \frac{C_{2'}}{C_2} = 0 \cdot 9$$

i.e.   Actual exit velocity, $C_{2'} = 0 \cdot 9 \times 986 = 887 \cdot 4$ m/s

Then using equation 10.4,

$$C_{2'} = 887 \cdot 4 = \sqrt{\{2(h_1 - h_{2'})\}}$$

$$\therefore \; h_1 - h_{2'} = \left(\frac{887 \cdot 4}{44 \cdot 72}\right)^2 = 393 \cdot 6 \text{ kJ/kg}$$

i.e. $h_{2'} = h_1 - 393 \cdot 6 = 3231 - 393 \cdot 6 = 2837 \cdot 4$ kJ/kg

State 2′ is now fixed at the point where the horizontal line at $h_{2'} = 2837 \cdot 4$ kJ/kg cuts the pressure line at exit, $p_2 = 4$ bar. From the chart, $t_{2'} = 189°C$, therefore from superheat tables, interpolating,

$$v_{2'} = 0 \cdot 4710 + \left(\frac{189 - 150}{200 - 150}\right)(0 \cdot 5345 - 0 \cdot 4710) = 0 \cdot 521 \text{ m}^3/\text{kg}$$

Then using equation 2.9,

$$A_2 = \frac{\dot{m}'v_{2'}}{C_{2'}} = \frac{0.2 \times 0.521 \times 10^6}{887.4} = 117.5 \text{ mm}^2$$

i.e.   Exit area $= 117.5$ mm^2

When an isentropic flow takes place entirely in the superheat region it is a reasonable approximation to use an isentropic relationship between the pressure and temperature,
i.e. using equation 4.21,

$$\frac{T_1}{T_2} = \left(\frac{p_1}{p_2}\right)^{(k-1)/k}$$

This equation must not be used for wet steam.
In example 10.7, $p_1 = 30$ bar, $p_2 = 4$ bar and $T_1 = 400 + 273 = 673$ K.
Therefore, using equation 4.21,

$$\frac{673}{T_2} = \left(\frac{30}{4}\right)^{0.3/1.3} = 1.593$$

i.e. $T_2 = \frac{673}{1.593} = 422$ K

$$\therefore \quad t_2 = 422 - 273 = 149°\text{C}$$

From the chart, for isentropic flow from $p_1 = 30$ bar, $t_1 = 400°$C to $p_2 = 4$ bar, then t_2 is read off as approximately 147°C. Thus using equation 4.21 to find t_2 gives a reasonable approximation in the case of superheated steam.

10.8 Supersaturation

When a superheated vapour expands isentropically and slowly, condensation within the vapour begins to form when the saturated vapour line is reached. As the expansion continues below this line into the wet region, then condensation proceeds gradually and the dryness fraction of the steam becomes progressively less. This is illustrated on a *T-s* and an *h-s* diagram in figs. 10.17a and 10.17b respectively. Point A represents the point at which condensation within the vapour just begins.

It is found that the expansion through a nozzle takes place so quickly that condensation within the vapour does not occur. The vapour expands as a superheated vapour until some point at which condensation occurs suddenly and irreversibly. The point at which condensation occurs may be within the nozzle or after the vapour leaves the nozzle.

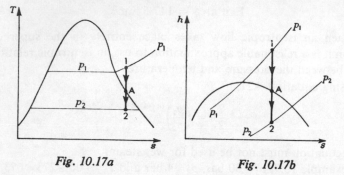

Fig. 10.17a   Fig. 10.17b

Up to the point at which condensation occurs the state of the steam is not one of stable equilibrium, yet it is not one of unstable equilibrium, since a small disturbance will not cause condensation to commence. The steam in this condition is said to be in a *metastable state*; the introduction of a large object (e.g. a measuring instrument) will cause condensation to occur immediately.

Such an expansion is called a *supersaturated expansion*.

Assuming isentropic flow, as before, a supersaturated expansion in a nozzle is represented on a *T-s* and an *h-s* diagram in figs. 10.18a

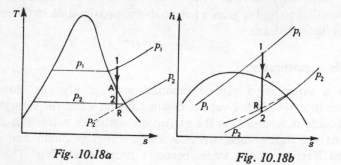

Fig. 10.18a   Fig. 10.18b

and 10.18b respectively. Line 1–2 on both diagrams represents the expansion with equilibrium throughout the expansion. Line 1–R represents supersaturated expansion. In supersaturated expansion the vapour expands as if the saturated vapour line did not exist, so that line 1–R intersects the pressure line p_2 produced from the superheat region (shown chain-dotted). It can be seen from fig. 10.18a that the temperature of the supersaturated vapour at p_2 is t_R, which is less than the saturation temperature t_2, corresponding to p_2. The vapour is said to be *supercooled* and the *degree of supercooling* is given by $(t_2 - t_R)$. Sometimes a *degree of supersaturation* is defined as the ratio of the actual pressure p_2, to the saturation pressure corresponding to the temperature t_R.

It can be seen from fig. 10.18b that the enthalpy drop in supersaturated flow, $(h_1 - h_R)$, is less than the enthalpy drop under equilibrium conditions. Since the velocity at exit, C_2, is given by equation 10.4, $C_2 = 44.72\sqrt{(h_1 - h_2)}$, it follows that the exit velocity for supersaturated flow is less than that for equilibrium flow. Nevertheless, the difference in the enthalpy drop is small, and since the square root of the enthalpy drop is used in equation 10.4, then the effect on the exit velocity is small.

If the approximations for isentropic flow are applied to the equilibrium expansion, then for the process illustrated in figs. 10.18a and 10.18b, the expansion from 1 to A obeys the law $pv^{1.3} = \text{constant}$, and the expansion from A to 2 obeys the law $pv^{1.135} = \text{constant}$. The equilibrium expansion and the supersaturated expansion are shown on a p-v diagram in fig. 10.19, using the same symbols as in fig. 10.18. It can be seen from fig. 10.19 that the specific volume at exit with

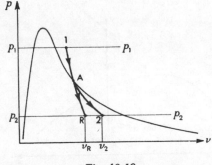

Fig. 10.19

supersaturated flow, v_R, is considerably less than the specific volume at exit with equilibrium flow, v_2. Now the mass flow through a given exit area, A_2, is given by equation 2.9, i.e. for equilibrium flow,

$$\dot{m} = \frac{A_2 C_2}{v_2}$$

and for supersaturated flow,

$$\dot{m}_s = \frac{A_2 C_R}{v_R}$$

It has been pointed out that C_2 and C_R are very nearly equal, therefore, since $v_R < v_2$, it follows that the mass flow with super-saturated flow is greater than the mass flow with equilibrium flow. It was this fact, proved experimentally, that led to the discovery of the phenomenon of supersaturation.

Example 10.8

A convergent-divergent nozzle receives steam at 7 bar and 200°C and expands it isentropically into a space at 3 bar. Neglecting the inlet velocity, calculate the exit area required for a mass flow of 0·1 kg/s

(a) When the flow is in equilibrium throughout,
(b) When the flow is supersaturated with $pv^{1\cdot3} = $ constant.

Calculate also for part (b) the degree of supercooling and the degree of supersaturation.

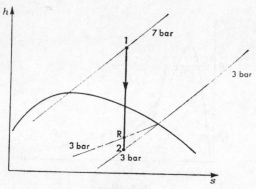

Fig. 10.20

A sketch of both processes (a) and (b) is shown on an h-s chart in fig. 10.20.

Part (a) is most conveniently solved by using the h-s chart.

From the chart at 7 bar, 200°C, $h_1 = 2846$ kJ/kg.

Also, $h_2 = 2682$ kJ/kg   and $x_{2'} = 0.98$

$\therefore v_2 = x_2 v_{g2} = 0.98 \times 0.6057 = 0.594$ m³/kg

From equation 10.4,

$$C_2 = \sqrt{\{2(h_1 - h_2)\}} = 44.72\sqrt{(2846 - 2682)} = 573 \text{ m/s}$$

Then using equation 2.9,

$$A_2 = \frac{mv_2}{C_2} = \frac{0.1 \times 0.594 \times 10^6}{573} = 103.7 \text{ mm}^2$$

i.e.   Exit area $= 103.7$ mm²

(b) Although the process is represented on an *h-s* diagram as 1–R in fig. 10.20, nevertheless the *h-s* chart cannot be used to find h_R, since the chain-dotted line representing 3 bar, produced from the superheat region, cannot be located easily on the chart. The enthalpy drop, $(h_1 - h_R)$, and hence the velocity, C_R, can be found using equation 10.18,

i.e. $$\frac{C_R^2}{2} = \frac{k}{k-1}(p_1 v_1 - p_2 v_R)$$

From tables at 7 bar and 200°C, $v_1 = 0.3001$ m³/kg.

Then, since $p_1 v_1^k = p_2 v_R^k$, we have,

$$\frac{v_R}{v_1} = \left(\frac{p_1}{p_2}\right)^{1/k} = \left(\frac{7}{3}\right)^{1/1.3} = 1.918$$

i.e. $v_R = 1.918 \times 0.3001 = 0.576$ m³/kg

Then substituting,

$$\frac{C_R^2}{2} = \frac{1.3 \times 10^5}{0.3}(7 \times 0.3001 - 3 \times 0.576) = \frac{1.3 \times 10^5 \times 0.3727}{0.3}$$

i.e. $$C_R = \sqrt{\left(\frac{2 \times 1.3 \times 10^5 \times 0.3727}{0.3}\right)} = 568 \text{ m/s}$$

Then using equation 2.9,

$$A_2 = \frac{\dot{m}v_R}{C_R} = \frac{0.1 \times 0.576 \times 10^6}{568} = 101.5 \text{ mm}^2$$

To find the degree of supercooling it is necessary to find T_R using equation 4.21.

i.e.
$$\frac{T_1}{T_R} = \left(\frac{p_1}{p_2}\right)^{(k-1)/k} = \left(\frac{7}{3}\right)^{0.3/1.3} = 1.216$$

$$\therefore T_R = \frac{200+273}{1.216} = 389 \text{ K}$$

i.e.
$$t_R = 389 - 273 = 116°C$$

The saturation temperature, t_2, corresponding to $p_2 = 3$ bar, is 133.5°C.

Therefore,

$$\text{Degree of supercooling} = (133.5 - 116) = 17.5 \text{ K}$$

The saturation pressure corresponding to $t_R = 116°C$ is found by interpolating as,

$$p_{sat} \text{ (at } t_R) = 1.7 + \left(\frac{116 - 115.2}{116.9 - 115.2}\right) \times 0.1 = 1.75 \text{ bar}$$

Then,
$$\text{Degree of supersaturation} = \frac{3}{1.75} = 1.713$$

When there is friction in the nozzle the flow is still supersaturated. This is illustrated in fig. 10.21, in which 1–2 and 1–2' represent the

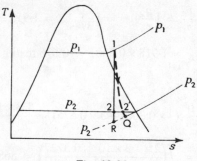

Fig. 10.21

ideal equilibrium process and the actual equilibrium process respectively, and 1–R and 1–Q represent the ideal supersaturated process and the actual supersaturated process respectively.

When supersaturation continues to the exit of the nozzle, then the nozzle efficiency can be taken from equation 10.11 as,

$$\text{Nozzle efficiency} = \frac{h_1 - h_Q}{h_1 - h_2}$$

Similarly the coefficient of velocity is given by equation 10.12 as,

$$\text{Coefficient of velocity} = \frac{C_Q}{C_2}$$

Applying a nozzle efficiency or a coefficient of velocity to a supersaturated flow with friction it is possible to evaluate C_Q, after having calculated $(h_1 - h_2)$. In order to calculate the exit area required for a given mass flow it is necessary to calculate v_Q, and this presents a difficulty. The h-s chart cannot be used since the flow is supersaturated, and no simple relationship between p, v, and T is known between states 1 and Q, since the process is irreversible. One approximation used is to assume a mean value of c_p for supersaturated steam and then to write,

$$h_1 - h_Q = c_p(t_1 - t_Q)$$

Then v_Q is given approximately by,

$$\frac{p_Q v_Q}{T_Q} = \frac{p_1 v_1}{T_1} \quad \text{or} \quad v_Q = \frac{p_1}{p_2} \times \frac{T_Q}{T_1} \times v_1$$

Perhaps a more accurate approximation is to use an approximate formula for the specific volume of superheated and supersaturated steam, given by,

$$\frac{p_Q}{\text{bar}} \times \frac{v_Q}{\text{m}^3/\text{kg}} \times 10^2 = \frac{0.3}{1.3}\left(\frac{h}{\text{kJ/kg}} - 1943\right) \qquad 10.20$$

Example 10.9

Recalculate the exit area in example 10.8b, assuming a nozzle efficiency of 0.92, and a mean value of $c_p = 1.925$ kJ/kg K for supersaturated steam. Check the answer obtained by assuming that equation 10.20 may be used.

From example 10.8a the enthalpy drop in equilibrium flow is $(2846 - 2682) = 164$ kJ/kg.

Then using equation 10.11,

$$\text{Nozzle efficiency} = \frac{h_1 - h_Q}{h_1 - h_2} = 0.92$$

$$\therefore\ h_1 - h_Q = 0.92 \times 164 = 150.9 \text{ kJ/kg}$$

Therefore from equation 10.4,

$$C_Q = \sqrt{\{2(h_1 - h_Q)\}} = 44.72\sqrt{150.9} = 549 \text{ m/s}$$

Also, $h_1 - h_Q = c_p(t_1 - t_Q)$, hence,

$$t_Q = t_1 - \frac{150.9}{1.925} = 200 - 78.3 = 121.7°C$$

Then,
$$v_Q = \frac{7}{3} \times \frac{394.7}{473} \times 0.3001 = 0.584 \text{ m}^3/\text{kg}$$

(where $T_Q = 121.7 + 273 = 394.7$ K, and $T_1 = 200 + 273 = 473$ K).

Hence from equation 2.9,

$$A_2 = \frac{\dot{m}v_Q}{C_Q} = \frac{0.1 \times 0.584 \times 10^6}{549} = 106.4 \text{ mm}^2$$

i.e.
$$\text{Exit area} = 106.4 \text{ mm}^2$$

Using equation 10.20 for the specific volume, we have

$$v_Q = \frac{0.3(h_Q - 1943)}{10^2 \times p_Q \times 1.3} = \frac{0.3(2846 - 150.9 - 1943)}{10^2 \times 3 \times 1.3} = 0.578 \text{ m}^3/\text{kg}$$

(where $(h_1 - h_Q) = 150.9$ kJ/kg, therefore $h_Q = h_1 - 150.9 = 2846 - 150.9$).

Then,
$$A_2 = \frac{0.1 \times 0.578 \times 10^6}{549} = 105.2 \text{ mm}^2$$

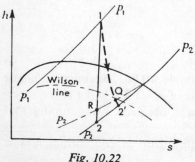

Fig. 10.22

It has been found experimentally that supersaturation does not continue indefinitely, and the vapour always reverts to its equilibrium condition at a dryness fraction between 0·97 and 0·96. A line can be plotted which gives the approximate limit of supersaturated flow; this is known as the *Wilson line,* shown in fig. 10.22. At point Q the vapour will revert to the equilibrium state.

Condensation takes place irreversibly although adiabatically, and since the condensation process is irreversible there is an increase in the entropy of the vapour. In a convergent-divergent nozzle supersaturation usually ends in the divergent portion. It is found that up to the throat the flow remains supersaturated, and hence a coefficient of discharge can always be used based on a law $pv^k = $ constant up to the throat. When supersaturated flow ends with irreversible condensation within the nozzle it is usually possible to use a nozzle efficiency which will take into account the effects of friction and of supersaturation.

10.9 Rocket propulsion

One very important use for the nozzle is as a means of propulsion. Since the fluid flowing through the nozzle is accelerated relative to the nozzle, then by Newton's third law it follows that the fluid exerts a thrust on the nozzle in the opposite direction to the fluid flow. In the jet aeroplane and the ram-jet the atmospheric air is drawn in, compressed, heated, and allowed to expand through a nozzle, leaving the aircraft at high velocity; the rate of change of momentum of the air backwards relative to the aircraft gives a reactive forward thrust to the aircraft. In order to achieve jet-propelled flight in space, where there is no atmosphere to be drawn into the vehicle, it is necessary that the fuel plus its oxidant should be carried. This is known as rocket propulsion.

A rocket operating on a chemical fuel consists of tanks containing the chemical propellant, and a *rocket motor* (or a rocket engine) which consists of a combustion chamber and a convergent-divergent nozzle. Some way of introducing the propellant from the tanks to the combustion chamber is also necessary, and this can be done using a pump or by having an additional tank of compressed nitrogen. When a pump is used it can be driven by a small turbine using the propellant as fuel. A simple line diagram of a rocket is shown in fig. 10.23.

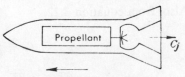

Fig. 10.23

Let the velocity of the jet leaving the rocket be C_j relative to the rocket, and let the mass flow of fluid be $\dot{m}$. Then, for a correctly designed nozzle,

Thrust = Rate of change of momentum of fluid

= (mass flow) × (change of velocity)

Initially the fluid is at rest relative to the nozzle, and finally the fluid velocity is C_j relative to the rocket, hence,

$$\text{Thrust} = \dot{m}C_j \qquad 10.21$$

The thrust per unit mass flow is of importance in rocket propulsion and is known as the *specific impulse*,

i.e. $$\text{Specific impulse} = C_j \qquad 10.22$$

The units of specific impulse are N per kg/s, or since 1 N is 1 kg m/s², the units are m/s.

Let the combustion pressure and temperature be p_1 and T_1 and let the exit pressure and temperature be p_2 and T_2, then applying the flow equation we have

$$\frac{C_j^2}{2} = (h_1 - h_2) = c_p(T_1 - T_2) \quad \text{for a perfect gas}$$

i.e. $$\frac{C_j^2}{2} = c_p T_1 \left(1 - \frac{T_2}{T_1}\right)$$

From equation 4.21, assuming isentropic flow in the nozzle,

$$\frac{T_2}{T_1} = \left(\frac{p_2}{p_1}\right)^{(\gamma-1)/\gamma}$$

Therefore, $$\frac{C_j^2}{2} = c_p T_1 \left\{1 - \left(\frac{p_2}{p_1}\right)^{(\gamma-1)/\gamma}\right\}$$

Also, from equation 3.22,

$$c_p = \frac{\gamma R}{(\gamma - 1)}$$

$$\therefore \frac{C_j^2}{2} = \frac{\gamma R T_1}{(\gamma - 1)} \left\{1 - \left(\frac{p_2}{p_1}\right)^{(\gamma-1)/\gamma}\right\}$$

From equation 3.9, $R = R_0/M$, therefore, the specific impulse is given by,

$$C_1 = \sqrt{\left[\frac{2\gamma R_0 T_1}{(\gamma-1)M}\left\{1 - \left(\frac{p_2}{p_1}\right)^{(\gamma-1)/\gamma}\right\}\right]} \qquad 10.23$$

The mass flow, $\dot{m}$, through the nozzle is fixed by the throat conditions, since the flow is choked. From equation 2.9, for the throat, we have

$$\dot{m} = \frac{A_c C_c}{v_c}$$

From equation 10.9, $C_c = \sqrt{(\gamma R T_c)}$, and from equation 3.5, $v = RT/p$, hence,

$$\dot{m} = \frac{p_c A_c \sqrt{(\gamma R T_c)}}{R T_c} = p_c A_c \sqrt{\left(\frac{\gamma}{R T_c}\right)}$$

Now, from equations 10.7 and 10.8 we have,

$$p_c = p_1\left(\frac{2}{\gamma+1}\right)^{\gamma/(\gamma-1)} \quad \text{and} \quad T_c = T_1\left(\frac{2}{\gamma+1}\right)$$

Hence substituting,

$$\dot{m} = p_1 A_c\left(\frac{2}{\gamma+1}\right)^{\gamma/(\gamma-1)} \times \sqrt{\left\{\frac{\gamma}{R T_1}\left(\frac{\gamma+1}{2}\right)\right\}}$$

i.e.

$$\dot{m} = p_1 A_c \sqrt{\left\{\frac{\gamma}{R T_1}\left(\frac{2}{\gamma+1}\right)^{(\gamma+1)/(\gamma-1)}\right\}} \qquad 10.24$$

Now from equation 10.21,

$$\text{Thrust} = \dot{m} C_1$$

Hence substituting for $\dot{m}$ and C_1 from equations 10.24 and 10.23, we have,

$$\text{Thrust} = p_1 A_c \sqrt{\left\{\frac{\gamma}{R T_1}\left(\frac{2}{\gamma+1}\right)^{(\gamma+1)/(\gamma-1)}\right\}}$$

$$\times \sqrt{\left[\frac{2\gamma R_0 T_1}{(\gamma-1)M}\left\{1 - \left(\frac{p_2}{p_1}\right)^{(\gamma-1)/\gamma}\right\}\right]}$$

i.e.

$$\text{Thrust} = p_1 A_c \sqrt{\left[\left(\frac{2\gamma^2}{\gamma-1}\right)\left(\frac{2}{\gamma+1}\right)^{(\gamma+1)/(\gamma-1)}\right.}$$

$$\left. \times \left\{1 - \left(\frac{p_2}{p_1}\right)^{(\gamma-1)/\gamma}\right\}\right] \qquad 10.25$$

It should be noted that equations 10.23, 10.24, and 10.25 are for ideal isentropic flow.

For a given value of γ for the gases flowing through the nozzle, and for a fixed nozzle throat area, A_c, and a fixed atmospheric pressure p_2, it can be seen from equation 10.25 that the thrust depends only on the combustion pressure p_1. However, from equation 10.23 for the specific impulse, it can be seen that the performance of the rocket depends mainly on the combustion temperature T_1. It should also be noted from equation 10.22 that gases with a low value of molar mass, M, will give a better performance than gases with a high molar mass.

When atmospheric conditions are such that the nozzle under-expands (i.e. when the pressure outside the nozzle is below the design value), then the gases expand outside the nozzle giving an additional *pressure thrust* given by,

$$\text{Pressure thrust} = (p_2 - p_a)A_2$$

(where p_2 and p_a are the exit pressure and the atmospheric pressure respectively, and A_2 is the nozzle exit area).

A more detailed discussion of pressure thrust with particular reference to jet propulsion is given in section 12.5.

The combustion temperature is usually high (of the order of 4000°C), and a means of cooling the walls of the combustion chamber and nozzle is required. This is achieved by allowing the liquid propellant to circulate round the walls of the nozzle and combustion space before it is injected into the chamber.

The fuel, and the oxygen required for its combustion, can be combined in one liquid (such as hydrogen peroxide), and this is known as a *monopropellant* system. More common is the *bi-propellant* system in which a fuel such as kerosene is burnt with a supply of oxygen such as liquid oxygen. Hydrogen peroxide and nitric acid are also used as sources of oxygen in bi-propellant systems.

There are certain fuels which can be made to release energy without using oxygen. For example, hydrazine can be made to decompose into nitrogen and hydrogen, and certain fuels can be made to react with fluorine or fluorine compounds.

Solid propellants are also used in a combination of an oxidiser and a binder (e.g. ammonia nitrate and polyester; nitroglycerine and nitrocellulose; magnesium metal components and an oxidant).

To increase the thrust of a rocket more than one rocket motor can be used. For space exploration or for earth satellites more than one rocket stage is necessary; each stage may have a different propellant system to suit the different thrust requirements.

The mass of the rocket is clearly of great importance, and the mass of the propellant tanks, the turbo-pumps and the nozzle, must be kept to a minimum. The *mass ratio* is defined as the ratio of the initial mass at take-off to the final mass at the end of the final rocket stage.

Other means of rocket propulsion are at present being investigated. These include *ion propulsion* in which a plasma jet of ionized gases is produced either by electric arc heating or by the use of strong magnetic fields, or by an electro-static field as in a linear accelerator. The working fluid may be a gas such as argon or hydrogen, and the electrical energy required could be provided by a nuclear reactor. Such rockets have a high specific impulse but a much lower thrust than those using chemical fuels.

The sun can be used as a source of power for rockets in space, use being made of solar heating devices to produce the high temperature of a gas such as hydrogen which then expands through a nozzle.

At present only chemical propellants are suitable for take-off, or for use as decelerating stages for landing on a planet.

It should be remembered that the velocity required in order to 'escape' from the earth's gravitational field is approximately 11·2 km/s; the escape velocity from the moon is about 2·4 km/s. Rockets developing a thrust of over ninety million newtons are now in use for spaceships capable of reaching the moon and returning.

10.10 Total head or stagnation conditions

Throughout this chapter it has been assumed that the inlet velocity to the nozzle is negligible. When this is not the case the concept of total head or stagnation conditions can be used.

Let a gas moving with velocity, C, at a temperature T, be brought to rest adiabatically, finally reaching a temperature T_t, when at rest. Then, applying the flow equation, for a perfect gas, we have,

$$c_p T + \frac{C^2}{2} = c_p T_t$$

or
$$T_t = T + \frac{C^2}{2c_p} \qquad\qquad 10.26$$

The temperature T_t is called the *total temperature* (or stagnation temperature) of the moving gas.

When a thermometer is inserted in a moving gas stream, the gas around the bulb is brought to rest adiabatically and hence the thermometer measures the total temperature. In order to measure the ordinary or static temperature, T, the thermometer would have to move at the gas velocity.

The term $C^2/(2c_p)$ in equation 10.26 is sometimes called the *temperature equivalent of velocity*. The error in the absolute temperature by neglecting this term is less than 1% for velocities up to about 75 m/s, for a gas at atmospheric temperature.

The total pressure, or stagnation pressure, p_t, of a gas stream is defined as the pressure the gas would attain if brought to rest isentropically.

From equation 4.21,

$$\frac{p_t}{p} = \left(\frac{T_t}{T}\right)^{\gamma/(\gamma-1)} \qquad 10.27$$

Using equation 10.26, we have,

$$\frac{p_t}{p} = \left(1 + \frac{C^2}{2c_p T}\right)^{\gamma/(\gamma-1)}$$

Also, from equation 3.22, $c_p = \gamma R/(\gamma-1)$, hence substituting,

$$\frac{p_t}{p} = \left\{1 + \frac{(\gamma-1)C^2}{2\gamma RT}\right\}^{\gamma/(\gamma-1)}$$

From equation 10.10, the velocity of sound in a gas, a, is equal to $\sqrt{(\gamma RT)}$, hence,

$$\frac{p_t}{p} = \left\{1 + \frac{(\gamma-1)C^2}{2a^2}\right\}^{\gamma/(\gamma-1)} = \left\{1 + \frac{(\gamma-1)(Ma)^2}{2}\right\}^{\gamma/(\gamma-1)} \qquad 10.28$$

$\left(\text{where } (Ma) = \dfrac{C}{a} \text{ is the } Mach\ number\right)$.

If the right-hand side of equation 10.28 is expanded by the binomial theorem we have,

$$\frac{p_t}{p} = 1 + \frac{\gamma(\gamma-1)(Ma)^2}{(\gamma-1)\times 2} + \left(\frac{\gamma}{\gamma-1}\right)\left\{\left(\frac{\gamma}{\gamma-1}\right) - 1\right\}\frac{1}{2}\frac{(\gamma-1)^2(Ma)^4}{4} + . \text{ etc.,}$$

i.e.
$$\frac{p_t}{p} = 1 + \frac{\gamma (Ma)^2}{2} + \frac{\gamma (Ma)^4}{8} + \cdots \quad \text{etc.}$$

When the velocity of the gas is low, and Ma is therefore small (say $Ma < 0.2$), then it is a good approximation to write,

$$\frac{p_t}{p} = 1 + \frac{\gamma (Ma)^2}{2} = 1 + \frac{\gamma C^2}{2\gamma RT} = 1 + \frac{C^2}{2RT}$$

or
$$p_t = p + \frac{C^2}{2} \frac{p}{RT}$$

Now the density, ρ, is the reciprocal of the specific volume,

i.e.
$$\rho = \frac{1}{v} = \frac{p}{RT}$$

Hence,   Total pressure, $p_t = p + \dfrac{\rho C^2}{2}$ 10.29

The term $\rho C^2/2$ in equation 10.29 is called the *velocity head*.

Applying total head conditions to flow through a nozzle we have, at inlet,

$$\frac{C_1^2}{2} + c_p T_1 = c_p T_{t_1}$$

At any other section of the nozzle where the velocity is C, and the temperature is T, we have

$$\frac{C^2}{2} + c_p T = c_p T_t$$

$$\therefore \frac{C_1^2}{2} + c_p T_1 = \frac{C^2}{2} + c_p T = c_p T_{t_1} = c_p T_t$$

Therefore the total temperature remains constant throughout the nozzle for adiabatic flow.

(The total pressure remains constant throughout the nozzle for isentropic flow, but not for irreversible adiabatic flow, since there is a pressure loss due to friction.)

The nozzle inlet velocity, C_1, can be treated by imagining the nozzle extrapolated back from the inlet to a section where the

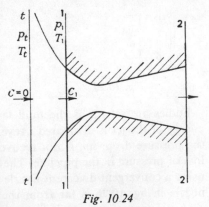

Fig. 10.24

velocity is zero. The conditions at this imaginary section are the total head conditions. This is illustrated in fig. 10.24. At section t the cross-sectional area is infinite.

The equations derived previously can be used with p_{t_1} and T_{t_1} substituted for p_1 and T_1. Therefore, equation 10.4 can be written

$$C = \sqrt{\{2c_p(T_{t_1} - T)\}}$$

Also, from equation 10.7 and 10.8, we have

$$\frac{p_c}{p_{t_1}} = \left(\frac{2}{\gamma+1}\right)^{\gamma/(\gamma-1)} \quad \text{and} \quad \frac{T_c}{T_{t_1}} = \frac{2}{\gamma+1}$$

10.11 Flow measurement

A nozzle is used frequently as a flow meter by inserting it into a pipe-line and measuring the pressure drop or differential between the inlet and the throat. This pressure drop must be kept small, and is measured by a water or mercury manometer.

A convergent nozzle can be used as shown in fig. 10.25. The difference in levels in the manometer is $\Delta p/w$, where Δp is the pressure difference between sections 1 and 2, and w is the specific weight of the manometer liquid.

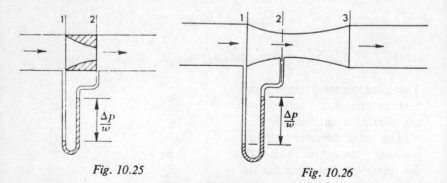

Fig. 10.25   Fig. 10.26

Eddies are set up as the fluid leaves the nozzle and the kinetic energy of the jet is dissipated irreversibly. This means that some of the pressure drop, Δp, is not recovered, and so the nozzle causes a loss of pressure in the pipeline. The pressure loss can be reduced by using a convergent-divergent nozzle as shown in fig. 10.26. Since the nozzle in fig. 10.26 is far from the choked condition, it acts as a *venturimeter* (see Section 10.4). The flow is expanded down to the throat at section 2, and then diffused from 2 to 3. In this way the pressure drop to the throat, Δp, is almost completely recovered in the diffuser portion, and the pressure loss in the pipeline due to the venturimeter is much smaller than that due to a convergent nozzle.

For both the nozzle and the venturimeter, equation 2.9 can be applied to sections 1 and 2,

i.e. $$\dot{m} = \frac{A_1 C_1}{v_1} = \frac{A_2 C_2}{v_2} \quad \therefore \quad C_1 = C_2 \frac{A_2 v_1}{A_1 v_2}$$

Also, applying the flow equation,

$$\frac{C_1^2}{2} + h_1 = \frac{C_2^2}{2} + h_2$$

i.e. $$\frac{C_2^2 - C_1^2}{2} = \frac{C_2^2 - C_2^2 \left(\frac{A_2 v_1}{A_1 v_2}\right)^2}{2} = (h_1 - h_2)$$

or $$C_2 = \sqrt{\left\{ \frac{2(h_1 - h_2)}{1 - \left(\frac{A_2 v_1}{A_1 v_2}\right)^2} \right\}} = E\sqrt{2} \times \sqrt{(h_1 - h_2)}$$

$$\left(\text{where } E = \frac{1}{\sqrt{\left\{ 1 - \left(\frac{A_2 v_1}{A_1 v_2}\right)^2 \right\}}} \right)$$

Then,

$$\dot{m} = \frac{A_2 C_2}{v_2} = \frac{A_2 E}{v_2} \sqrt{(h_1 - h_2)} \times \sqrt{2}$$

i.e. $$\dot{m} = \frac{A_2 E}{v_2} \sqrt{2} \times \sqrt{\{c_p (T_1 - T_2)\}}$$

$$= \frac{A_2 E}{v_2} \sqrt{2} \sqrt{\left\{ c_p T_1 \left(1 - \frac{T_2}{T_1}\right) \right\}}$$

From equation 4.21, assuming isentropic flow, we have

$$\frac{T_2}{T_1} = \left(\frac{p_2}{p_1}\right)^{(\gamma - 1)/\gamma}$$

i.e. $$\dot{m} = \frac{A_2 E}{v_2} \sqrt{2} \sqrt{\left[c_p T_1 \left\{ 1 - \left(\frac{p_2}{p_1}\right)^{(\gamma - 1)/\gamma} \right\} \right]} \qquad 10.29$$

Expressing v_2 in terms of the specific volume at section 1,

i.e. $$v_2 = v_1\left(\frac{p_2}{p_1}\right)^{-1/\gamma}$$

Substituting, in equation 10.30,

$$\dot{m} = \frac{A_2E}{v_1}\sqrt{2}\Bigg/\left[c_pT_1\left\{\left(\frac{p_2}{p_1}\right)^{2/\gamma} - \left(\frac{p_2}{p_1}\right)^{(\gamma+1)/\gamma}\right\}\right] \qquad 10.31$$

Equation 10.31 is clumsy to use, and an approximation can be made to equation 10.30 as follows:

$$\left(\frac{p_2}{p_1}\right)^{(\gamma-1)/\gamma} = \left(1 - \frac{p_1-p_2}{p_1}\right)^{(\gamma-1)/\gamma} = \left(1 - \frac{\Delta p}{p_1}\right)^{(\gamma-1)/\gamma}$$

(where Δp is equal to $(p_1 - p_2)$).

Expanding this expression using the binomial theorem we have

$$\left(\frac{p_2}{p_1}\right)^{(\gamma-1)/\gamma} = 1 - \left(\frac{\gamma-1}{\gamma}\right)\frac{\Delta p}{p_1} - \left(\frac{\gamma-1}{\gamma}\right)\left\{\left(\frac{\gamma-1}{\gamma}\right) - 1\right\}\frac{1}{2}\left(\frac{\Delta p}{p_1}\right)^2 \quad \text{etc.}$$

When Δp is small then terms such as $(\Delta p)^2$ and higher orders can be neglected,

i.e. $$\left(\frac{p_2}{p_1}\right)^{(\gamma-1)/\gamma} = 1 - \left(\frac{\gamma-1}{\gamma}\right)\frac{\Delta p}{p_1}$$

Substituting in equation 10.30,

$$\dot{m} = \frac{A_2E}{v_2}\sqrt{2}\Bigg/\left[c_pT_1\left\{1 - 1 + \left(\frac{\gamma-1}{\gamma}\right)\frac{\Delta p}{p_1}\right\}\right]$$

i.e. $$\dot{m} = \frac{A_2E}{v_2}\sqrt{2}\Bigg/\left\{\frac{c_pT_1(\gamma-1)\Delta p}{\gamma p_1}\right\}$$

Now from equation 3.22,

$$c_p = \frac{\gamma R}{(\gamma-1)} \quad \text{or} \quad \frac{c_p(\gamma-1)}{\gamma} = R$$

$$\therefore \ \dot{m} = \frac{A_2E}{v_2}\sqrt{2}\Bigg/\left(\frac{RT_1}{p_1}\Delta p\right)$$

i.e. $$\dot{m} = \frac{A_2E}{v_2}\sqrt{2}\sqrt{v_1\,\Delta p}$$

Or in terms of densities, $\rho_1 = 1/v_1$, and $\rho_2 = 1/v_2$, then,

$$\dot{m} = A_2E\rho_2\sqrt{2}\sqrt{\frac{\Delta p}{\rho_1}} \qquad 10.32$$

The effects of friction and eddies are allowed for by introducing a nozzle coefficient, C. Therefore equation 10.31 becomes

$$\dot{m} = CA_2E\rho_1\sqrt{\left[2c_pT_1\left\{\left(\frac{p_2}{p_1}\right)^{2/\gamma} - \left(\frac{p_2}{p_1}\right)^{(\gamma+1)/\gamma}\right\}\right]} \qquad 10.33$$

Equation 10.33 must be used when an accurate value of the mass flow is required, particularly when the pressure drop is not small.

PROBLEMS

10.1 Calculate the throat and exit areas of a nozzle to expand air at the rate of 4·5 kg/s from 8·3 bar, 327°C into a space at 1·38 bar. Neglect the inlet velocity and assume isentropic flow.

(3290 mm²; 4850 mm²)

10.2 It is required to produce a stream of helium at the rate of 0·1 kg/s travelling at sonic velocity at a temperature of 15°C. Calculate the inlet pressure and temperature required assuming a back pressure of 1·013 bar and a negligible inlet velocity. Calculate also the exit area of the nozzle. Assume isentropic flow, and assume also that helium is a perfect gas of molar mass 4 kg/kmol, and $\gamma = 1·66$.

(2·077 bar; 110°C; 592 mm²)

10.3 Re-calculate problem 10.1 assuming a coefficient of discharge of 0·96, and a nozzle efficiency of 0·92.   (3430 mm²; 5320 mm²)

10.4 A convergent-divergent nozzle expands air at 6·89 bar and 427°C into a space at 1 bar. The throat area of the nozzle is 650 mm² and the exit area 975 mm². Assuming that the inlet velocity is negligible, and the exit velocity is 680 m/s, calculate the mass flow through the nozzle and state whether the nozzle is underexpanding or overexpanding. Assume that friction in the convergent portion is negligible. Calculate also the nozzle efficiency and the coefficient of velocity.

(0·684 kg/s; underexpanding: $p_2 = 1·392$ bar; 0·892; 0·945)

10.5 Dry saturated steam at 7 bar expands isentropically in a convergent nozzle into a space at 2 bar. Estimate graphically the mass flow per m² of exit area. The inlet velocity may be assumed to be negligible. Assume that the flow is in equilibrium throughout the expansion

(1024 kg/s per m²)

10.6 Steam enters a convergent-divergent nozzle at 11 bar, dry saturated and expands isentropically to 2·7 bar. Calculate the area of the nozzle throat for a mass flow of 0·75 kg/s by a graphical method. Compare the value of the critical pressure thus obtained with the value found by assuming a law of expansion $pv^{1·135} = $ constant. Assume equilibrium flow conditions. Using the *h-s* chart calculate the nozzle exit area and compare this with the value obtained using the law $pv^{1·135} = $ constant for the complete expansion.

(474 mm^2; 6·3 bar; 6·35 bar; 647 mm^2; 645 mm^2)

10.7 Steam at 7 bar and 290°C expands in a convergent-divergent nozzle to a pressure of 0·55 bar. Assuming a negligible inlet velocity, calculate the mass flow and the nozzle exit area. The area of the nozzle throat is 970 mm^2 and the coefficient of discharge is 0·95. Take a coefficient of velocity of 0·92. Assume for the isentropic flow of superheated steam that $p_c/p_1 = 0·546$, and take the specific volume of superheated steam from equation 10.20.

Use the *h-s* chart throughout.   (0·875 kg/s; 2790 mm^2)

10.8 Steam at 20 bar and 240°C expands isentropically to a pressure of 3 bar in a convergent-divergent nozzle. Calculate the mass flow per m^2 exit area,

(a) assuming equilibrium flow.

(b) assuming supersaturated flow.

For supersaturated flow assume that $pv^{1·3} = $ constant.

(1540 kg/s; 1750 kg/s)

10.9 A convergent nozzle receives steam at 4 bar and 150°C and negligible inlet velocity, and expands it into a space at atmospheric pressure. Assuming supersaturated expansion and a nozzle efficiency of 0·9, calculate the nozzle throat area required for a mass flow of 1·2 kg/s. For supersaturated flow use equation 10.20.

(2010 mm^2)

10.10 A thermometer inserted into an air stream flowing at 33·5 m/s records a temperature of 15°C. Calculate the true static temperature of the air, assuming that the air round the thermometer bulb is brought to rest adiabatically. Calculate also the total pressure of the air if the static pressure in the duct is 1·01 bar.

(14·44°C; 1·017 bar)

REFERENCES

10.1 SHAPIRO, A. H., *The Dynamics and Thermodynamics of Compressible Fluid Flow*, vols 1 and 2 (Kreiger, 1983).

10.2 DOUGLAS, J. F., GASIOREK, J. M., and SWAFFIELD, J. A., *Fluid Mechanics* (Pitman, 1985).

11. Steam Turbines

The steam turbine is a power unit which produces power from a continuous supply of steam, the steam being delivered to the turbine at a high pressure and exhausted to the condenser at a low pressure. The way in which the overall pressure drop in the turbine occurs is a

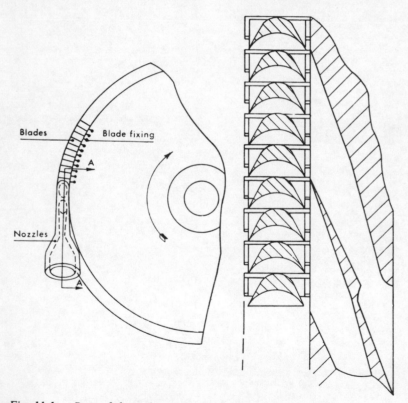

Blades   Blade fixing

Nozzles

Fig. 11.1a   Part of the turbine wheel

Fig. 11.1b

characteristic of the particular design, but each incremental drop in pressure is accompanied by an increase in the kinetic energy of the steam.

A jet of steam has momentum (mass × velocity), which is a vector quantity. A change in the magnitude of either the mass or the velocity, or a change in direction of the velocity, produces a change in momentum, and the force necessary to cause this change is, by Newton's second law, proportional to the rate of change of momentum. The blading on a turbine wheel causes a change in direction of the impinging jet of steam. The force required to produce the change of momentum of the jet is provided by the blades which are attached to the wheel in such a way that a tangential force is applied continuously to the wheel and a constant output torque is obtained. Fig. 11.1a shows a simple steam turbine of the impulse type; fig. 11.1b is an enlarged view of the section on A–A.

11.1 Classification of steam turbines

The steam can be made to flow across the blades in a direction which is either axial (i.e. parallel to the turbine shaft), or radial (i.e. in the plane perpendicular to the turbine shaft). The great majority of steam turbines are of the axial type, and these will now be considered. The radial flow turbine will be considered briefly in Section 11.10.

Axial flow turbines are classified as (a) impulse turbines, or (b) impulse-reaction turbines.

The basic principles of each will be considered.

(a) *Impulse turbine*

Fig. 11.2 shows a single, fixed curved blade on which a jet of steam of velocity *C* impinges. The jet is shown to be reversed in its direction as it passes over the blade without change in its speed.

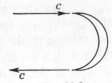

Fig. 11.2

By Newton's second law:

> Force ∝ rate of change of momentum;
> ∝ rate of change of (mass × velocity);

i.e.   Force ∝ mass flow × change in velocity.

Therefore for a coherent system of units,

$$\text{Force} = \dot{m} \times \text{change of velocity} \qquad 11.1$$

(where $\dot{m}$ is the mass flow rate).

In the simple case shown in fig. 11.2 the velocity of the steam at entry to the blade is $+C$ (assuming that the direction from left to right is positive), and the velocity of the steam leaving the blade is $-C$. Therefore the change in velocity in the positive direction is $-C-C=-2C$, and the force acting *on the jet*, using equation 11.1, is given by

$$F = -\dot{m}2C$$

The negative sign shows that the force acts from right to left. There is a reactive force acting on the blade which is equal in magnitude and opposite in direction to that acting on the jet, and tends to push the blade in the same direction as that of the impinging jet. If the blade is attached to a wheel which is free to rotate then the wheel will be rotated by the force of the jet on the blade. Consider the moving

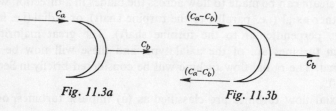

Fig. 11.3a   *Fig. 11.3b*

blade in fig. 11.3a. The blade moves with a velocity C_b as shown and the jet velocity is C_a. It is clear that if $C_b > C_a$, then the jet will not catch up with the blade and can have no effect on its motion. If $C_a > C_b$, then the steam impinges on the blade with a relative velocity of $(C_a - C_b)$ as in fig. 11.3b, and if there is no reduction of this velocity relative to the blade due to friction, then the steam leaves the blade with a relative velocity of $-(C_a - C_b)$. Hence the change in velocity of the jet is $-(C_a - C_b) - (C_a - C_b) = -2(C_a - C_b)$, and the force on the jet, using equation 11.1, is given by,

$$F = -2\dot{m}(C_a - C_b)$$

The impulsive force on the blade in the direction of motion is

$$F = 2\dot{m}(C_a - C_b)$$

As stated previously the turbine wheel consists of a series of blades, and if each blade is to receive steam from a series of nozzles in succession, then the steam cannot be supplied to the wheel in the direction shown in fig. 11.3. The nozzles must be inclined to the blade wheel at an angle α_i, (where i denotes inlet), as shown in fig. 11.4.

This permits the steam to enter the blades, and the steam leaves the wheel with an absolute velocity C_{a_e} in the direction defined by the angle α_e (where e denotes exit). The selection of the angle α_i, is one of compromise since an increase in α_i reduces the value of the velocity $C_{a_i} \cos \alpha_i$, and increases the axial, or flow component $C_{a_i} \sin \alpha_i$. The flow component must be sufficient to allow the steam to flow across the

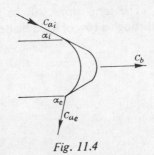

Fig. 11.4

wheel, and its value influences the size of the wheel annulus for a given rate of steam flow, since the volume flow is equal to the effective area of the blade annulus multiplied by the flow velocity.

(b) *Impulse-reaction turbine*

Before considering the impulse-reaction turbine it is necessary to mention the pure-reaction turbine. The first known turbine of this type is attributed to Hero of Alexandria about the first or second century B.C. A reaction turbine built on the principle of Hero's was demonstrated in 1883 by Gustaf de Laval, one of the most prominent pioneers of the steam turbine. Fig. 11.5 shows the reaction type; the radial tubes, which are in connection with the vertical supply tube, are free to rotate. The end of each tube is shaped as a nozzle and the steam from the supply tube passes along the radial tubes and then expands through the nozzles to atmosphere in a tangential direction. There is an increase in velocity of the steam relative to the rotating tube, and hence there is a reaction on the tube which makes it rotate. The turbine of de Laval had a speed of rotation of 42 000 rev/min, with a tip speed of 180 m/s. The turbine was inefficient and did not become of commercial interest, although it was an achievement of its time and indicated the possibilities of high speeds in turbine application.

The impulse-reaction turbine has an axial steam flow in the majority of cases, and receives its driving force partly as an impulsive force and partly as a reaction force due to an expansion of the steam in the blade passages. The impulse-reaction turbine is more often referred to simply as the reaction turbine. Both types of turbine, impulse and reaction, will be analysed further in the following sections.

The term *stage* will be used, and this refers to one expansion through a row of fixed blades or nozzles, and a row of moving blades.

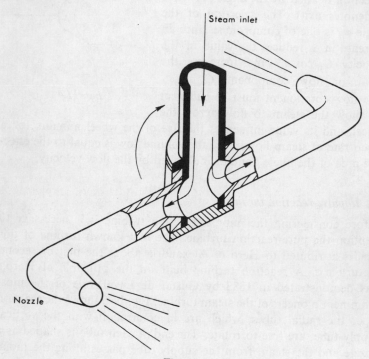

Fig. 11.5

11.2 The impulse turbine

The steam supplied to a single wheel impulse turbine expands completely in the nozzles and leaves with a high absolute velocity. This is the absolute inlet velocity to the blade, and it will be denoted for the moment by C_{a_i}, as shown in fig. 11.6a. The steam is delivered to the wheel at an angle α_i. The absolute velocity can be considered as having two components: C_b, which is equal to the blade velocity;

and C_r, the velocity of the steam relative to the blade. C_r is tangential to the blade profile at any point on it. The two points of particular interest are those at the inlet and exit of the blades. The velocities

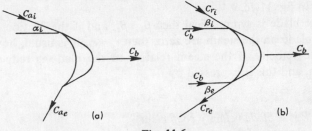

Fig. 11.6

at these points are shown in fig. 11.6b as C_{r_i} and C_{r_e} respectively, and the directions are defined by the angles β_i and β_e as shown. If the steam is to enter and leave the blades smoothly without shock, then β_i is the angle of the blade at inlet, and β_e the angle of the blade at exit.

The analysis of the velocity changes will be simplified by using the subscripts a_i, r_i, etc., to denote the velocities (e.g. a_i denotes absolute

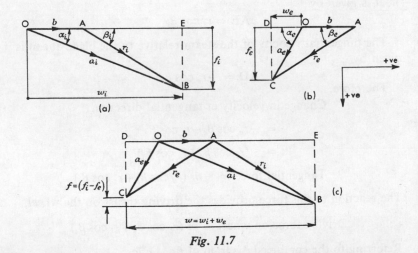

Fig. 11.7

velocity at inlet, and r_i denotes relative velocity at inlet). The velocity triangle for the inlet condition is drawn in fig. 11.7a from the information of fig. 11.6. The steam leaves the blade with a velocity r_e

relative to the blade, and at the blade exit angle of β_e. The absolute velocity at exit a_e is determined from the velocity triangle of fig. 11.7b, and its direction is α_e. Since both triangles have the common side $OA = b$, the triangles can be combined to give a single diagram as shown in fig. 11.7c.

If the blade is symmetrical then $\beta_i = \beta_e$, and if the friction effects of the blade on the steam are zero, then $r_e = r_i$. It is usual, however, that the velocity of the steam relative to the blade is reduced by friction, and this is expressed by

$$r_e = kr_i \tag{11.2}$$

(where k is a *blade velocity coefficient*).

The velocities of flow across the blade at inlet and exit are given by f_i and f_e (i.e. EB and DC respectively). There may be a difference between these values which means that there is a change in velocity in the axial direction and an associated *axial thrust*. The horizontal components of the absolute velocities at inlet and exit are called the *whirl velocities*, w_i and w_e, as shown in fig. 11.7c.

Using equation 11.1, the tangential force acting on the jet is given by $F = \dot{m} \times$ (change of velocity in the tangential direction).

The tangential velocity of the steam relative to the blade at inlet is given by

$$AE = r_i \cos \beta_i$$

The tangential velocity of the steam relative to the blade at exit is given by

$$AD = -r_e \cos \beta_e$$

Therefore,

Change in velocity in tangential direction

$$= -r_e \cos \beta_e - r_i \cos \beta_i$$
$$= -(r_e \cos \beta_e + r_i \cos \beta_i)$$

$\therefore$ Tangential force $= -\dot{m}\,(r_e \cos \beta_e + r_i \cos \beta_i)$

The reaction to this force provides the driving thrust on the wheel,

i.e. Driving force on wheel $= \dot{m}\,(r_e \cos \beta_e + r_i \cos \beta_i)$

Referring to the combined diagram of fig. 11.7c,

$$(r_e \cos \beta_e + r_i \cos \beta_i) = AD + AE = DE = w$$

(where w is the change in the velocity of whirl).

$$\therefore \text{ Driving force on wheel} = \dot{m}w \qquad 11.3$$

The rate at which work is done on the wheel is given by the product of the driving force and the blade velocity. Therefore, using equation 11.3,

$$\text{Rate at which work is done} = \dot{m}wb \qquad 11.4$$

The power output is then given by,

$$\text{Power output} = \dot{m}wb \qquad 11.5$$

It is usual to construct the blade velocity diagram to scale and to determine the value of w from it by measurement. Referring to fig. 11.7c it can be seen that the above result can also be obtained by considering the changes in absolute velocity of the steam between inlet and exit.

A part of this diagram is repeated in fig. 11.8a as defined by the lines OB and OC. The change in velocity of the jet is given by BC and the resultant force on the jet is equal to $\dot{m} \times \text{BC}$. The reaction to this is the force on the wheel (i.e. $\dot{m} \times \text{CB}$). This force can be expressed as its components $\dot{m} \times \text{CF}$ and $\dot{m} \times \text{FB}$ as in fig. 11.8b. From fig. 11.7c, $\text{FB} = \text{DE} = w$, and $\dot{m} \times \text{FB}$ is the tangential driving force as deter-

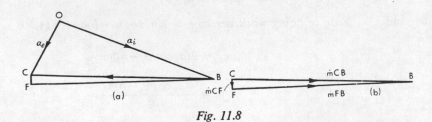

Fig. 11.8

mined previously. The axial component of the force, $\dot{m} \times \text{CF}$, is the axial thrust on the wheel, which must be taken up by the bearings in which the shaft is mounted. From fig. 11.7c it can be seen that $\text{CF} = \text{EB} - \text{DC} = f_i - f_e = f,$

i.e. $\qquad\qquad\qquad \text{Axial thrust} = \dot{m}f \qquad 11.6$

If the enthalpy of the steam at entry to the nozzle is h_0, and at the nozzle exit is h_i, then the maximum velocity of the steam impinging on the blades is given by equation 10.4 as

$$a_i = \sqrt{\{2(h_0 - h_i)\}} \text{ m/s} \qquad 11.7$$

(assuming negligible velocity at inlet to the nozzles).

The energy supplied to the blades is the kinetic energy of the jet, $a_i^2/2$, and the blading efficiency or *diagram efficiency* is defined by

$$\text{Diagram efficiency, } \eta_d = \frac{\text{rate of doing work per kg of steam}}{\text{energy supplied per kg of steam}}$$

The numerator is obtained from equation 11.4 as, wb per kg of steam per second, therefore,

$$\eta_d = wb \times \frac{2}{a_1^2} = \frac{2wb}{a_1^2} \qquad 11.8$$

For purpose of analysis, w can be expressed as

$$w = r_e \cos \beta_e + r_i \cos \beta_i \qquad 11.9$$

Also, if the blade velocity coefficient, $k = 1$ (i.e. $r_e = r_i$), and if the blades are symmetrical (i.e. $\beta_e = \beta_i$), then we have

$$w = 2r_i \cos \beta_i$$

But, $r_i \cos \beta_i = a_i \cos \alpha_i - b$ (see fig. 11.7a)

Hence, $w = 2(a_i \cos \alpha_i - b)$

and   Rate of doing work per kg/s $= 2(a_i \cos \alpha_i - b)b \qquad 11.10$

$$\therefore \eta_d = 2(a_i \cos \alpha_i - b)b \times \frac{2}{a_i^2}$$

$$= \frac{4(a_i \cos \alpha_i - b)b}{a_i^2}$$

i.e.   Diagram efficiency $= \dfrac{4b}{a_i}\left(\cos \alpha_i - \dfrac{b}{a_i}\right) \qquad 11.11$

(where b/a_i is called the *blade speed ratio*).

The simple impulse turbine is called the de Laval turbine, since it was invented by Dr Gustaf de Laval and patented by him in 1888. The first turbine produced by de Laval was a 11 kW marine turbine in 1892; this turbine ran at 16 000 rev/min with an output shaft speed of 330 rev/min using a double reduction in a double helical gear. It is a tribute to de Laval that so much of his work is either current practice or has received little modification up to the present day. He pioneered the use of high pressures and high speeds for steam turbines. His mechanical design from the point of view of blading shape and blade attachment compare with the best in modern practice. He also developed a form of double helical reduction gearing which is similar to that used in ships of modern construction.

Example 11.1

The velocity of steam leaving the nozzles of an impulse turbine is 900 m/s and the nozzle angle is 20°. The blade velocity is 300 m/s and the blade velocity coefficient is 0·7. Calculate for a mass flow of 1 kg/s, and symmetrical blading:

- (a) The blade inlet angle.
- (b) The driving force on the wheel.
- (c) The axial thrust.
- (d) The diagram power.
- (e) The diagram efficiency.

The information given is indicated on fig. 11.9, and the required quantities can be obtained either by construction or by calculation. In many cases calculation becomes tedious and long, and a graphical construction is recommended for all but the simplest problems.

(a) The blade inlet angle can be measured directly from fig. 11.9

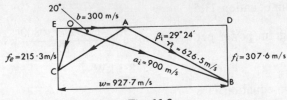

Fig. 11.9

if this figure is drawn to scale, but the angle will be found by an analytical method to illustrate the procedure.

Applying the cosine rule to triangle OAB,

$$r_i^2 = a_i^2 + b^2 - 2ba_i \cos \alpha_i$$
$$= 900^2 + 300^2 - 2 \times 300 \times 900 \times \cos 20° = 39\cdot25 \times 10^4$$
$$\therefore r_i = 626\cdot5 \text{ m/s}$$

Then using the sine rule in triangle OAB,

$$\frac{a_i}{\sin \text{OAB}} = \frac{r_i}{\sin \alpha_i}$$

also

$$\sin \text{OAB} = \sin (180 - \beta_i) = \sin \beta_i$$

$$\therefore \sin \beta_i = \frac{a_i \sin \alpha_i}{r_i} = \frac{900 \times \sin 20°}{626 \cdot 5} = 0 \cdot 491$$

$$\therefore \beta_i = 29° \, 24' = \beta_e$$

(b) Using equation 11.2,

$$r_e = kr_i = 626 \cdot 5 \times 0 \cdot 7 = 438 \cdot 5 \text{ m/s}$$
$$\text{AD} = r_i \cos \beta_i = 626 \cdot 5 \cos 29° \, 24' = 545 \cdot 8 \text{ m/s}$$
$$\text{AE} = r_e \cos \beta_e = 438 \cdot 5 \cos 29° \, 24' = 381 \cdot 9 \text{ m/s}$$
$$\therefore w = 545 \cdot 8 + 381 \cdot 9 = 927 \cdot 7 \text{ m/s}$$

From equation 11.3,

Driving force on wheel $= \dot{m}w = 1 \times 927 \cdot 7 = 927 \cdot 7$ N per kg/s

(c) $\quad f_i = r_i \sin \beta_i = 626 \cdot 5 \sin 29° \, 24' = 307 \cdot 6$ m/s
$$f_e = r_e \sin \beta_e = 438 \cdot 5 \sin 29° \, 24' = 215 \cdot 3 \text{ m/s}$$
$$\therefore f = 307 \cdot 6 - 215 \cdot 3 = 92 \cdot 3 \text{ m/s}$$

Then, from equation 11.6,

Axial thrust $= \dot{m}f = 1 \times 92 \cdot 3 = 92 \cdot 3$ N per kg/s

(d) From equation 11.5,

Diagram power per kg/s $= \dot{m}wb = \dfrac{1 \times 927 \cdot 7 \times 300}{10^3}$ kW

$$= 278 \cdot 3 \text{ kW}$$

(e) From equation 11.8,

Diagram efficiency, $\eta_d = \dfrac{2wb}{a_i^2} = \dfrac{2 \times 927 \cdot 7 \times 300}{900^2}$

$$= 0 \cdot 687 \quad \text{or} \quad 68 \cdot 7\%$$

Optimum operating conditions from the blade velocity diagrams
From equation 11.10 the rate of doing work on the blade wheel per kg/s steam flow is given by $2b(a_i \cos \alpha_i - b)$. For a given steam

velocity a_i, and a given blade velocity b, it can be seen that the rate of doing work is a maximum when $\cos \alpha_i = 1$ (i.e. when $\alpha_i = 0$). For this value of α_i the axial flow component would be zero. As mentioned before it is necessary to have an axial flow component to allow the steam to reach the blades properly and to clear the blades on leaving. As α_i is increased the rate of working on the blades is reduced, but the blade annulus area required for a given mass flow is reduced since the axial flow component of the velocity is increased. Further, the surface area of the blades will be reduced at higher values of α_i, and this means friction losses will be less. A selection of α_i must be made based on these conflicting requirements, and the usual values of α_i lie between 15° and 30°. For a fixed value of α_i the optimum blade speed ratio for maximum diagram efficiency can be obtained by differentiating equation 11.11 and putting the result equal to zero.

i.e. Diagram efficiency, $\eta_d = 4\dfrac{b}{a_i}\left(\cos \alpha_i - \dfrac{b}{a_i}\right)$

$$\therefore \frac{d(\eta_d)}{d\left(\dfrac{b}{a_i}\right)} = 4\cos \alpha_i - 8\frac{b}{a_i} = 0$$

$$\therefore \frac{b}{a_i} = \frac{\cos \alpha_i}{2} \qquad 11.12$$

i.e. Maximum diagram efficiency $= \dfrac{4\cos \alpha_i}{2}\left(\cos \alpha_i - \dfrac{\cos \alpha_i}{2}\right)$

$$= \cos^2 \alpha_i \qquad 11.13$$

The rate of doing work corresponding to the maximum diagram efficiency is then given by substituting in equation 11.10,

i.e. Maximum efficiency rate of doing work per kg/s $= 2b^2$ 11.14

The variation in η_d with b/a_i is shown in fig. 11.10.

The single stage impulse turbine is used only as a small power machine. The steam velocities may be as high as 1070 m/s, and for $\alpha_i = 20°$ the optimum blade speed ratio would be about 0·47, giving the maximum blade speed as 500 m/s. In practice the blade speed is limited to about 420 m/s. This value of velocity used in small machines would give high speeds of rotation of the order of 30 000 rev/min. Smaller diameter rotors mean a more economic construction, but high rotational speeds mean high stresses. Further, the high turbine speeds are not directly applicable, and a reduction gear is required to give output speeds which are in the useful range.

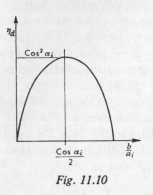

Fig. 11.10

The expansion of steam in the simple impulse turbine is carried out in a single stage, so the steam velocity at inlet to the wheel is high. There is no drop in pressure in the wheel casing. The blade velocity must be limited for mechanical reasons of strength and operating speed. From these considerations, and an inspection of the velocity diagrams, it is evident that the steam leaves the blade wheel with a high velocity. This constitutes a loss in the work available on the wheel, although a moderate velocity must be accepted in order to take the steam to the condenser. This *leaving loss* may amount to 11% of the input energy. The leaving velocity in the velocity diagram of fig. 11.7c is a_e, and the leaving loss is given by $a_e^2/2$.

Methods of improving the efficiency of the simple impulse turbine will be considered in the following section, and are known as compounding.

11.3 Pressure and velocity compounded impulse turbines

(i) *Pressure compounding (The Rateau turbine)*

The pressure drop available to the turbine is used in a series of small increments, each increment being associated with one stage of the turbine. The physical arrangement is shown in fig. 11.11. The nozzles are carried in diaphragms which separate each stage from the next. The steam pressure in the space between each pair of diaphragms is constant, but there is a pressure drop across each diaphragm as required by the nozzles. Precaution must be taken to prevent leakage of the steam from one section to the next at the shaft and outer casing. The steam speeds, and hence the blade speeds, are low if the number of stages is high. In fig. 11.11 the variations in pressure and velocity through the turbine are shown, the final pressure being that of the condenser, and the final velocity that required for the steam to leave the turbine. In fig. 11.11 only one set of wheels is shown, but these may be followed by another set with a larger mean radius. Each of the stages can be analysed by the method used previously for the single stage. A turbine with a series of simple impulse stages is called a Rateau turbine.

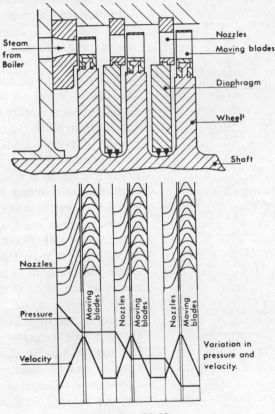

Fig. 11.11

(ii) *Velocity compounding (The Curtis turbine)*

From previous considerations it is seen that in the simple impulse stage the optimum condition of blade speed is hardly practicable, and with the speeds actually used only a small amount of the kinetic energy of the steam can be utilized. The velocity compounded stage, called the Curtis stage after its designer, is used to employ lower blade

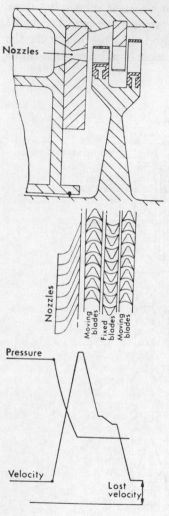

Fig. 11.12

speeds and a higher utilization of the kinetic energy of the steam. This design was patented by an American, C. G. Curtis, in 1895, but there were turbines of this type being used by the de Laval company some years previously.

In this type all the expansion takes place in a single set of nozzles, and the steam then passes through a series of blades attached to a single wheel or rotor. Since the blades move in the same direction it is necessary to change the direction of the steam between one set of moving blades and the next. For this purpose a stationary ring of blades is fitted between each pair of moving blades. A *two-row wheel* version of this turbine is shown in fig. 11.12. The steam velocity is high but the blade velocity is less than that for the single row turbine. The kinetic energy of the jet is thus utilized in the multiple stages. The inlet velocity to the fixed blades is the absolute exit velocity from the first row of moving blades. The absolute inlet velocity to the second row of moving blades is the exit velocity from the fixed blades. The velocity diagrams are shown in fig. 11.13.

Work done in the first row $= \dot{m}w_1 b$

Work done in the second row $= \dot{m}w_2 b$

$\therefore$ Total work done on the wheel $= \dot{m}b\,(w_1 + w_2)$ 11.15

If the moving and stationary blades are symmetrical and the relative velocities are unchanged on passing over a blade, then by the

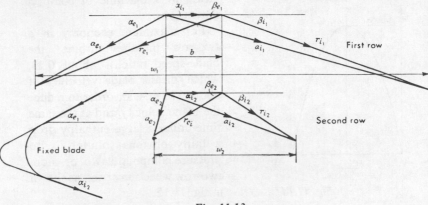

Fig. 11.13

procedure used for the single row impulse turbine it can be shown that the diagram efficiency is a maximum when

$$\frac{b}{a_i} = \frac{\cos \alpha_i}{4} \qquad 11.16$$

and for this condition the absolute velocity at exit is in the axial direction.

The maximum diagram efficiency is then

$$\eta_d = \cos^2 \alpha_i \qquad 11.17$$

The corresponding rate at which work is done is given by

$$\text{Rate of doing work per kg/s} = 8b^2 \qquad 11.18$$

Comparing equations 11.18 and 11.14 it can be seen that the enthalpy drop used in the two-row stage is four times that of the single-row stage. The variation of η_d with b/a_i is shown in fig. 11.14.

The *stage efficiency* of a turbine is defined in Section 11.8 by equation 11.32, and is a measure of the useful enthalpy drop for a given available enthalpy drop. The single-row, two-row, and three-row velocity compounded wheels show maximum stage efficiencies of approximately 0·8, 0·67 and 0·52 respectively, at blade-speed ratios of 0·46, 0·23, and 0·13 respectively. Hence the steam consumption increases with the increase in the number of rows of blades.

The three-row wheel is used usually for small turbines employed on auxiliary work. The Curtis stage permits a large expansion of the steam in a machine of compact dimensions.

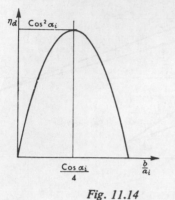

Fig. 11.14

For maximum economy in a two-row impulse turbine, the blade-speed ratio is about 0·23, which means blade velocities of about 275 m/s. In order to reduce the blade velocity and at the same time utilize a large enthalpy drop a fairly obvious solution is to pressure compound two or more two-row wheels in series, as shown in fig. 11.15.

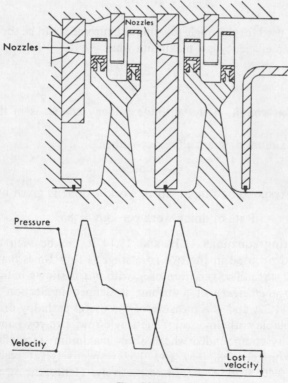

Fig. 11.15

Example 11.2

The first stage of a turbine is a two-row velocity compounded impulse wheel. The steam velocity at inlet is 600 m/s and the mean blade velocity is 120 m/s. The nozzle angle is 16° and the exit angles for the first row of moving blades, the fixed blades, and the second row of moving blades are 18°, 21°, and 35° respectively. Calculate the blade inlet angles for each row. Calculate also for each row of moving blades, the driving force and the axial thrust on the wheel for a mass flow of 1 kg/s. Calculate the diagram efficiency for the wheel and the diagram power per kg/s steam flow. What would be the maximum possible diagram efficiency for the given steam inlet velocity and nozzle angle? Take the blade velocity coefficient as 0·9 for all blades.

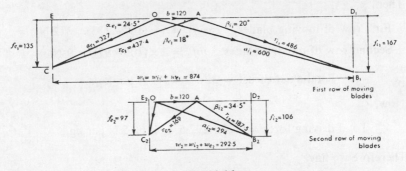

Fig. 11.16

The velocity diagrams are drawn to scale, as shown in fig. 11.16, and the relative velocities, r_{e_1} and r_{e_2} are calculated using equation 11.2.

i.e.
$$r_{e_1} = k_1 r_{t_1} = 0·9 \times 486 = 437·4 \text{ m/s}$$

$$r_{e_2} = k_2 r_{i_2} = 0·9 \times 187·5 = 169 \text{ m/s}$$

The absolute velocity at inlet to the second row of moving blades, a_{i_2}, is equal to the velocity of the steam leaving the fixed row of blades.

i.e.
$$a_{i_2} = k a_{e_1} = 0·9 \times 327 = 294 \text{ m/s}$$

The blade inlet angles are measured from the velocity diagram as,

Inlet blade angle, first row of moving blades, $\beta_{i_1} = 20°$

Inlet blade angle, fixed blades, $\alpha_{e_1} = 24·5°$

Inlet blade angle, second row of moving blades, $\beta_{i_2} = 34·5°$

Using equation 11.3,

$$\text{Driving force} = \dot{m}w$$

Then,

First row of moving blades, $\quad \dot{m}w_1 = 1 \times 874 = 874 \text{ N}$

Second row of moving blades, $\quad \dot{m}w_2 = 1 \times 292 \cdot 5 = 292 \cdot 5 \text{ N}$

(where w_1 and w_2 are scaled from the velocity diagram).

Using equation 11.6,

$$\text{Axial thrust} = \dot{m}f = \dot{m}\,(f_i - f_e)$$

Then,

First row of moving blades, $\quad \dot{m}f_1 = 1 \times (167 - 135) = 32 \text{ N}$

Second row of moving blades, $\quad \dot{m}f_2 = 1 \times (106 - 97) = 9 \text{ N}$

i.e. $\qquad$ Total axial thrust $= 32 + 9 = 41 \text{ N per kg/s}$

Now,

$$\text{Total driving force} = 874 + 292 \cdot 5 = 1166 \cdot 5 \text{ N per kg/s}$$

Therefore we have

$$\text{Power output} = \text{driving force} \times \text{blade velocity}$$

i.e. $\quad$ Power developed $= \dfrac{1166 \cdot 5 \times 120}{10^3} = 140 \text{ kW per kg/s}$

$$\text{Energy supplied to the wheel} = \frac{\dot{m}a_{i1}^2}{2} = \frac{1 \times 600^2}{2 \times 10^3} \text{ kJ/s}$$

$$\therefore \text{ Diagram efficiency} = \frac{140 \times 10^3 \times 2}{600^2} = 0 \cdot 779$$

i.e. $\qquad$ Diagram efficiency $= 77 \cdot 9\%$

From equation 11.17,

$$\text{Maximum diagram efficiency} = \cos^2 \alpha_i = \cos^2 16°$$
$$= 0 \cdot 923 \quad \text{or} \quad 92 \cdot 3\%$$

11.4 Turbine blade height

In the impulse turbine the nozzles do not occupy the complete circumference leading into the blade annulus, and this is referred to as *partial admission*.

The total nozzle exit area must be such as to satisfy the conditions of continuous mass flow of steam. Referring to fig. 11.17, if n is the length of the arc covered by the nozzles and the nozzle height is l, then the nozzle area in the exit plane is nl. If the specific volume of the steam at the exit condition is v, and the mass flow is $\dot{m}$, then the volume flow rate is $\dot{m}v$. The component of the steam velocity at exit and perpendicular to the area nl is $a_i \sin \alpha_i$. Therefore we have,

$$\dot{m}v = a_i \sin \alpha_i \times nl \qquad\qquad 11.19$$

The mass flow of steam, $\dot{m}$ passes through the blade channels of the first moving row and, due to friction, the relative velocity of the

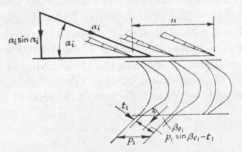

Fig. 11.17a

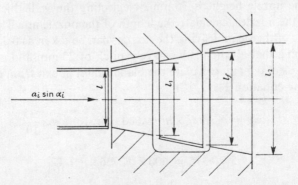

Fig. 11.17b

steam at exit is r_{e_1}. If the blade pitch at exit is p_1 and the blade thickness is t_1, then each blade channel exit area is $(p_1 \sin \beta_{e_1} - t_1)l_1$, where l_1 is the height of the blades at exit. These quantities are shown in fig. 11.17a and fig. 11.17b, where $p_1 \sin \beta_{e_1}$ is shown to be the effective width of the channel perpendicular to the direction of the relative velocity, and t_1 is the blade thickness measured in the same direction. The arc covered by the nozzles is of length n, therefore the number of blade channels accepting steam is given by n/p_1, and the total blade channel exit area by $(n/p_1)(p_1 \sin \beta_{e_1} - t_1)l_1$. As before, for the condition of continuity of mass flow of steam, we have,

$$\dot{m}v_{e_1} = \frac{n}{p_1}(p_1 \sin \beta_{e_1} - t_1)l_1 r_{e_1} \qquad 11.20$$

Fig. 11.17a shows only the nozzles and the first moving row of blades, and fig. 11.17b shows the nozzles as supplying steam to a two-row, velocity compounded wheel. The blade height is increased progressively, and for each row of blades, fixed and moving, an expression similar to that of equation 11.20 can be established.

Due to friction effects as the steam passes over the blades, the enthalpy at exit is higher than it would be if friction were absent. There is an associated increase in the specific volume of the steam compared with the frictionless case, but this is small enough to be negligible compared with the reduction in the relative velocity of the steam.

Example 11.3

For the nozzles and wheel of example 11.2 the steam flow is 5 kg/s and the nozzle height is 25 mm. Neglecting the wall thickness between the nozzles, calculate the length of the nozzle arc. The specific volume of the steam leaving the nozzles can be taken as 0.375 m³/kg. Assuming that all the blades have a pitch of 25 mm and an exit tip thickness of 0.5 mm calculate the blade height at exit from each row.

Using equation 11.19,

$$\dot{m}v = a_i \sin \alpha_i \times nl = 600 \times \sin 16° \times n \times \frac{25}{10^3}$$

i.e. $$5 \times 0.375 = 600 \times 0.2756 \times n \times \frac{25}{10^3}$$

$$\therefore \ n = 0.454 \text{ m}$$

i.e.   Length of nozzle arc $= 0.454$ m

Using equation 11.20,

$$\dot{m}v_e = \frac{n}{p}(p \sin \beta_e - t)lr_e$$

applied to each row of blades, we have

For the first row:

$$5 \times 0.375 = \frac{0.454}{0.025}(0.025 \times \sin 18° - 0.0005)l_1 \times 437.4$$

$$\therefore l_1 = \frac{5 \times 0.375 \times 0.025}{0.454 \times 0.00723 \times 437.4} = 0.0327 \text{ m}$$

i.e.   Blade height at exit $= 32.7$ mm

For the fixed row:

$$5 \times 0.375 = \frac{0.454}{0.025}(0.025 \times \sin 21° - 0.0005)l_f \times 294$$

$$\therefore l_f = \frac{5 \times 0.375 \times 0.025}{0.454 \times 0.00846 \times 294} = 0.0415 \text{ m}$$

i.e.   Blade height at exit $= 41.5$ mm

For the second row:

$$5 \times 0.375 = \frac{0.454}{0.025}(0.025 \times \sin 35° - 0.0005)l_2 \times 169$$

$$\therefore l_2 = \frac{5 \times 0.375 \times 0.025}{0.454 \times 0.01384 \times 169} = 0.0442 \text{ m}$$

i.e.   Blade height at exit $= 44.2$ mm

11.5 Impulse-reaction turbine

The reaction turbine applies the principle of both the pure impulse and the pure reaction turbine. Each stage of the reaction turbine consists of a fixed row of blades over the whole of the circumferential annulus, and an equal number of blades on a wheel. Admission of steam in the reaction turbine takes place over the complete annulus, and so there is *full admission*. The fixed blade channels are of nozzle shape and there is a comparatively small drop in pressure accompanied by an increase in velocity. The steam then passes over the moving blades and, as in the pure impulse turbine, a force is exerted on the blades by the steam. There is a further drop in pressure as the steam passes through the moving blades, since the moving blade channels are also of nozzle shape, and therefore there is an increase in the steam velocity relative to the blades. This is illustrated in the velocity diagram of fig. 11.18a. With a simple impulse type the value of r_e would be given by AD, but in the reaction turbine this velocity is increased to AC by further expansion of the steam in the blade channels. The net change in velocity of the steam is given by BC and the resultant force on the blades by mCB, as shown in fig. 11.18b. This force can be resolved into the tangential and axial thrusts, mCE and mEB as shown in fig. 11.18b.

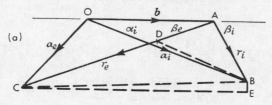

Fig. 11.18a

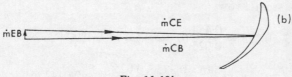

Fig. 11.18b

The reaction turbine was invented by Sir Charles Parsons who produced a 7·5 kW turbine running at 17 000 rev/min in 1884. Parsons was a contemporary of de Laval.

The reaction effect is one of degree since the impulse effect is always there. The degree of reaction, R, is defined as,

$$R = \frac{\text{The enthalpy drop in the moving blades}}{\text{The enthalpy drop in the stage}} = \frac{h_1 - h_2}{h_0 - h_2} \quad 11.21$$

(where h_0, h_1, and h_2 are the enthalpies of the steam at inlet to the fixed blades, at entry to the moving blades, and at exit from the moving blades respectively).

There are only two cases of practical interest. One is for $R=0$, which is the simple impulse turbine. The other is for the Parsons type of blading which has the same section for both the fixed and moving blades. This arrangement has the practical advantage that most of the blades used in a turbine of this type can be extruded from

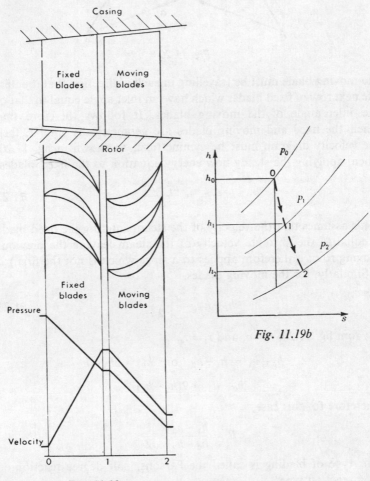

Fig. 11.19b

Fig. 11.19a

one set of dies. The two sets of blades, fixed and moving, are mounted in the relationship shown in fig. 11.19a, and the h-s diagram for the expansion is shown in fig. 11.19b. The moving blade exit angle

β_e is the same as the fixed blade exit angle α_i, and the velocity diagram for this blade arrangement is shown in fig. 11.20. The steam leaving

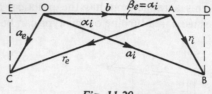

Fig. 11.20

the moving blade must be travelling in such a direction that it enters the next row of fixed blades which have an inlet angle equal to that of the inlet angle of the moving blades. It follows therefore that when the fixed and moving blades are geometrically similar, then the velocity diagram must be symmetrical, as shown in fig. 11.20. Then, applying the steady flow energy equation to the fixed blades,

$$h_0 - h_1 = \frac{a_i^2 - a_e^2}{2} \qquad 11.22$$

(This assumes that the velocity of the steam entering the fixed blade is equal to the absolute velocity of the steam leaving the previous moving row; it therefore applies to a stage which is not the first.)

Similarly, for the moving blades,

$$h_1 - h_2 = \frac{r_e^2 - r_i^2}{2} \qquad 11.23$$

From fig. 11.20, $r_e = a_i$ and $r_i = a_e$,

$$\therefore \quad h_0 - h_1 = h_1 - h_2 \quad \text{or} \quad h_0 = 2h_1 - h_2$$

i.e. $$h_0 - h_2 = 2(h_1 - h_2)$$

Therefore for this case,

$$R = \frac{h_1 - h_2}{h_0 - h_2} = \frac{1}{2}$$

This type of blading is called the Parsons, half degree reaction or 50% reaction type.

Blades are usually fitted in groups which have several rows of blades of the same mean radius and the same mean height. The enthalpy drop for each row is somewhat greater than that of the preceding row, and the degree of reaction is greater than 50%.

The energy input to the moving blade wheel can be written as

$$\frac{a_i^2}{2} + \frac{r_e^2 - r_i^2}{2}$$

Therefore, since $r_e = a_i$, this becomes

$$a_i^2 - \frac{r_i^2}{2}$$

From fig. 11.20,

$$r_i^2 = a_i^2 + b^2 - 2a_i b \cos \alpha_i$$

i.e.   Energy input $= a_i^2 - \left(\dfrac{a_i^2 + b^2 - 2a_i b \cos \alpha_i}{2} \right)$

$$= \frac{a_i^2 - b^2 + 2a_i b \cos \alpha_i}{2}$$

From equation 11.4,

Rate of doing work per kg/s $= wb$

Also, $w = ED = 2a_i \cos \alpha_i - b$

∴   Rate of doing work per kg/s $= b(2a_i \cos \alpha_i - b)$ 11.24

Therefore the diagram efficiency of the 50% reaction turbine is given by

$$\eta_d = \frac{\text{Rate of doing work}}{\text{Energy input}}$$

i.e. $\eta_d = \dfrac{2b(2a_i \cos \alpha_i - b)}{a_i^2 - b^2 + 2a_i b \cos \alpha_i}$

$$= \frac{2\dfrac{b}{a_i}\left(2 \cos \alpha_i - \dfrac{b}{a_i}\right)}{1 - \left(\dfrac{b}{a_i}\right)^2 + 2\left(\dfrac{b}{a_i}\right)\cos \alpha_i}$$ 11.25

(where b/a_i is the blade speed ratio).

Example 11.4

A stage of a turbine with Parsons blading delivers dry saturated steam at 2·7 bar from the fixed blades at 90 m/s. The mean blade height is 40 mm, and the moving blade exit angle is 20°. The axial velocity of the steam is 3/4 of the blade velocity at the mean radius. Steam is supplied to the stage at the rate of 9000 kg/h. The effect of the blade tip thickness on the annulus area can be neglected. Calculate,

(a) The wheel speed in rev/min.
(b) The diagram power.
(c) The diagram efficiency.
(d) The enthalpy drop of the steam in this stage.

The velocity diagram is shown in fig. 11.21a, and the blade wheel annulus is represented in fig. 11.21b.

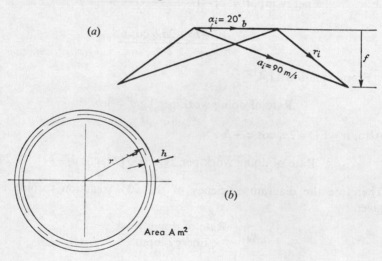

Fig. 11.21

(a) $$f = \frac{3}{4} b = a_i \sin \alpha_i = 90 \times \sin 20° = 30 \cdot 78 \text{ m/s}$$

$$\therefore b = \frac{4 \times 30 \cdot 78}{3} = 41 \cdot 04 \text{ m/s}$$

The mass flow of steam is given by

$$\dot{m} = \frac{fA}{v}$$

(where A is the annulus area, and v is the specific volume of the steam).

In this case, $v = v_g$ at 2·7 bar $= 0.6686$ m³/kg.

$$\therefore \quad \dot{m} = \frac{9000}{3600} = \frac{30.78A}{0.6686} \qquad \therefore \quad A = 0.054 \text{ m}^2$$

Now,   Annulus area, $A = 2\pi rh$

(where r is the mean radius, and h is the mean blade height).

$$\therefore \quad 0.054 = 2\pi r\, 0.04 \qquad \therefore \quad r = 0.215 \text{ m}$$

Then,

$$\text{Wheel speed, } N = \frac{\text{Blade speed}}{2\pi r} = \frac{60 \times 41.04}{2\pi \times 0.215} = 1824 \text{ rev/min}$$

(b) Using equation 11.5,

$$\text{Diagram power} = \dot{m}wb$$

Now,

$$w = 2a_i \cos \alpha_i - b = 2 \times 90 \times \cos 20° - 41.04 = 128.1 \text{ m/s}$$

$$\therefore \quad \text{Diagram power} = \frac{9000 \times 128.1 \times 41.04}{3600 \times 10^3} = 13.14 \text{ kW}$$

(c)

Rate of doing work per kg/s $= wb = 128.1 \times 41.04$ N m/s

Also.

$$\text{Energy input to the moving blades per stage} = a_i^2 - \frac{r_i^2}{2}$$

Referring to fig. 11.21a,

$$r_i^2 = a_i^2 + b^2 - 2a_i b \cos \alpha_i$$

$$= 90^2 + 41.04^2 - 2 \times 90 \times 41.04 \times \cos 20°$$

$$\therefore \quad r_i = 53.32 \text{ m/s}$$

i.e.   Energy input $= 90^2 - \dfrac{53.32^2}{2} = 6679$ N m per kg/s

$$\therefore \quad \text{Diagram efficiency, } \eta_d = \frac{128.1 \times 41.04}{6679} = 0.787 \quad \text{or} \quad 78.7\%$$

(d)

$$\text{Enthalpy drop in the moving blades} = \frac{r_e^2 - r_i^2}{2} = \frac{90^2 - 53 \cdot 32^2}{2 \times 10^3}$$

$$= 2 \cdot 63 \text{ kJ/kg}$$

$$\therefore \text{ Total enthalpy drop per stage} = 2 \times 2 \cdot 63 = 5 \cdot 26 \text{ kJ/kg}$$

Optimum operating conditions from the blade velocity diagram

The diagram efficiency for the 50% reaction wheel is given by equation 11.25 as,

$$\eta_d = \frac{2 \dfrac{b}{a_i}\left(2 \cos \alpha_i - \dfrac{b}{a_i}\right)}{1 - \left(\dfrac{b}{a_i}\right)^2 + 2\left(\dfrac{b}{a_i}\right)\cos \alpha_i}$$

By equating $d\eta_d/d(b/a_i)$ to zero, the value of blade speed ratio for maximum diagram efficiency can be shown to be given by,

$$\frac{b}{a_i} = \cos \alpha_i \qquad 11.26$$

From equation 11.24,

$$\text{Rate of doing work} = b(2a_i \cos \alpha_i - b)$$

$$\therefore \text{ For } \eta_{d\text{max}} \text{ rate of doing work} = b\left(2a_i \frac{b}{a_i} - b\right) = b^2 \qquad 11.27$$

Also, from equation 11.25,

$$\text{Maximum diagram efficiency} = \frac{2 \cos^2 \alpha_i}{1 + \cos^2 \alpha_i} \qquad 11.28$$

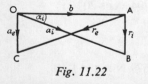

Fig. 11.22

For the optimum blade speed ratio a blade velocity diagram as shown in fig. 11.22 is obtained (i.e. $b = a_i \cos \alpha_i$).

The variation of η_d with blade speed ratio for the simple impulse and the reaction stage, are shown in fig. 11.23. It can be seen that for the reaction turbine the curve is reasonably flat in the region of the maximum value of diagram efficiency, so that a variation in cos α_i, and hence b/a_i, can be accepted without much variation in the diagram efficiency from the maximum value.

The variation of pressure and velocity through a reaction turbine is shown in fig. 11.24. The pressure falls continuously as the steam

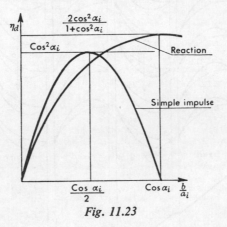

Fig. 11.23

passes over the fixed and moving blades of each stage. The steam velocities are low compared with those of the impulse turbine, and it can be seen from the diagram that the steam velocity is increased in each set of fixed blades. It is no longer convenient to talk of 'nozzles' and 'blades', since in the reaction turbine both fixed and moving blades act as nozzles. It is usual to refer to the two sets of blades as the stator blades and the rotor blades.

The pressure drop across the rotor produces an end thrust equal to the product of the pressure difference and the area of the annulus in contact with the steam. For the 50% reaction turbine the thrust due to the change in axial velocity is zero, but the side thrust is nevertheless greater than that of an equivalent impulse turbine, and larger thrust bearings are fitted. The net end thrust can be reduced by admitting the steam to the casing at the mid-section and allowing it to expand outwards to each end of the casing, passing over identical sets of blades. This has the additional feature of reducing the blade height at a given wheel for a given total mass flow of steam.

Turbines in practice very often consist of combinations of impulse and reaction types. A common combination is that of a Curtis stage followed by a series of reaction stages.

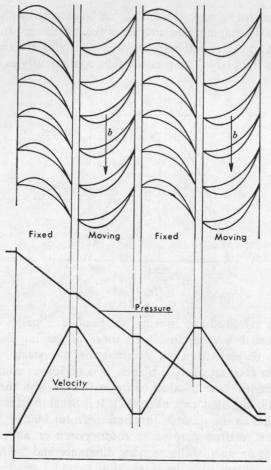

Fig. 11.24

11.6 Losses in steam turbines

In the previous sections it has been shown that the work done per
stage at optimum speed conditions for the simple impulse, two-row
impulse, and the 50% impulse-reaction turbines is given by,
$2\dot{m}b^2$, $8\dot{m}b^2$, and $\dot{m}b^2$, respectively. The arrangement of a particular
turbine to suit a specified purpose requires a consideration of other
factors. The most important of these is the losses involved. A detailed discussion of the experimental and associated theoretical
work which has been done on this topic will not be given here, and

reference 11.1 should be consulted for an authoritative treatment.

The losses which are of interest thermodynamically are the internal losses incurred as the steam passes through the blades. The losses may be classified in one of two groups:

(i) friction losses;   (ii) leakage losses

Group (i) includes friction losses in the nozzles, in the blades, and at the discs which rotate in the steam. Group (ii) includes losses at admission to the stages and leakage at glands, and the residual velocity loss, which has already been mentioned.

The friction loss in nozzles was mentioned in Section 10.5, and the effect of friction was allowed for by means of a nozzle efficiency defined by equation 10.11 as,

$$\text{Nozzle efficiency} = \frac{h_1 - h_{2'}}{h_1 - h_2}$$

(where $(h_1 - h_{2'})$ is the actual enthalpy drop in the nozzle, and $(h_1 - h_2)$ is the ideal isentropic enthalpy drop). Since the actual enthalpy drop is less than in the isentropic case, then the actual velocity of the steam leaving the nozzle is less than that obtained with isentropic expansion. The effect of friction in nozzle and blade passages is to cause losses which increase with the mean relative velocity through them, and with the surface area exposed to the steam. The losses are influenced by the nature of the flow, whether it is laminar or turbulent. The friction effects occur in the boundary layer on the surface and losses are higher in a turbulent boundary layer than in a laminar one. It is found that with curved blades the boundary layer is usually turbulent at the concave surface, and initially laminar at the convex surface. This laminar condition persists for some distance along the blade and then gives way to turbulence. In reaction stages, where a continuous pressure drop, and hence an acceleration of the steam, exists, the laminar condition persists over a greater length of the passage, so the friction loss is less than in the impulse stage. For a reaction turbine the enthalpy drop per stage is low and a large number of stages is required. This increases the blade surface area required, which increases the friction loss, but the average velocities are low and this helps to reduce friction losses.

In fig. 11.11 a blade wheel and diaphragm arrangement for a compounded impulse turbine is shown. It is evident that the wheel is rotating in a space full of steam. There is a loss at the surfaces of the wheel due to viscous friction of the steam on the wheel, and there is an admission loss as the steam passes from the nozzles to the wheel. Also, in impulse turbines the nozzles occupy only a part of the area opposite the blade annulus, and the blades pass areas in which they are not being served by nozzles. This tends to create eddies in the blade channels. The effects of this partial admission are referred to as blade windage losses.

The leakage loss in the impulse turbine between one stage and the next through the clearance space between the diaphragm and the shaft has been mentioned previously. Another leakage loss occurs at the external glands where the turbine shaft passes through the casing. At one end the tendency is for high pressure steam to escape into the atmosphere, and at the condenser end for air to leak in from the atmosphere. In diaphragm and external glands it is usual to use a form of labyrinth packing, examples of which are shown in figs. 11.25a and 11.25b. Fig. 11.25a shows a possible application to a diaphragm, and fig. 11.25b shows a section through part of an

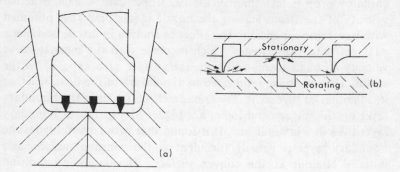

Fig. 11.25

external gland. At the low pressure gland it is usual to feed a supply of steam at low pressure to the centre of the gland. By this means there is a leakage of steam into the turbine, but the leakage of air into the turbine is prevented. For the diaphragm gland the labyrinth packing fits across the clearance space, but there is a small clearance at the tip through which some steam is throttled. For a more detailed description and discussion of the operation of this type of packing reference 11.1 should be consulted.

There is a leakage loss between the blade tips and the casing, and this is greater in the reaction turbine than in the impulse type due to the pressure difference across the clearance passage.

In reaction turbines the drum construction as shown in fig. 11.26 is usually preferred to the diaphragm and wheel construction of fig. 11.11.

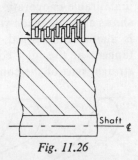

The effect of leakages is more important in the smaller turbines since, although the wheels, etc., can be scaled down, the working clearance cannot be reduced on the same scale. Leakages are highest where the pressure is high and are predominant over the friction losses in that

Fig. 11.26

case. It is advantageous to use impulse stages at the high pressure end of the turbine and reaction stages thereafter. At the lower pressures the friction losses become more important than the leakage losses.

11.7 Turbine blade profiles

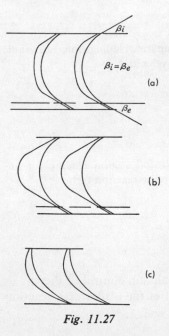

Fig. 11.27

The shape of the blades used in turbine construction is important from the point of view of design and performance, and also the ease of manufacture and attachment to the wheels or drum. Impulse blades are of the plate or profile types. Fig. 11.27a shows a somewhat refined plate type constructed from sheet metal with straight outlet edges to give a better control of the exit steam. This type of blade is easy to construct, is light, and is comparatively simple to mount. The channel section for steam flow increases to the centre and then decreases to the outlet. There is a break away of the steam at mid-section which creates eddies. This is avoided in the profile blade which is extended to take up the eddy space of the plate type. It is necessary to maintain a blade thickness at the edges to give the structure some rigidity to withstand the machining operations. This edge thickness is undesirable in operation as it disturbs the flow of steam. Fig. 11.27b shows a profile blade which is designed to give an inlet edge which is almost knife-edged. The inlet and exit angles are equal but the tangent at the inlet edge of the concave side diverges from the straight part of the convex side by a small amount (2 to 4 degrees).

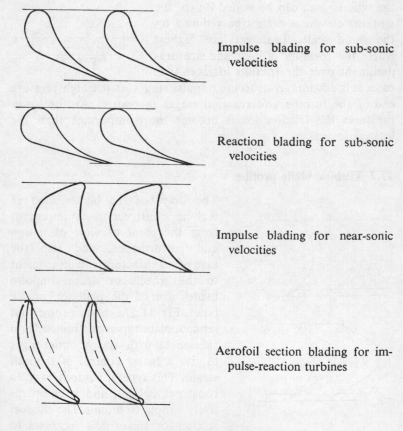

Impulse blading for sub-sonic velocities

Reaction blading for sub-sonic velocities

Impulse blading for near-sonic velocities

Aerofoil section blading for impulse-reaction turbines

Fig. 11.28

The blade has straight exit edges to maintain control of the jet.

Fig. 11.27c shows the typical section of the blades of a reaction turbine.

Recent developments in blade design are the result of experimental investigation into the effect of blade profile on the flow characteristics and friction losses. Details will not be given here, but fig. 11.28 shows the profile shapes proposed for different turbines and particular velocity ranges.

The mean blade speed varies from the value at the blade root to that at the blade tip. For short blades it is satisfactory to use a mean blade speed in design calculations. For long blades it may be necessary to account for this variation in blade speed in the design of the blade section. This increases the complexity and cost of manufacture of the blades.

11.8 Stage efficiency, overall efficiency, reheat factor, and the condition curve

It has been shown that as the steam expands through the turbine there are friction effects between the steam and the enclosing boundary surfaces of the nozzles and blade passages. Further losses are produced by leakage, which amount to a throttling of the steam. Both of these are irreversibilities in the expansion process and there is a reduction in the useful enthalpy drop which provides the work

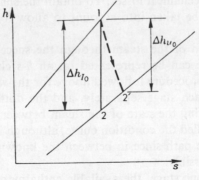

Fig. 11.29

output. The effect on the state of the steam is shown on an *h-s* chart in fig. 11.29. From state 1 an isentropic enthalpy drop would bring the steam to state 2. Due to friction and throttling effects the enthalpy drop is less, and there is an increase in entropy to give the actual state $2'$. The useful overall enthalpy drop is then given by $(h_1 - h_{2'})$, which will be denoted by Δh_{u_0}.

Applying the steady flow energy equation between the inlet and outlet of the turbine we have

$$h_1 + \frac{C_1^2}{2} + Q = h_{2'} + \frac{C_{2'}^2}{2} + W$$

It can be assumed that the flow is approximately adiabatic (i.e. $Q = 0$). Also the changes in velocity overall are usually negligible,

$$\therefore \quad W = h_1 - h_{2'} \qquad 11.29$$

Similarly, for an isentropic expansion through the turbine we have

$$W = h_1 - h_2 \qquad 11.30$$

Then the isentropic efficiency of the turbine is defined as

$$\text{Overall efficiency, } \eta_0 = \frac{h_1 - h_{2'}}{h_1 - h_2} = \frac{\Delta h_{u_0}}{\Delta h_{I_0}} \qquad 11.31$$

(where Δh_{I_0} is the isentropic overall enthalpy drop for the turbine between p_1 and p_2).

The overall efficiency so defined depends only on the change of properties of the steam during the expansion through the turbine. The work done at the turbine shaft will be less than $W = (h_1 - h_{2'})$, because of losses which do not affect the working fluid, and which can be called mechanical losses. To obtain the shaft work a further efficiency must be taken into account to allow for the mechanical losses.

The expansion of the steam through the successive stages of a reaction turbine can be represented on an h-s chart as shown in fig. 11.30. The procedure followed above for the whole turbine can be applied to each stage separately, and the dotted line joins the points representing the state of the steam between each stage. The dotted line is called the condition curve, although it does not give a continuous state path since in between the known points the processes are irreversible.

Considering one stage, the available enthalpy drop of the stage can be represented by Δh_u, and the isentropic enthalpy drop between the same pressures can be represented by Δh_I. Then a stage efficiency can be defined as,

$$\text{Stage efficiency, } \eta_s = \frac{\Delta h_u}{\Delta h_I} \qquad 11.32$$

From an inspection of fig. 11.30 it is seen that $BC < \Delta h_{I_2}$, $CD < \Delta h_{I_3}$, etc., since the lines of constant pressure diverge from left to right on the diagram,

i.e. $$\sum \Delta h_I > AB + BC + \text{ etc.}$$

i.e. $$\sum \Delta h_I > \Delta h_{I_0}$$

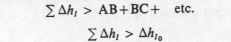

Fig. 11.30

From equation 11.32, $\Delta h_u = \eta_s \Delta h_I$, and if it can be assumed that the stage efficiency is the same for each stage, then,

$$\sum \Delta h_u = \eta_s \sum \Delta h_I$$

$$\therefore \ \Delta h_{u_0} = \eta_s \sum \Delta h_I$$

Dividing by Δh_{I_0} we have

$$\frac{\Delta h_{u_0}}{\Delta h_{I_0}} = \eta_s \frac{\sum \Delta h_I}{\Delta h_{I_0}}$$

or $$\eta_0 = \eta_s \times (\text{R.F.}) \qquad\qquad 11.33$$

(where R.F. is known as the *Reheat factor*; i.e. R.F. = $(\sum \Delta h_I)/(\Delta h_{I_0})$.

Since $\sum \Delta h_l$ is always greater than Δh_{l_0}, it follows that R.F. is always greater than unity. R.F. is usually of the order of 1·04.

It should be pointed out that if the inlet velocity to each stage is not equal to the outlet velocity from the stage, then the expression for stage efficiency can no longer apply.

Example 11.5

Steam at 15 bar and 350°C is expanded through a 50% reaction turbine to a pressure of 0·14 bar. The stage efficiency is 75% for each stage, and the reheat factor is 1·04. The expansion is to be carried out in twenty stages and the diagram power is required to be 12 000 kW. Calculate the flow of steam required, assuming that the stages all develop equal work. At one stage the pressure is 1 bar and the steam is dry saturated. The exit angle of the blades is 20°, and the blade speed ratio is 0·7. If the blade height is 1/12 of the blade mean diameter, calculate the value of the mean blade diameter and the rotor speed.

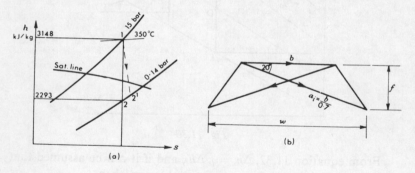

Fig. 11.31

The expansion is shown on an *h-s* chart in fig. 11.31a.
Using equation 11.33,

$$\eta_0 = \eta_s \times \text{R.F.} = 0.75 \times 1.04 = 0.78$$

From the chart, $h_1 = 3148$ kJ/kg, and state 1 is fixed.
Now $s_1 = s_2$, therefore state 2 is fixed, and $h_2 = 2293$ kJ/kg,

i.e.

Isentropic overall enthalpy drop $= 3148 - 2293 = 855$ kJ/kg

From equation 11.31,

$$\Delta h_{u_0} = \eta_0 \times \Delta h_{I_0} = 0.78 \times 855 = 667 \text{ kJ/kg}$$

i.e.   Enthalpy drop per stage $= \dfrac{667}{20} = 33.35 \text{ kJ/kg}$

Also,   Total diagram power $= \dot{m} \times \Delta h_{u_0}$

(where $\dot{m}$ is the mass flow in kg/s)

$$\therefore \dot{m} = \frac{12\,000}{667} = 17.99 \text{ kg/s}$$

i.e.   Steam mass flow $= 17.99 \times 3600 = 64\,770 \text{ kg/h}$

The blade velocity diagram for any one stage is shown in fig. 11.31b. From the diagram we have,

Work done per kg of steam $= wb = b(2a_i \cos 20° - b) \text{ N m}$

In this case $a_i = b/0.7 = 1.43b$

$$\therefore \text{ Work done per kg} = \frac{b}{10^3}(2 \times 1.43b \cos 20° - b) \text{ kJ}$$

$$= \frac{b^2}{10^3}(2 \times 1.43 \times \cos 20° - 1) = \frac{1.68b^2}{10^3} \text{ kJ}$$

Now work done per kg for one stage

$$= \text{enthalpy drop per stage}$$

$$= 33.35 \text{ kJ}$$

$$\therefore \frac{1.68b^2}{10^3} = 33.35 \quad \text{i.e.} \quad b = 141.4 \text{ m/s}$$

Also, from diagram,

$$f = a_i \sin 20° = 1.43b \sin 20° = 1.43 \times 141.4 \sin 20°$$

i.e. $f = 69.1 \text{ m/s}$

$$\therefore \text{ Volume flow per second at 1 bar} = f \times \pi D \times h$$

(where $D =$ blade mean diameter, and $h =$ blade height at this stage),

i.e.   Volume flow $= 69.1 \times \pi D \times \dfrac{D}{12} = 18.09 D^2 \text{ m}^3\text{/s}$

At 1 bar, $v_g = 1 \cdot 694$ m³/kg.

$$\therefore \text{ Mass flow} = \frac{18 \cdot 09 D^2}{1 \cdot 694} = 17 \cdot 99 \text{ kg/s}$$

i.e. $$D^2 = \frac{17 \cdot 99 \times 1 \cdot 694}{18 \cdot 09} \qquad \therefore \ D = 1 \cdot 298 \text{ m}$$

i.e.   Blade mean diameter $= 1 \cdot 298$ m

Then,   Blade speed, $b = \dfrac{\pi D N}{60}$

(where $N = $ rotor speed in rev/min).

$$\therefore \ 141 \cdot 4 = \frac{\pi \times 1 \cdot 298 \times N}{60} \qquad \therefore \ N = 2081 \text{ rev/min}$$

i.e.   Rotor speed $= 2081$ rev/min

11.9 Turbine governing and control

Turbines are designed to run at a particular load which is usually
the load to give maximum economy. The performance of the
turbine at other loads will be different from that at the design condition, and will depend to some extent on the means whereby the
speed of the turbine is governed to remain constant as the load is
changed.

The methods of governing used are as follows:

(a) *Throttle governing*

The simple throttle arrangement is shown in fig. 11.32. The stop

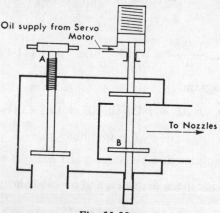

Fig. 11.32

valve A is opened for starting the turbine and for full load running. The double beat valve B is operated by an oil servo motor which is controlled by a centrifugal governor. As the speed of the turbine rises the valve B closes to throttle the steam, and reduces the supply to the nozzles.

The steam consumption plotted against the turbine load shows a linear relationship practically up to the maximum economic load. The steam rate can be used as a performance criterion for turbines working on the same cycles with similar conditions of steam pressure

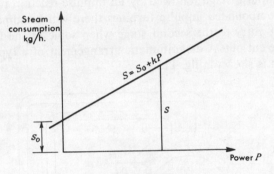

Fig. 11.33

and temperature. The graph of steam consumption against load is called the Willan's line for the turbine, and is shown in fig. 11.33.

Then, $$S = S_0 + kP \qquad 11.34$$

(where S_0 is the steam consumption when the turbine is giving zero output, S is the steam consumption when the turbine is developing the power, P, and k is a constant). S_0 is usually 0·10 to 0·14 of the full load steam consumption for a condensing turbine.

It is possible to allow for variations in back pressure on the turbine in the expression for Willan's law as follows:

$$S = k_1 p_b + k_2 P \qquad 11.35$$

(where k_1 and k_2 are constants, and p_b is the back pressure on the turbine).

(b) *Nozzle governing*

If throttling governing is carried out at low loads the efficiency of the turbine is considerably reduced. An alternative, and more efficient form of governing is by means of nozzle control. The nozzles are made up in sets, each set being controlled by a separate valve. The sets need not consist of equal numbers of nozzles, but are arranged to give the required degree of control. When a load lower than the design value is applied to the turbine the required number of nozzles can be shut off from the steam supply, and the rest will continue to receive steam at the supply pressure. There are several possible mechanical arrangements of nozzles and control valves. Nozzle control can only be applied to the first stage of a turbine. It is suitable for the simple impulse turbine and for larger units which have an impulse stage followed by an impulse-reaction turbine. In pressure compounded impulse turbines there will be some drop in pressure at entry to the second stage when some of the first stage nozzles are cut out. A diagrammatic arrangement of a typical control system is shown in fig. 11.34.

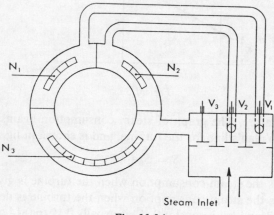

Fig. 11.34

With nozzle governing, the enthalpy drop available as the load is decreased is greater than with throttle governing.

(c) *By-pass governing*

With turbines designed to work at economic load it is desirable to have full admission of steam in the high-pressure stages. At the maximum load, which is greater than the economic load, the additional steam required could not pass through the first stage since additional nozzles are not available. By-pass regulation allows for this in a turbine which is throttle governed, by means of a second by-pass valve in the first stage nozzle chest (see fig. 11.35). This valve opens when the throttle valve has opened a definite amount. Steam

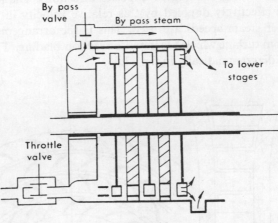

Fig. 11.35

is by-passed through the second valve to a lower stage in the turbine. When the by-pass valve operates it is under the control of the turbine governor. The secondary and tertiary supplies of steam in the lower stages increase the work output in these stages, but there is a loss in efficiency and a curving of the Willan's line.

With reaction turbines, because of the pressure drop required in the moving blades, nozzle control governing is not possible, and throttle governing plus by-pass governing, is used.

In modern power stations the turbines are designed to run at their maximum continuous rating, reduction of load being effected by shutting down a complete boiler-turbine set; the main throttle valves and reheat valves are automatically controlled to keep the speed within 4% of the rated value when going from full load to no load. The subject of governing and control has been considered in an elementary way and is a wide topic in itself. For further reading, reference 11.1 should be consulted.

11.10 The radial flow turbine (Ljungström turbine)

The turbines considered in the previous sections were of the axial flow type. In the radial flow turbine the steam is supplied near the

axis and expands radially outwards through a series of blades which
are fixed in concentric rings. This is called an outward flow radial
turbine, but inward flow turbines are also used. The concentric
rings of blades may be alternatively fixed and moving or all moving
with alternate rings moving in opposite directions. The blade velocities are effectively doubled by this relative velocity in the latter,
and higher steam speeds are used. This is the arrangement of the
Ljungström turbine which has impulse-reaction blading. The turbine
is shown diagrammatically in fig. 11.36.

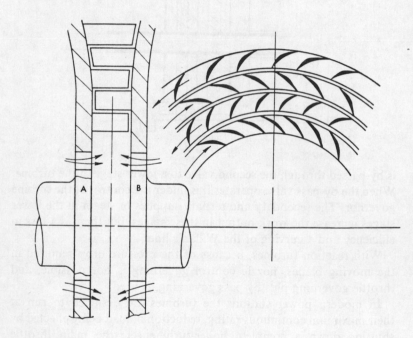

Fig. 11.36

The blade wheels A and B rotate in opposite directions and their
output shafts may be connected to different loads which operate at
the same speed. The steam is supplied to the centre of the turbine and
then expands radially outwards through the blades of wheels A and
B. The blades are of the 50% reaction type. For a more detailed
description and a theoretical analysis of this type of turbine, reference
11.1 should be consulted.

PROBLEMS

11.1 The velocity of steam at inlet to a simple impulse turbine is 1000 m/s, and the nozzle angle is 20°. The blade speed is 400 m/s and the blades are symmetrical. Determine the blade angles if the steam is to enter the blades without shock. If the friction effects on the blade are negligible, calculate the tangential force on the blades and the diagram power for a mass flow of 0·75 kg/s. What is the axial thrust and the diagram efficiency?

If the relative velocity at exit is reduced by friction to 80% of that at inlet, what is then the diagram power and the axial thrust? Calculate also the diagram efficiency in this case.

(32·5°; 810 N; 324 kW; 0; 86·2%; 291·5 kW; 51·29 N; 77·7%)

11.2 The steam from the nozzles of a single wheel impulse turbine discharges with a velocity of 600 m/s and at 20° to the plane of the wheel. The blade wheel rotates at 3000 rev/min and the mean blade radius is 590 mm. The axial velocity of the steam at exit from the blades is 164 m/s and the blades are symmetrical. Calculate,

(a) The blade angles.
(b) The diagram work per kg/s steam flow.
(c) The diagram efficiency.
(d) The blade velocity coefficient.

(28° 28′; 126·3 kJ per kg/s; 70%; 0·799)

11.3 The nozzles of the impulse stage of a turbine receive steam at 15 bar and 300°C and discharge it at 10 bar. The nozzle efficiency is 95% and the nozzle angle is 20°. The blade speed is that required for maximum work, and the inlet angle of the blades is that required for entry of the steam without shock. The blade exit angle is 5° less than the inlet angle. The blade velocity coefficient is 0·9. Calculate for a steam flow of 1350 kg/h.

(a) The diagram power.
(b) The diagram efficiency.   (30·3 kW; 86·3%)

11.4 The following particulars apply to a two-row velocity compounded impulse stage of a turbine:

Nozzle angle 17°; blade speed 125 m/s; exit angles of the first row moving blades, the fixed blades, and the second row moving blades, 22°, 26°, and 30° respectively. Take the blade velocity coefficient for each row of blades as 0·9, and assume that the absolute velocity of the steam leaving the stage is in the axial direction. Draw the velocity diagram for the stage and obtain,

(a) The absolute velocity of the steam leaving the stage.

(b) The diagram efficiency.

(72 m/s; 80%)

11.5 The first stage of a turbine is a two-row velocity compounded wheel. Steam at 40 bar and 400°C is expanded in the nozzles to 15 bar, and has a velocity at discharge of 700 m/s. The inlet velocity to the stage is negligible. The relevant exit angles are: nozzle 18°; first row blades 21°; fixed blades 26·5°; second row blades 35°. Take the blade velocity coefficient for all blades as 0·9. The mean diameter of the blading is 750 mm and the turbine shaft speed is 3000 rev/min. Draw the velocity diagram for this wheel and calculate,

(a) The diagram efficiency.

(b) The stage efficiency.

(70·8%; 67·4%)

11.6 For the turbine of problem 11.5 the mass flow of steam for each set of nozzles is 4·5 kg/s. Calculate the length of arc occupied by the nozzles if the nozzle height is 25 mm and the wall thickness between them is negligible. If the blades of the wheel have a pitch of 25 mm and the blade tip thickness at exit is 0·5 mm, calculate the blade exit height for each row.

(132·3 mm; 30·1 mm; 33·4 mm; 38·9 mm)

11.7 In a reaction stage of a steam turbine the nozzle angle is 20° and the absolute velocity of the steam at inlet to the moving blades is 240 m/s. The blade velocity is 210 m/s. If the blading is designed for 50% reaction, determine,

(a) The blade angle at inlet and exit.

(b) The enthalpy drop per kg of steam in the moving blades and in the complete stage.

(c) The diagram power for a steam flow of 1 kg/s.

(d) The diagram efficiency.

(77° 12′; 25·28 kJ/kg; 50·56 kJ/kg; 50·4 kW; 93·5%)

11.8 The speed of rotation of a blade group of a reaction turbine is 3000 rev/min. The mean blade velocity is 100 m/s. The blade speed ratio is 0·56 and the exit angle of the blades is 20°. If the mean specific volume of the steam is 0·65 m³/kg, and the mean height of the blades is 25 mm, calculate the mass flow of steam through the turbine in kg/h. Neglect the effect of blade thickness on the annulus area.

(16 900 kg/h)

11.9 In the blade group of problem 11.8 there are five pairs of blades. Calculate the useful enthalpy drop required for the group and the diagram power in kilowatts. (117·9 kJ/kg; 553 kW)

11.10 Calculate the optimum diagram efficiency for the reaction stage of problems 11.8 and 11.9, assuming that the nozzle angle and the axial velocity remain unchanged, the blade speed being adjusted. Calculate also the blade speed at optimum efficiency.

(94%; 167·7 m/s)

11.11 Ten stages of an ideal reaction turbine develop 3000 kW when the mass flow of steam is 18 000 kg/h. The mean value of the blade velocity is 0·8 of the steam velocity from the fixed blades. The exit angle of each blade is constant at 20° and the axial velocity is constant throughout the turbine. Calculate the inlet angle of the blades and the enthalpy drop in each moving row. (67·8°; 30 kJ/kg)

11.12 A reaction turbine is supplied with steam at 60 bar and 600°C. The condenser pressure is 0·07 bar. If the reheat factor can be assumed to be 1·04 and the stage efficiency is constant throughout at 80%, calculate the steam flow required in kg/h for a diagram power of 25 000 kW. (75 600 kg/h)

11.13 The steam consumption at full load for a 3000 kW turbine is 16 600 kg/h. At half load the steam consumption is 9100 kg/h. Applying Willan's line estimate the steam consumption at no load, and the consumption at a load of 2500 kW.

(1600 kg/h; 14 100 kg/h)

11.14 A reaction turbine expands 34 000 kg/h of steam from 20 bar, 400°C to a pressure of 0·2 bar. The turbine is designed such that the steam leaving is just dry saturated. The reheat factor is 1·05 and the isentropic efficiency of each stage is the same throughout. There are fourteen stages and the enthalpy drop is the same in each. All the blades have an exit angle of 22° and the mean value of the blade speed ratio is 0·82. Calculate the stage efficiency, the diagram power, the drum diameter, and the blade height for the last row of moving blades. The turbine speed is 2400 rev/min. Calculate also the pressure at the entry to the last stage and make a sketch on the *h-s* diagram showing the last stage expansion.

(67·7%; 6025 kW; 1·34 m; 175 mm; 0·32 bar)

REFERENCES

11.1 KEARTON, W. J., *Steam Turbine Theory and Practice* (Pitman & Sons, 1960).

11.2 DIXON, S. L., *Fluid Mechanics and Thermodynamics of Turbomachinery* (Pergamon, 1978).

12. Gas Turbines

The simple constant pressure cycle and the open and closed cycle gas turbine units are considered briefly in Chapter 6. In this chapter the various parts of the cycle will be considered in more detail and the practical limitations and modifications to the ideal cycle will be discussed.

The main use for the gas turbine at the present day is in the aircraft field, and in fact it is due to the work of Sir Frank Whittle, who invented the first jet engine in the years prior to the 1939–45 war, that the gas turbine unit has developed so rapidly. Large gas turbine units are used for electric power generation and for marine propulsion, but the oil engine and steam turbine are more frequently used in these fields. The gas turbine may have a future use in conjunction with the oil engine, and this possibility is discussed more fully in Chapter 18. The inefficiencies in the compression and expansion processes become greater for smaller gas turbine units and a heat exchanger is necessary in order to improve the thermal efficiency. A compact effective heat exchanger is necessary before the small gas turbine can compete for economy with the small oil engine or petrol engine.

The Holzwarth gas turbine, built from 1905 onwards, employed constant volume combustion, but this entails the use of valves and consequent intermittent operation. The use of constant pressure combustion with a rotary compressor driven by a rotary turbine, mounted on a common shaft, gives a combination which is ideal for conditions of steady mass flow over a wide operating range.

12.1 The practical gas turbine cycle

The most basic gas turbine unit is one operating on the open cycle in which a rotary compressor and a turbine are mounted on a common shaft, as shown diagrammatically in fig. 12.1. Air is drawn into the compressor, C, and after compression passes to a combustion chamber, CC. Energy is supplied in the combustion chamber by spraying fuel into the air stream, and the resulting hot gases expand through the turbine, T, to the atmosphere. In order to achieve net work output from the unit, the turbine must develop more gross work output than is required to drive the compressor and to overcome mechanical losses in the drive.

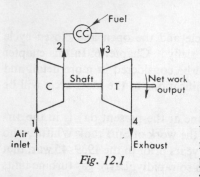

Fig. 12.1

The compressor used is either a centrifugal or an axial flow compressor (see Section 12.3), and the compression process is therefore irreversible but approximately adiabatic. Similarly the expansion process in the turbine is irreversible but adiabatic. Due to these irreversibilities, more work is required in the compression process for a given pressure ratio, and less work is developed in the expansion process. It is possible that the compressor and turbine may be so inefficient that the unit is not self-sustaining, and in fact it was the difficulties in improving the compressor and turbine design to cut down irreversibilities that retarded the development of the gas turbine unit.

As stated in Section 6.5 the open cycle gas turbine cannot be compared directly with the ideal constant pressure cycle. The actual cycle involves a chemical reaction in the combustion chamber which results in high temp. products which are chemically different from the reactants (see Section 15.9). During combustion there is no energy exchange with the surroundings, the effect being a gradual decrease in chemical energy with a corresponding increase in enthalpy of the working fluid. The combustion reaction will not be considered in detail here and a simplification will be made by assuming that the chemical energy released on combustion is equivalent to a transfer of heat at constant pressure to a working fluid of constant mean specific heat. This simplified approach allows the actual process to be compared with the ideal and to be represented on a *T-s* diagram.

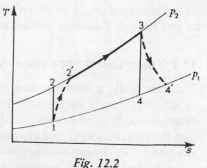

Fig. 12.2

Neglecting the pressure loss in the combustion chamber the cycle may be drawn on a T-s diagram as shown in fig. 12.2. Line 1–2′ represents irreversible adiabatic compression; line 2′–3 represents constant pressure heat supply in the combustion chamber; line 3–4′ represents irreversible adiabatic expansion. The process 1–2 represents the ideal isentropic process between the same pressures p_1 and p_2. Similarly the process 3–4 represents the ideal isentropic expansion process between the pressures p_2 and p_1. For the moment it will be assumed that the change in kinetic energy between the various points in the cycle is negligibly small compared with the enthalpy changes. Then applying the flow equation to each part of the cycle, we have, for unit mass:

For the compressor: Work input $= c_p(T_{2'} - T_1)$.
For the combustion chamber: Heat supplied $= c_p(T_3 - T_{2'})$.
For the turbine: Work output $= c_p(T_3 - T_{4'})$.

Then,   Net work output $= c_p(T_3 - T_{4'}) - c_p(T_{2'} - T_1)$

and   Thermal efficiency $= \dfrac{\text{Net work output}}{\text{heat supplied}}$

$$= \frac{c_p(T_3 - T_{4'}) - c_p(T_{2'} - T_1)}{c_p(T_3 - T_{2'})}$$

The value of the specific heat of a real gas varies with temperature; and also, in the open cycle, the specific heat of the gases in the combustion chamber and in the turbine is different from that in the compressor because fuel has been added and a chemical change has taken place. Curves showing the variation of c_p with temperature and air/fuel ratio can be used, and a suitable mean value of c_p and hence γ can be found. It is usual in gas turbine practice to assume fixed mean values of c_p and γ for the expansion process, and fixed mean values of c_p and γ for the compression process. For the combustion process curves as shown in fig. 12.25 (page 426) are used; for simple calculations a mean value of c_p can be assumed. In an open cycle gas turbine unit the mass flow of gases in the turbine is greater than that in the compressor due to the mass of fuel burned, but it is possible to neglect the mass of fuel, since the air/fuel ratios used are large. Also, in many cases, air is bled from the compressor for cooling purposes, or in the case of aircraft at high altitude, bleed air is used for de-icing and cabin air conditioning. This amount of air bleed is approximately the same as the mass of fuel injected.

The isentropic efficiency of the compressor is defined as the ratio of the work input required in isentropic compression between p_1 and p_2, to the actual work required.

Neglecting changes in kinetic energy, we have

$$\text{Compressor isentropic efficiency, } \eta_C = \frac{c_p(T_2 - T_1)}{c_p(T_{2'} - T_1)}$$

$$= \frac{T_2 - T_1}{T_{2'} - T_1} \qquad 12.1$$

Similarly the isentropic efficiency of the turbine is defined as the ratio of the actual work output, to the isentropic work output between the same pressures.

Neglecting kinetic energy changes,

$$\text{Turbine isentropic efficiency, } \eta_T = \frac{c_p(T_3 - T_{4'})}{c_p(T_3 - T_4)}$$

$$= \frac{T_3 - T_{4'}}{T_3 - T_4} \qquad 12.2$$

Example 12.1

A gas turbine unit has a pressure ratio of 6/1 and a maximum cycle temperature of 600°C. The isentropic efficiencies of the compressor and turbine are 0·82 and 0·85 respectively. Calculate the power output in kilowatts of an electric generator geared to the turbine when the air enters the compressor at 15°C at the rate of 15 kg/s. Take

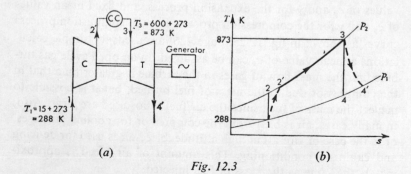

Fig. 12.3

$c_p = 1 \cdot 005$ kJ/kg K and $\gamma = 1 \cdot 4$ for the compression process, and take $c_p = 1 \cdot 11$ kJ/kg K and $\gamma = 1 \cdot 333$ for the expansion process.

A line diagram of the unit is shown in fig. 12.3a, and the cycle is shown on a T-s diagram in fig. 12.3b. In order to evaluate the net work output it is necessary to calculate the temperatures $T_{2'}$ and $T_{4'}$. To calculate $T_{2'}$ we must first calculate T_2 and then use the isentropic efficiency.

From equation 4.21, for an isentropic process,

$$\frac{T_2}{T_1} = \left(\frac{p_2}{p_1}\right)^{(\gamma - 1)/\gamma}$$

$$\therefore \ T_2 = 288 \times 6^{0 \cdot 4/1 \cdot 4} = 288 \times 1 \cdot 67 = 481 \text{ K}$$

Then using equation 12.1,

$$\eta_C = \frac{T_2 - T_1}{T_{2'} - T_1} = \frac{481 - 288}{T_{2'} - 288} = 0 \cdot 82$$

i.e.

$$(T_{2'} - 288) = \frac{193}{0 \cdot 82} = 235 \cdot 5 \text{ K}$$

$$\therefore \ T_{2'} = 288 + 235 \cdot 5 = 523 \cdot 5 \text{ K}$$

Similarly for the turbine,

$$\frac{T_3}{T_4} = \left(\frac{p_2}{p_1}\right)^{(\gamma - 1)/\gamma}$$

$$\therefore \ T_4 = \frac{873}{6^{0 \cdot 333/1 \cdot 333}} = \frac{873}{1 \cdot 564} = 558 \text{ K}$$

Then from equation 12.2,

$$\eta_T = \frac{T_3 - T_{4'}}{T_3 - T_4} = \frac{873 - T_{4'}}{873 - 558} = 0 \cdot 85$$

i.e.

$$(873 - T_{4'}) = 315 \times 0 \cdot 85 = 268 \text{ K}$$

$$\therefore \ T_{4'} = 873 - 268 = 605 \text{ K}$$

Hence,

$$\text{Compressor work input} = c_p(T_{2'} - T_1) = 1 \cdot 005 \times 235 \cdot 5$$
$$= 236 \cdot 2 \text{ kJ/kg}$$

$$\text{Turbine work output} = c_p(T_3 - T_{4'}) = 1 \cdot 11 \times 268$$
$$= 297 \cdot 5 \text{ kJ/kg}$$

$$\therefore \ \text{Net work output} = (297 \cdot 5 - 236 \cdot 2) = 61 \cdot 3 \text{ kJ/kg}$$

i.e.

$$\text{Power in kilowatts} = 61 \cdot 3 \times 15 = 920 \text{ kW}$$

Example 12.2

Calculate the thermal efficiency and the work ratio of the plant in example 12.1, assuming that c_p for the combustion process is $1 \cdot 11$ kJ/kg K.

$$\text{Heat supplied} = c_p(T_3 - T_{2'})$$

$$= 1 \cdot 11(873 - 523 \cdot 5) = 1 \cdot 11 \times 349 \cdot 5$$

i.e.   Heat supplied $= 388$ kJ/kg

Therefore,

$$\text{Thermal efficiency} = \frac{\text{net work output}}{\text{heat supplied}} = \frac{61 \cdot 3}{388}$$

i.e.   Thermal efficiency $= 0 \cdot 158 \quad \text{or} \quad 15 \cdot 8\%$

From the definition of work ratio given in Section 6.3, we have,

$$\text{Work ratio} = \frac{\text{net work output}}{\text{gross work output}} = \frac{61 \cdot 3}{297 \cdot 5} = 0 \cdot 206$$

In examples 12.1 and 12.2 the turbine is arranged to drive the compressor and to develop net work. It is sometimes more convenient to have two separate turbines, one of which drives the compressor

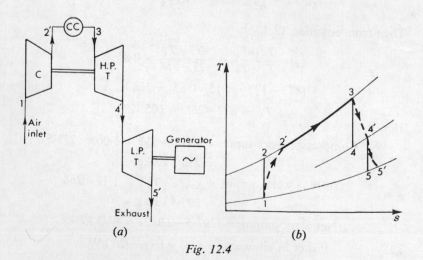

Fig. 12.4

while the other provides the power output. The first, or high-pressure turbine, is then known as the compressor turbine, and the second, or low-pressure turbine, is called the power turbine. The arrangement is shown in fig. 12.4a. Assuming that each turbine has its own isentropic efficiency, the cycle is as shown on a T-s diagram in fig. 12.4b. The numbers on fig. 12.4b correspond to those of fig. 12.4a. Neglecting kinetic energy changes, we have

Work from H.P. turbine = work input to compressor

i.e. $$c_{p_g}(T_3 - T_{4'}) = c_{p_a}(T_{2'} - T_1)$$

(where c_{p_g} and c_{p_a} are the specific heats at constant pressure of the gases in the turbine and the air in the compressor respectively).

The net work output is then given by the L.P. turbine,

i.e.   Net work output $= c_{p_g}(T_{4'} - T_{5'})$

Example 12.3

A gas turbine unit takes in air at 17°C and 1·01 bar and the pressure ratio is 8/1. The compressor is driven by the H.P. turbine and the L.P. turbine drives a separate power shaft. The isentropic efficiencies of the compressor, and the H.P. and L.P. turbines are 0·8, 0·85, and 0·83, respectively. Calculate the pressure and temperature of the gases entering the power turbine, the net power developed by the unit per kg/s mass flow, the work ratio and the thermal efficiency of the unit. The maximum cycle temperature is 650°C. For the compression process take $c_p = 1·005$ kJ/kg K and $\gamma = 1·4$; for the combustion process, and for the expansion process take $c_p = 1·15$ kJ/kg K and $\gamma = 1·333$. Neglect the mass of fuel.

The unit is as shown in figs. 12.4a and 12.4b.

From equation 4.21, for an isentropic process,

$$\frac{T_2}{T_1} = \left(\frac{p_2}{p_1}\right)^{(\gamma-1)/\gamma}$$

i.e. $T_2 = 290 \times 8^{0·4/1·4} = 290 \times 1·811 = 525$ K

Then, using equation 12.1,

$$\eta_C = \frac{T_2 - T_1}{T_{2'} - T_1} = \frac{525 - 290}{T_{2'} - 290} = 0·8$$

$$\therefore \ T_{2'} - 290 = \frac{235}{0·8}$$

i.e. $T_{2'} = 290 + 294 = 584$ K

Then, Work input to the compressor $= c_{p_a}(T_{2'} - T_1)$

$$= 1 \cdot 005 \times 294$$

$$= 295 \cdot 5 \text{ kJ/kg}$$

Now the work output from the H.P. turbine must be sufficient to drive the compressor,

i.e.

Work output from H.P. turbine $= c_{p_g}(T_3 - T_{4'}) = 295 \cdot 5 \text{ kJ/kg}$

$$\therefore \; T_3 - T_{4'} = \frac{295 \cdot 5}{1 \cdot 15} = 257 \text{ K}$$

$$\therefore \; T_{4'} = T_3 - 257 = 923 - 257 = 666 \text{ K}$$

Then, using equation 12.2,

$$\eta_{\text{T}} \text{ for H.P. turbine} = \frac{T_3 - T_{4'}}{T_3 - T_4} = \frac{923 - 666}{923 - T_4} = 0 \cdot 85$$

i.e.

$$923 - T_4 = \frac{257}{0 \cdot 85} = 302 \cdot 5 \text{ K}$$

$$\therefore \; T_4' = T_3 - 302 \cdot 5 = 923 - 302 \cdot 5 = 620 \cdot 5 \text{ K}$$

Then from equation 4.21 for an isentropic process,

$$\frac{T_3}{T_4} = \left(\frac{p_3}{p_4}\right)^{(\gamma - 1)/\gamma} \quad \text{or} \quad \frac{p_3}{p_4} = \left(\frac{T_3}{T_4}\right)^{\gamma/(\gamma - 1)}$$

$$= \left(\frac{923}{620 \cdot 5}\right)^{1 \cdot 333/0 \cdot 333} = 1 \cdot 488^4 = 4 \cdot 9$$

i.e.

$$p_4 = \frac{p_3}{4 \cdot 9} = \frac{8 \times 1 \cdot 01}{4 \cdot 9} = 1 \cdot 65 \text{ bar}$$

Hence the pressure and temperature at entry to the L.P. turbine are 1·65 bar and 393°C (where $t_{4'} = 666 - 273 = 393°C$).

To find the power output it is now necessary to evaluate $T_{5'}$. The pressure ratio, p_4/p_5, is given by, $(p_4/p_3) \times (p_3/p_5)$.

i.e.

$$\frac{p_4}{p_5} = \frac{p_4}{p_3} \times \frac{p_2}{p_1} \quad (\text{since } p_2 = p_3 \text{ and } p_5 = p_1)$$

$$\therefore \; \frac{p_4}{p_5} = \frac{8}{4 \cdot 9} = 1 \cdot 63$$

Then, $\dfrac{T_{4'}}{T_5} = \left(\dfrac{p_4}{p_5}\right)^{(\gamma-1)/\gamma} = 1{\cdot}63^{0{\cdot}333/1{\cdot}333} = 1{\cdot}131$

$$\therefore\ T_5 = \dfrac{666}{1{\cdot}131} = 588 \text{ K}$$

Then, using equation 12.2,

$$\eta_T \text{ for the L.P. turbine} = \dfrac{T_{4'}-T_{5'}}{T_{4'}-T_5}$$

i.e. $T_{4'} - T_{5'} = 0{\cdot}83(666-588) = 0{\cdot}83 \times 78 = 64{\cdot}8 \text{ K}$

Then,   Work output from L.P. turbine $= c_{p_g}(T_{4'}-T_{5'})$
$$= 1{\cdot}15 \times 64{\cdot}8$$
$$= 74{\cdot}5 \text{ kJ/kg}$$

Hence,   Net power output $= 74{\cdot}5 \times 1 = 74{\cdot}5$ kW

Work ratio $= \dfrac{\text{net work output}}{\text{gross work output}} = \dfrac{74{\cdot}5}{74{\cdot}5+295{\cdot}5} = \dfrac{74{\cdot}5}{370} = 0{\cdot}201$

Heat supplied $= c_{p_g}(T_3 - T_{2'}) = 1{\cdot}15(923-584)$

i.e.   Heat supplied $= 1{\cdot}15 \times 339 = 390$ kJ/kg

Then,   Thermal efficiency $= \dfrac{\text{net work output}}{\text{heat supplied}} = \dfrac{74{\cdot}5}{390}$

$$= 0{\cdot}191 \quad \text{or} \quad 19{\cdot}1\%$$

In a *jet engine* the propulsion nozzle takes the place of the low-pressure stage turbine, as shown diagrammatically in fig. 12.5a. The cycle is shown on a *T-s* diagram in fig. 12.5b, and it can be seen to be identical with fig. 12.4b. The aircraft is powered by the reactive thrust of the jet of gases leaving the nozzle, and this high velocity jet is obtained at the expense of the enthalpy drop from 4' to 5'. The turbine develops just enough work to drive the compressor and overcome mechanical losses. The jet engine will be dealt with more fully in Section 12.5.

In a *turbo-prop* engine the turbine drives the compressor and also the airscrew, or propeller, as shown in figs. 12.6a, and 12.6b. The net work available to drive the propeller is given by,

$$\text{Net work} = c_p(T_3 - T_{4'}) - c_p(T_{2'} - T_1)$$

(neglecting mechanical losses).

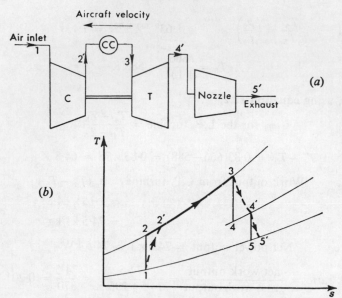

Fig. 12.5

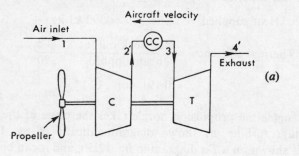

Fig. 12.6

In practice there is also a small jet thrust developed in a turbo-prop aircraft.

In some industrial and marine gas turbine units, the air flow is split into two streams after the compression process is completed. Some air is then passed to a combustion chamber which supplies hot gases to the turbine driving the compressor, while the rest of the air is passed to a second combustion chamber and from thence to the power turbine. The system is shown diagrammatically in fig. 12.7,

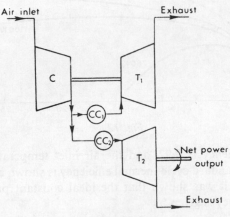

Fig. 12.7

and is called a *parallel flow* unit. In this system each turbine expands the gases received by it through the full pressure ratio. Parallel flow turbines are not so common as the more conventional series flow, and will not be considered in any detail.

12.2 Modifications to the basic cycle

It can be seen from examples 12.1, 12.2, and 12.3 that the work ratio and the thermal efficiency of the basic gas turbine cycle are low. These can be improved by increasing the isentropic efficiencies of the compressor and turbine, and this is a matter of blade design and manufacture.

In a practical cycle with irreversibilities in the compression and expansion processes the thermal efficiency depends on the maximum cycle temperature as well as on the pressure ratio. For fixed values of the isentropic efficiencies of the compressor and turbine, the thermal efficiency can be plotted against pressure ratio for various values of maximum temperature. This is illustrated in fig. 12.8, for a cycle in which the compressor isentropic efficiency is 0·89, the turbine

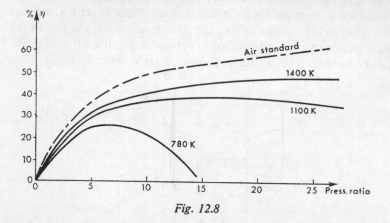

Fig. 12.8

isentropic efficiency is 0·92, and the air inlet temperature is 20°C. The ideal air standard cycle thermal efficiency is shown chain-dotted. In Section 6.4 it was shown that the ideal constant pressure cycle thermal efficiency is given by

$$\eta = 1 - \left(\frac{1}{r_p}\right)^{(\gamma - 1)/\gamma}$$

(where r_p is the pressure ratio) and is independent of the maximum cycle temperature.

It can be seen from fig. 12.8 that at any one fixed maximum cycle temperature there is a value of pressure ratio which will give maximum thermal efficiency.

The net work output also depends on the pressure ratio and on the maximum cycle temperature, and curves of specific power output in kW per kg/s against pressure ratio for various maximum temperatures are shown in fig. 12.9. The isentropic efficiencies of the compressor and turbine, and the air inlet temperature are the same as those used in deriving the curves of fig. 12.8. It can be seen that the thermal efficiency reaches a maximum at a different value of pressure ratio than the work output. The choice of pressure ratio is therefore a compromise.

The maximum cycle temperature is limited by metallurgical considerations. The blades of the turbine are under great mechanical stress and the temperature of the blade material must be kept to a safe working value. The temperature of the gases entering the turbine can be raised, provided a means of blade cooling is available.

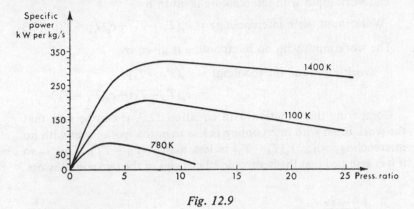

Fig. 12.9

Various methods of blade cooling have been investigated and a discussion of these will be found in reference 12.6. In order to have a basic gas turbine unit with a similar thermal efficiency to an oil engine of comparable power output, it would be necessary to have a maximum temperature of over 1100°C. In aircraft practice where the life expectancy of the engine is shorter, the maximum temperatures used are usually higher than those used in industrial and marine gas turbine units.

It is important to have as high a work ratio as possible, and methods of increasing the work ratio, such as intercooling between compressor stages, and reheating between turbine stages, will be considered in this section. Intercooling and reheating, while increasing the work ratio, can cause a decrease in the thermal efficiency, but when they are used in conjunction with a heat exchanger then intercooling and reheating increase both the work ratio and the thermal efficiency.

(a) *Intercooling*

When the compression is performed in two stages with an intercooler between the stages, then the work input for a given pressure ratio and mass flow is reduced. Consider a system as shown in fig. 12.10a; the *T-s* diagram for the unit is shown in fig. 12.10b. The actual cycle processes are 1–2' in the L.P. compressor, 2'–3 in the intercooler, 3–4' in the H.P. compressor, 4'–5 in the combustion chamber, and

5–6′ in the turbine. The ideal cycle for this arrangement is 1–2–3–4–5–6; the compression process without intercooling is shown as 1–A′ in the actual case, and 1–A in the ideal isentropic case.

The work input with intercooling is given by

Work input (with intercooling) $= c_p(T_{2'} - T_1) + c_p(T_{4'} - T_3)$ 12.3

The work input with no intercooling is given by

Work input (no intercooling) $= c_p(T_{A'} - T_1)$
$$= c_p(T_{2'} - T_1) + c_p(T_A - T_{2'})$$

Comparing this equation with equation 12.3, it can be seen that the work input with intercooling is less than the work input with no intercooling, when $c_p(T_{4'} - T_3)$ is less than $c_p(T_{A'} - T_{2'})$. This is so if it is assumed that the isentropic efficiencies of the two compressors,

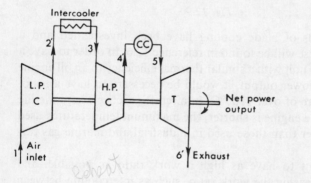

(a)

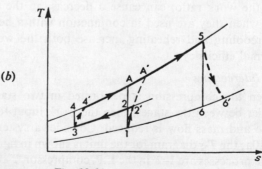

(b)

Fig. 12.10

operating separately, are each equal to the isentropic efficiency of the single compressor which would be required if no intercooling were used. Then, $(T_{4'} - T_3) < (T_{A'} - T_{2'})$ since the pressure lines diverge from left to right on the T-s diagram.

It can be shown that the best interstage pressure is the one which gives equal pressure ratios in each stage of compression; referring to fig. 12.10b this means that, $p_2/p_1 = p_4/p_3$. The work input required is a minimum when the pressure ratio in each stage is the same, and when the temperature of the air is cooled in the intercooler, back to the value at inlet to the unit (i.e. referring to fig. 12.10b, $T_3 = T_1$).

Now,

$$\text{Work ratio} = \frac{\text{net work output}}{\text{gross work output}}$$

$$= \frac{\text{work of expansion} - \text{work of compression}}{\text{work of expansion}}$$

It follows therefore, that when the compressor work input is reduced then the work ratio is increased. However, referring to fig. 12.10b, the heat supplied in the combustion chamber when intercooling is used in the cycle is given by,

$$\text{Heat supplied (with intercooling)} = c_p(T_5 - T_{4'})$$

Whereas, the heat supplied when intercooling is not used, with the same maximum cycle temperature T_5, is given by,

$$\text{Heat supplied (no intercooling)} = c_p(T_5 - T_{A'})$$

Hence the heat supplied when intercooling is used is greater than with no intercooling. Although the net work output is increased by intercooling it is found in general that the increase in the heat to be supplied causes the thermal efficiency to decrease. It will be shown later that this disadvantage is offset when a heat exchanger is also used.

When intercooling is used a supply of cooling water must be readily available. The additional bulk of the unit may offset the advantage to be gained by increasing the work ratio.

(b) *Reheat*

As stated earlier, the expansion process is very frequently performed in two separate turbine stages, the H.P. turbine driving the compressor and the L.P. turbine providing the useful power output. The work output of the L.P. turbine can be increased by raising the temperature at inlet to this stage. This can be done by placing a second combustion chamber between the two turbine stages in order to heat the gases leaving the H.P. turbine. The system is shown diagrammatically in fig. 12.11a, and the cycle is represented on a

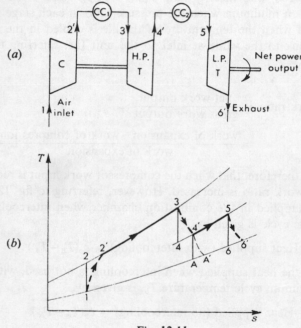

Fig. 12.11

T-s diagram in fig. 12.11b. The line 4′–A′ represents the expansion in the L.P. turbine if reheating is not used.

As before, the work output of the H.P. turbine must be exactly equal to the work input required for the compressor (neglecting mechanical losses),

i.e. $$c_{p_a}(T_{2'} - T_1) = c_{p_g}(T_3 - T_{4'})$$

The net work output, which is the work output of the L.P. turbine, is given by

$$\text{Net work output} = c_{p_g}(T_5 - T_{6'})$$

If reheating is not used, then the work of the L.P. turbine is given by

$$\text{Net work output (no reheat)} = c_{p_g}(T_{4'} - T_{A'})$$

Since pressure lines diverge to the right on the T-s diagram, it can be seen that the temperature difference $(T_5 - T_{6'})$, is always greater than $(T_{4'} - T_{A'})$, so that reheating increases the net work output.

Also,

$$\text{Work ratio} = \frac{\text{work of expansion} - \text{work of compression}}{\text{work of expansion}}$$

i.e. $$\text{Work ratio} = 1 - \frac{\text{work of compression}}{\text{work of expansion}}$$

Therefore, when the work of expansion is increased and the work of compression is unchanged, then the work ratio is increased.

Although the net work is increased by reheating, the heat to be supplied is also increased, and the net effect can be to reduce the thermal efficiency

i.e. $$\text{Heat supplied} = c_{p_g}(T_3 - T_{2'}) + c_{p_g}(T_5 - T_{4'})$$

However, the exhaust temperature of the gases leaving the L.P. turbine is much higher when reheating is used (i.e. $T_{6'}$ as compared with $T_{A'}$), and a heat exchanger can be used to enable some of the energy of the exhaust gases to be used.

(c) *Heat exchanger*

The exhaust gases leaving the turbine at the end of expansion are still at a high temperature, and therefore a high enthalpy (e.g. in

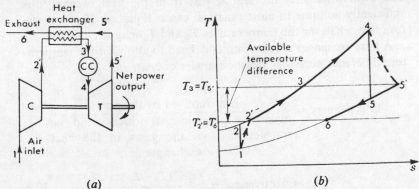

(a)

(b)

Fig. 12.12

example 12.3, $t_{5'} = 328 \cdot 2°C$). If these gases are allowed to pass into the atmosphere, then this represents a loss of available energy. Some of this energy can be recovered by passing the gases from the turbine through a heat exchanger, where the heat transferred from the gases is used to heat the air leaving the compressor. The simple unit with a heat exchanger added is shown diagrammatically in fig. 12.12a, and the cycle is represented on a T-s diagram in fig. 12.12b. In the ideal heat exchanger the air would be heated from $T_{2'}$ to $T_3 = T_{5'}$, and the gases would be cooled from $T_{5'}$ to $T_6 = T_{2'}$. This ideal case is shown in fig. 12.12b. In practice this is impossible, since a finite temperature difference is required at all points in the heat exchanger in order to overcome the resistance to the heat transfer. Referring to fig. 12.13, the required temperature difference between the gases and the air entering the heat exchanger is $(T_6 - T_{2'})$, and the required temperature difference between the gases and the air leaving the heat exchanger is $(T_{5'} - T_3)$.

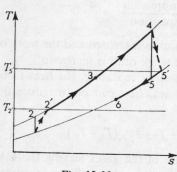

Fig. 12.13

If no heat is lost from the heat exchanger to the atmosphere, then the heat given up by the gases must be exactly equal to the heat taken up by the air,

i.e. $$\dot{m}_a c_{p_a}(T_3 - T_{2'}) = \dot{m}_g c_{p_g}(T_{5'} - T_6) \qquad 12.4$$

The assumption that no heat is lost from the heat exchanger is sufficiently accurate in most practical cases. Equation 12.4 is therefore true whatever the temperatures T_3 and T_6 may be.

A heat exchanger *effectiveness* is defined to allow for the temperature difference necessary for the transfer of heat,

i.e.

$$\text{Effectiveness} = \frac{\text{heat received by the air}}{\text{maximum possible heat which could be transferred from the gases in the heat exchanger}}$$

$$\therefore \text{Effectiveness} = \frac{\dot{m}_a c_{p_a}(T_3 - T_{2'})}{\dot{m}_g c_{p_g}(T_{5'} - T_{2'})} \qquad 12.5$$

A more convenient way of assessing the performance of the heat exchanger is to use a *thermal ratio*, defined as

$$\text{Thermal ratio} = \frac{\text{temperature rise of the air}}{\text{maximum temperature difference available}}$$

i.e. $$\text{Thermal ratio} = \frac{T_3 - T_{2'}}{T_{5'} - T_{2'}} \qquad 12.6$$

Comparing equations 12.5 and 12.6 it can be seen that the thermal ratio is equal to the effectiveness when the product, $\dot{m}_a c_{p_a}$, is equal to the product, $\dot{m}_g c_{p_g}$.

When a heat exchanger is used then the heat to be supplied in the combustion chamber is reduced, assuming that the maximum cycle temperature is unchanged. The net work output is unchanged and hence the thermal efficiency is increased.

Referring to fig. 12.13,

Heat supplied by the fuel (without heat exchanger) = $c_{p_g}(T_4 - T_{2'})$

Heat supplied by the fuel (with heat exchanger)   = $c_{p_g}(T_4 - T_3)$

A heat exchanger can only be used if there is a sufficiently large temperature difference between the gases leaving the turbine, and the air leaving the compressor.

For example, in the cycle shown in fig. 12.14 a heat exchanger could not possibly be used because the temperature of the exhaust gases, $T_{4'}$, is lower than the temperature of the air leaving the compressor, $T_{2'}$. In practice, although the gas temperature may be higher than the temperature of the air leaving the compressor, the difference in temperature may not be sufficiently large to warrant

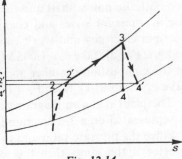

Fig. 12.14

the additional capital cost and subsequent maintenance required for a heat exchanger. Also, when the temperature difference is small in a heat exchanger, then the surface areas for the heat transfer must be made large in order to achieve a reasonably high value of the thermal ratio. In aircraft gas turbines, where the power/weight ratio is much more important than thermal efficiency or long life, the heat exchanger is not used; additional power is obtained by using higher pressure ratios, higher maximum temperatures, and by reheating. For small gas turbine units (e.g. for pumping sets or for motor cars) a compact heat exchanger must be designed before such units can hope to become competitive for economy with conventional internal combustion engines of equivalent power. In large gas turbine units for marine propulsion or industrial power, a heat exchanger is usually used.

Example 12.4

A 5000 kW gas turbine generating set operates with two compressor stages with intercooling between stages; the overall pressure ratio is 9/1. A high-pressure turbine is used to drive the compressors, and a low-pressure turbine drives the generator. The temperature of the gases at entry to the high-pressure turbine is 650°C and the gases are reheated to 650°C after expansion in the first turbine. The exhaust gases leaving the low-pressure turbine are passed through a heat exchanger to heat the air leaving the high-pressure stage compressor. The compressors have equal pressure ratios and intercooling is complete between stages. The air inlet temperature to the unit is 15°C. The isentropic efficiency of each compressor stage is 0·8, and the isentropic efficiency of each turbine stage is 0·85; the heat exchanger thermal ratio is 0·75. A mechanical efficiency of 98% can be assumed for both the power shaft and the compressor turbine shaft. Neglecting all pressure losses and changes in kinetic energy, calculate the thermal efficiency and work ratio of the plant, and the mass flow in kg/s. For air take $c_p = 1·005$ kJ/kg K and $\gamma = 1·4$, and for the gases in the combustion chamber and in the turbines and heat exchanger take $c_p = 1·15$ kJ/kg K and $\gamma = 1·333$. Neglect the mass of fuel.

The plant is shown diagrammatically in fig. 12.15a, and the cycle is represented on a T-s diagram in fig. 12.15b.

Since the pressure ratio and the isentropic efficiency of each compressor is the same, then the work input required for each compressor is the same since both compressors have the same air inlet temperature, i.e. $T_1 = T_3$ and $T_{2'} = T_{4'}$.

From equation 4.21,

$$\frac{T_2}{T_1} = \left(\frac{p_2}{p_1}\right)^{(\gamma-1)/\gamma} \quad \text{and} \quad \frac{p_2}{p_1} = \sqrt{9} = 3$$

$$\therefore \; T_2 = 288 \times 3^{0·4/1·4} = 394 \text{ K}$$

Then from equation 12.1,

$$\eta_C, \text{ L.P. compressor} = \frac{T_2 - T_1}{T_{2'} - T_1} = 0 \cdot 8$$

$$\therefore T_{2'} - T_1 = \frac{394 - 288}{0 \cdot 8} = \frac{106}{0 \cdot 8} = 132 \cdot 5 \text{ K}$$

i.e.

$$T_{2'} = 288 + 132 \cdot 5 = 420 \cdot 5 \text{ K}$$

Also, Work input per compressor stage $= c_{p_a}(T_{2'} - T_1)$

$$= 1 \cdot 005 \times 132 \cdot 5$$

$$= 133 \cdot 1 \text{ kJ/kg}$$

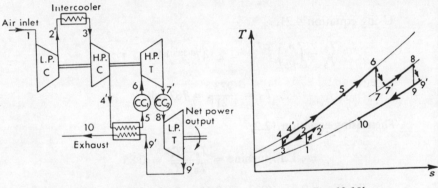

Fig. 12.15a   Fig. 12.15b

The H.P. turbine is required to drive both compressors and to over-come mechanical friction,

i.e.

$$\text{Work output of H.P. turbine} = \frac{2 \times 133 \cdot 1}{0 \cdot 98} = 272 \text{ kJ/kg}$$

$$\therefore c_{p_g}(T_6 - T_{7'}) = 272$$

i.e.

$$1 \cdot 15(923 - T_{7'}) = 272$$

$$\therefore 923 - T_{7'} = \frac{272}{1 \cdot 15} = 236 \cdot 5 \text{ K}$$

i.e.

$$T_{7'} = 923 - 236 \cdot 5 = 686 \cdot 5 \text{ K}$$

From equation 12.2,

$$\eta_T, \text{ H.P. turbine} = \frac{T_6 - T_{7'}}{T_6 - T_7} = 0.85$$

$$\therefore T_6 - T_7 = \frac{236.5}{0.85} = 278 \text{ K}$$

i.e. $$T_7 = 923 - 278 = 645 \text{ K}$$

Then using equation 4.21,

$$\frac{p_6}{p_7} = \left(\frac{T_6}{T_7}\right)^{\gamma/(\gamma-1)} = \left(\frac{923}{645}\right)^{1.333/0.333} = 4.19$$

Then, $$\frac{p_8}{p_9} = \frac{9}{4.19} = 2.147$$

Using equation 4.21,

$$\frac{T_8}{T_9} = \left(\frac{p_8}{p_9}\right)^{(\gamma-1)/\gamma} = 2.147^{0.333/1.333} = 1.211$$

$$\therefore T_9 = \frac{923}{1.211} = 762.6 \text{ K}$$

Then using equation 12.2,

$$\eta_T, \text{ L.P. turbine} = \frac{T_8 - T_{9'}}{T_8 - T_9} = 0.85$$

$$\therefore T_8 - T_{9'} = 0.85 \times (923 - 762.6) = 0.85 \times 160.4 = 136.3 \text{ K}$$

i.e. $$T_{9'} = 923 - 136.3 = 786.7 \text{ K}$$

Therefore,

$$\text{Net work output} = c_{p_g}(T_8 - T_{9'}) \times 0.98$$
$$= 1.15 \times 136.3 \times 0.98 = 153.7 \text{ kJ/kg}$$

From equation 12.6,

$$\text{Thermal ratio of heat exchanger} = \frac{T_5 - T_{4'}}{T_{9'} - T_{4'}} = 0.75$$

i.e.

$$T_5 - 420.5 = 0.75(786.7 - 420.5) = 0.75 \times 366.2 = 274.7 \text{ K}$$

$$\therefore T_5 = 420.5 + 274.7 = 695.2 \text{ K}$$

Now,

$$\text{Heat supplied} = c_{p_g}(T_6 - T_5) + c_{p_g}(T_8 - T_{7'})$$
$$= 1 \cdot 15(923 - 695 \cdot 2) + 1 \cdot 15(923 - 686 \cdot 5)$$

i.e.   Heat supplied $= 1 \cdot 15(227 \cdot 8 + 236 \cdot 5) = 534 \text{ kJ/kg}$

Then, from equation 5.2,

$$\text{Thermal efficiency} = \frac{W}{Q} = \frac{153 \cdot 7}{534} = 0 \cdot 288 \quad \text{or} \quad 28 \cdot 8\%$$

Gross work of the plant
$$= \text{work of H.P. turbine} + \text{work of L.P. turbine}$$

i.e. $\text{Gross work} = 272 + \dfrac{153 \cdot 7}{0 \cdot 98} = 429 \text{ kJ/kg}$

Therefore,

$$\text{Work ratio} = \frac{\text{net work output}}{\text{gross work output}} = \frac{153 \cdot 7}{429} = 0 \cdot 358$$

The electrical output is 5000 kW. Let the mass flow be $\dot{m}$ kg/s. Then

$$5000 = \dot{m} \times 153 \cdot 7$$

i.e. $\dot{m} = \dfrac{5000}{153 \cdot 7} = 32 \cdot 6 \text{ kg/s}$

i.e.   Rate of flow of air $= 32 \cdot 6 \text{ kg/s}$

12.3 Centrifugal and axial flow compressors

In the earliest gas turbine units for aircraft the centrifugal type of compressor was used. For low pressure ratios (no greater than about 4/1) the centrifugal compressor is lighter and is able to operate effectively over a wider range of mass flows at any one speed, than its axial flow counterpart.

For larger units with higher pressure ratios the axial flow compressor is more efficient and is usually preferred. For industrial and large marine gas turbine plants axial compressors are usually used, although some units may employ two or more centrifugal compressors with intercooling between stages. For aircraft the trend has been to higher pressure ratios, and the compressor is usually of the axial flow type. In aircraft units the advantage of the smaller diameter axial flow compressor can offset the disadvantage of the increased length and weight compared with an equivalent centrifugal compressor. However centrifugal compressors are cheaper to produce, more robust, less prone to icing troubles at high altitudes, and have a wider operating range than the axial flow types.

All design is a compromise, and it may be that under certain circumstances two centrifugal compressors in series may be preferred to an axial compressor. Perhaps the best example of this was in the series of Rolls-Royce Dart turbo-prop engines which used pressure ratios from 5·4/1 to 6·35/1 with two centrifugal compressor stages.

Each type of compressor will be described briefly. The design of the blading of a rotary compressor is more the province of the aerodynamicist particularly in the case of axial flow compressors for high speed aircraft. For a more extensive theoretical treatment references 12.1 and 12.2 should be consulted.

(a) *Centrifugal compressor*

A centrifugal compressor consists of an impeller with a series of curved radial vanes as shown in fig. 12.16. Air is drawn in near the

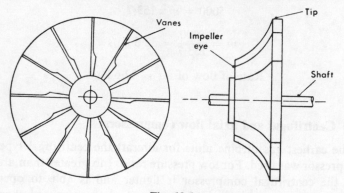

Fig. 12.16

hub, called the impeller eye, and is whirled round at high speed by the vanes on the impeller as the impeller rotates at high rotational speed. The static pressure of the air increases from the eye to the tip of the impeller in order to provide the centripetal force on the air. As the air leaves the impeller tip it is passed through diffuser passages which convert most of the kinetic energy of the air into an increase in enthalpy, and hence the pressure of the air is further increased.

The complete compressor is shown diagrammatically in fig. 12.17. The air from the discharge scroll passes to the combustion chamber.

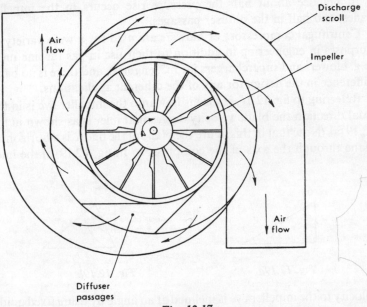

Fig. 12.17

The impeller may be double-sided, having an eye on either side of the compressor, so that air is drawn in on both sides, as shown

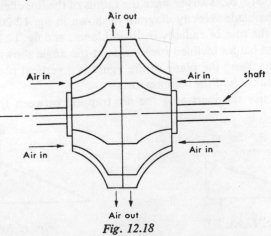

Fig. 12.18

diagrammatically in fig. 12.18. The advantage of this type is that the impeller is subjected to approximately equal forces in an axial direction.

In practice about half the pressure rise occurs in the impeller vanes, and half in the diffuser passages.

Centrifugal compressors or blowers are used for a wide variety of purposes in engineering in addition to their use in gas turbine units (e.g. blowers and superchargers for I.C. engines), and there is no basic difference in the design for any of the different applications.

Referring to fig. 12.16, if the air flow into the impeller eye is in the axial direction the blade velocity diagram at inlet is as shown in fig. 12.19 (a). (Note that in this figure the plane of the figure is a horizontal plane through the axis of the compressor.) In fig. 12.19 (b) the inlet

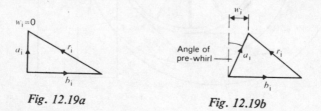

Fig. 12.19a

Fig. 12.19b

velocity to the impeller eye is inclined at an angle by using fixed guide vanes; this is known as pre-whirl. The symbols used in this section are as defined in Chapter 11, page 350.

At exit from the impeller the flow is in the radial direction and the blade velocity, b_e, is larger since the radius of the impeller is larger at outlet. The blade velocity diagram is shown in fig. 12.20; fig. 12.20 (a) is for the case of radially inclined blades, and fig. 12.20 (b) is for the case of blades inclined backwards at the angle shown, β_e. (Note that in this figure the plane of the figure is a vertical plane at right angles to the axis of the compressor.)

In practice the inertia of the air trapped between the impeller

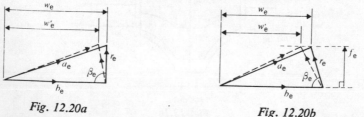

Fig. 12.20a

Fig. 12.20b

blades causes the actual whirl velocity at exit, w'_e, to be less than w_e; this phenomenon is known as *slip*.

$$\text{Slip factor} = \frac{w'_e}{w_e} = \frac{w_e'}{b_e - f_e \cot \beta_e}$$

Also, using equation 11.5,

$$\text{Power input} = \dot{m}\,(b_e\,w'_e - b_i\,w_i)$$

Example 12.5

A centrifugal compressor has a pressure ratio of 4/1 with an isentropic efficiency of 80% when running at 15 000 rev/min and inducing air at 20°C. Guide vanes at inlet give the air a pre-whirl of 25° to the axial direction at all radii and the mean diameter of the eye is 250 mm; the absolute air velocity at inlet is 150 m/s. At exit the blades are radially inclined and the impeller tip diameter is 590 mm. Calculate the slip factor of the compressor.

Temperature after isentropic compression

$$= (20 + 273) \times 4^{0\cdot286}$$

$$= 293 \times 1\cdot486 = 435\cdot4\,\text{K}$$

i.e. Isentropic temperature rise $= 435\cdot4 - 293 = 142\cdot4\,\text{K}$

$\therefore \quad$ Actual temperature rise $= \dfrac{142\cdot4}{0\cdot8} = 178\,\text{K}$

i.e. Power input per unit mass flow rate

$$= c_p \times \text{Actual temperature rise}$$

$$= 1\cdot005 \times 178 = 178\cdot9\,\text{kJ}$$

Referring to fig. 12.19(b):

$a_i = 150$ m/s (given), and the angle of pre-whirl is given as 25°.

$$b_i = \frac{15\,000 \times \pi \times 250}{60 \times 10^3}\,\text{m/s} = 196\cdot4\,\text{m/s}$$

and, $\quad w_i = a_i \sin 25 = 150 \times \sin 25 = 63\cdot4$ m/s

At exit, referring to fig. 12.20 (a):

$$b_e = \frac{15\,000 \times \pi \times 590}{60 \times 10^3} \text{ m/s} = 463 \cdot 4 \text{ m/s}$$

i.e. $w_e = 463 \cdot 4$ m/s, since the blades are radial.

Also,

Power input per unit mass flow rate $= b_e w'_e - b_i w_i$

$$= 178 \cdot 9 \text{ kJ (see above)}$$

i.e. $\qquad 178 \cdot 9 \times 10^3 = 463 \cdot 4 w'_e - 196 \cdot 4 \times 63 \cdot 4$

$\therefore \qquad\qquad\qquad w'_e = 412 \cdot 9$ m/s

Hence,

$$\text{Slip factor} = \frac{w'_e}{w_e} = \frac{412 \cdot 9}{463 \cdot 4} = 0 \cdot 89$$

(b) *Axial compressor*

An axial flow compression stage consists of a row of moving blades arranged round the circumference of a rotor, and a row of fixed blades arranged round the circumference of a stator. The air flows axially through the moving and fixed blades in turn; stationary guide vanes are provided at entry to the first row of moving blades. Basically, the compression is performed in a similar manner to that of the centrifugal type. The work input to the rotor shaft is transferred by the moving blades to the air, thus accelerating it. The blades are arranged so that the spaces between blades form diffuser passages, and hence the velocity of the air relative to the blades is decreased as the air passes through them, and there is a rise in pressure. The air is then further diffused in the stator blades, which are also arranged to form diffuser passages. In the fixed stator blades the air is turned through an angle so that its direction is such that it can be allowed to pass to a second row of moving rotor blades. It is usual to have a relatively large number of stages and to maintain a constant work input per stage (e.g. from 5 to 14 stages have been used).

The necessary reduction in volume may be allowed for by flaring the stator or by flaring the rotor. It is more common to use a flared rotor, and this type is shown diagrammatically in fig. 12.21. The rotor is built up of discs of steel or light alloy and the blades are fitted into tee-shaped, or dove-tailed, slots in the periphery of the disc. The stator blades are normally spot welded onto a ring at one end of the blade, and loosely fitted to a ring at the other end, to allow for expansion of the blade. This annulus of blades is then fixed to the casing by set screws. There are many possible alternative methods of blade fixing, and rotor and stator design; the above brief description has been included to give a general impression only.

It is usually arranged to have an equal temperature rise in the moving and the fixed blades, and to keep the axial velocity of the air constant throughout the compressor. Thus each stage of the compression is exactly similar with regard to air velocity and blade inlet and outlet angles.

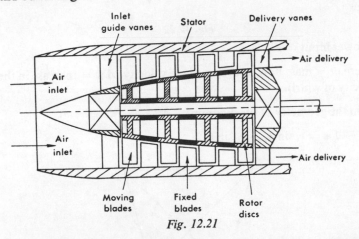

Fig. 12.21

A diffusing flow is less stable than a converging flow, and for this reason the blade shape and profile is much more important for a compressor than for a reaction turbine. The design of compressor blades is based on aerodynamic theory and an aerofoil shape is used.

The design of turbine blading is similar to that of steam turbine blading dealt with in Chapter 11. In aircraft practice the size of the turbine must be kept to a minimum, and high gas velocities are used so that the gas flow can pass through smaller annulus areas. Thus a larger power output per stage is obtained in aircraft practice.

Typical blade sections are shown in fig. 12.22; note that the convention is to measure the blade angles from the axial direction and not the tangential direction as in the case of steam turbines. The corresponding blade velocity diagram is shown in fig. 12.23.

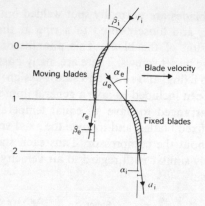

Fig. 12.22

Power input $= \dot{m} \, b \, \Delta w$

(where $\dot{m} =$ mass flow rate; $b =$ blade velocity; $\Delta w =$ increase in the velocity of whirl).

From the geometry of the diagram,

$$\Delta w = r_i \sin \beta_i - r_e \sin \beta_e$$

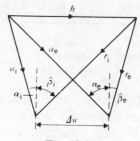

Fig. 12.23

As with the impulse-reaction steam turbine a degree of reaction is defined,

i.e.   Degree of reaction $= \dfrac{\text{Enthalpy rise in rotor}}{\text{Enthalpy rise in the stage}}$

$$= \frac{h_1 - h_0}{h_2 - h_0}$$

$$= \frac{r_i^2 - r_e^2}{(r_i^2 - r_e^2) + (a_e^2 - a_i^2)}$$

In practice the blade angles are made geometrically similar in the fixed and moving blade rows and hence the degree of reaction is 50% as in the case of the Parsons turbine (see Chapter 11).

Due to the non-uniformity of the velocity profile in the blade passages the work that can be put into a given blade passage is less than that given by the ideal diagram. A work done factor is introduced, defined by:

$$\text{Work done factor} = \frac{\text{Actual power input}}{\dot{m} \, b \, \Delta w}$$

This is usually about 0·85 for a compressor stage; for a turbine stage the velocity profile is much more uniform and this effect can be neglected.

Example 12.6

In an axial flow air compressor producing a pressure ratio of 6/1 with air entering at 20°C the mean velocity of the rotor blades is 200 m/s and the inlet and exit angles of both the moving and the fixed blades are 45° and 15° respectively. The degree of reaction is 50%, the work done factor is 0·86 throughout, there are twelve stages, and the axial velocity may be taken as constant through the compressor.

Calculate the isentropic efficiency of the compressor.

Explain why the pressure ratio for any stage is *not* equal to the overall pressure ratio to the power of $1/n$ where n is the number of stages.

By drawing the blade diagram to scale (see fig. 12.23), the value of Δw may be found,

i.e. $\Delta w = 115 \text{ m/s}$

Specific power input per stage $= b \, \Delta w \times$ work done factor

$$= 200 \times 115 \times 0.86 = 19\,780 \text{ J}$$

$$= 19.78 \text{ kJ}$$

i.e.   Compressor specific power input
$$= 12 \times 19.78 = 237.4 \text{ kJ}$$

For isentropic compression,

Exit temperature $= (20 + 273) \times 6^{0.286} = 293 \times 1.669 = 489.1 \text{ K}$

i.e. Isentropic specific power input $= 1 \cdot 005 (489 \cdot 1 - 293) = 197 \cdot 1$ kJ

$\therefore$ Compressor isentropic efficiency $= \dfrac{197 \cdot 1}{237 \cdot 4} = 0 \cdot 83$ or 83%

Each stage has the same rise in temperature since the specific work input to each stage is the same (i.e. $b \, \Delta w$). It is also a reasonable assumption that the stage efficiency is the same for all stages,

i.e. Isentropic temperature rise for any stage $= \eta_s \times \dfrac{b \, \Delta w}{c_p} \times Y$

(where $\eta_s =$ stage efficiency; $Y =$ work done factor)
Let the temperature at inlet to the compressor be T_1.
Then at the yth stage of the compressor,

$$\text{Actual temperature at exit} = T_1 + y \frac{(b \, \Delta w)}{c_p} Y$$

$$\text{Actual temperature at inlet} = T_1 + (y - 1) \frac{(b \, \Delta w)}{c_p} \, Y$$

$$\frac{\text{Isentropic temperature rise}}{\text{Temperature at inlet}} = \frac{\dfrac{\eta_s \, b \, \Delta w \, Y}{c_p}}{T_1 + (y - 1) \dfrac{(b \, \Delta w)}{c_p} \, Y}$$

$$\frac{\text{Isentropic temperature at exit}}{\text{Temperature at inlet}} = \frac{\eta_s \, b \, \Delta w \, Y}{c_p \, T_1 + (y - 1) \, b \, \Delta w \, Y} + 1$$

$$\text{Pressure ratio for the } y\text{th stage} = \left\{ \frac{\eta_s \, b \, \Delta w \, Y}{c_p \, T_1 + (y - 1) \, b \, \Delta w \, Y} + 1 \right\}^{\gamma/\gamma - 1}$$

Since η_s, b, Y and Δw are constant throughout the compressor, and T_1 is fixed, it follows that the pressure ratio for any stage varies according to the number of the stage, y. As y increases the pressure ratio decreases. The overall pressure ratio for the compressor must be equal to the product of the stage pressure ratios but since these are not equal then the overall pressure ratio is *not* equal to the pressure ratio for a stage raised to the power of n, where n is the number of stages.

12.4 Combustion

In the closed cycle gas turbine unit heat is transferred to the air in a heat exchanger, but in the open cycle unit the fuel must be sprayed into the air continuously, and combustion is a continuous process unlike the cyclic combustion of the I.C. engine.

There are two main combustion systems for open cycles; one in which the air leaving the compressor is split into several streams and each stream is supplied to a separate cylindrical 'can' type combustion chamber; and the other in which the air flows from the compressor through an annular combustion chamber. The annular type would appear to be more suitable for a unit using an axial flow compressor, but it is difficult to obtain good fuel/air distribution and research and development work on this type is harder than with the simpler can type. The annular type can be modified by having a series of interconnected cans placed in a ring; this is known as the cannular type. In aircraft practice at present the majority of engines use either the cannular or the can type of combustion chamber.

In industrial plants where space is not important, the combustion may be arranged to take place in one or two large cylindrical combustion chambers with ducting to convey the hot gases to the turbine; this sytem gives better control over the combustion process.

In all types of combustion chamber, combustion is initiated by electrical ignition, and once the fuel starts burning, a flame is stabilized in the chamber. In the can type it is usual to have inter-

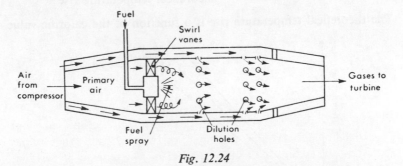

Fig. 12.24

connecting pipes between cans, to stabilize the pressure and to allow combustion to be initiated by a spark in one chamber on starting up. A typical can type chamber is shown diagrammatically in fig. 12.24. Some of the air from the compressor is introduced directly to the fuel burner; this is called primary air, and represents about 25% of the total air flow. The remaining air enters the annulus round the flame tube, thus cooling the upper portion of the flame tube, and then enters the combustion zone through dilution holes as shown in fig. 12.24. The primary air forms a comparatively rich mixture and the temperature is high in this zone. The air entering the dilution holes completes the combustion and helps to stabilize the flame in the high-temperature region of the chamber. In some combustion chambers the fuel is injected upstream into the air flow, and a sheet metal cone and perforated baffle plate ensure the necessary mixing of the fuel and air.

The air/fuel ratio overall is of the order of 60/1 to 120/1, and the air velocity at entry to the combustion chamber is usually not more than 75 m/s. There is a rich and a weak limit for flame stability, and the limit is usually taken at flame blow-out. Instability of the flame results in rough running with consequent effect on the life of the combustion chamber.

It should be noted that because of the high air/fuel ratios used, the gases entering the H.P. turbine contain a high percentage of oxygen, and therefore if reheating is performed between turbine stages, the additional fuel can be burned satisfactorily in the exhaust gas from the H.P. turbine.

A combustion efficiency may be defined as follows,

$$\text{Combustion efficiency} = \frac{\text{actual temperature rise}}{\text{theoretical temperature rise}} \qquad 12.7$$

The theoretical temperature rise is a function of the calorific value

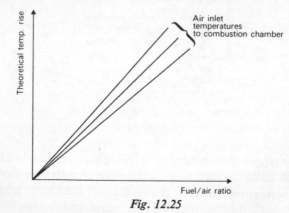

Fig. 12.25

of the fuel used, the fuel/air ratio, and the initial temperature of the air. The theoretical temperature rise for any one fuel of known calorific value can be plotted against the fuel/air ratio for various values of air inlet temperature to the chamber, and curves of the form shown in fig. 12.25 are obtained. The combustion efficiency can be evaluated by testing the chamber, and traversing the inlet and outlet sections to obtain true mean readings of the inlet and outlet total temperature. The fuel most commonly used in British gas turbine practice is kerosene which has a calorific value of approximately 43 300 kJ/kg.

In order to give a comparison of combustion chambers of different size operating under different ambient conditions, a *combustion intensity* is defined,

i.e.   Combustion intensity

$$= \frac{\text{rate of heat supply by fuel}}{\text{volume of combustion chamber}} \quad 12.8$$
$$\times \text{inlet pressure in atmospheres}$$

There is a pressure loss in the combustion chamber which is mainly due to friction and turbulence. There is also a small drop in pressure due to non-adiabatic flow in a duct of approximately constant cross-sectional area. The loss due to friction can be found experimentally by blowing air through the combustion chamber without initiating combustion and measuring the change in total pressure. This friction loss in pressure is therefore called the *cold loss*. The loss due to the heating process alone is called the *fundamental loss*. For a more extensive treatment of the combustion process reference 12.3 should be consulted.

12.5 Jet propulsion

Aircraft propulsion may be achieved by using a heat engine to drive an air-screw or propeller, or by allowing a high-energy fluid to expand and leave the aircraft in a rearward direction as a high-velocity jet. In the propeller type of aircraft engine the propeller takes a large mass flow and gives it a moderate velocity backwards relative to the aircraft. In the jet engine the aircraft induces a comparatively small air flow and gives it a high velocity backwards relative to the aircraft. In both cases the rate of change of momentum of the air provides a reactive forward thrust which propels the aircraft.

The propeller-type engine can be driven by a petrol engine or by a gas-turbine unit. By the end of the 1939–45 war the piston-engined aircraft had reached the limit of its development, and the gas-turbine unit with its higher power/weight ratio has now completely taken over for all but the very smallest aeroplanes.

If the velocity of the jet (from propeller or jet engine), backwards relative to the aircraft, is C_j, and the velocity of the aircraft is C_a, then the atmospheric air, initially at rest, is given a velocity of $(C_j - C_a)$. This is illustrated in fig. 12.26. Assuming for the moment

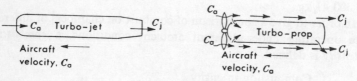

Fig. 12.26

that the jet leaves the aircraft in the case of the jet engine at atmospheric pressure, then there is no thrust due to pressure forces. The thrust available for propulsion is solely due to the rate of change of momentum of the stream,

i.e. $$\text{Thrust} = C_j - C_a \quad \text{N per kg/s flow} \qquad 12.9$$

The propulsive power is then given by

$$\text{Thrust power} = C_a(C_j - C_a) \quad \text{W per kg/s} \qquad 12.10$$

This is the rate at which work must be done in order to keep the aircraft moving at the constant velocity C_a against the frictional resistance or drag.

The net work output from the engine is given by the increase in kinetic energy, $(C_j^2 - C_a^2)/2$. This work output is used in two ways; it provides the thrust work as given by equation 12.10, and it gives the air a kinetic energy of $(C_j - C_a)^2/2$, i.e. the air previously at rest is given an absolute velocity of $(C_j - C_a)$,

i.e. $$C_a(C_j - C_a) + \frac{(C_j - C_a)^2}{2} = C_aC_j - C_a^2 + \frac{C_j^2}{2} + \frac{C_a^2}{2} - \frac{2C_jC_a}{2}$$

i.e. $$\text{Work output from engine} = \frac{C_j^2 - C_a^2}{2} \qquad 12.11$$

The propulsive efficiency, η_P, is defined as the thrust work divided by the rate at which work is done on the air in the aircraft. Therefore from equations 12.10 and 12.11 we have,

$$\eta_P = \frac{2C_a(C_j - C_a)}{C_j^2 - C_a^2}$$

$$\therefore \ \eta_P = \frac{2C_a}{C_j + C_a} \qquad\qquad 12.12$$

It can be seen from equation 12.12 that as the aircraft velocity, C_a, increases, then the propulsive efficiency increases. For a propeller driven aircraft the change of η_P is greater initially, but at speeds at which the propeller tip approaches sonic velocity the efficiency of the propeller falls off rapidly, and equation 12.12 is no longer applicable. Curves of the form shown in fig. 12.27 are obtained. It can be seen

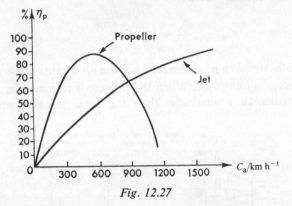

Fig. 12.27

that for aircraft speeds up to about 850 km/h the propeller is the more efficient means of propulsion, but for speeds above this the jet engine is superior.

The simplest form of jet engine is the *ram jet* (or athodyd). In the ram jet the air is compressed by the conversion of the kinetic energy of the atmospheric air relative to the aircraft; this is known as the *ram effect*. Fuel is then burned in the compressed air stream at approximately constant pressure, and the hot gases are allowed to expand through a nozzle, reaching a high velocity backwards relative to the aircraft. The ram jet is shown diagrammatically in fig. 12.28a, and the cycle is represented on a *T-s* diagram in fig. 12.28b.

If the ram jet velocity is C_a, then the air enters the diffuser with a kinetic energy of $C_a^2/2$ per unit mass of air. The velocity after diffusion can be allowed for by using the total temperature after diffusion, as follows:

Using the flow equation, and assuming isentropic flow,

$$h_1 + \frac{C_a^2}{2} = h_1 + \frac{C_1^2}{2}$$

$$\therefore \quad c_p T_1 + \frac{C_a^2}{2} = c_p\left(T_1 + \frac{C_1^2}{2c_p}\right) = c_p T_{t_1}$$

(using the equation for total temperature, 10.25),

i.e. $$\frac{C_a^2}{2} = c_p(T_{t_1} - T_1)$$

or $$T_{t_1} - T_1 = \frac{C_a^2}{2c_p} \qquad\qquad 12.13$$

Then, using equation 10.26,

$$\frac{T_{t_1}}{T_1} = \left(\frac{p_{t_1}}{p_1}\right)^{(\gamma-1)/\gamma}$$

The total pressure p_{t_1} is the pressure the air attains when the diffusion process is isentropic. When the process is irreversible, although still approximately adiabatic, then the total pressure attained is

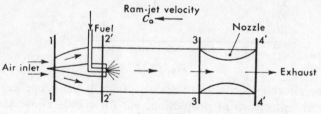

Fig. 12.28a

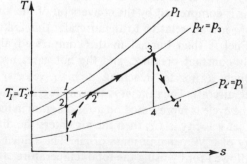

Fig. 12.28b

$p_{t_1'}$, which is less than p_{t_1}, as seen from fig. 12.28b. Since the kinetic energy available, $C_a^2/2$, is the same whether the process is reversible or not, then the temperature change remains the same (i.e. $T_{t_{2'}} = T_{t_1}$),

i.e. $$T_{t_{2'}} - T_1 = \frac{C_a^2}{2c_p} \qquad\qquad 12.14$$

Then the intake isentropic efficiency is defined as follows,

$$\text{Intake efficiency} = \frac{T_{t_2} - T_1}{T_{t_{2'}} - T_1} \qquad\qquad 12.15$$

An aircraft powered by ram jet requires an auxiliary power supply for starting in order to attain the velocity necessary to give a large enough ram compression. Some missiles and short-range fighter aircraft have been built using rockets for take-off and ram jets when in flight. Another use for the ram jet is as a means of driving the propeller of a helicopter; ram jets can be mounted on the propeller blades, placed so that they give a tangential thrust to the propeller. The difficulties in this latter application lie in supplying fuel to the ram jets.

In supersonic flight in an aircraft using ram jets a shock wave forms on the ram jet entry, and part of the pressure rise is through this shock wave.

In a jet engine or turbo-jet the kinetic energy of the incoming air can be used to obtain a ram compression in the intake duct, thus raising the overall efficiency of the unit. The layout of the unit has been considered briefly in Section 12.2. In aircraft gas turbine work it becomes imperative to use total head conditions, since velocity

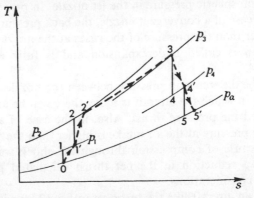

Fig. 12.29

changes through the unit are no longer negligible. Also, in general, temperature measuring instruments such as thermocouples, measure total head temperature and not static temperature. Using total head conditions, the isentropic efficiencies of the compressor and turbine can be re-defined, and an intake and jet pipe efficiency can be introduced.

Referring to fig. 12.29 for a typical jet engine, we have

$$\text{Isentropic efficiency of intake duct} = \frac{T_{t_1} - T_0}{T_{t_{1'}} - T_0} \qquad 12.16$$

$$\text{Isentropic efficiency of compressor} = \frac{T_{t_2} - T_{t_{1'}}}{T_{t_{2'}} - T_{t_{1'}}} \qquad 12.17$$

$$\text{Isentropic efficiency of turbine} = \frac{T_{t_3} - T_{t_{4'}}}{T_{t_3} - T_{t_4}} \qquad 12.18$$

and $$\text{Jet pipe efficiency} = \frac{T_{t_{4'}} - T_{5'}}{T_{t_{4'}} - T_5} \qquad 12.19$$

For adiabatic flow, the total temperature remains constant, and therefore $T_{t_0} = T_{t_{1'}}$, and $T_{t_{4'}} = T_{t_{5'}}$, for the intake duct and the jet pipe respectively. Note that fig. 12.29 is a diagram of static temperature against entropy.

In a practical unit there is a loss of pressure in the combustion chamber from 2' to 3.

Pressure Thrust

It has been assumed in the foregoing analysis that the gases expand down to atmospheric pressure in the jet nozzle. In practice, particularly in the case of a convergent nozzle, the back pressure will normally be lower than the pressure of the gases at the nozzle outlet; this phenomenon is called underexpansion and is fully explained in Section 10.4.

Due to the difference in pressure between the nozzle exit and the atmosphere in which the aircraft is flying there will be an additional thrust, called the pressure thrust. Also, in the case of a supersonic aircraft, the pressure at the air intake is higher than the atmospheric pressure because of compression through the shock wave formed; this causes a reduction in the net thrust calculated purely from momentum considerations.

Consider an aircraft like the turbo-jet in fig. 12.30 with an air in-

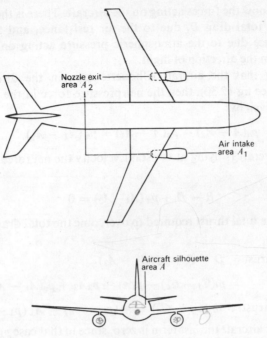

Nozzle exit area A_2

Air intake area A_1

Aircraft silhouette area A

Fig. 12.30

take of area A_1, inlet air pressure p_1, and a nozzle exit area A_2, exit pressure p_2; let the atmospheric pressure be p_a. For a control volume round the working fluid in the aircraft engine we have, using Newton's second law:

$$F + p_1 A_1 - p_2 A_2 = \text{rate of change of momentum of working fluid}$$
$$\text{in the direction of motion of the fluid.}$$

where F is the net force due to hydrostatic pressure and friction exerted by the inside of the aircraft on the working fluid in the direction of its motion,

i.e. $F + p_1 A_1 - p_2 A_2 = \dot{m}(C_J - C_a)$

$\therefore$ $\qquad\qquad\qquad F = \dot{m}(C_J - C_a) - p_1 A_1 + p_2 A_2$

There is an equal and opposite force, R, exerted by the working fluid on the inside of the aircraft engine,

i.e. $\qquad\qquad\qquad R = \dot{m}(C_J - C_a) - p_1 A_1 + p_2 A_2$

in the direction of motion of the aircraft.

Consider now the forces acting on the aircraft. There is the force R, there is the total drag D, due to the air resistance, and there is a pressure force due to the atmospheric pressure acting on the projected area in the direction of flight.*

Assuming that the aircraft silhouette area in the direction of flight is A (see fig 12.30), then the net pressure force in the direction of flight is given by:

$$p_a(A - A_2) - p_a(A - A_1) = p_a(A_1 - A_2)$$

Since the aircraft is flying at constant velocity the net force acting is zero,

i.e. $$R - D + p_a(A_1 - A_2) = 0$$

Therefore the total thrust required to overcome the total drag force is given by:

$$\text{Total thrust} = D = R + p_a(A_1 - A_2)$$

$$= \dot{m}(C_j - C_a) - p_1A_1 + p_2A_2 + p_a(A_1 - A_2)$$

$\therefore$ Total thrust $= \dot{m}(C_j - C_a) + A_2(p_2 - p_a) - A_1(p_1 - p_a)$

For subsonic aircraft the last term is zero, since in that case $p_1 = p_a$,

i.e. $$\text{Total thrust} = \underset{\substack{\text{(momentum} \\ \text{thrust)}}}{\dot{m}(C_j - C_a)} + \underset{\substack{\text{(pressure} \\ \text{thrust)}}}{A_2(p_2 - p_a)} \qquad 12.20$$

Example 12.7

A turbo-jet aircraft is flying at 800 km/h at 10 700 m where the pressure and temperature of the atmosphere are 0·24 bar and $-50°C$ respectively. The compressor pressure ratio is 10/1 and the maximum cycle temperature is 820°C. Calculate the thrust developed and the specific fuel consumption in kg/kN thrust s, using the following information: Entry duct efficiency 0·9; isentropic efficiency of compressor 0·9; total head pressure loss in the combustion chamber 0·14 bar; calorific value of fuel 43 300 kJ/kg; combustion efficiency 98%; isentropic efficiency of turbine 0·92; mechanical efficiency

---

* It may be easier to understand this force if the aircraft is imagined to be fixed to a test bed such that there is no flow of air over it. In flight there is considerable pressure variation over the aircraft surfaces, which is the cause of the lift and drag forces; the total drag force, D, incorporates all such effects and also includes form drag due to the vortices formed.

of drive 98%; jet pipe efficiency 0·92; nozzle outlet area 0·08 m²; c_p and γ for the compression process 1·005 kJ/kg K and 1·4; c_p and γ for the combustion and expansion processes 1·15 kJ/kg K and 1·333; assume that the nozzle is convergent.

The cycle is shown on a T-s diagram in fig. 12.31. The exhaust condition of the gases leaving the nozzle is not known until it is ascertained whether the nozzle is choked or not.

Kinetic energy of air at inlet

$$= \tfrac{1}{2} \times \left(\frac{800 \times 1000}{3600}\right)^2 = \tfrac{1}{2} \times (222 \cdot 2)^2 \text{ N m/kg}$$

$$= 24 \cdot 7 \text{ kJ/kg}$$

Therefore,

$$T_{t_{1'}} - T_0 = \frac{24 \cdot 7}{c_p} = \frac{24 \cdot 7}{1 \cdot 005} = 24 \cdot 6 \text{ K}$$

$$\therefore \; T_{t_{1'}} = (-50 + 273) + 24 \cdot 6 = 247 \cdot 6 \text{ K}$$

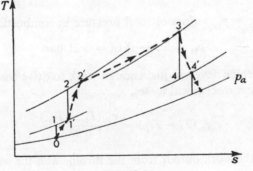

Fig. 12.31

Now from equation 12.16,

$$\text{Intake efficiency} = \frac{T_{t_1} - T_0}{T_{t_{1'}} - T_0} = 0 \cdot 9$$

$$\therefore \; T_{t_1} - T_0 = 0 \cdot 9 \times 24 \cdot 7 = 22 \cdot 2 \text{ K}$$

i.e.

$$T_{t_1} = (-50 + 273) + 22 \cdot 2 = 245 \cdot 2 \text{ K}$$

Then,

$$\frac{p_{t_1}}{p_a} = \left(\frac{T_{t_1}}{T_0}\right)^{\gamma/(\gamma - 1)} = \left(\frac{245 \cdot 2}{223}\right)^{1 \cdot 4/0 \cdot 4} = 1 \cdot 1^{3 \cdot 5} = 1 \cdot 396$$

$$\therefore \; p_{t_1} = p_{t_1} = 1 \cdot 396 \times 0 \cdot 24 = 0 \cdot 335 \text{ bar}$$

For the compressor, we have

$$\frac{T_{t_2}}{T_{t_1}} = \left(\frac{p_{t_2}}{p_{t_1}}\right)^{(\gamma-1)/\gamma} = 10^{0.4/1.4} = 1.932$$

$$\therefore \ T_{t_2} = 1.932 \times 247.7$$

i.e. $$T_{t_2} = 478 \ K$$

Then using equation 12.17,

$$\text{Isentropic efficiency} = \frac{T_{t_2} - T_{t_1}}{T_{t_2'} - T_{t_1'}} = 0.9$$

$$\therefore \ T_{t_2'} - T_{t_1'} = \frac{478 - 247.7}{0.9} = 255.8 \ K$$

i.e. $$T_{t_2'} = 247.7 + 255.8 = 503.5 \ K$$

Also, $$p_{t_2} = 10 \times p_{t_1'} = 10 \times 0.335 = 3.35 \ \text{bar}$$

Hence,

$$p_{t_3} = p_{t_2'} - (\text{loss of total pressure in combustion})$$

i.e. $$p_{t_3} = 3.35 - 0.14 = 3.21 \ \text{bar}$$

Now the turbine develops just enough work to drive the compressor and overcome mechanical losses,

i.e. $$c_{p_g}(T_{t_3} - T_{t_4'}) = \frac{c_{p_a}(T_{t_2'} - T_{t_1'})}{0.98}$$

[Note that the work output from the turbine and the work input to the compressor are given by the product of c_p and the difference in total temperature, when the flow in each is adiabatic,

e.g. for the turbine, using the flow equation we have

$$c_{p_g}T_3 + \frac{C_3^2}{2} = c_{p_g}T_{4'} + \frac{C_{4'}^2}{2} + W$$

$$\therefore \ W = c_{p_g}(T_{t_3} - T_{t_4'})].$$

$$\therefore \ T_{t_3} - T_{t_4'} = \frac{1.005(503.5 - 247.7)}{1.15 \times 0.98} = 228 \ K$$

i.e. $$T_{t_4'} = (820 + 273) - 228 = 865 \ K$$

Then using equation 12.18,

$$\text{Isentropic efficiency} = \frac{T_{t_3} - T_{t_4'}}{T_{t_3} - T_{t_4}} = 0.92$$

$$\therefore T_{t_3} - T_{t_4} = \frac{228}{0.92} = 248 \text{ K}$$

i.e. $$T_{t_4} = (820 + 273) - 248 = 845 \text{ K}$$

Then, $$\frac{p_{t_3}}{p_{t_4}} = \left(\frac{T_{t_3}}{T_{t_4}}\right)^{\gamma/(\gamma-1)} = \left(\frac{1093}{845}\right)^{1.333/0.333} = 2.815$$

$$\therefore p_{t_4} = \frac{3.21}{2.815} = 1.141 \text{ bar}$$

For choked flow in the nozzle the critical pressure ratio is given by equation 10.7,

i.e. $$\frac{p_c}{p_{t_4}} = \left(\frac{2}{\gamma+1}\right)^{\gamma/(\gamma-1)} = \left(\frac{2}{2.333}\right)^{1.333/0.333} = 0.54$$

$$\therefore p_c = 0.54 \times 1.141 = 0.616 \text{ bar}$$

Since the atmospheric pressure is 0.24 bar it follows that the nozzle is choking and hence the actual velocity of the gas at exit is sonic. The expansion in the nozzle is shown on a T-s diagram in fig. 12.32.

The temperature at exit, T_5', is given by equation 10.8 since the velocity at exit is sonic,

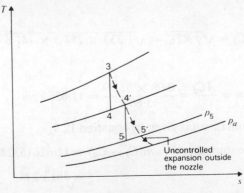

Fig. 12.32

i.e. $\qquad \dfrac{T_5'}{T_{t_4'}} = \dfrac{2}{\gamma + 1} = \dfrac{2}{1 \cdot 333 + 1}$

$$T_5' = \dfrac{865}{1 \cdot 1665} = 741 \cdot 5 \, \text{K}$$

Using equation 12.19:

$$\text{Jet pipe efficiency} = \dfrac{T_{t_4'} - T_5'}{T_{t_4'} - T_5} = 0 \cdot 92$$

$\therefore \qquad\qquad T_5 = 865 - \dfrac{(865 - 741 \cdot 5)}{0 \cdot 92} = 730 \cdot 8 \, \text{K}$

From equation 10.26:

$$\dfrac{p_{t_4'}}{p_5} = \left(\dfrac{T_{t_4'}}{T_5} \right)^{\gamma/\gamma - 1} = \left(\dfrac{865}{730 \cdot 8} \right)^{1 \cdot 333/0 \cdot 333} = 1 \cdot 963$$

$\therefore \qquad\qquad p_5 = \dfrac{1 \cdot 141}{1 \cdot 963} = 0 \cdot 581 \, \text{bar}$

Now, from equation 3.22,

$$R = \dfrac{c_p(\gamma - 1)}{\gamma} = \dfrac{1 \cdot 15 \times 0 \cdot 333}{1 \cdot 333}$$

i.e. $\qquad R = 0 \cdot 2873 \, \text{kJ/kg K}$

Hence,

$$v_5' = \dfrac{RT_5'}{p_5'} = \dfrac{287 \cdot 3 \times 741 \cdot 5}{0 \cdot 581 \times 10^5} = 3 \cdot 667 \, \text{m}^3/\text{kg}$$

Also,

Jet velocity, $C_J = \sqrt{\gamma RT_5'} = \sqrt{1 \cdot 333 \times 287 \cdot 3 \times 741 \cdot 5} = 532 \cdot 9 \, \text{m/s}$

Then,

$$\text{Mass flow} = \dfrac{AC_J}{v_5'} = \dfrac{0 \cdot 08 \times 532 \cdot 9}{3 \cdot 667} = 11 \cdot 626 \, \text{kg/s}$$

The momentum thrust is given by equation 12.9

i.e. $\qquad \text{Momentum thrust} = \dot{m}(C_J - C_a) = 11 \cdot 626 \, (532 \cdot 9 - 222 \cdot 2)$

$$= 3612 \cdot 2 \, \text{N}$$

The pressure thrust is given by equation 12.20,

i.e.   Pressure thrust $= (p_5' - p_a) A = (0 \cdot 581 - 0 \cdot 24) \times 0 \cdot 08 \times 10^5$

$$= 2728 \, \text{N}$$

$\therefore$ Total thrust $= 3612 \cdot 2 + 2728 = 6340 \, \text{N}$

Also,

Heat supplied $= \dot{m}c_{p_g}(T_{t_3} - T_{t_{2'}})$

$$= 11 \cdot 62 \times 1 \cdot 15(1093 - 503 \cdot 5) = 7870 \, \text{kJ/s}$$

If curves of theoretical total temperature rise against fuel/air ratio were available for the fuel used, then the fuel consumption could be found in this way. In this case it is sufficient to write

$$\text{Heat supplied} = \dot{m}_f \times \text{calorific value} = \frac{7870}{0 \cdot 98} \, \text{kJ/s}$$

(where $\dot{m}_f$ is the mass of fuel supplied in kg/s),

i.e. $$\dot{m}_f = \frac{7870}{43 \, 300 \times 0 \cdot 98} = 0 \cdot 186 \, \text{kg/s}$$

Hence,

$$\text{Specific fuel consumption} = \frac{0 \cdot 186 \times 10^3}{6340} = 0 \cdot 0293 \, \text{kg/kN s}$$

At sea level conditions with the same exit area for the propulsion nozzle, the thrust produced will be much higher since the mass flow through the unit will be considerably increased. It is possible to have a variable area nozzle which can be adjusted to give maximum thrust at any given altitude. When a convergent-divergent nozzle is used, the jet pipe efficiency is less because of the increased friction of the divergent portion, and the jet pipe is bulkier. When the pressure ratio across the jet pipe is not large a convergent nozzle is preferred, since the pressure thrust obtained due to underexpansion makes up for the loss of momentum thrust. When the pressure ratio across the jet pipe becomes large, a convergent-divergent nozzle is required to make full use of the energy available. The nozzle throat area must be made variable to avoid the losses which would occur if the nozzle were underexpanding.

Another means of obtaining a thrust boost is by using *after-burning*, which is thermodynamically equivalent to reheat. Fuel is sprayed into the gases leaving the turbine and this increases the jet velocity leaving the nozzle. A 50% increase in thrust can be obtained in this way, but it is very wasteful on fuel. Afterburning can be used on starting and as a reserve power source for thrust augmentation over short periods. It is usually arranged to keep the same pressure ratio across the nozzle when afterburning is used. Therefore, since jet velocity changes approximately directly with the square root of the exit temperature, but specific volume changes directly with the temperature, it follows that the exit area of the nozzle must be increased proportionately with the increase of jet velocity (i.e. $A = \dot{m}v/C =$ constant $\times$ $T/\sqrt{T} =$ constant $\times$ $\sqrt{T} =$ constant $\times$ C). Thus a variable area nozzle is necessary when afterburning is used in order to keep the mass flow through the unit constant, and therefore to allow the compressor and turbine to operate efficiently. The volume of the afterburner is large compared with a normal combustion chamber, since it operates at a lower pressure. This additional bulk and weight must be set against the increase in thrust available.

In a turbo-prop aircraft the ram effect of the incoming air relative to the aircraft can be used as in the turbo-jet. At exit from the turbo-prop engine, the gases should theoretically have a velocity relative to the aircraft, just high enough to carry the exhaust clear of the aircraft. In practice the whole pressure drop available is not used by the turbine, and the gases leaving the turbine expand in the jet pipe, thus leaving with a velocity relative to the aircraft which is higher than the relative velocity of the air entering the engine. This provides a momentum thrust which complements the thrust from the propeller. For basic calculations this additional thrust will be neglected, and it will be assumed that the gases leave the turbine at the ambient pressure. However, in order to make use of the total head isentropic efficiency of the turbine (given by equation 12.18), it is necessary to know the total temperature at turbine exhaust. This can be found if the exhaust velocity is known.

Example 12.8

A turbo-prop aircraft is flying at 650 km/h at an altitude where the ambient temperature is $-18°C$. The compressor pressure ratio is 9/1 and the maximum cycle temperature is 850°C. The intake duct efficiency is 0·9, and the total head isentropic efficiencies of the compressor and turbine are 0·89 and 0·93 respectively. Calculate the specific power output in kW per kg/s, and the thermal efficiency, taking a mechanical efficiency of 98% and neglecting the pressure loss in the combustion chamber. Assume that the exhaust gases leave the aircraft at 650 km/h relative to the aircraft, and take c_p and γ as in example 12.7

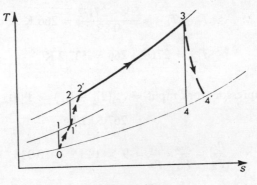

Fig. 12.33

The cycle is shown on a T-s diagram in fig. 12.33.

Kinetic energy of the air at inlet

$$= \frac{1}{2} \times \left(\frac{650 \times 10^3}{3600}\right)^2 = \frac{180 \cdot 5^2}{2} \text{ N m/kg}$$

Therefore,

$$T_{t_1'} - T_0 = \frac{180 \cdot 5^2}{2c_p} = \frac{180 \cdot 5^2}{2 \times 10^3 \times 1 \cdot 005} = 16 \cdot 2 \text{ K}$$

i.e. $$T_{t_1'} = (-18 + 273) + 16 \cdot 2 = 271 \cdot 2 \text{ K}$$

Now from equation 12.16,

$$\text{Intake efficiency} = \frac{T_{t_1} - T_0}{T_{t_1'} - T_0} = 0 \cdot 9$$

$$\therefore \ T_{t_1} - T_0 = 0 \cdot 9 \times 16 \cdot 2 = 14 \cdot 6 \text{ K}$$

i.e. $$T_{t_1} = 269 \cdot 6 \text{ K}$$

Then, $$\frac{p_{t_1}}{p_0} = \left(\frac{T_{t_1}}{T_0}\right)^{\gamma/(\gamma-1)} = \left(\frac{269 \cdot 6}{255}\right)^{1 \cdot 4/0 \cdot 4} = 1 \cdot 215$$

For the compressor, we have

$$T_{t_2} = T_{t_1} \left(\frac{p_{t_2}}{p_{t_1}}\right)^{(\gamma-1)/\gamma} = 271 \cdot 2 \times 9^{0 \cdot 286} = 508 \text{ K}$$

Then, using equation 12.17,

$$\text{Isentropic efficiency} = \frac{T_{t_2} - T_{t_{1'}}}{T_{t_{2'}} - T_{t_{1'}}} = 0.89$$

$$\therefore \ T_{t_{2'}} - T_{t_{1'}} = \frac{508 - 271.2}{0.89} = 266 \text{ K}$$

i.e. $$T_{t_{2'}} = 271.2 + 266 = 537.2 \text{ K}$$

Also,

$$\text{Compressor work input} = c_{p_a}(T_{t_{2'}} - T_{t_{1'}}) = 1.005 \times 266$$
$$= 267.5 \text{ kJ/kg}$$

Now, $$\frac{p_{t_3}}{p_4} = \frac{p_{t_2}}{p_{t_1}} \times \frac{p_{t_1}}{p_0} = 9 \times 1.215 = 10.935$$

Therefore,

$$T_4 = T_{t_3}\left(\frac{p_4}{p_{t_3}}\right)^{(\gamma-1)/\gamma} = \frac{1123}{10.935^{0.25}} = 617.6$$

Then, using equation 10.25,

$$T_{t_4} = T_4 + \frac{C_4^2}{2c_{p_g}} = 617.6 + \frac{180.5^2}{2 \times 10^3 \times 1.15} = 617.6 + 14.2$$

i.e. $$T_{t_4} = 631.8 \text{ K}$$

Then, using equation 12.18,

$$\text{Isentropic efficiency} = \frac{T_{t_3} - T_{t_{4'}}}{T_{t_3} - T_{t_4}} = 0.93$$

$$\therefore \ T_{t_3} - T_{t_{4'}} = 0.93(1123 - 631.8) = 0.93 \times 491.2 = 456.8 \text{ K}$$

Then,

$$\text{Turbine work output} = c_{p_g}(T_{t_3} - T_{t_{4'}}) = 1.15 \times 456.8$$
$$= 525.3 \text{ kJ/kg}$$

$$\therefore \ \text{Net work output} = (525.3 - 267.5)0.98 = 252.7 \text{ kJ/kg}$$

i.e.   Specific power output $= 252.7 \text{ kW per kg/s}$

Also,   Heat supplied $= c_{p_g}(T_{t_3} - T_{t_{2'}}) = 1.15(1123 - 537.2)$
$$= 675 \text{ kJ/kg}$$

Then,   Thermal efficiency $= \dfrac{252.7}{675} = 0.374$ or 37.4%

For aircraft flying at low speeds the amount of thrust from the jet pipe of a turbo-prop engine is usually very small, but as the aircraft speed is raised it becomes an advantage to use a portion of the available energy to obtain thrust in a nozzle, since the propulsion efficiency for a jet increases as shown in fig. 12.27 (see reference 12.5).

It was stated at the beginning of this section that the turbo-jet is more efficient than the turbo-prop at speeds of about 850 km/h and above. At lower speeds the propulsive efficiency can be increased by using a *ducted fan* engine. In the ducted fan engine a turbine drives a fan which draws air through a duct surrounding the engine, and delivers it either to the main gas stream leaving the turbine, or to atmosphere where it surrounds the main jet. Some of the energy available is used in driving the fan, but the decreased jet velocity at a higher mass flow, gives a higher propulsive efficiency and a similar thrust to that of the turbo-jet engine. A ducted fan engine is shown diagrammatically in fig. 12.34. Against the advantage of increased

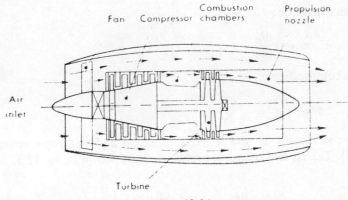

Fig. 12.34

propulsive efficiency must be set the disadvantage of increased size and weight due to the ducting required to deal with the comparatively large volumes of air.

An alternative to the ducted fan engine is the *by-pass* engine, in which some of the air flow is by-passed at an intermediate stage in the compression and passed directly to the main jet. In this engine and the ducted fan, fuel may be injected into the secondary air stream to give thrust boost over short periods.

Another important feature of the ducted fan and by-pass engines (sometimes called turbo-fan engines), is the great reduction in noise level compared with the turbo-jet engine. This is of great importance for aircraft flying on the commercial airlines, taking off and landing at airports in congested areas.

In considering the performance of gas turbine units many variables are involved. It is usual to express the variables in non-dimensional form. The non-dimensional characteristics of the compressor and turbine must then be *matched* (together with the propulsion nozzle, if a jet engine is considered), and a set of *equilibrium running* curves obtained for the unit. A good introductory treatment of equilibrium running is given in reference 12.5.

12.6 Polytropic or small-stage efficiency

The polytropic efficiency of an expansion or compression process is the isentropic efficiency of an infinitely small stage. It is a useful concept since, unlike isentropic efficiency, it is independent of the pressure ratio.

For an expansion:

$$\eta_{\infty e} = \frac{\mathrm{d}h_a}{\mathrm{d}h_\mathrm{I}}$$

For a compression:

$$\eta_{\infty c} = \frac{\mathrm{d}h_\mathrm{I}}{\mathrm{d}h_a}$$

(where subscripts 'a' and 'I' refer to actual and isentropic respectively). Taking an expansion process 1–2′ as shown in fig. 12.35 we have:

$$\eta_{\infty e} = \frac{\mathrm{d}h_a}{\mathrm{d}h_\mathrm{I}} = \frac{\mathrm{d}h_a}{v\,\mathrm{d}p}$$

(since $\mathrm{d}h = v\,\mathrm{d}p$ for an isentropic process, see page 68).
Also, for a perfect gas,

$$\mathrm{d}h_a = c_p\,\mathrm{d}T \text{ and } v = \frac{RT}{p}$$

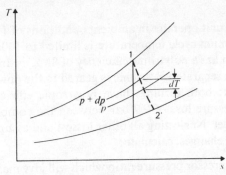

Fig. 12.35

$$\therefore \qquad \eta_{\infty e} = \frac{c_p}{R} \frac{\mathrm{d}T}{T} \frac{p}{\mathrm{d}p}$$

i.e.

$$\eta_{\infty e} \int_{p2}^{p} \frac{\mathrm{d}p}{p} = \frac{c_p}{R} \int_{T2'}^{T1} \frac{\mathrm{d}T}{T}$$

$$\eta_{\infty e} \log_e \frac{p_1}{p_2} = \frac{\gamma}{\gamma - 1} \log_e \frac{T_1}{T_2'}$$

or

$$\frac{T_1}{T_2'} = \left(\frac{p_1}{p_2}\right)^{(\gamma - 1)\eta_{\infty e}/\gamma}$$

For a compression process it can be shown that:

$$\frac{T_2'}{T_1} = \left(\frac{p_2}{p_1}\right)^{(\gamma - 1)/\gamma \eta_{\infty c}} \qquad\qquad 12.21$$

For a gas turbine with pressure ratio, r, and polytropic efficiency, $\eta_{\infty e}$, we have:

Isentropic efficiency, $\eta_T = \dfrac{T_1 - T_2'}{T_1 - T_2} = \dfrac{1 - \dfrac{1}{r^{(\gamma-1)\eta_{\infty e}/\gamma}}}{1 - \dfrac{1}{r^{(\gamma-1)/\gamma}}}$

Similarly for a compressor,

Isentropic efficiency, $\eta_c = \dfrac{r^{\gamma-1/\gamma} - 1}{r^{\gamma-1/\gamma\eta_{\infty c}} - 1}$

Example 12.9

A gas turbine unit operates in ambient conditions of 1·012 bar, 17°C, and the maximum cycle temperature is limited to 1000 K. The compressor, which has a polytropic efficiency of 88 %, is driven by the HP turbine, and a separate LP turbine is geared to the power output on a separate shaft; both turbines have polytropic efficiencies of 90 %. There is a pressure loss of 0·2 bar between the compressor and the HP turbine inlet. Neglecting all other losses, and assuming negligible kinetic energy changes, calculate:

(a) the compressor pressure ratio which will give maximum specific power output;

(b) the isentropic efficiency of the power turbine.

For the gases in both turbines, take $c_p = 1·15$ kJ/kg K and $\gamma = 4/3$.

For air take $c_p = 1·005$ kJ/kg K and $\gamma = 1·4$

(a) The cycle is shown on a T-s diagram in fig. 12.36.

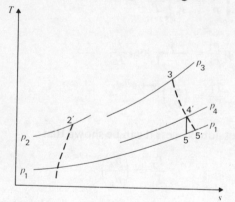

Fig. 12.36

Let $\dfrac{p_2}{p_1} = r$

From equation 12.21,

$$T_2{}' = T_1 r^{\gamma-1/\gamma}\eta_{\infty c} = (17 + 273) \times r^{0·4/1·4\times0·88} = 290r^{0·325}$$

Now,

$$p_3 = p_2 - 0·2 = (p_1 \times r) - 0·2 = 1·012r - 0·2$$
$$p_5 = p_1 = 1·012 \text{ bar}$$

i.e. $\dfrac{p_3}{p_5} = \dfrac{1\cdot012r - 0\cdot2}{1\cdot012} = r - 0\cdot198$

Since the polytropic efficiency of both turbines is the same then,

$$\frac{T_3}{T_5{}'} = \frac{(p_3)}{(p_1)}{}^{(\gamma-1)\eta\infty e/\gamma} = (r - 0\cdot198)^{\,0\cdot333 \times 0\cdot9/1\cdot333}$$

i.e. $\qquad T_5{}' = \dfrac{1000}{(r - 0\cdot198)^{0\cdot225}}$

Turbine specific power output $= c_p(T_3 - T_5{}')$

$$= 1\cdot15\left(1000 - \frac{1000}{(r - 0\cdot198)^{0\cdot225}}\right)$$

$$= 1150\,[1 - (r - 0\cdot198)^{-0\cdot225}]$$

Compressor specific power input

$$= c_p(T_2{}' - T_1)$$

$$= 1\cdot005\,(290r^{0\cdot325} - 290)$$

$$= 291\cdot5\,(r^{0\cdot325} - 1)$$

Net specific power output,

$$P = 1150\,[1 - (r - 0\cdot198)^{-0\cdot225}] - 291\cdot5\,(r^{0\cdot325} - 1)$$

This is a maximum when $\dfrac{dP}{dr} = 0$

i.e. when

$$0\cdot225 \times 1150 \times (r - 0\cdot198)^{-1\cdot225} = 0\cdot325 \times 291\cdot5 \times r^{-0\cdot675}$$

Trial and error, or graphical, solution gives, $r = 6\cdot65$,

i.e. Compressor pressure ratio for maximum specific power output $= 6\cdot65$.

(b) $\qquad T_2{}' = 290r^{0\cdot325} = 290(6\cdot65)^{0\cdot325} = 536\cdot8\ \text{K}$

Now,

HP turbine power output $=$ Compressor power input

$\therefore \qquad 1\cdot15\,(1000 - T_4{}') = 1\cdot005\,(536\cdot8 - 290)$

i.e.
$$T_4' = 784 \cdot 3 \text{ K}$$

Then,

$$\frac{p_3}{p_4} = \frac{(T_3)}{(T_4')}^{\gamma/(\gamma-1)\eta_{\infty e}} = \frac{(1000)}{(784 \cdot 3)}^{1 \cdot 333/0 \cdot 333 \times 0.9} = 2 \cdot 944$$

Also,

$$p_3 = 6 \cdot 65 p_1 - 0 \cdot 2 = (6 \cdot 65 \times 1 \cdot 012) - 0 \cdot 2 = 4 \cdot 73 \text{ bar}$$

$\therefore$
$$p_4 = \frac{4 \cdot 73}{2 \cdot 944} = 1 \cdot 607 \text{ bar}$$

Then,

$$\frac{T_4'}{T_5'} = \frac{(p_4)}{(p_5)}^{(\gamma-1)\eta_{\infty e}/\gamma} = \frac{(1 \cdot 607)}{(1 \cdot 012)}^{0 \cdot 333 \times 0 \cdot 9/1 \cdot 333} = 1 \cdot 110$$

and
$$\frac{T_4'}{T_5} = \frac{(p_4)}{(p_5)}^{\gamma-1/\gamma} = \frac{(1 \cdot 607)}{(1 \cdot 012)}^{0 \cdot 333/1 \cdot 333} = 1 \cdot 123$$

Using equation 12.2,

Turbine isentropic efficiency, $\eta_T = \dfrac{T_4' - T_5'}{T_4' - T_5} = \dfrac{1 - \dfrac{T_5'}{T_4'}}{1 - \dfrac{T_5}{T_4'}}$

i.e.
$$\eta_T = \frac{1 - \dfrac{1}{1 \cdot 11}}{1 - \dfrac{1}{1 \cdot 123}} = 0 \cdot 905 \text{ or } 90 \cdot 5\%$$

PROBLEMS

(For all problems c_p and γ may be taken as $1 \cdot 005$ kJ/kg K and $1 \cdot 4$ for all compression processes, and as $1 \cdot 15$ kJ/kg K and $1 \cdot 333$ for all combustion and expansion processes.)

12.1 A gas turbine has an overall pressure ratio of 5/1 and a maximum cycle temperature of 550°C. The turbine drives the compressor and an electric generator, the mechanical efficiency of the drive being 97%. The ambient temperature is 20°C and the isentropic efficiencies of the compressor and turbine are 0·8 and 0·83 respectively. Calculate the power output in kilowatts for an air flow of 15 kg/s. Calculate also the thermal efficiency and the work ratio. Neglect changes in kinetic energy, and the loss of pressure in the combustion chamber.

(655 kW; 12%; 0·168)

12.2 In a marine gas turbine unit a high-pressure stage turbine drives the compressor, and a low-pressure stage turbine drives the propeller through suitable gearing. The overall pressure ratio is 4/1, and the maximum temperature is 650°C. The isentropic efficiencies of the compressor, H.P. turbine, and L.P. turbine, are 0·8, 0·83, and 0·85 respectively, and the mechanical efficiency of both shafts is 98%. Calculate the pressure between turbine stages when the air intake conditions are 1·01 bar and 25°C. Calculate also the thermal efficiency and the shaft power when the mass flow is 60 kg/s. Neglect kinetic energy changes, and the pressure loss in combustion.

(1·57 bar; 14·9%; 4560 kW)

12.3 For the unit of problem 12.2, calculate the thermal efficiency obtainable when a heat exchanger is fitted. Assume a thermal ratio of 0·75. (23·4%)

12.4 In a gas turbine generating set two stages of compression are used with an intercooler between stages. The H.P. turbine drives the H.P. compressor, and the L.P. turbine drives the L.P. compressor and the alternator. The exhaust from the L.P. turbine passes through a heat exchanger which transfers heat to the air leaving the H.P. compressor. There is a reheat combustion chamber between turbine stages which raises the gas temperature to 600°C, which is also the gas temperature at entry to the H.P. turbine. The overall pressure ratio is 10/1, each compressor having the same pressure ratio, and the air temperature at entry to the unit is 20°C. Assuming isentropic efficiencies of 0·8 for both compressor stages, and 0·85 for both turbine stages, and that 2% of the work of each turbine is used in overcoming friction, calculate the power output in kilowatts for a mass flow of 115 kg/s. The heat exchanger thermal ratio may be taken as 0·7, and intercooling is complete between compressor stages. Neglect all losses in pressure, and assume that velocity changes are negligibly small. Calculate also the overall thermal efficiency of the plant.

(14 460 kW; 25·7%)

12.5 A motor-car gas-turbine unit has two centrifugal compressors in series giving an overall pressure ratio of 6/1. The air leaving the

H.P. compressor passes through a heat exchanger before entering the combustion chamber. The expansion is in two turbine stages, the first stage driving the compressors, and the second stage driving the car through gearing. The gases leaving the L.P. turbine pass through the heat exchanger before exhausting to atmosphere. The H.P. turbine inlet temperature is 800°C and the air inlet temperature to the unit is 15°C. The isentropic efficiency of the compression is 0·8, and that of each turbine is 0·85; the mechanical efficiency of each shaft is 98%. Neglecting pressure losses and changes in kinetic energy, calculate the overall thermal efficiency and the power developed when the air mass flow is 0·7 kg/s. The heat exchanger thermal ratio may be assumed to be 0·65. Calculate also the specific fuel consumption when the calorific value of the fuel used is 42 600 kJ/kg, and the combustion efficiency is 97%.

(28·7%; 94·3 kW; 0·303 kg/kW h)

12.6 In a gas turbine generating station the overall compression ratio is 12/1, performed in three stages with pressure ratios of 2·5/1, 2·4/1, and 2/1 respectively. The air inlet temperature to the plant is 25°C and intercooling between stages reduces the temperature to 40°C. The H.P. turbine drives the high-pressure and intermediate-pressure compressor stages; the L.P. turbine drives the low-pressure compressor and the alternator. The gases leaving the L.P. turbine are passed through a heat exchanger which heats the air leaving the high-pressure compressor. The temperature at inlet to the H.P. turbine is 650°C, and reheating between turbine stages raises the temperature to 650°C. The gases leave the heat exchanger at a temperature of 200°C. The isentropic efficiency of each compressor stage is 0·83, and the isentropic efficiencies of the H.P. and L.P. turbines are 0·85 and 0·88, respectively. Take the mechanical efficiency of each shaft as 98%. The air mass flow is 140 kg/s. Calculate the power output in kilowatts, the thermal efficiency, and the flow of cooling water required for the intercoolers when the rise in water temperature must not exceed 30 K. Neglect pressure losses and changes in kinetic energy, and take the specific heat of water as 4·19 kJ/kg K. Calculate also the heat exchanger thermal ratio.

(25 300 kW; 33·7%; 223 kg/s; 0·825)

12.7 In a turbo-prop engine the compressor pressure ratio is 6/1 and the maximum cycle temperature is 760°C. The total head isentropic efficiencies of the compressor and turbine are 0·85 and 0·88 respectively, and the mechanical efficiency is 99%. The intake duct efficiency is 0·9. Calculate the specific power output in kW per kg/s and the thermal efficiency when the aircraft is travelling at 725 km/h at an altitude where the ambient temperature is $-7°C$. Neglect the pressure loss in the combustion chamber, and assume that the gases in the turbine expand down to atmospheric pressure, and leave the aircraft at 725 km/h relative to the aircraft.

(166·7 kW per kg/s; 27·8%)

12.8 The loss of total pressure in the combustion process in problem 12.7 is 3% of the inlet total pressure to the chamber. Calculate the specific power output and the thermal efficiency when this is not neglected.   (164 kW per kg/s; 27·3%)

12.9 A turbo-jet aircraft is flying at 925 km/h and the entry duct efficiency is 0·9. The atmospheric conditions are 0·45 bar and $-26°C$, and the maximum allowable temperature is 800°C. The compressor pressure ratio is 6/1, and the total head isentropic efficiency is 0·85. The turbine total head isentropic efficiency is 0·89, and the mechanical efficiency is 98%. Calculate the required nozzle exit area and the net thrust developed when the air mass flow is 45 kg/s. Assume a loss of total pressure in the combustion chamber of 0·07 bar. Calculate also the air/fuel ratio, and the specific fuel consumption, assuming a combustion efficiency of 99% and a calorific value of the fuel of 43 300 kJ/kg. Assume a convergent propulsion nozzle with an efficiency of 0·9.   (0·229 m²; 20 450 N; 65·2/1; 0·0338 kg/kN s)

12.10 Afterburning is used in the aircraft of problem 12.9 to obtain an increase in the thrust. The temperature after the afterburner is 700°C and the afterburning process may be assumed to take place at constant pressure. Calculate the nozzle exit area now required to pass the same mass flow as in problem 12.9. Calculate also the new net thurst.   (0·242 m²; 22 200 N)

12.11 A centrifugal compressor running at 16 000 rev/min takes in air at 17°C and compresses it through a pressure ratio of 4/1 with an isentropic efficiency of 82%. The blades are radially inclined and the slip factor is 0·85. Guide vanes at inlet give the air an angle of pre-whirl of 20° to the axial direction; take the mean diameter of the impeller eye as 200 mm and the absolute air velocity at inlet as 120 m/s. Calculate the impeller tip diameter.   (549 mm)

12.12 Show that for an axial compressor of 50% reaction design with blade velocity, b, axial flow component of air flow, f, and inlet blade angle, β_i, that:

Specific power input per stage $= b^2[2\dfrac{f}{b}\tan\beta_i - 1] \times$ (work done factor)

A ten-stage axial flow compressor of 50% reaction design has a mean blade velocity of 250 m/s and the blade inlet angle for each row is 45°. The ratio of flow velocity to blade velocity is 0·75, the work done factor for each stage is 0·87 and the isentropic efficiency of the compressor is 0·85. Assuming an air inlet temperature of 20°C, calculate:

(a) the exit angle of the blades;
(b) the pressure ratio of the compressor;
(c) the pressure ratio for the first stage.

(Hint: for part (c) as a first approximation take the stage efficiency to be equal to the compressor polytropic efficiency.)

$$(11\cdot3°; \ 7\cdot6/1; \ 1\cdot317/1)$$

12.13 In a gas turbine plant air enters a compressor at atmospheric conditions of 15°C, 1·0133 bar and is compressed through a pressure ratio of 10/1. The air leaving the compressor passes through a heat exchanger before entering the combustion chamber. The hot gases leave the combustion chamber at 800°C and expand through an HP turbine which drives the compressor. On leaving the HP turbine the gases pass through a reheat combustion chamber which raises the temperature of the gases to 800°C before they expand through the power turbine, and thence to the heat exchanger where they flow in counter-flow to the air leaving the compressor. Sketch the plant layout and the T-s diagram for the various processes, and using the data given below, calculate:

(a) the air flow required for a net power output of 10 MW;
(b) the work ratio of the cycle;
(c) the temperature of the air entering the combustion chamber.

Data:
Polytropic efficiency of compressor 88%; polytropic efficiencies of HP and power turbines 82% and 80% respectively; mechanical

efficiency of HP turbine–compressor drive 92%; mechanical efficiency of power turbine drive 94%; R for hot gases in both turbines 0·2875 kJ/kg K; c_p for hot gases in both turbines 1·15 kJ/kg K; thermal ratio of heat exchanger 0·75; pressure drop on air side of heat exchanger 0·125 bar; pressure drop in combustion chamber 0·1 bar; pressure drop in reheat combustion chamber 0·08 bar; pressure drop on gas side of heat exchanger 0·1 bar. Neglect the mass of fuel added, and neglect changes of velocity throughout.

(93·4 kg/s; 0·234; 882·8 K)

REFERENCES

12.1 DIXON, S. L., *Fluid Mechanics and Thermodynamics of Turbomachinery* (Pergamon, 1978).

12.2 TURTON, R. K., *Principles of Turbomachinery* (E. & F. N. Spon, 1984).

12.3 SPALDING, D. B., 'Performance Criteria of Gas Turbine Combustion Chambers,' *Aircraft Engineering*, vol. 28, 1956.

12.4 MCKINLEY, B., *Aircraft Powerplants* (McGraw-Hill, 1978).

12.5 COHEN, H., ROGERS, G. F. C., and SARAVANAMUTTO, H. I. H., *Gas Turbine Theory* (Halstead Press, 1979).

12.6 HARTMAN, R. T. C., *Gas Turbine Engineering* (Macmillan, 1981).

13. Mixtures

A pure substance is defined as a substance having a constant and uniform chemical composition, and this definition can be extended to include a homogeneous mixture of gases when there is no chemical reaction taking place. The thermodynamic properties of a mixture of gases can be determined in the same way as for a single gas. The most common example of this is dry air, which is a mixture of oxygen, nitrogen, a small percentage of argon, and traces of other gases. The properties of air have been determined and it is considered as a single substance.

The mixtures to be considered in this chapter are those composed of perfect gases, and perfect gases and vapours. The properties of such mixtures are important in combustion calculations. Air and water vapour mixtures are considered later in the chapter with reference to surface condensers, but for moist atmospheric air there is a special nomenclature and this is considered in a separate chapter on Psychrometry.

For the purpose of this chapter some knowledge of relative atomic and molecular masses will be assumed, but in Section 15.1 will be found a brief treatment of the principles involved. Where necessary the relative molecular mass will be given in brackets after the gas—e.g. nitrogen (28).

13.1 Dalton's law and the Gibbs-Dalton law

Consider a closed vessel of volume V at temperature T, which contains a mixture of perfect gases at a known pressure. If some of the mixture were removed, then the pressure would be less than the initial value. If the gas removed were the full amount of one of the constituents then the reduction in pressure would be equal to the contribution of that constituent to the initial total pressure. Each constituent contributes to the total pressure by an amount which is known as the *partial pressure* of the constituent. The relationship between the partial pressures of the constituents is expressed by Dalton's law, as follows:

The pressure of a mixture of gases is equal to the sum of the partial pressures of the constituents.

The partial pressure of each constituent is that pressure which the gas would exert if it occupied alone that volume occupied by the mixture at the same temperature.

This is expressed diagrammatically in fig. 13.1. The gases A and B, originally occupying volume V at temperature T are mixed in the

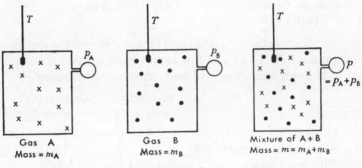

Fig. 13.1

third vessel which is of the same volume and is at the same temperature.

By the conservation of mass,

$$m = m_A + m_B \qquad 13.1$$

By Dalton's law,

$$p = p_A + p_B \qquad 13.2$$

Dalton's law is based on experiment and is found to be obeyed more accurately by gas mixtures at low pressures. As shown in fig. 13.1 each constituent occupies the whole vessel. The example given in fig. 13.1 and the relationships in equations 13.1 and 13.2 refer to a mixture of two gases, but the law can be extended to any number of gases,

i.e. $$m = m_A + m_B + m_C + \text{etc.} \quad \text{or} \quad m = \sum m_i \qquad 13.3$$

(where m_i is the mass of a constituent).

Similarly $$p = p_A + p_B + p_C + \text{etc.} \quad \text{or} \quad p = \sum p_i \qquad 13.4$$

(where p_i is the partial pressure of a constituent).

Air is the most common mixture and since it will be referred to frequently, its properties are given as follows:

Constituent (Relative Molecular Mass)	Chemical Symbol	Volumetric analysis %	Gravimetric analysis %
Oxygen (31·999)	O_2	20·95	23·14
Nitrogen (28·013)	N_2	78·09	75·53
Argon (39·948)	Ar	0·93	1·28
Carbon Dioxide (44·01)	CO_2	0·03	0·05

The mean relative molecular mass of air is 28·96, and the specific gas constant R is 0·2871 kJ/kg K. For approximate calculations the air is said to be composed of oxygen and 'atmospheric nitrogen'.

	Volumetric analysis %	Gravimetric analysis %
Oxygen (32)	21	23·3
Atmospheric nitrogen (28)	79	76·7
Nitrogen/Oxygen	3·76:1	3·29:1

(Note: volumetric analysis is the analysis by volume; gravimetric analysis is the analysis by weight or mass.)

Example 13.1

A vessel of volume 0·4 m³ contains 0·45 kg of carbon monoxide (28) and 1 kg of air, at 15°C. Calculate the partial pressure of each constituent and the total pressure in the vessel. The gravimetric analysis of air is to be taken as 23·3% oxygen (32), and 76·7% nitrogen (28).

$$\text{Mass of oxygen present} = \frac{23·3}{100} \times 1 = 0·233 \text{ kg}$$

$$\text{Mass of nitrogen present} = \frac{76·7}{100} \times 1 = 0·767 \text{ kg}$$

From equation 3.9,

$$R = \frac{R_0}{M}$$

and from equation 3.6,

$$pV = mRT$$

Hence,

$$p = \frac{mR_0T}{MV}$$

or for a constituent,

$$p_i = \frac{m_i R_0 T}{M_i V}$$

The volume V is 0.4 m^3 and the temperature T is $(15+273)=288$ K.
Therefore we have,

for O$_2$, $p_{O_2} = \dfrac{0.233 \times 8.314 \times 288}{32 \times 0.4} = 43.59$ kN/m^2

$$= \frac{43.59 \times 10^3}{10^5} = 0.4359 \text{ bar}$$

for N$_2$, $p_{N_2} = \dfrac{0.767 \times 8.314 \times 288}{28 \times 0.4} = 164$ kN/m^2

$$= \frac{164 \times 10^3}{10^5} = 1.64 \text{ bar}$$

for CO, $p_{CO} = \dfrac{0.45 \times 8.314 \times 288}{28 \times 0.4} = 96.2$ kN/m^2

$$= \frac{96.2 \times 10^3}{10^5} = 0.962 \text{ bar}$$

The total pressure in the vessel is given by equation 13.4,

$$p = \sum p_i = 0.436 + 1.64 + 0.962 = 3.038 \text{ bar}$$

i.e.   Pressure in vessel = 3.038 bar

Dalton's law was re-formulated by Gibbs to include a second statement on the properties of mixtures. The combined statement is known as the Gibbs-Dalton law, and is as follows:

The internal energy, enthalpy, and entropy, of a gaseous mixture are respectively equal to the sums of the internal energies, enthalpies, and entropies, of the constituents.

Each constituent has that internal energy, enthalpy, and entropy, which it would have if it occupied alone that volume occupied by the mixture at the temperature of the mixture.

This statement leads to the equations

$$mu = m_A u_A + m_B u_B + \text{etc.} \quad \text{or} \quad mu = \sum m_i u_i \qquad 13.5$$
and $$mh = m_A h_A + m_B h_B + \text{etc.} \quad \text{or} \quad mh = \sum m_i h_i \qquad 13.6$$
and $$ms = m_A s_A + m_B s_B + \text{etc.} \quad \text{or} \quad ms = \sum m_i s_i \qquad 13.7$$

13.2 Volumetric analysis of a gas mixture

The analysis of a mixture of gases is often quoted by volume as this is the most convenient for practical determinations. In Section 15.6

the analysis of exhaust or flue gases by means of the Orsat apparatus is described. The volume of the gas sample is measured at atmospheric pressure, and the temperature is held constant by means of a water jacket round the gas sample. The constituents are absorbed chemically one by one, and the remainder of the sample is measured after each absorption; the difference in volume gives the partial volume occupied by the constituent in the mixture.

Consider a volume V of a gaseous mixture at a temperature T, consisting of three constituents A, B, and C as in fig. 13.2a. Let each of the constituents be compressed to a pressure p equal to the total

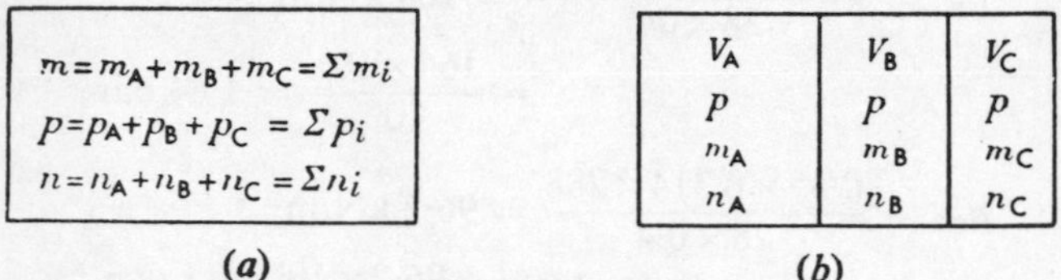

$$m = m_A + m_B + m_C = \Sigma m_i$$
$$p = p_A + p_B + p_C = \Sigma p_i$$
$$n = n_A + n_B + n_C = \Sigma n_i$$

V_A	V_B	V_C
p	p	p
m_A	m_B	m_C
n_A	n_B	n_C

(a)   (b)

Fig. 13.2

pressure of the mixture, and let the temperature remain constant. The partial volumes then occupied by the constituents will be V_A, V_B, and V_C. From equation 3.6 $pV = mRT$, therefore, referring to fig. 13.2a,

$$m_A = \frac{p_A V}{R_A T}$$

and referring to fig. 13.2b, $\quad m_A = \dfrac{p V_A}{R_A T}$

Equating the two values for m_A, we have

$$\frac{p_A V}{R_A T} = \frac{p V_A}{R_A T}$$

i.e. $\qquad p_A V = p V_A \quad \text{or} \quad V_A = \dfrac{p_A}{p} V$

In general therefore,

$$V_i = \frac{p_i}{p} V \qquad\qquad 13.8$$

i.e. $\qquad \sum V_i = \sum \dfrac{p_i V}{p} = \dfrac{V}{p} \sum p_i$

Now from equation 13.4, $p = \sum p_i$, therefore,

$$\sum V_i = V \qquad\qquad 13.9$$

Therefore, the volume of a mixture of gases is equal to the sum of the volumes of the individual constituents when each exists alone at the pressure and temperature of the mixture.

This is the statement of another empirical law, the law of partial volumes (sometimes called Amagat's law or Leduc's law).

Very often the analysis of mixtures is simplified if it is carried out in moles. The mole was defined in Section 3.3 and is given by equation 3.7 as $n = m/M$. By Avogadro's law, the number of moles of any gas is proportional to the volume of the gas at a given pressure and temperature. Referring to fig. 13.2a, the volume V contains n moles of the mixture at p and T. In fig. 13.2b, the gas A occupies a volume V_A at p and T, and this volume contains n_A moles. Similarly there are n_B moles of gas B in volume V_B, and n_C moles of gas C in volume V_C. Now from equation 13.9, $\sum V_i = V$, or, $V_A + V_B + V_C = V$. Therefore the total number of moles in the vessel must equal the sum of the moles of the individual constituents,

i.e. $$n_A + n_B + n_C = n \quad \text{or} \quad n = \sum n_i \qquad 13.10$$

13.3 The relative molecular mass and specific gas constant

For any gas in a gas mixture occupying a total volume of V at a temperature T, from equation 3.8 $pV = nR_0T$, and the definition of partial pressure, we have

$$p_i V = n_i R_0 T \qquad 13.11$$

$$\therefore \sum p_i V = \sum n_i R_0 T$$

i.e. $$V \sum p_i = R_0 T \sum n_i$$

From equation 13.4, $p = \sum p_i$, hence,

$$pV = R_0 T \sum n_i$$

Also from equation 13.10, $n = \sum n_i$, therefore,

$$pV = nR_0T$$

The mixture therefore acts as a perfect gas, and this is the characteristic equation for the mixture. A molar mass is defined by the equation, $M = m/n$, where m is the mass of the mixture and n is the number of moles of mixture. Similarly a specific gas constant is defined by the equation, $R = R_0/M$. It can be assumed that a mixture of perfect gases obeys all the perfect gas laws.

To find the specific gas constant for the mixture in terms of the specific gas constants of the constituents, consider equation 3.6 both for the mixture and for a constituent,

i.e. $$pV = mRT \quad \text{and} \quad p_iV = m_iR_iT$$

Then, $$\sum p_iV = \sum m_iR_iT$$

$$\therefore \quad V\sum p_i = T\sum m_iR_i$$

Now from equation 13.4, $p = \sum p_i$, therefore,

$$pV = T\sum m_iR_i \quad \text{or} \quad pV = mRT = T\sum m_iR_i$$

i.e. $$mR = \sum m_iR_i \quad \text{or} \quad R = \sum \frac{m_i}{m} R_i \qquad 13.12$$

(where m_i/m is the mass fraction of a constituent).

Example 13.2

The gravimetric analysis of air is 23.14% oxygen, 75.53% nitrogen, 1.28% argon, 0.05% carbon dioxide. Calculate the specific gas constant for air and the relative molecular mass. Take the relative molecular masses from the table on p. 484.

From equation 3.9, $R = R_0/M$, therefore,

$$R_{O_2} = \frac{8.314}{31.999} = 0.2598 \text{ kJ/kg K}$$

$$R_{N_2} = \frac{8.314}{28.013} = 0.2968 \text{ kJ/kg K}$$

$$R_{A_r} = \frac{8.314}{39.948} = 0.2081 \text{ kJ/kg K}$$

$$R_{CO_2} = \frac{8.314}{44.01} = 0.1889 \text{ kJ/kg K}$$

Then using equation 13.12, $R = \sum \frac{m_i}{m} R_i$, we have,

$$R = 0.2314 \times 0.2598 + 0.7553 \times 0.2968 + 0.0128 \times 0.2081$$
$$+ 0.0005 \times 0.1889 = 0.2871 \text{ kJ/kg K}$$

i.e. Specific gas constant for air $= 0.2871 \text{ kJ/kg K}$

From equation 3.9, $M = R_0/R$, therefore,

$$M = \frac{8 \cdot 3143}{0 \cdot 2871} = 28 \cdot 96 \text{ kg/kmol}$$

i.e. Relative molecular mass $= 28 \cdot 96$

When the approximate analysis for air is used (i.e. $23 \cdot 3\%$ O_2 and $76 \cdot 7\%$ N_2 by mass), the value obtained for M by the same method is $28 \cdot 84$ kg/kmol.

From equation 13.11, $p_i V = n_i R_0 T$, and combining this with equation 3.8 applied to the mixture (i.e. $pV = nR_0T$), we have

$$\frac{p_i V}{pV} = \frac{n_i R_0 T}{n R_0 T}$$

i.e. $$\frac{p_i}{p} = \frac{n_i}{n} \qquad\qquad 13.13$$

This can be combined with equation 13.8, to give

$$\frac{p_i}{p} = \frac{n_i}{n} = \frac{V_i}{V} \qquad\qquad 13.14$$

This is an important result which means that the molar analysis is identical with the volumetric analysis, and both are equal to the ratio of the partial pressure to the total pressure.

Another method of determining the relative molecular mass is as follows. Applying the characteristic equation, 3.6, to each constituent and to the mixture we have, $m_i = p_i V/R_i T$, and $m = pV/RT$.

From equation 13.3, $m = \sum m_i$, therefore,

$$\frac{pV}{RT} = \sum \frac{p_i V}{R_i T} \qquad \therefore \; \frac{p}{R} = \sum \frac{p_i}{R_i}$$

Using equation 3.9, $R = R_0/M$, and substituting, we have

$$\frac{pM}{R_0} = \sum \frac{p_i M_i}{R_0} \quad \text{or} \quad pM = \sum p_i M_i$$

i.e. $$M = \sum \frac{p_i}{p} M_i \qquad\qquad 13.15$$

Also using equation 13.14,

$$M = \sum \frac{V_i}{V} M_i \qquad\qquad 13.16$$

and $$M = \sum \frac{n_i}{n} M_i \qquad\qquad 13.17$$

Example 13.3

The gravimetric analysis of air is 23·14% oxygen, 75·53% nitrogen, 1·28% argon, and 0·05% carbon dioxide. Calculate the analysis by volume and the partial pressure of each constituent when the total pressure is 1 bar.

From equation 13.14, the analysis by volume, V_i/V, is the same as the mole fraction n_i/n. Also from equation 3.7, $n_i = m_i/M_i$; therefore considering 1 kg of mixture and using a tabular method, we have

Constituent	$\dfrac{m_i}{kg}$	$\dfrac{M_i}{kg/kmol}$	$\dfrac{n_i}{kmol} = \dfrac{m_i}{M_i}$	$\dfrac{n_i}{n} \times 100\% = \dfrac{V_i}{V} \times 100\%$
Oxygen	0·2314	31·999	0·007 23	$\dfrac{0{\cdot}007\,23 \times 100}{0{\cdot}034\,52} = 20{\cdot}95\%$
Nitrogen	0·7553	28·013	0·026 96	$\dfrac{0{\cdot}026\,96 \times 100}{0{\cdot}034\,52} = 78{\cdot}09\%$
Argon	0·0128	39·948	0·000 32	$\dfrac{0{\cdot}000\,32 \times 100}{0{\cdot}034\,52} = 0{\cdot}93\%$
Carbon Dioxide	0·0005	44·01	0·000 01	$\dfrac{0{\cdot}000\,01 \times 100}{0{\cdot}03452} = 0{\cdot}03\%$

$$n = \sum n_i = \overline{0{\cdot}034\,52}\,\text{kmol}$$

From equation 13.14, $\dfrac{p_i}{p} = \dfrac{V_i}{V} = \dfrac{n_i}{n}$, therefore, $p_i = \dfrac{n_i}{n}p$, hence,

for O_2	$p_{O_2} = 0{\cdot}2095 \times 1 = 0{\cdot}2095$ bar
for N_2	$p_{N_2} = 0{\cdot}7809 \times 1 = 0{\cdot}7809$ bar
for Ar	$p_{A_r} = 0{\cdot}0093 \times 1 = 0{\cdot}0093$ bar
for CO_2	$p_{CO_2} = 0{\cdot}0003 \times 1 = 0{\cdot}0003$ bar

Example 13.4

A mixture of 1 kmol CO_2 (44) and 3·5 kmol of air is contained in a vessel at 1 bar and 15°C. The volumetric analysis of air can be taken as 21% oxygen and 79% nitrogen. Calculate for the mixture,

(a) The masses of CO_2, O_2, and N_2, and the total mass;
(b) The percentage carbon content by mass;
(c) The relative molecular mass and the specific gas constant for the mixture;
(d) The specific volume of the mixture.

(a) From equation 13.14, $n_i = (V_i/V)n$, we have

$$n_{O_2} = 0.21 \times 3.5 = 0.735 \text{ kmol}$$

and

$$n_{N_2} = 0.79 \times 3.5 = 2.765 \text{ kmol}$$

From equation 3.7, $m_i = n_i M_i$, therefore,

$$m_{CO_2} = 1 \times 44 = 44 \text{ kg}$$

$$m_{O_2} = 0.735 \times 32 = 23.55 \text{ kg}$$

and

$$m_{N_2} = 2.765 \times 28 = 77.5 \text{ kg}$$

The total mass, $m = m_{O_2} + m_{N_2} + m_{CO_2}$

$$= 23.55 + 77.5 + 44 = 145.05 \text{ kg}$$

(b) The relative molecular mass of carbon is 12, therefore there are 12 kg of carbon present for every mole of carbon dioxide,

i.e.

$$\text{Percentage carbon in mixture} = \frac{12 \times 100}{145.05} = 8.27\% \quad \text{by mass}$$

(c) From equation 13.10, $n = \sum n_i$, then,

$$n = n_{CO_2} + n_{O_2} + n_{N_2} = 1 + 0.735 + 2.765 = 4.5 \text{ kmol}$$

Then using equation 13.17, $M = \sum \dfrac{n_i}{n} M_i$, we have

$$M = \frac{1}{4.5} \times 44 + \frac{0.735}{4.5} \times 32 + \frac{2.765}{4.5} \times 28 = 32.2 \text{ kg/kmol}$$

i.e. Relative molecular mass = 32.2

From equation 3.9, $R = R_0/M$, we have

$$R = \frac{8.314}{32.2} = 0.2581 \text{ kJ/kg K}$$

i.e. Specific gas constant for the mixture = 0.2581 kJ/kg K

(d) From equation 3.5, $pv = RT$,

$$\therefore v = \frac{RT}{p} = \frac{0.2581 \times 288 \times 10^3}{1 \times 10^5} = 0.7435 \text{ m}^3/\text{kg}$$

(where $T = 15 + 273 = 288$ K),

i.e. Specific volume of the mixture at 1 bar and 15°C is 0·7435 m³/kg.

Example 13.5

A mixture of H_2 (2), and O_2 (32), is to be made so that the ratio of H_2 to O_2 is 2 to 1 by volume. Calculate the mass of O_2 required and the volume of the container, per kg of H_2, if the pressure and temperature are 1 bar and 15°C respectively.

Let the mass of O_2 per kg of H_2 be x.

From equation 3.7, $n_i = m_i/M_i$,

$$\therefore n_{H_2} = \frac{1}{2} = 0·5 \text{ kmol} \quad \text{and} \quad n_{O_2} = \frac{x}{32} \text{ kmol}$$

From equation 13.14, $V_i/V = n_i/n$,

$$\therefore \frac{V_{H_2}}{V_{O_2}} = \frac{n_{H_2}}{n_{O_2}} \quad \text{and} \quad \frac{V_{H_2}}{V_{O_2}} = 2 \quad \text{(given)}$$

i.e. $$\frac{0·5}{x/32} = 2 \quad \therefore x = \frac{32 \times 0·5}{2} = 8 \text{ kg}$$

i.e. Mass of oxygen per kg of hydrogen = 8 kg

The total number of kilomoles in the vessel per kg of H_2 is

$$n = n_{H_2} + n_{O_2} = 0·5 + \frac{x}{32} = 0·5 + \frac{8}{32} = 0·5 + 0·25 = 0·75 \text{ kmol}$$

Then from equation 3.8,

$$pV = nR_0T$$

$$\therefore V = \frac{0·75 \times 8·314 \times 288 \times 10^3}{1 \times 10^5} = 17·96 \text{ m}^3$$

Example 13.6

A vessel contains a gaseous mixture of composition by volume, 80% H_2 (2), and 20% CO (28). It is desired that the mixture should be made in the proportion 50% H_2 and 50% CO by removing some of the mixture and adding some CO. Calculate per kmol of mixture the mass of mixture to be removed, and the mass of CO to be added. The pressure and temperature in the vessel remain constant during the procedure.

Since the pressure and temperature remain constant, then the number of kmol in the vessel remain the same throughout. Therefore,

kmol of mixture removed = kmol of CO added

Let x kg of mixture be removed and y kg of CO be added. For the mixture, from equation 13.16,

$$M = \sum \frac{V_i}{V} M_i$$

therefore, $M = 0.8 \times 2 + 0.2 \times 28 = 7.2$ kg/kmol

Then using equation 3.7, $n = m/M$, we have

kmol of mixture removed $= \dfrac{x}{7.2}$ — kmol of CO added $= \dfrac{y}{28}$ kmol

From equation 13.14, $V_i/V = n_i/n$, therefore,

kmol of H_2 in the mixture removed $= 0.8 \times \dfrac{x}{7.2} = \dfrac{x}{9}$ kmol

and kmol of H_2 initially $= 0.8 \times 1 = 0.8$ kmol

Hence, kmol of H_2 remaining in vessel $= 0.8 - \dfrac{x}{9}$ kmol

But 1 kmol of the new mixture is 50% H_2 and 50% CO, therefore

$$0.8 - \frac{x}{9} = 0.5$$

i.e. $x = (0.8 - 0.5) \times 9 = 2.7$ kg

i.e. Mass of mixture removed = 2.7 kg

Also since $x/7.2 = y/28$,

$$\therefore \ y = \frac{28}{7.2} \times x = \frac{28 \times 2.7}{7.2} = 10.5 \text{ kg}$$

i.e. Mass of CO added = 10.5 kg

13.4 Specific heats of a gas mixture

It was shown in Section 13.1 that as a consequence of the Gibbs-Dalton law, the internal energy of a mixture of gases is given by equation 13.5, $mu = \sum m_i u_i$.

Also for a perfect gas from equation 3.14, $u = c_v T$.
Hence substituting we have

$$mc_v T = \sum m_i c_{v_i} T$$

$$\therefore \ mc_v = \sum m_i c_{v_i}$$

or $$c_v = \sum \frac{m_i}{m} c_{v_i}$$ 13.18

Similarly from equation 13.6, $mh = \sum m_i h_i$, and from equation 3.18, $h = c_p T$, therefore,

$$mc_p T = \sum m_i c_{p_i} T$$

$$\therefore \ mc_p = \sum m_i c_{p_i}$$

or $$c_p = \sum \frac{m_i}{m} c_{p_i}$$ 13.19

From equations 13.18 and 13.19,

$$c_p - c_v = \sum \frac{m_i}{m} c_{p_i} - \sum \frac{m_i}{m} c_{v_i} = \sum \frac{m_i}{m} (c_{p_i} - c_{v_i})$$

Using equation 3.17, $c_{p_i} - c_{v_i} = R_i$, therefore,

$$c_p - c_v = \sum \frac{m_i}{m} R_i$$

Also from equation 13.12, $R = \sum \frac{m_i}{m} R_i$, therefore for the mixture

$$c_p - c_v = R$$

The equations 3.20, 3.21, and 3.22, can be applied to a mixture of gases,

i.e. $$\gamma = \frac{c_p}{c_v}; \quad c_v = \frac{R}{\gamma - 1}; \quad c_p = \frac{\gamma R}{\gamma - 1}$$

It should be noted that γ must be determined from equation 3.20; there is no weighted mean expression as there is for R, c_v, and c_p.

Example 13.7

The gas in an engine cylinder has a volumetric analysis of 12% CO_2, 11·5% O_2, and 76·5% N_2 The temperature at the beginning of expansion is 1000°C and the gas mixture expands reversibly through a volume ratio of 7 to 1, according to a law $pv^{1 \cdot 25} = \text{constant}$. Calculate the work done and the heat flow per kg of gas. The values of c_p for the constituents averaged over the temperature are as follows:

c_p for $CO_2 = 1 \cdot 271$ kJ/kg K; c_p for $O_2 = 1 \cdot 110$ kJ/kg K; c_p for $N_2 = 1 \cdot 196$ kJ/kg K.

From equation 3.7, $m_i = n_i M_i$, therefore a conversion from volume fraction to mass fraction is as follows:

Consider 1 mole of the mixture.

Constituent	$\dfrac{n_i}{\text{kmol}}$	$\dfrac{M_i}{\text{kg/kmol}}$	$\dfrac{m_i}{\text{kg}} = n_i M_i$	m_i/m = fraction by mass
CO_2	0·12	44	5·28	$\dfrac{5 \cdot 28}{30 \cdot 36} = 0 \cdot 174$
O_2	0·115	32	3·68	$\dfrac{3 \cdot 68}{30 \cdot 36} = 0 \cdot 121$
N_2	0·765	28	21·40	$\dfrac{21 \cdot 40}{30 \cdot 36} = 0 \cdot 705$
			$\sum m_i = 30 \cdot 36$	

Then using equation 13.19,

$$c_p = \sum \frac{m_i}{m} c_{p_i}$$

$$\therefore \quad c_p = 0 \cdot 174 \times 1 \cdot 271 + 0 \cdot 121 \times 1 \cdot 110 + 0 \cdot 705 \times 1 \cdot 196$$

$$= 1 \cdot 199 \text{ kJ/kg K}$$

From equation 13.12, $R = \sum \dfrac{m_i}{m} R_i$, and from equation 3.9,

$R_i = R_0/M_i$, therefore,

$$R = 0 \cdot 174 \times \frac{8 \cdot 314}{44} + 0 \cdot 121 \times \frac{8 \cdot 314}{32} + 0 \cdot 705 \times \frac{8 \cdot 314}{28}$$

$$= 0 \cdot 2739 \text{ kJ/kg K}$$

Then from equation 3.17, $c_p - c_v = R$, we have,

$$c_v = 1 \cdot 199 - 0 \cdot 2739 = 0 \cdot 925 \text{ kJ/kg K}$$

The work done per kg of gas can be obtained from equation 4.30,

$$W = \frac{R(T_1 - T_2)}{n - 1}$$

T_2 can be found using equation 4.28,

$$\frac{T_2}{T_1} = \left(\frac{v_1}{v_2}\right)^{n-1} = \left(\frac{1}{7}\right)^{0\cdot 25}$$

i.e.
$$T_2 = \frac{T_1}{7^{0\cdot 25}} = \frac{1273}{1\cdot 625} = 783\cdot 2 \text{ K}$$

(where $T_1 = 1000 + 273 = 1273$ K).

$$\therefore \ W = \frac{0\cdot 2739(1273 - 783\cdot 2)}{1\cdot 25 - 1} = 536\cdot 3 \text{ kJ/kg}$$

Also from equation 3.16, for 1 kg, $u_2 - u_1 = c_v(T_2 - T_1)$, therefore,

$$u_2 - u_1 = 0\cdot 925(783\cdot 2 - 1273) = -453\cdot 1 \text{ kJ/kg}$$

Finally from the non-flow energy equation, 2.2,

$$Q = (u_2 - u_1) + W = -453\cdot 1 + 536\cdot 3 = 83\cdot 2 \text{ kJ/kg}$$

i.e.
$$\text{Heat supplied} = 83\cdot 2 \text{ kJ/kg}$$

Example 13.8

Calculate for the data of example 13·7 the change of entropy per kg of mixture.

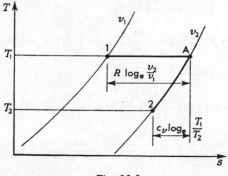

Fig. 13.3

Referring to fig. 13.3, the change of entropy between state 1 and state 2 can be found by imagining the process replaced by two other processes, 1 to A and A to 2. This method is described in Section 5.4.

For isothermal process 1 to A, from equation 5.12,

$$s_A - s_1 = R \log_e \frac{v_2}{v_1} = 0.2739 \times \log_e 7 = 0.533 \text{ kJ/kg K}$$

For the constant volume process A to 2,

$$s_A - s_2 = c_v \int_2^A \frac{dT}{T} = c_v \log_e \frac{T_1}{T_2} = 0.925 \times \log_e \frac{1273}{783.2}$$

i.e. $$s_A - s_2 = 0.449 \text{ kJ/kg K}$$

Then by subtraction,

$$s_2 - s_1 = 0.533 - 0.449 = 0.084 \text{ kJ/kg K}$$

It is often convenient to use kilomoles in problems on mixtures, and the specific heats may be expressed in terms of the kilomole. These are known as *molar heats*, and are denoted by C_p and C_v. Molar heats are defined as follows:

$$C_p = Mc_p \quad \text{and} \quad C_v = Mc_v \qquad \qquad 13.20$$

From equation 3.17, $c_p - c_v = R$, therefore,

$$C_p - C_v = Mc_p - Mc_v = MR$$

Also from equation 3.9, $MR = R_0$, hence,

$$C_p - C_v = R_0 \qquad \qquad 13.21$$

From equation 3.15,

$$U = mc_vT = \frac{mMc_vT}{M}$$

Also from equation 3.7, $m/M = n$, and from equation 13.20, $Mc_v = C_v$, therefore,

$$U = nC_vT \qquad \qquad 13.22$$

Similarly, $$H = nC_pT \qquad \qquad 13.23$$

By the Gibbs-Dalton law,

$$U = \sum U_i \quad \text{and} \quad H = \sum H_i$$

$$\therefore \quad nC_vT = \sum n_i C_{v_i} T \quad \text{and} \quad nC_pT = \sum n_i C_{p_i} T$$

i.e.
$$C_v = \sum \frac{n_i}{n} C_{v_i} \qquad \qquad 13.24$$

and
$$C_p = \sum \frac{n_i}{n} C_{p_i} \qquad \qquad 13.25$$

Example 13.9

A producer gas has the following volumetric analysis: 29% CO (28), 12% H_2 (2), 3% CH_4 (16), 4% CO_2 (44), 52% N_2 (28). Calculate the values of C_p, C_v, c_p, and c_v for the mixture. The values of C_p for the constituents are as follows: for CO $C_p = 29 \cdot 27$ kJ/kmol K; for H_2 $C_p = 28 \cdot 89$ kJ/kmol K; for CH_4 $C_p = 35 \cdot 8$ kJ/kmol K; for CO_2 $C_p = 37 \cdot 22$ kJ/kmol K; for N_2 $C_p = 29 \cdot 14$ kJ/kmol K.

From equation 13.25,

$$C_p = \sum \frac{n_i}{n} C_{pi}$$

Therefore,

$$C_p = 0 \cdot 29 \times 29 \cdot 27 + 0 \cdot 12 \times 28 \cdot 89 + 0 \cdot 03 \times 35 \cdot 8 + 0 \cdot 04 \times 37 \cdot 22$$
$$+ 0 \cdot 52 \times 29 \cdot 14$$

i.e.
$$C_p = 29 \cdot 676 \text{ kJ/kmol K}$$

From equation 13.21,

$$C_p - C_v = R_0$$

$$\therefore \quad C_v = C_p - R_0 = 29 \cdot 676 - 8 \cdot 314 = 21 \cdot 362 \text{ kJ/kmol K}$$

i.e.
$$C_v = 21 \cdot 362 \text{ kJ/kmol K}$$

The molar mass can be found from equation 13.17,

i.e. $\quad M = \sum \frac{n_i}{n} M_i$

$$= 0 \cdot 29 \times 28 + 0 \cdot 12 \times 2 + 0 \cdot 03 \times 16 + 0 \cdot 04 \times 44 + 0 \cdot 52 \times 28$$

i.e.
$$M = 25 \cdot 2 \text{ kg/kmol}$$

Then from equation 13.20,

$$c_p = \frac{C_p}{M} = \frac{29 \cdot 676}{25 \cdot 2} = 1 \cdot 178 \text{ kJ/kg K}$$

and

$$c_v = \frac{C_v}{M} = \frac{21 \cdot 362}{25 \cdot 2} = 0 \cdot 8476 \text{ kJ/kg K}$$

Values of γ, c_p, c_v, C_p, C_v, M, and R at 300 K for some of the more common gases are shown in Table 13.1. The relative molecular masses normally used in calculations are also included.

Table 13.1

Gas	c_p c_v kJ/kg K		γ	C_p C_v kJ/kmol K		M kg/kmol	R kJ/kg K	Rel mol mass
DIATOMIC								
Carbon Monoxide (CO)	1·041	0·7449	1·398	29·265	20·867	28·01	0·2968	28
Hydrogen (H$_2$)	14·323	10·1965	1·405	28·889	20·574	2·016	4·124	2
Nitrogen (N$_2$)	1·040	0·7436	1·40	29·140	20·825	28·013	0·2968	28
Oxygen (O$_2$)	0·9182	0·6586	1·394	29·391	21·076	32	0·2598	32
MONATOMIC								
Argon (Ar)	0·5203	0·3122	1·666	20·786	12·470	39·95	0·2081	40
Helium (He)	5·193	3·1160	1·666	20·787	12·473	4·003	2·077	4
TRIATOMIC								
Carbon Dioxide (CO$_2$)	0·8457	0·6573	1·29	37·220	28·906	44·01	0·1889	44
Sulphur Dioxide (SO$_2$)	0·6448	0·5150	1·25	41·324	33·009	64·06	0·1298	64
HYDROCARBONS								
Ethane (C$_2$H$_6$)	1·7668	1·4947	1·18	53·130	44·8155	30·07	0·2765	30
Methane (CH$_4$)	2·2316	1·7124	1·30	35·797	27·482	16·04	0·5183	16
Propane (C$_3$H$_8$)	1·6915	1·507	1·12	74·567	66·252	44·09	0·1886	44

13.5 Adiabatic mixing of perfect gases

Consider two gases A and B separated from each other in a closed vessel by a thin diaphragm, as shown in fig. 13.4. If the diaphragm

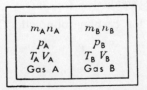

Fig. 13.4

is punctured or removed then the gases mix and each then occupies the total volume, behaving as if the other gas were not present. This process is equivalent to a free expansion of each gas, and is irreversible. The process can be simplified by the assumption that it is adiabatic; this means that the vessel is perfectly thermally insulated and there will therefore be an increase in entropy of the system. In Section 5.5 it was shown that there is always an increase in entropy of a thermally isolated system which undergoes an irreversible process.

It was shown in Section 4.4 that in a free expansion process the internal energy initially is equal to the internal energy finally. In this case, from equation 13.22,

$$U_1 = n_A C_{v_A} T_A + n_B C_{v_B} T_B$$

and

$$U_2 = (n_A C_{v_A} + n_B C_{v_B})T$$

Extending this result to any number of gases,

$$U_1 = \sum n_i C_{v_i} T_i \quad \text{and} \quad U_2 = T \sum n_i C_v$$

Then,

$$U_1 = U_2$$

i.e.

$$\sum n_i C_{v_i} T_i = T \sum n_i C_{v_i}$$

i.e.

$$T = \frac{\sum n_i C_{v_i} T_i}{\sum n_i C_{v_i}} \qquad 13.26$$

Example 13.10

A vessel of $1 \cdot 5 \text{ m}^3$ capacity contains oxygen at 7 bar and 40°C. The vessel is connected to another vessel of 3 m^3 capacity containing carbon monoxide at 1 bar and 15°C. A connecting valve is opened and the gases mix adiabatically.

Calculate,

(a) The final temperature and pressure of the mixture,
(b) The change in entropy of the system.

For oxygen $C_v = 21 \cdot 07 \text{ kJ/kmol K}$; for carbon monoxide $C_v = 20 \cdot 86 \text{ kJ/kmol K}$.

(a) From equation 3.8,

$$n = \frac{pV}{R_0 T}$$

Therefore,

$$n_{O_2} = \frac{7 \times 10^5 \times 1 \cdot 5}{8 \cdot 314 \times 313 \times 10^3} = 0 \cdot 4035 \quad \text{(where } T_{O_2} = 40 + 273 = 313 \text{ K)}$$

and

$$n_{CO} = \frac{1 \times 10^5 \times 3}{8 \cdot 314 \times 288 \times 10^3} = 0 \cdot 1253 \quad \text{(where } T_{CO} = 15 + 273 = 288 \text{ K)}$$

Before mixing,

$$U_1 = \sum n_i C_{v_i} T_i = 0 \cdot 4035 \times 21 \cdot 07 \times 313 + 0 \cdot 1253 \times 20 \cdot 86 \times 288$$

i.e. $$U_1 = 3413 \cdot 8 \text{ kJ}$$

After mixing,

$$U_2 = T \sum n_i C_{v_i} = T(0 \cdot 4035 \times 21 \cdot 07 + 0 \cdot 1253 \times 20 \cdot 86)$$

i.e. $$U_2 = 11 \cdot 118 \times T$$

For adiabatic mixing, $U_1 = U_2$, therefore,

$$3413 \cdot 8 = 11 \cdot 118 \times T \qquad \therefore \ T = \frac{3413 \cdot 8}{11 \cdot 118} = 307 \text{ K}$$

i.e.   Temperature of mixture $= 307 - 273 = 34°C$

From equation 3.8,

$$p = \frac{nR_0 T}{V}$$

Therefore,

$$p = \frac{(0 \cdot 4035 + 0 \cdot 1253) \times 8 \cdot 314 \times 307 \times 10^3}{(1 \cdot 5 + 3 \cdot 0) \times 10^5}$$

$$= \frac{0 \cdot 5288 \times 8 \cdot 314 \times 307 \times 10^3}{4 \cdot 5 \times 10^5}$$

i.e.   Pressure after mixing $= 3$ bar

(b) The change of entropy of the system is equal to the change of entropy of the oxygen plus the change of entropy of the carbon monoxide; this follows from the Gibbs-Dalton law.

Referring to fig. 13.5, the change of entropy of the oxygen can be calculated by replacing the process undergone by the oxygen by the two processes 1 to A and A to 2.

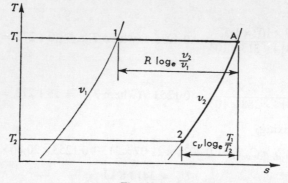

Fig. 13.5

For an isothermal process from 1 to A, from equation 5.12, we have

$$s_A - s_1 = R \log_e \frac{V_A}{V_1} \quad \text{or} \quad S_A - S_1 = mR \log_e \frac{V_A}{V_1}$$

i.e.

$$S_A - S_1 = nR_0 \log_e \frac{V_A}{V_1} = 0.4035 \times 8.314 \times \log_e \frac{4.5}{1.5} = 3.686 \text{ kJ/K}$$

At constant volume from A to 2,

$$s_A - s_2 = c_v \int_2^A \frac{dT}{T} = c_v \log_e \frac{T_1}{T_2} \quad \text{or} \quad S_A - S_2 = mc_v \log_e \frac{T_1}{T_2}$$

i.e.

$$S_A - S_2 = nC_v \log_e \frac{T_1}{T_2} = 0.4035 \times 21.07 \times \log_e \frac{313}{307} = 0.1683 \text{ kJ/K}$$

$$\therefore \quad S_2 - S_1 = 3.686 - 0.168 = 3.518 \text{ kJ/K}$$

Referring to fig. 13.6, the change of entropy of the carbon monoxide can be found in a similar way to the above,

i.e.

$$S_2 - S_1 = (S_B - S_1) + (S_2 - S_B)$$

$$\therefore \quad S_2 - S_1 = nR_0 \log_e \frac{V_B}{V_1} + nC_v \log_e \frac{T_2}{T_1}$$

$$= 0.1253 \times 8.314 \times \log_e \frac{4.5}{3} + 0.1253 \times 20.86 \times \log_e \frac{307}{288}$$

$$\therefore \quad S_2 - S_1 = 0.5897 \text{ kJ/K}$$

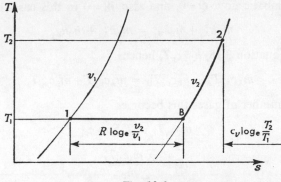

Fig. 13.6

Hence the change of entropy of the whole system is given by

$$(S_2 - S_1)_{\text{system}} = (S_2 - S_1)_{O_2} + (S_2 - S_1)_{CO}$$

i.e.

Change of entropy of system $= 3 \cdot 518 + 0 \cdot 590 = 4 \cdot 108$ kJ/K

Another form of mixing is that which occurs when two streams of fluid meet to form a common stream in steady flow. This is shown diagrammatically in fig. 13.7. The steady-flow energy equation can

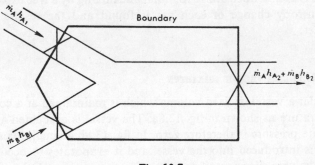

Fig. 13.7

be applied to the mixing section, and changes in kinetic and potential energy are usually negligible,

i.e. $$\dot{m}_A h_{A_1} + \dot{m}_B h_{B_1} + Q = \dot{m}_A h_{A_2} + \dot{m}_B h_{B_2} + W$$

For adiabatic flow $Q=0$, and also $W=0$ in this case, therefore,

$$\dot{m}_A h_{A_1} + \dot{m}_B h_{B_1} = \dot{m}_A h_{A_2} + \dot{m}_B h_{B_2}$$

From equation 3.18, $h=c_p T$, hence,

$$\dot{m}_A c_{p_A} T_A + \dot{m}_B c_{p_B} T_B = \dot{m}_A c_{p_A} T + \dot{m}_B c_{p_B} T$$

For any number of gases this becomes

$$\sum \dot{m}_i c_{p_i} T_i = T \sum \dot{m}_i c_{p_i}$$

i.e.
$$T = \frac{\sum \dot{m}_i c_{p_i} T_i}{\sum \dot{m}_i c_{p_i}} \qquad 13.27$$

Also, since from equation 13.20, $C_p = M c_p$, and $M = m/n$ from equation 3.7, then.

$$n C_p = m c_p$$

Hence,
$$T = \frac{\sum n_i C_{p_i} T_i}{\sum n_i C_{p_i}} \qquad 13.28$$

Equation 13.27 or 13.28 represents one condition which must be satisfied in an adiabatic mixing process of perfect gases in steady flow. In a particular problem some other information must be known (e.g. the final pressure or specific volume) before a complete solution is possible. To find the change of entropy in such a process the procedure is as described above for adiabatic mixing by a free expansion. The entropy change of each gas is found and the results added together.

13.6 Gas and vapour mixtures

Consider a vessel of fixed volume which is maintained at a constant temperature as shown in fig. 13.8a. The vessel is evacuated and the absolute pressure is therefore zero. In fig. 13.8b a small quantity of water is introduced into the vessel and it evaporates to occupy the whole volume. For a small quantity of water introduced, the pressure in the vessel will be less than the saturation pressure corresponding to the temperature of the vessel. At this condition of pressure and temperature the vessel will be occupied by superheated vapour. As more water is introduced the pressure increases and the water continues to evaporate until such a condition is reached that the volume can hold no more vapour. Any additional water introduced into the vessel after this will not evaporate but will exist as water, the condition being as in fig. 13.8c, which shows the vapour in contact with

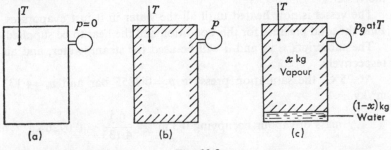

Fig. 13.8

its liquid. Per kg of water introduced, the vessel can be thought of as containing either $(1-x)$ kg of water plus x kg of dry saturated vapour, or as containing 1 kg of wet steam of dryness fraction x.

During the entire process of evaporation the temperature remains constant. If the temperature is now raised by the addition of heat, then more vapour will evaporate and the pressure in the vessel will increase. Eventually the vessel will contain a superheated vapour as before, but at a higher pressure and temperature.

The vessel in fig. 13.8 is considered to be initially evacuated, but the water would evaporate in exactly the same way if the vessel contained a gas or a mixture of gases. As stated in the Gibbs-Dalton law, each constituent behaves as if it occupies the whole vessel at the temperature of the vessel. When a little water is sprayed into a vessel containing a gas mixture, then the vapour formed will exert the saturation pressure corresponding to the temperature of the vessel, and this is the partial pressure of the vapour in the mixture. (It must be remembered that the vapour is only saturated when it is in contact with its liquid.)

When a mixture contains a saturated vapour, then the partial pressure of the vapour can be found from tables at the temperature of the mixture. This assumes that a saturated vapour obeys the Gibbs-Dalton law; this is only a good approximation at low values of the total pressure.

Example 13.11

A vessel of 0.3 m³ capacity contains air at 0.7 bar and 75°C. The vessel is maintained at this temperature as water is injected into it.

Calculate the mass of water to be injected so that the vessel is just filled with saturated vapour. If injection now continues until a total mass of 0·7 kg of water is introduced, find the new total pressure in the vessel.

The vessel is now heated until all the water in it just evaporates. Find the total pressure for this condition and the heat to be supplied.

The subscripts s, w, and a will be used for steam, water, and air respectively.

At 75°C, the saturation pressure $p_g = 0·3855$ bar and $v_g = 4·133$ m³/kg.

$$\therefore \text{ mass of vapour occupying } 0·3 \text{ m}^3 = \frac{0·3}{4·133} = 0·0726 \text{ kg}$$

i.e.   Mass of water to be injected $= 0·0726$ kg

By Dalton's law, equation 13.2, $p = p_a + p_s$,

i.e.   Total pressure in vessel $= 0·7 + 0·3855 = 1·0855$ bar

Note that the dry vapour is assumed to act as a perfect gas, hence the vapour and the air are assumed to occupy the same volume while each exerts its partial pressure.

When a total mass of 0·7 kg of water has been injected into the vessel it will exist partly as dry saturated vapour (say m_s, kg) and partly as water (say m_w kg; where $m_w = (0·7 - m_s)$) in such proportions that the mixture occupies the total volume of 0·3 m³

$$\therefore \quad m_s \times 4·133 + (0·7 - m_s) \times 0·001\ 026 = 0·3$$

(where 0·001026 m³/kg is the specific volume of water)

$$\therefore \quad m_s(4·133 - 0·001\ 026) = 0·3 - (0·7 \times 0·001\ 026)$$

i.e. $4·132 \times m_s = 0·2993$ $\therefore \quad m_s = \dfrac{0·2993}{4·132} = 0·0724$ kg

(Note that the volume of water is negligibly small compared to the volume of the air-vapour mixture.)

$m_w = 0·7 - 0·0724 = 0·6276$ kg
The volume occupied by the dry vapour $= 0·0724 \times 4·133 = 0·2993$ m³. The vessel may be assumed to contain air, dry saturated steam, and water, as shown in fig. 13.9.

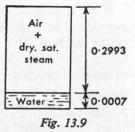

Fig. 13.9

Since $T_1 = T_2$, we can write

$$p_{a_1} V_{a_1} = p_{a_2} V_{a_2}$$

$$\therefore \ p_{a_2} = 0.7 \times \frac{0.3}{0.2993} = 0.7017 \text{ bar}$$

i.e.   Total pressure $= p_a + p_s = 0.7017 + 0.3855 = 1.0872$ bar

The water can be completely evaporated by raising the temperature to a value such that the total volume is occupied by saturated steam and air. This condition is reached when the steam has a specific volume v_g, such that, $0.7 \times v_g = 0.3$,

i.e. $$v_g = \frac{0.3}{0.7} = 0.4286 \text{ m}^3/\text{kg}$$

From tables the saturation pressure at $v_g = 0.4286 \text{ m}^3/\text{kg}$ is, by interpolating,

$$p = 4 + \left(\frac{0.4623 - 0.4286}{0.4623 - 0.4139}\right) \times (4.5 - 4.0) = 4.35 \text{ bar}$$

The air now occupies the volume of 0.3 m^3 while exerting its partial pressure p_{a_3} at the new temperature. The new temperature is that saturation temperature corresponding to the pressure of 4.35 bar.

From tables by interpolation at 4.35 bar,

$$t = 143.6 + \frac{0.35}{0.5} \times 4.3 = 146.6°\text{C} \ \therefore \ T = 146.6 + 273 = 419.6 \text{ K}$$

Then for the air,

$$\frac{p_{a_3}}{T_{a_3}} = \frac{p_{a_1}}{T_{a_1}} \quad \therefore \ p_{a_3} = 0.7 \times \frac{419.6}{348} = 0.8439 \text{ bar}$$

(where $T_{a_1} = 75 + 273 = 348$ K),

i.e.   Total pressure in vessel $= 4.35 + 0.8439 = 5.194$ bar

From the non-flow energy equation,

$$Q = (U_2 - U_1) + W$$

In this case $W = 0$, therefore $Q = (U_2 - U_1)$.

Then, $U_1 = m_{w_1} u_{w_1} + m_a u_{a_1} + m_{s_1} u_{s_1}$

and $U_2 = m_a u_{a_2} + m_{s_2} u_{s_2}$

For a perfect gas, from equation 3.15, $U = mc_v T$, therefore,

$$Q = m_{s_2} u_{s_2} - m_{s_1} u_{s_1} - m_{w_1} u_{w_1} + m_a c_v (T_2 - T_1)$$

Then taking u_s and u_w from tables, and substituting for

$$m_a = \frac{p_a V}{R_a T} = \frac{0.7 \times 10^5 \times 0.3}{0.287 \times 348 \times 10^3} = 0.2102 \text{ kg}$$

we have

$$Q = 0.7 \times 2556.8 - 0.0724 \times 2475.3 - 0.6276 \times 313.5 + 0.2102$$
$$\times 0.718(419.6 - 348)$$

i.e.   Heat supplied $= 1789.7 - 179.2 - 196.7 + 10.8 = 1424.6 \text{ kJ}$

Example 13.12

The products of combustion of a coal gas have a volumetric analysis of CO_2 8%, H_2O 15%, O_2 5.5%, and N_2 71.5%. If the total pressure is 1.4 bar, calculate the temperature to which the gas must be cooled at constant pressure for condensation of the H_2O just to commence.

From equation 13.14,

$$\text{Partial pressure of } H_2O = \frac{n_i}{n} \times p = 0.15 \times 1.4 = 0.21 \text{ bar}$$

The saturation temperature corresponding to 0.21 bar is 61.15°C, i.e., the gas must be cooled to 61.15°C for condensation of the H_2O to commence.

13.7 The steam condenser

The condenser was shown in Section 7.1 to be an essential part of a steam plant. The temperature at which condensation occurs is in the order of 27 to 38°C, the corresponding saturation pressures being 0.03564 bar and 0.06624 bar. The shell and tube type condenser is a vessel in which this low pressure is maintained by a pump, and the steam condenses on the outside of tubes through which cold water is flowing. This type is called a surface condenser. There will be some leakage of air into the condenser, both through the glands and from air dissolved in the feed water which comes out of solution and is carried into the condenser by the steam. This air impairs the condenser performance since it reduces the heat transfer from the steam to the cooling water.

The condenser contains a mixture of steam, air, and water. The air must be pumped out of the condenser continually to maintain the vacuum, and the air which is pumped out carries with it some of the steam. This results in a loss of feed water to the boiler. This loss has to be made up by the addition of cold water. Another effect of the presence of air is that the condensate is undercooled (i.e. cooled to a temperature below the saturation temperature), which means that more heat has to be supplied to the water in the boiler than if no undercooling had occurred.

The pressure in the condenser is approximately constant throughout and steam and air enter the condenser in fixed proportions when steady conditions prevail. As some of the steam is condensed, the partial pressure of the remaining steam decreases, and hence the partial pressure of the air increases to maintain the same total pressure. At reduced partial pressures the steam has a saturation temperature which is below that of the incoming steam. Hence condensation proceeds at progressively lower temperatures.

Some condensers are designed to make up for the deficiencies of the simple type. Two of these are indicated in figs. 13.10a and 13.10b. In fig. 13.10a most of the condensation is carried out on the main

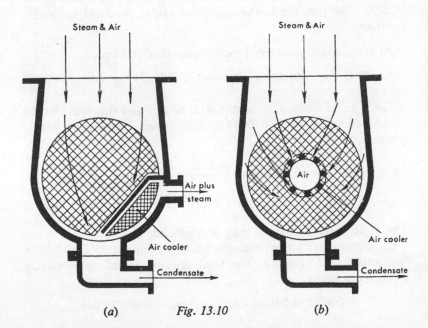

(a)   Fig. 13.10   (b)

bank of tubes and the air is drawn over another, smaller, bank which
is shielded from the main bank and is called the air cooler. Here
further condensation takes place at a lower temperature with a
subsequent saving in feed water, and a smaller pump is required for
the condenser. In fig. 13.10b the air cooling tubes are in the centre of
the condenser and the air is pumped away from this region. The
incoming steam passes all round the bank of tubes and some is
drawn upwards to the centre. In doing so it meets the undercooled
condensate which has been formed and reheats it, hence reducing the
amount of undercooling.

Example 13.13

A surface condenser is required to deal with 20 000 kg of steam per
hour, and the air leakage is estimated at 0·3 kg per 1000 kg of steam.
The steam enters the condenser dry saturated at 38°C. The condensate is extracted at the lowest point of the condenser at a temperature
of 36°C. The condensate loss is made up with water at 7°C. It is required to find the saving in condensate and the saving in heat supplied in the boiler, by fitting a separate air extraction pump which
draws the air over an air cooler. Assume that the air leaves the cooler
at 27°C. The pressure in the condenser can be assumed to remain
constant.

At entry, mass of air per kg of steam = 0·3/1000 kg.

At 38°C the saturation pressure is 0·06624 bar and $v_g = 21.63$
m^3/kg.

For 1 kg of steam the volume is 21·63 m^3, and this must be the
volume occupied by 0·3/1000 kg of air when exerting its partial
pressure,

i.e.   Partial pressure of air $= \dfrac{m_a R_a T}{V} = \dfrac{0.3 \times 0.287 \times 311 \times 10^3}{1000 \times 21.63 \times 10^5}$

$$= 1.2 \times 10^{-5} \text{ bar}$$

This is negligibly small and may be neglected.

Condensate extraction: the saturation pressure at 36°C is 0·0594
bar, and $v_g = 23.97$ m^3/kg. The total pressure in the condenser is
0·06624 bar, hence,

$$0.066\,24 = 0.0594 + p_a \qquad \therefore \ p_a = 0.006\,84 \text{ bar}$$

The mass of air removed per hour is

$$\frac{20\,000 \times 03}{1000} = 6 \text{ kg/h}$$

Hence the volume of air removed per hour is

$$\frac{mRT}{p} = \frac{6 \times 0.287 \times 309 \times 10^3}{0.006\,84 \times 10^5} = 778 \text{ m}^3/\text{h}$$

The mass of steam associated with the air removed is therefore given by

$$\frac{778}{23.97} = 32.45 \text{ kg/h}$$

Separate extraction: the saturation pressure at $27°\text{C}$ is 0.03564 bar and $v_g = 38.81$ m^3/kg.

The air partial pressure is $0.066\,24 - 0.035\,64 = 0.0306$ bar.

Therefore the volume of air removed is

$$\frac{mRT}{p} = \frac{6 \times 0.287 \times 300 \times 10^3}{0.0306 \times 10^5} = 168.9 \text{ m}^3/\text{h}$$

$$\therefore \text{ Steam removed} = \frac{168.9}{38.81} = 4.35 \text{ kg/h}$$

Hence the saving in condensate by using the separate extraction method is given by $32.45 - 4.35 = 28.1$ kg/h.

Also, the saving in heat to be supplied in the boiler is $28.1 \times 4.186(36 - 7) = 3411$ kJ/h.

Example 13.14

For the data of example 13.13 calculate the percentage reduction in air pump capacity by using the separate extraction method. If the temperature rise of the cooling water is 5.5 K, calculate the mass flow of cooling water required.

Air pump capacity without air cooler = 778 m^3/h
Air pump capacity with the air cooler = 168.9 m^3/h

Therefore,

$$\text{Percentage reduction in capacity} = \left(\frac{778 - 168.9}{778}\right) \times 100$$
$$= 78.3\%$$

The system to be analysed is shown in fig. 13.11. Let suffixes s, a, and c denote steam, air, and condensate respectively. Applying the steady flow energy equation and neglecting changes in kinetic energy, we have

$$Q = \dot{m}_{s_1}h_{s_1} + \dot{m}_{a_1}h_{a_1} - (\dot{m}_{s_2}h_{s_2} + \dot{m}_{a_2}h_{a_2}) - \dot{m}_c h_c$$
$$\dot{m}_{a_1} = \dot{m}_{a_2} = 6 \text{ kg/h}; \quad \dot{m}_{s_2} = 4\cdot35 \text{ kg/h}$$
$$\dot{m}_c = 20\,000 - 4\cdot35 = 20\,000 \text{ kg/h appr.}$$
$$h_{a_1} - h_{a_2} = c_p(T_1 - T_2) \quad \text{(from equation 3.18)}$$
$$\therefore \quad Q = 20\,000 \times 2570\cdot1 + 6 \times 1\cdot005(38 - 27)$$
$$- 4\cdot35 \times 2550\cdot3 - 20\,000 \times 150\cdot7$$

(where $h_c = h_f$ at 36°C $= 150\cdot7$ kJ/kg),

i.e. 　　　　　　　Heat rejected $= 48\cdot38 \times 10^6$ kJ/h

The mass of cooling water required for a 5·5 K rise in temperature is $48\cdot38 \times 10^6/(5\cdot5 \times 4\cdot187)$ $= 2\cdot1 \times 10^6$ kg/h, approximately.

Unless a very large natural supply of cooling water is available for large steam plants, means must be found to cool the cooling water after use. This can be done by passing the cooling water through a cooling tower; cooling towers are considered in Section 14.5.

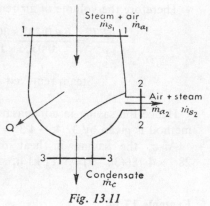

Fig. 13.11

PROBLEMS

(For values of M, R, C_p, C_v, etc., which are necessary in the following problems, refer to Table 13.1 on page 471.)

13.1 A mixture of carbon monoxide and oxygen is to be prepared in the proportion of 7 kg to 4 kg in a vessel of 0·3 m³ capacity. If the temperature of the mixture is 15°C, determine the pressure to which the vessel is subject. If the temperature is raised to 40°C, what will then be the pressure in the vessel? 　　　(29·9 bar; 32·5 bar)

13.2 For the mixture of problem 13.1 calculate the volumetric analysis, the relative molecular mass and the characteristic gas constant. Calculate also the total number of kmol in the mixture.

$$(33.3\% \ O_2; \ 66.7\% \ CO; \ 29.3; \ 0.283 \ kJ/kg \ K; \ 0.375)$$

13.3 An exhaust gas is analysed and is found to contain, by volume, 78% N_2, 12% CO_2, and 10% O_2. What is the corresponding gravimetric analysis? Calculate the mass of mixture per kmol, and the density if the temperature is 550°C and the total pressure is 1 bar.

$$(72.2\% \ N_2, \ 17.3\% \ CO_2, \ 10.6\% \ O_2; \ 30.28 \ kg/kmol; \ 0.442 \ kg/m^3)$$

13.4 A vessel of 3 m³ capacity contains a mixture of nitrogen and carbon dioxide, the analysis by volume showing equal quantities of each. The temperature is 15°C and the total pressure is 3.5 bar. Determine the mass of each constituent. (6.14 kg N_2; 9.65 kg CO_2)

13.5 The mixture of problem 13.4 is to be changed so that it is 70% CO_2 and 30% N_2 by volume. Calculate the mass of mixture to be removed and the mass of CO_2 to be added to give the required mixture at the same temperature and pressure as before.

$$(6.32 \ kg; \ 7.72 \ kg \ CO_2)$$

13.6 In a mixture of methane and air there are three moles of oxygen to one mole of methane. Calculate the values of c_p, c_v, C_p, C_v, R, and γ for the mixture. Assume that air contains only oxygen and nitrogen.

From initial conditions of 1 bar and 95°C, the gas is compressed reversibly and adiabatically through a volume ratio of 5 to 1. Calculate the final pressure and temperature and the work done per kg of mixture. Calculate also the change of entropy per kg of mixture and the change of internal energy per kg of mixture.

$$(1.051, \ 0.754 \ kJ/kg \ K; \ 29.52, \ 21.18 \ kJ/kmol \ K; \ 0.2954$$
$$kJ/kg \ K; \ 1.39; \ 9.4 \ bar; \ 415°C; \ 241.2 \ kJ/kg; \ 0; \ 241.2 \ kJ/kg)$$

13.7 A mixture is made up of 25% N_2, 35% O_2, 20% CO_2, and 20% CO by volume. Calculate,

(a) The relative molecular mass of the mixture.
(b) C_p and C_v for the mixture.
(c) γ for the mixture.
(d) The partial pressure of each constituent when the total pressure is 1.5 bar.

(e) The density of the mixture at 1·5 bar and 15°C.

$\qquad$ (32·6; 30·9, 22·53 kJ/kmol K; 1·37; 0·375, 0·525, 0·3, 0·3 bar;
$\qquad$ 2·04 kg/m^3)

13.8 Two vessels are connected by a pipe in which there is a valve. One vessel of 0·3 m^3 contains air at 7 bar and 32°C, and the other of 0·03 m^3 contains oxygen at 21 bar and 15°C. The valve is opened and the two gases are allowed to mix. Assuming that the system is well-lagged, calculate,

(a) The final temperature of the mixture,
(b) The final pressure of the mixture,
(c) The partial pressure of each constituent,
(d) The volumetric analysis of the mixture,
(e) The values of c_p, c_v, R, M, and γ for the mixture,
(f) The increase of entropy of the system per kg of mixture,
(g) The change in internal energy and enthalpy of the mixture per kg if the vessel is cooled to 10°C.

Assume that air consists only of oxygen and nitrogen.

$\qquad$ (27·7°C; 8·26 bar; 3·30, 4·96 bar; 60% N$_2$, 40% O$_2$;
$\qquad$ 0·982, 0·703 kJ/kg K; 0·281 kJ/kg K; 29·6 kg/kmol; 1·4;
$\qquad$ 0·182 kJ/kg K; 12·4, 17·4 kJ/kg)

13·9 Air and carbon monoxide are mixed in the proportion 3 to 1 by mass. The CO is supplied at 4 bar and 15°C, and the air is supplied at 7 bar and 32°C. The two constituents are passed in steady flow through non-return valves to mix adiabatically at a pressure of 1 bar. Calculate,

(a) The final temperature of the mixture,
(b) The partial pressure of each constituent of the mixture,
(c) The increase of entropy per kg of mixture,
(d) The volume flow of mixture for a flow of 1 kg/min of CO,
(e) The velocity of the mixture if the area of the pipe downstream of the mixing section is 0·1 m^2.

$\qquad$ (27·6°C; 0·255, 0·156, 0·589 bar; 0·687 kJ/kg K;
$\qquad$ 3·48 m^3/min; 0·581 m/s)

13.10 Ammonia in air is a toxic mixture when the ammonia is 0·55% by volume. Calculate how much leakage in kg from an ammonia compressor can be tolerated per 1000 m^3 of space. The pressure is 1 bar and the temperature is 15°C. The relative molecular mass of ammonia (NH_3) is 17, and it may be assumed to act as a perfect gas for the purposes of this problem. (3·88 kg)

13.11 A vessel of 0·3 m³ capacity contains a mixture of air and steam which is 0·75 dry. If the pressure is 7 bar and the temperature is 116·9°C, calculate the mass of water present, the mass of dry saturated vapour, and the mass of air. (0·102 kg; 0·307 kg; 1·39 kg)

13.12 If the vessel of problem 13.11 is cooled to 100°C calculate,

    (a) The mass of vapour condensed,

    (b) The final pressure in the vessel,

    (c) The heat removed. (0·13 kg; 5·99 bar; 297 kJ)

13.13 A closed vessel of volume 3 m³ contains air saturated with water vapour at 38°C and a vacuum pressure of 660 mm of mercury. The vacuum falls to 560 mm of mercury and the temperature falls to 26·7°C. Calculate the mass of air that has leaked in and the quantity of vapour that has condensed. Take the barometric pressure as 760 mm Hg. (0·58 kg; 0·063 kg)

13.14 The air in a cylinder fitted with a piston is saturated with water vapour. The volume is 0·3 m³, the pressure is 3·5 bar and the temperature is 60·1°C. The mixture is compressed to 5·5 bar, the temperature remaining constant. Find,

    (a) The masses of air and vapour present initially,

    (b) The mass of vapour condensed on compression.

(1·035 kg; 0·0392 kg; 0·0148 kg)

13.15 The temperature in a vessel is 36°C and the proportion of air to dry saturated steam is 0·1 kg/kg. What is the pressure in the vessel in bar and in mm of mercury vacuum? The barometric pressure is 760 mm Hg. (0·0631 bar; 712·5 mm Hg)

13.16 A surface condenser is fitted with separate air and condensate outlets. A portion of the cooling surface is screened from the incoming steam and the air passes over these screened tubes to the air extraction and becomes cooled below the condensate temperature. The condenser receives 20 000 kg/h of steam dry saturated at 36·2°C. At the condensate outlet the temperature is 34·6°C, and at the air extraction the temperature is 29°C. The volume of air plus vapour leaving the condenser is 3·8 m³/min. Assuming constant pressure throughout the condenser calculate,

(a) The mass of air removed per 10 000 kg of steam,
(b) The mass of steam condensed in the air cooler per minute,
(c) The heat to be removed per minute by the cooling water.

Neglect the partial pressure of the air at inlet to the condenser.

(2·63 kg; 0·492 kg; 807 050 kJ)

14. Psychrometry

Mixtures of air and water vapour are considered in Chapter 13; in this chapter moist atmospheric air (i.e. a mixture of dry air and water vapour) is considered as a separate topic.

It is often necessary to provide a controlled atmosphere in buildings where industrial processes are to be carried out, or to provide air conditioning in private and public buildings. The properties of atmospheric air have to be considered in these problems, and this is a subject which is receiving an increasing amount of attention and application. Another topic which will be considered is that of the cooling tower by means of which large quantities of cooling water are cooled for recirculation. These topics come under the title of Psychrometry (sometimes called Hygrometry).

14.1 Psychrometric mixtures

In Section 13.6 the evaporation of water into an evacuated space or into a volume occupied by a gas was described, and it was seen that before the saturated condition was reached the vapour existed in the mixture as a superheated vapour. At the saturation condition the partial pressure of the vapour can be obtained from steam tables as that pressure corresponding to the temperature of the mixture. If the space or gas is not saturated at a particular temperature, then the partial pressure of the vapour will be less than the saturation pressure corresponding to that temperature.

Consider atmospheric air at 1·013 bar and 15°C. The saturation pressure of water vapour corresponding to 15°C is 0·017 04 bar. Unless the water vapour is in contact with its liquid it will not be saturated, and its pressure will be below the saturation value of 0·017 04 bar. In normal applications the atmosphere is well removed from the saturated condition. At such low vapour pressures (i.e. well below one atmosphere) the vapour can be considered to act as a perfect gas, and the properties of the mixture can be found using the Gibbs-Dalton law. The properties of the mixture depend on its pressure and temperature, and are determined for a particular state with reference to the properties of saturated vapour.

✓ Assume that in a quantity of atmospheric air the vapour pressure is 0·010 01 bar at 15°C and the total pressure is 1·013 bar.

From equation 13.2,

$$p = p_a + p_s$$

(where p_a = partial pressure of the dry air, and p_s = partial pressure of the superheated vapour),

i.e. $p_a = p - p_s = 1·013 - 0·010\ 01 = 1·003$ bar

The saturation temperature corresponding to 0·010 01 bar is 7°C, hence the vapour in atmospheric air under these conditions has a degree of superheat of $(15 - 7) = 8$ K. This state is indicated by point 1 on a T-s diagram in fig. 14.1. Suppose a metal beaker containing

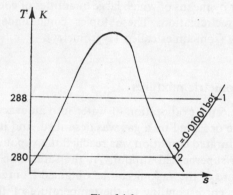

Fig. 14.1

water is placed in this atmosphere, and the water is progressively cooled by adding ice. At a particular temperature of the water it will be noticed that condensation begins to appear on the outside surface of the beaker. The vapour in contact with the beaker has been cooled at constant pressure to 7°C, as indicated by point 2 in fig. 14.1. This is the condition of saturation, and further cooling causes condensation of the water vapour. This temperature is called the *dew point* of the mixture, and it is the temperature to which an unsaturated mixture must be cooled in order to become just saturated. The dew point temperature is denoted by t_d.

If a room is warm and the outside atmosphere is cold, then a window which is colder than the walls of the room can produce condensation on its inside surface. A person wearing spectacles entering a warm room after a time spent in a cold atmosphere very often finds the vapour in the air condensing on the lenses as the vapour is cooled to its dew point. Condensation is noticed on cold water pipes which are situated in an atmosphere which is at a higher temperature and which is sufficiently humid.

14.2 Specific humidity, percentage saturation and relative humidity

The *specific humidity* (or *absolute humidity*) is the ratio of the mass of water vapour to the mass of dry air in a given volume of the mixture, and is denoted by the symbol ω,

i.e.
$$\omega = \frac{m_s}{m_a} \qquad 14.1$$

(where the subscript '*s*' denotes the vapour, and the subscript '*a*' denotes the dry air).

Since both masses occupy the volume V then

$$\omega = \frac{m_s/V}{m_a/V} = \frac{1/v_s}{1/v_a} = \frac{v_a}{v_s} \qquad 14.2$$

(where v_a and v_s are the specific volumes of the dry air and vapour respectively).

Since both the vapour and the dry air are considered as perfect gases then

$$m_s = \frac{p_s V}{R_s T} \quad \text{and} \quad m_a = \frac{p_a V}{R_a T}$$

Also,
$$R_s = \frac{R_0}{M_s} \quad \text{and} \quad R_a = \frac{R_0}{M_a}$$

Therefore,
$$m_s = \frac{p_s V M_s}{R_0 T} \quad \text{and} \quad m_a = \frac{p_a V M_a}{R_0 T}$$

Then, substituting in equation 14.1,

$$\omega = \frac{p_s V M_s}{R_0 T} \times \frac{R_0 T}{p_a V M_a}$$

i.e.
$$\omega = \frac{M_s}{M_a} \times \frac{p_s}{p_a}$$

$$\therefore \qquad \omega = \frac{18}{28 \cdot 96} \times \frac{p_s}{p_a} = 0 \cdot 622 \frac{p_s}{p_a}$$

If the total pressure is p, then from equation 13.2, $p = p_a + p_s$,

$$\therefore \qquad \omega = 0 \cdot 622 \times \left(\frac{p_s}{p - p_s} \right) \qquad 14.3$$

(The total pressure p is usually the barometric pressure.)

The *relative humidity* of the atmosphere is the ratio of the actual mass of the water vapour in a given volume to that which it would have if it were saturated at the same temperature.

i.e.   Relative humidity, $\phi = \dfrac{m_s}{(m_s)_{sat}}$

Now, $\qquad m_s = \dfrac{p_s V}{R_s T} \quad$ and $\quad (m_s)_{sat} = \dfrac{p_g V}{R_s T}$

(where p_g is the saturation pressure at the temperature of the mixture).

i.e $\qquad\qquad\qquad \phi = \dfrac{p_s}{p_g} \qquad\qquad\qquad 14.4$

The term *percentage saturation* is also used, defined as the ratio of the specific humidity of a mixture to the specific humidity of the mixture when saturated at the same temperature,

i.e. $\qquad\qquad$ Percentage saturation, $\psi = \dfrac{\omega}{\omega_g} \qquad 14.5$

From equations 14.3, 14.4 and 14.5 it can be seen that,

$$\text{Percentage saturation, } \psi = \phi \times \frac{(p - p_g)}{(p - p_s)} \qquad 14.6$$

In air conditioning practice the percentage difference between ψ and ϕ is in the range $0 \cdot 5\%$ to 2%, approximately.

Example 14.1

The air supplied to a room of a building in winter is to be at $17°C$ and have a relative humidity of 60%. If the barometric pressure is $1 \cdot 013\ 25$ bar, calculate the specific humidity. What would be the dew point under these conditions?

At 17°C, $p_g = 0.01936$ bar, hence using equation 14.4,

$$0.6 = \frac{p_s}{0.019\,36} \qquad \therefore \; p_s = 0.6 \times 0.019\,36 = 0.011\,616$$

Using equation 14.3, $\omega = 0.622 \times p_s/(p - p_s)$, we have

$$\omega = 0.622 \times \frac{0.011\,616}{1.013\,25 - 0.011\,616} = 0.007\,213$$

i.e., the atmosphere contains $0.007\,213$ kg of vapour per kg of dry air.

If the air is cooled at constant pressure the vapour will begin to condense at the saturation temperature corresponding to $0.011\,616$ bar. By interpolation from tables, the dew point temperature t_d is then

$$t_d = 9 + (10 - 9) \times \left(\frac{0.011\,616 - 0.011\,47}{0.012\,27 - 0.011\,47} \right) = 9.18°C$$

Example 14.2

If air at the condition of example 14.1 is passed at the rate of 0.5 m³/s over a cooling coil which is at a temperature of 6°C, calculate the amount of vapour which will be condensed. Assume that the barometric pressure is the same as in example 14.1, and that the air leaving the coil is saturated.

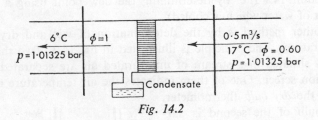

Fig. 14.2

The system is shown in fig. 14.2. The mass flow rate of dry air, $\dot{m}_a$, is given by

$$\dot{m}_a = \frac{p_a \dot{V}}{R_a T}$$

From equation 13.2, $p_a = p - p_s$, therefore,

$$p_a = 1.013\,25 - 0.011\,616 = 1.001\,63 \text{ bar}$$

$$\therefore \; \dot{m}_a = \frac{10^5 \times 1.001\,63 \times 0.5}{10^3 \times 0.287 \times 290} = 0.6017 \text{ kg/s}$$

This mass of air is constant throughout the process.

From equation 14.1, $\omega = m_s/m_a$, and ω has been determined as 0·007 213, therefore,

$$\dot{m}_{s_1} = 0.007\ 213 \times \dot{m}_a$$

After passing the cooling coil, $\phi = 1$, since the air is saturated. From equation 14.4, $p_s = p_g$ for this condition, and at 6°C, $p_g = 0.009\ 346$ bar, therefore, from equation 14.3,

$$\omega_2 = 0.622 \times \left(\frac{0.009\ 346}{1.013\ 25 - 0.009\ 346}\right) = 0.005\ 79$$

$$\therefore \dot{m}_{s_2} = 0.005\ 79 \times \dot{m}_a$$

Hence,

$$\text{Mass of condensate} = \dot{m}_{s_1} - \dot{m}_{s_2} = (0.007\ 213 - 0.005\ 79) \times m_a$$

$$= 0.001\ 423 \times 0.6017 \times 3600 = 3.082\ \text{kg/h}$$

Measurement of relative humidity

An instrument used to measure relative humidity is called a psychrometer, or a hygrometer. A simple psychrometer has been referred to in Section 14.1 (i.e. by determining the dew point using a metal beaker of water which is cooled).

Another method is by the determination of wet and dry bulb temperatures. The principle is illustrated in fig. 14.3. Two thermometers situated in a stream of unsaturated air are separated by a radiation screen. One of them indicates the air temperature and is called the *dry bulb* thermometer. The bulb of the second is surrounded by a wick which dips into a small reservoir of water and the temperature indicated is called the *wet bulb* temperature. As the air stream passes the wet wick, some of the water evaporates and this produces a cooling effect at the bulb. Heat is transferred from the air to the wick and an equilibrium condition is reached at which the

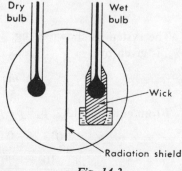

Fig. 14.3

wet bulb indicates a lower temperature than the dry bulb. The amount of this *wet bulb depression* depends on the relative humidity of the air. If the relative humidity is low, then the rate of evaporation at the wick is high, and hence the wet bulb depression is high.

The instrument may be made for use in stationary air, but satis-factory results are obtained only when the air velocity past the bulbs is between 1·85 m/s and 40 m/s. Over this range the results are fairly constant and enable the relative humidity to be calculated from the temperatures obtained. The air current can be produced by a small fan driving the air over the thermometer bulbs, or by mounting the thermometers on a frame which is whirled round by hand. This latter instrument is called a *sling psychrometer*. Another portable instrument has a fan which has a battery or clockwork drive. The wet and dry bulb temperatures are measured by thermocouples and read off an indicator. The advantages claimed are compactness and rapid response. The measurement of humidity is dealt with fully in reference 14.1.

Instruments are available which will give a continuous reading of humidity in the form of an electrical signal which may then be used as part of a control system. A common form of sensor is a thin polymer film which absorbs and desorbs moisture thus changing the dielectric constant and hence the capacitance. By measuring temperature and humidity simultaneously the enthalpy can be obtained and hence used as the control. This type of approach has been made possible using microchips which convert the readings from the sensors directly into relative humidity and/or enthalpy; calibration of the sensors against wet and dry bulb, or dew point, standard instruments can be programmed into the instrument.

A more accurate type of instrument uses an opto-electronic detection of vapour condensation on an electrically chilled solid gold mirror; the chilling is done using a thermoelectric solid state device.

Psychrometric chart

The properties of moist air can be obtained from tables, reference 14.2, but the specific humidity and percentage saturation are most conveniently obtained from a psychrometric chart. A reduced size copy of the CIBSE chart is shown in fig. 14.4. An ordinate is erected at the known dry bulb temperature and the point of intersection between it and the diagonal line representing the known wet bulb

CIBS
PSYCHROMETRIC CHART

BASED ON A BAROMETRIC
PRESSURE OF 101·325 kPa

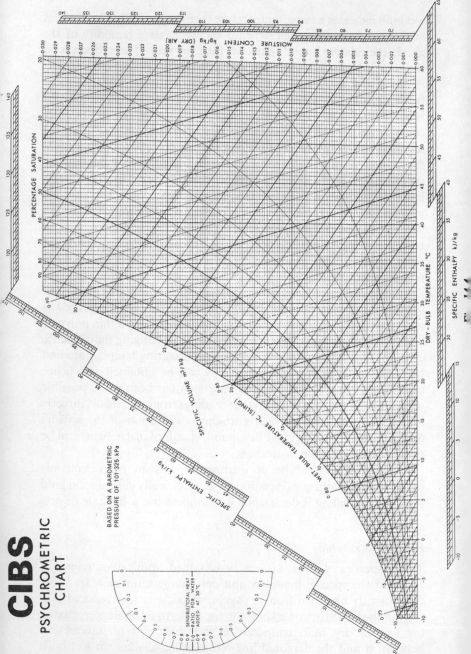

temperature is found. The percentage saturation is then found from the curve of constant percentage saturation which passes through this point. The specific humidity is read off the ordinate scale in kg of vapour per kg of dry air. The enthalpy of the mixture in kJ per kg of dry air can be read off the diagonal scale of enthalpy. The zero enthalpy for the vapour is always taken at 0°C. For the dry air the zero for enthalpy is also taken at 0°C.

From equation 14.3,

$$\omega_g = \frac{0 \cdot 622 p_g}{(p - p_g)}$$

Combining this with equation 14.5, we have

$$\psi = \frac{\omega}{\omega_g} = \frac{\omega(p - p_g)}{0 \cdot 622 p_g} \qquad 14.7$$

For a given barometric pressure, p, the percentage saturation, is a function of ω and p_g. Also p_g corresponds to the dry bulb temperature, t. The chart is prepared for a given barometric pressure and ω and h are the independent variables. A particular chart can be used for a small range of pressure (approximately $\pm 0 \cdot 1$ bar of the stated value).

14.3 Specific enthalpy, specific heat and specific volume of moist air

The enthalpy of a mixture is the sum of the enthalpies of the individual constituents (see equation 13.6),

i.e. $$mh = m_a h_a + m_s h_s$$

(where m = mass of mixture; h = enthalpy of mixture per unit mass of mixture; m_a = mass of dry air in the mixture; h_a = enthalpy of dry air per unit mass of dry air; m_s = mass of water vapour in the mixture; h_s = enthalpy of water vapour per unit mass of water vapour.)

∴ Enthalpy of mixture per unit mass of dry air $= mh/m_a$

$$= h_a + \frac{m_s h_s}{m_a}$$

$$= h_a + \omega h_s$$

At low partial pressures the specific enthalpy of water vapour can be expressed as:

$$h_s = (h_g \text{ at } p_s) + c_{ps}(t - t_g \text{ at } p_s) \qquad 14.8$$

(where the mean specific heat of superheated water vapour, c_{ps}, may be taken as approximately $1 \cdot 88$ kJ/kg K.)

Since the specific enthalpy of a vapour, from steam tables, is expressed above a datum of approximately $0°C$ then the specific enthalpy of dry air in the mixture is also expressed above the same datum,

i.e. $h_a = c_{pa}t$ 14.9

(where the specific heat of dry air, c_{pa}, may be taken as $1 \cdot 005$ kJ/kg K).
Then,

Enthalpy of mixture per unit mass of dry air

$$= c_{pa}t + \{(h_g \text{ at } p_s) + c_{ps}(t - t_g \text{ at } p_s)\}\omega$$

It can be shown that over the temperature range encountered in air conditioning the term: $\{(h_g \text{ at } p_s) - c_{ps}(t_g \text{ at } p_s)\}$ may be taken as a constant, C. Therefore we can write:

Enthalpy of mixture per unit mass of dry air

$$= c_{pa}t + \omega(C + c_{ps}t) \qquad 14.10$$

(where $C = 2500$ kJ/kg).

Alternatively, since for low pressures the enthalpy of superheated vapour is approximately equal to the saturation value at the same *temperature*, then we have:

Enthalpy of mixture per unit mass of dry air

$$= c_{pa}t + \omega(h_g \text{ at } t) \qquad 14.11$$

(where $c_{pa} = 1 \cdot 005$ kJ/kg K, as before).

For example, at $5°C$ and $\omega = 0.002\,82$, $h = 12 \cdot 11$ kJ/kg dry air.
From equation 14.10,

$$h = 1 \cdot 005 \times 5 + 0 \cdot 002\,82\,(2500 + 1 \cdot 88 \times 5) = 12 \cdot 10 \text{ kJ/kg dry air.}$$

From equation 14.11,

$$h = 1 \cdot 005 \times 5 + 0.002\,82 \times 2509 \cdot 9 = 12 \cdot 10 \text{ kJ/kg dry air.}$$

Similarly, at $30°C$ and $\omega = 0.0142$, $h = 66 \cdot 48$ kJ/kg dry air.

From equation 14.10,

$$h = 1{\cdot}005 \times 30 + 0{\cdot}0142 \,(2500 + 1{\cdot}88 \times 30)$$

$$= 66{\cdot}45 \text{ kJ/kg dry air.}$$

From equation 14.11,

$$h = 1{\cdot}005 \times 30 + 0{\cdot}0142 \times 2555{\cdot}7$$

$$= 66{\cdot}44 \text{ kJ/kg dry air.}$$

It can be seen that the error is negligible in both cases for both equations 14.10 and 14.11. Equation 14.11 is easier to use than equation 14.10 except for cases where the temperature, t, is the unknown term.

Specific heat of moist air

Assuming that the superheated water vapour acts as a perfect gas, then using equation 13.19,

Specific heat of mixture per unit mass of mixture:

$$c_p = \frac{m_a c_{pa}}{m} + \frac{m_s c_{ps}}{m}$$

Then,

Specific heat of mixture per unit mass of dry air, c_{pma}, is given by

$$c_{pma} = c_{pa} + \frac{m_s c_{ps}}{m_a}$$

$$c_{pma} = c_{pa} + \omega c_{ps} \qquad \qquad 14.12$$

(where $c_{pa} = 1{\cdot}005$ kJ/kg K, and $c_{ps} = 1{\cdot}88$ kJ/kg K, as before).

Specific volume

Since the enthalpy of the mixture is expressed per unit mass of dry air it is convenient to use the specific volume of the dry air. Therefore when the volume flow of the mixture is known the rate of mass flow of dry air can be found directly,

i.e.   Specific volume of dry air, $v_a = \dfrac{R_a T}{p_a}$

and,

$$\dot{m}_a = \frac{\dot{V}}{v_a} \qquad\qquad 14.13$$

The specific volume of dry air is plotted on the psychrometric chart. By reference to the chart it may be noted that over the normal range of room temperatures and humidities the density expressed as kg dry air per m^3 of mixture is approximately $1 \cdot 2$ (i.e. $v_a = 1/1 \cdot 2 = 0 \cdot 833$ m^3/kg dry air); this is a useful approximation for many practical problems.

14.4 Air conditioning systems

In the United Kingdom air conditioning is used mainly for industrial purposes and to supply a controlled atmosphere to public buildings such as offices, cinemas, halls, etc. In tropical and sub-tropical countries cooling by means of air conditioning is a necessary feature of modern development.

The following classification of air conditioning systems can be made:

(a) *Conventional:* the air is processed in a central plant and is distributed to the conditioned spaces via ducts;

(b) *Terminal reheat:* air supply to the units in the room is from ducting as in (a) and provides the cooling and dehumidification load; the units in the room, provided with water coils, supply the necessary reheat.

(c) *Induction:* similar to (b) but only a small quantity of primary conditioned air is supplied to each unit in the rooms where it expands through nozzles thus inducing a large volume of secondary room air into the unit; the secondary air passes over a coil before mixing with the primary air, the mixture then being delivered to the room.

(d) *Fan-coil:* air is drawn from the room and from outside the building into the room units and passed over coils as necessary, the coils carrying chilled or hot water supplied from a central plant.

(e) *Dual-duct:* twin ducts of high velocity air, one with hot air the other with cold air, are supplied to room units from a central plant as in (a); mixing at the units gives the required condition of the room air.

(f) *Variable air volume* (VAV): a high velocity flow of cooled air is supplied in a single duct and the change of load is met by varying the air volume while maintaining the same supply temperature; winter conditions can be catered for using a terminal reheat unit as (b), or by adding a second duct carrying hot air with a mixing box as in (e) or by providing a perimeter system of low pressure hot water heating.

(g) *Panel air systems:* chilled water is circulated through radiant ceiling panels at a temperature above the room dew point.

(h) *Integrated environmental design:* this is an all-embracing term to cover complex systems incorporating some of the features described above but with the emphasis on energy recovery, incorporating air flow through luminaires, the use of thermal wheels, and air-to-air or air-to-water heat pumps.

References 14.3, 14.4 and 14.6 should be consulted for a more detailed discussion.

Summer air conditioning

The air conditioning load on a room or space may be considered in two parts: the sensible heat load which is defined as the energy added per unit time which increases the dry bulb temperature; and the latent heat load which is defined as the energy added per unit time due to the enthalpy of the moisture added plus the heat required to evaporate the moisture added.

The sensible heat gains are due to heat transfer through the fabric, including solar radiation, plus internal gains from people, lighting, machinery, etc. The latent heat gains are mainly due to the occupants of the room.

Fig. 14.5 shows a typical room condition line on the psychrometric chart; point 1 represents the moist air from the air conditioning plant entering the room; point 2 represents the moist air leaving the room. It may be assumed that the air at state 2 is at the design conditions for the room.

Point X is such that:

$$\omega_1 = \omega_X \quad \text{and} \quad t_2 = t_X$$

Sensible heat load $= \dot{m}_a(h_X - h_1) = \dot{m}_a c_{pma}(t_X - t_1)$

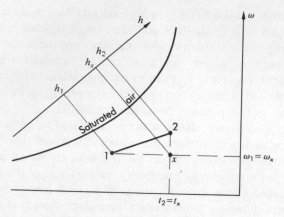

Fig. 14.5

Or using equation 14.12,

$$\text{Sensible heat load} = \dot{m}_a(c_{pa} + \omega c_{ps})(t_X - t_1) \qquad 14.14$$

Also,

$$\text{Latent heat load} = \dot{m}_a(h_2 - h_X)$$

Or using equation 14.11 with $t_2 = t_X$,

$$\text{Latent heat load} = \dot{m}_a(\omega_2 - \omega_1)(h_g \text{ at } t_2) \qquad 14.15$$

$$\text{Total heat load} = \text{sensible heat load} + \text{latent heat load}$$

$$= \dot{m}_a(h_2 - h_1) \qquad 14.16$$

The *room ratio line* 1–2 is given by:

$$\frac{h_X - h_1}{h_2 - h_1} = \frac{\text{sensible heat load}}{\text{total heat load}}$$

where, for zero latent heat load the ratio is unity and the line on the chart is horizontal, and for zero sensible heat load the ratio is zero and the line is vertical.

The ratio of sensible heat load to total heat load, and hence the slope of the room condition line on the chart, is given by a protractor in the top left hand corner of the chart (see fig. 14.4).

A typical conventional air conditioning system is shown diagrammatically in fig. 14.6a. Some of the air is recirculated and mixed with a

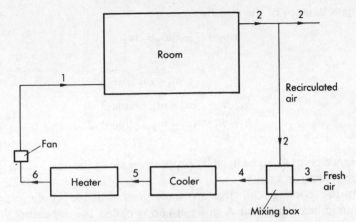

Fig. 14.6a

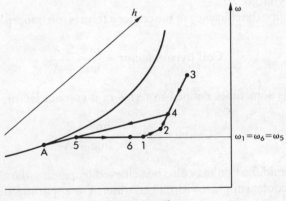

Fig. 14.6b

quantity of fresh air. Assuming adiabatic mixing, we have:

$$rh_2 + (1-r)h_3 = h_4$$

(where $r =$ mass flow of dry recirculated air per unit mass flow of dry air supplied to the room).

$$\therefore \quad r = \frac{h_3 - h_4}{h_3 - h_2}$$

A mass balance of the moisture gives,

$$r\omega_2 + (1-r)\omega_3 = \omega_4$$

$$\therefore \; r = \frac{\omega_3 - \omega_4}{\omega_3 - \omega_2}$$

i.e.
$$r = \frac{h_3 - h_4}{h_3 - h_2} = \frac{\omega_3 - \omega_4}{\omega_3 - \omega_2} = \frac{\text{line } 3\text{--}4}{\text{line } 3\text{--}2} \qquad 14.17$$

Hence point 4 can be fixed by proportion along the line 3–2 on fig. 14.6b when r is known.

In the cooling coil the air undergoes sensible cooling and dehumidification. Point A in fig. 14.6b is called the *apparatus dew point*. The moist air leaving the coil is at some intermediate state 5, and points 5 and A would only coincide if the coil surface were infinitely large.

To define the efficiency of the cooler a term is introduced as follows:

$$\text{Coil bypass factor} = \frac{\text{line } 5\text{--}A}{\text{line } 4\text{--}A} \qquad 14.18$$

This is sometimes defined in terms of a contact factor,

$$\text{Contact factor} = \frac{\text{line } 4\text{--}5}{\text{line } 4\text{--}A} \qquad 14.19$$

Dehumidification may also be achieved by passing the air through a spray cooler supplied with chilled water. The apparatus dew point is then the water temperature. In this case the contact factor given by equation 14.19 is usually renamed the spray cooler, or washer, efficiency and is expressed as a percentage.

The actual condition line of the moist air in both a coil type and a spray type cooler is not straight on the psychrometric chart; the exact path can be plotted using the theory of combined heat and mass transfer (see reference 14.5 or 14.7).

The reheat coil provides sensible heating of the air to bring the air intake to the room to state 1 such that the slope of the line 1–2 will match the required sensible and latent heat loads.

The fan provides a small amount of sensible heating which can usually be neglected.

Winter air conditioning

In winter the fabric heat losses are partially compensated by solar radiation and internal heat gains from people, lighting, machinery, etc.; the latent heat gains from people are again the main source of moisture addition.

A typical conventional type air conditioning system for winter use is shown in fig. 14.7a and the corresponding state points are shown on fig. 14.7b. The various parts of the system are similar to those of fig. 14.6 except for the humidifier. The humidification process 5–6 in the case shown is assumed to be adiabatic and takes place at constant wet bulb temperature if pumped recirculation of the water is used as shown in fig. 14.7a.

As before, a washer efficiency is defined as (line 5–6)/(line 5–A).

In general, direct contact air washers and humidifiers may be classified as follows:

(a) pumped recirculation;
(b) (i) no recirculation, with a water spray which is continuously evaporated;
     (ii) no recirculation, with steam blown into the air stream.

It can be shown for (a), assuming an adiabatic process, that the process occurs at a constant thermodynamic wet bulb temperature (see reference 14.5).

For cases (b)(i) and (b)(ii), assuming that the process changes from state 5 to state 6, we have:

$$\text{Mass of water or steam added, } \dot{m}_s = \dot{m}_a(\omega_6 - \omega_5)$$

Also,

$$\dot{m}_a(h_6 - h_5) = \dot{m}_s h_s$$

i.e.

$$\dot{m}_a(h_6 - h_5) = \dot{m}_a(\omega_6 - \omega_5)h_s$$

or,

$$h_s = \frac{h_6 - h_5}{\omega_6 - \omega_5} \qquad 14.20$$

It can be seen from equation 14.20 that the slope of the process line 5–6 depends on the enthalpy of the water or steam added. When water is added it can be shown that over the range of possible water temperatures the process line can be approximated to a line of constant wet bulb temperature as in (a). For example for a change in

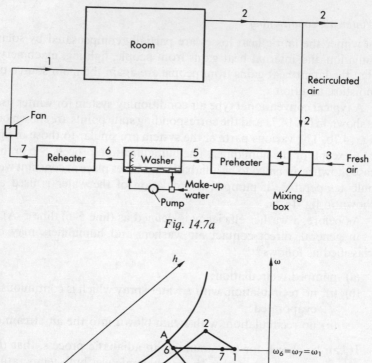

Fig. 14.7a

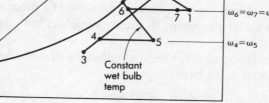

Fig. 14.7b

moisture content of 0·01, say, and water at 100°C then the enthalpy change, $(h_6 - h_5)$, is only $0·01 \times 419·1 = 4·2$ kJ/kg; for water at 20°C the enthalpy change is only 0·8 kJ/kg; it can be seen from the chart that these enthalpy changes are of the order found in following a line of constant wet bulb over this sort of moisture content range.

When steam is injected the enthalpy, h_s, is much larger than for water, and heating and humidification can be obtained. For example, from equations 14.15, 14.16 and 14.20 it can be seen that for a value of $h_s = (h_g$ at $t_5)$ then there will be no sensible heating or cooling and the process will be such that $t_6 = t_5$.

Example 14.3

An air conditioned room is to be maintained at 18°C, percentage saturation 40%. The fabric heat gains are 3000 W and there are a maximum of 20 people in the room at any time. Neglecting all other heat gains or losses calculate the required volume flow rate of air to be supplied to the room and its percentage saturation when the air supply temperature is 10°C.

Data:

Sensible heat gains per person = 100 W; latent heat gains per person = 30 W; barometric pressure = 1·013 25 bar.

$$\text{Sensible heat gain} = 3000 + (20 \times 100) = 5000 \text{ W}$$

$$\text{Latent heat gain} = 20 \times 30 = 600 \text{ W}$$

The process is shown on a sketch of the psychrometric chart in fig. 14.5, where point 2 represents the design state and point 1 the state of the supply air to the room.

At 18°C, from tables $p_{g2} = 0·020\,63$ bar.

Using equation 14.7,

$$\omega_2 = \frac{0·4 \times 0·622 \times 0·020\,63}{1·013\,25 - 0·020\,63} = 0·005\,17$$

From equation 14.15,

$$\text{Latent heat load} = \dot{m}_a(\omega_2 - \omega_1)(h_g \text{ at } t_2)$$

$$\therefore \omega_2 - \omega_1 = \frac{600}{\dot{m}_a \times 2533·9}$$

$$\therefore \omega_1 = 0·005\,17 - \frac{0·2368}{\dot{m}_a} \tag{a}$$

Also, using equation 14.14,

$$\text{Sensible heat load} = \dot{m}_a(c_{pa} + \omega_1 c_{ps})(t_2 - t_1)$$

$$\therefore 5000 = \dot{m}_a(1·005 + \omega_1 \times 1·88)(18 - 10)$$

$$\therefore \dot{m}_a = \frac{5000}{8·04 + 15·04\omega_1} \tag{b}$$

Substituting from (b) into (a),

$$\omega_1 = 0.005\,17 - \frac{0.2368}{5000}(8.04 + 15.04\omega_1)$$

$$\therefore \omega_1 = 0.004\,79$$

Substituting in (b),

$$\dot{m}_a = 616.4 \text{ kg/s}$$

From equation 14.3,

$$\frac{p_{s1}}{p_{a1}} = \frac{0.004\,79}{0.622} = 0.0077$$

i.e. $\dfrac{p - p_{a1}}{p_{a1}} = 0.0077$ $\therefore\ p_{a1} = 1.005\,51$

Density, $\rho_{a1} = \dfrac{1.005\,51 \times 10^5}{287 \times 283} = 1.238 \text{ kg/m}^3$

i.e.   Volume flow rate of supply air $= \dfrac{616.4}{1.238} = 498 \text{ m}^3/\text{s}$

Using equation 14.7,

$$\psi_1 = \frac{0.004\,79(1.013\,25 - 0.012\,27)}{0.622 \times 0.012\,27} = 0.628$$

$$\therefore \text{ Percentage saturation of supply air} = 62.8\%$$

This problem can be solved very quickly, if less accurately, using the psychrometric chart. Point 2 can be located on the chart from the information given, then we have:

$$\frac{\text{Sensible heat load}}{\text{Total heat load}} = \frac{5000}{5600} = 0.893$$

Using the chart protractor and drawing a line of slope 0·893 from point 2 to where it cuts the 10°C dry bulb line fixes point 1. Hence the percentage saturation at point 1 and h_1 can be read from the chart. Also,

$$\text{Total heat load} = 5600 = \dot{m}_a(h_2 - h_1)$$

Hence $\dot{m}_a$ can be found; the specific volume at state 1 can also be read from the chart and hence the volume flow rate may be calculated.

Example 14.4

Air at 1°C dry bulb and 80% percentage saturation mixes adiabatically with air at 18°C dry bulb, 40% percentage saturation, in the ratio of 1 to 3 by volume. Calculate the temperature and percentage saturation of the mixture. Take the barometric pressure as 1·013 25 bar.

The process is shown on a sketch of the chart in fig. 14.8, where point 3 represents the condition of the mixture.

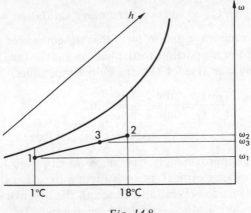

Fig. 14.8

The problem can be solved very easily using the psychrometric chart. From equation 14.17,

$$r = \frac{\text{line } 3\text{-}2}{\text{line } 1\text{-}2} = \frac{\dot{m}_{a1}}{\dot{m}_{a1} + \dot{m}_{a2}}$$

From the chart:

$$v_{a1} = 0.78 \text{ m}^3/\text{kg} \quad \text{and} \quad v_{a2} = 0.831 \text{ m}^3/\text{kg}$$

And from the information given,

$$\frac{\dot{V}_2}{\dot{V}_1} = 3$$

Then using equation 14.13,

$$\frac{\dot{m}_{a2}}{\dot{m}_{a1}} = \frac{\dot{V}_2}{v_{a2}} \times \frac{v_{a1}}{V_1} = \frac{3 \times 0.78}{0.831} = 2.82$$

$$r = \frac{1}{1 + 2.82} = 0.262$$

Then by measurement from the chart:

$$\text{line } 1\text{--}2 = 7.9 \text{ mm}$$

$$\text{line } 3\text{--}2 = 0.262 \times 7.9 = 2.07 \text{ mm}$$

Hence point 3 can be located on the line 1–2.

From the chart:

$$t_3 = 13.6°\text{C} \qquad \text{and} \qquad \text{Percentage saturation} = 48\%$$

It can be seen from equation 14·17 that the enthalpies and specific humidities are in proportion to the lengths on the line 1–2. This is approximately true also for the dry bulb temperature,

i.e. $$\frac{t_2 - t_3}{t_2 - t_1} \simeq \frac{\text{line } 3\text{--}2}{\text{line } 1\text{--}2} = 0.262$$

$$\therefore \; t_3 \simeq 18 - 0.262(18 - 1) = 13.6°\text{C}$$

This is only approximately true since the dry bulb temperature lines are not exactly vertical. In constructing the chart the dry bulb line at 30°C is made vertical and hence all the other dry bulb lines have a slight slope to the vertical (see reference 14.2).

Example 14.5

An air conditioning plant is designed to maintain a room at a condition of 20°C dry bulb and specific humidity 0·0079 when the outside condition is 30°C dry bulb and 40% percentage saturation and the corresponding heat gains are 18 000 W (sensible), and 3600 W (latent). The supply air contains one-third outside air by mass and the supply temperature is to be 15°C dry bulb.

The plant consists of a mixing chamber for fresh and recirculated air, an air washer with chilled spray water with an efficiency of 80%, an after heater battery and supply fan.

Neglecting temperature changes in fan and ducting, calculate:

(i)   the mass flow rate of supply air necessary;
(ii) the moisture content of the supply air;
(iii) the cooling duty of the washer;
(iv) the heating duty of the after heater.

Use the psychrometric chart assuming the barometric pressure is 1·013 25 bar.

The plant is shown in fig. 14.9a and the processes are shown on the chart in fig. 14.9b. Points 2 and 3 can be fixed since the conditions are known. Fresh air is to be one-third by mass of the total air to the room, hence point 4 is fixed one-third of the way from 2 to 3.

$$\frac{\text{Sensible heat load}}{\text{Total heat load}} = \frac{18\,000}{18\,000 + 3600} = 0\cdot833$$

Using the chart protractor a line of slope 0·833 is drawn from point 2 and where it cuts the dry bulb line of 15°C gives point 1. Neglecting the fan work then points 6 and 1 are coincident.

The washer efficiency is 80% and point 5 must lie on the horizontal line through point 1 since there is no change in moisture content across the heater,

i.e.
$$\frac{\text{line } 4\text{–}5}{\text{line } 4\text{–}A} = 0\cdot8$$

or,
$$\frac{\omega_4 - \omega_5}{\omega_4 - \omega_A} = 0\cdot8 \qquad \therefore \ \omega_A = \omega_4 - \frac{(\omega_4 - \omega_1)}{0\cdot8}$$

i.e.
$$\omega_A = 0\cdot0089 - \frac{(0\cdot0089 - 0\cdot0075)}{0\cdot8} = 0\cdot007\,15$$

Point 5 is fixed by joining points 4 and A; where this line cuts the horizontal line through 1 fixes point 5 at $t_5 = 12$°C dry bulb.

(i)   From the chart: $h_1 = 33\cdot9$ kJ/kg, $h_2 = 40\cdot2$ kJ/kg.

$$\text{Total heat load} = 18\,000 + 3600 = 21\cdot6 \text{ kW}$$

$$\therefore \ \dot{m}_{a1} = \frac{21\cdot6}{(40\cdot2 - 33\cdot9)} = 3\cdot43 \text{ kg/s}$$

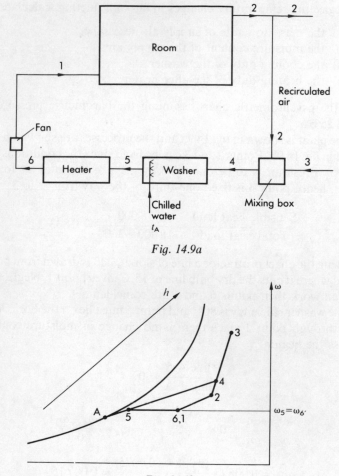

Fig. 14.9a

Fig. 14.9b

Also, $\omega_1 = 0.0075$, therefore,

Mass flow rate of supply air $= 3.43(1 + 0.0075)$

$= 3.45 \text{ kg/s}$

(ii) From the chart,

Moisture content of supply air $= 0.007\,45$

(iii) From the chart: $h_4 = 46 \cdot 2$ kJ/kg, $h_5 = 31 \cdot 1$ kJ/kg.

$$\text{Cooling load on washer} = \dot{m}_{a1}(h_4 - h_5)$$
$$= 3 \cdot 43(46 \cdot 2 - 31 \cdot 1)$$
$$= 51 \cdot 8 \text{ kW}$$

(iv) $\text{Heating load} = \dot{m}_{a1}(h_6 - h_5) = 3 \cdot 43(33 \cdot 9 - 31 \cdot 1)$
$$= 9 \cdot 6 \text{ kW}$$

14.5 Cooling towers

Some industrial processes require large quantities of cooling water. The position of the plant may be such that a convenient supply of water (e.g. from the sea or a river) is not available and a recirculatory system is necessary. A necessary part of this system is a cooler which cools down the cooling water. A convenient cooling medium is necessary and this is inevitably the atmosphere. It would be possible to produce some cooling by means of a heat exchanger, the cooling water passing through it and the air passing over it. A more satisfactory method employs the cooling effect produced when water evaporates. This is done by spraying the water into the air over a pond, or into the air passing through a cooling tower. A current of air rises, in the latter method, by means of a natural or forced draught, through a cooling tower and the hot water enters at some point and is sprayed into the air. The cooling effect is greater with a forced draught due to the increased flow of air.

As the water falls, some of it evaporates and to assist this process the tower contains packing which breaks up the stream. The warm water is cooled, the temperature of the air is raised and it becomes almost completely saturated with water vapour. The cooling water can theoretically be cooled to the wet bulb temperature of the incoming air, but a compromise is reached between the amount of cooling obtained and the size of the tower, and the figure used in design for the cooling water leaving the tower is about 8 K above the wet bulb temperature. Induced and natural draught cooling towers are shown in fig. 14.10 and fig. 14.11, respectively. The packing of the tower is usually formed from wooden slats. A modern design of tower employs a plastic impregnated cellulose material which has high water-absorption qualities and a long working life. For a given

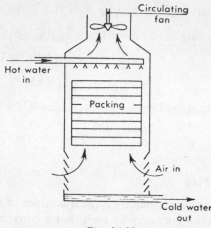

Fig. 14.10

duty the size of tower using this type of packing is about one-fifth of that using wooden packing, and it is of much lighter construction. The compact design means that the tower could possibly be situated on the top of a building without a special structure being required. This design which is of the induced draught type, delivers the warm water over the packing from a rotating header.

Some of the cooling water is lost to the atmosphere in the evaporation process in all cooling towers, and so a small amount of make-up water is required.

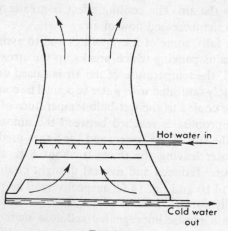

Fig. 14.11

Example 14.6

A small-size cooling tower is designed to cool 5·5 litres of water per second, the inlet temperature of which is 44°C. The motor-driven fan induces 9 m³/s of air through the tower and the power absorbed is 4·75 kW. The air entering the tower is at 18°C, and has a relative humidity of 60%. The air leaving the tower can be assumed to be saturated and its temperature is 26°C. Calculate the final temperature of the water and the amount of cooling water make-up required per second. Assume that the pressure remains constant throughout the tower at 1·013 bar.

The cooling tower is shown diagrammatically in fig. 14.12. At inlet, using equation 14.4,

$$\phi = p_s/p_g \quad \text{and} \quad p_g \text{ at } 18°C = 0.020\,63 \text{ bar}$$
$$\therefore \ p_{s_1} = 0.6 \times 0.020\,63 = 0.012\,38 \text{ bar}$$

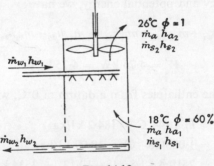

Fig. 14.12

From equation 13.2,

$$p_{a_1} = 1.013 - 0.012\,38 = 1.0006 \text{ bar}$$

Then,
$$\dot{m}_a = \frac{10^5 \times 1.0006 \times 9}{10^3 \times 0.287 \times 291} = 10.78 \text{ kg/s}$$

and
$$\dot{m}_{s_1} = \frac{10^5 \times 0.012\,38 \times 9}{10^3 \times 0.4618 \times 291} = 0.0829 \text{ kg/s}$$

At exit at 26°C,

$$p_g = 0.033\,60 \text{ bar} \quad \text{and} \quad \phi = 1$$
$$\therefore \ p_{s_2} = 0.033\,60 \text{ bar}$$

Using equation 14.3,

$$\omega_2 = 0.622\left(\frac{p_{s2}}{p-p_{s2}}\right) = \left(\frac{0.622 \times 0.0336}{1.013 - 0.0336}\right) = 0.02133$$

Then using equation 14.1,

$$\dot{m}_{s2} = 10.78 \times 0.021\ 33 = 0.23 \text{ kg/s}$$

Hence,

Make-up water required $= 0.23 - 0.0829 = 0.1471$ kg/s

Also,

$$\dot{m}_{w_1} = 5.5 \times 1 = 5.5 \text{ kg/s}$$

and

$$\dot{m}_{w_2} = \dot{m}_{w_1} - \text{(make-up water)} = 5.5 - 0.1471 = 5.353 \text{ kg/s}$$

Applying the steady flow energy equation and neglecting changes in kinetic energy and potential energy, we have

$$W + \dot{m}_{w_1}h_{w_1} + \dot{m}_{a_1}h_{a_1} + \dot{m}_{s_1}h_{s_1} = \dot{m}_{a_2}h_{a_2} + \dot{m}_{s_2}h_{s_2} + \dot{m}_{w_2}h_{w_2}$$

Now, $\qquad\qquad W = 4.75 \text{ kW} = 4.75 \text{ kJ/s}$

Evaluating the enthalpies from a datum of $0°C$, we have

$$h_{w_1} = h_f \text{ at } 44°C = 184.2 \text{ kJ/kg};$$
$$h_{a_1} = 1.005(18 - 0) = 18.09 \text{ kJ/kg};$$
$$h_{s_1} = 2519.4 + 1.86(18 - 10.13) = 2534 \text{ kJ/kg}$$
$$h_{s_2} = h_g \text{ at } 26°C = 2548.4 \text{ kJ/kg};$$
$$h_{a_2} = 1.005(26 - 0) = 26.13 \text{ kJ/kg}$$

The vapour is superheated at 1, being above $10.13°C$, the saturation temperature corresponding to 0.01238 bar.

Then, substituting,

$$4.75 + 5.5 \times 184.2 + 10.78 \times 18.09 + 0.0829 \times 2534$$
$$= 10.78 \times 26.13 + 0.23 \times 2548.4 + 5.353\ h_{w_2}$$

i.e. $\qquad 5.353\ h_{w_2} = 556.3 \qquad \therefore\ h_{w_2} = 104 \text{ kJ/kg}$

i.e. $\qquad$ by interpolation, $h_f = 104$ kJ/kg at $24.8°C$

14.6 Mixtures of gases and a vapour other than water vapour

The methods of the previous sections can be applied to mixtures of vapours other than water vapour, in air or other gases. One such mixture is that induced into the cylinder of a petrol engine from the inlet manifold, consisting of atmospheric air and petrol vapour. Between the carburettor and the inlet valve the mixture receives heat from the hot manifold, and hence its condition varies during the induction process. With similar data available the problems can be solved in the same way as for dry air-water vapour mixtures, and the same terminology is used. This is mentioned again in Section 15.10 and example 15.13 illustrates the procedure.

PROBLEMS

14.1 Air at 32°C is saturated with water vapour and the barometric pressure is 1·013 bar. Determine the partial pressure of the vapour and of the dry air. What volume of the mixture would contain 1 kg of vapour? Calculate also the mass of air associated with this amount of vapour and the specific and relative humidities of the mixture.

(0·047 54 bar; 0·9655 bar; 29·6 m³; 32·6 kg; 0·031; 1)

14.2 The pressure of the water vapour in an atmosphere which is at 32°C and 1·013 bar is 0·020 63 bar. By how much is the water vapour superheated? What are the specific and relative humidities of the air? To what temperature would the air have to be cooled for it to be just saturated with water vapour? If the air is cooled to 10°C from its original condition, how much condensate is formed per kg of dry air? (14 K; 0·012 93; 43·4% 18°C; 0·0053 kg)

14.3 An air and water vapour mixture at 1 bar and 26·7°C has a specific humidity of 0·0085. Determine the percentage saturation.

(37·7%)

14.4 A mixture of air and water vapour at 1·013 bar and 16°C has a dew point of 5°C. Determine the relative and specific humidities.

(48%, 0·0054 kg/kg dry air)

14.5 Atmospheric air at a pressure of 760 mm Hg has a temperature of 32°C and a percentage saturation as determined from a psychrometric chart of 52%. Calculate,

(a) The partial pressures of the vapour and the dry air,

(b) The specific humidity,
(c) The dew point,
(d) The density of the mixture.

$$(0.024\ 72\ \text{bar};\ 0.988\ 53\ \text{bar};\ 0.015\ 56;\ 20.9°\text{C};\ 1.147\ \text{kg/m}^3)$$

14.6 Compare the characteristic constant for dry air ($R=0.287$ kJ/ kg K), with the value for dry air saturated with water vapour at 16°C and 1.013 bar. $\qquad$ (0.2889 kJ/kg K)

14.7 The temperature in a room of volume 38 m³ is 25°C and the pressure is 1.013 bar; the dew point of the air in the room is 14°C. If a vessel containing water is placed in the room estimate the maximum amount which may be lost by evaporation. Assume that the pressure in the room remains constant. $\qquad$ (0.447 kg)

14.8 An air conditioned room is maintained at a temperature of 21°C and a relative humidity of 55% when the barometric pressure is 740 mm Hg. Calculate the specific humidity of the air-water vapour mixture. Calculate also the temperature of the inside of the windows in the room if moisture is just beginning to form on them.

What mass of water vapour per kg of dry air in the room must be removed from the mixture in order to prevent condensation on the windows when their temperature drops to 4°C? Calculate the initial relative humidity to satisfy this condition if the temperature remains at 21°C. The barometric pressure remains constant.

$$(0.008\ 76;\ 11.62°\text{C};\ 0.0036\ \text{kg};\ 32.7\%)$$

14.9 A mixture of air and steam at 50°C and 1.013 bar has a gravimetric analysis of 4% moisture and 96% dry air. Calculate,

(a) The analysis by volume of the mixture,
(b) The partial pressures of the vapour and the dry air,
(c) The relative humidity of the mixture,

$$(6.27\%,\ 93.73\%;\ 0.0636\ \text{bar},\ 0.949\ \text{bar};\ 51.5\%)$$

14.10 For the mixture of problem 14.9 calculate the enthalpy per kg of the mixture reckoned from 0°C. Calculate also the heat to be removed at a constant pressure of 1.013 bar for condensation to begin. $\qquad$ (151.8 kJ/kg; 13.3 kJ/kg)

14.11 For the mixture of problem 14.8 calculate the specific volume of the vapour and the heat to be removed per kg of mixture if it is cooled at constant volume to the condition where condensation just begins. What is the dew point and the pressure in the room at this condition?   (98·7 m³/kg; 7·1 kJ/kg; 11·2°C; 0·953 bar)

14.12 The readings taken in a room from a sling psychrometer gave a dry and a wet bulb temperature of 25°C and 19·7°C respectively. Using a psychrometric chart determine,

(a) The specific humidity,
(b) The percentage saturation,
(c) The dew point,
(d) The specific volume of the mixture,
(e) The enthalpy per kg of dry air.

Take the atmospheric pressure as 1·013 25 bar.
(12·4 g/kg; 62%; 17·2°C; 0·86 m³/kg; 57 kJ/kg)

14.13 If the atmosphere of problem 14.12 is cooled to 5°C and then heated until the dry bulb temperature is 17·5°C, both processes being at constant pressure, determine, from the chart, assuming that the air leaving the cooler is saturated,

(a) the final percentage saturation,
(b) the final specific humidity,
(c) the final wet bulb temperature,
(d) the amount of condensate collected at the cooler per kg of dry air,
(e) the heat removed in the cooling process and that supplied in the heating process per kg of dry air.
(43·6%; 0·0055; 10·9°C; 6·9 g; 38·2 kJ/kg; 12·7 kg)

14.14 In an air conditioning system air at 28°C and 1·013 bar is drawn into a building with a percentage saturation of 50%. It is required to maintain the air in the building at 20°C and 40% percentage saturation. The total heat gains to the room (sensible and latent) are 15 kW, and the latent heat gains are 3 kW. Calculate the temperature to which the inlet air should be cooled at the cooling coil (assuming a coil by-pass factor of 0·2), the refrigerating load, and the heat input to the heater. The system requires 5 m³ of fresh air per second and no air is recirculated.   (6·5°C; 220 kW; 67·1 kW)

14.15 A room in summer is to be maintained at 18°C, 50% percentage saturation when the outside conditions are 30°C, 80% percentage saturation. The sensible heat gains and latent heat gains are 4·4 kW and 1·89 kW respectively.

The conditioned air is supplied through ducts from a central station consisting of a cooler battery, a reheat battery and a fan. Fresh air is supplied to a mixing unit where it mixes with a certain percentage of air recirculated from the room, the remainder of the room air being expelled to atmosphere.

The air entering the room is at 12·5°C, the air temperature rise in the fan and duct work is 1 K, the air leaving the cooler battery and entering the reheat battery is at 7°C, and the apparatus dew point of the cooler is 1·5°C.

Draw a sketch of the plant, numbering the relevant points, and calculate:

(i)   the ratio of the mass rate of flow of recirculated air to the mass rate of air supplied to the room;
(ii) the cooler battery load;
(iii) the reheater battery load;
(iv) the cooler battery by-pass factor.

Use the psychrometric chart, taking the barometric pressure as 1·01325 bar.   (0·88, 15·3 kW, 3·5 kW, 0·3)

14.16 Air enters a natural draught cooling tower at 1·013 bar and 13°C and relative humidity 50%. Water at 60°C from turbine condensers is sprayed into the tower at the rate of 22·5 kg/s and leaves at 27°C. The air leaves the tower at 38°C, 1·013 bar and is saturated. Calculate,

(a) The air flow required in m^3/s,
(b) The make-up water required in kg/s.

(21 m^3/s; 1 kg/s)

REFERENCES

14.1 *Measurement of Humidity*, National Physical Laboratory (H.M.S.O., 1970).

14.2 CIBSE Guide C1 and 2: *Properties of humid air, water and steam* (CIBSE, 1975).

14.3 KELL, J. R., and MARTIN, P. L., *Faber and Kell's Heating and Air Conditioning of Buildings* (Architectural Press, 1984).

14.4 JONES, W. P., *Air Conditioning Engineering* (Edward Arnold, 1985).

14.5 THRELKELD, J. L., *Thermal Environmental Engineering* (Prentice Hall, 1970).

14.6 CROOME, D. J., and ROBERTS, B. M., *Air Conditioning and Ventilation of Buildings* (Pergamon, 1981).

14.7 EASTOP, T. D., and GASIOREK, J. M. *Air Conditioning through Worked Examples* (Longman, 1968).

15. Combustion

The ideal cycles previously considered use fluids which remain unchanged chemically as they pass through the various processes of the cycle. In practical engines and power plants the source of heat is the chemical energy of substances called fuels. This energy is released during the chemical reaction of the fuel with oxygen. The fuel elements combine with oxygen in an oxidation process which is rapid and is accompanied by the evolution of heat.

The combustion process takes place in a controlled manner in some form of combustion chamber after initiation of combustion by some means (e.g. in a petrol engine the combustion is started by an electric spark). The most convenient source of oxygen supply is that of the atmosphere which contains oxygen and nitrogen and traces of other gases. Normally no attempt is made to separate out the oxygen from the atmosphere, and the nitrogen, etc. accompanies the oxygen into the combustion chamber.*

Nitrogen does not oxidize easily and is inert as far as the combustion process is concerned, but it acts as a moderator in that it absorbs some of the heat of combustion and so limits the maximum temperature reached. As combustion proceeds the oxygen is progressively used up and the proportion of nitrogen plus products of combustion to the available oxygen, increases. For a given amount of fuel there is a definite amount of oxygen, and therefore air, which is required for the complete combustion of a given fuel. To ensure complete combustion it is usual to supply air in excess of the amount required for chemically correct combustion. The oxygen not consumed in the reaction passes into the exhaust with the products of combustion.

Internal-combustion engines are run on liquid fuels which are grouped as petrols and diesel oils; gas turbines are run mainly on kerosene. Engines burning solid fuels have been built but are mainly experimental; engines using gaseous fuels are being used to a decreasing extent. In the many and diverse applications in industry,

* Rocket motors make no use of the atmospheric oxygen but carry a liquid oxidant (e.g. hydrogen peroxide or nitric acid).

solid, liquid, and gaseous fuels are used. Generalization is not possible on the selection of fuels, since the fuel used and its necessary firing equipment depend on the particular application, the practical circumstances and economic considerations.

15.1 Basic chemistry

It is necessary to understand the construction and use of chemical formulae, before combustion problems can be considered. This involves elementary concepts which have been met before by most students, but a brief explanation will be given here.

Atoms: Chemical elements cannot be divided indefinitely and the smallest particle which can take part in a chemical change is called an atom. If an atom is split as in a nuclear reaction, the divided atom does not retain the original chemical properties.

Molecules: Elements are seldom found to exist naturally as single atoms. Some elements have atoms which exist in pairs, each pair forming a molecule (e.g. oxygen), and the atoms of each molecule are held together by strong inter-atomic forces. The isolation of a molecule of oxygen would be tedious, but possible; the isolation of an atom of oxygen would be a different prospect.

The molecules of some substances are formed by the mating up of atoms of different elements. For example, water (which is chemically the same as ice and steam) has a molecule which consists of two atoms of hydrogen and one atom of oxygen.

The atoms of different elements have different masses and these values are important when a quantitative analysis is required. The actual masses are infinitesimally small, and the ratios of the masses of atoms are used. These ratios are given by the relative atomic masses quoted on a scale which defines the atomic mass of isotope 12 of carbon as 12. The *relative atomic mass* of a substance is the mass of a single entity of the substance relative to a single entity of carbon 12. Table 15.1 gives the relative atomic masses of some common elements rounded off to give values accurate enough for most purposes.

Relative molecular masses are based on the relative masses of the atoms which constitute the molecule. In chemical formulae one atom of an element is represented by the symbol for the element, i.e. an atom of hydrogen is written as H, and other examples are given in Table

Table 15.1

Element	Oxygen	Hydrogen	Carbon	Sulphur	Nitrogen
Atomic symbol	O	H	C	S	N
Relative atomic mass	16	1	12	32	14
Molecular grouping	O_2	H_2	C	S	N_2
Relative molecular mass (rounded)	32	2	12	32	28
Accurate values	31·999	2·016	12	32·030	28·013

15.1. If a substance exists as a molecule containing say two atoms, as for hydrogen, it is written as H_2. Two molecules of hydrogen is written as $2H_2$, etc. Table 15.1 includes relative molecular masses, rounded off, and, for comparison, the accurate values.

Some of the other substances met in combustion work are given in Table 15.2 to illustrate the calculations of the relative molecular mass from the relative atomic masses of the elements.

Table 15.2   Compounds and their Relative Molecular Masses

Compound	Formula	Relative molecular mass
Water, steam	H_2O	$2 \times 1 + 1 \times 16 = 18$
Carbon monoxide	CO	$1 \times 12 + 1 \times 16 = 28$
Carbon dioxide	CO_2	$1 \times 12 + 2 \times 16 = 44$
Sulphur dioxide	SO_2	$1 \times 32 + 2 \times 16 = 64$
Methane	CH_4	$1 \times 12 + 4 \times 1 = 16$
Ethane	C_2H_6	$2 \times 12 + 6 \times 1 = 30$
Propane	C_3H_8	$3 \times 12 + 8 \times 1 = 44$
n-Butane	C_4H_{10}	$4 \times 12 + 10 \times 1 = 58$
Ethylene	C_2H_4	$2 \times 12 + 4 \times 1 = 28$
Propylene	C_3H_6	$3 \times 12 + 6 \times 1 = 42$
n-Pentane	C_5H_{12}	$5 \times 12 + 12 \times 1 = 72$
Benzene	C_6H_6	$6 \times 12 + 6 \times 1 = 78$
Toluene	C_7H_8	$7 \times 12 + 8 \times 1 = 92$
n-Octane	C_8H_{18}	$8 \times 12 + 18 \times 1 = 114$

15.2 Fuels

The most important fuel elements are carbon and hydrogen, and most fuels consist of these and sometimes a small amount of sulphur.

The fuel may contain some oxygen and a small quantity of incombustibles (e.g. water vapour, nitrogen, or ash).

Coal is the most important solid fuel and the various types are divided into groups according to their chemical and physical properties. An accurate chemical analysis by mass of the important elements in the fuel is called the *ultimate analysis*, the elements usually included being carbon, hydrogen, nitrogen, and sulphur. The main groups are shown in Table 15.3, and their ultimate analyses are given.

Table 15.3   Analyses of Solid Fuels

| Fuel | Moisture content % by mass | Ultimate analysis % by mass in dry fuel | | | | | Volatile matter % by mass in dry fuel |
		Carbon	Hydrogen	Oxygen	Nitrogen	Ash	
Anthracite	1	90·27	3·00	2·32	1·44	2·97	4
Bituminous coal	2	81·93	4·87	5·98	2·32	4·90	25
Lignite	15	56·52	5·72	31·89	1·62	4·25	50
Peat	20	43·70	6·42	44·36	1·52	4·00	65

The analyses are typical but may vary from one sample to another within the group, and hence can be taken only as a guide. Another analysis of coal, called the *proximate analysis*, gives the percentages of moisture, volatile matter, combustible solid (called fixed carbon), and ash. The fixed carbon is found as a remainder by deducting the percentages of the other quantities. The volatile matter includes the water derived from the chemical decomposition of the coal (not to be confused with free, or inherent moisture), the combustible gases (e.g. hydrogen, methane, ethane, etc.), and tar (i.e. a complex mixture of hydrocarbons and other organic compounds). The procedures for both analyses are given in reference 15.1.

Most liquid fuels are hydro-carbons which exist in the liquid phase at atmospheric conditions. Petroleum oils are complex mixtures of sometimes hundreds of different fuels, but the necessary information to the engineer is the relative proportions of carbon, hydrogen, etc., as given by the ultimate analysis. Table 15.4 gives the ultimate analyses of some liquid fuels.

Gaseous fuels are chemically the simplest of the three groups. Some gaseous fuels exist naturally at atmospheric conditions (e.g. methane (CH_4), which is a paraffin). Other gaseous fuels are manufactured by the various treatments of coal. Carbon monoxide is an important gaseous fuel which is a constituent of other gas mixtures, and is also a product of the incomplete combustion of carbon.

Table 15.4   Analyses of Liquid Fuels

Fuel	Carbon	Hydrogen	Sulphur	Ash etc.
100 octane petrol	85·1	14·9	0·01	—
Motor petrol	85·5	14·4	0·1	—
Benzole	91·7	8·0	0·3	—
Kerosene (paraffin)	86·3	13·6	0·1	—
Diesel oil	86·3	12·8	0·9	—
Light fuel oil	86·2	12·4	1·4	—
Heavy fuel oil	86·1	11·8	2·1	—
Residual fuel oil	88·3	9·5	1·2	1·0

Table 15.5 gives the typical analyses of different gaseous fuels. The table gives the analyses by volume, each constituent having been

Table 15.5   Volumetric and Molar Analyses of Some Gaseous Fuels

Fuel	H_2	CO	CH_4	C_2H_4	C_2H_6	C_4H_8	O_2	CO_2	N_2
Coal gas	53·6	9·0	25	—	—	3	0·4	3	6
Producer gas	12	29	2·6	0·4	—	—	—	4	52
Natural gas	—	1	93	—	3	—	—	—	3
Blast furnace gas	2	27	—	—	—	—	—	11	60

measured by volume at atmospheric pressure and temperature. The volumetric analysis is the same as the molar analysis (see equation 13.14).

Fuels are tested according to standardized procedures and for further information reference 15.1 should be consulted.

15.3 Combustion equations

Proportionate masses of air and fuel enter the combustion chamber where the chemical reaction takes place, and from which the products of combustion pass to the exhaust. By the conservation of mass the mass flow remains constant (i.e. total mass of products equals total mass of reactants), but the reactants are chemically different from the products, and the products leave at a higher temperature. The total number of atoms of each element concerned in the combustion remains constant, but the atoms are rearranged into groups having different chemical properties. This information is expressed in the chemical equation which shows:

(a) The reactants and the products of combustion,
(b) The relative quantities of the reactants and products.

The two sides of the equation must be consistent, each having the same number of atoms of each element involved. It should not be assumed that if an equation can be written, that the reaction it represents is inevitable or even possible. For possibility and direction the reaction has to be considered with reference to the Second Law of Thermodynamics. For the present the only concern is known combustion equations.

The equation shows the number of molecules of each reactant and product. The mole, introduced in Section 3.3, is proportional to the number of molecules, hence the relative numbers of molecules of the reactants and the products gives the molar, and therefore the volumetric, analysis of the gaseous constituents.

As stated earlier the oxygen supplied for combustion is usually provided by atmospheric air, and it is necessary to use accurate and consistent analyses of air by mass and by volume. It is usual in combustion calculations to take air as $23\cdot3\%$ O_2, $76\cdot7\%$ N_2 by mass, and 21% O_2, 79% N_2 by volume. The small traces of other gases in dry air are included in the nitrogen, which is sometimes called 'atmospheric nitrogen'.

Consider the combustion equation for hydrogen:

$$2H_2 + O_2 \rightarrow 2H_2O \qquad \text{15.1}$$

This tells us that

(a) Hydrogen reacts with oxygen to form steam or water,
(b) Two molecules of hydrogen react with one molecule of oxygen to give two molecules of steam or water,

i.e.   2 volumes H_2 + 1 volume $O_2 \rightarrow$ 2 volumes H_2O

The H_2O may be a liquid or a vapour depending on whether the product has been cooled sufficiently to cause condensation. The proportions by mass are obtained by using relative atomic masses,

i.e. $2H_2 + O_2 \rightarrow 2H_2O$

$\therefore\ 2\times(2\times1) + 2\times16 \rightarrow 2\times(2\times1+16)$

i.e.   4 kg H_2 + 32 kg $O_2 \rightarrow$ 36 kg H_2O

or   1 kg H_2 + 8 kg $O_2 \rightarrow$ 9 kg H_2O

The same proportions are obtained by writing the equation 15.1 as $H_2 + \frac{1}{2}O_2 \rightarrow H_2O$, and this is sometimes done.

It will be noted from equation 15.1 that the total volume of the reactants is 2 volumes $H_2 + 1$ volume $O_2 = 3$ volumes. The total volume of the product is only 2 volumes. There is therefore a volumetric contraction on combustion.

Since oxygen is accompanied by nitrogen if air is supplied for the combustion, then this nitrogen should be included in the equation. As nitrogen is inert as far as the chemical reaction is concerned, it will appear on both sides of the equation.

With one mole of oxygen there are 79/21 moles of nitrogen, hence equation 15.1 becomes,

$$2H_2 + O_2 + \frac{79}{21}N_2 \rightarrow 2H_2O + \frac{79}{21}N_2 \qquad 15.2$$

Similar equations can be found for the combustion of carbon. There are two possibilities to consider:

(a) The complete combustion of carbon to carbon dioxide,

$$C + O_2 \rightarrow CO_2 \qquad 15.3$$

and including the nitrogen,

$$C + O_2 + \frac{79}{21}N_2 \rightarrow CO_2 + \frac{79}{21}N_2 \qquad 15.4$$

Considering the volumes of reactants and products,

$$0 \text{ volume } C + 1 \text{ volume } O_2 + \frac{79}{21} \text{ volumes } N_2$$

$$\rightarrow 1 \text{ volume } CO_2 + \frac{79}{21} \text{ volumes } N_2$$

The volume of carbon is written as zero since the volume of a solid is negligible in comparison with that of a gas.

By mass,

$$12 \text{ kg } C + (2 \times 16) \text{ kg } O_2 + \frac{79}{21}(2 \times 14) \text{ kg } N_2$$

$$\rightarrow (12 + 2 \times 16) \text{ kg } CO_2 + \frac{79}{21}(2 \times 14) \text{ kg } N_2$$

i.e. $12 \text{ kg } C + 32 \text{ kg } O_2 + 105 \cdot 3 \text{ kg } N_2 \rightarrow 44 \text{ kg } CO_2 + 105 \cdot 3 \text{ kg } N_2$

or $1 \text{ kg } C + \frac{8}{3} \text{ kg } O_2 + \frac{105 \cdot 3}{12} \text{ kg } N_2 \rightarrow \frac{11}{3} \text{ kg } CO_2 + \frac{105 \cdot 3}{12} \text{ kg } N_2$

(b) The incomplete combustion of carbon.

This occurs when there is an insufficient supply of oxygen to burn the carbon completely to carbon dioxide.

$$2C + O_2 \rightarrow 2CO \qquad\qquad 15.5$$

and including the nitrogen,

$$2C + O_2 + \frac{79}{21} N_2 \rightarrow 2CO + \frac{79}{21} N_2 \qquad\qquad 15.6$$

By mass,

$$(2 \times 12) \text{ kg C} + (2 \times 16) \text{ kg O}_2 + \frac{79}{21} (2 \times 14) \text{ kg N}_2$$

$$\rightarrow 2(12 + 16) \text{ kg CO} + \frac{79}{21}(2 \times 14) \text{ kg N}_2$$

i.e. $24 \text{ kg C} + 32 \text{ kg O}_2 + 105 \cdot 3 \text{ kg N}_2 \rightarrow 56 \text{ kg CO} + 105 \cdot 3 \text{ kg N}_2$

or $1 \text{ kg C} + \frac{4}{3} \text{ kg O}_2 + \frac{105 \cdot 3}{24} \text{ kg N}_2 \rightarrow \frac{7}{3} \text{ kg CO} + \frac{105 \cdot 3}{24} \text{ kg N}_2$

If a further supply of oxygen is available then the combustion can continue to completion,

$$2CO + O_2 + \frac{79}{21} N_2 \rightarrow 2CO_2 + \frac{79}{21} N_2 \qquad\qquad 15.7$$

By mass,

$$56 \text{ kg CO} + 32 \text{ kg O}_2 + \frac{79 \times 28}{21} \text{ kg N}_2 \rightarrow 88 \text{ kg CO}_2 + \frac{79 \times 28}{21} \text{ kg N}_2$$

or

$$1 \text{ kg CO} + \frac{4}{7} \text{ kg O}_2 + \frac{105 \cdot 3}{56} \text{ kg N}_2 \rightarrow \frac{11}{7} \text{ kg CO}_2 + \frac{105 \cdot 3}{56} \text{ kg N}_2$$

15.4 Stoichiometric, or chemically correct, air fuel ratio

A stoichiometric mixture of air and fuel is one that contains just sufficient oxygen for the complete combustion of the fuel. A mixture which has an excess of air is termed a *weak mixture*, and one which has a deficiency of air is termed a *rich mixture*. The percentage of excess air is given by the following,

Percentage excess air

$$= \frac{\text{actual A/F ratio} - \text{stoichiometric A/F ratio}}{\text{stoichiometric A/F ratio}} \qquad 15.8$$

(where A denotes air and F denotes fuel).

For gaseous fuels the ratios are expressed by volume and for solid and liquid fuels the ratios are expressed by mass. The equation 15.8 gives a positive result when the mixture is weak, and a negative result when the mixture is rich. For boiler plant the mixture is usually greater than 20% weak; for gas turbines it can be as much as 300% weak. Petrol engines have to meet various conditions of load and speed, and operate over a wide range of mixture strengths. The following definition is used,

$$\text{Mixture strength} = \frac{\text{stoichiometric A/F ratio}}{\text{actual A/F ratio}} \qquad 15.9$$

The working values range between 80% (weak) and 120% (rich) (see Section 18.6).

Where fuels contain some oxygen (e.g. ethyl alcohol C_2H_6O) this oxygen is available for the combustion process, and so the fuel requires a smaller supply of air.

15.5 Exhaust and flue gas analysis

The products of combustion are mainly gaseous. When a sample is taken for analysis it is usually cooled down to a temperature which is below the saturation temperature of the steam present. The steam content is therefore not included in the analysis, which is then quoted as the analysis of the dry products. Since the products are gaseous, it is usual to quote the analysis by volume. An analysis which includes the steam in the exhaust is called a wet analysis. The following examples illustrate the principles covered in this chapter up to this point.

Example 15.1

Calculate the stoichiometric A/F ratio for the combustion of a sample of dry anthracite of the following composition by mass:

C 90%; H 3%; O 2·5%; N 1%; S 0·5%; ash 3%

Determine the A/F ratio and the dry and wet analysis of the products of combustion by volume, when 20% excess air is supplied.

In the case of a fuel with several constituents a tabular method is advisable, as shown below. Each constituent is taken separately and the amount of oxygen required for complete combustion is found from the chemical equation. The oxygen in the fuel is included in the column headed 'oxygen required' as a negative quantity.

	Mass per kg coal	Combustion equation	Oxygen required per kg of coal	Products per kg of coal
C	0·9	$C + O_2 \rightarrow CO_2$ $12\ kg + 32\ kg$ $\rightarrow 44\ kg$	$0·9 \times \frac{32}{12} = 2·4\ kg$	$0·9 \times \frac{44}{12} = 3·3\ kg\ CO_2$
H	0·03	$2H_2 + O_2 \rightarrow 2H_2O$ $1\ kg + 8\ kg \rightarrow 9\ kg$	$0·03 \times 8 = 0·24\ kg$	$0·03 \times 9 = 0·27\ kg\ H_2O$
O	0·025	—	$-0·025\ kg$	—
N	0·01	—	—	$0·01\ kg\ N_2$
S	0·005	$S + O_2 \rightarrow SO_2$ $32\ kg + 32\ kg \rightarrow 64\ kg$	$0·005 \times \frac{32}{32} = 0·005\ kg$	$0·005 \times \frac{64}{32} = 0·01\ kg\ SO_2$
ash	0·03	—	—	—
			Total = 2·620 kg	

From table:

$$O_2 \text{ required per kg of coal} = 2·62\ kg$$

$$\therefore\ \text{air required per kg of coal} = \frac{2·62}{0·233} = 11·25\ kg$$

(where air is assumed to contain 23·3% O_2 by mass).

$$N_2 \text{ associated with this air} = 0·767 \times 11·25 = 8·63\ kg$$

$$\therefore\ \text{total } N_2 \text{ in products} = 8·63 + 0·01 = 8·64\ kg$$

The stoichiometric A/F ratio = 11·25/1

For an air supply which is 20% in excess, using equation 15.8,

$$\text{Actual A/F ratio} = 11·25 + \frac{20}{100} \times 11·25 = 1·2 \times 11·25 = 13·5/1$$

Therefore N_2 supplied $= 0·767 \times 13·5 = 10·36$ kg.
Also O_2 supplied $= 0·233 \times 13·5 = 3·144$ kg.
In the products then, we have

$$N_2 = 10·36 + 0·01 = 10·37\ kg$$

$$\text{and excess } O_2 = 3·144 - 2·62 = 0·524\ kg$$

The products are entered in the following table and the analysis by volume is obtained. In column 3 the percentage by mass is given by the mass of each product divided by the total mass of 14·47 kg. In column 5 the kmol per kg of coal are given by equation 3.7, $n = m/M$. The total of 0·4764 in column 5 gives the total kmol of wet products per kg of coal, and by subtracting the kmol of H_2O from this total, the total kmol of the dry products is obtained as 0·4614. Column 6 gives the proportion of each constituent of column 5 expressed as a percentage of the total kmol of the wet products. Similarly column 7 gives the percentage by volume of the dry products.

Product	Mass/kg coal	% by mass	$\dfrac{M}{\text{kg/kmol}}$	kmol/kg coal	% by vol. wet	% by vol. dry
1	2	3	4	5	6	7
CO_2	3·3	22·8	44	0·075	15·77	16·3
H_2O	0·27	1·87	18	0·015	3·16	—
SO_2	0·01	0·07	64	0·0002	0·03	0·03
O_2	0·52	3·6	32	0·0162	3·4	3·51
N_2	10·37	71·65	28	0·37	77·8	80·3
	14·47 kg		Total wet =	0·4764	100·16	100·14
			−H_2O =	0·015		
			Total dry =	0·4614		

Example 15.2

The analysis of a supply of coal gas is: H_2 49·4%; CO 18%; CH_4 20%; C_4H_8 2%; O_2 0·4%; N_2 6·2%; CO_2 4%. Calculate the stoichiometric A/F ratio. Find also the wet and dry analysis of the products of combustion if the actual mixture is 20% weak.

The example is again solved by a tabular method; a specimen calculation is shown more fully as follows:

For CH_4,

$$CH_4 + 2O_2 \rightarrow CO_2 + 2H_2O$$

i.e.   1 kmol CH_4 + 2 kmol O_2 → 1 kmol CO_2 + 2 kmol H_2O

There are 0·2 kmol of CH_4 per kmol of the coal gas, hence,

  0·2 kmol CH_4 + 0·2 × 2 kmol O_2

           → 0·2 kmol CO_2 + 0·2 × 2 kmol H_2O

Therefore the oxygen required for the CH_4 in the coal gas is 0·4 kmol per kmol of coal gas.

The oxygen in the fuel (0·004 kmol) is included in column 4 as a negative quantity.

1	kmol/kmol fuel	Combustion equation	O_2 kmol/ kmol fuel	Products CO_2	H_2O
	2	3	4	5	6
H_2	0·494	$2H_2 + O_2 \rightarrow 2H_2O$	0·247	—	0·494
CO	0·18	$2CO + O_2 \rightarrow 2CO_2$	0·09	0·18	—
CH_4	0·2	$CH_4 + 2O_2 \rightarrow CO_2 + 2H_2O$	0·4	0·2	0·4
C_4H_8	0·02	$C_4H_8 + 6O_2 \rightarrow 4CO_2 + 4H_2O$	0·12	0·08	0·08
O_2	0·004	—	−0·004	—	—
N_2	0·062	—	—	—	—
CO_2	0·04	—	—	0·04	—
		Total =	$\overline{0·853}$	$\overline{0·50}$	$\overline{0·974}$

$$\text{Air required} = \frac{0·853}{0·21} = 4·06 \text{ kmol/kmol of fuel}$$

(where air is assumed to contain 21% O_2 by volume),

i.e.   Stoichiometric A/F ratio = 4·06/1 by volume

For a mixture which is 20% weak, using equation 15.8,

$$\text{Actual A/F ratio} = 4·06 + \frac{20}{100} \times 4·06 = 1·2 \times 4·06 = 4·872/1$$

Associated $N_2 = 0·79 \times 4·872 = 3·85$ kmol/kmol fuel

Excess oxygen $= 0·21 \times 4·872 - 0·853 = 0·1706$ kmol/kmol fuel

Total kmol of N_2 in products
$$= 3·85 + 0·062 = 3·912 \text{ kmol/kmol fuel}$$

The analysis by volume of the wet and dry products is shown in the following table:

Product	kmol/kmol fuel	% by vol. (dry)	% by vol. (wet)
CO_2	0·50	10·90	9·0
H_2O	0·974	—	17·5
O_2	0·171	3·72	3·08
N_2	3·912	85·4	70·4
Total wet =	5·557	100·02	99·98
−H_2O =	0·974		
Total dry =	4·583		

In examples 15.1 and 15.2 it will be noted that the analyses do not add up to exactly 100%. However, the accuracy shown is quite sufficient and no time should be wasted in attempting to get exact analyses.

Example 15.3

Find the stoichiometric A/F ratio for the combustion of ethyl alcohol (C_2H_6O), in a petrol engine. Calculate the A/F ratios for the extreme mixture strengths of 90% and 120%, by the definition of equation 15.9. Determine the wet and dry analyses by volume of the exhaust gas for each mixture strength.

The equation for the combustion of ethyl alcohol is as follows:

$$C_2H_6O + 3O_2 \rightarrow 2CO_2 + 3H_2O$$

Since there are two atoms of carbon in each mole of C_2H_6O then there must be two moles of CO_2 in the products, giving two atoms of carbon on each side of the equation. Similarly, since there are six atoms of hydrogen in each mole of ethyl alcohol then there must be three moles of H_2O in the products, giving six atoms of hydrogen on each side of the equation. Then balancing the atoms of oxygen, it is seen that there are $(2 \times 2 + 3) = 7$ atoms on the right-hand side of the equation, hence 7 atoms must appear on the left-hand side of the equation. There is one atom of oxygen in the ethyl alcohol, therefore a further six atoms of oxygen must be supplied, and hence three moles of oxygen are required as shown. Since the O_2 is supplied as air, the associated N_2 must appear in the equation,

i.e. $$C_2H_6O + 3O_2 + 3 \times \frac{79}{21} N_2 \rightarrow 2CO_2 + 3H_2O + 3 \times \frac{79}{21} N_2$$

1 kmol of fuel has a mass of $(2 \times 12 + 6 + 16) = 46$ kg. 3 kmol of oxygen have a mass of $(3 \times 32) = 96$ kg.

Therefore,

$$O_2 \text{ required per kg of fuel} = \frac{96}{46} = 2 \cdot 09 \text{ kg}$$

$$\therefore \text{ Stoichiometric A/F ratio} = \frac{2 \cdot 09}{0 \cdot 233} = 8 \cdot 96/1$$

Considering a mixture strength of 90% then, from equation 15.9,

$$0 \cdot 9 = \frac{\text{stoichiometric A/F ratio}}{\text{actual A/F ratio}}$$

Therefore, $$\text{Actual A/F ratio} = \frac{8 \cdot 96}{0 \cdot 9} = 9 \cdot 95/1$$

This means that $1/0.9$, or 1.11, times as much air is supplied as is necessary for complete combustion. The exhaust will therefore contain 0.11 of the stoichiometric oxygen,

i.e.

$$C_2H_6O + 1.11\left(3O_2 + 3 \times \frac{79}{21} N_2\right)$$

$$\rightarrow 2CO_2 + 3H_2O + 0.11 \times 3O_2 + 1.11 \times 3 \times \frac{79}{21} N_2$$

i.e. the products are:

$$2 \,\text{kmol}\, CO_2 + 3 \,\text{kmol}\, H_2O + 0.33 \,\text{kmol}\, O_2 + 12.54 \,\text{kmol}\, N_2$$

The total kilomoles $= 2 + 3 + 0.33 + 12.54 = 17.87 \,\text{kmol}$

Hence wet analysis is

$$\frac{2}{17.87} \times 100 = 11.20\% \, CO_2; \qquad \frac{3}{17.87} \times 100 = 16.8\% \, H_2O;$$

$$\frac{0.33}{17.87} \times 100 = 1.85\% \, O_2; \qquad \frac{12.54}{17.87} \times 100 = 70.2\% \, N_2$$

The total dry kilomoles $= 2 + 0.33 + 12.54 = 14.87 \,\text{kmol}$

Hence dry analysis is

$$\frac{2}{14.87} \times 100 = 13.45\% \, CO_2; \qquad \frac{0.33}{14.87} \times 100 = 2.22\% \, O_2;$$

$$\frac{12.54}{14.87} \times 100 = 84.4\% \, N_2$$

Considering a mixture strength of 120%, then from equation 15.9,

$$1.2 = \frac{\text{stoichiometric A/F ratio}}{\text{actual A/F ratio}}$$

Therefore,   Actual A/F ratio $= \dfrac{8.96}{1.2} = 7.47/1$

This means that $1/1.2$, or 0.834 of the stoichiometric air is supplied. The combustion cannot be complete, as the necessary oxygen is not available. It is usual to assume that all the hydrogen is burned to H_2O, since hydrogen atoms have a greater affinity for oxygen than carbon atoms. The carbon in the fuel will burn to CO and CO_2, but the relative proportions have to be determined. Let the number of kilomoles of CO_2 in the products be a, and let the number of kilomoles of CO in the products be b. Then the combustion equation is as follows:

$$C_2H_6O + 0.834\left(3O_2 + 3 \times \frac{79}{21} N_2\right)$$
$$\rightarrow aCO_2 + bCO + 3H_2O + 0.834 \times 3 \times \frac{79}{21} N_2$$

To find a and b a balance of the carbon and the oxygen atoms can be made,

i.e.   Carbon balance: $2 = a+b$

            Oxygen balance: $1 + 2 \times 0.834 \times 3 = 2a + b + 3$

Subtracting the equations gives

$$a = 1.004 \quad \text{and then} \quad b = 2 - 1.004 = 0.996$$

i.e. the products are: 1.004 kmol $CO_2 + 0.996$ kmol $CO + 3$ kmol $H_2 + 9.41$ kmol N_2.

The total kilomoles $= 1.004 + 0.996 + 3 + 9.41 = 14.41$ kmol

Hence wet analysis is

$$\frac{1.004}{14.41} \times 100 = 6.94\% \ CO_2; \qquad \frac{0.996}{14.41} \times 100 = 6.94\% \ CO;$$

$$\frac{3}{14.41} \times 100 = 20.8\% \ H_2; \qquad \frac{9.41}{14.41} \times 100 = 65.3\% \ N_2$$

The total dry kilomoles $= 1.004 + 0.996 + 9.41 = 11.41$ kmol.

Hence dry analysis is

$$\frac{1.004}{11.41} \times 100 = 8.7\% \ CO_2; \qquad \frac{0.996}{11.41} \times 100 = 8.7\% \ CO;$$

$$\frac{9.41}{11.41} \times 100 = 82.5\% \ N_2$$

Example 15.4

Calculate for the stoichiometric mixture of example 15.3, the volume of the mixture per kg of fuel at a temperature of 65°C and a pressure of 1.013 bar. Calculate also the volume of the products of combustion per kg of fuel after cooling to a temperature of 120°C at a pressure of 1 bar.

As before,

$$C_2H_6O + 3O_2 + 3 \times \frac{79}{21} N_2 \rightarrow 2CO_2 + 3H_2O + 3 \times \frac{79}{21} N_2$$

Therefore,

Total kilomoles reactants $= 1 + 3 + 3 \times \frac{79}{21} = 15 \cdot 3$ kmol

From equation 3.8, $pV = nR_0T$,

$$\therefore V = \frac{15 \cdot 3 \times 10^3 \times 8 \cdot 314 \times 338}{10^5 \times 1 \cdot 013} = 424 \cdot 4 \text{ m}^3/\text{kmol of fuel}$$

(where $T = 65 + 273 = 338$ K).

In 1 kmol of fuel there are $(2 \times 12 + 6 + 16) = 46$ kg.

$$\therefore \text{ Volume of reactants per kg of fuel} = \frac{424 \cdot 4}{46} = 9 \cdot 226 \text{ m}^3$$

When the products are cooled to 120°C the H_2O exists as steam, since the temperature is well above the saturation temperature corresponding to the partial pressure of the H_2O. (This must be so since the saturation temperature corresponding to the *total* pressure is 99·6°C, and saturation temperature decreases with pressure.) The total kilomoles of the products is

$$\left(2 + 3 + 3 \times \frac{79}{21} \right) = 16 \cdot 3 \text{ kmol}$$

From equation 3.8, $pV = nR_0T$,

$$\therefore V = \frac{16 \cdot 3 \times 10^3 \times 8 \cdot 314 \times 393}{10^5 \times 1} = 533 \cdot 8 \text{ m}^3/\text{kmol of fuel}$$

(where $T = 120 + 273 = 393$ K).

Then,

Volume of products per kg of fuel $= \dfrac{533 \cdot 8}{46} = 11 \cdot 58 \text{ m}^3$

Example 15.5

If the products in example 15.4 are cooled to 15°C at constant pressure, calculate the amount of water which will condense per kg of fuel.

At 15°C, since some condensation has taken place, the steam remaining is dry saturated, being in contact with its liquid. The saturation pressure at 15°C is 0·017 04 bar, and this is the partial pressure of the dry saturated steam.

Then using equation 13.14,

$$\frac{V_i}{V} = \frac{n_i}{n} = \frac{p_i}{p}$$

For the steam

$$\frac{n_s}{n} = \frac{0·017\ 04}{1} = 0·017\ 04$$

From example 15·4 the total kilomoles of dry products is 13·3 kmol, therefore,

$$\frac{n_s}{n_s + 13·3} = 0·017\ 04 \qquad \therefore\ n_s = \left(\frac{0·017\ 04 \times 13·3}{1 - 0·017\ 04}\right) = 0·2305$$

i.e.   kilomoles of dry saturated steam remaining at 15°C = 0·2305

$$\therefore\ \text{kilomoles of water condensed} = 3 - 0·2305 = 2·77$$

1 kmol of H_2O contains $(2+16) = 18$ kg. Therefore mass of water condensed is $2·77 \times 18$ kg for every kmol of fuel, i.e.

$$\text{Mass of water condensed per kg of fuel} = \frac{2·77 \times 18}{46} = 1·084\ \text{kg}$$

Any problem in combustion can be solved using the kilomole. The following examples illustrate the method for a solid and for a gaseous fuel; these examples should be compared with the method used in examples 15.1 and 15.2.

Example 15.6

The gravimetric analysis of a sample of coal is given as 80% C, 12% H, and 8% ash. Calculate the stoichiometric A/F ratio and the analysis of the products by volume.

1 kg of coal contains 0·8 kg C and 0·12 kg H

$$\therefore\ 1\ \text{kg of coal contains}\ \frac{0·8}{12}\ \text{kmol C and } 0·12\ \text{kmol H}$$

Let the oxygen required for complete combustion be x kmol, the nitrogen supplied with the oxygen is then $x \times \dfrac{79}{21} = 3·76x$ kmol.

For 1 kg of coal the combustion equation is therefore as follows:

$$\frac{0\cdot8}{12}C + 0\cdot12H + xO_2 + 3\cdot76xN_2 \rightarrow aCO_2 + bH_2O + 3\cdot76xN_2$$

Then,

Carbon balance: $\dfrac{0\cdot8}{12} = a$ $\therefore$ $a = 0\cdot067$ kmol

Hydrogen balance: $0\cdot12 = 2b$ $\therefore$ $b = 0\cdot06$ kmol

Oxygen balance: $2x = 2a + b$ $\therefore$ $x = 0\cdot097$ kmol

The mass of 1 kmol of oxygen is 32 kg, therefore the mass of O_2 supplies per kg of coal is $32 \times 0\cdot097$ kg,

i.e.   Stoichiometric A/F ratio $= \dfrac{32 \times 0\cdot097}{0\cdot233} = 13\cdot3/1$

Total kilomoles of products $= a + b + 3\cdot76x$

$$= 0\cdot067 + 0\cdot06 + 3\cdot76 \times 0\cdot97$$

$$= 0\cdot492 \text{ kmol}$$

Hence wet analysis is,

$\dfrac{0\cdot067}{0\cdot492} \times 100 = 13\cdot6\%$ CO_2; $\dfrac{0\cdot06}{0\cdot492} \times 100 = 12\cdot2\%$ H_2;

$$\dfrac{0\cdot365}{0\cdot492} \times 100 = 74\cdot2\% \text{ } N_2$$

Example 15.7

A gas engine is supplied with coal gas of the following composition: 53·6% H_2; 9% CO; 25% CH_4; 3% C_4H_8; 0·4% O_2; 3% CO_2; 6% N_2. If the air fuel ratio is 6·5/1 by volume, calculate the analysis of the dry products of combustion. It can be assumed that the stoichiometric A/F ratio is less than 6·5/1.

Since we are told that the actual A/F ratio is greater than the stoichiometric, then it follows that excess air has been supplied. The products will therefore consist of CO_2, H_2O, O_2, and N_2. The combustion equation can be written as follows:

$0\cdot536H_2 + 0\cdot09CO + 0\cdot25CH_4 + 0\cdot03C_1H_8$

$\qquad\qquad\qquad + 0\cdot004O_2 + 0\cdot03CO_2 + 0\cdot06N_2$

$\qquad\qquad\qquad + 0\cdot21 \times 6\cdot5O_2 + 0\cdot79 \times 6\cdot5N_2$

$\qquad\rightarrow aCO_2 + bH_2O + cO_2 + dN_2$

Then,

Carbon balance: $0.09+0.25+0.12+0.03 = a$ $\therefore$ $a = 0.49$

Hydrogen balance: $0.536 \times 2+0.25 \times 4+0.03 \times 8 = b$

$\therefore$ $b = 1.26$

Oxygen balance: $0.09+0.004 \times 2+0.03 \times 2+0.21 \times 6.5 \times 2$
$$= 2a+b+2c \quad \therefore \quad c = 0.378$$

Nitrogen balance: $0.06 \times 2+0.79 \times 6.5 \times 2 = 2d$ $\therefore$ $d = 5.2$

Therefore,

Total kilomoles of dry products $= 0.49+0.378+5.2 = 6.068$

Then analysis by volume is,

$$\frac{0.49}{6.068} \times 100 = 8.08\% \ CO_2; \qquad \frac{0.378}{6.068} \times 100 = 6.22\% \ O_2;$$

$$\frac{5.2}{6.068} \times 100 = 85.7\% \ N_2$$

15.6 Practical analysis of combustion products

The experimental investigation of a combustion process requires the analysis of the products of combustion. It is essential that representative samples of the products, from a sufficient number of points in the plant, are obtained. Having obtained a sample it then remains to analyse it. If the analysis is by chemical means, then suitable reagents are used, each one of which will preferably absorb only one of the constituents. As each constituent is absorbed the volume of the remainder is measured at the same pressure and temperature as the initial sample. The volume of the constituent absorbed is obtained by taking the difference between volume readings before and after the absorption. Some of the requirements are now evident: a means of taking a representative sample; a means of measuring the volume at constant pressure and temperature; suitable reagents for the absorption process. The most common means of analysis is the Orsat apparatus, see fig. 15.1.

The gas sample and reagents are manipulated by raising or lowering the levelling bottle. With the three-way valve open to atmosphere the air in the capillary tube is exhausted. The three-way valve is then closed and the levelling bottle lowered, thus producing a pressure in the capillary which is below atmospheric. By opening the taps on the reagent vessels each of the levels can be brought to the mark on the stem. With the reagent levels correct and the taps closed the three-way valve is opened to atmosphere. The measuring burette is then

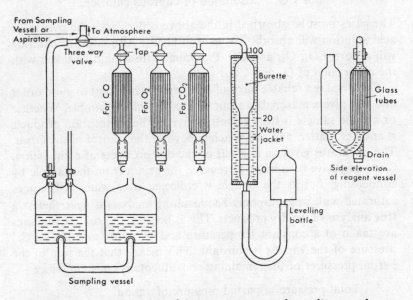

Fig. 15.1 Arrangement of Orsat apparatus and sampling vessels

emptied of air by raising the levelling bottle. The three-way valve is turned to connect the sample bottle or aspirator to the burette, and the gas sample is induced into the burette by lowering the levelling bottle. When a sample has been drawn in, the three-way valve is closed and the pressure in the burette is brought to atmospheric by equalizing the liquid levels in the burette and in the bottle. The volume is read off the scale of the burette, the temperature being maintained constant by a water jacket surrounding the burette. The sample is now passed into each one of the reagent vessels A, B, and C in turn. After each absorption in a reagent bottle the sample is returned to the burette, and the reagent is restored to its original level. This is repeated for each reagent bottle until the same burette reading is obtained for two or more successive readings.

In the normal version of the Orsat apparatus three reagents are used for the absorption of CO_2, O_2, and CO respectively. The unabsorbed portion of the sample is always taken to be nitrogen. The reagents used are as follows:

> For CO_2:   a solution of caustic potash,
>
> For O_2:   a pyrogallic acid solution,
>
> For CO: a solution of cuprous chloride.

The gases must be absorbed in the above order, since the pyrogallic acid solution will absorb CO_2 as well as O_2. Also cuprous chloride will absorb CO_2, O_2, and CO. Particular care should be taken with the absorption of O_2 and CO.

The Orsat is a reliable piece of equipment when kept in good order and will give a reasonably accurate analysis of the sample. Whether or not the sample is truly representative of the combustion products is another matter, and one which is beyond the control of the Orsat. The apparatus gives an analysis of the dry products of combustion. Steps may have been taken to remove the steam from the sample by condensing, but as the sample is collected over water it becomes saturated with water vapour. The resulting analysis is nevertheless a true analysis of the dry products. This is because the volume readings are taken at a constant temperature and pressure, and the partial pressure of the vapour is constant. This means that the sum of the partial pressures of the remaining constituents is constant, since

$$\text{Total pressure} = \text{partial pressure of vapour} + \text{partial pressure of dry constituents}$$

i.e.
$$p = p_s + p_c$$

The vapour then occupies the same proportion of the total volume at each measurement. Using equation 13.14, $V_s = (p_s/p)V$. Hence the vapour does not affect the result of the analysis. A more formal proof of this is given in reference 15.2.

The analysis will show whether or not the combustion is complete. For instance the presence of CO will indicate that the combustion is not complete, and if an oxygen reading is obtained this will mean that excess air has been supplied. Both CO and O_2 may appear in the analysis as a result of incomplete combustion and dissociation (see Section 15.8). Quantitatively the dry product analysis can be used to calculate the A/F ratio. This method of obtaining the A/F ratio is not so reliable as direct measurement of the air consumption and fuel consumption of the engine. More caution is required when analysing the products of combustion of a solid fuel since some of the products do not appear in the flue gases (e.g. ash and unburnt carbon). The residual solid must be analysed as well in order to determine the carbon content, if any. With an engine using petrol or diesel fuel the exhaust may include unburnt particles of carbon and this quantity will not appear in the analysis. The exhaust from internal combustion engines may contain also some CH_4 and H_2 due to incomplete combustion. The CH_4 content is approximately 0·22% of the dry products and the H_2 content is of the order of half the CO content. Another piece of equipment called the Haldane apparatus measures the CH_4 content as well as CO_2, O_2, and CO. Other methods of exhaust gas analysis will be mentioned later in this section.

Considerable care is required in combustion calculations as inaccuracies are caused due to taking small differences between quantities and then using these as ratios. The use which can be made of the dry analysis is illustrated by the following problems. Alternative methods are given for some of the problems, but this does not mean that these are the only methods available. There are other methods, but in general the mole method is the most convenient, since it is applicable to solid, liquid, and gaseous fuels.

Example 15.8

An Orsat analysis of the exhaust from an engine running on benzole showed a CO_2 content of 15%, but no CO. Assuming that the remainder of the exhaust contains only oxygen and nitrogen, calculate the A/F ratio of the engine. The ultimate analysis of benzole is 90% C and 10% H.

Method 1

Consider 1 kilomole of dry exhaust gas (D.E.G.).

Let the D.E.G. contain a kmol of O_2. The kmol of CO_2 in 1 kmol of D.E.G. are 0·15. Therefore the D.E.G. contains $(1-a-0.15) = (0.85-a)$ kmol of N_2.

Carbon content per kmol D.E.G. $= 12 \times 0.15 = 1.8$ kg

Carbon content per kg fuel $= 0.9$ kg

$\therefore$ Moles D.E.G. per kg fuel $= \dfrac{0.9}{1.8} = 0.5$

Mass of N_2 in exhaust $= 28 \times (0.85-a)$ kg/kmol D.E.G.

$$\therefore \text{ Total } O_2 \text{ supplied} = \frac{0.233}{0.767} \times 28 \times (0.85 - a)$$

$$= (7.24 - 8.5a) \quad \text{kg/kmol D.E.G.}$$

O_2 accounted for in the D.E.G. is

$$32 \times (0.15 + a) = (4.8 + 32a) \text{ kg/kmol D.E.G.}$$

Therefore O_2 associated with H_2O is

$$(7.24 - 8.5a) - (4.8 + 32a) = (2.44 - 40.5a) \text{ kg/kmol D.E.G.}$$

i.e. $\quad O_2$ burned to $H_2O = 0.5 \times (2.44 - 40.5a) \quad \text{kg/kg fuel}$

$$= (1.22 - 20.25a) \quad \text{kg/kg fuel}$$

Now 0·1 kg H in the fuel requires $0.1 \times 8 = 0.8$ kg O_2 for complete combustion (see equation 15.1),

i.e. $\quad\quad 0.8 = 1.22 - 20.25a \quad\quad \therefore \ a = 0.0208$ kmol

Therefore oxygen in products is 0·0208 kmol/kmol D.E.G. The nitrogen is then

$$(0.85 - 0.0208) = 0.8292 \text{ kmol/kmol D.E.G.}$$

$$\therefore N_2 \text{ per kg fuel} = 0.5 \times 0.8292 = 0.4146 \text{ kmol/kg}$$

i.e. $\quad\quad\quad$ A/F ratio $= \dfrac{28 \times 0.4146}{0.767} = 15.1/1$

Method 2

1 kg of fuel, consisting of 0·9 kg C and 0·1 kg H, can be written as 0·9/12 kmol C and 0·1 kmol H. Therefore considering 1 kmol of D.E.G. we can write the combustion equation as follows:

$$X \left(\frac{0.9}{12} C + 0.1H \right) + YO_2 + \frac{79}{21} YN_2$$

$$\rightarrow 0.15CO_2 + aO_2 + (0.85 - a)N_2 + bH_2O$$

where X is the mass of fuel per kilomole D.E.G.

$\quad\quad Y$ is the kilomoles of O_2 per kilomole D.E.G.

$\quad\quad a$ is the kilomole of excess O_2 per kilomole D.E.G.

$\quad\quad b$ is the kilomoles of H_2O per kilomole D.E.G.

Then,

Carbon balance: $\dfrac{0.9}{12} X = 0.15$ $\therefore$ $X = 2.0$

Hydrogen balance: $0.1X = 2b$ $\therefore$ $b = 0.1$

Oxygen balance: $2Y = 2 \times 0.15 + 2a + b$ $\therefore$ $Y = 0.2 + a$

Nitrogen balance: $3.76 \times 2 \times Y = 2 \times (0.85 - a)$
$$\therefore \ Y = 0.226 - 0.266a$$

Equating the expressions for Y gives

$$0.226 - 0.266a = 0.2 + a \qquad \therefore \ a = 0.0206$$

i.e. $$Y = 0.2 + 0.0206 = 0.221 \text{ kmol}$$

$$\therefore \ O_2 \text{ supplied} = 0.221 \times 32 \text{ kg/kmol D.E.G.}$$

i.e. $$\text{Air supplied} = \frac{0.221 \times 32}{0.233} = 30.4 \text{ kg/kmol D.E.G.}$$

Since $X = 2$, then, the fuel supplied per kmol D.E.G. is 2 kg.

$$\therefore \ \text{A/F ratio} = \frac{30.4}{2} = 15.2/1$$

Example 15.9

The analysis of the dry exhaust from an internal-combustion engine gave: 12% CO_2; 2% CO; 4% CH_4; 1% H_2; 4.5% O_2; and the remainder nitrogen. Calculate the proportions by mass of carbon to hydrogen in the fuel, assuming it to be a pure hydrocarbon.

Method 1

$$N_2 \text{ in D.E.G.} = 1 - 0.12 - 0.02 - 0.04 - 0.01 - 0.045$$

$$= 0.765 \text{ kmol/kmol D.E.G.}$$

$$\therefore \ N_2 \text{ in D.E.G.} = 28 \times 0.765 = 21.4 \text{ kg/kmol D.E.G.}$$

$$O_2 \text{ associated with this nitrogen} = \frac{0.233 \times 21.4}{0.767}$$
$$= 6.5 \text{ kg/kmol D.E.G.}$$

$$O_2 \text{ accounted for in the D.E.G.} = 32 \left(0.12 + \frac{0.02}{2} + 0.045 \right)$$

$$= 5.6 \text{ kg/kmol D.E.G.}$$

$$\therefore \quad O_2 \text{ burned to } H_2O = 6\cdot5 - 5\cdot6 = 0\cdot9 \text{ kg/kmol D.E.G.}$$

Therefore since 1 kg of H_2 requires 8 kg of O_2 for complete combustion we have

$$H_2 \text{ burned to } H_2O = \frac{0\cdot9}{8} = 0\cdot1125 \text{ kg/kmol D.E.G.}$$

$$H_2 \text{ accounted for in the D.E.G.} = (0\cdot04 \times 4 + 0\cdot01 \times 2)$$

$$= 0\cdot18 \text{ kg/kmol D.E.G.}$$

$$\therefore \quad H_2 \text{ in fuel} = 0\cdot18 + 0\cdot1125 = 0\cdot2925 \text{ kg/kmol D.E.G.}$$

$$\text{Mass of carbon in fuel} = 12(0\cdot12 + 0\cdot02 + 0\cdot04)$$

$$= 2\cdot16 \text{ kg/kmol D.E.G.}$$

i.e. $$\text{Ratio of C to } H_2 \text{ in fuel} = \frac{0\cdot2925}{2\cdot16} = 7\cdot38/1$$

Method 2

Let 1 kg of fuel contain x kg C and y kg H. Then, considering 1 kmol of D.E.G. and introducing X and Y as defined in example 15.8, we can write

$$X\left(\frac{xC}{12} + yH\right) + YO_2 + \frac{79}{21}YN_2 \rightarrow 0\cdot12CO_2 + 0\cdot02CO$$

$$+ 0\cdot04CH_4 + 0\cdot01H_2 + 0\cdot045O_2 + aH_2O + 0\cdot765N_2$$

Then,

Nitrogen balance: $\quad 3\cdot76Y = 0\cdot765 \quad \therefore Y = 0\cdot2035$

Oxygen balance: $\quad 0\cdot2035 = 0\cdot12 + \dfrac{0\cdot02}{2} + 0\cdot045 + \dfrac{a}{2}$

$$\therefore \quad a = 0\cdot057$$

Carbon balance: $\quad \dfrac{Xx}{12} = 0\cdot12 + 0\cdot02 + 0\cdot04$

$$\therefore \quad Xx = 2\cdot16 \quad (1)$$

Hydrogen balance: $\quad Xy = 4 \times 0\cdot04 + 2 \times 0\cdot01 + 2 \times 0\cdot057$

$$\therefore \quad Xy = 0\cdot294 \quad (2)$$

Dividing equations (1) and (2) gives,

$$\frac{Xx}{Xy} = \frac{2 \cdot 16}{0 \cdot 294} = 7 \cdot 35$$

i.e. Ratio of C to H in fuel $= \dfrac{x}{y} = 7 \cdot 35/1$

The data in the above problems has been presented in such a way as to be consistent. In a real analysis the CO_2, CO and O_2 components would be measured and the N_2 component calculated. If the fuel analysis is known then the combustion equation formed will enable the air–fuel ratio to be calculated by two methods. Close agreement between the two answers will tend to validate the data obtained for the analysis. Disagreement will indicate a source of error and, assuming the analysis to have been made correctly and accurately, one may suspect the presence of unburned carbon, or other products of combustion, in the exhaust gas.

The analytical procedure is indicated by the following method.

Writing the combustion equation as before as

$$X\left(\frac{x}{12}C + yH\right) + YO_2 + 3 \cdot 76YN_2$$
$$\rightarrow aCO_2 + bCO + cO_2 + dH_2O + eN_2$$

where x, y, a, b, c and e are known and X, Y and d are unknown.

Method 1

C balance: $\dfrac{Xx}{12} = a + b$ $\therefore$ X is determined

N_2 balance: $3 \cdot 76Y = e$ $\therefore$ Y is determined

$$\text{A/F ratio, by mass,} = \frac{32Y}{0 \cdot 233X}$$

Method 2

Y is obtained from the N_2 balance as before.

O_2 balance: $Y = a + \dfrac{b}{2} + c + \dfrac{d}{2}$ $\therefore$ d is determined

H_2 balance: $Xy = 2d$ $\therefore$ X is determined

the A/F ratio is calculated as before and agreement between Methods (1) and (2) should be obtained.

15.7 Further methods for the analysis of combustion products

For research purposes more accurate analyses of the combustion products are required than are given by the Orsat apparatus. The chemical methods can be extended to give the required accuracy, but the apparatus becomes more complex. Analysis by chemical means is a slow process and other methods have been developed to give an accurate analysis more quickly. Another desirable feature is that analysis should be continuous if possible.

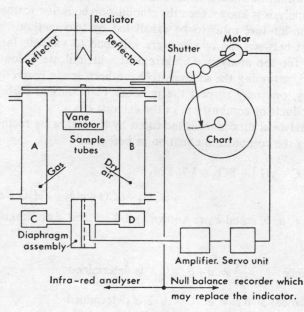

Fig. 15.2

Gas analysis by infra-red spectra

The concentration of the constituent gas is measured by recording its 'optical' absorption in the infra-red spectrum. The mixture is continuously examined by being drawn through a tube, the inside of which is subject to radiation from an infra-red source, each gas absorbing on a particular wave-band of the radiation.

Referring to fig. 15.2, the gas being analysed is passed through tube A and dry air is passed through tube B. The chambers C and D contain pure samples of the constituent to be detected. Radiation from nichrome elements is passed through tubes A and B and thence to chambers C and D where it is absorbed. The subsequent heating of the gases in C and D causes an increase in pressure in the two chambers. C and D are separated by a thin metal diaphragm which, together with an insulated, perforated plate, forms a capacitor.

If the constituent being sought is not present in tube A then equal amounts of radiation will be absorbed in C and D, which will be heated equally and there will be no pressure difference across the diaphragm. If the constituent is present in A, it will absorb some of the radiation admitted, and the rest will pass on to the chamber C. The radiation absorbed in C will then be less than in D so that a greater pressure will be reached in D than in C, and the diaphragm will be displaced. This displacement produces a change in capacitance of the condenser and a current is produced which is amplified to give a reading on a micro-ammeter. The micro-ammeter scale is calibrated to give the corresponding concentration of the constituent in the gas being analysed.

To avoid zero errors the radiation is cut off from both tubes simultaneously and allowed to fall on them simultaneously, by means of a vane which rotates at a low frequency. The pressure changes are then related to the temperature changes produced by the differential absorption in C and D.

To provide a continuously recording instrument as shown in the right-hand part of fig. 15.2, a 'null' balance recorder is employed. The principle is that the pressure difference created should be nullified by cutting off from the vessel B a sufficient amount of radiation to balance that absorbed in A. This is done by means of a shutter driven by a servo-motor which receives a signal from the detector unit. A recording pen is linked to the balancing shutter mechanism and records on a circular chart.

The instrument is calibrated against accurately prepared samples of gas mixtures. This method is suitable for carbon monoxide, carbon dioxide, sulphur and nitrogen compounds, methane and other hydro carbons, and organic vapours. The following lower limits are quoted in parts per million for the most common applications;

$$CO_2 \text{ and } N_2 \ldots 3; \qquad CO \text{ and } C_2H_4 \ldots 10;$$
$$CH_4, C_2H_2, \text{ and } C_2H_6 \ldots 20$$

Oxygen, hydrogen, nitrogen, argon, chlorine, and helium, do not absorb infra-red radiation and so will not be detected by this type of instrument.

CO_2 and O_2 recorders

Boiler house engineers require a continuous indication of the quality of the combustion process in the plant under their control. This enables comparisons to be made and a falling off in efficiency becomes immediately apparent. For continuous firing a continuous record is required, and digital instruments are available with in-built microchips which enable boiler efficiency to be obtained directly from individual readings of temperature, and percentage by volume of CO_2 and O_2. The variations and applications of these instruments are many, and are made to suit particular requirements. The general principles only will be dealt with here.

(a) CO_2 measurement by thermal conductivity variations

The CO_2 content of a flue gas is an important criterion of efficient and economic combustion, and is important in observing the regulations governing smoke emission.

When a heated wire is placed in a gaseous atmosphere it loses heat by radiation, convection and conduction. If the losses by radiation and convection are kept constant, the total heat loss is dependent on the heat loss by conduction, which varies with the constituents of the gas since each has a different and characteristic thermal conductivity. If a constant heat input is supplied to the wire there is an equilibrium temperature for each mixture, and if the CO_2 content alone is varied, then its concentration will be indicated by a measurement of the temperature of the wire. In an actual instrument the heat loss from the wire is mainly by conduction, the other means (e.g. convection, radiation, end cooling, diffusion) account for about 1% each of the total loss. The convection loss is reduced by mounting the wire vertically. The instrument is calibrated against mixtures of known composition.

The sample of gas is passed over an electrically heated platinum wire in a cell which forms one arm of a Wheatstone bridge. In another arm of the bridge is a similar cell containing air. A difference in CO_2 content between the two cells causes a difference in temperature between the two wires and hence a difference in resistance. The out-of-balance potential of the bridge is measured by a recording potentiometer, calibrated to give the CO_2 content directly.

(b) Oxygen measurement by magnetic means

Gases may be classified in two groups:

(a) **Diamagnetic gases** which seek the weakest part of a magnetic field.

(b) **Paramagnetic gases** which seek the strongest part of a magnetic field.

Most gases are diamagnetic, but oxygen is paramagnetic, and this property of oxygen can be utilized in measuring the oxygen content of gas mixtures. Referring to fig. 15.3 the gas sample is introduced into

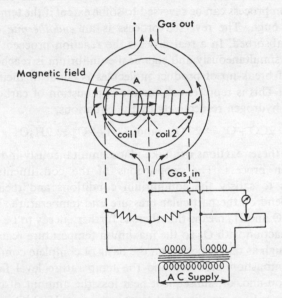

Fig. 15.3

the analysis cell and passes through the annulus as shown. The horizontal cross-tube carries two identical platinum windings, coils 1 and 2, which are connected in adjacent arms of a Wheatstone bridge, and are heated by the applied voltage. The winding at A is traversed by a magnetic field of high intensity from a large permanent magnet. When the gas sample passes the end of the tube the oxygen is drawn into the cross-tube. It is heated and its paramagnetic property decreases due to the increase in temperature as it passes through the tube. The induction of fresh cool oxygen continues, hence a continuous flow is established (sometimes called the 'magnetic breeze'). The result is that the two windings are cooled by different amounts, the resistance changes and the bridge goes out of balance. The resulting e.m.f. is measured by a potentiometer, and since this is proportional to the oxygen content, the reading gives the oxygen content of the mixture.

15.8 Dissociation

It is found that during adiabatic combustion the maximum temperature reached is lower than that expected on the basis of elementary calculation. One important reason for this is that the *exothermic* combustion process can be reversed to some extent if the temperature is high enough. The reversed process is an *endothermic* one, i.e. energy is absorbed. In a real process the reaction proceeds in both directions simultaneously and chemical equilibrium is reached when the rate of break-up of product molecules is equal to their rate of formation. This is represented, for the combustion of carbon monoxide and hydrogen respectively by the equations,

$$2CO + O_2 \rightleftharpoons 2CO_2 \quad \text{and} \quad 2H_2 + O_2 \rightleftharpoons 2H_2O$$

Both of these reactions can take place simultaneously in the same combustion process. The proportions of the constituents adjust themselves to satisfy the equilibrium conditions and their actual values depend on the particular pressure and temperature. The presence of CO and H_2 means that there is further energy to be released on their reaction with O_2 so the maximum temperature reached can not be as high as that expected on the basis of complete combustion. As the combustion proceeds and the temperature level falls, due to expansion and, or, subsequent heat loss the amount of dissociation decreases (it is significant at temperatures > 1500 K) and combustion proceeds to completion. However, since the energy is released at lower temperatures and, in a positive expansion cylinder, at a lower effective compression ratio the efficiency of the process is reduced. The condition of equilibrium during a reversible combustion process can be studied by means of a conceptual device known as the 'van't Hoff equilibrium box' as shown in fig. 15.4.

Consider the general reversible combustion process

$$a \text{ kmol } A + b \text{ kmol } B \rightleftharpoons c \text{ kmol } C + d \text{ kmol } D$$

which occurs at a fixed temperature T and a pressure, p, in the equilibrium box. The reactants A and B are each initially at p_1 and T and the products C and D are each finally at p_1 and T. As the process is reversible some energy transfer will take place in the form of work

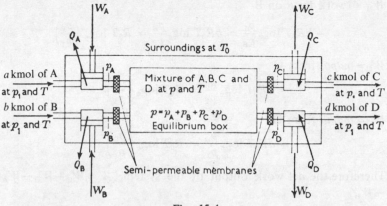

Fig. 15.4

and this is allowed for by the inclusion of isothermal compressors and expanders. The equilibrium box contains a mixture of gases A, B, C and D at total pressure p and temperature T and to allow reversible mass flow of the constituents the pressure of each constituent at entry to the box must be equal to its partial pressure in the box. The pressure adjustments are made by the isothermal expanders and compressors and each constituent enters or leaves through a *semi-permeable membrane**.

The process may proceed equally well in either direction but it is illustrated here as going from left to right in the equation and in fig. 15.4. With a reversal of the process the heat and work transfers would be reversed in direction.

The work done during an isothermal expansion by a perfect gas between states 1 and 2 is given by

$$W = mRT \log_e \frac{p_1}{p_2} = nR_0 T \log_e \frac{p_1}{p_2}$$

where $n =$ the number of kilomoles, and this can be applied to each of the compressors and expanders in the system of fig. 15.4

$W_A =$ work *input* on A

$$= -aR_0 T \log_e \frac{p_1}{p_A} = aR_0 T \log_e \frac{p_A}{p_1} = R_0 T \log_e \left(\frac{p_A}{p_1}\right)^a$$

---

* A semi-permeable membrane in this sense is a concept based on reality. Some substances permit one gas to pass through but prevent other gases, e.g. a glowing aluminium sheet allows hydrogen to pass through but not other gases. It is assumed here that such substances are available for gases A, B, C and D.

W_B = work *input* on B

$$= -bR_0T\log_e\frac{p_1}{p_B} = bR_0T\log_e\frac{p_B}{p_1} = R_0T\log_e\left(\frac{p_B}{p_1}\right)^b$$

W_C = *output* from C

$$= cR_0T\log_e\frac{p_C}{p_1} = R_0T\log_e\left(\frac{p_C}{p_1}\right)^c$$

W_D = *output* from D

$$= dR_0T\log_e\frac{p_D}{p_1} = R_0T\log_e\left(\frac{p_D}{p_1}\right)^d$$

Therefore the net work output of the system, $W = W_C + W_D - W_A - W_B$

i.e.

$$W = R_0T\left\{\log_e\left(\frac{p_C}{p_1}\right)^c + \log_e\left(\frac{p_D}{p_1}\right)^d - \log_e\left(\frac{p_A}{p_1}\right)^a - \log_e\left(\frac{p_B}{p_1}\right)^b\right\}$$

$$= R_0T\left\{\log_e\frac{p_C^c p_D^d}{p_A^a p_B^b} + \log_e p_1^{a+b-c-d}\right\}$$

Suppose that in a second similar system in the same surroundings the pressure in the equilibrium box is p' then it will have a net work output W' given by

$$W' = R_0T\left\{\log_e\frac{(p_C')^c(p_D')^d}{(p_A')^a(p_B')^b} + \log_e p_1^{a+b-c-d}\right\}$$

where $\qquad\qquad p' = p_A' + p_B' + p_C' + p_D'$

It is proposed that $W \neq W'$ and this statement is to be investigated. Suppose $W > W'$, then the second system can be reversed, as shown in fig. 15.5, and a single system formed by using the work output from the first system W to provide the work input for the second system reversed.

The result is a single system giving a net output of work $(W - W')$ whilst exchanging heat with a single source at temperature T. This is a contradiction of the Second Law of Thermodynamics thus the proposition that $W \neq W'$ is not true so that $W = W'$

$$\therefore \frac{p_C^c p_D^d}{p_A^a p_B^b} = \frac{(p_C')^c(p_D')^d}{(p_A')^a(p_B')^b} = K \qquad\qquad 15.10$$

where K is the *equilibrium* or *dissociation* constant and has been shown to be independent of the pressure in the equilibrium box.

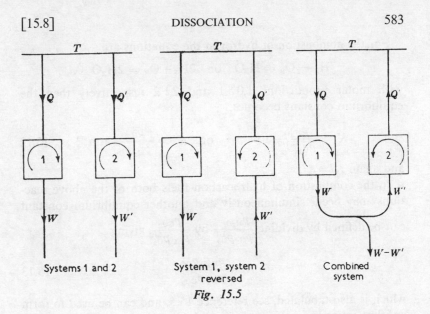

Fig. 15.5

It is a function of temperature and values of K are tabulated against temperature for each reaction *equation*, see reference 15.5. As the partial pressures of the constituents are proportional to the molar proportions then K is an indication of the ratio of products to reactants and so is a measure of the amount of dissociation. If K is large then the proportion of product is high and the amount of dissociation is small.

The above general expression for K, 15.10, will now be applied to particular important reactions and as the chemical equation can be written in alternative ways it will be seen that alternative values of K are defined and evaluated, e.g. the combustion of carbon monoxide to carbon dioxide can be written as

$$CO + \tfrac{1}{2}O_2 \rightleftharpoons CO_2 \quad \text{or} \quad 2CO + O_2 \rightleftharpoons 2CO_2$$

with mole proportions for CO, O_2 and CO_2 of 1,0·5,1 and 2,1,2 respectively thus K can be written as

$$K = \frac{p_{CO_2}}{p_{CO} p_{O_2}^{1/2}} \text{ atm}^{-1/2} \quad \text{or} \quad K' = \frac{p_{CO_2}^2}{p_{CO}^2 p_{O_2}} \text{ atm}^{-1} \qquad 15.11$$

with units as stated if the partial pressures are measured in atmospheres. It can be seen that $K' = K^2$ and for calculation purposes the corresponding equation and equilibrium constant must be used.

For the combustion of hydrogen the equations are

$$H_2 + \tfrac{1}{2}O_2 \rightleftharpoons H_2O \quad \text{or} \quad 2H_2 + O_2 \rightleftharpoons 2H_2O$$

with molar proportions 1,0·5,1 and 2,1,2 respectively then the equilibrium constant becomes

$$K = \frac{p_{H_2O}}{p_{H_2}p_{O_2}^{1/2}} \text{ atm}^{-1/2} \quad \text{or} \quad K' = \frac{p_{H_2O}^2}{p_{H_2}^2 p_{O_2}} \text{ atm}^{-1} \qquad 15.12$$

and again $K' = K^2$.

In the combustion of hydrocarbon fuels both of the above reactions may occur simultaneously and another equilibrium constant can be defined by dividing $\dfrac{p_{H_2O}}{p_{H_2}p_{O_2}^{1/2}}$ by $\dfrac{p_{CO_2}}{p_{CO}p_{O_2}^{1/2}}$ giving

$$K = \frac{p_{H_2O}p_{CO}}{p_{H_2}p_{CO_2}} \qquad 15.13$$

which is also tabulated, see reference 15.5, and can be used to form another equation in the analysis. Values of the various equilibrium constants discussed are shown plotted against temperature in fig. 15.6.

It is readily seen that dissociation as described introduces an added complexity into the analysis of the combustion process but the complication does not end there. The conditions of equilibrium must be satisfied at any particular temperature and also the energy balance for the process must be satisfied. The temperature reached will depend on the amount of fuel burned, the proportions of the constituents and their thermodynamic properties all of which are also dependent on pressure or temperature. Thus several conditions have to be satisfied before any particular state in the process can be determined and it is necessary to establish a sufficient number of equations to complete the analysis. This is discussed more fully in Section 15.9 and for the immediate purpose it will be assumed that the energy requirements have been met in the process and only the dissociation effects are required.

The analysis of combustion may be extended to include the formation of nitric oxide, $\tfrac{1}{2}N_2 + \tfrac{1}{2}O_2 \rightleftharpoons NO$, which occurs at high temperature and the dissociation of H_2O vapour into hydrogen and hydroxyl, $\tfrac{1}{2}H_2 + OH \rightleftharpoons H_2O$, as well as into its constituent gases. There may also be the dissociation of molecules of oxygen, hydrogen and nitrogen into atoms. These aspects of combustion are detailed

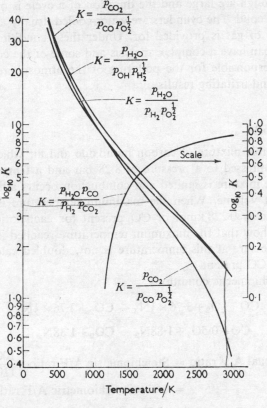

Fig. 15.6

and involve small proportions of the charge. It has been stated previously that recombination occurs as the temperature falls so that combustion is completed with a loss in efficiency of the cycle such that less work output is obtained than expected. The completion of combustion requires the absence of low temperature quenching conditions, which may arrest the process, and a sufficiency of time for the reaction to be completed.

The importance of the analysis of combustion increases with advancing engine technology as typified in the petrol engine. This engine includes the extremes of combustion conditions starting from a complex chemical fuel and a mixture of air, water vapour and residual exhaust gas. The pressure and temperature levels passed through are large and the duration of a cycle is only a fraction of a second. The cylinders are water-cooled usually and sudden exhausting of gas is provided for. Under these conditions the exhaust gas can have a complex analysis and some of the constituents are held responsible for the polluting of the atmosphere with undesirable and irritating results.

Example 15.10

A combustible mixture of carbon monoxide and air which is 10% rich is compressed to a pressure of 8·28 bar and a temperature of 282°C. The mixture is ignited and combustion occurs adiabatically at constant volume. When the maximum temperature is attained analysis shows 0·228 kmol of CO present for each kmol of CO supplied. Show that the maximum temperature reached is 2677°C.

If the pressure at this temperature is now doubled calculate the amount of CO present.

For stoichiometric conditions

$$CO + \tfrac{1}{2}O_2 + 3·76 \times \tfrac{1}{2}N_2 \rightarrow CO_2 + 3·76 \times \tfrac{1}{2}N_2$$

$$CO + 0·5O_2 + 1·88N_2 \rightarrow CO_2 + 1·88N_2$$

$$\text{Actual A/F ratio} = \text{Stoichiometric A/F ratio} \times \frac{100}{110}$$

$$= 0·91 \times \text{Stoichiometric A/F ratio}$$

Therefore, for the actual conditions

$$CO + 0·91(0·5O_2 + 1·88N_2) = aCO_2 + bCO + 0·91 \times 1·88N_2$$

With dissociation there will be some break up of CO_2 giving CO and O_2 in the products such that $CO + 0·91(0·5O_2 + 1·88N_2) = aCO_2 + bCO + cO_2 + 0·91 \times 1·88N_2$ (here a and b have different values from the previous equation).

The question states that $b = 0·228$

$$\therefore CO + 0·455\,O_2 + 1·71N_2 \rightarrow aCO_2 + 0·228CO + cO_2 + 1·71N_2$$

$\therefore$ for C balance $\quad a + 0·228 = 1 \quad \therefore a = 0·772$

$\therefore$ for O balance $\quad 1 + 0·91 = 2a + 0·228 + 2c$

$$1·91 = 2 \times 0·772 + 0·228 + 2c$$

$$\therefore c = 0·069$$

For the reaction $CO + \frac{1}{2}O_2 \rightleftharpoons CO_2$,

$$K = \frac{p_{CO_2}}{p_{CO} p_{O_2}^{1/2}}$$

and

$$p_{CO_2} = \frac{a}{n_2} p_2, \quad p_{CO} = \frac{b}{n_2} p_2, \quad p_{O_2} = \frac{c}{n_2} p_2$$

$$\therefore K = \frac{a}{b} \frac{1}{\sqrt{\dfrac{c}{n_2} p_2}} \tag{1}$$

where p_2 = the total pressure at the required temperature

and n_2 = total kilomoles of products = $a + b + c + 1 \cdot 71$
$$= 0 \cdot 772 + 0 \cdot 228 + 0 \cdot 069 + 1 \cdot 71$$
$$= 2 \cdot 779 \text{ kmol}$$

At ignition

$$p_1 = \frac{8 \cdot 28}{1 \cdot 013} = 8 \cdot 175 \text{ atm} \quad \text{and} \quad T_2 = 273 + 282$$
$$= 555 \text{ K}$$

$$p_1 V = n_1 R_0 T_1 \quad \text{and} \quad p_2 V = n_2 R_0 T_2 \quad \text{and} \quad V = \text{constant}$$

$$\therefore p_2 = p_1 \frac{n_2}{n_1} \frac{T_2}{T_1}$$

where n_1 = kilomoles of reactants = $1 + 0 \cdot 455 + 1 \cdot 71 = 3 \cdot 165$

$$\therefore p_2 = 8 \cdot 175 \times \frac{2 \cdot 779}{3 \cdot 165} \times \frac{2950}{555} = 38 \cdot 1 \text{ atm}$$

assuming that

$$T_2 = 273 + 2677 = 2950 \text{ K}$$

substituting in equation (1),

$$K = \frac{0 \cdot 772}{0 \cdot 228} \frac{1}{\sqrt{\left(\dfrac{0 \cdot 069}{2 \cdot 779} \times 38 \cdot 1 \right)}} = 3 \cdot 48 \text{ atm}^{-1/2}$$

From fig. 15.6 or tables, reference 15.5, it is seen that $K = 3 \cdot 5$ for this reaction at 2950 K showing the assumed value to be true. The corresponding pressure was calculated at 38·1 atm.

If the pressure is now doubled it becomes 76·2 atm and the

previous combustion equation must be used to find the new molar proportions

$$CO + 0.455\,O_2 + 1.71\,N_2 \rightarrow aCO_2 + bCO + cO_2 + 1.71\,N_2$$

for C balance $\quad a+b = 1 \quad \therefore b = 1-a$

for O balance $\quad 1+0.91 = 2a+b+2c$

$$1.91 = 2a+(1-a)+2c$$

$$\therefore c = 0.455-0.5a$$

and

$$n_2 = a+b+c+1.71 = 1+0.455-0.5a+1.71 = 3.165-0.5a$$

$$\therefore K = \frac{a}{b}\frac{1}{\sqrt{\dfrac{c}{n_2}}\,p_2}$$

and as the temperature is 2950 K then $K = 3.5\ \text{atm}^{-1/2}$

$$\therefore 3.5 = \frac{a}{1-a}\frac{1}{\sqrt{\left\{\left(\dfrac{0.455-0.5a}{3.165-0.5a}\right)\times 76.2\right\}}}$$

which on simplification gives

$$a^3 - 2.9a^2 + 2.82a - 0.91 = 0$$

This can be solved sufficiently accurately to indicate $0.99 < a < 1$, and hence since $b = 1-a$ there is a negligible amount of CO at this condition.

Example 15.11

A mixture of Heptane (C_7H_{16}) and air which is 10% rich is initially at a pressure of 1 atmosphere and temperature 100°C, and is compressed through a volumetric ratio of 6 to 1. It is ignited and adiabatic combustion proceeds at constant volume. The maximum temperature reached is 2627°C and at this temperature the equilibrium constants are

$$\frac{p_{H_2O}\,p_{CO}}{p_{CO_2}\,p_{H_2}} = 6.72 \quad \text{and} \quad \frac{p_{CO}^2\,p_{O_2}}{p_{CO_2}^2} = 0.054\ \text{atm}$$

If the constituents of the gas are CO_2, CO, H_2O, H_2, O_2 and N_2 show that approximately 30.2% of the carbon has burned incompletely. Pressures are measured in atmospheres.

The processes are shown in fig. 15.7 and for a 10% rich mixture

$$C_7H_{16} + \frac{10}{11}(11\,O_2 + 41 \cdot 36\,N_2)$$

$$\rightarrow aCO_2 + bCO + cH_2O + dH_2 + eO_2 + 37 \cdot 6N_2$$

$$C_7H_{16} + 10O_2 + 37 \cdot 6N_2 \rightarrow aCO_2 + bCO + cH_2O + dH_2 + eO_2 + 37 \cdot 6N_2$$

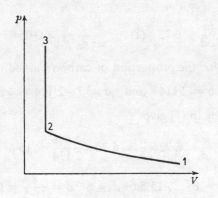

Fig. 15.7

and by an atomic balance

$$\text{for } O_2 \qquad a + \frac{b}{2} + \frac{c}{2} + e = 10$$

$$\text{for } C \qquad a + b = 7$$

$$\text{for } H_2 \qquad c + d = 8$$

number of kmol initially, $n_1 = 1 + 10 + 37 \cdot 6 = 48 \cdot 6$; number of kmol finally, $n_3 = a + b + c + d + e + 37 \cdot 6$; and the characteristic equation for initial and final conditions can be written

$$p_1 V_1 = n_1 R_0 T_1 \quad \text{and} \quad p_3 V_3 = n_3 R_0 T_3$$

and combining these gives

$$\frac{p_3}{n_3} = \frac{p_1}{n_1} \frac{V_1}{V_3} \frac{T_3}{T_1} = \frac{1}{48 \cdot 6} \times \frac{6}{1} \times \frac{2900}{373} = 0 \cdot 958 \text{ atm/kmol}$$

(where $T_1 = 273 + 100 = 373$ K, and $T_2 = 273 + 2627 = 2900$ K)

$$p_{CO_2} = \frac{a}{n_3}p_3, \quad p_{CO} = \frac{b}{n_3}p_3,$$

$$p_{H_2O} = \frac{c}{n_3}p_3, \quad p_{H_2} = \frac{d}{n_3}p_3, \quad p_{O_2} = \frac{e}{n_3}p_3;$$

$$\frac{p_{CO}p_{H_2O}}{p_{CO_2}p_{H_2}} = 6\cdot72 \qquad \frac{p_{CO}^2 p_{O_2}}{p_{CO_2}^2} = 0\cdot054$$

$$\therefore \frac{bc}{ad} = 6\cdot72 \quad (1) \qquad \frac{b^2}{a^2}\frac{e}{n_3}p_3 = 0\cdot054 \quad (2)$$

Put $b/7 = 0\cdot302$, the proportion of carbon burned incompletely

$$\therefore b = 2\cdot114 \quad \text{and} \quad a = 7 - 2\cdot114 = 4\cdot886$$

$\therefore$ substituting in (1) gives

$$\frac{c}{d} = 6\cdot72 \times \frac{a}{b} = 6\cdot72 \times \frac{4\cdot886}{2\cdot114} = 15\cdot5$$

$$c + d = 8 \quad \therefore 15\cdot5d + d = 8 \quad d = \frac{8}{16\cdot5} = 0\cdot485$$

$$c = 15\cdot5 \times 0\cdot485 = 7\cdot515$$

$$e = 10 - a - \frac{b}{2} - \frac{c}{2} = 10 - 4\cdot886 - \frac{2\cdot114}{2} - \frac{7\cdot515}{2}$$

$$= 0\cdot297$$

substituting in (2) gives

$$\frac{b^2}{a^2}\frac{e}{n_3}p_3 = \left(\frac{2\cdot114}{4\cdot886}\right)^2 \times 0\cdot297 \times 0\cdot958 = 0\cdot0531 \text{ atm}$$

which gives sufficient agreement to the $0\cdot054$ quoted showing that approximately $30\cdot2\%$ of carbon was burned to CO.

15.9 Internal energy and enthalpy of combustion

Previous consideration of the combustion process has not included the energy released during the process and final temperatures attained. It is evident however that such a process must obey the First Law of Thermodynamics. Applications of this law to other processes have been for pure substances, or those that can be considered to be so, with the stipulation that their thermodynamic state is identified by two independent properties. In the type of process now considered there is the potential chemical energy of the fuel to be included which is released during the change from reactants to products.

It is an experimental fact that the energy released on the complete combustion of unit mass of a fuel depends on the temperature at which the process is carried out. Thus such quantities quoted are related to temperature. It will be shown that if the energy release is known for a fuel at one temperature that at other temperatures can be calculated.

The combustion process is defined as taking place from reactants at a state identified by the reference temperature T_0 and another property, either pressure or volume, to products at the same state. If the process is carried out at constant volume then the non-flow energy equation, $Q = (U_2 - U_1) + W$, can be applied to give

$$Q = (U_{P_0} - U_{R_0}),\qquad\qquad 15.14$$

(where $W = 0$ for constant volume combustion, $U_1 = U_{R_0}$ the internal energy of the reactants which is a mixture of fuel and air at T_0, $U_2 = U_{P_0}$ the internal energy of the products of combustion at T_0.)

The change in internal energy does not depend on the path between the two states but only on the initial and final values and is given by the quantity Q, the heat transferred to the surroundings during the process. This is illustrated in fig. 15.8 and also the property diagram of fig. 15.9.

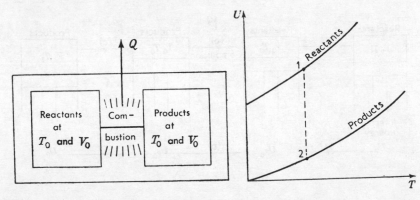

Fig. 15.8   Fig. 15.9

The heat so transferred is called the *internal energy of combustion at T_0* or the *constant volume heat of combustion* and is denoted by ΔU_0

i.e.
$$\Delta U_0 = U_{P_0} - U_{R_0} \qquad\qquad 15.15$$

As the internal energy of the reactants includes the potential chemical energy, and since heat is transferred from the system, it is evident that as defined ΔU_0 is a negative quantity.

For real constant volume combustion processes the initial and final temperatures will be different from the reference temperature T_0. For analytical purposes the change in internal energy between reactants at state 1 to products at state 2 can be considered in three stages

(a) the change for the reactants from state 1 to the reference temperature T_0

(b) the constant volume combustion process from reactants to products at T_0

(c) the change for the products from T_0 to state 2.

The complete process can be conceived as taking place in a piston-cylinder device as indicated in fig. 15.10.

Thus the change in internal energy between states 1 and 2, $(U_2 - U_1)$ can be written more explicitly as $(U_{P_2} - U_{R_1})$ to show the chemical

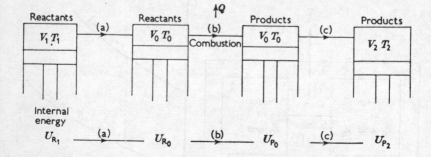

Fig. 15.10

change involved and this can be further expanded for analytical purposes

$$U_{P_2} - U_{R_1} = (U_{P_2} - U_{P_0}) + (U_{P_0} - U_{R_0}) + (U_{R_0} - U_{R_1})$$
$$\therefore U_{P_2} - U_{R_1} = (U_{P_2} - U_{P_0}) + \Delta U_0 + (U_{R_0} - U_{R_1}) \qquad 15.16$$

$$\underset{\text{Products}}{} \qquad \underset{\text{Reactants}}{}$$
$$\underset{\text{(c)}}{} \qquad \underset{\text{(b)}}{} \qquad \underset{\text{(a)}}{}$$

The non-flow energy equation applied to a process involving combustion and work gives

$$Q = (U_{P_2} - U_{R_1}) + W$$

It can be seen that the expression for $U_{P_2} - U_{R_1}$ has been conveniently split up to give a term $U_{P_2} - U_{P_0}$ which requires the product mixture to be regarded as a single substance or a summation of single substance constituents, a similar term for the reactants and the quantity ΔU_0 previously defined. The values of internal energy for the constituents of the mixtures remain to be determined. These are functions of temperature and the most accurate method is to use tabulated values such as those of reference 15.5. It is permissible to calculate changes in internal energy assuming the gaseous constituents to be perfect using an average value of c_v for the temperature range involved. If the temperature range is T_1 to T_2 then the value of c_v at $T = (T_1 + T_2)/2$ can be used, this assumes a linear change in c_v with temperature, but if the temperature range is large the result may not be accurate enough and tabulated values of the properties are required. The tables available take the values of u and h as zero at the absolute zero of temperature and do not include chemical changes and so are only used for processes which do not include chemical changes. The analysis presented has reduced the procedure to this set of circumstances with chemical consistency for the reactants and products.

The changes in internal energy for the reactants $(U_{R_0} - U_{R_1})$ and for the products $(U_{P_2} - U_{P_0})$ can be calculated from the following expressions

$$U_{R_0} - U_{R_1} = \sum_R n_i (u_{i_0} - u_{i_1}) \qquad 15.17$$

(where $\sum_R$ denotes the summation for all the constituents of the reactants denoted by i, u_i is the tabulated value of the internal energy for the constituent at the required temperature T_0 or T_1 in heat units per kmol and n_i is the number of kmol of the constituent).

Alternatively if a mass base is used for the tabulated values or calculation

$$U_{R_0} - U_{R_1} = \sum_R m_i(u_{i_0} - u_{i_1}) \qquad 15.18$$

where u_i is now the internal energy per unit mass.

In terms of the specific heats which are average values for the required temperature range

$$U_{R_0} - U_{R_1} = \sum_R m_i c_{v_i}(T_0 - T_1) = (T_0 - T_1) \sum_R m_i c_{v_i} \qquad 15.19$$

and similar expressions for the products are

$$U_{P_2} - U_{P_0} = \sum_P n_i(u_{i_2} - u_{i_0}) \quad \text{to a kmol basis} \qquad 15.20$$

$$U_{P_2} - U_{P_0} = \sum_P m_i(u_{i_2} - u_{i_0}) \quad \text{to a mass basis} \qquad 15.21$$

$$U_{P_2} - U_{P_0} = \sum_P m_i c_{v_i}(T_2 - T_0) = (T_2 - T_0) \sum_P m_i c_{v_i} \qquad 15.22$$

Note that $n_i C_{v_i} = m_i c_{v_i}$.

The process has been analysed on the basis of a non flow process which involves combustion at constant volume. A similar analysis can be made for a steady flow or constant pressure combustion process in which the changes in enthalpy are important

$$H_{P_2} - H_{R_1} = \underbrace{(H_{P_2} - H_{P_0})}_{\text{Products}} + \Delta H_0 + \underbrace{(H_{R_0} - H_{R_1})}_{\text{Reactants}} \qquad 15.23$$

where $\Delta H_0 = $ *enthalpy of combustion* at T_0 or the *constant pressure heat of combustion* at T_0 and

$$\Delta H_0 = H_{P_0} - H_{R_0}, \text{ and is always negative.} \qquad 15.24$$

The expressions for the change in enthalpy of reactants and products are

$$H_{R_0} - H_{R_1} = \sum_R n_i(h_{i_0} - h_{i_1}) \qquad \text{to a kmol basis} \qquad 15.25$$

$$H_{R_0} - H_{R_1} = \sum_R m_i(h_{i_0} - h_{i_1}) \qquad \text{to a mass basis} \qquad 15.26$$

and if mean specific heats are used

$$H_{R_0} - H_{R_1} = \sum_R m_i c_{p_i}(T_0 - T_1) = (T_0 - T_1 \sum_R m_i c_{p_i} \qquad 15.27$$

$$H_{P_2} - H_{P_0} = \sum_P n_i(h_{i_2} - h_{i_0}) \qquad \text{to a kmol basis} \qquad 15.28$$

$$H_{P_2} - H_{P_0} = \sum_P m_i(h_{i_2} - h_{i_0}) \qquad \text{to a mass basis} \qquad 15.29$$

and if mean specific heats are used

$$H_{P_2} - H_{P_0} = \sum_P m_i c_{p_i}(T_2 - T_0) = (T_2 - T_0)\sum_P m_i c_{p_i} \qquad 15.30$$

Note that $n_i C_{p_i} = m_i c_{p_i}$,

Values of ΔH_0 are quoted in tables such as reference 15.6 as:

Reaction	ΔH_0 kJ/kmol at 25°C (298·15 K)
$C(sol) + O_2 \rightarrow CO_2$	$-393\,520$
$CO + \frac{1}{2}O_2 \rightarrow CO_2$	$-282\,990$
$C_6H_6(vap) + 7\frac{1}{2}O_2 \rightarrow 6CO_2 + 3H_2O(liq)$	$-3\,301\,397$

It will be noted that the state of the fuel is given if it is solid (sol), liquid (liq) or vapour (vap) if this is required. If H_2O is a product of the combustion then it is necessary to know the state, liquid or vapour, at the end of the process by which ΔH_0 was determined. If ΔH_0 is known for a particular fuel with the H_2O formed in the liquid state the value of ΔH_0 with the H_2O in the vapour state can be calculated. ΔH_0 may be quoted per kmol of fuel or per unit mass.

Example 15.12

ΔH_0 for Benzene vapour (C_6H_6), at 25°C is $-3\,301\,397$ kJ/kmol with the H_2O in the liquid phase. Calculate ΔH_0 for the H_2O in the vapour phase.

If the H_2O remains as a vapour the heat transferred to the surroundings will be less than that when the vapour condenses by the amount due to the change in enthalpy of the vapour during condensation at the reference temperature.

$$\Delta H_0(vap) = \Delta H_0(liq) + m_s h_{fg_0}$$

where m_s = mass of H_2O formed

h_{fg_0} = change in enthalpy of steam between saturated liquid and saturated vapour at the reference temperature T_0

= 2441·8 kJ/kg at 25°C

for the reaction

$$C_6H_6 + 7\tfrac{1}{2}O_2 \rightarrow 6CO_2 + 3H_2O$$

3 kmol of H_2O are formed on combustion of 1 kmol of C_6H_6; 3 kmol of $H_2O = 3 \times 18 = 54$ kg H_2O

$$\Delta H_0(\text{vap}) = -3\,301\,397 + 54 \times 2441 \cdot 8$$

$$= -3\,169\,540 \text{ kJ/kmol}$$

In the equations for the change in internal energy and enthalpy for reactants and products, and for ΔU_0 and ΔH_0, if a change in state takes place (e.g. from liquid fuel to vapour), a term describing this process must be included.

Nothing has been said of the air-fuel ratio for combustion during the determination of ΔU_0 or ΔH_0. Consideration will show that this does not matter provided there is sufficient air to ensure complete combustion. Excess oxygen, like the nitrogen present, starts and finishes at the reference temperature T_0 and so suffers no change in internal energy or enthalpy, thus not affecting ΔU_0 or ΔH_0.

From the definition of the enthalpy of a perfect gas

$$H = U + pV = U + nR_0T$$

So if we are concerned only with gaseous mixtures in the reaction then for products and reactants

$$H_{P_0} = U_{P_0} + n_P R_0 T_0 \quad \text{and} \quad H_{R_0} = U_{R_0} + n_R R_0 T_0$$

where n_P and n_R are the kilomoles of products and reactants respectively and the temperature is the reference temperature T_0.

Then, using equations 15.15 and 15.24 we have:

$$\Delta H_0 = \Delta U_0 + (n_P - n_R)R_0T_0 \qquad 15.31$$

If there is no change in the number of kilomoles during the reaction, or if the reference temperature is absolute zero, then ΔH_0 and ΔU_0 will be equal.

Example 15.13

Calculate ΔU_0 in kJ/kg for the combustion of Benzene (C_6H_6) vapour at 25°C given that $\Delta H_0 = -3\,169\,540$ kJ/kmol and the H_2O is in the vapour phase.

The combustion equation is

$$C_6H_6 + 7\tfrac{1}{2}O_2 \rightarrow 6CO_2 + 3H_2O(\text{vap})$$

$$n_R = 1 + 7 \cdot 5 = 8 \cdot 5, \quad n_P = 6 + 3 = 9$$

From equation 15.31

$$\Delta U_0 = \Delta H_0 - (n_P - n_R)R_0 T_0$$
$$= -3\,169\,540 - (9 - 8\cdot 5) \times 8\cdot 314 \times 298$$

(where $T_0 = 273 + 25 = 298$ K)

$$= -3\,169\,540 - 1239 = -3\,170\,779 \text{ kJ/kmol}$$

(note that ΔU_0 is negligibly different from ΔH_0)

$$1 \text{ kmol of } C_6H_6 = 6 \times 12 + 6 \times 1 = 78 \text{ kg}$$

$$\therefore \quad \Delta U_0 = -\frac{3\,170\,779}{78} = -40\,651 \text{ kJ/kg}$$

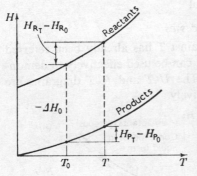

Fig. 15.11

Change in reference temperature

It has already been mentioned that the internal energy and enthalpy of combustion depend on the temperature at which the reaction occurs. This is due to the change in enthalpy and internal energy of the reactants and products with temperature.

It can be seen from the property diagram of fig. 15.11 that the enthalpy of combustion at temperature T, ΔH_T can be obtained from ΔH_0 at T_0 by the relationship

$$-\Delta H_T = -\Delta H_0 + (H_{R_T} - H_{R_0}) - (H_{P_T} - H_{P_0}) \qquad 15.32$$

where $H_{R_T} - H_{R_0} =$ increase in enthalpy of the reactants from T_0 to T

and $\quad H_{P_T} - H_{P_0} =$ increase in enthalpy of the products from T_0 to T

Example 15.14

ΔH_0 for carbon monoxide at 60°C is given as $-285\,200$ kJ/kmol. Calculate ΔH_0 at 2500°C given that the enthalpies of the gases concerned in kJ/kmol, are as follows

Gas	60°C	2500°C
CO	9705	94 080
O	9696	99 790
CO	10 760	149 100

The reaction equation is $CO + \frac{1}{2}O_2 \rightarrow CO_2$

$$H_{R_0} = 1 \times 9705 + \tfrac{1}{2} \times 9696 = 14\,553 \text{ kJ}$$

$$H_{R_T} = 1 \times 94\,080 + \tfrac{1}{2} \times 99\,790 = 143\,975 \text{ kJ}$$

$$H_{P_0} = 1 \times 10\,760 \text{ kJ} \qquad H_{P_T} = 149\,100 \text{ kJ}$$

Then using equation 15.32,

$$-\Delta H_T = -\Delta H_0 + (H_{R_T} - H_{R_0}) - (H_{P_T} - H_{P_0})$$

$$-\Delta H_T = 285\,200 + (143\,975 - 14\,553) - (149\,100 - 10\,760)$$

$$= 285\,200 + 129\,422 - 138\,340 = 276\,282$$

$$\therefore \ \Delta H_T = -276\,282 \text{ kJ/kmol}$$

The property diagram and real processes

The property diagram of U against T has already been referred to and is shown in fig. 15.12 and it can be used effectively to demonstrate real processes of interest. The U–T and H–T diagrams are shown in fig. 15.12a and b respectively.

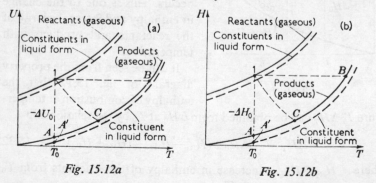

Fig. 15.12a   *Fig. 15.12b*

The solid lines indicate the property variations with T if the constituents are gaseous. If reactants or products contain a liquid component then the property lines will be modified as shown by the dotted lines. It can be seen by inspection that the effect of condensation of the H_2O in the products is to increase ΔU_0 and ΔH_0.

In processes 1–A, $T_A = T_0$ and there is a maximum energy transfer to the surroundings ΔU_T or ΔH_T.

In processes 1–B the internal energy, or enthalpy, initially and finally is the same so that the increase in temperature is a maximum and the combustion process is adiabatic.

In 1–C the processes are general with heat transfer and possibly work transfer.

Process 1–A' in fig. 15.12a corresponds to the constant volume Bomb Calorimeter test and in fig. 15.12b, 1–A' corresponds to the

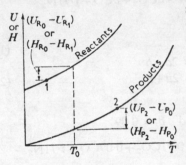

Fig. 15.13

steady flow combustion Boy's Calorimeter test. Both of these tests are discussed later in Section 15.10.

In a general non-flow or steady flow process the initial state (1) and final state (2) will be different and neither will be at the reference temperature T_0. The quantities of interest are $U_2 - U_1$, and $H_2 - H_1$ for the respective processes. It can be seen readily from fig. 15.13 that

$$U_2 - U_1 = -(U_1 - U_2) = -\{-\Delta U_0 - (U_{R_0} - U_{R_1}) - (U_{P_2} - U_{P_0})\}$$
$$U_2 - U_1 = U_{P_2} - U_{R_1} = (U_{P_2} - U_{P_0}) + \Delta U_0 + (U_{R_0} - U_{R_1})$$

This is equation 15.16 repeated.

A corresponding expression is obtained for $H_2 - H_1$.

The principles dealt with in this section will now be applied to typical problems.

Example 15.15

A combustible mixture of carbon monoxide and air which is 10% rich is compressed to a pressure of 8·28 bar and a temperature of 282°C. The mixture is ignited and combustion occurs adiabatically at constant volume. Calculate the maximum temperature and pressure reached assuming that no dissociation takes place. Take average specific heats for the temperature ranges considered. At the reference temperature of 25°C, ΔH_0 for CO is $-282\,990$ kJ/kmol.

(Note that this example was the data of example 15.10.)

For stoichiometric conditions

$$CO + \tfrac{1}{2}O_2 + 3 \cdot 76 \times \tfrac{1}{2}N_2 \rightarrow CO_2 + 3 \cdot 76 \times \tfrac{1}{2}N_2$$

$$CO + 0 \cdot 5O_2 + 1 \cdot 88N_2 \rightarrow CO_2 + 1 \cdot 88N_2$$

and for a mixture strength of 110%

$$\text{Actual A/F ratio} = \text{Stoichiometric A/F ratio} \times \frac{100}{110}$$

$$= 0 \cdot 91 \text{ Stoichiometric A/F ratio}$$

Therefore, for the actual conditions

$$CO + 0 \cdot 91(0 \cdot 5O_2 + 1 \cdot 88N_2) \rightarrow aCO_2 + bCO + 1 \cdot 71N_2$$

and by an atomic balance

for C $1 = a + b$ (1)

for O $1 + (0 \cdot 91 \times 0 \cdot 5 \times 2) = 2a + b$

$$1 \cdot 91 = 2a + b \qquad (2)$$

From (1) and (2)

$$a = 0 \cdot 91 \quad \text{and} \quad b = 0 \cdot 09$$

$$\therefore \ CO + 0 \cdot 455O_2 + 1 \cdot 71N_2 \rightarrow 0 \cdot 91CO_2 + 0 \cdot 09CO + 1 \cdot 71N_2$$

$$\Delta H_0 = \Delta U_0 + (n_P - n_R)R_0T$$

Also for the reaction

$$CO + \tfrac{1}{2}O_2 \rightarrow CO_2$$

$$n_R = 1 + 0 \cdot 5 = 1 \cdot 5 \quad \text{and} \quad n_P = 1$$

$$\therefore \ \Delta H_0 = \Delta H_0 - (n_P - n_R)R_0T_0 = -282\,990 - (1 - 1 \cdot 5) \times 8 \cdot 314 \times 298$$

$$= -282\,990 + 1239 = -281\,751 \text{ kJ/kmol}$$

The non-flow process is defined by

$$Q = W + (U_2 - U_1)$$

Also $Q = 0$; $W = 0$ at constant volume; $U_1 = U_R$, $U_2 = U_{P_2}$

$$\therefore \ 0 = 0 + (U_{P_2} - U_{R_1})$$

$$0 = (U_{P_2} - U_{P_0}) + (U_{P_0} - U_{R_0}) + (U_{R_0} - U_{R_1})$$

$$0 = (U_{P_2} - U_{P_0}) + \Delta U_0 + (U_{R_0} - U_{R_1}) \qquad (3)$$

$$U_{P_2} - U_{P_0} = (T_2 - T_0) \sum_P n_i C_{v_i}$$
$$U_{R_0} - U_{R_1} = (T_0 - T_1) \sum_R n_i C_{v_i}$$

$$(4)$$

The temperature range for the reactants is known, from $T_0 = 298$ K to $T_1 = 273 + 282 = 555$ K and average value of C_{v_i} at $T = \dfrac{298 + 555}{2} = 426 \cdot 5$ K are found from tables of C_v, reference 15.6, and calculated in kJ/kmol K as

$$CO \ 21 \cdot 35; \quad O_2 \ 22 \cdot 23; \quad N_2 \ 21 \cdot 14$$

The temperature range for the products is unknown as T_2 is to be determined. A value of T_2 is assumed and in this case can be taken as a high value, say 3000 K. The values of C_{v_i} are then calculated in kJ/kmol K for an average temperature of $\dfrac{298 + 3000}{2} = 1649$ K as

$$CO_2 \ 45 \cdot 93; \quad CO \ 25 \cdot 33; \quad N_2 \ 25 \cdot 13$$

Hence substituting in equation (3) above using equations (4) gives, for the combustion of $1 - 0 \cdot 09 = 0 \cdot 91$ kmol of CO

$$0 = (T_2 - 298)(0 \cdot 91 \times 45 \cdot 93 + 0 \cdot 09 \times 25 \cdot 33 + 1 \cdot 71 \times 25 \cdot 13) - 0 \cdot 91$$
$$\times 281 \, 751 - (555 - 298)(1 \times 21 \cdot 35 + 0 \cdot 455 \times 22 \cdot 23 + 1 \cdot 71 \times 21 \cdot 14)$$

$$0 = (T_2 - 298) \times 87 \cdot 02 - 256 \, 393 - 257 \times 67 \cdot 71$$

$$\therefore \ T_2 = \frac{256 \, 393 + 17 \, 400}{87 \cdot 02} + 298 = 3444 \text{ K} (3171^\circ \text{C})$$

The calculation could be repeated using this value of T_2 to give a more accurate answer.

The temperature calculated in example 15.15 is that which would be obtained by adiabatic combustion of the mixture. This temperature would not be attained in practice due to the effect of dissociation, this was discussed in Section 15.8 and the inclusion of this effect is illustrated in the next example.

Example 15.16

Calculate the final temperature for the mixture defined in example 15.15 including the effect of dissociation.

The proportions of the constituents depend on the temperature so the combustion equation is written as

$$CO + 0 \cdot 455 O_2 + 1 \cdot 71 N_2 \rightarrow a CO_2 + b CO + c O_2 + 1 \cdot 71 N_2$$

and b and c can be expressed in terms of a by an atomic balance.

for C $\quad\quad 1 = a+b \quad \therefore b = 1-a$

for O $\quad\quad 1+0\cdot455\times2 = 2a+b+2c$

$\quad\quad\quad\quad\quad\quad\quad 1\cdot91 = 2a+b+2c$

$\quad\quad\quad\quad\quad\quad\quad\quad 2c = 1\cdot91-2a-(1-a)$

$\quad\quad\quad\quad\quad\quad \therefore c = 0\cdot455-0\cdot5a$

$\therefore CO+0\cdot455O_2+1\cdot71N_2$

$\quad\quad \rightarrow aCO_2+(1-a)CO+(0\cdot455-0\cdot5a)O_2+1\cdot71N_2$

The energy equation can be written as previously in terms of a and using the value for the reactants calculated in example 15.15.

$$0 = (T_2-298)\{aC_{vCO_2}+(1-a)C_{vCO}$$
$$+(0\cdot455-0\cdot5a)C_{vO_2}+1\cdot71C_{vN_2}\}-a\times281\,751-17\,400$$

$$\therefore a\left(C_{vCO_2}-C_{vCO}-0\cdot5C_{vO_2}-\frac{281\,751}{(T_2-298)}\right)$$

$$= \frac{17\,400}{(T_2-298)}-(C_{vCO}+0\cdot455C_{vO_2}+1\cdot71C_{vN_2}) \quad (1)$$

A second equation is derived using the definition of the equilibrium constant K for the reaction $CO+\tfrac{1}{2}O_2\rightleftharpoons CO_2$ as

$$K = \frac{p_{CO_2}}{p_{CO}p_{O_2}^{1/2}} \text{ atm}^{-1/2}$$

and the partial pressures are given by

$$p_{CO_2} = \frac{a}{n_2}p_2; \quad p_{CO} = \frac{1-a}{n_2}p_2; \quad p_{O_2} = \frac{(0\cdot455-0\cdot5a)}{n_2}p_2$$

p_2 = the pressure of the mixture at T_2.

For the constant volume process 1–2

$$p_1V = n_1R_0T_1 \quad \text{and} \quad p_2V = n_2R_0T_2$$

which combine to give

$$\frac{p_2}{n_2} = \frac{T_2}{T_1}\frac{p_1}{n_1}$$

Substituting in the expression for K gives

$$K = \frac{a}{1-a}\frac{1}{\sqrt{\left(\frac{(0\cdot455-0\cdot5a)p_2}{n_2}\right)}}$$

and in this for $\frac{p_2}{n_2}$ gives

$$K = \frac{a}{1-a} \frac{1}{\sqrt{\left((0\cdot455-0\cdot5a)\dfrac{T_2 p_1}{T_1 n_1}\right)}}$$

$$T_1 = 555 \text{ K}, \quad p_1 = \frac{8\cdot28}{1\cdot013} = 8\cdot175 \text{ atm}$$

$$n_1 = 1+0\cdot455+1\cdot71 = 3\cdot165$$

$$\therefore \quad K = \frac{a}{1-a} \frac{1}{\sqrt{\left(\dfrac{0\cdot455-0\cdot5a)\times T_2 \times 8\cdot175}{555 \times 3\cdot165}\right)}} \tag{2}$$

Equations (1) and (2) have to be solved by a trial and error method using corresponding values of K and T_2 from tables and also values of C_v for the constituents.

The technique is to select a value of T_2, calculate a from (1) and obtain a calculated value of K called K_{calc} from (2). Agreement between K and K_{calc} gives the true value of T_2. If the results are plotted on a graph the solution will be simplified somewhat. This type of calculation is most conveniently carried out by digital computer. This was the method used in this case and the resulting curves are shown in fig. 15.14 with the final values $T_2 = 2949$ K, $K = 3\cdot5$ and $a = 0\cdot772$.

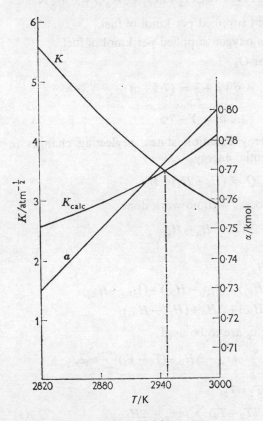

Fig. 15.14

Example 15.17

Benzene vapour (C_6H_6) is to be used as a fuel in a steady flow combustion process. The vapour is mixed with air both being at the reference temperature of 25°C and the final temperature of the products after adiabatic combustion is required to be 370°C. Calculate the air/fuel ratio required. Neglect changes in kinetic energy, and the effects of dissociation.

It was calculated in example 15.12 that for the combustion of benzene with the H_2O in the vapour phase $\Delta H_0 = -3\,169\,540$ kJ/kmol at $T_0 = 298$ K (25°C). The combustion equation, including the nitrogen which will act as a moderator and limit the maximum temperature reached, is given by

$$C_6H_6 + YO_2 + 3 \cdot 76 YN_2 \rightarrow 6CO_2 + aO_2 + 3H_2O + 3 \cdot 76 YN_2$$

Y = kmol of oxygen supplied per kmol of fuel.

a = kmol of excess oxygen supplied per kmol of fuel.

By atomic balance for O_2

$$Y = 6 + a + \frac{3}{2} = (7 \cdot 5 + a)$$

$$\therefore a = Y - 7 \cdot 5$$

The steady flow energy equation states, neglecting changes in kinetic energy and potential energy,

$$Q = (H_2 - H_1) + W$$

and for an adiabatic process with no work done

$$0 = H_2 - H_1$$

Therefore, in this case,

$$0 = H_{P_2} - H_{R_1}$$
$$0 = (H_{P_2} - H_{P_0}) + (H_{P_0} - H_{R_0}) + (H_{R_0} - H_{R_1})$$

i.e. $$0 = (H_{P_2} - H_{P_0}) + \Delta H_0 + (H_{R_0} - H_{R_1})$$

and if average values of c_p are to be used

$$0 = (T_2 - T_0) \sum_P m_i c_{p_i} + \Delta H_0 + (T_0 - T_1) \sum_R m_i c_p$$

Then, since in this case $T_1 = T_0$,

$$0 = (T_2 - T_0) \sum_P m_i c_{p_i} + \Delta H_0 \qquad (1)$$

The average temperature of the products is

$$\frac{25+360}{2} = 192\cdot5°C \ (= 465\cdot5\ K)$$

so values of c_{p_i} in kJ/kg K will be taken at 465 K from the tables of reference 15.6 as

$$CO_2\ 0\cdot9965;\quad O_2\ 0\cdot9629;\quad H_2O\ 1\cdot938;\quad N_2\ 1\cdot055$$

Considering the combustion of 1 kmol of C_6H_6 the mass proportion of the products are

$$m_{CO_2} = 6\times44 = 264\ kg$$
$$m_{O_2} = 32(Y-7\cdot5)\ kg$$
$$m_{H_2O} = 3\times18 = 54\ kg$$
$$m_{N_2} = 3\cdot76\ Y\times28 = 105\cdot3\ Y\ kg$$
$$\text{and } T_2 = 273+370 = 643\ K$$

Substituting in equation (1) gives

$$0 = (643-298)\{264\times0\cdot9965+32(Y-7\cdot5)\times0\cdot9629+54\times1\cdot938$$
$$+105\cdot3Y\times1\cdot055\}-3\ 169\ 540$$

$$\therefore\ 263\cdot4+30\cdot82(Y-7\cdot5)+104\cdot7+110\cdot6Y = \frac{3\ 169\ 540}{345} = 9187$$

$$\therefore\ 141\cdot42Y = 9050; \text{ i.e. } Y = 64\ \text{kmol of } O_2/\text{kmol fuel.}$$

Therefore,

$$\text{air supplied/kmol fuel} = \frac{32\times64}{0\cdot233} = 8790\ kg$$

$$1\ \text{kmol of } C_6H_6 = 6\times12+6 = 78\ kg$$

$$\therefore\ \text{A/F ratio} = \frac{8790}{78} = 112\cdot7 \text{ to } 1$$

15.10 Enthalpy of formation ΔH_{fo}

The combustion reaction is a particular kind of chemical reaction in which products are formed from reactants with the release or absorption of energy as heat is transferred to or from the surroundings. As some substances, for instance hydrocarbon fuels, may be many in number and complex in structure the heat of reaction (or combustion) may be calculated on the basis of known values of the enthalpy of formation, ΔH_{fo} of the constituents of the reactants and products at the reference temperature T_0.

The enthalpy of formation ΔH_{fo} is the increase in enthalpy when a compound is formed from its constituent elements in their natural form and in a standard state. Something needs to be said about the standard form. The normal forms of oxygen (O_2) and hydrogen (H_2) are gaseous, so ΔH_{fo} for these can be put equal to zero. The normal form of carbon (C) is graphite, so ΔH_{fo} for solid carbon is put to zero. Carbon in another form, e.g. diamond or gas, is not 'normal' and ΔH_{fo} is quoted. The standard state is 25°C, and 1 atm pressure, but it must be borne in mind that not all substances can exist in the natural form, e.g. H_2O cannot be a vapour at 1 atm and 25°C.

For calculation purposes, for a particular reaction

$$\Delta H_0 = \sum_P n_i \Delta H_{fo_i} - \sum_R n_i \Delta H_{fo_i} \qquad 15.33$$

Typical values of ΔH_{fo} are quoted for different substances in Table 15.6.

Table 15.6 Typical Heats of Formation ΔH_{fo} of Various Species at 25°C (298 K) in kJ/kmol

Substance	Formula	State	ΔH_{fo}
	O	gas	249 190
Oxygen	O_2	gas	0
Water	H_2O	liqu.	−285 820
Water	H_2O	vap	−241 830
Carbon	C	gas	714 990
Carbon	C	diamond	1900
Carbon	C	graphite	0
Carbon monoxide	CO	gas	−110 530
Carbon dioxide	CO_2	gas	−393 520
Methane	CH_4	gas	−74 870
Methyl alcohol	CH_3OH	vap.	−240 532
Ethyl alcohol	C_2H_5OH	vap.	−281 102
Ethane	C_2H_6	gas	−83 870
Ethene	C_2H_4	gas	51 780
Propane	C_3H_8	gas	−102 900
Butane	C_4H_{10}	gas	−125 000
Octane	C_8H_{18}	liq.	−247 600

Example 15.18

Calculate the enthalpy of combustion at 25°C of ethyl alcohol, C_2H_5OH, using the data of Table 15.6.

Using equation 15.33

$$\Delta H_0 = \sum_P n_i \Delta H_{fo_i} - \sum_R n_i \Delta H_{fo_i}$$

and the combustion equation

$$C_2H_5OH + 3O_2 \rightarrow 2CO_2 + 3H_2O$$

$$\sum_R n_i \Delta H_{fo_i} = 1 \times (-281\,102) + 3 \times 0 = -281\,102$$

$$\sum_P n_i \Delta H_{fo_i} = 2 \times (-393\,520) + 3 \times (-241\,830) = -1\,512\,530$$

$$\therefore \quad \Delta H_0 = -1\,512\,530 - (-281\,102)$$

$$= -1\,231\,428 \text{ kJ/kmol}$$

15.11 Calorific value of fuels

The quantities ΔH_0 and ΔU_0 are approximated to in fuel specifications by quantities called *calorific values* which are obtained by the combustion of the fuels in suitable apparatus. This may be of the constant volume type (e.g. bomb calorimeter) or constant pressure, steady flow type (e.g. Boys' calorimeter). Both of these are described in Section 15.13. The various calorific values quoted are discussed fully in reference 15.7 and definitions only will be given here.

(1) The energy transferred as heat to the surroundings (cooling water) per unit quantity of fuel when burned at constant volume with the H_2O product of combustion in the liquid phase is called the Higher (or Gross) Calorific Value (H.C.V.) at Constant Volume $Q_{gr,\,v}$. This approximates to $-\Delta U_0$ at a reference temperature of 25°C with the H_2O in the liquid phase.

If the H_2O products are in the vapour phase the energy released per unit quantity is called the Lower (or Net) Calorific Value (L.C.V.) at Constant Volume, $Q_{net,\,v}$. This approximates to $-\Delta U_0$ at 25°C with the H_2O in the vapour phase.

(2) The energy transferred as heat to the surroundings (cooling water) per unit quantity of fuel when burned at constant pressure with the H_2O products of combustion in the liquid phase is called the Higher (or Gross) Calorific Value (H.C.V.) at Constant Pressure, $Q_{gr,\,p}$. This approximates to $-\Delta H_0$ at a reference temperature of 25°C with the H_2O in the liquid phase.

If the H_2O products are in the vapour phase the energy released is called the Lower (or Net) Calorific Value (L.C.V.) at Constant

Pressure, $Q_{net. p}$. This approximates to $-\Delta H_0$ at 25°C with the H_2O in the vapour phase.

Contrary to the definition of ΔU_0 and ΔH_0 it is usual to quote calorific values as positive quantities. If $Q_{gr. p}$ and the fuel composition is known, the other quantities can be calculated. The above quantities are related as follows:

For the constant volume process

$$Q_{gr. v} = Q_{net. v} + m_c u_{fg} \qquad 15.34$$

For the constant pressure process

$$Q_{gr. p} = Q_{net. p} + m_c h_{fg} \qquad 15.35$$

where $m_c =$ mass of condensate per unit quantity of fuel.

$$u_{fg} = u_{fg} \text{ at 25°C for } H_2O = 2304{\cdot}4 \text{ kJ/kg}$$
$$h_{fg} = h_{fg} \text{ at 25°C for } H_2O = 2441{\cdot}8 \text{ kJ/kg}$$

The difference between u_{fg} and h_{fg} is small.

The calorific values differ from ΔH_0 and ΔU_0 due to the departure of experimental conditions from ideal with regard to temperatures of products and reactants and also heat transfer conditions. This topic is discussed further in Section 15.13.

15.12 Power plant thermal efficiency

The purpose of any power plant is the power output which should be obtained as economically as possible consistent with capital cost and running conditions. It is necessary to assess the overall performance of a plant for comparison purposes and an important criterion is the *overall thermal efficiency* η_0. This is defined as

$$\eta_0 = \frac{\text{Work output (W)}}{\text{Fuel energy supplied}}$$

It is necessary to decide on the denominator for this definition. It is desirable, if we consider a steady flow process, to relate the plant conditions to those for the steady flow calorimeter in which the products are cooled to atmospheric temperature giving $Q_{gr. p}$. However it is not possible, or desirable, to cool the products of combustion in a real plant to atmospheric temperature so there is a substantial energy loss to atmosphere. Complete cooling of the products would require large heat transfer surfaces which are expensive and the condensate produced would form corrosive acids. As achieving these conditions is not even attempted it seems that the use of $Q_{gr, p}$ is unsatisfactory and $Q_{net, p}$ is more appropriate and this is often preferred. It does not matter really for comparison purposes except that both values are in use making the definition of η_0 somewhat arbitrary as

$$\eta_0 = \frac{W}{Q_{gr, p}} \quad \text{or} \quad \frac{W}{Q_{net, p}} \qquad 15.36$$

It is necessary to state with the definition whether $Q_{gr, p}$ or $Q_{net, p}$ is being used.

In a plant it may be required to assess the boiler or steam generator performance only, in which case the *boiler efficiency*, given similarly in equation 7.18, is defined as

$$\text{Boiler Efficiency} = \frac{\text{Heat transferred to working fluid}}{\text{Fuel energy supplied}}$$

$$\therefore \eta_B = \frac{\text{Heat transferred to working fluid}}{Q_{gr, p} \quad \text{or} \quad Q_{net, p}} \qquad 15.37$$

Example 15.19

A medium size steam boiler required to supply a generator of output 25 000 kW has a performance specification as follows:

Steam output	31·6 kg/s
Steam pressure	60 bar
Steam temperature	500°C
Feed water temperature	100°C
Fuel, Natural gas	(96·5% CH_4, 0·5% C_2H_6, remainder incombustible)
Higher Calorific Value	38 700 kJ/m³ at 1·013 bar and 15°C.
Fuel consumption	2·85 m³/s

Calculate the boiler efficiency and the overall thermal efficiency based on the Lower Calorific Value of the fuel.

1 kmol of CH_4 burns to give 2 kmol H_2O

$\therefore$ 0·965 kmol of CH_4 burns to give $2 \times 0.965 \times 18 = 34.8$ kg H_2O

1 kmol of C_2H_6 burns to give 3 kmol H_2O

0.005 kmol of C_2H_6 burns to give $3 \times 0.005 \times 18 = 0.28$ kg H_2O

$\therefore$ 1 kmol of gas produces $34.8 + 0.28 = 35.08$ kg H_2O

1 kmol of gas is 23.65 m^3 at 1.013 bar and $15°C$

$\therefore$ steam formed per m^3 of gas $= \dfrac{35.08}{23.65} = 1.483$ kg

Using equation 15.35

$$Q_{gr, \, p} = Q_{net, \, p} + m_c h_{fg}$$

$\therefore Q_{net, \, p} = 38\,700 - 1.483 \times 2441.8 = 35\,074$ kJ/m^3

Heat to working fluid $= h_{supply\,steam} - h_{feed\,water}$
$$= 3421 - 419.1$$
$$= 3001.9 \text{ kJ/kg}$$

Using equation 15.37

$$\therefore \eta_B = \frac{3001.9 \times 31.6}{2.85 \times 35\,074} = 0.95 \ (95\%)$$

and using equation 15.36

$$\eta_0 = \frac{W}{Q_{net, \, p}} = \frac{25\,000}{2.85 \times 35\,074} = 0.25 \ (25\%)$$

15.13 Practical determination of calorific values

The methods for determining calorific value depend on the type of fuel; solid and liquid fuels are usually tested in a bomb calorimeter, and gaseous fuels in a continuous flow apparatus such as the Boys' calorimeter. The apparatus in every case is required to meet with a standard specification and the procedure to be adopted is also laid down.

In the bomb calorimeter combustion occurs at constant volume and is a non-flow process; in the Boys' calorimeter the gas is burnt continuously under steady flow conditions.

The bomb is a small stainless steel vessel in which a small mass of the fuel is held in a crucible (see fig. 15.15). If the fuel is solid, it is usually crushed, passed through a sieve, and then pressed into the form of a pellet in a special press. The size of pellet is estimated from the expected heat release, and is such that the temperature rise to be measured does not exceed 2 to 3 K. The pellet is ignited by fusing a piece of platinum or nichrome wire which is in contact with it. The wire forms part of an electric circuit which can be completed by a firing button, which is situated in a position remote from the bomb.

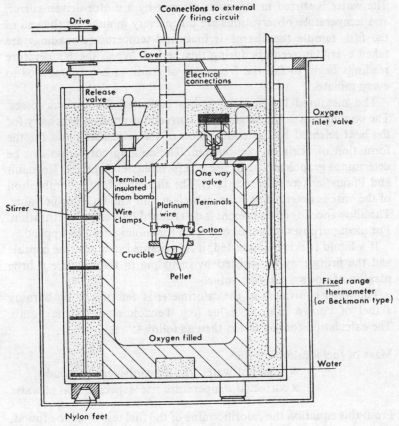

Fig. 15.15

With a special form of press the pellet can be formed with the fuse wire passing through it. This facilitates the firing, particularly with some of the more difficult fuels. The crucible carrying the pellet is located in the bomb, a small quantity of distilled water is put into the bomb to absorb the vapours formed by combustion and to ensure that the water vapour produced is condensed, and the top of the bomb is screwed down. Oxygen is then admitted slowly until the pressure is above 23 atmospheres. The bomb is located in the calorimeter and a measured quantity of water is poured into the calorimeter. The calorimeter is closed, the external connections to the circuit are made, and an accurate thermometer of the fixed range or the Beckman type is immersed to the proper depth in the water. The water is stirred in a regular manner by a motor-driven stirrer, and temperature observations are taken every minute. At the end of the fifth minute the charge is fired and temperature readings are taken every ten seconds during this period. When the temperature readings begin to fall the frequency of readings can be reduced to every minute.

The measured temperature rise is corrected for various losses. The cooling loss is the largest, but corrections are also necessary for the heat released by the combustion of the wire itself, and for the formation of acids on combustion. The cooling correction can be determined graphically or by use of a formula developed by Regnault and Pfaundler (see reference 15.1). The allowance for the combustion of the wire is determined from its weight and known calorific value. The allowance for acids present is determined by a chemical titration. For most purposes only the correction for cooling need be applied.

If a liquid fuel is being tested, it is contained in a gelatine capsule and the firing may be assisted by including in the crucible a little paraffin of known calorific value.

The water equivalent of the calorimeter is determined by burning a fuel of known calorific value (e.g. benzoic acid) in the bomb. The calculation for the test is then as follows:

Mass of fuel × calorific value

$$= \text{(mass of water + water equivalent of bomb)}$$
$$\times \text{ corrected temperature rise} \times \text{specific heat of water}$$

From this equation the calorific value of the fuel tested can be found.

Example 15.20

Table 15.7 gives the results of a bomb calorimeter test on a sample of coal. The mass of coal burned was 0·825 g and the total water equivalent of the apparatus was determined as 2500 g. Calculate the calorific value of the coal in kJ/kg. The temperature rise is to be corrected according to the formula by Regnault and Pfaundler, but no correction need be made for the acids formed.

Table 15.7

Pre Firing Period		Heating		Cooling Period	
Time/min	Temp/°C	Time/min	Temp/°C	Time/min	Temp/°C
0	25·730	t_1 6	27·340	t_n 10	27·880
1	25·732	t_2 7	27·880	11	27·878
2	25·734	t_3 8	27·883	12	27·876
3	25·736	t_4 9	27·885	13	27·874
4	25·738			14	27·872
t_0 5	25·740			15	27·870

From reference 15.1, the cooling correction is obtained as,

$$\text{Correction} = nv + \left(\frac{v_1 - v}{t_1 - t}\right)\left\{\sum_1^{(n-1)}(t) + \tfrac{1}{2}(t_0 + t_n) - nt\right\}$$

(where n is the number of minutes between the time of firing and the first reading after the temperature begins to fall from the maximum; v is the rate of fall of temperature per minute during the pre-firing period; v_1 is the rate of fall of temperature per minute after the maximum temperature; t and t_1 are the average temperatures during the pre-firing and final periods respectively; $\sum_1^{(n-1)}(t)$ is the sum of the readings during the period between firing and the start of cooling; $\tfrac{1}{2}(t_0 + t_n)$ is the mean of the temperature at the moment of firing and the first temperature after the rate of change of temperature becomes constant. The pre-firing and final periods are of the same duration).

In this example,

$$n = 10 - 5 = 5 \text{ min}$$

$$v = -\left(\frac{25 \cdot 740 - 25 \cdot 730}{5}\right) = -0 \cdot 002 \text{ K/min}$$

(the negative sign indicates that the temperature was rising in the pre-firing period).

$$v_1 = \frac{27 \cdot 880 - 27 \cdot 870}{5} = 0 \cdot 002 \text{ K/min}$$

$$t = 25 \cdot 735°C \quad \text{and} \quad t_1 = 27 \cdot 875°C$$

$$\sum_1^{(n-1)}(t) = 110 \cdot 988°C \quad \text{and} \quad \tfrac{1}{2}(t_0 + t_n) = \frac{25 \cdot 740 + 27 \cdot 880}{2} = 26 \cdot 81°C$$

Substituting the values in the equation gives

$$\text{Correction} = -5 \times 0.002 + \left(\frac{0.002 + 0.002}{27.875 - 25.735}\right)$$
$$\times (110.988 + 26.81 - 5 \times 25.735)$$

i.e. $$\text{Correction} = 0.00705 \text{ K}$$

Uncorrected temperature rise $= t_n - t_0$
$$= 27.880 - 25.740 = 2.14 \text{ K}$$

∴ Corrected temperature rise $= 2.14 + 0.00705 = 2.147 \text{ K}$

Then,

Heat released from 0.825 g coal $= 2.147 \times 2500 \times 4.187 \times 10^{-3}$
$$= 22.5 \text{ kJ}$$

i.e. Calorific value $= \dfrac{22.5}{0.825 \times 10^{-3}} = 27\,250 \text{ kJ/kg}$

i.e. Calorific value of fuel $= 27\,250 \text{ kJ/kg}$

For a gaseous fuel a continuous supply of the gas is metered and passed at constant pressure into the calorimeter, where it is burned in an ample supply of air (see fig. 15.16). The products of combustion are cooled as nearly as possible to the initial temperature of the reactants by a continuously circulating supply of cooling water. The rates of flow are adjusted to burn gas at about 2.36 l/min, with water circulating at 2100 l/min for a gas of declared H.C.V. 18 600 kJ/m³. The gas pressure and temperature are measured and the amount of gas burned is referred to 1013 mbar and 15°C. The temperature rise of the circulating water is measured, and the condensate from the products of combustion is collected. A test is carried out over a fixed time period. The water flow rate is measured and the condensate is weighed. Then we have

(Volume of fuel at 1013 m bar and 15°C) × (calorific value)

$\qquad$ = (mass of water circulated) × specific heat

$\qquad$ × (temperature rise of water)

The calorific value of the fuel is obtained in MJ/m³ of gas. The correct procedure for this determination is given in reference 15.4.

In both the bomb calorimeter test and the Boys' calorimeter test the steam formed on combustion is condensed, and so the heat released by the steam on condensing is transferred to the cooling

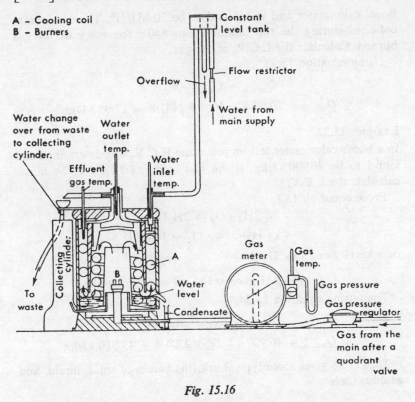

A – Cooling coil
B – Burners

Fig. 15.16

water. This condensation does not occur in practice and would be undesirable if it did, since the acids formed as a consequence would corrode metal surfaces. (If there was sulphur in the fuel the SO_2 formed would combine with O_2 and the water to form sulphuric acid.) If the calorific value is quoted for the H_2O in the vapour phase it is called the Lower Calorific Value (L.C.V.), and if it is quoted with the H_2O in the liquid phase (as in the calorimeter tests) it is called the Higher Calorific Value (H.C.V.). To calculate the L.C.V. from the H.C.V. a standard allowance is made for the condensate formed, as discussed in Section 15.11 and to the relationships given by equations 15.34 and 15.35 for the constant volume and constant pressure processes respectively.

Example 15.21

The calorific value of a supply of town gas was determined using a

Boys' calorimeter and was found to be 20 MJ/m³. The condensate collected during the test amounted to 840 g for every m³ of gas burned. Calculate the L.C.V. of the gas.

Using equation 15.35,

$$Q_{net, p} = Q_{gr, p} - m_c \times h_{fg}$$

$$\therefore Q_{net, p} = 20 - 840 \times 10^{-6} \times 2441 \cdot 8 = 17 \cdot 95 \text{ MJ/m}^3$$

Example 15.22

In a bomb calorimeter test on petrol the H.C.V. was determined and found to be 46 900 kJ/kg. If the fuel contains 14·4% H by mass, calculate the L.C.V.

From equation 15.1,

$$2H_2 + O_2 \rightarrow 2H_2O$$

$$\therefore 4 \text{ kg H} + 32 \text{ kg O}_2 \rightarrow 36 \text{ kg H}_2O$$

or 1 kg H gives 9 kg H_2O

$$\therefore 0 \cdot 144 \text{ kg H gives } 0 \cdot 144 \times 9 = 1 \cdot 296 \text{ kg H}_2O$$

Then using equation 15.34,

$$Q_{net, v} = Q_{gr, v} - m_c \times u_{fg}$$

$$\therefore Q_{net, v} = 46 \ 900 - 1 \cdot 296 \times 2304 \cdot 4 = 43 \ 910 \text{ kJ/kg}$$

Table 15.8 gives some typical calorific values of solid, liquid, and gaseous fuels.

Table 15.8   Typical Calorific Values of Fuels (see also Table 8.6)

Fuel	Calorific value	
	Higher	Lower
SOLID kJ/kg at 15°C		
Anthracite	34 600	33 900
Bituminous coal	33 500	32 450
Coke	30 750	30 500
Lignite	21 650	20 400
Peat	15 900	14 500
LIQUID kJ/kg at 15°C		
100 octane petrol	47 300	44 000
Motor petrol	46 800	43 700
Benzole	42 000	40 200
Kerosene	46 250	43 250
Diesel oil	46 000	43 250
Light fuel oil	44 800	42 100
Heavy fuel oil	44 000	41 300
Residual fuel oil	42 100	40 000

Table 15.8—cont.

Fuel	Calorific value	
	Higher	Lower
GAS MJ/m³ at 15°C and 1013 mbar		
Coal gas	20·00	17·85
Producer gas	6·04	6·00
Natural gas	36·20	32·60
Blast furnace gas	3·41	3·37
Carbon monoxide	11·79	11·79
Hydrogen	11·85	10·00

15.14 Air and fuel-vapour mixtures

The mixture supplied to an engine fitted with a carburettor is one of air and fuel vapour, and the quality of the mixture is controlled by the carburettor. If the mixture is saturated with fuel vapour then the relative proportions of fuel to air can be determined from a knowledge of the temperature/pressure relationship for the saturated fuel. Such values are given for ethyl alcohol in Table 15.9.

Table 15.9 Approximate Saturation Temperatures and Pressures for Ethyl Alcohol (C_2H_6O)

Temperature °C	Saturation pressure bar
0	0·0162
10	0·0314
20	0·0584
30	0·1049
40	0·1800
50	0·2960
60	0·4690

Example 15.23

In example 15.3 the stoichiometric A/F ratio for ethyl alcohol, C_2H_6O, was found to be 8·96/1. If the L.C.V. of ethyl alcohol is 27 800 kJ/kg, calculate the calorific value of the combustion mixture per m³ at 1·013 bar and 15°C.

The molar mass of ethyl alcohol is 46 kg/kmol, therefore, we have

$$\text{kmol of fuel per kg of fuel} = \frac{1}{46} = 0\cdot021\ 75 \text{ kmol/kg}$$

The molar mass of air is 28·96 kg/kmol, therefore,

$$\text{kmol of air per kg of fuel} = \frac{8·96}{28·96} = 0·31 \text{ kmol/kg}$$

i.e.

$$\text{Total kmol of mixture} = 0·021\,75 + 0·31 = 0·3318 \text{ kmol/kg}$$

From equation 3.8,

$$pV = nR_0T$$

$$\therefore V = \frac{0·3318 \times 8314 \times 288}{1·013 \times 10^5} = 7·83 \text{ m}^3 \text{ per kg of fuel}$$

Now,   L.C.V. of fuel $= Q_{net, v} = 27·8$ MJ/kg

$$\therefore \text{L.C.V. of mixture} = \frac{27·8}{7·83} = 3·55 \text{ MJ/m}^3 \text{ of mixture}$$

Example 15.24

For the stoichiometric mixture of ethyl alcohol and air calculate the temperature above which there will be no liquid fuel in the mixture. The pressure of the mixture is 1·013 bar.

Using the results of example 15.23, we have

$$\text{kmol of ethyl alcohol} = 0·021\,75$$

and   Total number of kmol $= 0·3318$

Then using equation 13.14, $p_i = (n_i/n)p$, we have

$$\text{Partial pressure of ethyl alcohol vapour} = \frac{0·021\,75}{0·3318} \times 1·013$$

$$= 0·0665 \text{ bar}$$

From table 15.9, the saturation temperature corresponding to this pressure lies between 20°C and 30°C. Therefore, interpolating,

$$t = 20 + \left(\frac{0·0665 - 0·0584}{0·1049 - 0·0584}\right) \times (30 - 20) = 21·74°C$$

Hence the minimum temperature of the mixture is 21·74°C for complete evaporation of the liquid fuel.

PROBLEMS

15.1 A sample of bituminous coal gave the following ultimate analysis by mass: C 81·9%; H 4·9%; O 6%; N 2·3%; ash 4·9%. Calculate,

(a) The stoichiometric A/F ratio.

(b) The analysis by volume of the wet and dry products of combustion when the air supplied is 25% in excess of that required for complete combustion.

(10·8/1; CO_2 14·16%; H_2O 5·07%; O_2 4·04%; N_2 76·7%; CO_2 14·9%; O_2 4·24%; N_2 80·9%)

15.2 The analysis of coal gas gave the following values: H_2 54%; CO 9%; CH_4 25%; C_4H_8 3%; CO_2 3%; N_2 8%. Determine the stoichiometric A/F ratio and the analysis of the wet and dry products of combustion when the A/F ratio is 7/1.

(4·75/1; CO_2 6·0%; H_2O 15·1%; O_2 6·1%; N_2 72·8%; CO_2 7·0%; O_2 7·21%; N_2 85·9%)

15.3 Calculate the stoichiometric A/F ratio for benzene (C_6H_6), and the wet and dry analysis of the combustion products.

(13·2/1; CO_2 16%; H_2O 8·05%; N_2 75·8%; CO_2 17·5%; N_2 82·5%)

15.4 In the actual combustion of benzene in an engine the A/F ratio was 12/1. Calculate the analysis of the wet products of combustion.

(CO_2 13·34%; CO 3·98%; H_2O 8·63%; N_2 74·1%)

15.5 The ultimate analysis of a sample of petrol was 85·5% C and 14·5% H. Calculate,

(a) The stoichiometric A/F ratio.

(b) The A/F ratio when the mixture strength is 90%.

(c) The A/F ratio when the mixture strength is 120%.

For (b) and (c) determine the product analysis expected using an Orsat apparatus. Calculate also for (c) the volume flow of the products through the engine exhaust per kg of fuel per minute if the pressure is 1·013 bar and the temperature is 110°C.

(14·8/1; 16·5/1; 12·35/1; CO_2 13·3%; O_2 2·33%; N_2 84·4%; CO_2 8·87%; CO 8·55%; N_2 82·6%; 15·1 m³/min.)

15.6 The ultimate analysis of a sample of petrol was C 85·5% and H 14·5%. The analysis of the dry products gave 14% CO_2 and some O_2. Calculate the A/F ratio supplied to the engine, and the mixture strength.

(15·71/1; 94%)

15.7 In an engine test the dry product analysis was CO_2 15·5%; O_2 2·3% and the remainder N_2. Assuming that the fuel burned was a pure hydrocarbon, calculate the ratio of carbon to hydrogen in the fuel, the A/F ratio used, and the mixture strength.

(92% C; 8% H; 14·88/1; 89·6%)

15.8 The ultimate analysis of a sample of petrol was 85% C and 15% H. The analysis of the dry products showed 13·5% CO_2, some CO and the remainder N_2. Determine the actual A/F ratio and the mixture strength. Calculate the mass of H_2O vapour carried by the exhaust gas per kg of total exhaust gas. Calculate also the temperature to which the gas must be cooled before condensation of the H_2O vapour begins, if the pressure in the exhaust pipe is 1·013 bar.

(14·4/1; 104%; 0·088 kg/kg; 52·8°C)

15.9 A quantity of coal used in a boiler had the following analysis: 82% C; 5% H; 6% O; 2% N; 5% ash. The dry flue gas analysis showed 14% CO_2 and some oxygen. Calculate,

(a) The oxygen content of the dry flue gas,
(b) The A/F ratio and the excess air supplied.

(5·25%; 14·4/1; 32·4%)

15.10 For the mixture of problem 15.4 calculate the calorific value per m^3 of mixture at 1·013 bar and 38°C. The calorific value of benzene is 40 700 kJ/kg. (3·82 MJ/m^3)

15.11 The lower explosive limit of ethyl alcohol in air is 3·56% by volume at a pressure of 1013 mbar. If the pressure in a room is 1013 mbar calculate the lowest temperature at which the explosive mixture would be formed. What quantity of ethyl alcohol in litres would be needed in a room of volume 115 m^3 to produce this mixture. The specific gravity of liquid ethyl alcohol is 0·794. Use the data of table 15.9 for this problem. (11·72°C; 10·14 l)

15.12 The products of combustion of a hydrocarbon fuel, of carbon to hydrogen ratio 0·85:0·15, were analysed by an Orsat apparatus as CO_2 8%, CO 1%, O_2 8·5%. Calculate the air to fuel ratio for the process by two methods and hence check the consistency of the data. (23·70, 23·52)

15.13 At a particular point in the combustion of a gaseous fuel in air the constituents are in the following molar proportions

CO 0·0002; CO_2 0·057; H_2O 0·127; H_2O 0·00008;
OH 0·0015; O_2 0·072;   NO 0·0068; N_2 0·695;

Define equilibrium constants for the following reactions $CO + \frac{1}{2}O_2$ $\rightleftharpoons CO_2$; $OH + \frac{1}{2}H_2 \rightleftharpoons H_2O$; $\frac{1}{2}N_2 + \frac{1}{2}O \rightleftharpoons NO$; and also that given by K_{H_2O}/K_{CO_2} where K_{H_2O} is for the reaction $H_2 + \frac{1}{2}O_2 \rightleftharpoons H_2O$.

The pressure is 21·75 atm and the temperature is 1878°C. Show that the analysis given is consistent with these conditions using the following values of K.

Temp K	REACTION			
	$CO + \frac{1}{2}O_2 \rightleftharpoons CO_2$	$OH + \frac{1}{2}H_2 \rightleftharpoons H_2O$	$\frac{1}{2}N_2 + \frac{1}{2}O_2 \rightleftharpoons NO$	K_{H_2O}/K_{CO_2}
2110	304·09	2763	0·02687	5·167
2166	202·77	1808	0·03068	5·365

(Values of K, 229, 2032, 0·0301, 5·27)

15.14 A stoichiometric mixture of CO and air was burned adiabatically at constant volume and at the peak pressure of 7·2 atm the temperature was 2454°C and analysis showed the amount of CO_2 to be dissociated per kmol of CO supplied as 0·192. Show by calculating the equilibrium constant for $CO + \frac{1}{2}O_2 \rightleftharpoons CO_2$ that the data is consistent. The value of K for this reaction at 2722 K is 8·9557 and at 2777 K it is 7·0259.

15.15 A reciprocating engine compresses a stoichiometric mixture of carbon monoxide and air from initial conditions of 1 atm and 60°C through a compression ratio of 5 to 1 and then burns this adiabatically at constant volume. Show that the equilibrium constant for the reaction $CO + \frac{1}{2}O_2 \rightleftharpoons CO_2$ is given by

$$K = \frac{(1-x)}{x^{3/2}} \left(\frac{2nT_1}{rp_1T_2} \right)^{1/2}$$

where x = number of CO_2 kmol dissociated per kmol of CO supplied

p_1 = initial pressure, atm.

T_1 = initial temperature, K.

r = compression ratio.

T_2 = temperature at the end of combustion, K.

n = number of kmol of the combustion reactants per kmol of CO supplied.

For the initial condition given the amount of CO_2 dissociated is 0·177 kmol per kmol of CO at the maximum temperature reached, calculate the maximum temperature by comparing calculated values of K with values from the following table:

Temperature/K	K
2777	7·026
2833	5·571
2889	4·458
2944	3·598

Calculate the corresponding pressure. State the other condition to be met for the calculated temperature to be a true one.

(2894 K, 38·1 atm)

15.16 Carbon dioxide is heated in a constant pressure process at 1 atm from 25°C to 1949°C at which temperature the equilibrium constant for the reaction $CO + \frac{1}{2}O_2 \rightleftharpoons CO_2$ is 138·36 atm$^{-1/2}$. Calculate, in kmol per kmol of CO supplied, the composition of the mixture.

If the pressure is increased to 9 atm at this temperature calculate the composition.

(CO_2 0·954, CO 0·046, O_2 0·023; CO_2 0·978, CO 0·022, O_2 0·011)

15.17 A stoichiometric mixture of benzene (C_6H_6) and air is induced into an engine of volumetric compression ratio 5 to 1. The pressure and temperature at the beginning of compression are 1 atm and 100°C. The estimated maximum temperature reached, allowing for dissociation, after adiabatic combustion at constant volume is 2704°C and at this temperature the dissociation constants are

$$\frac{(p_{CO})^2 p_{O_2}}{(p_{CO_2})^2} = 0·098 \qquad \frac{p_{CO} p_{H_2O}}{p_{CO_2} p_{H_2}} = 6·876$$

where the partial pressures are in atmospheres.

Show that about 75% of the carbon in the fuel is burned to CO_2 and calculate the maximum pressure reached. (41·3 atm)

15.18 The enthalpy of combustion of propane gas, C_3H_8, at 25°C with the H_2O in the products in the liquid phase is $-50\,360$ kJ/kg. Calculate the enthalpy of combustion with the H_2O in the vapour phase per kg of fuel and per kmol of fuel.

($-46\,350$ kJ/kg, $-2\,039$ kJ/kmol)

15.19 Calculate, for propane liquid, C_3H_8, at 25°C the enthalpy of combustion with the H_2O in the products in the vapour phase in kJ/kg of fuel. Use the data of question 15.18 and take h_{fg} at 25°C for propane as 372 kJ/kg. ($-45\,980$ kJ/kg)

15.20 Calculate the internal energy of combustion of propane vapour, C_3H_8, at 25°C in kJ/kg of fuel with the H_2O in the vapour phase from the corresponding value of $\Delta H_0 = -46\ 350$ kJ/kg.

(−48 828 kJ/kg)

15.21 Calculate the internal energy of combustion for gaseous propane, C_3H_8, at 25°C in kJ/kg with the H_2O of combustion in the liquid phase from the corresponding value of $\Delta H_0 = -50\ 360$ kJ/kg.

(−42 910 kJ/kg)

15.22 Calculate the internal energy of combustion for liquid propane, C_3H_8, at 25°C in kJ/kg with the H_2O of combustion in the vapour phase from the corresponding value of $\Delta H_0 = -45\ 980$ kJ/kg.

(−50 930 kJ/kg)

15.23 ΔH_0 for hydrogen at 60°C is given as $-242\ 400$ kJ/kmol. Calculate ΔH_0 at 1950°C given that the enthalpies of the gases concerned in kJ/kmol are as follows:

Gas	60°C	1950°C
H_2	9492	69 250
O_2	9697	76 500
H_2O	11 147	94 620

(−252 000 kJ/kmol)

15.24 A stoichiometric mixture of hydrogen and air at 25°C is ignited and combustion takes place adiabatically at a constant pressure of 1 atm. ΔH_0 for hydrogen at 25°C with the H_2O in the liquid phase is $-286\ 000$ kJ/kmol. Calculate, neglecting changes in kinetic energy, and using the specific heats in reference 15.6.

(a) the temperature reached if the process is assumed to be adiabatic and dissociation is neglected.

(b) the temperature reached after adiabatic combustion if the constituents are H_2, O_2, H_2O and N_2. Assume the temperature to be between 2500 and 2600 K.
   At 25°C h_{fg} for H_2O is 2441·8 kJ/kg.   (2480°C, 2255°C)

15.25 Octane vapour, C_8H_{18}, is to be burned in air in a steady flow process. Both the fuel and air are supplied at 25°C and the product temperature is to be 760°C. Dissociation is negligible and the heat loss is to be taken as 10% of the increase in enthalpy of the products above the reference temperature.

ΔH_0 for octane vapour is $-5\,510\,294$ kJ/kmol with the products in the liquid phase. Calculate the air to fuel ratio required. h_{fg} at 25°C for H_2O is 2441·8 kJ/kg. Use the specific heats in reference 15.6.

(49·3)

15.26 Calculate the enthalpy of combustion of methane CH_4, at 25°C and 1 atm pressure. Use the data of Table 15.6 and assume the H_2O of combustion to be in the liquid phase.

($-890\,290$ kJ/kmol)

15.27 Calculate the enthalpy of combustion in kJ/kmol at 25°C and 1 atm pressure of Producer gas of volumetric analysis 12% H_2, 29% CO, 2·6% CH_4, 0·4% C_2H_4, 4% CO_2 and 52% N_2 for the H_2O in the vapour phase and in the liquid phase.

($-153\,000$, $-160\,900$ kJ/kmol)

REFERENCES

15.1 HIMUS, G. W., *Elements of Fuel Technology* (Leonard Hill, 1958).

15.2 ROGERS, G. F. C., and MAYHEW, Y. R., *Engineering Thermodynamics, Work and Heat Transfer* (Longmans, 1980).

15.3 DRYDEN, I. G. C., *The efficient use of energy* (Butterworth Scientific, 1982).

15.4 *The Gas Examiners' General Directions* (1956).

15.5 GEYER, E. M., and BRUGES, E. A., *Tables of Properties of Gases* (Longmans, 1948).

15.6 ROGERS, G. F. C., and MAYHEW, Y. R., *Thermodynamic Properties of Fluids and Other Data* (Basil Blackwell, 1980).

15.7 British Standards Institution, *Definition of the Calorific Values of Fuels* B.S. 526: 1961

15.8 Institution of Mechanical Engineers Conference: Combustion in Engineering Vol. 2 (MEP, 1983).

16. Refrigeration

The purpose of a refrigerator is to transfer heat from a cold chamber which is at a temperature lower than that of its surroundings. A temperature gradient is thus established from the surroundings to the chamber and heat will flow naturally in this direction. The heat flow can be resisted by insulating the chamber from the surroundings by the use of suitable insulating materials, but practical requirements and conditions make necessary a continuous means of transfer of heat from the chamber.

Elementary refrigerators have been used which utilize the melting of ice or the sublimation of solid carbon dioxide at atmospheric pressure to provide the cooling effect. The continuous consumption of the refrigerating substance with the means of replenishment required from another source of supply make these methods inconvenient. The temperatures attainable by these methods are limited, but although inefficient for continuous refrigeration, these methods are sometimes convenient forms of cooling in the laboratory and workshop.

The nature of the problem suggests a means of refrigeration which consists of a cycle of processes with the same quantity of working fluid, called the refrigerant, in continuous circulation. If the refrigerant receives energy in the cold chamber at a temperature below that of the surroundings, then this energy must be rejected before the refrigerant can return to the cold chamber in its initial state. This energy rejection must be carried out at a temperature above that of the surroundings. The energy at rejection is of a higher quality, because of its higher temperature, than that received in the cold chamber. This energy can be used for heating purposes and refrigerating plants designed entirely for this purpose are called heat pumps. The term heat pump is appropriate to the action of the plant since energy is transferred against the natural temperature gradient from a low temperature to a higher one. It is analogous to the pumping of water from a low level to a higher one against the natural gradient of gravitational force. Both actions require an input of energy for their accomplishment. There is no difference in operation between a refrigerator and a heat pump. With the refrigerator the important quantity is the energy removed from the cold chamber called the *refrigerating effect*, and with the heat pump it is the energy to be rejected by the refrigerant for heating purposes. The machine can be used for both purposes and one particular domestic unit provides for the cooling of a larder and the heating of water.

The refrigerating plant chosen depends on the particular purpose since each application has to meet specific requirements. A number of substances are utilized as refrigerants and most methods use the refrigerants in the liquid-vapour states. The reasons for this will be discussed later (see Section 16.2). Air is used as the refrigerant in air conditioning plant where it is also the conditioning medium (see Section 16.8).

16.1 Reversed heat engine cycles

In Section 2.1 the First Law of Thermodynamics was stated as follows:

When a system undergoes a thermodynamic cycle then the net heat supplied to the system from its surroundings is equal to the net work done by the system on its surroundings.

This is illustrated in fig. 16.1a in which the heat Q_1 is supplied at

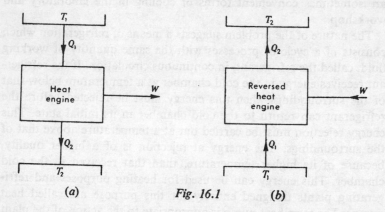

(a)   Fig. 16.1   (b)

temperature T_1 to the system which does work W and rejects the heat Q_2 at temperature T_2. The First Law of Thermodynamics as expressed by equation 2.1 applied to this system gives

$$\sum \mathrm{d}Q = \sum \mathrm{d}W$$

or $$Q_1 - Q_2 = W \qquad\qquad 16.1$$

Fig. 16.1b shows the heat-engine cycle in reverse, quantities Q_1, Q_2, and W having the opposite directions to those in fig. 16.1a. The effect of the reversed heat engine is to transfer a quantity of heat Q_1 from a cold source at temperature T_1 and reject a quantity Q_2 at temperature T_2. It will be noted that in fig. 16.1a T_1 and T_2 are the higher and lower temperatures of the cycle respectively, and in fig. 16.1b T_1 and T_2 are the lower and higher temperatures of the cycle respectively.

The reversed heat engine fulfils the requirements of a refrigerator and the First Law of Thermodynamics applied to the system of fig. 16.1b gives

$$\sum dQ = \sum dW$$

or $$Q_1 - Q_2 = -W$$

$$\therefore \ W = Q_2 - Q_1 \qquad\qquad 16.2$$

For a refrigerator the important quantity is Q_1, and for a heat pump it is the quantity Q_2. The power input is W and is important because it is the quantity which has to be paid for and constitutes the main item of the running cost.

Refrigerator and heat-pump performances are defined by means of the Coefficient of Performance (C.O.P.) which is given by

$$\text{C.O.P. refrigerator} = \frac{Q_1}{W} \qquad\qquad 16.3$$

$$\text{C.O.P. heat pump} = \frac{Q_2}{W} \qquad\qquad 16.4$$

(C.O.P. heat pump is sometimes called the Performance Ratio.) The above terms will be abbreviated to C.O.P.$_r$ and C.O.P.$_{hp}$, respectively.

The best C.O.P. will be given by a cycle which is a Carnot cycle operating between the given temperature conditions. Such a cycle using a wet vapour as the working substance is shown diagrammatically in fig. 16.2a. Wet vapour is used as the example, since the processes of constant pressure heat supply and heat rejection are made at

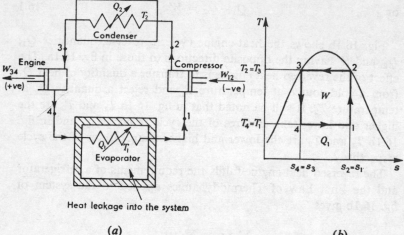

(a)   *(b)*

Fig. 16.2

constant temperature, a necessary requirement of the Carnot cycle and one which is not fulfilled by using a superheated vapour. The changes in the thermodynamic properties of the refrigerant throughout the cycle are indicated on the *T-s* diagram of fig. 16.2b. The cycle events are as follows:

1–2. Wet vapour at state 1 enters the compressor and is compressed isentropically to state 2. The work input for this process is represented by W_{1-2}.

2–3. The vapour enters the condenser at state 2 and is condensed at constant pressure and temperature to state 3 when it is completely liquid. The heat rejected by the refrigerant is Q_2.

3–4. The liquid expands isentropically behind the piston of the engine doing work of amount W_{3-4}.

4–1. At the lower pressure and temperature of state 4 the refrigerant enters the evaporator where the heat necessary for evaporation Q_1 is taken from the cold source.

The boundaries of the system are as shown in fig. 16.2a and therefore,

The net *work input to* the system $W = W_{1-2} - W_{3-4}$

The net *heat rejected by* the system $Q = Q_2 - Q_1$

and from equation 16.2

$$W = W_{1-2} - W_{3-4} = Q_2 - Q_1$$

and from equations 16.3 and 16.4

$$\text{C.O.P.}_r = \frac{Q_1}{W} = \frac{Q_1}{Q_2 - Q_1}$$

and

$$\text{C.O.P.}_{hp} = \frac{Q_2}{W} = \frac{Q_2}{Q_2 - Q_1}$$

From the T-s diagram, fig. 16.2b, since the areas on the T-s diagram are proportional to the heat quantities, then

$$Q_1 = T_1(s_1 - s_4)$$

and

$$Q_2 = T_2(s_2 - s_3)$$
$$= T_2(s_1 - s_4)$$

$$\therefore \; \text{C.O.P.}_r = \frac{T_1(s_1 - s_4)}{(T_2 - T_1)(s_1 - s_4)}$$

and

$$\text{C.O.P.}_{hp} = \frac{T_2(s_1 - s_4)}{(T_2 - T_1)(s_1 - s_4)}$$

i.e.

$$\text{C.O.P.}_r = \frac{T_1}{T_2 - T_1} \qquad\qquad 16.5$$

and

$$\text{C.O.P.}_{hp} = \frac{T_2}{T_2 - T_1} \qquad\qquad 16.6$$

Equations 16.5 and 16.6 give the maximum possible values of C.O.P.$_r$ and C.O.P.$_{hp}$ between given values of T_1 and T_2, the temperatures of the refrigerant in the evaporator and condenser coils respectively.

Example 16.1

A refrigerator has working temperatures in the evaporator and condenser coils of $-30°C$ and $32°C$ respectively. What is the maximum C.O.P. possible? If the actual refrigerator has a C.O.P. of 0·75 of the maximum calculate the refrigerating effect in kW per kW power input.

$$T_1 = -30 + 273 = 243 \text{ K} \quad \text{and} \quad T_2 = 32 + 273 = 305 \text{ K}$$

From equation 16.5,

$$\text{C.O.P.}_r = \frac{T_1}{T_2 - T_1} = \frac{243}{305 - 243} = 3.92$$

$$\text{Actual C.O.P.}_r = 0.75 \times 3.92 = 2.94$$

Using equation 16.3, where W is 1 kW, we have

$$\text{C.O.P.}_r = \frac{Q_1}{W} = 2.94$$

$$\therefore Q_1 = 2.94W = 2.94 \times 1 = 2.94 \text{ kW}$$

i.e.   Refrigerating effect per kW power input = 2.94 kW

The areas representing the quantities Q_1 and W are shown in fig. 16.2b. A consideration of these areas and equations 16.3, and 16.4 shows the relationship between C.O.P.$_r$ and C.O.P.$_{hp}$.

From equation 16.2,

$$Q_2 = Q_1 + W$$

Dividing through by W gives,

$$\frac{Q_2}{W} = \frac{Q_1}{W} + 1$$

Then using the definitions of equations 16.3 and 16.4,

$$\text{C.O.P.}_{hp} = \text{C.O.P.}_r + 1 \qquad 16.7$$

Equation 16.7 indicates that ideally the heating effect of a heat pump is greater than the work put in, and this suggests that it would provide an effective heater. However a heating effect can always be obtained easily by direct heating and the application of a heat pump would require a thorough analysis of all the factors concerned.

The definitions of equations 16.5 and 16.6 show that the values of C.O.P.$_r$ and C.O.P.$_{hp}$ decrease with increased temperature difference $(T_2 - T_1)$. In a practical unit this temperature difference is increased above that between the source and receiver because of the temperature differences required for the purpose of heat transfer. Fig. 16.3a shows a heat pump which takes low-grade energy from a large river and utilizes it to heat a building. The temperature differences required are indicated on the T-s diagram in fig. 16.3b. It can be seen that $(T_2 - T_1) > (T_b - T_a)$. A secondary working fluid is required to take the heat rejected by the refrigerant in the condenser and reject

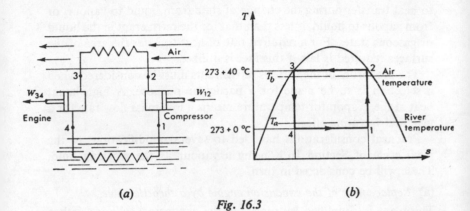

(a)   *(b)*

Fig. 16.3

it in the space to be heated. If the temperature limits of the refrigerant are 0°C and 40°C, as indicated in fig. 16.3b, then, using equation 16.6,

$$\text{C.O.P.}_{\text{hp}} = \frac{313}{313 - 273} = \frac{313}{40} = 7 \cdot 825$$

(without using a secondary fluid).

If it was proposed to use hot water as a secondary fluid at a temperature of 60°C and heat the space using radiators, then the C.O.P.$_{\text{hp}}$ becomes $333/60 = 5 \cdot 55$.

Practical values of the C.O.P.$_{\text{hp}}$ for these temperature limits have been quoted as $5 \cdot 7$ and $3 \cdot 5$ (see reference 16.1). It may be more advisable to use water at 40°C to 60°C for panel heating than to use the higher temperature water with radiators. The values calculated using equation 16.6 are ideal values and the difference between these and those for the actual plant are due to the modifications made to the ideal cycle, and to irreversibilities.

16.2 Vapour-compression refrigeration cycles

The most widely used refrigerators are those which use a liquefiable vapour as the refrigerant. The evaporation and condensation processes take place when the fluid is receiving and rejecting latent heat, and these are constant temperature and constant pressure processes. The cycle is one in which these two processes correspond to those of the reversed Carnot cycle for a vapour, and this enables the temperature range for a given duty to be kept low. The resistance to heat transfer during the change of state from liquid to vapour, or from vapour to liquid, is less than that for the refrigerant in the liquid or gaseous states. For a required rate of heat transfer the area of the surfaces required is less if this fact is utilized.

The properties of the various refrigerants must be considered when a selection is to be made for a particular purpose. A high latent heat at the evaporator temperature means a low mass flow rate for a given refrigerating effect.

Practical considerations have led to several modifications to the ideal cycle of Section 16.1, using a vapour as the working fluid. These will be considered in turn.

(a) Replacement of the expansion engine by a throttle valve

The plant is simplified by replacing the expansion cylinder with a simple throttle valve. Throttling was discussed in Section 4.4 and the process was shown to occur such that the initial enthalpy equals the final enthalpy. The process is highly irreversible so that the whole cycle becomes irreversible. The process is represented by the dotted line 3–4 on fig. 16.4a. A comparison of figs. 16.4a and 16.2b shows

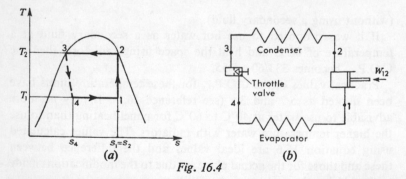

Fig. 16.4

that the refrigerating effect, $Q_1 = T_1(s_1 - s_4)$ is reduced by using a throttle valve instead of the expansion cylinder.

(b) Condition at the compressor inlet

To make complete use of the latent heat of the refrigerant in the evaporator it is desirable to continue the process until the vapour is dry saturated. In a practical unit this process is extended to give the vapour a definite amount of superheat as it leaves the evaporator. This is really undesirable, since the work to be done by the compressor is increased, as will be shown later. It is a practical necessity to allow the refrigerant to become superheated in this way in order to prevent the carry-over of liquid refrigerant into the compressor, where it interferes with the lubrication. The amount of superheat should be kept to a minimum. The compression process under these conditions is shown in fig. 16.5 and it is seen that the isentropic compression takes the refrigerant well into the superheat region. The rejection of heat in the condenser cannot now be carried out at constant temperature, and this represents another departure from the ideal reversed Carnot cycle.

(c) *Undercooling of the condensed vapour*

The condensed vapour can be cooled at constant pressure to a temperature below that of the saturation temperature corresponding to the condenser pressure. This effect is shown in fig. 16.6, in which the constant pressure line is shown further from the liquid line than

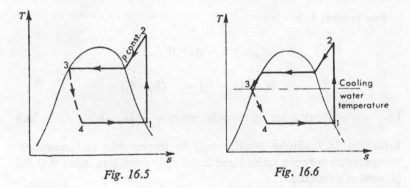

Fig. 16.5   *Fig. 16.6*

it would actually appear, in order to illustrate the point. The effect of undercooling is to move the line 3–4, representing the throttling process, to the left on the diagram. The result of this is that the refrigerating effect in process 4–1 is increased. The amount of undercooling is limited by the temperature of the cooling water and the essential temperature difference required for the transfer of heat.

Fig. 16.6 includes all the modifications of this section, and this can be taken as showing the practical ideal cycle. The term 'ideal' in this context means that all the individual processes behave exactly as specified; the compression process is isentropic and the refrigerant passes through the evaporator and condenser coils at constant pressure, requirements which are not met with in practice.

The cycle consists of a number of flow processes and can be analysed by the application of the steady flow energy equation, 2.8,

$$h_1 + \frac{C_1^2}{2} + Q = h_2 + \frac{C_2^2}{2} + W$$

If changes in kinetic energy are neglected, then

$$h_1 + Q = h_2 + W$$

Applying this equation to each of the processes in fig. 16.6, we have

For process 4–1:

$$h_4 + Q_1 = h_1 + 0$$
$$\therefore\ Q_1 = (h_1 - h_4)$$

i.e.   Refrigerating effect $Q_1 = (h_1 - h_4)$ 16.8

For process 1–2:

$$h_1 + 0 = h_2 + W$$

$$\therefore\ W = (h_1 - h_2) = -(h_2 - h_1)$$

i.e.   Work done *on* the refrigerant $= +(h_2 - h_1)$ 16.9

Equation 16.9 applies equally well to irreversible and reversible compression between states 1 and 2, the only condition being that the process is adiabatic.

If the process is reversible and adiabatic, then it is isentropic (i.e. $s_1 = s_2$).

For process 2–3:

$$h_2 + Q_2 = h_3 + 0$$
$$\therefore\ Q_2 = (h_2 - h_3) = -(h_3 - h_2)$$

i.e.   Heat *rejected* by the refrigerant $= +(h_3 - h_2)$ 16.10

For process 3–4:

$$h_3 + 0 = h_4 + 0$$

i.e. $h_3 = h_4$ in throttling process

The solution to numerical problems depends on the means of obtaining the enthalpies h_1, h_2, and h_3. Three methods are available:

(a) Using the tabulated values of the thermodynamic properties of the refrigerant.

(b) Using known values of latent heats and approximate values of specific heats for the refrigerant and making use of the fact that areas on the T-s diagram represent heat quantities.

(c) Using a chart which gives the properties of the refrigerant, the most useful chart for this purpose being the pressure-enthalpy, p-h, chart. This method is dealt with in Section 16.4; charts are available for most common refrigerants.

Example 16.2

The pressure in the evaporator of an ammonia refrigerator is 1·902 bar and the pressure in the condenser is 12·37 bar. What is the ideal coefficient of performance for a machine working between the corresponding saturation temperatures, and what is the ideal refrigerating effect? Calculate the refrigerating effect per kg of refrigerant and the C.O.P.$_r$ for the practical cycle working between the same pressures, when,

(a) dry saturated vapour is delivered to the condenser after isentropic compression, and there is no undercooling of the condensed liquid;

(b) dry saturated vapour is delivered to the compressor where it is compressed isentropically, and there is no undercooling of the condensed liquid;

(c) dry saturated vapour is delivered to the compressor, and the liquid after condensation is undercooled by 10 K.

The ideal C.O.P.$_r$ is that given by a reversed Carnot cycle and is given by equation 16.5. The cycle is shown in fig. 16.7.

$$\text{C.O.P.}_r = \frac{T_1}{T_2 - T_1} = \frac{-20 + 273}{32 - (-20)} = \frac{253}{52} = 4 \cdot 86$$

(where $t_1 = -20°C$ is the saturation temperature corresponding to the evaporator pressure of 1·902 bar, and $t_2 = 32°C$ is the saturation temperature corresponding to the condenser pressure of 12·37 bar. Note that $(t_2 - t_1) = (T_2 - T_1)$).

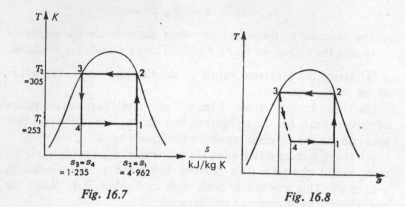

Fig. 16.7   Fig. 16.8

Refrigerating effect, Q_1, is given by

$$Q_1 = T_1(s_1 - s_4) = T_1(s_2 - s_3)$$

From tables

$$s_2 = s_g \text{ at } 12·37 \text{ bar} = 4·962 \text{ kJ/kg K}$$

and

$$s_3 = s_f \text{ at } 12·37 \text{ bar} = 1·235 \text{ kJ/kg K}$$

$$\therefore Q_1 = 253(4·962 - 1·235) = 942·8 \text{ kJ/kg}$$

i.e.   Ideal refrigerating effect = 942·8 kJ/kg

(a) The cycle is shown in fig. 16.8.

At 12·37 bar,

$$h_2 = h_g = 1469·9 \text{ kJ/kg} \quad \text{and} \quad h_3 = h_f = 332·8 \text{ kJ/kg}$$

At 1·902 bar,

$$h_4 = h_3 = 332·8 \text{ kJ/kg}$$

To find h_1 use the fact that process 1–2 is isentropic,

i.e. $s_1 = s_2 = s_g \text{ at } 12·37 \text{ bar} = 4·962 \text{ kJ/kg K}$

$$\therefore 0·368 + x_1(5·623 - 0·368) = 4·962$$

$$\therefore x_1 = \frac{4·594}{5·255} = 0·874$$

Then from equation 3.2,

$$h_1 = h_{f_1} + x_1 h_{f_{g_1}}$$

i.e. $h_1 = 89{\cdot}8 + 0{\cdot}874 \times (1420 - 89{\cdot}8) = 1251{\cdot}8 \text{ kJ/kg}$

From equation 16.8,

Refrigerating effect $= (h_1 - h_4) = 1251{\cdot}8 - 332{\cdot}8 = 919 \text{ kJ/kg}$

From equation 16.9,

$$\text{Work done } on \text{ refrigerant} = +(h_2 - h_1)$$

i.e.

Work done on refrigerant $= (1469{\cdot}9 - 1251{\cdot}8) = 218{\cdot}1 \text{ kJ/kg}$

Then from equation 16.3,

$$\text{C.O.P.}_r = \frac{Q_1}{W} = \frac{919}{218{\cdot}1} = 4{\cdot}2$$

(b) The cycle is shown in fig. 16.9.

At 1·902 bar

$$h_1 = h_g = 1420 \text{ kJ/kg}$$

as before, $h_4 = h_3 = 332{\cdot}8 \text{ kJ/kg}$

Also at 1·902 bar,

$$s_1 = s_g = 5{\cdot}623 \text{ kJ/kg K} = s_2$$

At 12·37 bar, $s_g = 4{\cdot}962 \text{ kJ/kg K}$, hence the refrigerant is super-heated at state 2.

Interpolating,

$$h_2 = 1613 + \left(\frac{5{\cdot}623 - 5{\cdot}397}{5{\cdot}731 - 5{\cdot}397}\right) \times (1739{\cdot}3 - 1613) = 1698{\cdot}5 \text{ kJ/kg}$$

From equation 16.8,

$$\text{Refrigerating effect, } Q_1 = (h_1 - h_4) = 1420 - 332{\cdot}8$$
$$= 1087{\cdot}2 \text{ kJ/kg}$$

From equation 16.9,

Work done on refrigerant $= (h_2 - h_1) = (1698{\cdot}5 - 1420)$
$$= 278{\cdot}5 \text{ kJ/kg}$$

Then from equation 16.3,

$$\text{C.O.P.}_r = \frac{Q_1}{W} = \frac{1087{\cdot}2}{278{\cdot}5} = 3{\cdot}9$$

(c) The cycle is shown in fig. 16.10. The values of h_1 and h_2 are as determined for part (b). The value of $h_3 = h_4$ can be found by assuming that the undercooling takes place along the saturated liquid line, and therefore $h_3 = h_f$ at t_3. This is a good approximation for

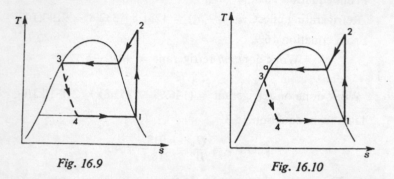

Fig. 16.9   Fig. 16.10

most refrigerants. Another way of obtaining h_3 is by assuming a constant specific heat, c, for the ammonia liquid, and then,

$$h_3 = (h_f \text{ at } t_a) - c(t_a - t_3)$$

It is usually more convenient to use the first approximation,

i.e. $h_3 = h_f \text{ at } t_3 = 284 \cdot 6 \text{ kJ/kg}$

(where $t_3 = 32 - 10 = 22°C$).

Then, from equation 16.8,

Refrigerating effect, $Q_1 = (h_1 - h_4) = 1420 - 284 \cdot 6$
$$= 1135 \cdot 4 \text{ kJ/kg}$$

Also from equation 16.3,

$$\text{C.O.P.}_r = \frac{Q_1}{W} = \frac{1135 \cdot 4}{278 \cdot 5} = 4 \cdot 08$$

(where W is the same as in part (b)).

Example 16.3

An ammonia vapour-compression refrigerator operates between an evaporator temperature of $-12°C$ and a condenser temperature of $32°C$. The vapour is delivered to the condenser in a dry saturated condition after isentropic compression. There is no undercooling in the condenser. The latent heat of ammonia at 32°C is 1137·1 kJ/kg and the specific heat of the liquid ammonia may be taken as constant at 4·815 kJ/kg K. Calculate the coefficient of performance. The tables are not to be used for this example.

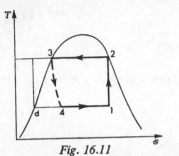

Fig. 16.11

The cycle is shown in fig. 16.11. The enthalpies h_1, h_2, and $h_3 = h_4$ can be found with reference to a datum point using the theory of Section 5.3a. Let the datum point be point d in fig. 16.11.

Then,

$$h_3 - h_d = c(t_3 - t_d) = 4·815 \times (32 - (-12)) = 211·8 \text{ kJ/kg}$$

also $h_2 - h_3 = h_{f_g} = 1137·1 \text{ kJ/kg}$ (given)

$$\therefore \ h_2 - h_d = 211·8 + 1137·1 = 1348·9 \text{ kJ/kg}$$

Using equation 5.7,

$$s_3 - s_d = \int_d^3 \frac{c \, dT}{T} = c \log_e \frac{T_3}{T_d}$$

$$= 4·815 \times \log_e \frac{305}{261} = 0·7521 \text{ kJ/kg K}$$

(where $T_3 = 32 + 273 = 305$ K, and $T_d = -12 + 273 = 261$ K).

Also,
$$s_{f_g} = \frac{h_{f_g}}{T}$$

$$\therefore \ s_2 - s_3 = \frac{1137·1}{305} = 3·728 \text{ kJ/kg K}$$

i.e.
$$s_2 - s_d = 0·7521 + 3·728 = 4·48 \text{ kJ/kg K}$$

Now $s_2 = s_1$, hence,

$$s_1 - s_d = s_2 - s_d = 4·48 \text{ kJ/kg K}$$

and
$$h_1 - h_d = T_d(s_1 - s_d) = 261 \times 4·48 = 1169 \text{ kJ/kg}$$

Then from equation 16.8,

$$Q_1 = (h_1 - h_4) \quad \text{and} \quad h_4 = h_3$$

i.e.

Refrigerating effect $= (h_1 - h_d) - (h_4 - h_d) = 1169 - 211\cdot8$

$$= 957\cdot2 \text{ kJ/kg}$$

From equation 16.9,

$$W = (h_2 - h_1)$$

$\therefore \quad W = (h_2 - h_d) - (h_1 - h_d) = 1348\cdot9 - 1169 = 179\cdot9 \text{ kJ/kg}$

Therefore from equation 16.3,

$$\text{C.O.P.}_r = \frac{Q_1}{W} = \frac{957\cdot2}{179\cdot9} = 5\cdot32$$

In most systems the space to be cooled is not cooled directly by the refrigerant in the evaporator. The space is encircled by pipes carrying brine (a solution of sodium or calcium chloride in water). The brine is cooled in the evaporator of the refrigerator before being pumped through pipes passing through the cold chamber. The temperature rise of the brine is limited to a value which is less than 3 K. This introduces a further complication to the system, but when large distances have to be covered by the pipes carrying the cold fluid, the refrigerant leakage problem is reduced by using brine. In some cases (e.g. in storage warehouses) there may be damage to goods due to the leakage of refrigerant. The cost of the plant is increased by introducing brine circulation, and the cost of pumping it may be an important factor in the running costs. The calcium chloride brine solution is used more often than the sodium chloride solution since it can be applied to cases where the temperature is to be below $-18°\text{C}$.

16.3 Refrigerating load and the unit of refrigeration

The most important quantity in the application of a refrigerator is the amount of heat which must be transferred per unit time from the cold chamber and in SI units this is expressed in kilowatts. The American unit of refrigeration is called the *ton* and is defined as a rate of heat transfer of 200 Btu/min, based on the cooling rate required to produce 2000 lb of ice at 32°F from water at 32°F in a time of 24 hours, i.e. 1 ton = 200 Btu/min = 3·516 kW.

The rate at which heat must be removed decides the mass flow of a given refrigerant when working under specified conditions,

i.e.

$$\text{Mass flow of refrigerant} = \frac{\text{Heat to be removed per unit time}}{\text{Refrigerating effect per unit mass}} \qquad 16.11$$

Example 16.4

Calculate for the data of example 16.2b the mass flow of ammonia required and the indicated power of the compressor per kW of refrigeration.

From example 16.2b we have

$$\text{Refrigerating effect} = 1087 \cdot 2 \text{ kJ/kg}$$

Therefore from equation 16.11,

$$\text{Mass flow required} = \frac{1 \times 3600}{1087 \cdot 2} = 3 \cdot 31 \text{ kg/h}$$

Also from example 16.2b,

$$\text{Work done on refrigerant} = 278 \cdot 5 \text{ kJ/kg}$$

$$\therefore \text{Work input per second} = \frac{278 \cdot 5 \times 3 \cdot 31}{3600} = 0 \cdot 256 \text{ kJ/s}$$

i.e. $$\text{Indicated power} = 0 \cdot 256 \text{ kW}$$

16.4 The pressure-enthalpy diagram

Up to this point refrigeration cycles have been represented on a T-s diagram. The pressure-enthalpy diagram is more convenient for refrigeration cycles since the enthalpies required for the calculation can be read off direct. The essential features of the diagram are given in fig. 16.12, and a typical refrigeration cycle is shown on a p-h diagram in fig. 16.13. The points 1, 2, 3, and 4 represent the same positions in the cycle as they did in the previous sections; a cycle with under-cooling is shown. For the rest of this chapter all refrigeration cycles will be shown on p-h diagrams.

Example 16.5

A vapour-compression heat pump using ammonia is to be used to heat 0·5 m³ of air per second measured at 30°C from 5°C to 30°C, at a constant pressure of 1·013 bar. The temperature in the evaporator is $-8°C$ and the pressure in the condenser is 14·7 bar. The vapour

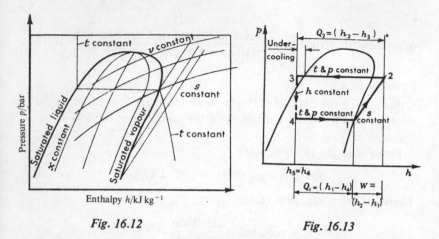

Fig. 16.12   Fig. 16.13

is dry saturated at the inlet to the compressor, and the liquid enters the throttle valve at 26°C. Calculate, assuming isentropic compression,

(a) The rate of flow of refrigerant in kg/s,
(b) The power input to the compressor,
(c) The coefficient of performance,
(d) The equivalent power which would be required to achieve the same purpose by direct heating.

The values required should be taken from a *p-h* chart for ammonia.

For ammonia the *p-h* chart appears as shown in fig. 16.14. Only a portion of the wet loop is plotted, since the critical pressure of ammonia is high (114·2 bar) and pressures above about 20 bar are seldom used in refrigeration. Further, as the parts of the diagram of most interest are in the region of the liquid line or the saturated vapour line, it is usual to plot the chart with a scale change on the enthalpy axis. The evaporator pressure line is found by the position at which the $-8°C$ constant temperature line in the liquid region intersects the liquid line. The corresponding pressure is 3·153 bar. The point 1 is given by the intersection of the constant pressure line at 3·153 bar and the dry saturated vapour line. The value of h_1 is read from the enthalpy axis as 1435 kJ/kg. From state 1 to state 2 the path follows a line of constant entropy until the pressure line

14·7 bar is reached. The value of h_2 is then read off the enthalpy axis as 1661 kJ/kg. The value of $h_3 = h_4$ is given by the intersection of the 26°C line in the liquid region and the 14·7 bar pressure line, and is

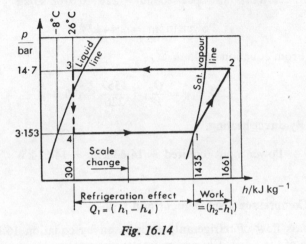

Fig. 16.14

read off the enthalpy axis as 304 kJ/kg. Hence all the required values of enthalpy are obtained from the chart.

Heat required = mass flow of air $\times c_p \times$ temperature rise of air

The mass flow of air is given by

$$\dot{m} = \frac{p\dot{V}}{RT} = \frac{1·013 \times 10^5 \times 0·5}{0·287 \times 10^3 \times 303} = 0·5825 \text{ kg/s}$$

(where $T = 30 + 273 = 303$ K).

$\therefore$ Heat required $= 0·5825 \times 1·005 \times (30 - 5) = 14·63$ kW

(a) From equation 16.10,

Heat rejected $= (h_2 - h_3) = 1661 - 304 = 1357$ kJ/kg

$\therefore$ Mass flow of refrigerant $= \dfrac{\text{Heat required per second}}{\text{Heat rejected per kg}}$

$$= \frac{14·63}{1357} = 0·0108 \text{ kg/s}$$

(b) From equation 16.9,

Work done on refrigerant $= (h_2 - h_1) = 1661 - 1435 = 226 \text{ kJ/kg}$

i.e.   Work input per second $= 226 \times 0.0108 \text{ kJ/s}$

i.e.   Power input $= 2.44 \text{ kW}$

(c) From equation 16.4,

$$\text{C.O.P.}_{\text{hp}} = \frac{Q_2}{W} = \frac{1357}{226} = 6.0$$

(d) By direct heating,

Power input required $= 14.63 \text{ kJ/s} = 14.63 \text{ kW}$

16.5 Compressor displacement and type

The mass flow of refrigerant, $\dot{m}$, is given by equation 16.11. The volume of the refrigerant drawn into the compressor is given by

$$\dot{V} = \dot{m}v_1 \qquad 16.12$$

(where v_1 is the specific volume of the refrigerant at state 1).

If the compressor is of the reciprocating type and has a volumetric efficiency of η_V (which is usually between 65% and 85%), then the swept volume per unit time of the compressor cylinder is given by

$$V_s = \frac{\dot{V}}{\eta_V}$$

or $$V_s = \frac{\dot{m}v_1}{\eta_V} \qquad 16.13$$

Example 16.6

Calculate the swept volume per unit time of the compressor in example 16.5, if the volumetric efficiency is 80%.

The value of v_1 can be read from the p–h chart, but in this case the refrigerant is dry saturated at compressor inlet, hence,

$$v_1 = v_g \text{ at } 3.153 \text{ bar} = 0.3879 \text{ m}^3/\text{kg}$$

Therefore from equation 16.13,

$$V_s = \frac{\dot{m}v_1}{\eta_v} = \frac{60 \times 0.0108 \times 0.3879}{0.8} = 0.314 \text{ m}^3/\text{min}$$

Household refrigerators of the vapour-compression type usually employ a hermetically sealed positive displacement rotary compressor. Commercial refrigerators use reciprocating compressors or centrifugal compressors if the system is large and in continuous operation. The compression process with the latter type of machine is irreversible although approximately adiabatic. The state of the refrigerant before compression can be determined and the state after compression is known if the pressure and temperature are known, since the vapour after compression is invariably superheated.

Example 16.7

A plant using refrigerant-12 has an evaporator saturation temperature of −25°C and a condenser saturation temperature of 35°C. The vapour is dry saturated at entry to the compressor and has a temperature of 65°C after compression to the condenser pressure. Calculate the work done per kW of refrigeration and the C.O.P.$_r$. Compare this result with that obtained when the compression is isentropic. There is no undercooling of the liquid in the condenser.

Use of the chart is encouraged but the values used in the examples are taken from tables.

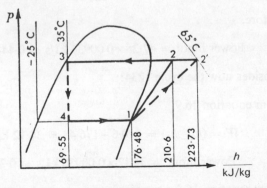

Fig. 16.15

A sketch of a *p-h* chart for refrigerant-12 is shown in fig. 16.15 with enthalpy values as shown. Line 1 to 2 represents isentropic compression, and line 1 to 2' represents irreversible adiabatic compression. Consider first the cycle 12'34:

From equation 16.9,

$$W = (h_{2'} - h_1) = 223 \cdot 73 - 176 \cdot 48 = 47 \cdot 25 \text{ kJ/kg}$$

(Note that equation 16.9 is true for irreversible adiabatic processes as well as for isentropic processes.)

From equation 16.8,

Refrigerating effect, $Q_1 = (h_1 - h_4) = 176 \cdot 48 - 69 \cdot 55 = 106 \cdot 93$ kJ/kg

Then from equation 16.3,

$$\text{C.O.P.}_{\cdot r} = \frac{Q_1}{W} = \frac{106 \cdot 93}{47 \cdot 25} = 2 \cdot 26$$

Also using equation 16.11, taking the heat to be removed as 1 kW of refrigeration, we have

$$\text{Mass flow of refrigerant} = \frac{\text{Heat to be removed per second}}{\text{Refrigerating effect per kg}}$$

$$= \frac{1}{106 \cdot 93} = 0 \cdot 00936 \text{ kg/s}$$

Therefore,

$$\text{Power input} = 47 \cdot 25 \times 0 \cdot 00936 \text{ kJ/s} = 0 \cdot 442 \text{ kW}$$

Consider now the cycle 1234:

From equation 16.9,

$$W = (h_2 - h_1) = 210 \cdot 6 - 176 \cdot 48 = 34 \cdot 12 \text{ kJ/kg}$$

$$\therefore \text{ Power input} = 34 \cdot 12 \times 0 \cdot 00936 \text{ kJ/s} = 0 \cdot 319 \text{ kW}$$

From equation 16.3,

$$\text{C.O.P.}_{\cdot r} = \frac{Q_1}{W} = \frac{106 \cdot 93}{34 \cdot 12} = 3 \cdot 13$$

16.6 The use of the flash chamber

Fig. 16.16 shows the cycle which was discussed in Section 16.2(c). Consider the throttling process 3 to 4 which shows the refrigerant

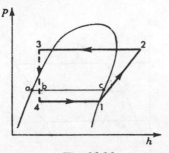

as an undercooled liquid at state 3, and a wet vapour at 4. Vapour begins to form at the pressure at which the line 3–4 crosses the liquid line, and as expansion proceeds more liquid becomes vapour. In an actual process the change from liquid to vapour is not gradual as might be suggested by the p-h diagram, but the liquid refrigerant is immediately exposed

Fig. 16.16

to the evaporator pressure in passing through the valve and some of it 'flashes' into vapour. The vapour is a wet mixture (i.e. a mixture of dry saturated vapour and saturated liquid), and the proportions of dry vapour to liquid are given by the ratio ab/bc at any pressure as shown on fig. 16.16.

The basic practical cycle of Section 16.2 can be improved if an increase in the complexity of the plant is acceptable. It was shown that undercooling of the condensate improved the refrigerating effect per kg of refrigerant, and hence increased the C.O.P.$_r$. It would be an advantage if the dry saturated vapour formed by flashing as the liquid expands through the throttle valve did not pass through the evaporator, since it can make no contribution to the refrigerating effect.

Using a flash chamber at some intermediate pressure, the flash vapour at this pressure can be bled off and fed back to the compression process. The throttling process is then carried out in two stages, each one starting with a liquid. Suitable valving is required and the compression process is best done in two separate compressor stages, the flash vapour at the intermediate pressure mixing with the refrigerant after the first stage of compression. Fig. 16.17a shows a diagrammatic arrangement of the plant, and fig. 16.17b shows the cycle plotted on a p-h diagram. Some care is necessary in making the calculations since the mass flow is not the same in all parts of the circuit.

Consider 1 kg of refrigerant flowing through the condenser, and use the subscript i to denote the interstage and flash chamber conditions. At the flash chamber x kg of dry saturated vapour at pressure p_i and enthalpy h_{g_i} are bled off to the interstage of the compressor. The remaining mass of $(1-x)$ kg of liquid of enthalpy h_{f_i} passes

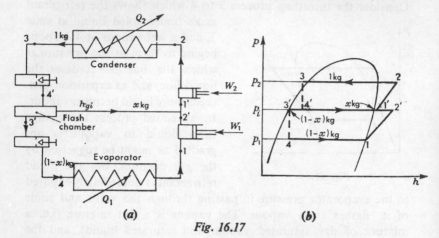

Fig. 16.17

through the second throttle valve to the evaporator. In the first stage of compression $(1-x)$ kg of fluid are compressed from p_1 to p_i (i.e. from state 1 to state 2′). At the interstage pressure, $(1-x)$ kg of vapour at state 2′ are mixed adiabatically with x kg of flash vapour of enthalpy h_{g_i}. The resultant mixture at some state 1′ is compressed in the second stage compressor from p_i to p_2 at which state it is delivered to the condenser.

The amount of dry saturated vapour bled off is given by the dryness fraction at state 4′. The mixture at 4′ consists of x kg of dry saturated vapour and $(1-x)$ kg of liquid for every kg of refrigerant in the condenser. The value of x can be determined by equating the enthalpy before throttling at 3 to the enthalpy after throttling at 4′,

$$\therefore \ h_{f_i} + x h_{g_i} = h_3 \quad \text{or} \quad x = \frac{h_3 - h_{f_i}}{h_{g_i}} \qquad 16.14$$

Using equation 16.9,

$$\text{Total work input} = W_1 + W_2$$

$$= (h_{2'} - h_1)(1-x) + (h_2 - h_{1'}) \qquad 16.15$$

Using equation 16.8,

$$\text{Refrigerating effect} = (h_1 - h_4)(1-x) \qquad 16.16$$

The heat rejected in the condenser is given by equation 16.10,

i.e.   Heat rejected in condenser $= (h_2 - h_3)$

Note that the work and heat quantities given by equations 16.15 and 16.16 are for 1 kg of refrigerant in the condenser.

A consideration of fig. 16.17 shows that if the compression process were carried out in an infinite number of stages then the compression line would be coincident with the saturated vapour line. The throttling process would show a decreasing amount of liquid in a succession of states from p_2 to p_1 along the liquid line. This cycle is the reverse of that used in steam power plant for regenerative feed heating (see Section 7.7).

Example 16.8

A vapour compression plant uses refrigerant-12 and has a suction pressure of 2·61 bar and a condenser pressure of 12·19 bar. The vapour is dry saturated on entering the compressor and there is no undercooling of the condensate. The compression is carried out isentropically in two stages and a flash chamber is employed at an interstage pressure of 4·914 bar. Calculate,

(a) The amount of vapour bled off at the flash chamber,
(b) The state of the vapour at the inlet to the second stage of compression,
(c) The refrigerating effect per kg of refrigerant in the condenser,
(d) The work done per kg of refrigerant in the condenser,
(e) The coefficient of performance.

The cycle is shown on fig. 16.18 and the enthalpy values are as indicated.

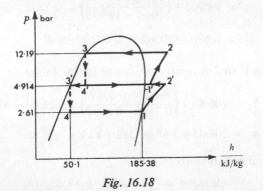

Fig. 16.18

(a) From equation 16.14,

$$x = \frac{h_3 - h_{f_1}}{h_{f_{g_1}}} = \frac{84 \cdot 94 - 50 \cdot 1}{193 \cdot 78 - 50 \cdot 1} = \frac{34 \cdot 84}{143 \cdot 68} = 0 \cdot 242$$

i.e.   Vapour bled off $= 0 \cdot 242$ kg per kg in the condenser

(b) Adiabatic mixing was dealt with in Section 4.4. In this case x kg of dry saturated vapour at 4·914 bar are mixed with $(1-x)$ kg of superheated vapour at 4·914 bar. Since $s_1 = s_{2'} = 0 \cdot 6991$ kJ/kg K then the value of $h_{2'}$ can be found by interpolating from superheat tables,

i.e. $h_{2'} = 193 \cdot 78 + \left(\dfrac{0 \cdot 6991 - 0 \cdot 6901}{0 \cdot 7251 - 0 \cdot 6901} \right) \times (204 \cdot 1 - 193 \cdot 78)$

$= 196 \cdot 45$ kJ/kg

From equation 4.33 for adiabatic mixing,

$1 \times h_{1'} = (1-x) \times h_{2'} + x h_{g_1}$

$= (1 - 0 \cdot 242) \times 196 \cdot 45 + 0 \cdot 242 \times 193 \cdot 78 = 195 \cdot 8$ kJ/kg

Therefore the vapour at inlet to the second stage compressor is still superheated.

(c) From equation 16.16,

Refrigerating effect $= (1-x)(h_1 - h_4)$

$= (1 - 0 \cdot 242) \times (185 \cdot 38 - 50 \cdot 1) = 102 \cdot 6$ kJ/kg

(d) The entropy $s_{1'}$ is found by interpolation at 4·914 bar and $h_1 = 195 \cdot 8$ kJ/kg,

i.e.

$$s_{1'} = 0 \cdot 6901 + \left(\frac{195 \cdot 8 - 193 \cdot 78}{204 \cdot 1 - 193 \cdot 78} \right) \times (0 \cdot 7251 - 0 \cdot 6901)$$

$\therefore \ s_{1'} = 0 \cdot 690 + 0 \cdot 0069 = 0 \cdot 697$ kJ/kg K

The process 1′ to 2 is isentropic, hence interpolating,

$$h_2 = 206 \cdot 45 + \left(\frac{0 \cdot 697 - 0 \cdot 6797}{0 \cdot 7166 - 0 \cdot 6797} \right) \times (218 \cdot 64 - 206 \cdot 45)$$

$\therefore \ h_2 = 206 \cdot 45 + 5 \cdot 68 = 212 \cdot 1$ kJ/kg

Then from equation 16.15,

Work input $= (1-x)(h_{2'} - h_1) + (h_2 - h_{1'})$

i.e.

Work input $= \{(1-0\cdot242)(196\cdot45-185\cdot38)+(212\cdot1-195\cdot8)\}$

$\qquad\qquad\quad = 8\cdot39+16\cdot3 = 24\cdot69$ kJ/kg

(e) From equation 16.3,

$$\text{C.O.P.}_{\cdot r} = \frac{Q_1}{W} = \frac{102\cdot6}{24\cdot69} = 4\cdot16$$

It is interesting to compare these results with those obtained with the basic cycle as shown in fig. 16.19.

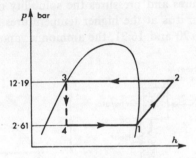

Fig. 16.19

As before, $s_1 = 0\cdot6991$ kJ/kg K, therefore, since $s_1 = s_2$, interpolating,

$$h_2 = 206\cdot45 + \left(\frac{0\cdot6991-0\cdot6797}{0\cdot7166-0\cdot6797}\right) \times (218\cdot64-206\cdot45)$$

$$= 206\cdot45 + 6\cdot4 = 212\cdot85 \text{ kJ/kg}$$

Then from equation 16.9,

$$W = (h_2-h_1) = (212\cdot85-185\cdot38) = 27\cdot47 \text{ kJ/kg}$$

From equation 16.8,

Refrigerating effect, $Q_1 = (h_1-h_4) = 185\cdot38-84\cdot94 = 100\cdot44$ kJ/kg

Also from equation 16.3,

$$\text{C.O.P.}_{\cdot r} = \frac{Q_1}{W} = \frac{100\cdot44}{27\cdot47} = 3\cdot66$$

16.7 Absorption refrigerators

The absorption refrigerator is one which has been in operation over a long period of years. The refrigerant is absorbed on leaving the evaporator, the absorbing medium being a solid or a liquid. In order that the sequence of events should be continuous it is necessary for the refrigerant to be separated from the absorbent and subsequently condensed before being returned to the evaporator. The separation is accomplished by the application of direct heat in a 'generator'. The solubility of the refrigerant and absorbent must be suitable and the plant which uses ammonia as the refrigerant and water as the absorbent will be described.

At low temperatures and pressures the solubility of ammonia in water is higher than it is at the higher temperatures and pressures. Referring to fig. 16.20 and 16.21, the ammonia vapour leaving the

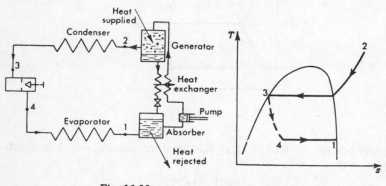

Fig. 16.20   *Fig. 16.21*

evaporator at point 1 is readily absorbed in the low temperature water in the absorber. This process is accompanied by the rejection of heat. The ammonia in water solution is pumped to the higher pressure and is heated in the generator. Due to the reduced solubility of ammonia in water at the higher pressure and temperature, the vapour is removed from the solution. The vapour then passes to the condenser and the weakened ammonia in water solution is returned to the absorber. To reduce the heat input required in the generator the returning weak solution passes through one side of a heat exchanger and helps to heat the strong solution as it goes from the pump to the generator.

The work done on compression is less than in the vapour-compression cycle (since pumping a liquid requires much less work than compressing a vapour between the same pressures), but a heat input to the generator is required. This heat may be supplied by any convenient form (e.g. gas, steam, or electricity).

This system will not be analysed thermodynamically, since such an analysis is complex and depends on the properties of ammonia-water solutions. A further treatment can be found in references 16.2 and 16.3.

In what is perhaps the best known absorption type of refrigerator, the domestic Electrolux, the pump is dispensed with. A diagrammatic representation of the system is shown in fig. 16.22.

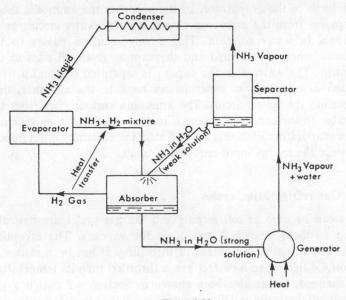

Fig. 16.22

The small energy supply is by means of a heater which may be electric or gas. The principle involved makes use of the properties of gas-vapour mixtures. If a liquid is exposed to an inert atmosphere, it will evaporate until the atmosphere is saturated with the vapour of the liquid. This evaporation requires heat which is taken from the surroundings in which the evaporation takes place. A cooling effect is thus produced. The partial pressure of the refrigerant vapour (in this case ammonia), must be low in the evaporator, and higher in the condenser. The total pressure throughout the circuit must be constant so that the only movement of the working fluid is by convection currents. The partial pressure of the ammonia is kept low in the requisite parts of the circuit by concentrating hydrogen in those parts. The action of the refrigerator is described briefly in the following paragraph. The actual plant includes refinements and practical modifications which will not be included here.

The ammonia liquid leaving the condenser enters the evaporator and evaporates into the hydrogen at the low temperature corresponding to its low partial pressure. The ammonia+hydrogen mixture passes to the absorber into which is also admitted water from the separator. The water absorbs the ammonia and the hydrogen returns to the evaporator. In the absorber the ammonia therefore passes from the ammonia circuit into the water circuit as an ammonia in water solution. This strong solution passes to the generator where it is heated and the vapour given off rises to the separator. The water with the vapour is separated out and a weak solution of ammonia in water passes back to the absorber, thus completing the water circuit. The ammonia vapour rises from the separator to the condenser where it is condensed and then returned to the evaporator. With this type of machine efficiency is not important since the energy input required is small.

16.8 Gas refrigeration cycles

Gases can be used as refrigerants and the gas cycles are basically similar to those described previously for vapours. The exception made here is that of the means of throttling. It has been shown in Section 4.4 that when a perfect gas is throttled then its temperature is unchanged. It has also been shown in Section 4.2 that if a gas expands adiabatically and does external work then there is a reduction in temperature. If the expansion process is also reversible and therefore isentropic then the relationship between the initial and final temperature is given by equation 4.21,

$$\frac{T_1}{T_2} = \left(\frac{p_1}{p_2}\right)^{(\gamma-1)/\gamma}$$

It is necessary, in the case of a gaseous refrigerant, to provide an expansion cylinder to replace the throttling process. The plant arrangement is shown in fig. 16.23a, and the cycle is represented on a T-s diagram in fig. 16.23b. It will be noticed that the cycle is a reversed

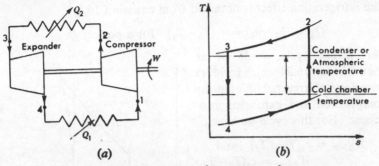

Fig. 16.23.   Gas refrigeration cycle

constant pressure or Joule cycle. The gas does not receive and reject heat at constant temperature, and this shows a deficiency of gas as a refrigerant when compared with a liquefiable vapour. The temperature range for the wet vapour is seen from fig. 16.23b to be t_1 to t_3, and the temperature range for the gas is seen to be t_4 to t_2. The significance of this is that the gas cycle will be less efficient than the vapour cycle for given evaporator and condenser temperatures. The volume of refrigerant to be dealt with is much greater for a gas and larger surfaces are required for heat transfer.

The use of a gas as a refrigerant becomes more attractive when a double purpose is to be met. This is so in the case of air conditioning when the air can be both the refrigerating and the conditioning medium. Another advantage of using air is that it is safe as a refrigerant. The reversed constant pressure cycle was used in the early days of refrigeration for this reason, the refrigerator using this cycle being known as the Bell-Coleman refrigerator. In modern air refrigeration cycles the large displacement volumes can be handled best by the rotary type compressor and expander, as these are more compact for a given flow than the reciprocating machines.

Fig. 16.23b shows the ideal cycle for a plant using a rotary compressor and turbine. The work done by the air expanding is used to help to drive the compressor. The net input to the plant is given by $W = W_{1-2} - W_{3-4}$. Then applying the steady flow equation to the cycle and neglecting changes in kinetic energy

$$W_{1-2} = (h_2 - h_1) \quad \text{and} \quad W_{3-4} = (h_3 - h_4)$$

Therefore for a perfect gas,

$$W_{1-2} = c_p(T_2 - T_1) \quad \text{and} \quad W_{3-4} = c_p(T_3 - T_4)$$

The refrigerating effect is obtained from equation 16.8,

$$Q_1 = (h_1 - h_4) = c_p(T_1 - T_4) \quad \text{for a perfect gas.}$$

The actual cycle would be as shown in fig. 16.24 which includes the effects of irreversibilities in the compression and expansion processes. For this cycle we have,

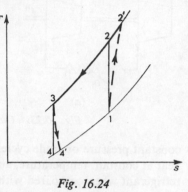

Fig. 16.24

$$W_{1-2'} = c_p(T_{2'} - T_1) \quad \text{and}$$
$$W_{3-4'} = c_p(T_3 - T_{4'})$$

and from the definition of isentropic efficiency in Section 12.1, and equations 12.1 and 12.2, we have

$$T_{2'} = T_1 + \frac{T_2 - T_1}{\eta_C} \quad \text{and} \quad T_{4'} = T_3 - \eta_T(T_3 - T_4)$$

(where η_C and η_T are the isentropic efficiencies of the compressor and the turbine respectively).

The refrigerating effect in the actual cycle is then reduced to,

$$Q_1 = c_p(T_1 - T_{4'})$$

One application of this system is that met with in modern aircraft practice in which the air delivered to the cabin must be conditioned. Air is bled from the compressor of the engine. The proportion of this air to be used for air conditioning is passed through a heat exchanger and is cooled by ram air at atmospheric temperature, which passes over the outside of the heat exchanger. This cooling takes place ideally at constant pressure and the cooled air then passes to the refrigerator turbine through which it expands and gives a work output. The air, after expanding approximately adiabatically, is passed to the cabin at a low temperature. The work output of the turbine is used to drive essential auxiliaries such as pumps, and perhaps a fan to draw air over the heat exchanger. As this turbine would not develop sufficient power to drive all the auxiliaries its effort is joined to that of another turbine which uses the rest of the bleed air from the main compressor. The arrangement is shown in fig. 16.25a, and the cycle is shown on a *T-s* diagram in fig. 16.25b. The process follows an open cycle, starting from the atmosphere and exhausting to the atmosphere.

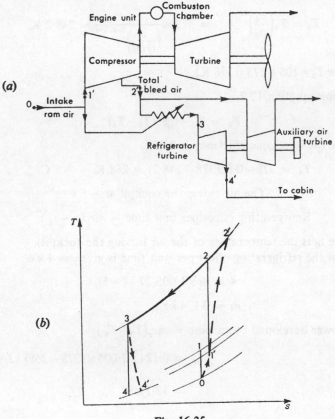

Fig. 16.25

Example 16.9

In the air-cooling system of a jet aircraft, air is bled from the engine compressor at 3 bar, and is cooled in a heat exchanger to 105°C. It is expanded to 0·69 bar in an air turbine, the isentropic efficiency of the process being 85%. The air is then delivered to the cockpit and leaves the aircraft at 27°C. Calculate the temperature at which the air enters the cockpit and the mass flow of air required for a refrigerating effect of 4 kW. If the air turbine is used to help to drive the auxiliaries, calculate its contribution in power.

The expansion process is shown by the line 3–4' on the T-s diagram of fig. 16.25b.

Using equation 4.21,

$$T_4 = T_3\left(\frac{p_4}{p_3}\right)^{(\gamma-1)/\gamma} = 378 \times \frac{1}{\left(\dfrac{3}{0\cdot69}\right)^{0\cdot4/1\cdot4}} = 248\cdot2 \text{ K}$$

(where $T_3 = 105 + 273 = 378$ K).

Using equation 12.2,

$$T_{4'} = T_3 - 0\cdot85(T_3 - T_4)$$

(where the isentropic efficiency is 0·85),

i.e. $T_{4'} = 378 - 0\cdot85(378 - 248\cdot2) = 268 \text{ K} = -5°\text{C}$

∴ The air enters the cockpit at $-5°$C

Refrigerating effect per unit time $= \dot{m}c_p(t_C - t_{4'})$

(where t_C is the temperature of the air leaving the cockpit).

Now the refrigerating effect per unit time is given as 4 kW,

i.e. $4 = \dot{m} \times 1\cdot005(27 - (-5))$

∴ $\dot{m} = 0\cdot124$ kg/s

Power developed by turbine $= \dot{m}c_p(T_3 - T_{4'})$

$$= 0\cdot124 \times 1\cdot005 \times (378 - 268) \text{ kJ/s}$$

$$= 13\cdot7 \text{ kW}$$

16.9 Liquefaction of gases

If a gas is to be liquefied, its temperature must be reduced to a value below its critical temperature (e.g. for nitrogen the critical temperature is $-147°$C). In order to reach such low temperatures various means are used. The arrangement described in Section 16.8 can be extended to employ four refrigerating systems in series, the refrigerants being ammonia, NH_3, ethylene, C_2H_4, methane, CH_4, and nitrogen, N_2, for the production of liquid nitrogen. The plant required for this would be complex and a simpler method is available for the liquefaction of some other gases. The arrangement is shown in fig. 16.26a, and the corresponding T-s diagram is shown in fig. 16.26b.

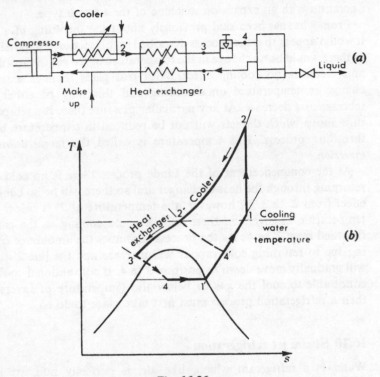

Fig. 16.26

This is called the *Linde process*. The corresponding state points are indicated on both diagrams. Substances which solidify at a temperature above that of the required liquid temperature must be removed from the gas before admission to the plant. The gas is compressed to a pressure of 100 to 200 atmospheres before delivery to the cooler. The gas is cooled in the cooler to a temperature which depends on the temperature of the cooling water available. The gas passes through a heat exchanger where heat is transferred from it to the returning low temperature vapour. It is cooled at 3 to a temperature which is in the region of its critical value and is then throttled to atmospheric pressure, at which pressure it exists as a wet vapour. The liquid is drawn off and the vapour is returned to the compressor. The quantity is made up from an external supply before induction into the compressor.

Lower temperatures could be obtained by replacing the throttling operation with an expansion machine of the turbine type.

From what has been said previously about the throttling of a gas it would appear that the process described in this section is impossible, as there would be no change in the temperature of the gas in throttling and therefore no cooling effect. With real gases there is a small change in temperature on throttling and this may be either an increase or a decrease. At any particular pressure there is a temperature above which the gas will not be reduced in temperature by a throttling process. This temperature is called the *temperature of inversion*.

At the commencement of the Linde process there is no cold gas returning through the heat exchanger and so there will be no cooling effect from 2' to 3. If, however, the temperature at 2' is below the temperature of inversion there will be some cooling as the gas is throttled from 2' to 4'. As the process continues the amount of cooling due to returning cold vapour will increase and the line 2' to 4' will gradually move down to position 3 to 4. If conventional cooling is not able to cool the gas to below the temperature of inversion then a refrigeration process must first take place to do so.

16.10 Steam jet refrigeration

Water is a refrigerant which, like air, is perfectly safe. At low temperatures the saturation pressures are low and the specific volumes high (e.g. at 4°C the saturation pressure is 0·008129 bar and the specific volume, v_g, is 157·3 m³/kg). The temperatures which can be attained using water as a refrigerant are not low enough for most applications of refrigeration but are in the range which may satisfy the conditions of air conditioning. The large volumes can be dealt with by means of a steam jet, and so this application is more attractive if a supply of medium pressure steam is available. The steam jet can be applied to high-temperature cooling and can be adjusted quickly to meet variations in the load. In the summer, when steam for heating is not required, a supply of steam is readily available for refrigeration, and the equipment of a steam-jet refrigerator can be installed for a low capital expenditure. Another essential is a liberal supply of cooling water for the steam condenser. Steam jets are not used when temperatures below 5°C are required.

The steam jet was applied successfully to refrigeration in the early years of this century. The lay-out of a modern steam jet plant is as shown in fig. 16.27. The secondary ejectors are for the removal of air and take approximately 10% of the total amount of steam required to produce the refrigerating effect. Steam expands through a nozzle to form a high-speed, low-pressure jet which entrains the vapour to be extracted from the vacuum chamber. The combined flow is diffused in the diverging part of the venturi to the exhaust pressure. After condensation some of the water can be returned to the evaporator, the rest is returned to the boiler feed.

16.11 Properties of refrigerants

Refrigerants used in vapour-compression systems can be divided into two main groups: halocarbons and inorganic compounds. The first group (halocarbons) were developed comparatively recently and are probably better known as 'freons'. Freon is a registered trade name of the DuPont Nemours company of the U.S.A.; the other trade names for the same family of compounds include 'arcton' which is the trade name used by I.C.I. The second main group of refrigerants (inorganic compounds) includes ammonia, carbon dioxide and sulphur dioxide, all of which have been used extensively in the past; only ammonia is still of commercial importance as a refrigerant. The American Society of Refrigerating Engineers (ASRE) has adopted a designation of refrigerants which is also used in this country. For halocarbon refrigerants a number is given which can be of three digits: the digit on the right represents the number of fluorine (F) atoms; the middle digit is one more than the number of hydrogen atoms; the digit on the left is one less than the number of carbon atoms. When the digit on the left is zero (i.e., only one atom of carbon present), then it is omitted. For example: dichlorodifluoromethane CCl_2F_2 is refrigerant 12; trichlorotrifluoroethane CCl_2FCClF_2 is refrigerant 113; dichloromonofluoromethane $CHCl_2F$ is refrigerant 21. For refrigerants other than halocarbons the designation is obtained by adding 700 to the approximate molecular weight. For example, ammonia (molecular weight 17) is refrigerant 717; water vapour (molecular weight 18) is refrigerant 718; air (molecular weight 28·96) is refrigerant 729.

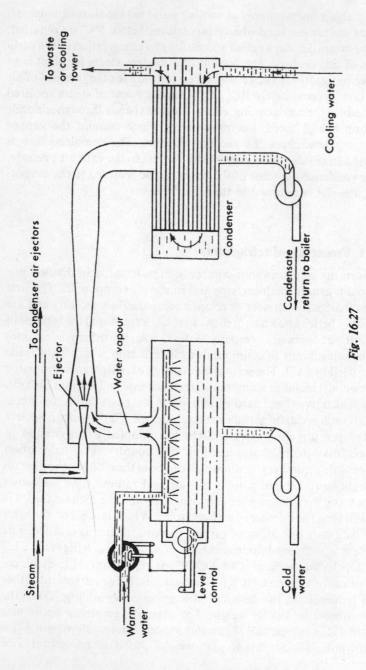

Fig. 16.27

Hydrocarbons and some other refrigerants are designated as for the halocarbons. The hydrocarbons (e.g., methane CH_4, refrigerant 50), are not used commercially, but one refrigerant that is used is methyl chloride CH_3Cl, and this is refrigerant 40.

No single refrigerant is applicable to every type of refrigeration. The following are the principal factors to be considered.

(a) Refrigerants should be non-toxic for the safety of personnel. Most refrigerants are toxic, if only to a small extent, the exceptions being the halocarbons, and carbon dioxide. Ammonia and sulphur are irritants; carbon dioxide, methyl chloride and methylene chloride are odourless; also methyl chloride, dichloroethylene, and ethyl chloride are anaesthetics. Chlorine refrigerants and the hydro-fluoric acids in fluorine compounds decompose when subject to an open flame, and form hydrochloric acid gas and traces of phosgene gas ($COCl_2$). Acrolein is added to methyl chloride as an irritant to make its detection easier on leakage from the plant.

(b) There should be no fire or explosion risk. Ammonia and methyl chloride are poor in this respect. Some possible refrigerants are ruled out completely because of this factor (e.g. ethyl chloride, butane, and propane).

(c) The pressures involved in the circuit are important as they influence the mechanical design of the plant with respect to size and sealing problems. Leakages are more difficult to prevent when the pressures are high. The pressures involved are highest with carbon dioxide.

(d) The volume of refrigerant to be dealt with per second per unit of refrigeration must be considered. This is important when deciding the size of compressor required. The power input to the compressor depends on the pressures involved and the rate of flow of refrigerant. If the pressures are low and the mass flow is high, then it may be possible to use a centrifugal type of compressor.

(e) The refrigerant should not react with the lubricating oil of the compressor and should not cause the oil to be carried over into the condenser and evaporator, where it reduces the heat transfer and causes fouling.

16.12 Dual refrigeration cycles

The different characteristics of refrigerants can be combined as in the two-stage cycle, which employs two refrigerants (e.g. carbon dioxide and ammonia). The carbon dioxide is used in the low temperature stage and this is topped by an ammonia cycle in the higher temperature range of the unit. The evaporator of the ammonia cycle is used to condense the vapour in the condenser of the carbon dioxide cycle. The carbon dioxide compression may be carried out in two stages as described in Section 16.6. The work done is high in the dual refrigerant cycle as the ammonia is liquefied at a temperature below that of the condensation of the carbon dioxide, and because of the refrigerating load handled by the ammonia which is greater than that of the carbon dioxide. The C.O.P.$_r$ for the dual cycle increases as the ammonia cycle occupies an increasing proportion of the temperature range.

16.13 Control of refrigerating capacity

The refrigerating load is seldom constant and the refrigerator must be controlled to meet the demand. It is desirable that automatic controls should be available and small refrigerating plants are operated on the on-off principle. The control in larger plants is obtained by regulating the mass flow of the refrigerant. This can be done by means of a manual speed control on the compressor, but this is uneconomical and inconvenient.

With the automatic on-off device a metal bellows, charged with a volatile liquid, is connected to a temperature sensitive element which is located at the evaporator coil. An increasing temperature at the evaporator causes the temperature and therefore the pressure in the bellows to increase. The bellows expand and the end of the bellows operates a switch which closes the compressor motor circuit. A decreasing temperature in the evaporator produces the reverse effect.

In a multi-cylinder reciprocating compressor the capacity can be controlled by regulating the number of cylinders which are effective at any time. When a reduction in refrigerant flow is required the suction valves of one or more cylinders are held open by a hydraulically operated mechanism. The control also applies for starting conditions when all suction valves are held open. The capacity of centrifugal compressors can be controlled by means of a speed control but this involves expensive equipment. There is a limit to the reduction in mass flow rate, which is determined by the surging characteristics of the compressor. With a constant speed machine, control can be achieved by,

(a) Reducing the cooling water flow to the condenser. The temperature and pressure in the condenser increase and this reduces the effective capacity of the machine. The compressor characteristics should be carefully studied.

(b) Throttling the inlet to the compressor. This reduces the inlet pressure and hence the density of the incoming charge. The delivery pressure is reduced because of the lower inlet pressure.

16.14 Insulating materials in refrigeration

Insulation is one of the main problems in a refrigerating plant. After the selection of the materials they have to be properly applied. Precautions in design must be taken, but service conditions produce cracking and distortion of the protective coating during the expansion and contraction of the pipe or container. Moisture is one of the worst difficulties to be overcome and is found on the cold surfaces. It increases the conductivity of the insulating material and tends to rot it and loosen it from its support.

The factors influencing the choice of material are: thermal conductivity; cost of erection; moisture absorbing properties; the yearly load factor; temperature and humidity of the air; rate of deterioration; bulk. Some of these could lead to conflicting requirements.

Corkboard is a common material used for insulation and is made in sheets for the insulation of flat walls, or in moulded sections for the insulation of pipes. Rock cork is a loose mineral wool combined with a waterproof binder, and this is also moulded into sheets.

The jointing of the insulating material to the surface to be insulated is important. The plasters used for finishing the wall may be of asphalt or cement. Cement is stronger but asphalt is waterproof and is also cheaper. Cement plaster does not bond well with cork. To insulate the brick or concrete walls of a warehouse or cold room the surface is primed with an emulsified asphalt primer to make an airtight joint. The insulating board is applied to the wall with hot asphalt; a second layer may be applied similarly to the first layer by means of wood skewers. A surface finish of cement plaster may then be applied.

When pipes are being insulated the spaces between the moulded sections and the pipe must be filled with a suitable insulating material.

Joints should be made up properly with cork fibre or cement. Hair felt is used on pipes as it can be wrapped round the pipe and held in position by metal bands.

Some insulating materials are used in the form of a loose fill, such as balsam wool, expanded rubber, paper, aluminium foil and fibre glass. The wall of the cold room or household refrigerator is of a double construction with the space between filled with loose insulating material.

PROBLEMS

16.1 The temperature in a refrigerator evaporator coil is $-6°C$ and that in the condenser coil is $22°C$. Assuming that the machine operates on the reversed Carnot cycle, calculate the C.O.P.$_r$, the refrigerating effect per kilowatt of input work, and the heat rejected to the condenser. (9·54; 9·54 kW; 10·54 kW)

16.2 An ammonia vapour-compression refrigerator operates between an evaporator pressure of 2·077 bar and a condenser pressure of 12·37 bar. The following cycles are to be compared; in each case there is no undercooling in the condenser, and isentropic compression may be assumed:

 (a) The vapour has a dryness fraction of 0·9 at entry to the compressor,
 (b) The vapour is dry saturated at entry to the compressor,
 (c) The vapour has 5 K of superheat at entry to the compressor.

In each case calculate the C.O.P.$_r$ and the refrigerating effect per kg. What would be the C.O.P.$_r$ of a reversed Carnot cycle operating between the same saturation temperatures?

(4·5; 957·5 kJ/kg; 4·13; 1089·9 kJ/kg; 4·1; 1101·4 kJ/kg; 5·1)

16.3 A plant using refrigerant-12 operates between saturation temperatures of $-10°C$ and $60°C$, at which temperatures the latent heats are 156·32 kJ/kg and 113·52 kJ/kg respectively. The refrigerant is dry saturated at entry to the compressor and the liquid is not undercooled in the condenser. The specific heat of the liquid is 0·970 kJ/kg K and that of the superheated vapour is 0·865 kJ/kg K. The vapour is compressed isentropically in the compressor. Using no other information than that given, calculate the temperature at the compressor delivery, and the refrigerating effect per kg of refrigerant.

(69·6°C; 88·42 kJ/kg)

16.4 A heat pump using ammonia as the refrigerant operates between saturation temperatures of 6°C and 38°C. The refrigerant is compressed isentropically from dry saturation and there is 6 K of undercooling in the condenser. Calculate, the C.O.P.$_{hp}$, the mass flow of refrigerant, and the heat available per kilowatt input.

(8·8; 25·06 kg/h; 8·8 kW)

16.5 An ammonia vapour-compression refrigerator has a single stage, single-acting reciprocating compressor which has a bore of 127 mm, a stroke of 152 mm and a speed of 240 rev/min. The pressure in the evaporator is 1·588 bar and that in the condenser is 13·89 bar. The volumetric efficiency of the compressor is 80% and its mechanical efficiency is 90%. The vapour is dry saturated on leaving the evaporator and the liquid leaves the condenser at 32°C. Calculate the mass flow of refrigerant, the refrigerating effect, and the power ideally required to drive the compressor.

(0·502 kg/min; 9·04 kW; 2·73 kW)

16.6 An ammonia refrigerator operates between evaporating and condensing temperatures of −16°C and 50°C respectively. The vapour is dry saturated at the compressor inlet, the compression process is isentropic and there is no undercooling of the condensate. Calculate the refrigerating effect per kg, the mass flow and power input per kW of refrigeration, and the C.O.P.$_r$.

A flash chamber is introduced with an interstage pressure of 5·346 bar, calculate,

(a) The mass flow rate of refrigerant through the condenser per kW of refrigeration,
(b) The power input per kW of refrigeration,
(c) The C.O.P.$_r$.

(1003·4 kJ/kg; 3·59 kg/h; 0·338 kW; 2·96; 3·57 kg/h;
0·3 kW; 3·33)

16.7 In a refrigerating plant using refrigerant-12 the vapour leaves the evaporator dry saturated, and is compressed adiabatically but not reversibly in a centrifugal compressor. The evaporator and condenser pressures are 1·826 bar and 7·449 bar respectively, and the temperature of the vapour leaving the compressor is 45°C. The liquid leaves the condenser at 25°C and is throttled to the evaporator pressure. Calculate the refrigerant effect, the work done per kg of refrigerant, and the C.O.P.$_r$ for this machine. What would be the C.O.P.$_r$ if the compression could be carried out reversibly and approximately adiabatically, in a reciprocating compressor? Under what circumstances would a centrifugal machine be preferable?

(121·27 kJ/kg; 29·66 kJ/kg; 4·09; 4·89)

16.8 A cold storage plant is used to cool 9000 litres of milk per hour from 27°C to 4°C, and the heat leakage into the plant is estimated at 3600 kJ/min. The refrigerant is to be ammonia and the temperature required in the evaporator is −6°C. The compressor delivery pressure is 10·34 bar and the condenser liquid is undercooled to 24°C before throttling. The plant has a brine circulating system and the rise in temperature of the brine is to be limited to 3 K. Determine, assuming that the vapour is dry saturated on leaving the evaporator and that the compression process is isentropic,

(a) the power input required in kW taking the mechanical efficiency of the compressor as 90%,

(b) the swept volume of each cylinder of the twin-cylinder, single-acting compressor, for which the volumetric efficiency can be taken as 85% (the rotational speed is 200 rev/min),

(c) the rate at which the brine must be circulated in litres per second.

For milk:

specific heat = 3·77 kJ/kg K: density = 1030 kg/m³

For brine:

specific heat = 2·93 kJ/kg K: density = 1190 kg/m³

(42·65 kW; 0·0157 m³; 2·71 l/s)

16.9 A combined domestic unit serves the dual purpose of cooling the kitchen larder and providing hot water. The motor driving the refrigerator operates for approximately one third of the day, and has an electrical input of 0·225 kW. The heat leakage into the larder from the kitchen is 0·29 kJ/s. The mechanical efficiency of the compressor is 85% and the compression process can be taken to be adiabatic. All the heat rejected by the refrigerant is taken by the water in the domestic hot water tank, which is heated from 10°C to 60°C. Calculate the amount of hot water which can be supplied by this plant in litres per hour.

(6·09 l/h)

16.10 An air refrigerator is to be designed consisting of a centrifugal compressor and an air turbine mounted coaxially such that the power output of the turbine contributes to the work required to drive the compressor. The temperature of the air at the compressor inlet is 15°C and the pressure ratio is 2·5/1. The air during its passage from the compressor to the turbine passes through an intercooler and enters the turbine at 40°C. The cold space temperature is required to be maintained at 15°C. Take the isentropic efficiencies of both compressor and turbine to be 84%. Calculate,

(a) The refrigerating effect per kg of refrigerant,
(b) The mass flow per kW of refrigeration,
(c) The driving power required per kW of refrigeration.

Take the mechanical efficiency of the turbine/compressor drive as 90%.   (35·8 kJ/kg; 1·675 kg/min; 1·317 kW)

16.11 In an aircraft refrigerating unit air is bled from the engine compressor at 3·5 bar and 270°C, and is passed through an air cooled heat exchanger. The refrigerant air bleed leaves the exchanger at 3·5 bar and 75°C and is expanded through a turbine to 0·76 bar. The isentropic efficiency of the turbine is 85%. The air is then delivered to the aircraft cabin and leaves the aircraft at 16°C. Calculate the refrigerating effect per kg of air and the power developed by the air turbine per kg of air per second.   (45·25 kJ/kg; 104·6 kW)

16.12 It is proposed to use a heat pump working on the ideal vapour compression cycle for the purpose of heating the air supply to a building. The supply of heat is taken from a river at 7°C. Air is required to be delivered into the building at 1·013 bar, and 32°C at the rate of 0·5 m³/s. The air is heated at constant pressure from 10°C as it passes over the condenser coils of the heat pump. The refrigerant is refrigerant-12 which is dry saturated leaving the evaporator; there is no undercooling in the condenser. A temperature difference of 17 K is necessary for the transfer of heat from the river to the freon in the evaporator. The delivery pressure of the compressor is 10·84 bar. Calculate,

(a) The mass flow of refrigerant in kg/s,
(b) The motor power required to drive the compressor if the mechanical efficiency is 87%,
(c) The C.O.P.$_{hp}$,
(d) The swept volume of the compressor which is single-acting and which runs at 240 rev/min. The volumetric efficiency can be taken as 85%.   (0·097 kg/s; 3·18 kW; 4·63; 2180 cm³)

REFERENCES

16.1 KELL, J. R., and MARTIN, P. L., *Faber and Kell's Heating and Air Conditioning of Buildings* (The Architectural Press, 1984).

16.2 DOSSAT, R. J., *Principles of Refrigeration* (Wiley, 1981).

16.3 GOSNEY, W. B., *Principles of Refrigeration* (Cambridge University Press, 1982).

16.4 ELONKA, S. M., and MINICH, Q. W., *Standard Refrigeration and Air Conditioning* (Tata McGraw-Hill, 1983).

16.5 HOLLAND, F. A., WATSON, F. A., and DEROTTA, S., *Thermodynamic Design Data for Heat Pump Systems* (Pergamon, 1982).

16.6 MEACOCK, H. M., *Refrigeration Processes* (Pergamon, 1979).

16.7 LUDWING VON CUBE, H., and STEIMLE, F., *Heat Pump Technology* (Butterworth, 1981).

16.8 PITA, E. G., *Refrigeration Principles and Systems* (Wiley, 1984).

17. Heat Transfer

The transfer of heat across the boundaries of a system, either to or from the system, is considered in previous chapters for non-flow and flow processes, but up to this point no mention is made of the actual mechanism of heat transfer; the definition of heat used throughout this book, and given in Section 1.1, simply states that heat is a form of energy which is transferred from one body to another body at a lower temperature, by virtue of the temperature difference between the bodies. When the mechanism of the transfer of heat is considered a slightly different approach is necessary compared with the approach of fundamental thermodynamics. For instance it becomes difficult to define a system. In order to illustrate this point consider a bar of metal being heated at one end and cooled at the other. Now a boundary may be put round the source of heat or round the sink for the rejection of heat, but a boundary encircling the metal bar encloses a body the temperature of which varies throughout its length. In order to apply the laws of thermodynamics to the system consisting of the metal bar, a mean temperature must be assumed.

In previous chapters many problems have been considered in which a certain quantity of heat has been transferred from one system to another. In this chapter we shall be concerned with the *rate* at which heat is transferred. The rate of heat transfer may be constant or variable, depending on whether conditions are such that the temperatures remain the same or change continually with time. Most problems in practice are concerned with *steady state* heat transfer, in which heat flows continuously at a uniform rate but there are many cases of *transient* heat transfer and some of these will also be considered.

In general there are three ways in which heat may be transferred:

(a) *By conduction*

Conduction is the transfer of heat from one part of a substance to another part of the same substance, or from one substance to another in physical contact with it, without appreciable displacement of the molecules forming the substance. For example, the heat transfer in the metal bar mentioned previously is by conduction.

(b) *By convection*

Convection is the transfer of heat within a fluid by the mixing of one portion of the fluid with another. The movement of the fluid may be caused by differences in density resulting from the temperature differences as in *natural convection* (or *free convection*), or the motion may be produced by mechanical means, as in *forced convection*. For example, the heat transferred from a hot-plate to the atmosphere is by natural convection, whereas the heat transferred by a domestic fan-heater, in which a fan blows air across an electric element, is by forced convection.

The transfer of heat through solid bodies is by conduction alone, whereas the heat transfer from a solid surface to a liquid or gas takes place partly by conduction and partly by convection. Whenever there is an appreciable movement of the gas or liquid, the heat transfer by conduction in the gas or liquid becomes negligibly small compared with the heat transfer by convection. However, there is always a thin boundary layer of fluid on a surface, and through this thin film the heat is transferred by conduction.

(c) *By radiation*

All matter continuously emits electro-magnetic radiation unless its temperature is absolute zero. It is found that the higher the temperature then the greater is the amount of energy radiated. If, therefore, two bodies at different temperatures are so placed that the radiation from each body is intercepted by the other, then the body at the lower temperature will receive more energy than it is radiating, and hence its internal energy will increase; similarly the internal energy of the body at the higher temperature will decrease. Thus there is a net transfer of energy from the high-temperature body to the low-temperature body by virtue of the temperature difference between the bodies. This form of energy transfer satisfies the definition of heat given in Section 1.1, and hence we may say that heat is transferred by radiation.

Radiant energy, being electro-magnetic radiation, requires no medium for its propagation, and will pass through a vacuum. Heat transfer by radiation is most frequent between solid surfaces, although radiation from gases also occurs. Certain gases emit and absorb radiation on certain wavelengths only, whereas most solids radiate over a wide range of wavelengths. The selective emitting and absorbing property of gases has already been mentioned in Section 15.7.

In any particular example in practice heat may be transferred by a combination of conduction, convection, and radiation, and it is usually possible to assess the effects of each mode of heat transfer separately and then to sum up the results. There are two main groups of problems; first, the desirable transfer of heat to or from a fluid as in a heat exchanger, boiler or condenser, and second, the prevention of heat losses from a fluid to its surroundings.

17.1 Fourier's law of conduction

Fourier's law states that the rate of flow of heat through a single homogeneous solid is directly proportional to the area A of the section at right angles to the direction of heat flow, and to the change of temperature with respect to the length of the path of the heat flow, dt/dx. This is an empirical law based on observation.

The law is illustrated in fig. 17.1a in which a thin slab of material of thickness dx and surface area A has one face at a temperature t and the other at a lower temperature $(t-dt)$. Heat flows from the high-temperature face to the low-temperature face, and the

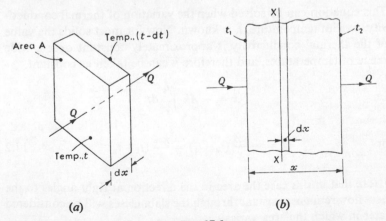

Fig. 17.1

temperature change in the direction of the heat flow is $-\mathrm{d}t$. Then applying Fourier's law we have

$$\text{Rate of heat flow, } Q \propto A \frac{\mathrm{d}t}{\mathrm{d}x}$$

or

$$Q = -kA \frac{\mathrm{d}t}{\mathrm{d}x} \qquad 17.1$$

The constant k is called the *thermal conductivity* of the material. The thermal conductivity of a substance can be defined as the heat flow per unit area per unit time when the temperature decreases by one degree in unit distance.

When Q is in W, A in m², and $\mathrm{d}t/\mathrm{d}x$ in K/m, then the units of k are $\dfrac{\mathrm{W}}{\mathrm{m}^2\,\mathrm{K/m}}$. This is usually written as W/m K.

Consider the transfer of heat through a slab of material as shown in fig. 17.1b. At section X–X, using equation 17.1,

$$Q = -kA \frac{\mathrm{d}t}{\mathrm{d}x} \quad \text{or} \quad Q\,\mathrm{d}x = -kA\,\mathrm{d}t$$

Integrating,

$$\int_0^z Q\,\mathrm{d}x = -\int_{t_1}^{t_2} kA\,\mathrm{d}t$$

i.e.

$$Qx = -A \int_{t_1}^{t_2} k\,\mathrm{d}t$$

This equation can be solved when the variation of thermal conductivity, k, with temperature, t, is known. Now for most solids the value of the thermal conductivity is approximately constant over a wide range of temperatures, and therefore k can be taken as constant,

i.e.

$$Qx = -Ak \int_{t_1}^{t_2} \mathrm{d}t$$

or

$$Q = -\frac{kA}{x}(t_2 - t_1) = \frac{kA}{x}(t_1 - t_2) \qquad 17.2$$

(Note that in this case the area in the direction at right angles to the heat flow remains constant through the slab. Cases will be considered later in which the area varies.)

The thermal conductivities of some materials encountered in engineering are shown in Table 17.1. It follows from equation 17.1

Table 17.1

Substance	Thermal conductivity W/m K
Pure copper	386
Pure aluminium	229
Duralumin	164
Cast iron	52
Mild steel	48·5
Lead	34·6
Concrete	0·85 to 1·4
Building brick	0·35 to 0·7
Wood (oak)	0·15 to 0·2
Rubber	0·15
Cork board	0·043

that materials with high thermal conductivities are good conductors of heat, whereas materials with low thermal conductivities are good thermal insulators. Conduction of heat occurs most readily in pure metals, less so in alloys, and much less readily in non-metals. The very low thermal conductivities of certain thermal insulators (e.g. cork) is due to their porosity, the air trapped within the material acting as an insulator. Gases and liquids are good insulators, but unless a completely stagnant layer of fluid is obtained, heat is transferred by convection currents.

Example 17.1

The inner surface of a plane brick wall is at 40°C and the outer surface is at 20°C. Calculate the rate of heat transfer per m² of surface area of the wall, which is 250 mm thick. The thermal conductivity of the brick is 0·52 W/m K.

From equation 17.2,

$$Q = \frac{kA}{x}(t_1 - t_2)$$

$$\therefore \frac{Q}{A} = q = \frac{10^3 \times 0·52}{250} \times (40-20) = 41·6 \text{ W/m}^2$$

(Note that the symbol q is used for the rate of heat transfer per unit area.)

17.2 Newton's law of cooling

In order to consider the rate at which heat is transferred from one fluid to another through a plane wall it is necessary to know something of the way in which heat is transferred from a solid surface to a fluid and vice versa.

Newton's law of cooling states that the heat transfer from a solid surface of area A, at a temperature t_w, to a fluid of temperature t, is given by,

$$Q = hA(t_w - t) \qquad 17.3$$

where h is called the *heat transfer coefficient*.

The units of h are seen to be W/m² K.

The heat transfer coefficient, h, depends on the properties of the fluid and on the fluid velocity; it is usually necessary to evaluate it by experiment. This will be discussed more fully in Section 17.6.

Equation 17.3 does not include the heat loss from the surface by radiation. This effect can be calculated separately (see Section 17.13), and in many cases is negligible compared with the heat transferred by conduction and convection from the surface to the fluid. When the surface temperature is high, or when the surface loses heat by natural convection, then the heat transfer due to radiation is of a similar magnitude to that lost by convection.

Consider the transfer of heat from a fluid A to a fluid B through a dividing wall of thickness x, and thermal conductivity k, as shown in

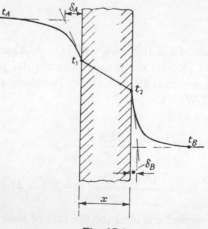

Fig. 17.2

fig. 17.2. The variation of temperature in the direction of the heat transfer is also shown. In fluid A the temperature decreases rapidly from t_A to t_1 in the region of the wall, and similarly in fluid B the temperature decreases rapidly from t_2 to t_B in the region of the wall. In most practical cases the fluid temperature is approximately constant throughout its bulk, apart from a thin film near the solid surface bounding the fluid. The dotted lines drawn on fig. 17.2 show that the thickness of this film of fluid is given by δ_A for fluid A and δ_B for fluid B. The heat transfer in these films is by conduction only, hence applying equation 17.2 we have, considering unit surface area,

from fluid A to the wall,

$$q = \frac{k_A}{\delta_A}(t_A - t_1) \tag{a}$$

from the wall to fluid B,

$$q = \frac{k_B}{\delta_B}(t_2 - t_B) \tag{b}$$

Also from equation 17.3,

from fluid A to the wall,

$$q = h_A(t_A - t_1) \tag{c}$$

from the wall to fluid B,

$$q = h_B(t_2 - t_B) \tag{d}$$

Comparing equations (a) and (c), and equations (b) and (d), it can be seen that,

$$h_A = \frac{k_A}{\delta_A} \quad \text{and} \quad h_B = \frac{k_B}{\delta_B}$$

In general, $h = k/\delta$, where δ is the thickness of the stagnant film of fluid on the surface.

The heat flow through the wall in fig. 17.2 is given by equation 17.2.

For unit surface area, $q = \dfrac{k}{x}(t_1 - t_2)$

For steady state heat transfer, the heat flowing from fluid A to the wall is equal to the heat flowing through the wall, which is also equal to the heat flowing from the wall to fluid B. If this were not so, then the temperatures t_A, t_1, t_2, and t_B would not remain constant but would change with time.

We therefore have

$$q = h_A(t_A - t_1) = \frac{k}{x}(t_1 - t_2) = h_B(t_2 - t_B)$$

Rewriting these equations in terms of the temperatures, then,

$$(t_A - t_1) = \frac{q}{h_A}; \qquad (t_1 - t_2) = \frac{qx}{k}; \qquad (t_2 - t_B) = \frac{q}{h_B}$$

Hence adding the corresponding sides of the three equations,

$$(t_A - t_1) + (t_1 - t_2) + (t_2 - t_B) = \frac{q}{h_A} + \frac{qx}{k} + \frac{q}{h_B}$$

$$\therefore (t_A - t_B) = q\left(\frac{1}{h_A} + \frac{x}{k} + \frac{1}{h_B}\right)$$

i.e.

$$q = \frac{(t_A - t_B)}{\left(\dfrac{1}{h_A} + \dfrac{x}{k} + \dfrac{1}{h_3}\right)}$$

By analogy with equation 17.3 this can be written as

$$q = U(t_A - t_B) \qquad\qquad 17.4$$

or

$$Q = UA(t_A - t_B) \qquad\qquad 17.5$$

where

$$\frac{1}{U} = \left(\frac{1}{h_A} + \frac{x}{k} + \frac{1}{h_B}\right) \qquad\qquad 17.6$$

U is called the *overall heat transfer coefficient*, and it has the same units as h.

Example 17.2

A mild steel tank of wall thickness 10 mm contains water at 90°C. Calculate the rate of heat loss per m² of tank surface area when the atmospheric temperature is 15°C. The thermal conductivity of mild steel is 50 W/m K, and the heat transfer coefficients for the inside and outside of the tank are 2800 and 11 W/m² K, respectively. Calculate also the temperature of the outside surface of the tank.

The wall of the tank is shown diagrammatically in fig. 17.3.

From equation 17.6,

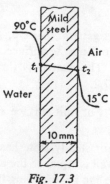

Fig. 17.3

$$\frac{1}{U} = \frac{1}{h_A} + \frac{x}{k} + \frac{1}{h_B} = \frac{1}{2800} + \frac{10}{10^3 \times 50} + \frac{1}{11}$$

$$= 0.000\ 357 + 0.0002 + 0.0909$$

i.e.

$$\frac{1}{U} = 0.0915$$

Then substituting in equation 17.4, $q = U(t_A - t_B)$,

i.e.

$$q = \left(\frac{90 - 15}{0.0915}\right) = 820 \text{ W/m}^2$$

i.e.   Rate of heat loss per m^2 of surface area $= 0.82$ kW

From equation 17.3,

$$q = h_B(t_2 - t_B) \qquad \therefore \quad 820 = 11 \times (t_2 - 15)$$

(where t_2 is the temperature of the outside surface of the tank as shown in fig. 17.3),

i.e.

$$t_2 = \frac{820}{11} + 15 = 89.6°C$$

i.e.   Temperature of outside surface of tank $= 89.6°C$

17.3 The composite wall and the electrical analogy

There are many cases in practice when different materials are constructed in layers to form a composite wall. An example of this is the wall of a building, which usually consists of a layer of plaster, a row of bricks, an air gap, a second row of bricks, and perhaps a cement rendering on the outside surface.

Consider the general case of a composite wall as shown in fig. 17.4.

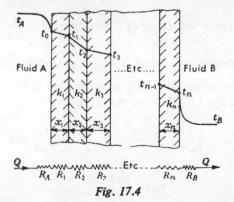

Fig. 17.4

There are n layers of material of thickness x_1, x_2, x_3, etc., and of thermal conductivity k_1, k_2, k_3, etc. On one side of the composite wall there is a fluid A at temperature t_A, and the heat transfer coefficient from fluid to wall is h_A; on the other side of the composite wall there is a fluid B, and the heat transfer coefficient from wall to fluid is h_B. Let the temperature of the wall in contact with fluid A be t_0 and the temperature of the wall in contact with fluid B be t_n; the interface temperatures are then t_1, t_2, t_3, etc., as shown. The most convenient method of solving such a problem is by making use of an electrical analogy. The flow of heat can be thought of as analogous to an electric current. The heat flow is caused by a temperature difference whereas the current flow is caused by a potential difference, hence it is possible to postulate a *thermal resistance* analogous to an electrical resistance. From Ohm's law we have,

$$V = IR \quad \text{or} \quad I = \frac{V}{R}$$

(where V is potential difference, I is current, and R is resistance).

Comparing this equation with equation 17.2, $Q = \{kA/x\}(t_1 - t_2)$, we have,

$$\text{Thermal resistance, } R = \frac{x}{kA} \qquad 17.7$$

(where Q is analogous to I, and $(t_1 - t_2)$ is analogous to V).

The composite wall is analogous to a series of resistances, as shown in fig. 17.4, and resistances in series can be added to give the total resistance. To find the resistance of a fluid film it is necessary to compare Ohm's law with equation 17.3, $Q = hA(t_w - t)$,

i.e. $$\text{Thermal resistance of a fluid film, } R = \frac{1}{hA} \qquad 17.8$$

(where Q is analogous to I and $(t_w - t)$ is analogous to V).

Note that the units of thermal resistance are K/W.
Referring to fig. 17.4 we therefore have

$$R_A = \frac{1}{h_A A}, \quad R_1 = \frac{x_1}{k_1 A}, \quad R_2 = \frac{x_2}{k_2 A}, \text{ etc.,}$$

$$R_n = \frac{x_n}{k_n A} \quad \text{and} \quad R_B = \frac{1}{h_B A}$$

The total resistance to heat flow is then

$$R_T = R_A + R_1 + R_2 + \ldots + R_n + R_B = \frac{1}{h_A A} + \frac{x_1}{k_1 A} + \text{etc.,} \ \frac{x_n}{k_n A} + \frac{1}{h_B A}$$

Or for any number of layers of material,

$$\text{Total resistance, } R_T = \frac{1}{h_A A} + \sum \frac{x}{kA} + \frac{1}{h_B A} \qquad 17.9$$

It can be seen from equation 17.9 that in this case the surface area, A, remains constant through the wall, and it is usual to calculate the total resistance for unit surface area in such problems. Cases in which the area varies through the various layers are considered in Section 17.4.

Using the electrical analogy for the overall heat transfer we have,

$$Q = \frac{t_A - t_B}{R_T} \qquad 17.10$$

(analogous to $I = V/R$).

In equation 17.6 the overall heat transfer coefficient, U, is defined as

$$\frac{1}{U} = \frac{1}{h_A} + \frac{x}{k} + \frac{1}{h_B}$$

For any number of walls we have

$$\frac{1}{U} = \frac{1}{h_A} + \sum \frac{x}{k} + \frac{1}{h_B}$$

It can be seen that the reciprocal of U is simply the thermal resistance for unit area,

i.e. $$\frac{1}{U} = R_T A \quad \text{or} \quad U = \frac{1}{R_T A} \qquad 17.11$$

If the inner and outer wall surface temperatures are known then the heat transfer can be found by calculating the thermal resistance of the composite wall only,

i.e. $$R = \sum \frac{x}{kA}$$

The overall heat transfer coefficient from one wall surface to the other is given by

$$\frac{1}{U} = \sum \frac{x}{k}$$

It should be noted that there may be an additional thermal resistance at the various interfaces of a composite wall, due to the small pockets of air trapped between the surfaces.

Example 17.3

A furnace wall consists of 125 mm wide refractory brick and 125 mm wide insulating firebrick separated by an air gap. The outside wall is covered with a 12 mm thickness of plaster. The inner surface of the wall is at 1100°C and the room temperature is 25°C. Calculate the rate at which heat is lost per m² of wall surface. The heat transfer coefficient from the outside wall surface to the air in the room is 17 W/m² K, and the resistance to heat flow of the air gap is 0.16 K/W. The thermal conductivity of refractory brick, insulating firebrick, and plaster are 1.6, 0.3, and 0.14 W/m K, respectively. Calculate also each interface temperature, and the temperature of the outside surface of the wall.

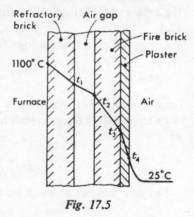

Fig. 17.5

The wall is shown in fig. 17.5. Consider 1 m² of surface area. Then using equation 17.7, $R = x/kA$, we have

$$\text{Resistance of refractory brick} = \frac{125}{10^3 \times 1.6} = 0.0781 \text{ K/W}$$

$$\text{Resistance of insulating firebrick} = \frac{125}{10^3 \times 0.3} = 0.417 \text{ K/W}$$

$$\text{Resistance of plaster} = \frac{12}{10^3 \times 0.14} = 0.0857 \text{ K/W}$$

Also using equation 17.8 for a fluid film, $R = 1/hA$, we have

$$\text{Resistance of air film on outside surface} = \frac{1}{17} \text{ K/W}$$

Hence,

$$\text{Total resistance, } R_T = 0.0781 + 0.417 + 0.0857 + \frac{1}{17} + 0.16$$

(where the resistance of the air gap is 0.16 K/W),

i.e. $$R_T = 0.8 \text{ K/W}$$

Then using equation 17.10,

$$Q = \frac{t_A - t_B}{R_T} = \frac{1100 - 25}{0.8} = 1344 \text{ W}$$

i.e.   Rate of heat loss per m² of surface area = 1·344 kW

Referring to fig. 17.5, the interface temperatures are t_1, t_2, and t_3; the outside surface is at t_4. Applying the electrical analogy to each layer and using the values of thermal resistance calculated above, we have

$$Q = 1344 = \frac{1100 - t_1}{0.0781}$$

i.e. $$t_1 = 1100 - 1344 \times 0.0781 = 995°C$$

Also, $$Q = 1344 = \frac{t_1 - t_2}{0.16}$$

i.e. $$t_2 = 995 - 0.16 \times 1344 = 780°C$$

$$Q = 1344 = \frac{t_2 - t_3}{0.417}$$

i.e. $$t_3 = 780 - 1344 \times 0.417 = 220°C$$

And, $$Q = 1344 = \frac{t_3 - t_4}{0.0857}$$

i.e. $$t_4 = 220 - 1344 \times 0.0857 = 104°C$$

The temperature t_4 can also be found by considering the air film,

i.e. $$Q = 1344 = \frac{t_4 - 25}{1/17}$$

i.e. $$t_4 = 1344 \times \frac{1}{17} + 25$$

$$\therefore t_4 = 104·1°C$$

17.4 Heat flow through a cylinder and a sphere

One of the most commonly occurring problems in practice is the case of heat being transferred through a pipe or cylinder. Less common is the case of heat being transferred through a spherical wall, but both cases will now be considered.

(a) *The cylinder*

Consider a cylinder of internal radius r_1, and external radius r_2 as shown in fig. 17.6. Let the inside and outside surface temperatures be t_1 and t_2, respectively. Consider the heat flow through a small element, thickness dr, at any radius r, where the temperature is t. Let the conductivity of the material be k. Then applying equation 17.1, for unit length in the axial direction, we have

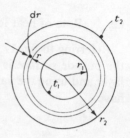

Fig. 17.6

$$Q = -kA \frac{dt}{dx} = -k(2\pi r \times 1) \frac{dt}{dr}$$

i.e.

$$Q \frac{dr}{r} = -2\pi k \, dt$$

Integrating between the inside and outside surfaces,

$$Q \int_{r_1}^{r_2} \frac{dr}{r} = -2\pi k \int_{t_1}^{t_2} dt$$

(where Q and k are both constant),

$$\therefore \quad Q \log_e \frac{r_2}{r_1} = -2\pi k(t_2 - t_1) = 2\pi k(t_1 - t_2)$$

i.e.

$$Q = \frac{2\pi k(t_1 - t_2)}{\log_e r_2/r_1} \qquad 17.12$$

Now from equation 17.2,

$$Q = \frac{kA}{x}(t_1 - t_2)$$

If we substitute a mean area A_m in this equation, and substitute also for the thickness $x = (r_2 - r_1)$, we have

$$Q = \frac{kA_m(t_1 - t_2)}{(r_2 - r_1)}$$

Comparing this equation with equation 17.12, then

$$Q = \frac{kA_m(t_1 - t_2)}{(r_2 - r_1)} = \frac{2\pi k(t_1 - t_2)}{\log_e r_2/r_1}$$

$$\therefore \frac{A_m}{r_2 - r_1} = \frac{2\pi}{\log_e r_2/r_1}$$

i.e.
$$A_m = \frac{2\pi(r_2 - r_1)}{\log_e r_2/r_1} = \frac{A_2 - A_1}{\log_e r_2/r_1}$$

A_m is called the logarithmic mean area, and using this area in equation 17.2 an exact solution is obtained. It can be seen from the above that there is also a logarithmic mean radius given by

$$r_m = \frac{r_2 - r_1}{\log_e r_2/r_1}$$

In the case of a composite cylinder (e.g. a metal pipe with several layers of lagging) the most convenient approach is again that of the electrical analogy; by using equation 17.7

$$R = \frac{x}{kA_m}$$

(where x is the thickness of a layer, and A_m is the logarithmic mean area for that layer).

From equation 17.12, applying the electrical analogy, $(I = V/R)$, it can be seen that,

$$R = \frac{\log_e (r_2/r_1)}{2\pi k} \qquad 17.13$$

The film of fluid on the inside and outside surfaces can be treated as before using equation 17.8,

i.e.
$$R_{outside} = \frac{1}{h_0 A_0}$$

(where A_0 is the outside surface area, $2\pi r_2$, referring to fig. 17.6, and h_0 is the heat transfer coefficient for the outside surface).

Also,
$$R_{inside} = \frac{1}{h_i A_i}$$

(where A_i is the inside surface area, $2\pi r_1$ and h_i is the heat transfer coefficient for the inside surface)

It can be seen from equation 17.12,

$$Q = \frac{2\pi k(t_1 - t_2)}{\log_e r_2/r_1}$$

that the heat transfer rate depends on the ratio of the radii, r_2/r_1, and not on the difference, $(r_2 - r_1)$. The smaller the ratio, r_2/r_1, then the higher is the heat flow for the same temperature difference. In many practical problems the ratio, r_2/r_1, tends towards unity since the pipe wall thickness or lagging thickness is usually small compared with the mean radius. In these cases it is a sufficiently close approximation to use the arithmetic mean radius,

i.e.   Arithmetic mean radius $= \dfrac{r_2 + r_1}{2}$

The error in the rate of heat transfer in using the arithmetic mean instead of the logarithmic mean is just over 4% for a ratio $r_2/r_1 = 2$. Most heat transfer experiments in practice cannot give better accuracy than about 4% or 5%, hence it is a good approximation to use the arithmetic mean area when r_2/r_1 is less than 2.

Example 17.4

A steel pipe of 100 mm bore and 7 mm wall thickness, carrying steam at 260°C, is insulated with 40 mm of a moulded high-temperature diatomaceous earth covering. This covering is in turn insulated with 60 mm of asbestos felt. If the atmospheric temperature is 15°C calculate the rate at which heat is lost by the steam per m length of pipe. The heat transfer coefficients for the inside and outside surfaces are 550 and 15 W/m² K, respectively, and the thermal conductivities of steel, diatomaceous earth, and asbestos felt are 50, 0·09, and 0·07 W/m K respectively. Calculate also the temperature of the outside surface.

A cross-section of the pipe is shown in fig. 17.7.

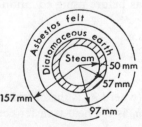

Fig. 17.7

Consider 1 m length of the pipe. From equation 17.8,

$$R = \frac{1}{hA}$$

i.e.

$$\text{Resistance of steam film} = \frac{10^3}{550 \times 2\pi \times 50 \times 1} = 0.005\ 79 \text{ K/W}$$

From equation 17.13, for the steel pipe

$$R = \frac{\log_e (r_2/r_1)}{2\pi k}$$

$$\therefore \ \text{Resistance of pipe} = \frac{\log_e (57/50)}{2\pi \times 50} = 0.000\ 417 \text{ K/W}$$

Similarly,

$$\text{Resistance of diatomaceous earth} = \frac{\log_e (97/57)}{2\pi \times 0.09}$$

$$= 0.94 \text{ K/W}$$

Similarly,

$$\text{Resistance of asbestos felt} = \frac{\log_e (157/97)}{2\pi \times 0.07}$$

$$= 1.095 \text{ K/W}$$

From equation 17.8, for the air film on the outside surface,

$$\text{Resistance of air film} = \frac{1}{hA} = \frac{10^3}{15 \times 2\pi \times 157 \times 1} = 0.0675 \text{ K/W}$$

Hence,

Total resistance, $R_T = 0.005\ 79 + 0.000\ 417 + 0.94 + 1.095 + 0.0675$

i.e. $\qquad\qquad R_T = 2.1087 \text{ K/W}$

(Note that the resistance to heat flow of the pipe metal is very small; also in this case the resistance of the film on the inside surface is very small because the heat transfer coefficient for steam is high.)

Then, using equation 17.10,

$$Q = \frac{t_A - t_B}{R_T} = \frac{260 - 15}{2 \cdot 1087} = 116 \text{ W}$$

i.e.   Rate of heat loss per m length of pipe = 116 W

Using the electrical analogy for the air film we have,

$$Q = 116 = \frac{t - 15}{0 \cdot 0675}$$

(where t is the temperature of the outside surface),

i.e. $t = 116 \times 0 \cdot 0675 + 15 = 22 \cdot 8°C$

i.e.   Temperature of outside surface = $22 \cdot 8°C$

(b) *The sphere*

Consider a hollow sphere of internal radius r_1 and external radius r_2, as shown in fig. 17.8. Let the inside and outside surface temperatures be t_1 and t_2, and let the thermal conductivity be k. Consider a small element of thickness dr at any radius r. It can be shown that the surface area of this spherical element is given by $4\pi r^2$. Then, using equation 17.1,

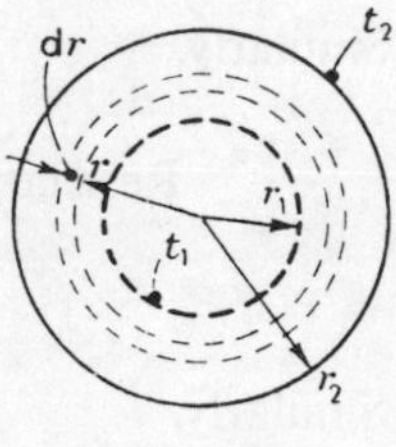

Fig. 17.8

$$Q = -kA \frac{dt}{dr} = -k4\pi r^2 \frac{dt}{dr}$$

Integrating,

$$Q \int_{r_1}^{r_2} \frac{dr}{r^2} = -4\pi k \int_{t_1}^{t_2} dt$$

$$\therefore \ -Q\left(\frac{1}{r_2} - \frac{1}{r_1}\right) = -4\pi k(t_2 - t_1)$$

i.e. $$\frac{Q(r_2 - r_1)}{r_1 r_2} = 4\pi k(t_1 - t_2)$$

or $$Q = \frac{4\pi k r_1 r_2 (t_1 - t_2)}{(r_2 - r_1)} \qquad (a)$$

Hence applying the electrical analogy, $(I = V/R)$, we have,

$$R = \frac{(r_2 - r_1)}{4\pi k r_1 r_2} \qquad 17.14$$

If a mean area, A_m, is introduced, then from equation 17.2,

$$Q = \frac{kA_m}{x}(t_1 - t_2) = \frac{kA_m(t_1 - t_2)}{(r_2 - r_1)} \qquad \text{(b)}$$

Comparing the equations (a) and (b) above, we have

$$A_m = 4\pi r_1 r_2$$

A mean radius, r_m, can be defined,

i.e. $$A_m = 4\pi r_m^2 = 4\pi r_1 r_2$$

$$\therefore \text{ Mean radius, } r_m = \sqrt{(r_1 r_2)}$$

It can be seen that r_m is a geometric mean radius.

Example 17.5

A small hemispherical oven is built of an inner layer of insulating firebrick 125 mm thick, and an outer covering of 85% magnesia 40 mm thick. The inner surface of the oven is at 800°C and the heat transfer coefficient for the outside surface is 10 W/m² K; the room temperature is 20°C. Calculate the heat loss through the hemisphere if the inside radius is 0·6 m. Take the thermal conductivities of firebrick and 85% magnesia as 0·31 and 0·05 W/m K, respectively.

For the insulating firebrick:

From equation 17.14, for a hemisphere,

$$\text{Resistance of firebrick} = \frac{0·125}{2\pi \times 0·31 \times 0·6 \times 0·725}$$

$$= 0·1478 \text{ K/W}$$

For the 85% magnesia:

$$\text{Resistance of 85\% magnesia} = \frac{0·04}{2\pi \times 0·05 \times 0·725 \times 0·765}$$

$$= 0·2295 \text{ K/W}$$

For the outside surface:

From equation 17.8,

$$\text{Resistance of outside air film} = \frac{1}{hA} = \frac{1}{10 \times 2\pi \times 0·765^2}$$

$$= 0·0272 \text{ K/W}$$

Hence,

$$\text{Total resistance, } R_T = 0.1478 + 0.2295 + 0.0272$$
$$= 0.4045 \text{ K/W}$$

Then using equation 17.10,

$$Q = \frac{t_A - t_B}{R_T} = \frac{800 - 20}{0.4045} = 1930 \text{ W}$$

i.e.   Rate of heat loss from the oven = 1.93 kW

17.5 General conduction equation

A general equation may be derived for a three-dimensional solid in which there is uniform internal heat generation (due to ohmic heating for example), and a change of temperature with time.

Consider an element at a temperature, t, at any instant of time, τ, within a homogeneous solid as shown in fig. 17.9. Let the rate of internal heat generation per unit volume be q_g. Let the density of the material be, ρ, the specific heat, c, and the thermal conductivity, k; assume that these properties are uniform and constant with time. By

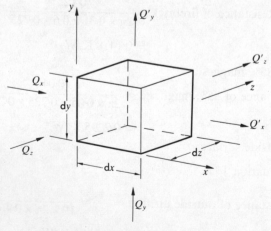

Fig. 17.9

Fourier's law:

$$Q_x = -k(\mathrm{d}y\,\mathrm{d}z)\frac{\partial t}{\partial x}$$

$$Q_y = -k(\mathrm{d}x\,\mathrm{d}z)\frac{\partial t}{\partial y}$$

$$Q_z = -k(\mathrm{d}x\,\mathrm{d}y)\frac{\partial t}{\partial z}$$

For the x-direction,

$$Q'_x - Q_x = \frac{\partial Q}{\partial x}\,\mathrm{d}x = -k\frac{\partial^2 t}{\partial x^2}\,\mathrm{d}y\,\mathrm{d}z\,\mathrm{d}x$$

Similarly, for the y and z directions.

$$\text{Rate of heat generation} = q_g(\mathrm{d}x\,\mathrm{d}y\,\mathrm{d}z)$$

Also,

Rate of increase of energy of the element = mass × specific heat × rate of change of temperature with time

$$= \rho(\mathrm{d}x\,\mathrm{d}y\,\mathrm{d}z)c\frac{\partial t}{\partial \tau}$$

Hence an energy balance on the element gives,

$$q_g(\mathrm{d}x\,\mathrm{d}y\,\mathrm{d}z) - \{(Q'_x - Q_x) + (Q'_y - Q_y) + (Q'_z - Q_z)\} = \rho c(\mathrm{d}x\,\mathrm{d}y\,\mathrm{d}z)\frac{\partial t}{\partial \tau}$$

$$\therefore\ q_g - \left(-k\frac{\partial^2 t}{\partial x^2} - k\frac{\partial^2 t}{\partial y^2} - k\frac{\partial^2 t}{\partial x^2}\right) = \rho c\frac{\partial t}{\partial \tau}$$

Dividing through by k and introducing the thermal diffusivity, $\alpha = k/\rho c$, we have,

$$\frac{\partial^2 t}{\partial x^2} + \frac{\partial^2 t}{\partial y^2} + \frac{\partial^2 t}{\partial x^2} + \frac{q_g}{k} = \frac{1}{\alpha}\frac{\partial t}{\partial \tau} \qquad 17.15$$

Equations using cylindrical or spherical coordinates may be derived in a similar way, or obtained from equation 17.15 by transforming the coordinates. For one-dimensional problems (e.g. an infinitely long cylinder or a sphere), it is simpler to derive the equations directly as in Section 17.4, and as below.

(a) Semi-infinite slab (see fig. 17.1b).
    From equation 17.15 we have:

$$\frac{\partial^2 t}{\partial x^2} + \frac{q_g}{k} = \frac{1}{\alpha}\frac{\partial t}{\partial \tau} \qquad 17.16$$

(b) Infinitely long cylinder (see fig. 17.6).
Applying an energy balance to an element of thickness, dr, we have:

$$q_g 2\pi r\, dr - \frac{\partial Q}{\partial r}\, dr = \rho c 2\pi r\, dr\, \frac{\partial t}{\partial \tau}$$

$$\therefore\ q_g 2\pi r\, dr - \frac{\partial}{\partial r}\left(-k2\pi r\, \frac{\partial t}{\partial r}\right) dr = \rho c 2\pi r\, dr\, \frac{\partial t}{\partial \tau}$$

$$\therefore\ q_g r + \left(kr\frac{\partial^2 t}{\partial r^2} + k\frac{\partial t}{\partial r}\right) = \rho c r\, \frac{\partial t}{\partial \tau}$$

i.e.
$$\frac{\partial^2 t}{\partial r^2} + \frac{1}{r}\frac{\partial t}{\partial r} + \frac{q_g}{k} = \frac{1}{\alpha}\frac{\partial t}{\partial \tau} \qquad 17.17$$

(c) sphere (see fig. 17.8).
Applying an energy balance,

$$q_g 4\pi r^2\, dr - \frac{\partial Q}{\partial r}\, dr = \rho c 4\pi r^2\, dr\, \frac{\partial t}{\partial \tau}$$

$$\therefore\ q_g 4\pi r^2\, dr - \frac{\partial}{\partial r}\left(-k4\pi r^2\, \frac{\partial t}{\partial r}\right) dr = \rho c 4\pi r^2\, dr\, \frac{\partial t}{\partial \tau}$$

i.e.
$$\frac{\partial^2 t}{\partial r^2} + \frac{2}{r}\frac{\partial t}{\partial r} + \frac{q_g}{k} = \frac{1}{\alpha}\frac{\partial t}{\partial \tau} \qquad 17.18$$

For steady state cases the right-hand side of equations 17.16, 17.17 and 17.18 becomes zero, and the equations become ordinary differential equations.

Example 17.6

A hollow cylindrical copper conductor of 30 mm outside diameter and 14 mm inside diameter has a current density of 40 A/mm^2. The external surface is covered with a uniform layer of insulation of thickness 10 mm, and the ambient temperature is 10°C. Neglecting axial conduction and assuming that the temperature of the insulation must not exceed 135°C at any point, calculate:

(i) the heat required to be removed by forced cooling from the inside of the conductor;

(ii) the temperature at the inside surface of the conductor.

Data: thermal conductivity of copper $= 380$ W/m K; thermal conductivity of insulating material $= 0.3$ W/m K; heat transfer coefficient at outside surface $= 40$ W/m^2 K; electrical resistivity of copper $= 2 \times 10^{-5}$ Ω mm.

From equation 17.17, for the steady state:

$$\frac{d^2 t}{dr^2} + \frac{1}{r}\frac{dt}{dr} + \frac{q_g}{k} = 0$$

or

$$\frac{1}{r}\frac{d}{dr}\left(r\frac{dt}{dr}\right) = -\frac{q_g}{k}$$

Hence integrating,

$$r\frac{dt}{dr} = -\frac{q_g r^2}{2k} + C_1$$

$$\therefore \frac{dt}{dr} = -\frac{q_g r}{2k} + \frac{C_1}{r} \tag{a}$$

Integrating,

$$t = -\frac{q_g r^2}{4k} + C_1 \log_e r + C_2 \tag{b}$$

(C_1 and C_2 are integration constants).

The heat generated per unit volume due to the current flowing is given by:

$$q_g = \frac{I^2 R}{AL}$$

(where $I =$ current; $R =$ electrical resistance; $A =$ cross-sectional area; $L =$ length)

Current density, $J = I/A$ and Resistance, $R = sL/A$

(where $s =$ electrical resistivity of the conductor material)

i.e.

$$q_g = \frac{J^2 A^2}{AL}\frac{sL}{A} = J^2 s \tag{17.19}$$

$$\therefore q_g = 40^2 \times 2 \times 10^{-5} \text{ W/mm}^3 = 32 \times 10^6 \text{ W/m}^3$$

(i) The maximum temperature of the insulation ($=135°C$) occurs at the interface between the insulation and the copper tube. Hence for the insulation, using equation 17.12,

$$\text{Heat transfer to the outside, } Q_o = \frac{2\pi \times 0.3(135-t)}{\log_e (50/30)}$$

(where t is the temperature of the outside surface of the insulation)

$$\therefore \quad 135-t = 0.271 Q_o \tag{c}$$

For the heat transferred from the outside surface of the insulation by convection, from equation 17.3,

$$Q_o = hA(t - t_{\text{fluid}})$$

i.e.

$$Q_o = 40 \times 2\pi \times 0.025(t - 10)$$

$$\therefore \quad t - 10 = 0.159 Q_o \tag{d}$$

Adding equations (c) and (d),

$$135 - 10 = 0.43 Q_o \quad \therefore \quad Q_o = 290.7 \text{ W}$$

Total heat generated internally $= q_g \times \text{volume}$

$$= 32 \times 10^6 \times \frac{\pi}{4}(0.03^2 - 0.014^2)$$

$$= 17\,693.5 \text{ W}$$

Hence,

Heat removed from inside of conductor $= (17\,693.5 - 290.7) \text{ W}$
$$= 17.4 \text{ kW}$$

(ii) Two boundary conditions are required to find the constants C_1 and C_2 and hence to obtain the solution of equation (b).
    At the inside surface of the conductor,

$$\text{Heat supplied to conductor} = -kA\left(\frac{dt}{dr}\right)_{r=0.007} = -17\,400 \text{ W}$$

$$\therefore \quad \left(\frac{dt}{dr}\right)_{r=0.007} = \frac{17\,400}{380 \times \pi \times 0.014} = 1041.1 \text{ K/m}$$

Substituting in equation (a),

$$1041 \cdot 1 = -\frac{32 \times 10^6 \times 0 \cdot 007}{2 \times 380} + \frac{C_1}{0 \cdot 007}$$

$$\therefore \ C_1 = 9 \cdot 351$$

At the outside surface of the conductor, $t = 135°C$, hence in equation (b),

$$135 = -\frac{32 \times 10^6 \times 0 \cdot 015^2}{4 \times 380} + 9 \cdot 351 \log_e 0 \cdot 015 + C_2$$

$$\therefore \ C_2 = 179$$

Therefore the complete solution for the temperature distribution in the conductor is,

$$t = -\left(\frac{32 \times 10^6}{4 \times 380}\right) r^2 + 9 \cdot 351 \log_e r + 179$$

Hence, at the inside surface, when $r = 0 \cdot 007$,

$$t = -\frac{32 \times 10^6 \times 0 \cdot 007^2}{4 \times 380} + 9 \cdot 351 \log_e 0 \cdot 007 + 179$$

$$= 131 \cdot 6°C$$

Transient conduction in one-dimension
The equations for one-dimensional transient conduction, 17.16, 17.17 and 17.18, can be solved using the separation of variables method. For example, from equation 17.16 for the case when there is no internal heat generation,

$$\frac{\partial^2 t}{\partial x^2} = \frac{1}{\alpha} \frac{\partial t}{\partial \tau} \qquad 17.20$$

it can be shown that:

$$t = e^{-\alpha \lambda^2 \tau} \{ C_1 \sin(\lambda x) + C_2 \cos(\lambda x) \}$$

(where $\lambda = $ eigenvalue; C_1 and C_2 are constants to be determined by the boundary conditions).

For a semi-infinite slab of half-thickness, L, initially at a uniform temperature, t_i, throughout, which is suddenly exposed to a fluid at a constant temperature, t_F, the temperature at any point, x, at time, τ, is given by:

$$\frac{t-t_F}{t_i-t_F} = 2 \sum_{n=1}^{\infty} e^{-(\lambda_n L)^2 Fo} \left\{ \frac{\sin(\lambda_n L)\cos(\lambda_n x)}{(\lambda_n L)+\sin(\lambda_n L)\cos(\lambda_n L)} \right\} \qquad 17.21$$

where
$$(\lambda_n L)\tan(\lambda_n L) = Bi \qquad\qquad 17.22$$

(Bi = Biot number, hL/k; Fo = Fourier number, $k\tau/\rho c L^2$ or $\alpha\tau/L^2$).

Similar equations can be derived for the cases of the semi-infinite cylinder and sphere, and graphs of non-dimensional temperature against Fourier number for various values of $1/Bi$ have been drawn (see reference 17.8).

Newtonian heating or cooling

This approach, which is sometimes known as *lumped capacity*, may be used when the temperature within a body does not vary appreciably as the body's average temperature changes with time due to exposure of the body to a fluid at a different temperature. This is the case when the surface thermal resistance is very much greater than the internal thermal resistance, and hence the heat transfer from the surface is the controlling factor.

For a body of surface area, A, volume, V, specific heat, c, and density, ρ, with an average temperature, $\bar{t}$, at any time, τ, we have:

$$hA(\bar{t}-t_F) = -\rho Vc \frac{d\bar{t}}{d\tau}$$

(where h = surface heat transfer coefficient; t_F = temperature of the fluid surrounding the body, assumed constant with time)

$$\therefore \int_{t_i}^{\bar{t}} \frac{d\bar{t}}{t-t_F} = -\int_0^\tau \frac{hA}{\rho Vc}\,d\tau$$

(where t_i = initial temperature of the body)

i.e.
$$\log_e\left(\frac{\bar{t}-t_F}{t_i-t_F}\right) = -\frac{hA\tau}{\rho Vc}$$

or
$$\frac{\bar{t}-t_F}{t_i-t_F} = e^{-(hA\tau/\rho Vc)} \qquad\qquad 17.23$$

Equation 17.23 can also be written as:

$$\frac{\bar{t}-t_F}{t_i-t_F} = e^{-(AL/V)BiFo} \qquad\qquad 17.24$$

The dimension of length, L, may be the half-thickness of a semi-infinite slab, or the radius of an infinite cylinder, or the radius of a sphere. The term AL/V for a semi-infinite slab, cylinder or sphere, may be shown to be 1, 2, or 3 respectively, e.g. for a sphere,

$$\frac{AL}{V} = \frac{(4\pi L^2)L}{(4\pi L^3/3)} = 3$$

In the previously considered exact solution, equation 17.21, it can be shown that when $Fo > 0.2$ then only the first term of the summation need be considered within engineering accuracy. Also, when Bi is small then in equation 17.22, $\tan(\lambda_1 L)$ approximates to $(\lambda_1 L)$, and hence Bi approximates to $(\lambda_1 L)^2$. Similarly, $\sin(\lambda_1 L)$ approximates to $(\lambda_1 L)$ and $\cos(\lambda_1 L)$ approaches unity. Therefore, substituting these approximations into equation 17.21, for the centre where $x = 0$, and hence $\cos(\lambda_1 x)$ is 1, we have:

$$\frac{t - t_F}{t_i - t_F} = 2\,e^{-BiFo}\left(\frac{\lambda_1 L}{\lambda_1 L + \lambda_1 L}\right) = e^{-BiFo}$$

Comparing this with equation 17.24 it can be seen to be equivalent to Newtonian cooling of a semi-infinite slab, i.e. when Bi is very small the problem approximates to Newtonian cooling.

Example 17.7

For transient conduction in a sphere when $Fo > 0.2$ it can be shown that the solution of equation 17.18 for the temperature at the centre of the sphere, t_c, when initially at t_i, and plunged into a fluid at t_F, is given by:

$$\frac{t_c - t_F}{t_i - t_F} = \frac{(\sin \lambda_1 L - \lambda_1 L \cos \lambda_1 L)}{(\lambda_1 L - \sin \lambda_1 L \cos \lambda_1 L)}\, 2\,e^{-(\lambda_1 L)^2 Fo}$$

where $$1 - \lambda_1 L \cot \lambda_1 L = Bi$$

Using the data below determine the temperature at the centre of a sphere, initially at a uniform temperature of 500°C, twenty minutes after it is plunged into a large bath of liquid at a temperature of 20°C:

(i) from the above equation;   (ii) assuming Newtonian cooling

Data: radius of sphere = 50 mm; density of sphere = 7600 kg/m³;

thermal conductivity of sphere = 40 W/m K; specific heat of sphere = 0·5 kJ/kg K; heat transfer coefficient from sphere surface to liquid = 88·8 W/m² K. It may be assumed that the heat transfer coefficient and the temperature of the liquid remain constant over the time period.

(i)
$$Bi = hL/k = (88\cdot8 \times 0\cdot05)/40 = 0\cdot111$$

$$\therefore \; 1 - \lambda_1 L \cot \lambda_1 L = 0\cdot111$$

or
$$\lambda_1 L \cot \lambda_1 L = 0\cdot889$$

This equation may be solved by trial and error,

i.e.

$\lambda_1 L$	0·6	0·7	0·5
$\lambda_1 L \cot \lambda_1 L$	0·877	0·831	0·915

By further trial and error, or by drawing a graph, it can be shown that $\lambda_1 L = 0\cdot57$.

Also,

$$Fo = \alpha\tau/R^2 = k\tau/\rho c R^2 = (40 \times 20 \times 60)/(7600 \times 0\cdot5 \times 10^3 \times 0\cdot05^2)$$
$$= 5\cdot053$$

$$\therefore \; \frac{t_c - t_F}{t_i - t_F} = \frac{(\sin 0\cdot57 - 0\cdot57 \cos 0\cdot57)}{(0\cdot57 - \sin 0\cdot57 \cos 0\cdot57)} \, 2 \, e^{-(0\cdot57)^2 \times 5\cdot053}$$

i.e.
$$\frac{t_c - t_F}{t_i - t_F} = 0\cdot5161 \times 2 \times 0\cdot1936 = 0\cdot2$$

$$\therefore \; t_c = 20 + 0\cdot2(500 - 20) = 116°C$$

(ii) For Newtonian cooling of a sphere, from equation 17.24,

$$\frac{\bar{t} - t_F}{t_i - t_F} = e^{-3BiFo} = 3^{-3 \times 0\cdot111 \times 5\cdot053}$$

$$\therefore \; \bar{t} = 20 + 0\cdot1859(500 - 20) = 109\cdot2°C$$

17.6 Numerical methods for conduction

The most commonly used numerical method is the finite difference method in which a differential equation is replaced by an approximate algebraic expression. The set of equations thus produced can be solved using a computer. The reader is recommended to consult the references given at the end of the chapter for a fuller treatment of numerical methods and their application in heat transfer.

A different method known as the finite element method is increasingly being used for heat transfer applications (see reference 17.6), but is not considered in this book.

In specialized texts the derivation of finite difference expressions is given in detail, using for example the Taylor series, but in this book only the following brief illustration will be given. Referring to a graph of t against x, fig. 17.10, three approximations to the true tangent to the curve, dt/dx, are illustrated,

i.e.
$$\frac{dt}{dx} = \frac{(t_x - t_{x-\delta x})}{\delta x} \qquad \text{backward difference} \qquad 17.25$$

$$\frac{dt}{dx} = \frac{(t_{x+\delta x} - t_x)}{\delta x} \qquad \text{forward difference} \qquad 17.26$$

$$\frac{dt}{dx} = \frac{(t_{x+\delta x} - t_{x-\delta x})}{2\delta x} \qquad \text{central difference} \qquad 17.27$$

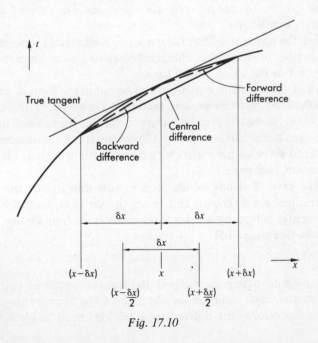

Fig. 17.10

It can be seen from the figure that the central difference approximation, equation 17.27, is a more accurate approximation to the true slope.

The second derivative, $\mathrm{d}^2t/\mathrm{d}x^2$, is the rate of change of slope at the point x. This may be approximated as the change of $\mathrm{d}t/\mathrm{d}x$ over the distance δx. From fig. 17.10 it can be seen that the slope at $x + (\delta x/2)$ is approximately $(t_{x+\delta x} - t_x)/\delta x$, and the slope at $x - (\delta x/2)$ is approximately $(t_x - t_{x-\delta x})/\delta x$. Hence the rate of change of slope over the distance δx is given by:

$$\frac{\mathrm{d}^2t}{\mathrm{d}x^2} = \left\{ \frac{(t_{x+\delta x} - t_x)}{\delta x} - \frac{(t_x - t_{x-\delta x})}{\delta x} \right\} \frac{1}{\delta x}$$

i.e.
$$\frac{\mathrm{d}^2t}{\mathrm{d}x^2} = \frac{t_{x+\delta x} - t_{x-\delta x} - 2t_x}{\delta x^2} \qquad 17.28$$

To solve a conduction problem by the finite difference method the relevant partial differential equation is replaced using expressions such as the above. The space and time dimensions are divided into a number of increments of finite size and the approximate expression which replaces the differential equation applies to every point in the grid of points, or nodes; separate equations are derived for the boundary conditions.

Hence the relevant differential equation is effectively replaced by a large number of identical algebraic expressions for the temperature at each point in the space at any time.

A set of simultaneous equations can be put in the form of a matrix and solved by *matrix inversion* methods. However, in the case of conduction the matrix of temperature coefficients has a small number of non-zero terms and hence matrix inversion is not recommended. It is better to solve such a matrix by a direct method such as Gaussian elimination, (see page 715).

In the case of steady conduction in two dimensions the initial temperatures are unknown and hence the set of equations is more conveniently solved by a relaxation method such as Gauss–Siedel iteration, (see page 705).

Errors

Using a finite difference method the answer obtained converges towards the exact solution as the size of the increments chosen approaches zero. Finite difference expressions must be chosen such that the computer solution converges towards the exact solution; in certain cases the solution will become unstable because errors generated are increasing in size as the solution proceeds, or are growing at a faster rate than the rate of convergence.

There are basically two types of error: round-off error and discretization error. Round-off error occurs when the answer is taken to a specific number of significant figures, and is cumulative; fortunately in modern computers this error is not usually important Discretization error is mainly due to the inaccuracy of the finite difference expression, see fig. 17.10, and can be reduced by reducing the size of the increments.

Notation
Referring to fig. 17.11, a two-dimensional space may be divided into a grid of nodes as shown. The temperature at any point may then be

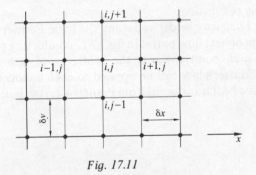

Fig. 17.11

designated as $t_{i,j}$. Note that i increases from left to right, and j from bottom to top, of the grid, following the normal x-direction and y-direction respectively.

For a problem in transient conduction the temperature at any time will be denoted by $t_{i,j}^{\tau}$; the next time is therefore $\tau+1$ and the temperature at that instant is $t_{i,j}^{\tau+1}$. For transient problems in one-dimension the j-direction will be omitted.

17.7 Two-dimensional steady conduction

From equation 17.15, for zero internal heat generation and for steady conduction in two dimensions the equation reduces to the Laplace equation:

$$\frac{\partial^2 t}{\partial x^2} + \frac{\partial^2 t}{\partial y^2} = 0 \qquad 17.29$$

This equation may be put into finite difference form using the central difference expression, equation 17.28. Using the notation outlined in section 17.6 (see fig. 17.11), we have:

$$\frac{(t_{i+1,j} + t_{i-1,j} - 2t_{i,j})}{\delta x^2} + \frac{(t_{i,j+1} + t_{i,j-1} - 2t_{i,j})}{\delta y^2} = 0$$

The grid may be chosen such that, $\delta x = \delta y$, then,

$$t_{i,j} = (t_{i,j-1} + t_{i+1,j} + t_{i,j+1} + t_{i-1,j})/4 \qquad 17.30$$

All the internal points within the boundaries of the two-dimensional space are represented by equation 17.30.

Conducting rod analogy

Equation 17.30 may be derived using the basic Fourier equation and the concept of heat flow paths. In fig. 17.12 conducting paths of width, δx, from each point towards the centre point, are shown cross-hatched. Fourier's law can be applied to each conducting path; for example, the heat transferred from point $(i+1, j)$ to point (i, j) is given

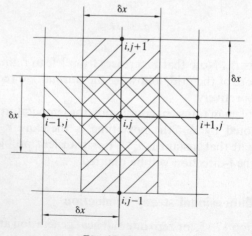

Fig. 17.12

by:

thermal conductivity × area × temperature gradient

$$= k(\delta x)\frac{(t_{i+1,j}-t_{i,j})}{\delta x}$$

Then a simple energy balance gives:

$$k(\delta x)\frac{(t_{i+1,j}-t_{i,j})}{\delta x} + k(\delta x)\frac{(t_{i,j+1}-t_{i,j})}{\delta x} + k(\delta x)\frac{(t_{i-1,j}-t_{i,j})}{\delta x}$$

$$+ k(\delta x)\frac{(t_{i,j-1}-t_{i,j})}{\delta x} = 0$$

This equation reduces to the same expression as in equation 17.30.

The conducting rod analogy can be used in more complex cases, including the case with internal heat generation or at a boundary convecting to a fluid; it may be found easier to apply since it relates to a simple physical model.

Boundary conditions
(a) Surface convecting to a fluid:

For a point (i,j) on the surface (see fig. 17.13):

$$-k\frac{dt}{dx} = h(t_F - t_{i,j})$$

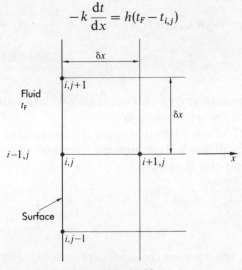

Fig. 17.13

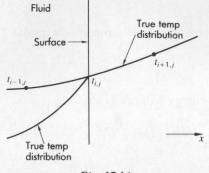

Fig. 17.14

Using a central difference expression for dt/dx, equation 17.27,

$$-k \frac{(t_{i+1,j}-t_{i-1,j})}{2\delta x} = h(t_F - t_{i,j}) \qquad 17.31$$

The point $(i-1, j)$ is fictitious and can be eliminated from the equation by assuming that it lies on the extrapolated temperature distribution line (see fig. 17.14), i.e. from equation 17.30:

$$t_{i-1,j} = 4t_{i,j} - t_{i+1,j} - t_{i,j+1} - t_{i,j-1}$$

Substituting in equation 17.31,

$$t_{i+1,j} - 4t_{i,j} + t_{i+1,j} + t_{i,j+1} + t_{i,j-1} = \frac{2h\,\delta x}{k}(t_{i,j} - t_F)$$

$$\therefore\ t_{i,j} = \frac{2t_{i+1,j} + t_{i,j+1} + t_{i,j-1} + (2h\,\delta x t_F/k)}{4 + (2h\,\delta x/k)} \qquad 17.32$$

Similar expressions may be obtained for a surface with the fluid on the right, or at the top or bottom.

Equation 17.32 can be derived using the conducting rod analogy with a rod of half-width, $\delta x/2$, running from point $(i, j+1)$ to point (i, j), and from point $(i, j-1)$ to (i, j),

i.e. $\quad k(\delta x)\dfrac{(t_{i+1,j}-t_{i,j})}{\delta x} + k\dfrac{(\delta x)}{2}\dfrac{(t_{i,j+1}-t_{i,j})}{\delta x} + k\dfrac{(\delta x)}{2}\dfrac{(t_{i,j-1}-t_{i,j})}{\delta x}$

$$= h\,\delta x(t_{i,j} - t_F)$$

Simplifying this equation the same expression as in equation 17.32 is obtained.

(b) Insulated surface.

At an insulated surface, $-k \, dt/dx = 0$, or $h = 0$, and hence for a left-hand surface which is insulated equation 17.32 reduces to,

$$t_{i,j} = \frac{2t_{i+1,j} + t_{i,j+1} + t_{i,j-1}}{4}$$

It should be noted that a line of thermal symmetry within a two-dimensional space will act as an insulated surface (see Example 17.8).

(c) corners.

Expressions can be derived for the temperatures at outside corners (top left, bottom right, etc.) and at inside corners.

For example, for a top left outside corner,

$$t_{i,j} = \frac{t_{i+1,j} + t_{i,j-1} + (2h \, \delta x t_F / k)}{2 + (2h \, \delta x / k)}$$

For a bottom right inside corner,

$$t_{i,j} = \frac{2t_{i+1,j} + 2t_{i,j-1} + t_{i,j+1} + t_{i-1,j} + (2h \, \delta x t_F / k)}{6 + (2h \, \delta x / k)}$$

The derivation of expressions for corner points such as the above is left as an exercise for the reader; the conducting rod analogy is the best method, particularly for inside corners.

Choice of numerical method

The Laplace equation is an elliptic equation and as stated in Section 17.6 the recommended numerical method in this case is *Gauss–Siedel iteration*.

In the Gauss–Siedel method the grid points are assigned initial values at every point, then, from the set of equations for the internal nodes and all the surface points, new values for each point are calculated point by point. The new value at a point is compared with the previous value and the process continued until the difference between successive values is small compared with the actual temperature, according to the accuracy required. Provided the solution converges then the complete temperature field is obtained.

This method is called a relaxation method because the values of temperature are modified, or relaxed, at each point in turn to satisfy the equation, which then alters the temperatures at adjacent points. In general, at any point (i, j) the value after one iteration $t_{i,j}^{(2)}$ in terms of the initial values is given by:

$$t_{i,j}^{(2)} = \omega \frac{(t_{i,j-1}^{(1)} + t_{i+1,j}^{(1)} + t_{i,j+1}^{(1)} + t_{i-1,j}^{(1)})}{4} + (1-\omega)t_{i,j}^{(1)}$$

where ω is a relaxation factor.

For $\omega = 0$, $t_{i,j}^{(2)} = t_{i,j}^{(1)}$, and no relaxation occurs.

For $\omega = 1$, equation 17.30 applies, which is Gauss–Siedel iteration.

Values of ω between 0 and 1 will slow down the relaxation process, whereas for $\omega > 1$ the solution will converge more rapidly. It can be shown that the solution will not converge for $\omega > 2$ and hence a value of ω between 1 and 2 is chosen, determined largely by experience. This method is known as *Successive Over Relaxation* (SOR).

Example 17.8

A long duct of square cross-section $0.5 \text{ m} \times 0.5 \text{ m}$ is buried in deep soil with one of its sides parallel to the surface of the soil as shown in fig. 17.15; the centre-line of the duct is at a depth of 1.25 m. The surface of the soil is at an equilibrium temperature of $0°C$, and at a soil depth of 2.5 m it may be assumed that the uniform equilibrium temperature is

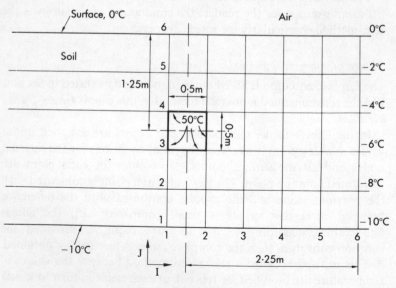

Fig. 17.15

− 10°C across the horizontal cross-section. The temperature of each side of the duct is 50°C and it may be assumed that at a vertical cross-section a horizontal distance of 2·25 m from the duct centre-line, the heat transfer vertically downwards from the surface is simple one-dimensional conduction.

Taking a square mesh of side 0·5 m, use a numerical method to obtain an approximation for the temperatures within the soil to the nearest degree, and estimate the heat loss per metre length of duct. The thermal conductivity of the soil is 1 W/m K.

At the vertical cross-section where the conduction is one-dimensional the temperature must vary linearly, hence the temperatures can be written in as shown on fig. 17.15.

At the centre line of the duct there is thermal symmetry, i.e. no heat can flow across the centre-line and hence the temperatures on either side of this line are equal.

Number the grid as shown with columns, I, from 1 to 6 horizontally, and rows, J, from 1 to 6 vertically.

The temperatures which are known are as follows: $I = 1$ to 6, $J = 6$, $t = 0$°C; $I = 1$, $J = 3$ and 4, $t = 50$°C; $I = 2$, $J = 3$ and 4, $t = 50$°C; $I = 1, 2, 3, 4, 5,$ and 6, $J = 1$, $t = -10$°C; $I = 6$, $J = 2$, $t = -8$°C; $I = 6$, $J = 3$, $t = -6$°C; $I = 6$, $J = 4$, $t = -4$°C; $I = 6$, $J = 5$, $t = -2$°C.

For all other points the temperature is given by equation 17.30,

$$t_{i,j} = (t_{i,j-1} + t_{i+1,j} + t_{i,j+1} + t_{i-1,j})/4$$

Also, from the condition of symmetry as stated above, the temperature at $I = 1$, $J = 2$ is equal to the temperature at $I = 2$, $J = 2$; similarly the temperature at $I = 1$, $J = 5$ is equal to the temperature at $I = 2$, $J = 5$.

To solve the problem we assign initial values to the temperatures at the internal points and then proceed by iteration, using equation 17.30 for internal points, and the values given above for the other points. This is best done by computer when the size of mesh and the required accuracy can be greater than those specified in the problem. To illustrate the method a procedure is outlined below using mental arithmetic only; the first column on the left side is the set of initial guesses, the second column is the first iteration, and so on.

The temperature values in brackets for each iteration are taken in the same order as in equation 17.30 at each point.

Temperature at (I,J)	First Iteration	Second Iteration	Third Iteration
$t(2,2)=10$	$\dfrac{(-10-5+50+10)}{4}=11$	$\dfrac{(-10+3+50+11)}{4}=14$	$\dfrac{(-10+5+50+14)}{4}=15$
$t(3,2)=-5$	$\dfrac{(-10-8+20+11)}{4}=3$	$\dfrac{(-10-4+21+14)}{4}=5$	$\dfrac{(-10-1+21+15)}{4}=6$
$t(4,2)=-8$	$\dfrac{(-10-8+0+3)}{4}=-4$	$\dfrac{(-10-4+4+5)}{4}=-1$	$\dfrac{(-10-5+6+6)}{4}=-1$
$t(5,2)=-8$	$\dfrac{(-10-8+0-4)}{4}=-6$	$\dfrac{(-10-8-2-1)}{4}=-5$	$\dfrac{(-10-8-1-1)}{4}=-5$
$t(3,3)=20$	$\dfrac{(3+0+30+50)}{4}=21$	$\dfrac{(5+4+25+50)}{4}=21$	$\dfrac{(6+16+23+50)}{4}=21$
$t(4,3)=0$	$\dfrac{(-4+0+0+21)}{4}=4$	$\dfrac{(-1-2+7+21)}{4}=6$	$\dfrac{(-1-1+9+21)}{4}=7$
$t(5,3)=0$	$\dfrac{(-6-6+10+4)}{4}=-2$	$\dfrac{(-5-6+0+6)}{4}=-1$	$\dfrac{(-5-6+1+7)}{4}=-1$
$t(3,4)=30$	$\dfrac{(21+0+30+50)}{4}=25$	$\dfrac{(21+7+13+50)}{4}=23$	$\dfrac{(21+9+13+50)}{4}=24$
$t(4,4)=0$	$\dfrac{(4+0+0+25)}{4}=7$	$\dfrac{(6+0+5+23)}{4}=9$	$\dfrac{(7+1+6+24)}{4}=10$
$t(5,4)=0$	$\dfrac{(-2-4+0+7)}{4}=0$	$\dfrac{(-1-4+1+9)}{4}=1$	$\dfrac{(-1-4+1+10)}{4}=2$
$t(2,5)=30$	$\dfrac{(50+30+0+30)}{4}=28$	$\dfrac{(50+13+0+28)}{4}=23$	$\dfrac{(50+13+0+23)}{4}=22$
$t(3,5)=30$	$\dfrac{(25+0+0+28)}{4}=13$	$\dfrac{(23+5+0+23)}{4}=13$	$\dfrac{(24+6+0+22)}{4}=13$
$t(4,5)=0$	$\dfrac{(7+0+0+13)}{4}=5$	$\dfrac{(9+1+0+13)}{4}=6$	$\dfrac{(10+1+0+13)}{4}=6$
$t(5,5)=0$	$\dfrac{(0-2+0+5)}{4}=1$	$\dfrac{(1-2+0+6)}{4}=1$	$\dfrac{(2-2+0+6)}{4}=2$

The temperatures at each point after the third iteration are within one degree of the previous value which is within the required accuracy. Note that in this approximate solution the temperatures at each point have been rounded up to the nearest degree and this error will be accumulative. It is left to the reader to write a simple program in BASIC or FORTRAN to obtain a more accurate solution using the same method.

From the conducting rod analogy the heat transfer from the duct per unit length is given by,

$$q = k \frac{(0 \cdot 5)}{2} \frac{(50-22)}{0 \cdot 5} \times 2 + k \frac{(0 \cdot 5)}{2} \frac{(50-24)}{0 \cdot 5} \times 2 + k \frac{(0 \cdot 5)}{2} \frac{(50-21)}{0 \cdot 5} \times 2$$

$$+ k \frac{(0 \cdot 5)}{2} \frac{(50-15)}{0 \cdot 5} \times 2$$

i.e. $$q = 118 \text{ W/m}$$

17.8 One-dimensional transient conduction by finite difference

The case of the semi-infinite slab will be considered as given in Section 17.5, equation 17.20. This equation is parabolic and therefore the solution starts from an initial boundary value and proceeds step by step in time.

In the finite difference method of solution a central difference expression is used as before for $\partial^2 t/\partial x^2$ but it is not possible to use a central difference expression for $\partial t/\partial \tau$ because this leads to instability. $\partial t/\partial \tau$ can either be replaced by a forward difference expression, equation 17.26 (leading to an *explicit solution*), or by a backward difference expression, equation 17.25 (leading to an *implicit solution*).

Explicit solution (or Euler solution)

For a semi-infinite slab, equation 17.20,

$$\frac{\partial^2 t}{\partial x^2} = \frac{1}{\alpha} \frac{\partial t}{\partial \tau}$$

is replaced by:

$$\frac{t_{i+1}^\tau + t_{i-1}^\tau - 2t_i^\tau}{\delta x^2} = \frac{1}{\alpha} \frac{(t_i^{\tau+1} - t_i^\tau)}{\delta \tau}$$

i.e. $$t_i^{\tau+1} = Fo(t_{i+1}^\tau + t_{i-1}^\tau - 2t_i^\tau) + t_i^\tau \qquad 17.33$$

(where $Fo = \alpha \, \delta\tau/\delta x^2$)

or, $$t_i^{\tau+1} = Fo\left\{ t_{i+1}^\tau + t_{i-1}^\tau + \left(\frac{1}{Fo} - 2\right)t_i^\tau \right\} \qquad 17.34$$

When the term $(1/Fo - 2)$ becomes negative the solution becomes unstable; hence the condition for stability is:

$$Fo \leqslant 1/2 \qquad 17.35$$

The size of the increments, δx, and $\delta \tau$, must be chosen such that Fo is less than or equal to 0·5, and then a set of equations such as equation 17.33 can be solved at each time step.

Note that when Fo is chosen to be equal to 0·5 then from equation 17.34 it can be seen that:

$$t_i^{\tau+1} = (t_{i+1}^{\tau} + t_{i-1}^{\tau})/2 \qquad 17.36$$

In this type of problem the initial temperatures are known at all the nodes and the values at the next time step can be found one by one using equation 17.33. This process is repeated for the next time step and so on until the required time is reached, or until the temperature within the slab reaches a certain value.

Boundary condition (*surface convecting to a fluid*)

At any instant of time the equation for heat transfer at the surface is the same as that previously considered in Section 17.7 for two-dimensional conduction. Referring to fig. 17.13 and equation 17.31, in this case the equation may be written:

$$-k \frac{(t_{i+1}^{\tau} - t_{i-1}^{\tau})}{2\delta x} = h(t_F^{\tau} - t_i^{\tau}) \qquad 17.37$$

As before the point $(i-1)$ is fictitious and can be eliminated from equation 17.37 using, in this case, equation 17.33, i.e. substituting for t_{i-1}^{τ} from equation 17.33 in equation 17.37 we have,

$$t_i^{\tau+1} = Fo\left\{2t_{i+1}^{\tau} + \left(\frac{1}{Fo} - 2 - \frac{2h\,\delta x}{k}\right)t_i^{\tau} + \frac{2h\,\delta x}{k}\,t_F^{\tau}\right\} \qquad 17.38$$

It can be seen from equation 17.38 that the condition for stability is now more restrictive,

i.e. $$Fo \leqslant \frac{1}{2 + (2h\,\delta x)/k} \qquad 17.39$$

By sacrificing some accuracy but retaining the condition for stability given by equation 17.35, the term $\partial t/\partial x$ can be replaced by a forward difference expression,

i.e. $$k\frac{(t_{i+1}^{\tau} - t_i^{\tau})}{\delta x} = h(t_i^{\tau} - t_F^{\tau})$$

$$\therefore \ t_i^{\tau} = \frac{t_{i+1}^{\tau} + (h\,\delta x t_F^{\tau}/k)}{1 + (h\,\delta x/k)} \qquad 17.40$$

Boundary condition (insulated face or line of thermal symmetry)

As considered previously in Section 17.7, at an insulated face $\partial t/\partial x$ is zero and hence the temperature at an insulated surface is obtained from equation 17.33 by putting $t_{i+1}^{\tau} = t_{i-1}^{\tau}$,

i.e. $$t_i^{\tau+1} = 2Fo(t_{i+1}^{\tau} - t_i^{\tau}) + t_i^{\tau}$$

for a left-hand insulated surface.

(Note that when $Fo = 0.5$ that $t_i^{\tau+1} = t_{i+1}^{\tau}$.)

In an important class of problem called *quenching*, a slab is suddenly subjected to surroundings at a different temperature. In such a case the centre-line of the slab is a line of thermal symmetry and hence only one half of the slab need be considered, and is treated as a slab with one face insulated.

Boundary condition (sudden increase in temperature)

When the fluid in contact with a surface, or the surface itself, is suddenly changed in temperature, a decision is required as to which temperature to use for the initial time step calculations. Provided the time taken for the change of temperature is very small compared with the time step $\delta\tau$, a suitable value to take is the arithmetic mean of the initial and final values.

For example, if the surface of a slab is suddenly raised from 20°C to 100°C, then for the first set of calculations use a surface temperature of 60°C, then for the next time step use a surface temperature of 100°C.

Example 17.9

A large steel plate, 300 mm thick, is heated to 800°C in a furnace and then allowed to cool in air at 25°C. Find the temperature distribution across the plate thickness 14 minutes after the start of cooling.

For the plate, $k = 18.75$ W/m K and $\alpha = 7.5 \times 10^{-6}$ m²/s. The heat transfer coefficient from the plate surfaces to the air may be taken as 125 W/m² K throughout the cooling process.

An explicit solution will be chosen with $Fo = 0.5$, hence equation 17.36 applies for the internal points. The less accurate expression for the surface is chosen to ensure stability with the chosen value of Fo, i.e. for the surface, equation 17.40 applies.

The plate half-thickness of 150 mm is divided into five increments of thickness, $x = 0.03$ m.

Then,

$$Fo = \alpha \, \delta\tau/\delta x^2 = 0.5$$

$$\therefore \ \delta\tau = \frac{0.5 \times 0.03^2}{7.5 \times 10^{-6}} = 60 \text{ s} = 1 \text{ min}$$

For the surface, from equation 17.40:

$$t_i^\tau = \frac{t_{i+1}^\tau + (125 \times 0.03 \times 25/18.75)}{1 + (125 \times 0.03/18.75)} = \frac{t_{i+1}^\tau + 5}{1.2}$$

For internal points from equation 17.36,

$$t_i^{\tau+1} = \frac{t_{i+1}^\tau + t_{i-1}^\tau}{2}$$

The problem can now be solved by calculating all the temperatures, one by one, at each time step until the time period of 14 minutes has elapsed. For a simple problem like this one the calculation can easily be done with sufficient accuracy using a hand calculator as in the table which follows.

To illustrate how the figures are calculated:

at time 6 min from the start, at the surface $t_o = (644.7 + 5)/1.2 = 541.4°C$; at internal section 2 at the same time, $t_2 = (663.1 + 777.2)/2 = 720.2°C$; at the centre-line at the same time, $t_5^{\tau+1} = t_4^\tau$ i.e. $t_5 = 792.0°C$

τ min	surface 0	1	2	3	4	centre-line 5
0	800	800	800	800	800	800
1	670.8	800	800	800	800	800
2	617.0	735.4	800	800	800	800
3	594.6	708.5	767.7	800	800	800
4	571.8	681.2	754.3	783.9	800	800
5	556.8	663.1	732.6	777.2	792.0	800
6	541.4	644.7	720.2	762.3	788.6	792.0
7	529.8	630.8	703.5	754.4	777.2	788.6
8	518.1	616.7	692.6	740.4	771.5	777.2
9	508.7	605.4	678.6	732.1	758.8	771.5
10	498.9	593.7	668.8	718.7	751.8	758.8
11	490.8	583.9	656.2	710.3	738.8	751.8
12	482.1	573.5	647.1	697.5	731.1	738.8
13	474.7	564.6	635.5	689.1	718.2	731.1
14	466.8	555.1	626.9	676.9	710.1	718.2

Implicit solution

In equation 17.20 a backward difference approximation is used by expressing the $\partial^2 t/\partial x^2$ term at the next time interval while retaining the same expression for $\partial t/\partial \tau$,

i.e.
$$\frac{t_{i+1}^{\tau+1} + t_{i-1}^{\tau+1} - 2t_i^{\tau+1}}{\delta x^2} = \frac{1}{\alpha} \frac{(t_i^{\tau+1} - t_i^{\tau})}{\delta \tau} \qquad 17.41$$

All the temperatures at $(\tau + 1)$ in equation 17.41 are unknown, hence a set of simultaneous equations such as the above must be solved at each time interval. This makes the method longer, with greater computer storage requirements than the explicit method, but there is not the limitations on the size of Fourier number because the solution is stable for all values of Fo. Also, the accuracy is not strongly dependent on the size of the time step and therefore some of the increased computer time and storage can be off-set by using a larger time interval.

From equation 17.41:

$$t_i^{\tau+1} - t_i^{\tau} = Fo(t_{i+1}^{\tau+1} + t_{i-1}^{\tau+1} - 2t_i^{\tau+1}) \qquad 17.42$$

This is the fully implicit method. In general we can write:

$$\left(\frac{\partial^2 t}{\partial x^2}\right) = \theta\left(\frac{\partial^2 t}{\partial x^2}\right)^{\tau+1} + (1-\theta)\left(\frac{\partial^2 t}{\partial x^2}\right)^{\tau}$$

When $\theta = 0$ the equation reduces to that for the explicit method; when $\theta = 1$ the equation reduces to the fully implicit method.

When θ is put equal to 0·5 the method is known as the *Crank–Nicholson* method,

i.e.
$$t_i^{\tau+1} - t_i^{\tau} = Fo\tfrac{1}{2}(t_{i+1}^{\tau+1} + t_{i-1}^{\tau+1} - 2t_i^{\tau+1})$$
$$+ (1 - \tfrac{1}{2})Fo(t_{i+1}^{\tau} + t_{i-1}^{\tau} - 2t_i^{\tau})$$

Since there is no stability restriction on the value of Fo it can be put equal to unity. Then we have:

$$4t_i^{\tau+1} - t_{i+1}^{\tau+1} - t_{i-1}^{\tau+1} = t_{i+1}^{\tau} + t_{i-1}^{\tau} \qquad 17.43$$

For a surface convecting to a fluid a central difference expression can be used with a fictitious point which is then eliminated by substitution from equation 17.42.

For example, for a left-hand boundary:

$$t_{i+1}^{\tau} - t_{i-1}^{\tau} = \frac{2h\,\delta x}{k}\,(t_i^{\tau} - t_F^{\tau})$$

and

$$t_{i+1}^{\tau+1} - t_{i-1}^{\tau+1} = \frac{2h\,\delta x}{k}\,(t_i^{\tau+1} - t_F^{\tau+1})$$

Adding these equations,

$$t_{i+1}^{\tau} + t_{i+1}^{\tau+1} - (t_{i-1}^{\tau} + t_{i-1}^{\tau+1}) = \frac{2h\,\delta x}{k}\,(t_i^{\tau} + t_i^{\tau+1} - t_F^{\tau} - t_F^{\tau+1}) \quad 17.44$$

From equation 17.43,

$$t_{i-1}^{\tau} + t_{i-1}^{\tau+1} = 4t_i^{\tau+1} - t_{i+1}^{\tau+1} - t_{i+1}^{\tau}$$

Substituting in equation 17.44 and rearranging,

$$-t_i^{\tau+1}\left(2 + \frac{h\,\delta x}{k}\right) + t_{i+1}^{\tau+1} + \frac{h\,\delta x}{k}\,t_F^{\tau+1}$$

$$= -\frac{h\,\delta x}{k}\,t_F^{\tau} - t_{i+1}^{\tau} + \frac{h\,\delta x}{k}\,t_i^{\tau} \quad 17.45$$

Equation 17.45, with equation 17.43 for the internal points, form a set of equations which must be solved at each time step.

As an illustration of the Crank–Nicholson method consider the previous Worked Example, 17.9.

Taking $\delta x = 0.03$ m as before, we then have, $\delta\tau = 2$ min since $Fo = 1$ instead of 0.5 as previously.

Numbering the points across the slab as in the previous table (page 712), then from equation 17.45 for the convecting surface:

$$-t_0^{\tau+1}\left(2 + \frac{125 \times 0.03}{18.75}\right) + t_1^{\tau+1}$$

$$= \frac{125 \times 0.03}{18.75}\,t_0^{\tau} - \frac{2 \times 125 \times 0.03 \times 25}{18.75} - t_1^{\tau}$$

$$\therefore \quad -2.2t_0^{\tau+1} + t_1^{\tau+1} = 0.2t_0^{\tau} - 10 - t_1^{\tau}$$

i.e. for the first time step,

$$-2.2t_0^{\tau+1} + t_1^{\tau+1} = 0.2 \times 800 - 10 - 800 = -650 \quad\quad \text{(a)}$$

For the internal points, using equation 17.43:

$$4t_1^{\tau+1} - t_2^{\tau+1} - t_0^{\tau+1} = t_2^{\tau} + t_0^{\tau} = 800 + 800 = 1600 \qquad (b)$$

$$4t_2^{\tau+1} - t_3^{\tau+1} - t_1^{\tau+1} = t_3^{\tau} + t_1^{\tau} = 1600 \qquad (c)$$

$$4t_3^{\tau+1} - t_4^{\tau+1} - t_2^{\tau+1} = t_4^{\tau} + t_2^{\tau} = 1600 \qquad (d)$$

$$4t_4^{\tau+1} - t_5^{\tau+1} - t_3^{\tau+1} = t_5^{\tau} + t_3^{\tau} = 1600 \qquad (e)$$

$$4t_5^{\tau+1} - t_4^{\tau+1} - t_4^{\tau+1} = t_4^{\tau} + t_4^{\tau}$$

(using the insulated surface condition)

i.e.
$$4t_5^{\tau+1} - 2t_4^{\tau+1} = 2t_4^{\tau} = 1600 \qquad (f)$$

In equations (a) to (f) the left-hand side remains the same for each time step, and only the right-hand side must be updated after each set of calculations. These equations form the following matrix for the time step, $\tau = 2$ min.

$$
\begin{vmatrix}
-2 \cdot 2 & +1 & 0 & 0 & 0 & 0 \\
-1 & +4 & -1 & 0 & 0 & 0 \\
0 & -1 & +4 & -1 & 0 & 0 \\
0 & 0 & -1 & +4 & -1 & 0 \\
0 & 0 & 0 & -1 & +4 & -1 \\
0 & 0 & 0 & 0 & -2 & +4
\end{vmatrix}
\begin{vmatrix}
t_0 \\ t_1 \\ t_2 \\ t_3 \\ t_4 \\ t_5
\end{vmatrix}
=
\begin{vmatrix}
-650 \\ +1600 \\ +1600 \\ +1600 \\ +1600 \\ +1600
\end{vmatrix}
$$

This matrix, which has a large number of zero values, is called tri-diagonal and may be solved particularly easily on a computer by *Gaussian elimination*. The temperatures t_0, t_1, t_2, t_3, t_4, and t_5 are thus obtained at time, $\tau = 2$ min. These values are then substituted in the right-hand sides of equations (a) to (f) and a new matrix obtained for the temperatures at time, $\tau = 4$ min, which is solved in the same way. The solution proceeds thus until the time, $\tau = 14$ min as specified in the problem.

Note: the Gaussian elimination method of solving a set of simultaneous equations is the traditional method taught in school for solving two or three simultaneous equations; unlike iterative methods it leads to an exact solution. In the above matrix the coefficient of t_0 in the first row is put equal to -1 by dividing the equation by $2 \cdot 2$; this row is then subtracted from the second row thus eliminating t_0 from the second equation. In the modified matrix t_1 is then eliminated from the third row in the same way, and so on until an equation for t_5 is obtained in the last row. The value of t_5 is then back substituted into the preceding row to give t_4, and t_4 then back substituted into the preceding row to give t_3, and so on until all the temperatures are found. Sub-routines are available for most computers for solving a tri-diagonal matrix by Gaussian elimination as described above.

17.9 Heat exchangers

One of the most important processes in engineering is the heat exchange between flowing fluids. In heat exchangers the temperature of each fluid changes as it passes through the exchanger, and hence the temperature of the dividing wall between the fluids also changes along the length of the exchanger. Examples in practice in which flowing fluids exchange heat are air intercoolers and preheaters, condensers and boilers in steam plant, condensers and evaporators in refrigeration units, and many other industrial processes in which a liquid or gas is required to be either cooled or heated.

There are three main types of heat exchanger; the most important type is the *recuperator* in which the flowing fluids exchanging heat are on either side of a dividing wall; the second type is the *regenerator* in which the hot and cold fluids pass alternately through a space containing solid particles, these particles providing alternately a sink and a source for heat flow; the third type is the *evaporative type* in which a liquid is cooled evaporatively and continuously in the same space as the coolant. Examples of the latter type are the cooling tower (see Section 14.5) and the jet condenser (see Section 7.5). Very often when the term 'heat exchanger' is used it refers to the recuperative type, which is by far the most commonly used in engineering practice. This section will deal almost entirely with the recuperative type.

Parallel-flow and counter-flow recuperators

Consider the simple case of a fluid flowing through a pipe and exchanging heat with a second fluid flowing through an annulus surrounding the pipe. When the fluids flow in the same direction along the pipe the system is known as *parallel-flow*, and when the fluids flow in opposite directions to each other the system is known as *counter-flow*. Parallel-flow is shown in fig. 17.16 and counter-flow is shown in fig. 17.16. Let the mean inlet and outlet temperatures of

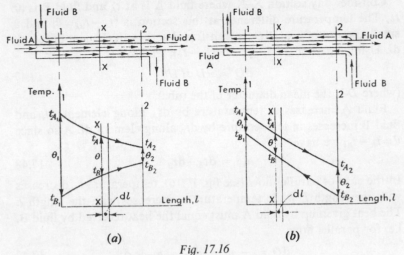

Fig. 17.16

fluid A be t_{A_1} and t_{A_2} respectively, and let the mean temperatures of fluid B at sections 1 and 2 be t_{B_1} and t_{B_2} respectively. Let the mass flow rates of fluid A and fluid B be $\dot{m}_A$ and $\dot{m}_B$, respectively, and let the specific heats of fluid A and fluid B be c_A and c_B respectively. The temperature difference at section 1 is $(t_{A_1} - t_{B_1}) = \theta_1$, and the temperature difference at section 2 is $(t_{A_2} - t_{B_2}) = \theta_2$.

Since the tube wall separating the fluids is thin it is possible to use an overall heat transfer coefficient, U, based on equation 17.11,

i.e. $$U = \frac{1}{R_T A}$$

(where A is the mean surface area of the tube).

Since the resistance of the tube wall is negligibly small we can write,

$$\frac{1}{U} = \frac{1}{h_A} + \frac{1}{h_B} \qquad 17.46$$

In practice the values of h_A and h_B will vary along the length of the tube, but suitable mean values can be found. A mean value of U along the tube will be assumed.

Consider any section X–X where fluid A is at t_A and fluid B is at t_B. The temperature difference at this section is $(t_A - t_B) = \theta$, and a small amount of heat, dQ, is transferred across an element of length dl. Using equation 17.5, $Q = UA(t_A - t_B)$, we have

$$dQ = \pi D \, dl \, U\theta \qquad 17.47$$

(where D is the mean diameter of the tube).

Fluid A increases in temperature by dt_A along element dl, and fluid B increases in temperature by dt_B along element dl. Also since $\theta = t_A - t_B$, we have

$$d\theta = dt_A - dt_B \qquad 17.48$$

In the case of parallel-flow (see fig. 17.16), temperature t_A decreases with the length l, while temperature t_B increases with the length l. The heat given up by fluid A must equal the heat received by fluid B, i.e. for parallel flow,

$$dQ = -\dot{m}_A c_A \, dt_A = \dot{m}_B c_B \, dt_B \qquad 17.49$$

$$\therefore \ dt_A = \frac{-dQ}{\dot{m}_A c_A} \quad \text{and} \quad dt_B = \frac{dQ}{\dot{m}_B c_B}$$

Substituting in equation 17.48,

$$d\theta = \frac{-dQ}{\dot{m}_A c_A} - \frac{dQ}{\dot{m}_B c_B} = -dQ\left(\frac{1}{\dot{m}_A c_A} + \frac{1}{\dot{m}_B c_B}\right) \qquad 17.50$$

Integrating equation 17.50 between sections 1 and 2,

$$\theta_2 - \theta_1 = -Q\left(\frac{1}{\dot{m}_A c_A} + \frac{1}{\dot{m}_B c_B}\right)$$

or

$$\theta_1 - \theta_2 = Q\left(\frac{1}{\dot{m}_A c_A} + \frac{1}{\dot{m}_B c_B}\right) \qquad 17.51$$

Also, from equation 17.50,

$$dQ = \frac{-d\theta}{\left(\dfrac{1}{\dot{m}_A c_A} + \dfrac{1}{\dot{m}_B c_B}\right)}$$

Substituting in equation 17.47,

$$\frac{-d\theta}{\left(\dfrac{1}{\dot{m}_A c_A} + \dfrac{1}{\dot{m}_B c_B}\right)} = \pi D \, dl \, U\theta$$

$$\therefore \ \frac{-d\theta}{\theta} = \pi D U\left(\frac{1}{\dot{m}_A c_A} + \frac{1}{\dot{m}_B c_B}\right) dl$$

Integrating between sections 1 and 2,

$$-\log_e \frac{\theta_2}{\theta_1} = \pi DlU\left(\frac{1}{\dot{m}_A c_A} + \frac{1}{\dot{m}_B c_B}\right) \qquad 17.52$$

(where l is the total length of the tube).

Now from equation 17.51,

$$\left(\frac{1}{\dot{m}_A c_A} + \frac{1}{\dot{m}_B c_B}\right) = \frac{\theta_1 - \theta_2}{Q}$$

Hence substituting in equation 17.52,

$$-\log_e \frac{\theta_2}{\theta_1} = \frac{\pi DlU(\theta_1 - \theta_2)}{Q}$$

or

$$Q = \frac{\pi DlU(\theta_1 - \theta_2)}{\log_e \theta_1/\theta_2} \qquad 17.53$$

In the case of counter flow (see fig. 17.16), both temperature t_A and temperature t_B decrease *in the direction of the length l.* In place of equation 17.49 we therefore have

$$dQ = -\dot{m}_A c_A \, dt_A = -\dot{m}_B c_B \, dt_B$$

When the same procedure as for parallel-flow is carried out equation 17.53 is again obtained; this procedure is left as an exercise for the reader.

Comparing equation 17.53 with equation 17.5, $Q = UA(t_A - t_B)$, it can be seen that the mean temperature difference, θ_m, is given by

$$\theta_m = \frac{\theta_1 - \theta_2}{\log_e \theta_1/\theta_2} \qquad 17.54$$

θ_m is known as the *logarithmic mean temperature difference*, frequently given the symbol *LMTD*.

Then we have

$$Q = UA \, LMTD \qquad 17.55$$

(where A is the mean surface area of the tube, πDl).

There are several important points which should be mentioned here:

(i) When one of the fluids is a wet vapour or a boiling liquid then its temperature remains constant. Assuming fluid A to be a wet vapour, then the temperature variations are as shown in fig. 17.17. It follows that under these circumstances the variation in temperature of fluid B is the same whether the flow is parallel-flow or counter-flow.

(ii) In counter-flow the temperature range possible is greater, since, in theory, the fluid being heated can be raised to a higher

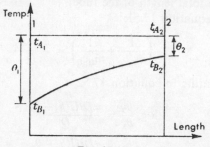

Fig. 17.17

temperature than that of the heating fluid at exit. In parallel-flow the final temperatures of the fluids must be somewhere between the initial values of each fluid. This should be clear from figs. 17.16a and 17.16b.

(iii) When the product $\dot{m}_A c_A$ is equal to $\dot{m}_B c_B$ then the temperature difference in counter-flow is the same all along the length of the tube. This must be the case since the heat given up by fluid A is equal to the heat received by fluid B. Referring to fig. 17.16b, we have

$$\dot{m}_A c_A(t_{A_1} - t_{A_2}) = \dot{m}_B c_B(t_{B_1} - t_{B_2})$$

$$\therefore (t_{A_1} - t_{A_2}) = (t_{B_1} - t_{B_2}) \quad \text{or} \quad (t_{A_1} - t_{B_1}) = (t_{A_2} - t_{B_2})$$

i.e. $$\theta_m = \theta_1 = \theta_2$$

Note that if we attempt to substitute in equation 17.54 under these circumstances, then the result is indeterminate,

i.e. $$\theta_m = \frac{\theta_1 - \theta_2}{\log_e \theta_1/\theta_2} = \frac{0}{\log_e 1} = \frac{0}{0}$$

The proof of the logarithmic mean temperature difference given previously is not valid when θ_1 is equal to θ_2, since $d\theta$ is then zero.

(iv) From equation 17.55, $Q = UA \, LMTD$, it can be seen that, for a given surface area, A, and a given mean value of the overall heat transfer coefficient, U, then the logarithmic mean temperature difference, $LMTD$, must be made as large as possible. It is found that for given temperature changes, $LMTD$ is always greater for counter-flow than it is for parallel-flow. The initial temperature difference, θ_1, is greater for parallel-flow, but the value of *LMTD* is always less. It follows that for given rates of mass flow of the two fluids, and for given temperature changes, the surface area required is less for counter-flow.

Example 17.10

Exhaust gases flowing through a tubular heat exchanger at the rate of 0·3 kg/s are cooled from 400°C to 120°C by water initially at 10°C. The specific heat of exhaust gases and water may be taken as 1·13 and 4·19 kJ/kg K respectively, and the overall heat transfer coefficient from gases to water is 140 W/m² K. Calculate the surface area required when the cooling water flow is 0·4 kg/s,

(a) for parallel-flow;

(b) for counter-flow.

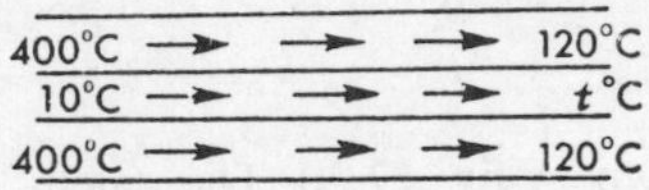

$$
\begin{array}{llll}
\overline{400°C \rightarrow} & \rightarrow & \rightarrow & 120°C \\
\overline{10°C \rightarrow} & \rightarrow & \rightarrow & t°C \\
\overline{400°C \rightarrow} & \rightarrow & \rightarrow & 120°C
\end{array}
$$

Fig. 17.18

(a) *Parallel-flow*

The heat exchanger is shown diagrammatically in fig. 17.18. The heat given up by the exhaust gases is equal to the heat taken up by the water,

i.e. $Q = 0·3 \times 1·13 \times (400 - 120) = 0·4 \times 4·19 \times (t - 10)$

i.e. $t = \dfrac{0·3 \times 1·13 \times 280}{0·4 \times 4·19} + 10 = 66·6°C$

Also, $Q = 0·3 \times 1·13 \times 280 = 95 \text{ kW}$

From equation 17.54,

$$\theta_m = \frac{\theta_1 - \theta_2}{\log_e \theta_1/\theta_2} = \frac{(400-10)-(120-66·6)}{\log_e \left(\dfrac{400-10}{120-66·6}\right)} = \frac{336·6}{1·99} = 169 \text{ K}$$

From equation 17.55,

$$Q = UA \; LMTD$$

i.e. $95 \times 10^3 = 140 \times A \times 169 \text{ W}$

i.e. $A = \dfrac{95 \times 10^3}{140 \times 169} = 4·01 \text{ m}^2$

i.e.   Surface area required $= 4·01 \text{ m}^2$

(b) *Counter-flow*

The heat exchanger is shown diagrammatically in fig. 17.19. The water temperature at outlet is 66·6°C and $Q = 95$ kW as calculated in part (a).

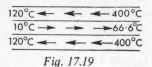

Fig. 17.19

From equation 17.54,

$$\theta_m = \frac{\theta_1 - \theta_2}{\log_e \theta_1/\theta_2} = \frac{(120-10)-(400-66\cdot6)}{\log_e\left(\dfrac{120-10}{400-66\cdot6}\right)} = \frac{223\cdot4}{1\cdot11} = 201 \text{ K}$$

From equation 17.55,

$$Q = UA\,LMTD \qquad \therefore \quad 95 \times 10^3 = 140 \times A \times 201 \text{ W}$$

i.e.
$$A = \frac{95 \times 10^3}{140 \times 201} = 3\cdot37 \text{ m}^2$$

i.e.   Surface area required $= 3\cdot37$ m^2

Cross-flow recuperator

A simple cross-flow recuperator is shown in fig. 17.20. The calculation of the mean temperature difference is much more difficult in

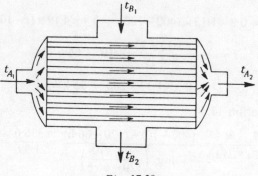

Fig. 17.20

this case. θ_m depends on the ratio of the product of the mass flow and specific heat of fluids A and B, as well as on the ratio of the temperature difference between the fluids at inlet and outlet. Tables are available of a correction factor for various values of the ratios

$$\frac{t_{B_2} - t_{A_2}}{t_{B_1} - t_{A_1}} \quad \text{and} \quad \frac{\dot{m}_A c_A}{\dot{m}_B c_B}$$

The correction factor is multiplied by the arithmetic mean temperature difference to give the true value of θ_m. When the temperature differences at inlet and outlet are not substantially different, it is a sufficiently good approximation to use the arithmetic mean temperature difference,

i.e. Arithmetic mean temperature difference

$$\theta_{AM} = \left(\frac{t_{A_1} + t_{A_2}}{2}\right) - \left(\frac{t_{B_1} + t_{B_2}}{2}\right) \qquad 17.56$$

It has been shown in example 17.10 that the surface area required for a given heat flow is smaller with counter-flow than with parallel-flow. For cross-flow the required surface area is between that for parallel-flow and counter-flow. As with counter-flow, the outlet temperature of the heated fluid in cross-flow can be raised to a higher temperature than the outlet temperature of the cooled fluid (e.g. in fig. 17.20, t_{A_2} can be higher than t_{B_2}); this is not possible in parallel-flow.

Multi-pass and mixed flow recuperators

The simple parallel-flow and counter-flow heat exchangers discussed above occur very rarely in practice. To obtain the necessary surface area with a simple tube and annulus arrangement the length of the tube may be too large for practical purposes. For instance, in example 17.10b, if the tube diameter were 150 mm, the length required would be,

$$l = \frac{A}{\pi D} = \frac{3 \cdot 37}{\pi \times 0 \cdot 15} = 7 \cdot 15 \text{ m}$$

In order to make the heat exchanger more compact, which is desirable from space considerations, and also to reduce the heat loss from the outside surface, it is necessary to have several tubes and perhaps several passes or bundles of tubes. The flow can be either cross-flow, or a mixture of parallel-flow, counter-flow, and cross-flow. The latter case is called mixed-flow. A typical example of a shell-type mixed flow heat exchanger is shown in fig. 17.21. The analysis of a mixed-flow heat exchanger is complex and correction factors have been plotted, which must be used to evaluate the mean temperature difference. The logarithmic mean temperature difference in counter-flow is evaluated and then multiplied by the correction factor. Correction factors for most types of mixed-flow heat exchangers are given in reference 17.2. Note that when one of the fluids is a condensing vapour or a boiling liquid then the mean temperature

Fig. 17.21

difference is the same whether the heat exchanger is parallel-flow, counter-flow, cross-flow, or mixed-flow.

In certain heat exchangers of the multi-pass type, the mean temperature difference for counter-flow or parallel-flow can still be used as a reasonable approximation. For example, the heat exchanger in fig. 17.22 is essentially a counter-flow type. The larger the number of passes made by fluid B then the nearer the heat exchanger is to pure counter-flow.

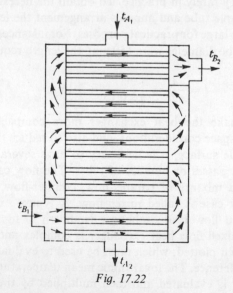

Fig. 17.22

In most heat exchangers the fluid flowing is not completely free from dirt, oil, grease, and chemical deposits, and a coating tends to collect on all metal surfaces. This increases the resistance to heat transfer and must be allowed for in design calculations. It is usual to allow for the effect of this coating of dirt by adding a *fouling resistance* to the total thermal resistance. Typical values of fouling resistance for 1 m² of surface area are: 1·8 K/kW for fuel oil; 0·6 K/kW for river water; and 0·2 K/kW for boiler feed water which has been treated. Facility must be provided for easy periodic cleaning of the tubes. For a comprehensive treatment of process heat exchangers see references 17.2 and 17.4.

Another form of recuperator which should be mentioned briefly is the *extended surface* type. The metal wall containing a fluid to be cooled can be extended on the outside in the form of fins, studs, or ribs. Examples of this type are the finned hot water space-heater (sometimes mis-named 'radiator') and the air-cooled cylinders of small air compressors and internal-combustion engines. The fins on the surface give a larger outside surface area for the same internal surface area, and hence increase the cooling effect for a given volume. Details of compact heat exchangers are given in reference 17.3.

Regenerators

In the various types of recuperator described above, the hot and cold fluids are separated at all times by a metal wall. The characteristic feature of a regenerator is that the fluids occupy the same space in turn. The fluids used in regenerators are nearly always gaseous. When the hot gas occupies the space, it gives up heat to the walls, or to solid matter distributed throughout the space, called a matrix. The hot gas is then withdrawn, the cold gas enters the space and is heated by the walls and the matrix; the process is a cyclic one and analysis is complex. In order to have continuous operation it is usual to use two regenerators with hot gas flowing through one, and cold gas flowing through the other at any instant. When the flows are switched from one regenerator to the other, the hot gas is cooled while the cold gas is heated. The period of time between the switching of the flows must be chosen to give the required heat transfer between the two gases. This type of regenerator is shown diagrammatically in fig. 17.23a.

Another method used is the *rotating matrix* type, in which a cylinder containing solid inserts is rotated so that it passes alternately through the cold and hot gas str3ams, which are sealed from each other. An example of this type of regenerator is the Ljungström air preheater for boiler furnaces, shown diagrammatically in fig. 17.23b.

There are many applications for regenerators, from air preheaters

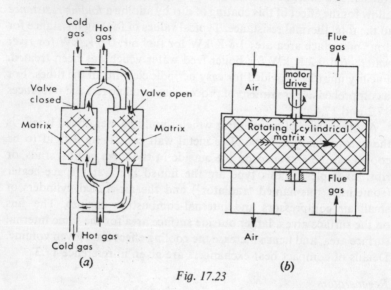

Fig. 17.23

in blast furnaces to heat exchangers in plants for gas liquefaction. One application of the rotating matrix type which is becoming important is in energy conservation (see Chapter 8: the energy wheel).

17.10 Forced convection

The study of forced convection is concerned with the transfer of heat between a moving fluid and a solid surface. In order to apply Newton's law of cooling, given by equation 17.3 ($Q = hA(t_w - t)$), it is necessary to find a value for the heat transfer coefficient, h. It has been stated in Section 17.2 that h is given by k/δ, where k is the thermal conductivity of the fluid and δ is the thickness of the fluid film on the surface. The problem is then to find a value for δ in terms of the fluid properties and the fluid velocity. The thickness of the fluid film, δ, depends on the type of fluid flow across the surface and this is governed by the Reynolds number, Re.

The Reynolds number is a dimensionless group given by

$$Re = \frac{\rho Cl}{\mu} \quad \text{or} \quad \frac{Cl}{\nu}$$

(where ρ = fluid density; C = fluid mean velocity; l = characteristic linear dimension; μ = dynamic viscosity of the fluid; ν = kinematic viscosity of the fluid = μ/ρ).

The various kinds of forced convection, such as flow in a tube, flow across a tube, flow across a flat plate, etc., can be solved mathematically when certain assumptions are made with regard to the boundary conditions. It is exceedingly difficult to obtain an exact mathematical solution to such problems, particularly in the case of turbulent flow, but approximate solutions can be obtained by making suitable assumptions.

It is not within the scope of this book to approach the subject of forced convection fundamentally. However, many of the results used in heat transfer are derived from experiment, and in fact for many problems no mathematical solution is available and empirical values are essential. These empirical values can be generalized using dimensional analysis, which will now be considered.

Dimensional analysis

In order to apply dimensional analysis it is necessary to know from experience all the variables upon which the desired function depends. The results must apply to geometrically similar bodies, therefore one of the variables must always be a characteristic linear dimension.

Consider the dimensional analysis for forced convection, assuming that the effects of free convection, due to differences in density, may be neglected. It is found that the heat transfer coefficient, h, depends on the fluid viscosity, μ, the fluid density, ρ, the thermal conductivity of the fluid, k, the specific heat of the fluid, c, the temperature difference between the surface and the fluid, θ, and the fluid velocity, C. Therefore we have

$$h = \mathrm{f}(\mu, \rho, k, c, \theta, C, l) \qquad 17.57$$

(where l is a characteristic linear dimension, and f is some function).

Equation 17.57 can be written as follows,

$$h = A\mu^{a_1}\rho^{b_1}k^{c_1}c^{d_1}\theta^{e_1}C^{f_1}l^{g_1} + B\mu^{a_2}\rho^{b_2}k^{c_2}c^{d_2}\theta^{e_2}C^{f_2}l^{g_2} + \text{etc.} \quad 17.58$$

(where A and B are constants, and a_1, b_1, c_1, etc., are arbitrary indices).

Each term on the right-hand side of the equation must have the same dimensions as the dimensions of h. Considering the first term only, we can write

Dimensions of h = Dimensions of $(\mu^{a_1} \rho^{b_1} k^{c_1} c^{d_1} \theta^{e_1} C^{f_1} l^{g_1})$

Each of the properties in the equation can be expressed in terms of five fundamental dimensions; these are, mass, M, length, L, time, T, temperature, t, and heat, Q.

For h the units are	$\dfrac{W}{m^2\,K}$	i.e.	$\dfrac{Q}{L^2 T t}$
For μ the units are	$\dfrac{kg}{m\,s}$	i.e.	$\dfrac{M}{LT}$
For k the units are	$\dfrac{W}{m\,K}$	i.e.	$\dfrac{Q}{LTt}$
For ρ the units are	kg/m^3	i.e.	$\dfrac{M}{L^3}$
For c the units are	$\dfrac{kJ}{kg\,K}$	i.e.	$\dfrac{Q}{Mt}$
For θ the units are	K	i.e.	t
For C the units are	m/s	i.e.	$\dfrac{L}{T}$
For l the units are	m	i.e.	L

Hence, substituting,

$$\frac{Q}{L^2 T t} = \left(\frac{M}{LT}\right)^a \times \left(\frac{M}{L^3}\right)^b \times \left(\frac{Q}{LTt}\right)^c \times \left(\frac{Q}{Mt}\right)^d \times (t)^e \times \left(\frac{L}{T}\right)^f \times (L)^g$$

i.e.

$$\frac{Q}{L^2 T t} = (M)^{a+b-d} \times (L)^{f+g-a-3b-c} \times (T)^{-a-c-f} \times (t)^{e-c-d} \times (Q)^{c+d}$$

For the dimensions of each side of the equation to be the same, the power to which each fundamental dimension is raised must be the same on both sides of the equation. Therefore, equating indices we have,

$$
\begin{aligned}
\text{For } Q: & \quad 1 = c + d \\
\text{For } L: & \quad -2 = f + g - a - 3b - c \\
\text{For } T: & \quad -1 = -a - c - f \\
\text{For } t: & \quad -1 = e - c - d \\
\text{For } M: & \quad 0 = a + b - d
\end{aligned}
$$

We have five equations and seven unknowns, therefore a solution can only be obtained in terms of two of the indices. It is most useful to express a, b, c, e, and g in terms of d and f. Then it can be shown that

$$a = (d-f); \quad b = f; \quad c = (1-d); \quad e = 0; \quad g = (f-1)$$

Substituting these values in equation 17.58, we have

$$h = A\mu^{(d_1-f_1)}\rho^{f_1}k^{(1-d_1)}c^{d_1}\theta^0 C^{f_1}l^{(f_1-1)}$$
$$+ B\mu^{(d_2-f_2)}\rho^{f_2}k^{(1-d_2)}c^{d_2}\theta^0 C^{f_2}l^{(f_2-1)} + \text{etc.},$$

i.e.
$$h = A\frac{k}{l}\left(\frac{c\mu}{k}\right)^{d_1}\left(\frac{\rho Cl}{\mu}\right)^{f_1} + B\frac{k}{l}\left(\frac{c\mu}{k}\right)^{d_2}\left(\frac{\rho Cl}{\mu}\right)^{f_2} + \text{etc.}$$

Hence it can be seen that

$$\frac{hl}{k} = K\mathbf{F}\left\{\left(\frac{c\mu}{k}\right), \left(\frac{\rho Cl}{\mu}\right)\right\}$$

(where K is a constant and $\mathbf{F}$ is some function).

The dimensionless group, hl/k, is called the Nusselt number, Nu; the dimensionless group, $c\mu/k$, is called the Prandtl number, Pr; and the dimensionless group, $\rho Cl/\mu$, is the Reynolds number, Re,

i.e.
$$Nu = K\mathbf{F}\{(Pr), (Re)\} \qquad 17.59$$

Experiments can be performed in order to evaluate K, and to determine the nature of the function $\mathbf{F}$.

When evaluating Nu, Pr, and Re it is necessary to take the fluid properties at a suitable mean temperature, since the properties vary with temperature. For cases in which the temperature of the bulk of the fluid is not very different from the temperature of the solid surface, then fluid properties are evaluated at the *mean bulk temperature* of the fluid. When the temperature difference is large, errors may be caused by using a mean bulk temperature, and a *mean film temperature* is sometimes used, defined by

$$t_f = \frac{t_b + t_w}{2} \qquad 17.60$$

(where t_b is the mean bulk temperature and t_w is the surface temperature).

When using an empirical equation it is essential to know at what reference temperature the properties have been evaluated by the experimenter.

It should be noted that the Prandtl number, $Pr = c\mu/k$, consists entirely of fluid properties and therefore is itself a property. Some values of μ, c, k, Pr, and ρ for various fluids are tabulated in reference 17.7.

Example 17.11

Calculate the heat transfer coefficient for water flowing through a 25 mm diameter tube at the rate of 1·5 kg/s, when the mean bulk temperature is 40°C. For turbulent flow of a liquid take

$$Nu = 0.0243 \ Re^{0.8} \times Pr^{0.4}$$

(where the characteristic dimension of length is the tube diameter and all properties are evaluated at mean bulk temperature).

First it is necessary to ascertain whether the flow is turbulent or laminar. For flow through a tube it can be assumed that the flow is turbulent when $Re > 2100$ approximately. The properties of water can be taken from reference 17.7.

Then,

$$\text{Volume flow} = 1.5 \times v_f = 1.5 \times 0.001 = 0.0015 \ \text{m}^3/\text{s}$$

i.e. $$\text{Velocity in tube, } C = \frac{0.0015}{\pi/4 \times 0.025^2} = 3.06 \ \text{m/s}$$

Now, $$Re = \frac{\rho C d}{\mu} = \frac{C d}{v_f \mu} = \frac{3.06 \times 0.025}{0.001 \times 651 \times 10^{-6}} = 117\,500$$

The flow is therefore well into the turbulent region and the formula given for turbulent flow can be applied.

From tables, $Pr = 4.3$, hence substituting,

$$Nu = 0.0243 \times (117\,500)^{0.8} \times (4.3)^{0.4}$$

$$= 0.0243 \times 11\,380 \times 1.792 = 496$$

i.e. $$Nu = \frac{hd}{k} = 496$$

$$\therefore \ h = \frac{496 \times 632 \times 10^{-6}}{0.025} = 12.55 \ \text{kW/m}^2 \ \text{K}$$

i.e.   Heat transfer coefficient $= 12.55 \ \text{kW/m}^2 \ \text{K}$

For laminar flow in a tube an exact mathematical solution has been found; this gives $Nu = 3.65$. It can be seen that, since $Nu = hd/k = 3.65$, the heat transfer coefficient, h, for any one tube, depends only on the thermal conductivity of the fluid.

In the foregoing dimensional analysis five fundamental dimensions, heat Q, length L, time T, temperature t, and mass M, were chosen. The units of work, or energy in general, are given by

$$(\text{force} \times \text{distance}) = (\text{mass} \times \text{acceleration} \times \text{distance})$$

$$= M \frac{L}{T^2} L = \frac{ML^2}{T^2}$$

Since heat is a form of energy it can be seen that there is no need to choose heat as one of the fundamental dimensions. If the dimension, Q, is omitted, and the units of heat are replaced by ML^2/T^2 whenever they occur, then four dimensionless groups are obtained,

i.e. $$Nu = K\text{F}\left\{(Pr), (Re), \left(\frac{C^2}{c\theta}\right)\right\}$$

Now if the group $C^2/c\theta$ is divided by $(\gamma - 1)$, which is a constant for any one gas, and if θ is replaced by the absolute bulk temperature of the gas, T, then we have

$$\frac{C^2}{cT(\gamma - 1)} = \frac{C^2}{\gamma RT} = \frac{C^2}{a^2} = (Ma)^2$$

(where a is the velocity of sound in the gas and Ma is the Mach number, see Section 10.10).

Hence, $$Nu = K'\text{F}\{(Pr), (Re), (Ma)^2\}$$ 17.61

(where K' is a constant).

The influence of the Mach number, Ma, on the heat transfer is negligible for most problems. For high-speed flow however, large amounts of kinetic energy are dissipated by friction in the boundary layer near the surface, and the Mach number becomes an important parameter.

Reynolds analogy

Reynolds postulated that the heat transfer from a solid surface is similar to the transfer of fluid momentum from the surface, and hence that it is possible to express the heat transfer in terms of the frictional resistance to the flow.

Consider turbulent flow:

It can be assumed that particles of mass, m, transport heat and momentum to and from the surface, moving perpendicular to the surface. Then, on the average,

$$\text{Heat transferred per unit area, } q = \dot{m}c\theta$$

(where c = specific heat of the fluid; θ = temperature difference between the surface and the bulk of the fluid).

Also, the rate of change of momentum across the stream is given by

$$\dot{m}(C - C_w) = \dot{m}C$$

(where C = velocity of the bulk of the fluid; C_w = fluid velocity at the surface = 0).

Then, $$\text{Force per unit area} = \tau_w = \dot{m}C$$

(where τ_w = shear stress in the fluid at the wall).

Combining the equations for heat flow and momentum transfer, then,

$$\frac{q}{c\theta} = \frac{\tau_w}{C}$$

or $$q = \frac{\tau_w c\theta}{C} \qquad 17.62$$

For turbulent flow in practice there is always a thin layer of fluid on the surface in which viscous effects predominate. This film is known as the laminar sub-layer. In this layer heat is transferred purely by conduction.

Therefore, from Fourier's law, for unit area

$$q = -k\left(\frac{d\theta}{dy}\right)_{y=0}$$

(where k = thermal conductivity of the fluid; y = distance from the surface perpendicular to the surface).

Also, for viscous flow,

$$\text{Shear stress, } \tau = \mu \times (\text{velocity gradient})$$

Hence the shear stress at the wall is given by

$$\tau_w = \mu \left(\frac{dC}{dy}\right)_{y=0}$$

(where μ = fluid viscosity; C = fluid velocity).

Now since the laminar sub-layer is very thin it may be assumed that the temperature and velocity vary linearly with the distance from the wall, y,

i.e.

$$q = -\frac{k\theta}{\delta_b} \quad \text{and} \quad \tau_w = \frac{\mu C}{\delta_b}$$

(where δ_b is the thickness of the laminar sub-layer).

Then eliminating δ_b, and neglecting the minus sign, we have

$$\frac{q}{k\theta} = \frac{\tau_w}{\mu C}$$

i.e.

$$q = \frac{\tau_w k\theta}{\mu C}$$

It can be seen that this equation is identical with equation 17.62 when

$$c = \frac{k}{\mu}$$

i.e. when

$$\frac{c\mu}{k} = 1 \quad \text{or} \quad Pr = 1$$

Therefore for fluids whose Prandtl number is approximately unity the simple Reynolds analogy can be applied, since the heat transferred across the laminar sub-layer can be considered in a similar way to the heat transferred from the sub-layer to the bulk of the fluid. For most gases, dry vapours, and superheated vapours Pr lies between about 0.65 and 1.2.

For unit surface area, $q = h\theta$, therefore substituting in equation 17.62, we have

$$\frac{h}{c} = \frac{\tau_w}{C}$$

Dividing through by ρC (where ρ is the mean density of the fluid) we have

$$\frac{h}{\rho c C} = \frac{\tau_w}{\rho C^2}$$

Both sides of this equation are dimensionless. The term on the left side is called the Stanton number, St,

i.e.
$$St = \frac{h}{\rho Cc} \qquad\qquad 17.63$$

A dimensionless friction factor, f, is defined by

$$f = \frac{\tau_w}{\left(\dfrac{\rho C^2}{2}\right)} \qquad\qquad 17.64$$

Therefore we have for Reynolds analogy

$$St = \frac{f}{2} \qquad\qquad 17.65$$

The Stanton number, St, can be written as

$$St = \frac{h}{\rho Cc} = \frac{hl}{k} \times \frac{\mu}{\rho Cl} \times \frac{k}{c\mu} = \frac{Nu}{RePr}$$

i.e.
$$St = \frac{Nu}{RePr} \qquad\qquad 17.66$$

The friction factor, f, can be derived mathematically for some cases, but in other cases a practical determination is necessary. For turbulent flow in a pipe, a simple measurement of the pressure drop gives f, and then, using equation 17.62 or equation 17.65, the approximate heat flow can be found.

For flow in a pipe of diameter, d, the resistance to flow over unit length is given by

$$\text{Resistance} = \tau_w \pi d = \Delta p \frac{\pi}{4} d^2$$

(where $\Delta p =$ the pressure drop in unit length).

$$\therefore \ \tau_w = \frac{\Delta p d}{4} \qquad\qquad 17.67$$

An important factor in heat exchanger design is the pumping power required. The pumping power is the rate at which work is done in overcoming the frictional resistance,

i.e. for flow in a pipe,

$$\text{Pumping power per unit length,} \ W = \tau_w \pi d C$$

Also, from equation 17.62,

$$\text{Heat flow per unit area, } q = \frac{\tau_w c \theta}{C}$$

i.e. $$\text{Heat flow per unit length, } Q = \frac{\tau_w c \theta \pi d}{C}$$

Then the ratio of the pumping power, W, to the heat flow, Q, can be expressed as

$$\frac{W}{Q} = \frac{\tau_w \pi d C C}{\tau_w c \theta \pi d} = \frac{C^2}{c\theta} \qquad 17.68$$

(for a heat exchanger, θ is the log mean temperature difference).

It can be seen from equation 17.68 that the power required for a given heat transfer rate can be reduced by decreasing the velocity of flow, C. However, a reduction in fluid velocity means that the required surface area must be increased, and hence a compromise must be made.

Example 17.12

Air is heated by passing it through a 25 mm bore copper tube which is maintained at 280°C. The air enters at 15°C and leaves at 270°C at a mean velocity of 30 m/s. Using the Reynolds analogy, calculate the length of the tube and the pumping power required. For turbulent flow in a tube take $f = 0.0791(Re)^{-1/4}$, and all properties at mean film temperature.

The mean film temperature can be found from equation 17.60,

i.e. $t_f = \dfrac{t_b + t_w}{2} = \dfrac{1}{2}\left(\dfrac{15 + 270}{2} + 280\right) = \dfrac{142.5 + 280}{2} = 211.25°C$

From tables at $t_f = 211.25°C = 484.4$ K the properties of air can be found.

Then, $$Re = \frac{Cd}{\nu} = \frac{30 \times 0.025}{3.591 \times 10^{-5}} = 20\,900$$

$$\therefore f = \frac{0.0791}{(20\,900)^{1/4}} = \frac{0.0791}{12.01} = 0.006\,58$$

From equation 17.65,

$$St = f/2$$

i.e. $$St = \frac{Nu}{RePr} = f/2 = \frac{0.006\,58}{2} = 0.003\,29$$

i.e. $Nu = 0.003\,29 \times 20\,900 \times 0.681 = 46.8$

$$\therefore \frac{hd}{k} = 46.8$$

i.e. $h = \dfrac{46.8 \times 3.938 \times 10^{-5}}{0.025} = 0.0737 \text{ kW/m}^2 \text{ K}$

Mass flow of air $= \dfrac{\pi}{4} \times 0.025^2 \times 30 \times 0.73 = 0.010\,75 \text{ kg/s}$

Hence,

Heat received by the air $= \dot{m}c(t_{a_2} - t_{a_1})$
$$= 0.010\,75 \times 1.027 \times (270 - 15)$$
$$= 2.815 \text{ kW}$$

Also, from equation 17.3,

$$Q = hA\theta = 2.815 \text{ kW}$$

Using equation 17.54,

$$\theta_m = \frac{\theta_1 - \theta_2}{\log_e \theta_1/\theta_2} = \frac{(280 - 15) - (280 - 270)}{\log_e\left(\dfrac{280 - 15}{280 - 270}\right)} = 77.9 \text{ K}$$

Then, $Q = 2.815 = 0.0737 \times 77.9 \times A$

$$\therefore A = \frac{2.815}{0.0737 \times 77.9} = 0.49 \text{ m}^2$$

Therefore,   Tube length $= \dfrac{0.49}{\pi \times 0.025} = 6.24 \text{ m}$

From equation 17.68,

$$\frac{W}{Q} = \frac{C^2}{c\theta}$$

$$\therefore W = \frac{2.815 \times 30 \times 30}{1.027 \times 77.9} = 31.7 \text{ W}$$

i.e.   Pumping power $= 31.7 \text{ W}$

Example 17.13

In a 25 mm diameter tube the pressure drop per m length is 0.0002
bar at a section where the mean velocity is 24 m/s, and the mean

specific heat of the gas is $1 \cdot 13$ kJ/kg K. Calculate the heat transfer coefficient.

For a 1 m length,

$$\Delta p = 0 \cdot 0002 \text{ bar}$$

From equation 17.67,

$$\tau_w = \frac{\Delta p d}{4} = \frac{10^5 \times 0 \cdot 0002 \times 25}{4 \times 10^3} = 0 \cdot 125 \text{ N/m}^2$$

Then from equation 17.64,

$$f = \frac{\tau_w}{\left(\dfrac{\rho C^2}{2}\right)} = \frac{2 \times 0 \cdot 125}{\rho C^2}$$

Also, from equation 17.65,

$$St = \frac{f}{2} \quad \text{i.e.} \quad \frac{h}{\rho c C} = \frac{2 \times 0 \cdot 125}{2 \rho C^2}$$

$$\therefore \quad h = \frac{0 \cdot 125 \rho c C}{\rho C^2} = \frac{0 \cdot 125 \times 1 \cdot 13}{24} \text{ kW/m}^2 \text{ K}$$

i.e.   Heat transfer coefficient, $h = 0 \cdot 005 \, 88$ kW/m^2 K

$$= 5 \cdot 88 \text{ W/m}^2 \text{ K}$$

Various modifications have been made to the simple Reynolds analogy in an attempt to obtain an equation which will give a solution for turbulent heat transfer over a wide range of Prandtl numbers. (For very viscous oil the Prandtl number is of the order of thousands, whereas for liquid metals it may be as low as $0 \cdot 01$.) Equations based on modern theories of turbulent flow give the Stanton number as a function of the Reynolds number, the Prandtl number, and the friction factor, and in general these equations reduce to, $St = f/2$, when the Prandtl number is put equal to unity (see references 17.1, 17.5 and 17.6).

Two further points should be mentioned here:

(i) When the temperature difference between the surface and the bulk of the fluid is very high, then the property variations become large enough to be taken into consideration. It is then no longer sufficient to use a mean film temperature to evaluate the properties, as given by equation 17.60. The variation of each property with temperature across the stream must be known; sometimes it is sufficiently accurate to use an equation of the form,

$$Nu = K\phi\left\{(Pr), (Re), \left(\frac{T_s}{T_w}\right)\right\}$$

(where T_s and T_w are the absolute temperatures at the axis of the pipe and at the pipe wall respectively, and fluid properties are taken at the mean film temperature).

(ii) The equations for flow in a pipe do not usually allow for the effects of the *entry length*. At the entry to a heated tube the hydrodynamic and thermal boundary layers start to build up on the wall, gradually thickening until the flow becomes *fully developed*. In this initial region of the tube the heat transfer coefficient is much larger since the resistance to heat flow of the boundary layer is less, and hence an equation which neglects this effect will give a low value for the calculated heat transfer. The effect is more marked for laminar flow than for turbulent flow, and is much more important for fluids with high Prandtl numbers. In most heat exchange processes the flow is turbulent and the tube length is sufficiently long to make the entry length effect negligibly small. In the case of oil coolers the flow is laminar, the Prandtl number is high, and hence the entry effect may be appreciable.

When flow across a flat plate is considered, the characteristic dimension of length is taken as the distance from the leading edge, and the heat transfer coefficient obtained is then the local value at that section of the plate. The average value of the heat transfer coefficient over the whole plate is the value to be used in calculating the heat transfer to or from the plate. It can be shown that the average heat transfer coefficient for a heated plate over a length l, is twice the local heat transfer coefficient at a distance l from the leading edge.

Example 17.14

Air at 20°C, flowing at 25 m/s, passes over a flat plate, the surface of which is maintained at 270°C. Calculate the rate at which heat is transferred per m width from both sides of the plate over a distance of 0·25 m from the leading edge. For heat transfer from a flat plate

$$Nu = 0.332(Pr)^{1/3} \times (Re)^{1/2}$$

(where the characteristic linear dimension is the distance from the leading edge, and all properties are evaluated at mean film temperature).

$$\text{Mean film temperature} = \frac{20+270}{2} = 145°C = 418 \text{ K}$$

Taking the values from tables of properties of air, we have

$$Pr = 0.687 \quad \text{and} \quad Re = \frac{Cl}{\nu} = \frac{25 \times 0.25 \times 10^5}{2.8} = 223\,000$$

Then, $Nu = 0.332 \times (0.687)^{1/3} \times (223\,000)^{1/2}$

$$= 0.332 \times 0.883 \times 472 = 138.5$$

i.e. $$Nu = \frac{hl}{k}$$

$$\therefore \ h = \frac{138.5 \times 3.49}{0.25 \times 10^5} = 0.0193 \text{ kW/m}^2 \text{ K}$$

Hence the average heat transfer coefficient is

$$0.0193 \times 10^3 \times 2 = 38.6 \text{ W/m}^2 \text{ K}$$

Then the heat transferred from both sides of the plate over the length of 0.25 m for 1 m width is given by

$$Q = hA\theta = 38.6 \times 0.25 \times 1 \times 2 \times (270 - 20) \doteqdot 4825 \text{ W}$$

i.e.   Heat transferred $= 4.825$ kW

The friction loss for the initial length of a flat plate where the boundary layer is still laminar is given by

$$f = 0.664(Re)^{-1/2}$$

Hence it can be seen that the simple Reynolds analogy, given by equation 17.65, $St = f/2$, gives for the initial length of a flat plate

$$St = 0.332(Re)^{-1/2}$$

or $$\frac{Nu}{PrRe} = 0.332(Re)^{-1/2}$$

i.e. $$Nu = 0.332(Pr)(Re)^{1/2}$$

This is the same as the equation given in example 17.14, if the Prandtl number is unity.

17.11 Heat exchanger effectiveness

In certain cases of heat exchanger design the efficiency of the heat transfer process becomes very important; for example, for compact heat exchangers, particularly in the aircraft industry where weight is also important. A method due to Nusselt and developed by Kays and London (see reference 17.3) is described in this section.

The *effectiveness*, ε, of a heat exchanger is defined as the ratio of the actual heat transferred to the maximum possible heat transfer.

For any heat exchanger with mass flow rates of hot and cold fluids, $\dot{m}_H$ and $\dot{m}_C$, with specific heats, c_H and c_C, let the overall temperature changes of each fluid be Δt_H and Δt_C.

Neglecting heat losses to the surroundings:

$$Q = \dot{m}_H c_H \, \Delta t_H = \dot{m}_C c_C \, \Delta t_C$$

or
$$Q = C_H \, \Delta t_H = C_C \, \Delta t_C \qquad 17.69$$

(where $C_H = \dot{m}_H c_H$ and $C_C = \dot{m}_C c_C$ are the *thermal capacities* of the hot and cold fluids).

From equation 17.69 it can be seen that the fluid with the smaller thermal capacity, C, has the greater temperature change, Δt. The maximum possible temperature change of one of the fluids is $(t_{H_{Max}} - t_{C_{Min}})$, and this ideal temperature change can only be aspired to by the fluid with the minimum thermal capacity,

i.e.
$$\varepsilon = \frac{Q}{C_{Min}(t_{H_{Max}} - t_{C_{Min}})} \qquad 17.70$$

The object of a well-designed heat exchanger is to obtain the maximum possible change of temperature of a fluid for a given driving force, that is for a given logarithmic mean temperature difference, LMTD. Hence another useful measure of the efficiency of a heat exchanger is the *number of transfer units*, NTU, defined as:

$$NTU = \frac{\text{Maximum temperature change of one fluid}, (\Delta t)_{Max}}{LMTD}$$

Now,

$$Q = UA \, LMTD = C_{Min}(\Delta t)_{Max}$$

$$\therefore \ NTU = \frac{(\Delta t)_{Max}}{LMTD} = \frac{UA}{C_{Min}} \qquad 17.71$$

The greater the number of transfer units the more effective is the heat exchanger.

The ratio of the minimum to the maximum thermal capacity is usually given the symbol, R,

i.e. $$R = C_{Min}/C_{Max} \qquad 17.72$$

Note that R may vary between 1 (when both fluids have the same thermal capacity) and 0 (when one of the fluids has an infinite thermal capacity, e.g. a condensing vapour or a boiling liquid).

Figure 17.24 shows a typical example of a graph of effectiveness, ε, against NTU for various values of the thermal capacity ratio, R.

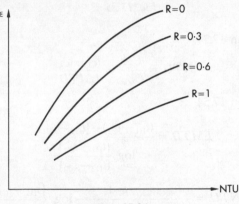

Fig. 17.24

Consider a counter-flow heat exchanger as shown in fig. 17.25. From the figure it can be seen that, $C_C = C_{Min}$, since $\Delta t_C > \Delta t_H$.

i.e. $$R = C_C/C_H$$

or, using equation 17.69,

$$R = \frac{t_{H1} - t_{H2}}{t_{C1} - t_{C2}} \qquad 17.73$$

From equation 17.70,

$$\varepsilon = \frac{Q}{C_{Min}(t_{H_{Max}} - t_{C_{Min}})} = \frac{C_{Min}(t_{C1} - t_{C2})}{C_{Min}(t_{H1} - t_{C2})} = \frac{t_{C1} - t_{C2}}{t_{H1} - t_{C2}} \qquad 17.74$$

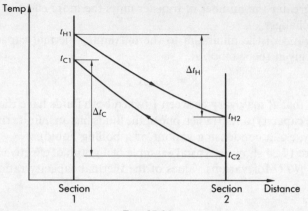

Fig. 17.25

From equation 17.71,

$$NTU = \frac{UA}{C_{\text{Min}}} = \frac{t_{C1} - t_{C2}}{LMTD}$$

From equation 17.54,

$$LMTD = \frac{(t_{H1} - t_{C1}) - (t_{H2} - t_{C2})}{\log_e \dfrac{(t_{H1} - t_{C1})}{(t_{H2} - t_{C2})}}$$

$$\therefore\ NTU = \frac{(t_{C1} - t_{C2})}{(t_{H1} - t_{C1}) - (t_{H2} - t_{C2})} \log_e \left\{ \frac{t_{H1} - t_{C1}}{t_{H2} - t_{C2}} \right\}$$

$$\text{or}\quad NTU \frac{(t_{H1} - t_{H2}) - (t_{C1} - t_{C2})}{(t_{C1} - t_{C2})} = \log_e \left\{ \frac{(t_{H1} - t_{C2}) - (t_{C1} - t_{C2})}{(t_{H1} - t_{C2}) - (t_{H1} - t_{H2})} \right\}$$

$$\therefore\ NTU(R - 1) = \log_e \left\{ \frac{[(t_{C1} - t_{C2})/\varepsilon] - (t_{C1} - t_{C2})}{[(t_{C1} - t_{C2})/\varepsilon] - R(t_{C1} - t_{C2})} \right\}$$

(using equations 17.73 and 17.74)

i.e.
$$NTU(R - 1) = \log_e \frac{(1 - \varepsilon)}{(1 - R\varepsilon)}$$

or
$$\varepsilon = \frac{1 - e^{-NTU(1 - R)}}{1 - R\,e^{-NTU(1 - R)}} \qquad\qquad 17.75$$

Note that for a counter-flow heat exchanger when $C_H = C_C$, i.e. $R = 1$ (say for a gas turbine heat exchanger), then the expression for effectiveness cannot be obtained by substituting $R = 1$ in equation 17.75. For this case the temperature change of each fluid is the same, since $C_H = C_C$, and hence the *LMTD* is equal to the temperature difference between the hot and cold fluids which remains constant throughout the heat exchanger. Equation 17.71 is therefore written as, $NTU = (t_{C1} - t_{C2})/(t_{H1} - t_{C1})$ and the derivation then proceeds as above, giving:

$$\varepsilon = \frac{NTU}{1 + NTU} \qquad 17.76$$

For a parallel-flow heat exchanger it can be shown that:

$$\varepsilon = \frac{1 - e^{-NTU(1 + R)}}{1 + R} \qquad 17.77$$

When $R = 0$ in the case of a condenser say, then it can be seen from equation 17.75 or equation 17.77 that the effectiveness is,

$$\varepsilon = 1 - e^{NTU} \qquad 17.78$$

Example 17.15

A single-pass shell and tube counter-flow heat exchanger uses waste gas on the shell side to heat a liquid in the tubes. The waste gas enters at a temperature of 400°C at a mass flow rate of 40 kg/s; the liquid enters at 100°C at a mass flow rate of 3 kg/s.

Assuming that the velocity of the liquid is not to exceed 1 m/s, calculate using the data below:

(i)   the required number of tubes;
(ii)   the effectiveness of the heat exchanger;
(iii) the exit temperature of the liquid.

Neglect fouling factors and the thermal resistance of the tube wall.
Data: tube inside diameter = 10 mm; tube outside diameter = 12·7 mm; tube length = 4 m; specific heat of waste gas = 1·04 kJ/kg K; specific heat of liquid = 1·5 kJ/kg K; density of liquid = 500 kg/m³; heat transfer coefficient on the shell side = 260 W/m² K; heat transfer coefficient on the tube side = 580 W/m² K.

(i)   Volume flow rate of liquid $= \dfrac{\dot{m}}{\rho} = \dfrac{3}{500} = 0.006\ \text{m}^3/\text{s}$

   $\therefore$ Total cross-sectional area for a velocity of 1 m/s
   $= 0.006\ \text{m}^2$,

i.e.   Number of tubes $= \dfrac{0.006 \times 4}{\pi \times 0.01^2} = 76.39$, say 77

(Note: the velocity in the tubes is then less than 1 m/s as required.)
(ii) From equation 17.46, taking into account the area difference,

$$\frac{1}{UA_o} = \frac{1}{h_o A_o} + \frac{1}{h_i A_i}$$

(where subscripts o and i refer to the outside and inside of the tube).
   The overall heat transfer coefficient, U, is referred to the outside
area which is the usual practice in heat exchanger design,

i.e. $\dfrac{1}{U} = \dfrac{1}{h_o} + \dfrac{A_o}{h_i A_i} = \dfrac{1}{260} + \dfrac{12.7}{580 \times 10} = 0.006\ 04$

(since $A_o/A_i = D_o/D_i$)

i.e. $U = 165.68\ \text{W/m}^2\ \text{K}$

Then from equation 17.71,

$$NTU = \frac{165.68 \times \pi \times 0.0127 \times 4 \times 77}{3 \times 1.5 \times 1000} = 0.452$$

Also,

$$R = \frac{3 \times 1.5}{40 \times 1.04} = 0.1082$$

Then from equation 17.75,

$$\varepsilon = \frac{1 - e^{-NTU(1-R)}}{1 - R\,e^{-NTU(1-R)}} = \frac{1 - e^{-0.452 \times 0.8918}}{1 - 0.1082\,e^{-0.452 \times 0.8918}}$$

i.e. $\varepsilon = 0.358$

(iii) From equation 17.74,

$$\varepsilon = 0.358 = \frac{t_{L2} - 100}{400 - 100}$$

(where t_{L2} is the exit temperature of the liquid)

$$\therefore \quad t_{L2} = 300 \times 0.358 + 100 = 207.4°C$$

17.12 Natural convection

As stated previously, heat transfer by free or natural convection is due to differences in density in the fluid causing a natural circulation, and hence a transfer of heat. For the majority of problems in which a fluid flows across a surface, the superimposed effect of natural convection is small enough to be neglected. When there is no forced velocity of the fluid then the heat is transferred entirely by natural convection (when radiation is negligible). The heat transfer in this case depends on the coefficient of cubical expansion, β, which is given by

$$\rho_1 = \rho_2(1 + \beta\theta) \quad \text{or} \quad (\rho_1 - \rho_2) = \rho_2\beta\theta$$

(where θ is the temperature difference between the two parts of the fluid of density ρ_1 and ρ_2).

The upthrust per unit volume of fluid is $(\rho_1 - \rho_2)g$, and the velocity of the convection current is dependent on the upthrust,

i.e. natural convection depends on

$$(\rho_1 - \rho_2)g = \rho_2\beta\theta g$$

The heat transfer also depends on the fluid viscosity, the thermal conductivity of the fluid, and a characteristic dimension of length. Since the coefficient of cubical expansion, β, and the local acceleration due to gravity, g, do not have a separate effect on the heat transfer, then only the product, βg, need be considered. For a dimensional analysis we then have

$$h = A\mu^{a_1}\rho^{b_1}k^{c_1}c^{d_1}\theta^{e_1}(\beta g)^{f_1}l^{g_1} + B\mu^{a_2}\rho^{b_2}k^{c_2}c^{d_2}\theta^{e_2}(\beta g)^{f_2}l^{g_2} + \text{etc.}$$

Then by the same procedure as in Section 17.6 it can be shown that

$$Nu = KF\left\{\left(\frac{c\mu}{k}\right), \left(\frac{\beta g\rho^2 l^3\theta}{\mu^2}\right)\right\}$$

or

$$Nu = KF\{(Pr), (Gr)\}$$

(where

$$Gr = \frac{\beta g\rho^2 l^3\theta}{\mu^2} = \frac{\beta g l^3\theta}{\nu^2}$$

is called the Grashof number).

In many cases of natural convection it is possible to use an approximate equation to evaluate the heat transfer coefficient, h. For example, for natural convection from a horizontal pipe

$$h = 1 \cdot 32\left(\frac{\theta}{d}\right)^{1/4} \quad \text{when} \quad 10^4 < Gr < 10^9$$

and $\qquad h = 1 \cdot 25\theta^{1/3} \quad \text{when} \quad 10^9 < Gr < 10^{12}$

(where h is in W/m^2 K, θ is in K, and d is in m).

Example 17.16

Calculate the heat loss by natural convection per m length from a horizontal pipe of 150 mm diameter, the surface of which is at 277°C. The room temperature is 17°C. It has been shown that for a horizontal cylinder (see reference 17.1)

$$Nu = 0 \cdot 527(Pr)^{1/2}(Pr + 0 \cdot 952)^{-1/4}(Gr)^{1/4}$$

(where the properties are evaluated at the surface temperature). Take the coefficient of cubical expansion, β, as $1/T$, where T K is the absolute temperature of the air.

From tables, at a surface temperature of $(277 + 273) = 550$ K, we have, $Pr = 0 \cdot 68$ and,

$$Gr = \frac{\beta g \theta d^3}{\nu^2} = \frac{9 \cdot 81 \times (277 - 17) \times 0 \cdot 15^3}{290 \times (4 \cdot 439 \times 10^{-5})^2}$$

$$\left(\text{Note that } g = 9 \cdot 81 \text{ m/s}^2, \text{ and } \beta = \frac{1}{17 + 273} = \frac{1}{290}\right)$$

i.e. $\qquad\qquad\qquad\qquad\qquad Gr = 15 \cdot 1 \times 10^6$

Substituting,

$$Nu = 0 \cdot 527(0 \cdot 68)^{1/2}(0 \cdot 68 + 0 \cdot 952)^{-1/4}(15 \cdot 1 \times 10^6)^{1/4}$$
$$= 0 \cdot 527 \times 0 \cdot 825 \times 0 \cdot 885 \times 62 \cdot 34 = 24$$

i.e. $\qquad\qquad\qquad\qquad\qquad Nu = \frac{hd}{k} = 24$

$$\therefore \; h = \frac{24 \times 4 \cdot 357 \times 10^{-5}}{0 \cdot 15} = 0 \cdot 00697 \text{ kW/m}^2 \text{ K}$$

Then from equation 17.3,

$$Q = hA(t_w - t) = 0 \cdot 00697 \times \pi \times 0 \cdot 15 \times 1 \times (277 - 17) = 0 \cdot 855 \text{ kW}$$

i.e. $\qquad\qquad$ Heat loss per m length = 855 W

Example 17.17

Recalculate the heat transfer coefficient for example 17.16 using the approximate equation,

$$h = 1 \cdot 32 \left(\frac{\theta}{d} \right)^{1/4} \quad \text{for} \quad 10^4 < Gr < 10^9$$

(where h is in W/m² K, θ is in K, and d is in m),

i.e.　　$h = 1 \cdot 32 \times \left(\frac{277 - 17}{0 \cdot 15} \right)^{1/4} = 1 \cdot 32 \times (1733)^{1/4} = 1 \cdot 32 \times 6 \cdot 45$

i.e.　　　　　　　Heat transfer coefficient = 8·52 W/m² K

For natural convection from a vertical wall the air in rising, due to the convection currents, builds up a boundary layer, starting from the bottom and thickening gradually up the wall. The heat transfer coefficient varies up the wall, and the formulae for heat transfer from a vertical wall give the local heat transfer coefficient at a distance, l, from the bottom of the wall, where the characteristic linear dimension to be used in the Grashof number is the length, l.

It can be shown that the average heat transfer coefficient for the wall from the bottom up to the distance, l, is given by

$$h_{av} = \frac{4}{3} h$$

(where h_{av} is the average heat transfer coefficient, and h is the heat transfer coefficient at the section distance, l, from the bottom of the wall).

Example 17.18

A vertical surface 1 m high is at a temperature of 327°C, and the atmospheric temperature is 30°C. Calculate the rate at which heat is lost by convection from the surface per m width. For natural convection from a vertical surface take

$$h = 1 \cdot 42 \left(\frac{\theta}{l} \right)^{1/4} \quad \text{for} \quad 10^4 < Gr < 10^9$$

or　　　　$h = 1 \cdot 31 \theta^{1/3} \quad \text{for} \quad 10^9 < Gr < 10^{12}$

(where all properties are at the surface temperature, and $\beta = 1/T$, where T is the absolute air temperature; h is in W/m² K, θ is in K, and l is in m).

The Grashof number in such problems has the same limiting function as the Reynolds number in fluid flow. For the lower range of Grashof numbers the flow of air due to natural convection remains laminar on the wall surface, whereas for the larger Grashof numbers the boundary layer on the wall is turbulent. It can be seen from the second equation above, $h = 1·31\theta^{1/3}$, that when the boundary layer is turbulent the heat transfer coefficient is assumed to be the same at all parts of the wall, since h no longer depends on the distance, l.

$$\text{Surface temperature} = (327+273) = 600 \text{ K}$$

Taking properties from tables, we have

$$Gr = \frac{\beta g l^3 \theta}{v^2} = \frac{1 \times 9·81 \times l^3 \times (327-30)}{303 \times (5·128 \times 10^{-5})^2} = 3·65 \times 10^9$$

$$\left(\text{where } \beta = \frac{1}{30+273} = \frac{1}{303}\right).$$

Hence,
$$h = 1·31\theta^{1/3} = 1·31(327-30)^{1/3}$$
$$= 1·31 \times 6·67 = 8·75 \text{ W/m}^2 \text{ K}$$
and
$$Q = hA\theta = 8·75 \times 1 \times 1 \times (327-30) = 2600 \text{ W}$$
i.e.
$$\text{Heat loss per m width} = 2·6 \text{ kW}$$

(Note: expressions for h as above give average values.)

17.13 Extended surfaces

From equation 17.55 it can be seen that for a given heat transfer coefficient and given fluid temperatures the heat transfer can be increased by increasing the heat transfer area. One way of doing this is to increase the area on one side of the heat exchanger by adding fins or studs which project into the fluid; the effective heat transfer area is then increased.

The thermal resistance on either side of a heat exchanger is $1/hA$, therefore when h is very large the resistance to heat transfer is low and hence there is no advantage in increasing the area. One of the fluids usually has a much lower value of h than the other and hence the resistance on this side of the heat exchanger controls the heat transfer. It is therefore on this side that the area can be extended with advantage. (The resistance of the separating wall is usually small compared with the resistance of the fluid film on either side.)

Heat transfer coefficients are very high for condensing and boiling fluids; they are generally higher for liquids than for gases, and generally higher for forced convection than for natural convection. A typical application for an extended surface would therefore be natural convection to air.

A simplified analysis now follows for an extended one-dimensional surface in the steady state; for a fuller treatment the reader is referred to references 17.6 and 17.8.

Consider an extended surface which has a constant cross-sectional area, A, which is small compared to its length, L, so that heat transfer along it is one-dimensional (see fig. 17.26).

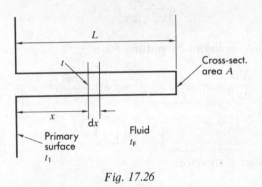

Fig. 17.26

The heat transfer at any section, distance x from the primary surface, where the temperature is t, is given by:

$$Q = -kA \frac{dt}{dx}$$

and at a section $(x+dx)$ from the primary surface is,

$$Q' = Q + \frac{\partial Q}{\partial x} dx = Q - kA \frac{d^2 t}{dx^2} dx \qquad 17.79$$

For a perimeter, P, and a heat transfer coefficient, h, assumed constant with temperature and uniform over the surface, then:

Heat loss from surface $= hP\, dx(t - t_F)$

(where t_F = temperature of the surrounding fluid, assumed uniform and constant). Therefore, for the steady state,

$$Q - Q' = hP \, dx(t - t_F)$$

Substituting from equation 17.79,

$$kA \frac{d^2t}{dx^2} = hP(t - t_F)$$

or,
$$\frac{d^2t}{dx^2} - \frac{hP}{kA}(t - t_F) = 0 \qquad\qquad 17.80$$

Writing $(t - t_F)$ as θ then,

$$\frac{d^2\theta}{dx^2} - \frac{hP}{kA}\theta = 0$$

The solution is found by putting $\theta = e^{mx}$,

i.e.
$$\frac{d^2\theta}{dx^2} = m^2 \, e^{mx} \qquad \therefore \ m^2 \, e^{mx} - \frac{hP}{kA} e^{mx} = 0$$

$$\therefore \ m = \pm\left(\frac{hP}{kA}\right)^{1/2}$$

The solution is therefore,

$$\theta = C_1 \, e^{mx} + C_2 \, e^{-mx} \qquad\qquad 17.81$$

(where C_1 and C_2 are constants).

C_1 and C_2 are found from the boundary conditions:

(i)   at $x = 0$, $\theta = \theta_1 = (t_1 - t_F)$

Therefore in equation 17.81),

$$\theta = C_1 + C_2 \qquad\qquad 17.82$$

(ii)   At the end of the extended surface the heat convected from the end is equal to the heat conducted at the section $x = L$.

Assuming the same heat transfer coefficient, h, for the end then,

$$-kA\left(\frac{d\theta}{dx}\right)_{x=L} = hA(\theta)_{x=L} \qquad\qquad 17.83$$

From equation 17.81,

$$\left(\frac{d\theta}{dx}\right)_{x=L} = mC_1\,e^{mL} - mC_2\,e^{-mL}$$

and

$$\theta_{x=L} = C_1\,e^{mL} + C_2\,e^{-mL}$$

Substituting in equation 17.83,

$$-kmC_1\,e^{mL} + kmC_2\,e^{-mL} = hC_1\,e^{mL} + hC_2\,e^{-mL} \qquad 17.84$$

From equations 17.82 and 17.84 the values of C_1 and C_2 can be found, and hence the solution to equation 17.80 is,

$$\frac{\theta}{\theta_1} = \frac{(e^{m(L-x)} + e^{-m(L-x)}) + \dfrac{h}{mk}(e^{m(L-x)} - e^{-m(L-x)})}{(e^{mL} + e^{-mL}) + \dfrac{h}{mk}(e^{mL} + e^{-mL})}$$

or,

$$\frac{\theta}{\theta_1} = \frac{\cosh m(L-x) + \dfrac{h}{mk}\sinh m(L-x)}{\cosh mL + \dfrac{h}{mk}\sinh mL} \qquad 17.85$$

(Note: $\cosh mL = (e^{mL} + e^{-mL})/2$ and $\sinh mL = (e^{mL} - e^{-mL})/2$.)

At the end of the extended surface, at $x=L$,

$$\frac{\theta_2}{\theta_1} = \frac{1}{\cosh mL + \dfrac{h}{mk}\sinh mL} \qquad 17.86$$

The heat transfer from the extended surface, which is the same as the heat leaving the primary surface to the fin, is given by:

$$Q_1 = -kA\left(\frac{d\theta}{dx}\right)_{x=0}$$

By differentiating equation 17.85 and putting $x=0$, it can be shown that,

$$Q_1 = hA\theta_1 \frac{\left(1 + \dfrac{\tanh mL}{h/mk}\right)}{(1 + (h/mk)\tanh mL)} \qquad 17.87$$

From equation 17.87 it can be seen that when $h/mk = 1$ then, $Q_1 = hA\theta_1$, which is the heat loss from the primary surface with no extended surface, i.e. when $h = mk$ an extended surface of whatever length, L, will not increase the heat transfer from the primary surface.

For $h/mk > 1$ then $Q_1 < hA\theta_1$ and hence adding secondary surface reduces the heat transfer; the surface added acts as an insulation.

For $h/mk < 1$ then $Q_1 > hA\theta_1$ and the extended surface will increase the heat transfer.

This is illustrated in fig. 17.27. Note:

$$\frac{h}{mk} = \left(\frac{h^2}{k^2}\frac{kA}{hP}\right)^{1/2} = \left(\frac{hA}{kP}\right)^{1/2}$$

Hence assuming $h/mk < 1$ then the heat transfer becomes more effective when h/k is low for a given geometry.

Approximate end condition

A simplified approach is possible by making the approximation that the loss of heat from the end of the extended surface is negligible,

i.e.   at $x = L$ $\qquad -kA\left(\frac{\mathrm{d}\theta}{\mathrm{d}x}\right)_{x=L} = 0$

Therefore in equation 17.81,

$$-kAm(C_1\, \mathrm{e}^{mL} - C_2\, \mathrm{e}^{-mL}) = 0 \qquad\qquad 17.88$$

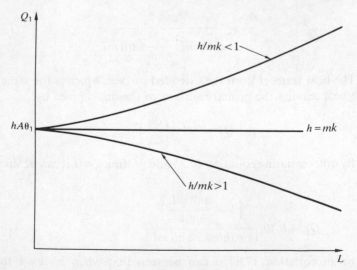

Fig. 17.27

Solving for C_1 and C_2 from equations 17.88 and 17.82 we have,

$$\frac{\theta}{\theta_1} = \frac{\cosh m(L-x)}{\cosh mL} \qquad 17.89$$

(compare this equation with equation 17.85).

Then at $x = L$,

$$\frac{\theta_2}{\theta_1} = \frac{1}{\cosh mL} \qquad 17.90$$

(compare with equation 17.86).

The heat transfer from the extended surface is then,

$$Q_1 = mkA\theta_1 \tanh mL \qquad 17.91$$

In most cases in practice the approximate expressions given by equations 17.89 to 17.91 can be used instead of the more accurate expressions given by equations 17.85 to 17.87.

In the important case of compact plate-fin heat exchangers where corrugated plates are sandwiched between flat plates (see for example fig. 17.28), then the expressions given in equations 17.89 to 18.91 are the accurate ones. In this case the fin bridges the two hot surfaces which may be assumed to be at the same temperature; at the mid-point of the fin the change of temperature with fin length is zero which

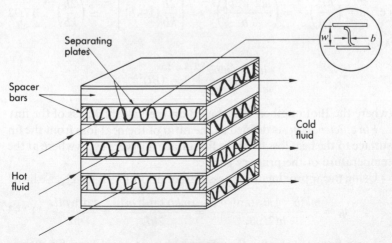

Fig. 17.28

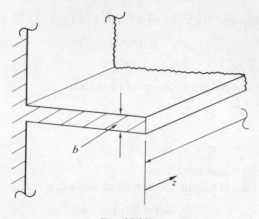

Fig. 17.29

corresponds to the condition of zero heat transfer. In the equations 17.89 to 17.91 the half width, $w/2$, is substituted for the length, L.

Rectangular section fins

For a fin of rectangular cross-section on a plane surface as shown in fig. 17.29, the perimeter, P, is given by $(2 + 2b)$ per unit length in the z direction, and the cross-sectional area, A, is given by b per unit length in the z direction,

i.e. $\quad m = \left(\dfrac{hP}{kA}\right)^{1/2} = \left(\dfrac{h(2+2b)}{kb}\right)^{1/2} = \left[\dfrac{2h}{kb}(1+b)\right]^{1/2} \simeq \left(\dfrac{2h}{kb}\right)^{1/2} \quad 17.92$

Also,

$$\frac{h}{mk} = \left(\frac{hb/2}{k}\right)^{1/2} = (Bi)^{1/2}$$

(where the Biot number, Bi, is based on the half thickness of the fin).

Fin efficiency, η_F, is defined as the ratio of the heat loss from the fin surface to the heat loss from the fin surface if it were everywhere at the temperature of the primary surface.

Using the approximate expression, equation 17.91, for Q_1, we have:

$$\eta_F = \frac{mk(b \times 1)\theta_1 \tanh mL}{h(2L)\theta_1} = \frac{mkb \tanh mL}{2hL} = \frac{\tanh mL}{mL} \quad 17.93$$

For a finned surface with unfinned area, A_b, and total fin surface area, A_F, then:

Heat loss from unit length in the z direction $= h\theta_1 (A_b + \eta_F A_F)$

$$17.94$$

Example 17.19

A typical plate-fin cross-flow heat exchanger is shown in fig. 17.28. For a point in one of the hot fluid channels where the fluid mean temperature within the channel is 200°C and the separating plates on either side are at 100°C, calculate:

(i)   the mean temperature of a fin at that point in the heat exchanger;

(ii) the fin efficiency.

Data: Height of flow channel $= 11.78$ mm; thickness of fin $= 0.203$ mm; heat transfer coefficient between the hot fluid and all surfaces $= 137$ W/m^2 K; thermal conductivity of fin material $= 168$ W/m K.

From equation 17.89,

$$\frac{\theta}{\theta_1} = \frac{\cosh m\left(\dfrac{w}{2} - x\right)}{\cosh (mw/2)}$$

Then,

$$\frac{\theta_{\text{Mean}}}{\theta_1} = \frac{\int_0^{w/2} \theta \, dx}{\theta_1 w/2} = \frac{\left[-\sinh m\left(\dfrac{w}{2} - x\right)\right]_0^{w/2}}{(mw/2)\cosh(mw/2)} = \frac{\sinh(mw/2)}{(mw/2)\cosh(mw/2)}$$

$$= \frac{\tanh(mw/2)}{(mw/2)}$$

It can be seen by reference to equation 17.93 that the ratio, $\theta_{\text{Mean}}/\theta_1$ is an alternative expression for fin efficiency.

(i) From equation 17.92,

$$m = \left(\frac{2h}{kb}\right)^{1/2} = \left(\frac{2 \times 137}{168 \times 0.203 \times 10^{-3}}\right)^{1/2} = 89.63$$

$$\therefore \quad mw/2 = 89.63 \times 11.78/2 \times 10^3 = 0.5279$$

$$\therefore \frac{\theta_{\text{Mean}}}{\theta_1} = \frac{\tanh 0 \cdot 5279}{0 \cdot 5279} = 0 \cdot 916$$

$$t_{\text{Mean}} = 200 - 0 \cdot 916 \times (200 - 100) = 108 \cdot 4°C$$

i.e.   Mean temperature of fin = 108·4°C

(ii)   Fin efficiency = $\theta_{\text{Mean}}/\theta_1 = 0 \cdot 916 = 91 \cdot 6\%$

17.14 Black body radiation

Thermal radiation consists of electro-magnetic waves emitted due to the agitation of the molecules of a substance. The waves are similar to light waves in that they are propagated in straight lines at the speed of light and they require no medium for propagation. Radiation striking a body can be absorbed by the body, reflected from the body, or transmitted through the body. The fractions of the radiation absorbed, reflected, and transmitted are called the absorptivity, α, the reflectivity, ρ, and the transmissivity, τ, respectively. Then we have

$$\alpha + \rho + \tau = 1$$

For most solids and liquids encountered in engineering the amount of radiation transmitted through the substance is negligible, and it is possible to write

$$\alpha + \rho = 1 \qquad\qquad 17.95$$

It is useful to define an ideal body which absorbs all the radiation which falls upon it; such a body is called a *black body*. For a black body, $\alpha = 1$ and $\rho = 0$. It should be noted that the term 'black' in this context does not necessarily imply black to the eye. A surface which is black to the eye is one which absorbs all the light incident upon it, but a surface can absorb all the thermal radiation incident upon it without necessarily absorbing all the light (e.g. snow is almost 'black' to thermal radiation, $\alpha = 0 \cdot 985$). Although no totally black body exists in practice, many surfaces approximate to the definition. For example, consider a small object radiating energy in a large space, as shown in fig. 17.30. The energy striking the surface surrounding the body is reflected and absorbed many times by the surface, and the fraction of energy reflected back and intercepted by the body is exceedingly small. Therefore, when a body is placed in large surroundings the surroundings are approximately black to thermal radiation. As a better example of a black body, consider a small hole in the surface of a wall, as shown in fig. 17.31. The hole leads into a

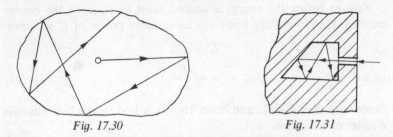

Fig. 17.30   Fig. 17.31

small chamber as shown. Rays of thermal radiation entering the hole are successively absorbed by the walls of the chamber such that only a negligible amount of radiation is emitted from the hole. Thus the hole acts as a black body. This is the closest approximation to a black body which can be devised in practice; the inside surfaces of the chamber can be made of a material with a high absorptivity (e.g. lampblack).

The energy which is radiated from a body per unit area per unit time is called the *emissive power*, E. It can be shown that a black body, as well as being the best possible absorber of radiation, is also the best possible emitter. Consider an enclosure at a uniform temperature, and let a black body be placed in the enclosure as shown in fig. 17.32. If the body is at the same temperature as the enclosure then it follows that all the energy radiated by the body and absorbed by the walls of the enclosure, must exactly equal the energy radiated by the enclosure and absorbed by the body. If this were not so then the body would gain or lose energy, and this is not possible in an isolated system, by the laws of thermodynamics. Let the emissive power of the black body be E_B. Therefore the rate at which energy impinges on unit surface of the black body is also E_B. Now replace the black body by any other body at the same temperature, and of the same shape and size. This body must receive exactly the same amount of energy from the enclosure as the black body received when it was in the same position in the enclosure. However, this body is not black and hence will only absorb a fraction of the energy it receives,

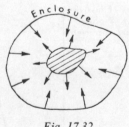

Fig. 17.32

i.e.   Energy absorbed $= \alpha E_B$

(where α is the absorptivity of the body).

Now as before the energy absorbed must be equal to the energy emitted, therefore if the body has an emissive power of E, we have

$$E = \alpha E_B$$

or
$$\alpha = \frac{E}{E_B}$$
17.96

Since $\alpha < 1$ then $E < E_B$, and hence the black body is the best possible emitter of radiation.

The ratio of the emissive power of a body to the emissive power of a black body is called the *emissitivy*, ε. From equation 17.95 it can be seen that when two bodies are at the same temperature, then the absorpitivity, α, equals the emissivity, ε. This is known as *Kirchhoff's law*, which may be stated as follows:

The emissivity of a body radiating energy at a temperature, T, is equal to the absorptivity of the body when receiving energy from a source at a temperature, T.

17.15 The grey body

In Section 17.14 it has been assumed that the energy emitted by thermal radiation is the same for all wavelengths of the radiation. In fact this is not the case, and fig. 17.33 shows the emissive power

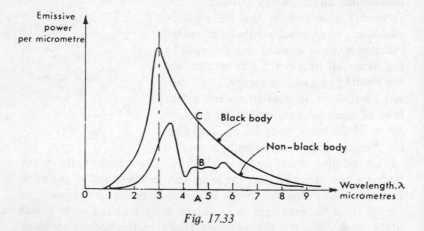

Fig. 17.33

per unit wavelength plotted against wavelength, λ, in micrometres for a black body at any one temperature. A corresponding curve at the same temperature is shown for a non-black body. The ratio of an ordinate of each curve at any wavelength gives the emissivity, and hence the absorptivity, at that wavelength. For example, at a wavelength of 4·5 micrometres, we have

$$\varepsilon_\lambda = \alpha_\lambda = \frac{AB}{AC}$$

The terms ε_λ and α_λ are called the *monochromatic emissivity* and the *monochromatic absorptivity*, respectively. It can be seen from fig. 17.33 that the monochromatic emissivity varies with wavelength. The variation is greater for some materials than for others, and there are certain materials for which the emissivity is practically constant over the entire waveband (e.g. slate). To simplify calculations, surfaces in practice are very often assumed to have a constant emissivity over all wavelengths and for all temperatures. Such an ideal surface is called a *grey body*. Then, for a grey body, $\alpha = \varepsilon$ at all temperatures, where α and ε are the total absorptivity and the total emissivity over all wavelengths.

It is an experimental fact that the emissive power of a body increases as the temperature of the body is increased. This is illustrated in fig. 17.34 in which the emissive power of a black body per

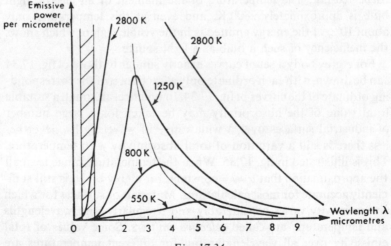

Fig. 17.34

unit wavelength is plotted against the wavelength in micrometres, for several temperatures. It can be seen that the wavelength which gives maximum emissive power becomes smaller as the temperature is increased, and hence more and more of the energy emitted is radiated over the shorter wavelengths as the temperature increases. The value of the wavelength for maximum emissive power is given by Wien's law,

$$\lambda_{max} = \frac{2900}{T} \qquad\qquad 17.97$$

(where λ_{max} is in micrometres and T is in K).

The limits of the visible spectrum are $\lambda = 0.4$ micrometres at the blue end and $\lambda = 0.8$ micrometres at the red end. Now the sun has a temperature of approximately 6000 K, hence using equation 17.97, the maximum wavelength of the radiation is

$$\lambda_{max} = \frac{2900}{6000} = 0.483 \text{ micrometres}$$

Therefore most of the thermal radiation from the sun is in the visible waveband. The waveband for light is shown shaded in fig. 17.34. At a temperature of 800 K a very small amount of the energy emitted is just within the red end of the visible spectrum. A surface at 800 K will appear as a dull red colour. At about 1250 K more of the energy emitted is in the visible range and the surface is then said to be red-hot. The temperature of the filament of an electric light bulb is approximately 2800 K, and even at this temperature only about 10% of the energy emitted is in the visible region, which shows the inefficiency of such a bulb as a light source.

For a grey body a set of curves exactly similar to those of fig. 17.34 can be drawn, with each ordinate only a fraction, ε, of the corresponding ordinate of the curves of fig. 17.34. In practice, although a suitable total value of the absorptivity may be taken for a large number of industrial surfaces over a wide range of wavelengths, nevertheless there is still a variation of total absorptivity with temperature. This is illustrated in fig. 17.35. When the temperature range is small the approximation that $\alpha = \varepsilon = $ constant, for a grey body, is still sufficiently accurate for most calculations. Materials or surfaces for which the emissivity varies considerably and irregularly with wavelengths and temperature, are called *selective emitters*. Some values of total emissivity over all wavelengths but for different temperatures are shown in table 17.2.

Surface finish plays a large part in determining the emissivity of a material. When the surface is very smooth it reflects radiation

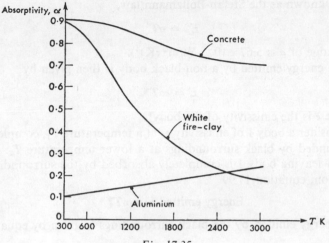

Fig. 17.35

Table 17.2

Surface	0–40°C	120°C	260°C	540°C
White paints	0·95	0·94	0·88	0·70
Black glossy paints	0·95	0·94	0·90	0·85
Lamp black	0·97	0·97	0·97	0·97
Building brick	0·93	0·93	0·79	0·74
Concrete	0·85	0·84	0·69	0·69
Polished steel	0·07	0·09	0·11	0·14

specularly; when the surface is rough, as in most practical cases, it reflects *diffusely*. Rough surfaces are much better absorbers, and hence much better emitters, of radiation than smooth surfaces. For mild steel, rough turned, ε at 15°C is 0·87; for mild steel, well-finished on a lathe, ε at 15°C is 0·39; and it can be seen from table 17.2 that when the steel is polished well the emissivity, ε, is reduced to 0·07.

17.16 The Stefan-Boltzmann law

It was found experimentally by Stefan, and proved theoretically by Boltzmann, that the emissive power of a black body is directly proportional to the fourth power of its absolute temperature, and this is known as the Stefan-Boltzmann law,

i.e. $$E_B = \sigma T^4 \qquad 17.98$$

(the value of σ is $5 \cdot 67 \times 10^{-8}$ W/m^2 (K)^4).

The energy emitted by a non-black body is then given by

$$E = \varepsilon \sigma T^4 \qquad 17.99$$

(where ε is the emissivity of the body).

Consider a body 1 of emissivity ε_1 at a temperature T_1, completely surrounded by black surroundings at a lower temperature T_2. The energy leaving body 1 is completely absorbed by the surroundings, and from equation 17.99

$$\text{Energy emitted} = \varepsilon_1 \sigma T_1^4$$

The energy emitted by the black surroundings is given by equation 17.98,

$$E_B = \sigma T_2^4$$

Now the fraction of this energy which is absorbed by body 1 depends on the absorptivity of body 1. For a grey body $\alpha = \varepsilon$ at all temperatures and hence,

$$\text{Energy absorbed} = \varepsilon \sigma T_2^4 = \alpha \sigma T_2^4$$

Then the heat transferred from the body to its surroundings per m^2 of the body is

$$q = \varepsilon \sigma T_1^4 - \varepsilon \sigma T_2^4$$

i.e. $$q = \varepsilon \sigma (T_1^4 - T_2^4) \qquad 17.100$$

If the emissivity of the body at T_1 is largely different from the emissivity of the body at T_2 then the approximation of the grey body may not be sufficiently accurate. In that case it is a good approximation to take the absorptivity of the body 1 when receiving radiation from a source at T_2 as being equal to the emissivity of body 1 when emitting radiation at T_2.

Then, $$q = \varepsilon_{T_1} \sigma T_1^4 - \varepsilon_{T_2} \sigma T_2^4 \qquad 17.101$$

The absorptivity, while depending mainly on the temperature of the source of radiation, also depends on the temperature of the surface itself. For most metals this factor can be important and it has been shown that the absorptivity of a metal surface at T_1 for radiation from a source at T_2 is approximately equal to the emissivity of the surface when at a temperature, T_3, given by

$$T_3 = \sqrt{(T_1 T_2)} \qquad 17.102$$

Example 17.20

A body at 1100°C in black surroundings at 550°C has an emissivity of 0·4 at 1100°C and an emissivity of 0·7 at 550°C. Calculate the rate of heat loss by radiation per m²,

   (a) when the body is assumed to be grey with $\varepsilon = 0·4$;

   (b) when the body is not grey.

Assume that the absorptivity is independent of the surface temperature.

   (a) Using equation 17.100,

$$q = \varepsilon\sigma(T_1^4 - T_2^4) = 0·4 \times \frac{5·67}{10^8} \times (1373^4 - 823^4)$$

(where $T_1 = 1100 + 273 = 1373$ K, and $T_2 = 550 + 273 = 823$ K),

i.e. $q = 0·4 \times 5·67 \times (13·73^4 - 8·23^4) = 0·4 \times 5·67 \times 30\,960$

i.e.   Heat loss per m² by radiation $= 70\,220$ W $= 70·22$ kW

   (b) When the body is not grey then,

Absorptivity when source is at 550°C = emissivity when body is at 550°C,

i.e. $\alpha = 0·7$

Then,   Energy emitted $= \varepsilon\sigma T_1^4 = 0·4 \times \dfrac{5·67}{10^8} \times 1373^4$

and   Energy absorbed $= \alpha\sigma T_2^4 = 0·7 \times \dfrac{5·67}{10^8} \times 823^4$

i.e. $q = 0·4 \times 5·67 \times 13·73^4 - 0·7 \times 5·67 \times 8·23^4$

$$= 80\,630 - 18\,210 = 62\,420 \text{ W}$$

i.e.   Heat loss per m² by radiation $= 62·42$ kW

It can be seen that the grey body assumption of part (a) over-estimates by

$$\left(\frac{70·22 - 62·42}{62·42}\right) \times 100 = 12·49\%$$

Example 17.21

Calculate the rate of heat loss per m² by radiation from a body at 1100°C in black surroundings at 40°C, when the emissivity at 40°C is 0·9, and the emissivity at 1100°C is as in example 17.20,

(a) when the body is grey with $\varepsilon = 0.4$;
(b) when the body is not grey.

Assume that the absorptivity is independent of the surface temperature.

(a) As in example 17.20,

$$q = \varepsilon\sigma(T_1^4 - T_2^4) = 0.4 \times 5.67 \times (13.73^4 - 3.13^4)$$

i.e. $\qquad q = 0.4 \times 5.67 \times 35\,454 = 80\,410$ W

i.e. $\qquad$ Heat loss per m² by radiation $= 80.41$ kW

(b) As in example 17.20,

$$\text{Energy emitted} = 0.4 \times 5.67 \times 13.73^4 = 80\,630 \text{ W/m}^2$$

and Energy absorbed $= 0.9 \times 5.67 \times 3.13^4 = 489.6$ W/m²

i.e. $\qquad q = (80\,630 - 490) = 80\,140$ W

i.e. $\qquad$ Heat loss per m² by radiation $= 80.14$ kW

Therefore the grey body assumption over-estimates by

$$\left(\frac{80.41 - 80.14}{80.14}\right) = 0.337\%$$

It can be seen from examples 17.20 and 17.21 that the grey body assumption gives a very accurate approximation when one of the temperatures is small compared with the other. The assumption also gives a very accurate approximation when both temperatures are small.

17.17 Lambert's law and the geometric factor

Most surfaces do not emit radiation strongly in all directions; the greater part of the energy emitted is in a direction normal to the surface. Before considering the interchange of energy between two bodies which receive only a part of the radiation emitted by each other, it is necessary to find out how the radiation is distributed in the various directions from the two surfaces. In order to do this the *intensity of radiation*, i, must be defined. The rate of energy emission from unit surface area through unit solid angle, along a normal to the surface is called the *intensity of normal radiation*, i_N. The intensity of radiation in any other direction at any angle ϕ to the normal, is denoted by i_ϕ. (Note: a surface subtends a solid angle at a point distance r from all points on the surface, equal to the surface area divided by r^2. The surface area of a sphere is $4\pi r^2$ and hence the solid angle subtended by the surface of the sphere at its centre is 4π.)

The variation in the intensity of radiation is given by *Lambert's Cosine law*,

$$i_\phi = i_N \cos \phi \qquad\qquad 17.103$$

The energy emitted from a surface of area dA is then given by,

$$\int i_\phi \, \mathrm{d}w \, \mathrm{d}A$$

(where dw is a small solid angle).

Consider a small area dA, and consider the radiation from dA

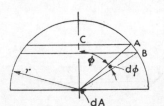

Fig. 17.36

which passes through a small element of the surface area of a hemisphere with dA at its centre, as shown in fig. 17.36. The element subtends an angle ϕ at the centre of the hemisphere and the small increase in angle over the width of the element is then dϕ. The width of the element is the length of the arc, of angle dϕ, and radius r (i.e. AB in fig. 17.36). Therefore,

$$\text{Width of element, AB} = r \, \mathrm{d}\phi$$

The radius of the element is CA $= r \sin \phi$. Hence the surface area of the element is given by,

$$\text{Surface area} = (\text{width} \times \text{circumference}) = r \, \mathrm{d}\phi \times 2\pi r \sin \phi$$

i.e.   Solid angle, dw, subtended at d$A = \dfrac{2\pi r^2 \sin \phi \, \mathrm{d}\phi}{r^2}$

i.e. $\mathrm{d}w = 2\pi \sin \phi \, \mathrm{d}\phi$

Hence the total energy emitted from dA is given by

$$E \, dA = \int_0^{\pi/2} i_\phi \, dw \, dA = \int_0^{\pi/2} dA \, i_\phi 2\pi \sin \phi \, d\phi$$

Substituting from equation 17.103, $i_\phi = i_N \cos \phi$, then

$$E \, dA = 2\pi \, dA \, i_N \int_0^{\pi/2} \cos \phi \sin \phi \, d\phi$$

or $\qquad E \, dA = 2\pi \, dA \, i_N \int_0^{\pi/2} \frac{\sin 2\phi}{2} \, d\phi = \pi i_N \, dA$

Now from equation 17.99,

$$E = \varepsilon \sigma T^4$$

Therefore, $\qquad \varepsilon \sigma T^4 \, dA = \pi i_N \, dA$

i.e. $\qquad\qquad i_N = \frac{\varepsilon \sigma T^4}{\pi}$ $\qquad\qquad$ 17.104

Consider two small black surfaces of area dA_1 and dA_2 at temperatures T_1 and T_2, and distance x apart. The angles of inclination of surfaces are as shown in fig. 17.37. This is a case where neither body receives all the radiation from the other. Let the surface dA_2 subtend a solid angle dw_1 at the centre of the surface dA_1. Then we have,

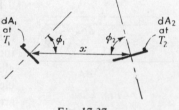

Fig. 17.37

Energy emitted by dA_1 and received by $dA_2 = i_{N_1} \cos \phi_1 \, dw_1 \, dA_1$

From equation 17.104, $i_N = \sigma T^4/\pi$, for a black surface, therefore,

$$\text{Energy received by } dA_2 = \frac{\cos \phi_1 \, dw_1 \, dA_1 \, \sigma T_1^4}{\pi}$$

Also, from the definition of solid angle,

$$dw_1 = \frac{dA_2 \cos \phi_2}{x^2}$$

Hence,

$$\text{Energy received by } dA_2 = \frac{\cos \phi_1 \cos \phi_2 \, dA_1 \, dA_2 \, \sigma T_1^4}{\pi x^2}$$

Now the total energy emitted by dA_1 is $\sigma \, dA_1 \, T_1^4$. The ratio of the energy received by the second body to the energy emitted by the first, is called the *geometric factor*, F_{1-2},

i.e.
$$F_{1-2} = \frac{\cos \phi_1 \cos \phi_2 \, dA_1 \, dA_2 \, \sigma T_1^4}{\pi x^2 \sigma \, dA_1 \, T_1^4}$$

$$\therefore F_{1-2} = \frac{\cos \phi_1 \cos \phi_2 \, dA_2}{\pi x^2} \qquad\qquad 17.105$$

In the same way it can be shown that the geometric factor for radiation from surface 2 to surface 1 is given by

$$F_{2-1} = \frac{\cos \phi_1 \cos \phi_2 \, dA_1}{\pi x^2} \qquad\qquad 17.106$$

The net energy interchange between the surfaces is given by

$$dQ_{1-2} = \frac{\sigma \cos \phi_1 \cos \phi_2 \, dA_1 \, dA_2}{\pi x^2} (T_1^4 - T_2^4)$$

This can be written as

$$dQ_{1-2} = F_{1-2} \, dA_1 \, \sigma(T_1^4 - T_2^4)$$

or
$$dQ_{1-2} = F_{2-1} \, dA_2 \, \sigma(T_1^4 - T_2^4)$$

The geometric factors F_{1-2} and F_{2-1} can be found by a double integration of equation 17.105 and 17.106; this can be done analytically or graphically. For a larger area made up of small surface areas dA_1 and dA_2, average geometric factors can be defined in the same way as above,

i.e.
$$Q_{1-2} = F_{1-2} A_1 \sigma(T_1^4 - T_2^4) \qquad\qquad 17.107$$

and
$$Q_{1-2} = F_{2-1} A_2 \sigma(T_1^4 - T_2^4) \qquad\qquad 17.108$$

From equations 17.107 and 17.108 it can be seen that:

$$A_1 F_{1-2} = A_2 F_{2-1} \qquad\qquad 17.109$$

This is known as the reciprocal relationship or theorem of reciprocity.

In practice calculating F can be a long and difficult process except for simple shapes; charts are available for some of the more common configurations (see reference 17.1).

When a body, 1, is completely enclosed by other surfaces then,

$$F_{1-surfaces} = 1$$

If the surfaces have separate elements, 2, 3, etc, it follows that,

$$F_{1-surfaces} = F_{1-1} + F_{1-2} + F_{1-3} +, etc, = 1 \qquad 17.110$$

The term F_{1-1} is necessary in cases where the body 1 can 'see' parts of itself, e.g. a concave body.

In Table 17.3 values of geometric factor, F_{1-2}, are given for some common configurations; for more complex geometries see references 17.1 and 17.8

Table 17.3

Configuration	Geometric factor, F_{1-2}
(a) Body 1 completely enclosed by body 2	1
(b) Parallel circular discs, radii r_1 and r_2, distance x apart on a common axis	$\dfrac{(x^2 + r_1^2 + r_2^2) - \sqrt{\{(x_1^2 + r_1^2 + r_2^2)^2 - 4r_1^2 r_2^2\}}}{2r_1^2}$
(c) Small disc opposite a parallel circular plate of radius R at a perpendicular distance L	$\dfrac{R^2}{R^2 + L^2}$
(d) Small sphere opposite a circular plate of radius R at a perpendicular distance L	$\dfrac{1}{2}\left\{1 - \dfrac{L}{\sqrt{(L^2 + R^2)}}\right\}$
(e) Small sphere at the centre of the axis of a cylinder of radius R and length $2L$	$\dfrac{L}{\sqrt{(L^2 + R^2)}}$

Example 17.22

A hemispherical cavity of 0·6 m radius is covered by a plate with a hole of 0·2m diameter drilled in its centre. The inner surface of the plate is maintained at 250°C by a heater embedded in the surface. The surfaces may be assumed to be black and the hemisphere may be assumed to be well insulated.

Calculate:

(a) the temperature of the surface of the hemisphere;

(b) the heat input to the heater.

State any other assumption made.

(a) Referring to fig. 17.38, let the inner surface of the plate be 1, the hemisphere surface 2, and hole projected surface 3, as shown.

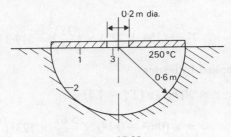

Fig. 17.38

Then, since surface 1 is completely surrounded, we have,

$$F_{1-2} + F_{1-3} = 1$$

or $F_{1-2} = 1$ since surface 1 cannot 'see' surface 3.

From equation 17.109,

$$F_{2-1} = \frac{A_1 F_{1-2}}{A_2} = \frac{\pi(0\cdot6^2 - 0\cdot1^2) \times 1}{2\pi \times 0\cdot6^2} = \frac{35}{72}$$

Similarly,

$$F_{3-2} = 1 \text{ and } A_2 F_{2-3} = A_3 F_{3-2}$$

$$\therefore \qquad F_{2-3} = \frac{\pi \times 0\cdot1^2 \times 1}{2\pi \times 0\cdot6^2} = \frac{1}{72}$$

Then,

Energy leaving surface $2 = A_2 F_{2-3}\, \sigma T_2^4 + A_2 F_{2-1}\, \sigma T_2^4$

$$= A_2 \sigma T_2^4 \left(\frac{1}{72} + \frac{35}{72}\right) = A_2 \sigma T_2^4 \times 0\cdot5$$

The energy falling on surface 2 may be taken as the energy leaving surface 1, since the energy entering the hole from outside will be negligible if the surroundings are large and at normal temperature,

i.e. Energy falling on surface $2 = A_1 F_{1-2}\, \sigma\, T_1^4 = A_1\, \sigma\, T_1^4$

Then, $\qquad\qquad A_1 \sigma T_1^4 = A_2 \sigma T_2^4 \times 0.5 \qquad$ for the steady state

i.e. $\qquad\qquad T_2^4 = \dfrac{T_1^4 \times 2 \times \pi(0.6^2 - 0.1^2)}{2\pi \times 0.6^2} = T_1^4 \times \dfrac{35}{36}$

$$T_2 = (250 + 273) \times \left(\frac{35}{36}\right)^{\frac{1}{4}} = 519.3\ \text{K} = 246.3°\text{C}$$

(b) Heat supplied by heater

$$= A_1 F_{1-2}\, \sigma\, (T_1^4 - T_2^4)$$

$$= \pi \times (0.6^2 - 0.1^2) \times \frac{5.67}{10^8} \times 523^4 \left(1 - \frac{35}{36}\right)$$

$$= 129.6\ \text{W}$$

17.18 Radiant interchange between grey bodies

Radiosity, J, is defined as the total radiant energy leaving a body per unit area per unit time.

Irradiation, G, is defined as the total radiant energy incident on a body per unit area per unit time.

Hence,

Net heat transfer from body, $Q = (J - G)A$

(where A is the area of the body surface)

For a black body, from equation 17.98,

$$J = \sigma T^4$$

For a grey body, the radiosity must include the fraction of energy which is reflected from the surface.

i.e. $\qquad\qquad J = \varepsilon\, \sigma\, T^4 + \rho G$

Also, for a grey body, $\varepsilon = \alpha = 1 - \rho$, neglecting transmissivity (see equation 17.95).

$\therefore \qquad\qquad J = \varepsilon\, \sigma\, T^4 + (1 - \varepsilon)G$

or

$$G = \frac{J - \varepsilon \sigma T^4}{1 - \varepsilon}$$

i.e.

$$\frac{Q}{A} = J - G = J - \frac{(J - \varepsilon \sigma T^4)}{1 - \varepsilon}$$

or

$$Q = \frac{\varepsilon A}{1 - \varepsilon}(\sigma T^4 - J) \qquad 17.111$$

For any two bodies 1 and 2, the geometric factor, F_{1-2}, is the fraction of radiation $A_1 J_1$ which is intercepted by body 2,

i.e.

$$Q_{1-2} = A_1 F_{1-2} J_1 - A_2 F_{2-1} J_2$$

Using equation 17.109,

$$Q_{1-2} = A_1 F_{1-2}(J_1 - J_2) \qquad 17.112$$

An electrical analogy can be used based on Ohm's law. For example, from equation 17.111,

$$\text{Resistance due to emissivity of surface} = \frac{1 - \varepsilon}{\varepsilon A} \qquad 17.113$$

(where Q is analogous to current and $(\sigma T^4 - J)$ is analogous to potential difference).

Similarly, from equation 17.112,

$$\text{Resistance due to geometry} = \frac{1}{A_1 F_{1-2}} \qquad 17.114$$

Take the simple case of a body 1, completely enclosed by a body 2. Fig. 17.39 shows the electrical analogy.

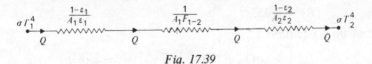

Fig. 17.39

Total resistance, $R_T = \dfrac{1 - \varepsilon_1}{A_1 \varepsilon_1} + \dfrac{1}{A_1 F_{1-2}} + \dfrac{1 - \varepsilon_2}{A_2 \varepsilon_2}$

Also in this case, $F_{1-2} = 1$,

$$\therefore R_T = \frac{1}{A_1}\left(\frac{1}{\varepsilon_1} - 1 + 1 + \frac{A_1}{A_2}\left\{\frac{1}{\varepsilon_2} - 1\right\}\right) = \left(\frac{1}{\varepsilon_1} + \frac{A_1}{A_2}\left\{\frac{1}{\varepsilon_2} - 1\right\}\right)\frac{1}{A_1}$$

i.e.
$$Q_{1-2} = \frac{\sigma(T_1^4 - T_2^4)}{R_T} = \frac{A_1\sigma(T_1^4 - T_2^4)}{\dfrac{1}{\varepsilon_1} + \dfrac{A_1}{A_2}\left\{\dfrac{1}{\varepsilon_2} - 1\right\}} \qquad 17.115$$

When the bodies are very close together then $A_1 \simeq A_2$.

i.e.
$$Q_{1-2} = \frac{A_1\sigma(T_1^4 - T_2^4)}{\dfrac{1}{\varepsilon_1} + \dfrac{1}{\varepsilon_2} - 1}$$

The latter expression for the heat transfer also applies to the case of two large flat parallel surfaces where the size of the surfaces is large compared with their distance apart, i.e. the radiant energy escaping to the surroundings is negligible.

Example 17.23

It is desired to cut down the radiation loss between two parallel surfaces by inserting a sheet of aluminium foil mid-way between them. The temperatures of the two surfaces are maintained at 40°C and 5°C, and the emissivity of both surfaces is 0·85. The emissivity of aluminium foil is 0·05. Calculate the percentage reduction in heat loss by radiation using the aluminium foil, assuming that the surface temperatures are the same in both cases and that all surfaces are grey. Neglect end effects.

(a) Without the aluminium foil.

For two long parallel surfaces, for 1 m² of surface,

$$R_T = 1/\varepsilon_1 + 1/\varepsilon_2 - 1 = 2/0\cdot85 - 1$$

i.e. $R_T = 1\cdot353$

Then from equation 17.115

$$Q_{1-2} = \frac{\sigma(T_1^4 - T_2^4)}{R_T} = \frac{5\cdot67 \times (3\cdot13^4 - 2\cdot78^4)}{1\cdot353}$$

(where $T_1 = 40 + 273 = 313$ K $= 3\cdot13 \times 10^2$ K; and $T_2 = 5 + 273 = 278$ K $= 2\cdot78 \times 10^2$ K),

i.e. $Q_{1-2} = 151\cdot8$ W/m²

(b) With the aluminium foil (see fig. 17.40).

Let the temperature of the foil be T K. Now from surface 1 to the foil, from equation 17.115,

$$Q_{1-F} = \sigma \frac{(T_1^4 - T^4)}{R_{T_{1-F}}}$$

and from the foil to surface 2,

$$Q_{F-2} = \sigma \frac{(T^4 - T_2^4)}{R_{T_{F-2}}}$$

Since both sides of the foil act in a similar way, and since $\epsilon_1 = \epsilon_2$, then $R_{T1-F} = R_{TF-2}$. Therefore,

$$(T_1^4 - T^4) = (T^4 - T_2^4) \quad \text{or} \quad T^4 = \frac{T_1^4 + T_2^4}{2} = \frac{(3 \cdot 13^4 + 2 \cdot 78^4) \times 10^8}{2}$$

$$T^4 = 77 \cdot 82 \times 10^8$$

Also, $R_{T1-F} = R_{TF-2} = 1/0 \cdot 85 + 1/0 \cdot 05 - 1 = 20 \cdot 176$

Then from equation 17.115

$$Q_{1-F} = \sigma \frac{(T_1^4 - T^4)}{R_T} = \frac{5 \cdot 67 \times (3 \cdot 13^4 - 77 \cdot 82)}{20 \cdot 176}$$

i.e. $Q_{1-F} = 5 \cdot 1 \text{ W/m}^2$

Therefore,

$$\text{Percentage reduction in heat loss} = \left(\frac{151 \cdot 8 - 5 \cdot 1}{151 \cdot 8} \right) \times 100$$

$$= 96 \cdot 5\%$$

It can be seen from this example that a material of low emissivity can act as a very efficient radiation shield. This is used to advantage in many cases in practice (e.g. radiation shields for thermocouples and thermometers).

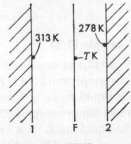

Fig. 17.40

For the case of body 1 small compared with body 2 then $\frac{A_1}{A_2} \to 0$, i.e.   from equation 17.115, $Q_{1-2} = \varepsilon_1 A_1 \sigma (T_1^4 - T_2^4)$

Note that this equation applies even if body 2 is not black, the reason being that a negligible amount of the energy reflected from body 2 is intercepted by body 1 because it is small compared with body 2.

When more than two surfaces exchange heat then an equivalent electric circuit can be drawn using the expressions for resistance given by equations 17.113 and 17.114. For the case shown in fig. 17.41 a

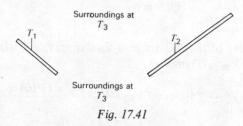

Surroundings at T_3

Surroundings at T_3

Fig. 17.41

body 1 exchanges heat with body 2, the surroundings 3 being at a different temperature.

The equivalent circuit is shown in fig. 17.42 with the resistances, potentials and currents as shown. Applying Ohm's law to each part of the net-work we obtain six equations,

i.e. $\sigma T_1^4 - J_1 = I_1 \dfrac{(1-\varepsilon_1)}{A_1 \, \varepsilon_1}, J_2 - \sigma T_2^4 = I_2 \dfrac{(1-\varepsilon_2)}{\varepsilon_2 A_2}$

$J_3 - \sigma T_3^4 = I_3 \dfrac{(1-\varepsilon_3)}{\varepsilon_3 A_3}, J_1 - J_2 = \dfrac{I_5}{A_1 F_{1-2}}$

$J_1 - J_3 = \dfrac{I_4}{A_1 F_{1-3}}, J_2 - J_3 = \dfrac{I_6}{A_2 F_{2-3}}$

Also, from Kirchhoff's law of electric circuits,

$I_1 = I_4 + I_5, I_2 = I_5 - I_6, \text{and } I_3 = I_4 + I_6$

Example 17.24

A cylindrical vessel of diameter 2 m contains molten metal whose surface temperature is 1327°C; the height of the vessel above the level of the liquid metal is 1 m and the vessel is in large surroundings at a mean temperature of 27°C. The cylindrical sides of the vessel above the liquid metal level are cooled so that the average inside temperature of the surface is 427°C.

Draw the radiation resistance net-work for this problem and hence

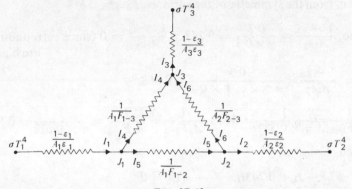

Fig. 17.42

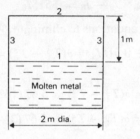

Fig. 17.43

calculate the rate of heat transfer from the liquid metal surface by radiation. Emissivity of liquid metal $= 0.3$; emissivity of inside surface of vessel $= 0.7$.

The vessel is shown in fig. 17.43; the molten metal is surface 1, the open top of the vessel is surface 2 at the temperature of the surroundings, and the sides of the vessel are surface 3. The equivalent circuit is as shown in fig. 17.42.

From Table 17.3 (b),

$$F_{1-2} = (1 + \tfrac{1}{2}) - \sqrt{\{(1 + \tfrac{1}{2})^2 - 1\}} = 0.382$$

From equation 17.110,

$$F_{1-2} + F_{1-3} = 1 \qquad \therefore F_{1-3} = 1 - 0.382 = 0.618$$

Then, from the symmetry of the surfaces, $F_{2-3} = 0.618$

Also, $\dfrac{1 - \varepsilon_1}{A_1\varepsilon_1} = \dfrac{0.7}{\pi \times 0.3} = 0.743 \quad \dfrac{1 - \varepsilon_2}{A_2\varepsilon_2} = 0$ (since surroundings are black)

$$\dfrac{1 - \varepsilon_3}{A_3\varepsilon_3} = \dfrac{0.3}{\pi \times 2 \times 1 \times 0.7} = 0.068$$

$$\dfrac{1}{A_1F_{1-2}} = \dfrac{1}{\pi \times 0.382} = 0.833 \quad \dfrac{1}{A_2F_{2-3}} = \dfrac{1}{A_1F_{1-3}} = \dfrac{1}{\pi \times 0.618} = 0.515$$

Then,

$$\sigma T_1^4 - J_1 = 0.743I_1, \quad J_2 - \sigma T_2^4 = 0$$

$$J_3 - \sigma T_3^4 = 0.068I_3, \quad J_1 - J_2 = 0.833I_5$$

$$J_1 - J_3 = 0.515I_4, \quad J_2 - J_3 = 0.515I_6$$

Combining the above equations to eliminate J_1, J_2 and J_3 we have,

$$0.743I_1 + 0.068I_3 + 0.515I_4 = \sigma(T_1^4 - T_3^4) = 5.67\,(16^4 - 7^4)$$
$$= 0.358 \times 10^6$$

$$0.743I_1 + 0.833I_5 = \sigma(T_1^4 - T_2^4) = 5.67\,(16^4 - 3^4) = 0.371 \times 10^6$$

$$0.515I_6 + 0.068I_3 = \sigma(T_2^4 - T_3^4) = 5.67\,(3^4 - 7^4) = -0.0132 \times 10^6$$

Also,

$$I_1 = I_4 + I_5, \quad I_2 = I_5 - I_6, \quad \text{and } I_3 = I_4 + I_6$$

Eliminating I_2, I_3, I_4, I_5 and I_6 we may calculate I_1,

i.e. $$I_1 = 0.336 \times 10^6$$

$\therefore$ Heat loss by radiation from molten metal $= 0.336 \times 10^6$ W
$$= 336 \text{ kW}$$

If in the above example the sides of the vessel were well-insulated rather than cooled, then the equivalent circuit would be as shown in fig. 17.44. This is a much simpler case; the equivalent resistance of the circuit can be calculated directly since there is no flow of 'current' outwards or inwards at T_3,

i.e. Equivalent resistance $= 0.743 + \dfrac{1}{\dfrac{1}{0.515 \times 2} + \dfrac{1}{0.833}}$

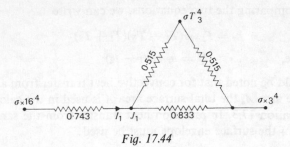

Fig. 17.44

$$= 0 \cdot 743 + 0 \cdot 461 = 1 \cdot 204$$

$$\therefore \quad I_1 = \frac{5 \cdot 67 \, (16^4 - 3^4)}{1 \cdot 204} = 0 \cdot 308 \times 10^6 \text{ W}$$

i.e. Heat loss by radiation from metal surface $= 308$ kW

In this case the temperature of surface 3 will not be 427°C as before. The 'current' flowing from 1 to 3 is given by,

$$308 \times \frac{0 \cdot 461}{2 \times 0 \cdot 515} = 137 \cdot 9 \text{ kW}$$

$$\therefore \quad \sigma(T_3^4 - T_2^4) = 137 \cdot 9 \times 10^3 \times 0 \cdot 515 = 0 \cdot 0710 \times 10^6$$

i.e.

$$\frac{T_3^4}{10^8} = 3^4 + \frac{0 \cdot 0710 \times 10^6}{5 \cdot 67}$$

$$\therefore \qquad T_3 = 1059 \cdot 5 \text{ K} = 786 \cdot 5 \text{ °C}$$

17.19 Heat transfer coefficient for radiation

A heat transfer coefficient for radiation, h_r, is sometimes defined analogously to the heat transfer coefficient, h, for convection.

From equation 17.107

$$Q_{1-2} = F_{1-2} A_1 \sigma (T_1^4 - T_2^4)$$
$$= F_{1-2} A_1 \sigma (T_1^2 + T_2^2)(T_1^2 - T_2^2)$$

i.e.

$$Q_{1-2} = F_{1-2} A_1 \sigma (T_1^2 + T_2^2)(T_1 + T_2)(T_1 - T_2)$$

For convection heat transfer, we have, from equation 17.3,

$$Q = hA(t_w - t)$$

Hence, comparing the two equations, we can write

$$h_r = F_{1-2}\sigma(T_1^2 + T_2^2)(T_1 + T_2) \qquad 17.116$$

Therefore, $Q = h_r A_1(t_1 - t_2)$ 17.117

It should be noted that for convective heat transfer from a surface of surface area A, the total surface area A is used in the calculation, as in equation 17.3. In radiation heat transfer from the same body the area of the surface envelope must be used.

Example 17.25

A ribbed cylinder of outside diameter 0·6 m is at a surface temperature of 260°C in large surroundings at 20°C. Calculate the heat transfer coefficient for radiation, and the total heat loss due to radiation and convection. The cylinder is 0·9 m long and is made of cast iron of emissivity 0·8. The surface area of the ribbed cylinder is 5 m², and the heat transfer coefficient for convection may be taken as 8·8 W/m² K. Neglect end effects.

Since the cylinder is small compared with the surroundings, then $F_{1-2} = 1$. Then, from equation 17.116,

$$h = \varepsilon_1 F_{1-2}\sigma(T_1^2 + T_2^2)(T_1 + T_2)$$

i.e. $h_r = 0·8 \times \dfrac{5·67}{10^8} \times (533^2 + 293^2)(533 + 293)$

(where $T_1 = 260 + 273 = 533$ K, and $T_2 = 20 + 273 = 293$ K),

i.e. $h_r = 0·8 \times \dfrac{5·67}{10^4} \times 37 \times 826 = 13·87$ W/m² K

Then, from equation 17.117

Heat loss by radiation $= h_r A_1(t_1 - t_2)$

$$= 13·87 \times \pi \times 0·6 \times 0·9 \times (260 - 20)$$

$$= 5650 \text{ W}$$

i.e.   Heat loss by radiation $= 5·65$ kW

From equation 17.3,

Heat loss by convection $= hA(t_w - t) = 8·8 \times 5 \times (260 - 20)$

i.e.   Heat loss by convection $= 10\,560$ W $= 10·56$ kW

Therefore,

$$\text{Total heat loss} = 5.65 + 10.56 = 16.21 \text{ kW}$$

17.20 Gas radiation

In the problems considered in the previous sections the effect of the transmission of radiation through the gaseous atmosphere has been neglected; some radiation will be absorbed by the atmosphere, but this is so small that it can be neglected. However, radiation from gases is of importance in furnaces and similar applications. There is one very important difference between radiation between solids and radiation from gases, and that is that gases have a high transmissivity so that radiation is no longer just a surface effect. Also, gases are selective emitters, only emitting and absorbing radiation in certain narrow wavebands; use is made of this fact in infra-red gas analysers, (see Section 15.8). In general, gases with symmetrical molecules (e.g. oxygen and nitrogen), neither emit nor absorb radiation appreciably, whereas gases with asymmetrical molecules (e.g. carbon dioxide, carbon monoxide, and water vapour), emit and absorb strongly in certain wavelengths.

Flames in combustion processes may be thought of as hot gases in which a chemical reaction is taking place. However, the radiation must be treated in a different way from gas radiation since the bulk of the radiation is due to suspended solid particles in the flame. The topic of gas and flame radiation is too complex to discuss concisely, and the reader is recommended to reference 17.1 for a simple treatment.

It is impossible to cover all aspects of heat transfer in even a cursory manner within the limits of one chapter. Many of the topics considered herein have been much simplified, and some major topics have not been mentioned, for example condensation and boiling, and combined heat and mass transfer. The references at the end of the chapter cover these topics, and deal much more completely with the other topics which are considered in this chapter.

PROBLEMS

17.1 A furnace wall consists of 250 mm firebrick, 125 mm insulating brick, and 250 mm building brick. The inside wall is at a temperature of 600°C and the atmospheric temperature is 20°C. Calculate the heat loss per m² of wall area and the temperature of the outside wall surface of the furnace. The heat transfer coefficient for the outside surface is 10 W/m² K, and the thermal conductivities of the firebrick, insulating brick, and building brick are 1·4, 0·2, and 0·7 W/m K, respectively. Neglect radiation.   (0·46 kW/m²; 66° C)

17.2 An electric hot-plate is maintained at a temperature of 350°C and is used to keep a solution just boiling at 95°C. The solution is contained in an enamelled cast-iron vessel of wall thickness 25 mm and enamel thickness 0·8 mm. The heat transfer coefficient for the boiling solution is 5·5 kW/m² K, and the thermal conductivities of cast iron and enamel are 50 and 1·05 W/m K, respectively. Calculate the resistance to the heat transfer for unit area, and the heat transferred per m² per second.   (1·44 m² K/kW; 176·5 kJ)

17.3 In problem 17.2 recalculate the heat transferred per m² per second if the base of the cast-iron vessel is not perfectly flat, and the resistance of the resultant air film is 35 K/kW.   (7 kJ)

17.4 The wall of a house consists of two 125 mm thick brick walls with an inner cavity. The inside wall has a 10 mm coating of plaster, and there is a cement rendering of 5 mm on the outside wall. In one room of the house the external wall is 4 m by 2·5 m, and contains a window of 1·8 m by 1·2 m of 1·5 mm thick glass. The heat transfer coefficients for the inside and outside surfaces of the wall and window are 8·5 and 31 W/m² K, respectively. The thermal conductivities of brick, plaster, cement, and glass are 0·43, 0·14, 0·86, and 0·76 W/m K, respectively. Calculate the proportion of the total heat transfer which is due to the heat loss through the window. Assume that the resistance of the air cavity is 0·16 m² K/W. Neglect all end effects, and neglect radiation.   (63·8%)

17.5 Water at 80°C flows through a 50 mm bore steel pipe of 6 mm thickness, and the atmospheric temperature is 15°C. Calculate the heat loss per m length of pipe. The thermal conductivity of steel is 48 W/m K and the inside and outside heat transfer coefficients are 2800 and 17 W/m² K, respectively. Neglect radiation.   (0·213 kW)

17.6 Calculate the percentage reduction in heat loss for the pipe in problem 17.5 when a layer of hair felt 12 mm thick, of thermal conductivity 0·03 W/m K, is wrapped round the outside surface. Assume that the heat transfer coefficient for the outside surface remains unchanged.   (86·2%)

17.7 A steam main of 150 mm outside diameter containing wet steam at 28 bar is insulated with an inner layer of diatomaceous earth, 40 mm thick, and an outer layer of 85% magnesia, 25 mm thick. Calculate the heat loss per m length of the pipe and the temperature of the outside surface of the lagging, when the room temperature is 20°C. Neglect the thermal resistance of the pipe itself. Assume that the inside surface of the pipe is at the steam temperature, and take the heat transfer coefficient for the outside surface of the lagging as 17 W/m² K. The thermal conductivities of diatomaceous earth and 85% magnesia are 0·09, and 0·06 W/m K, respectively. Neglect radiation.   (0·157 kW; 30·5°C)

17.8 A spherical pressure vessel of 1 m inside diameter is made of 20 mm steel plate. The vessel is lagged with a 25 mm thickness of vermiculite held in position by 10 mm thick asbestos. Calculate the heat loss from the sphere when the inside surface is at 500°C, and the room temperature is 20°C. The heat transfer coefficient for the outside surface is 20 W/m² K, and the thermal conductivities of steel, vermiculite, and asbestos are 48, 0·047, and 0·21 W/m K, respectively. Neglect radiation.   (2·74 kW)

17.9 A solid copper conductor of 13 mm diameter carries a current density of 5 A/mm². The conductor is electrically insulated with a thickness of rubber insulation such that the wire temperature is kept to the minimum possible. Assuming that the surrounding air is at 30°C, calculate:

(i)   the thickness of insulation;
(ii) the wire temperature at the axis;
(iii) the temperature of the outside surface of the insulation;
(iv) the wire temperature at the axis with the insulation removed and the new steady state reached.

Give a physical explanation why any larger or smaller thickness of insulation will lead to a higher wire temperature.

Data: heat transfer coefficient for outside surface of rubber or copper (assumed constant) = 20 W/m² K; thermal conductivities of copper and rubber $= 380$ and 0.2 W/m K; electrical resistivity of copper $= 2 \times 10^{-5} \ \Omega$ mm.

$$(3.5 \ \text{mm}; \ 105.6°C; \ 82.8°C; \ 111.3°C)$$

17.10 A gas-cooled nuclear reactor has solid fuel rods of radius, r_F, thermal conductivity, k_F, sheathed with zirconium of thickness, t, thermal conductivity, k_Z. The heat transfer coefficient from the sheath to the gas in the surrounding annulus is h.

Assuming that the uniform heat generation rate per unit volume within the fuel is q_g, show from first principles that the temperature difference between the axis of the fuel road and the bulk of the coolant at any cross-section is given by:

$$\frac{q_g r_F^2}{4 k_F} \left\{ 1 + \frac{2 k_F}{k_Z} \log_e (1 + t/r_F) + \frac{2 k_F}{h(r_F + t)} \right\}$$

At a particular cross-section in the reactor the gas mean bulk temperature is 220°C when the internal heat generation rate in the fuel rod is 30 MW/m³. Using the data below, calculate:

(i) the temperature at the axis of the fuel rod;
(ii) the temperature at the inner and outer surfaces of the sheath.

Data: thermal conductivities of reactor fuel and zirconium $= 33.5$ and 18.7 W/m K; radius of fuel rod $= 12$ mm; thickness of sheath $= 3.6$ mm; heat transfer coefficient from sheath to cooling gas $= 500$ W/m² K.

$$(559.1°C; \ 526.9°C; \ 496.6°C)$$

17.11 The concrete biological shield of a nuclear reactor is 2 m thick and can be considered to be an infinite flat plate of uniform thermal conductivity, $k = 2.0$ W/m K. The heat generated per unit volume due to the incident gamma radiation is given by:

$$q_g = H \, e^{-8.5x} \ \text{W/m}^3$$

where x is the distance in metres measured from the inside surface, and H is a constant dependent on the gamma radiation.

The maximum temperature difference in the concrete is to be limited to 4 K, and it may be assumed that the outer surface is well-insulated.

Calculate the maximum allowable value of the gamma radiation on the inside surface of the concrete in W/m^2.

(68 W/m^2)

17.12 (a) The temperature–time history of the centre of a large slab of material initially at a constant temperature which is suddenly plunged into a fluid at a different temperature can be shown to be given by:

$$\frac{\theta_c}{\theta_i} = 2 \sum_{n=1}^{\infty} e^{-(\lambda_n L)^2 Fo} \frac{\sin(\lambda_n L)}{(\lambda_n L) + \sin(\lambda_n L)\cos(\lambda_n L)}$$

where $(\lambda_n L)\tan(\lambda_n L) = Bi$.

(θ_c = temperature difference at time τ, between the centre of the slab thickness and the surrounding fluid; θ_i = temperature difference between slab and fluid initially; Fo = Fourier number, $\alpha\tau/L^2$; α = thermal diffusivity of slab; L = half-thickness of slab; Bi = Biot number, hL/k; h = heat transfer coefficient on slab surfaces; k = thermal conductivity of slab material.)

Show that for a case where the temperature of the slab surfaces is approximately equal to the fluid temperature (i.e. $h \to \infty$), and for a reasonably long time period:

$$\frac{\theta_c}{\theta_i} = \frac{4}{\pi} e^{-(\pi^2/4)Fo}$$

(b) A large slab of rubber of thickness 40 mm is vulcanized by heating the faces using steam at 330°C. The required temperature at the centre is 120°C and the rubber is initially at 20°C. Calculate the time required for the process,

(i) using the method of (a) above;
(ii) using a numerical method.

Take α for rubber as 64.5×10^{-9} m^2/s. (26.4 min)

17.13 A large metal plate of thickness 200 mm is initially at a uniform temperature of 20°C. One surface of the plate is in contact with ambient air at a constant temperature of 20°C while the other surface may be exposed to a constant net radiant heat flux of 100 kW/m^2 from an electric element.

(i) Assuming as an approximation that the plate temperature is uniform throughout at any instant, calculate the time taken for the plate to reach 70°C from the instant the electric element is switched on.

(ii) Estimate the temperature distribution through the plate thickness 9 minutes after the element is switched on using a numerical method with four space increments and a Fourier number of 0·5.

Data: thermal conductivity of plate $= 45$ W/m K; density of plate $=$ 7800 kg/m^3; specific heat of plate $= 0·5$ kJ/kg K; heat transfer coefficient from plate to air $= 200$ W/m^2 K.

(6·85 min; 41·9°C, 46·6°C, 82·6°C, 152·0°C, 263·1°C)

17.14 The wall of a large vessel is 50 mm thick and is initially at 12·5°C throughout. A hot fluid at a constant temperature of 500°C is suddenly pumped across the inside surface; the outside surface may be assumed to be perfectly insulated.

Using a numerical method, determine the time taken for the junction of wall and insulation to reach 110°C.

Data: thermal conductivity of wall $= 22$ W/m K; thermal diffusivity of wall $= 6·22 \times 10^{-6}$ m^2/s; heat transfer coefficient from fluid to wall $= 110$ W/m^2 K. (455 s)

17.15 A thick fin of rectangular cross-section, 1 m × 1 m, projects from a flat surface at 200°C into a fluid at 20°C. Using the data below, estimate the temperature distribution in the steady state assuming two-dimensional conduction, and hence calculate the heat loss from the fin surface per unit length.

Data: thermal conductivity of fin material $= 25$ W/m K; heat transfer coefficient for all parts of the fin surface $= 10$ W/m^2 K. (2·5 kW/m)

17.16 An exhaust pipe of 75 mm outside diameter is cooled by surrounding it by an annular space containing water. The exhaust gas enters the exhaust pipe at 350°C, and the water enters from the mains at 10°C. The heat transfer coefficients for the gases and water may be taken as 0·3 and 1·5 kW/m^2 K, respectively, and the pipe thickness may be taken to be negligible. The gases are required to be cooled to 100°C and the mean specific heat at constant pressure is 1·13 kJ/kg K. The gas flow is 200 kg/h and the water flow is 1400 kg/h. Calculate the required length of pipe, (a) for parallel-flow; (b) for counter-flow. Take the specific heat of water as 4·19 kJ/kg K.

(1·485 m; 1·438 m)

17.17 In a chemical plant a solution of density 1100 kg/m³ and specific heat 4·6 kJ/kg K is to be heated from 65°C to 100°C; the flow of solution required is 11·8 kg/s. It is desired to use a tubular heat exchanger, the solution flowing at about 1·2 m/s in 25 mm bore iron tubes, and being heated by wet steam at 115°C. The length of the tubes must not exceed 3·5 m. Estimate the number of tubes required and the number of tube passes required. Take the inside and outside heat transfer coefficients as 5 and 10 kW/m² K, respectively. Neglect the thermal resistance of the iron tube. (18; 4)

17.18 An oil engine develops 300 kW, and the specific fuel consumption is 0·21 kg/kW h. The exhaust from the engine is used in a tubular water heater, flowing through 25 mm diameter tubes, entering with a velocity of 12 m/s, at 340°C and leaving at 90°C. The water enters the heater at 10°C and leaves at 90°C, flowing in counter-flow to the hot gases. The air/fuel ratio of the engine is 20/1, and the exhaust pressure is 1·01 bar. Calculate the rate of water flow, the number of tubes, and the tube length. Take the overall heat transfer coefficient as 56 W/m² K, but allow for a fouling resistance of 0·5 K/kW for 1 m². For the gases take $c_p = 1·11$ kJ/kg K; and $R = 0·29$ kJ/kg K. Take the specific heat of water as 4·19 kJ/kg K.

(1096 kg/h; 110; 1·457 m)

17.19 In an oil cooler the oil enters 10 mm diameter tubes at 160°C and is cooled to 40°C. Calculate the heat transfer coefficient. For turbulent flow of a liquid being cooled take,

$$Nu = 0·0265(Re)^{0·8} \times (Pr)^{0·3}$$

and for laminar flow take $Nu = 3·65$.

(Take all properties at the mean bulk temperature.)

The following table gives some properties of engine oil:

$t\,°C$	ρ kg/m³	ν cSt	k W/m K	c kJ/kg K
40	878	251·0	0·144	1·96
100	839	20·4	0·137	2·22
160	806	5·7	0·131	2·48

(1 centistoke (cSt) $= 10^{-6}$ m²/s).

The mean velocity of the oil in the tubes is $1 \cdot 5$ m/s.

$$(50 \ W/m^2 \ K)$$

17.20 In problem 17.19 the length of each tube is $1 \cdot 2$ m. In order to allow for the entry length effect a more accurate expression for the Nusselt number for laminar flow is given by,

$$Nu = 3 \cdot 65 + \frac{0 \cdot 0668(d/L)(Re)(Pr)}{1 + 0 \cdot 04\{(d/L)(Re)(Pr)\}^{2/3}}$$

(where properties are taken at the mean bulk temperature).

Calculate the heat transfer coefficient using this formula.

$$(282 \ W/m^2 \ K)$$

17.21 $34 \cdot 2 \ m^3$/h of air at $15°C$ and 1 bar are to be heated to $285°C$ while flowing through a 25 mm diameter tube which is maintained at $455°C$. Calculate the length of tube required. Assume that the simple Reynolds analogy is valid and take $f = 0 \cdot 0791(Re)^{-1/4}$ for turbulent flow. Take all properties at the mean bulk temperature.

$$(1 \cdot 84 \ m)$$

17.22 Air flows through a 20 mm diameter tube 2 m long with a mean velocity of 40 m/s. The tube wall temperature is $150°C$ and the air temperature increases from $15°C$ to $100°C$. Using the simple Reynolds analogy with all properties at the mean bulk temperature, estimate the pressure loss in millimetres of water in the tube due to friction, and the pumping power required. Take the air mean pressure as 1 atmosphere. $(174 \ mm \ H_2O; \ 21 \cdot 45 \ W)$

17.23 In an air cooler the air is blown across a bank of tubes at the rate of 240 kg/h at a velocity of 24 m/s, the air entering at $97°C$ and leaving at $27°C$. The cooling water enters the tubes at $10°C$ and leaves at $20°C$, at a mean velocity of $0 \cdot 6$ m/s. The tubes are 6 mm diameter and the wall thickness may be neglected. The heat transfer coefficient from the air to the tubes may be calculated from

$$Nu = 0 \cdot 33(Re)^{0 \cdot 6} \times (Pr)^{0 \cdot 33}$$

with properties at mean bulk temperature.

The heat transfer coefficient from the water to the tubes is given by,

$$St = \frac{f/2}{1 + (Pr)^{-1/6}(Re)^{-1/8}(Pr - 1)}$$

where $f = 0 \cdot 0791(Re)^{-1/4}$ and properties are at mean bulk temperature. Assuming that the tubes are arranged in 6 passes, and that the logarithmic mean temperature difference for counter-flow can be used, calculate the number of tubes required in each pass and the necessary tube length. (7; 0·528 m)

17.24 Air at a temperature of 15°C is blown across a flat plate at a mean velocity of 6 m/s. Calculate the heat transferred per m width from both sides of the plate over the first 150 mm of the plate, when the surface temperature is 550°C. For heat transfer from a flat plate with a large temperature difference between the plate and the fluid take

$$Nu = 0.332(Pr)^{1/3} \times (Re)^{1/2} \times \left(\frac{T_w}{T_s}\right)^{0.117}$$

(where all properties are at mean film temperature, and T_w and T_s are the absolute temperatures of the plate and the free stream of the air respectively). Neglect radiation. (4·39 kW)

17.25 A two-pass shell-and-tube heat exchanger is used to condense a chemical on the shell side at a rate of 50 kg/s at a saturation temperature of 80°C. The chemical enters as a dry saturated vapour and is not undercooled during the process. Water at 10°C and a mass flow rate of 100 kg/s is available as a coolant; the velocity of the water is to be approximately 1·5 m/s.

Using the data below and taking a nominal tube diameter of 25 mm, neglecting tube wall thickness, determine:

  (i)   the number of tubes required;
  (ii) the tube length;
  (iii) the number of transfer units;
  (iv) the effectiveness of the heat exchanger.

Data: latent heat of vapourization of chemical = 417·8 kJ/kg; heat transfer coefficient for the shell side = 10 kW/m² K; fouling factor, shell side = 0·1 m² K/kW; fouling factor, tube side = 0·2 m² K/kW. For turbulent flow in a pipe,

$$Nu = 0.023Re^{0.8}Pr^{0.4}$$

(with properties at the mean bulk temperature).
(274; 10·61 m; 0·981; 62·5%)

17.26 An oil cooler consists of a single-pass, counter-flow shell-and-tube heat exchanger with 300 tubes of internal diameter 7·3 mm and length 8 m. The oil flows in the tube side entering at a mass flow rate of 8 kg/s at a temperature of 70°C. Cooling water in the shell side enters at a mass flow rate of 12 kg/s at a temperature of 15°C.

Using the data below, calculate:

(i) the number of transfer units;
(ii) the effectiveness of the heat exchanger;
(iii) the outlet temperature of the oil.

Neglect the thermal resistance of the tube wall and assume that the heat exchanger is clean.

Data: shell side heat transfer coefficient $= 1000$ W/m^2 K; heat transfer coefficient for the tube side given by, $Nu = 0.023 Re^{0.8} Pr^{0.4}$, with properties as follows: specific heat of oil $= 3.42$ kJ/kg K; density of oil $= 900$ kg/m^3; viscosity of oil $= 1.5 \times 10^{-3}$ kg/m s; thermal conductivity of oil $= 0.15$ kW/m K.

$$(1.1; \ 0.588; \ 37.7°C)$$

17.27 A double pipe heat exchanger has an effectiveness of 0.5 when the flow is counter-current and the thermal capacity of one fluid is twice that of the other fluid. Calculate the effectiveness of the heat exchanger if the direction of flow of one of the fluids is reversed with the same mass flow rates as before. $\qquad$ (0.469)

17.28 500 kg/h of oil at 120°C is to be cooled in the annulus of a double pipe counter-flow heat exchanger by water which enters the inside pipe at 10°C. The inner pipe has an inside diameter of 25 mm and a wall thickness of 2 mm, and the inside diameter of the outer pipe is 50 mm; the effective length is 12 m. Using the data below calculate the exit temperature of the oil.

Data:

oil   take $Nu = 30$, based on an equivalent diameter, d_e, given by $d_e = 4 \times$ (flow area)/(heat transfer area per unit length); specific heat $= 2.31$ kJ/kg K; thermal conductivity $= 0.135$ W/m K; fouling factor $= 0.001$ m^2 K/W.

water   assume the simple Reynolds analogy holds true, taking the velocity as 1 m/s and the friction factor, f, as 0.002; specific heat $= 4.18$ kJ/kg K; density $= 1000$ kg/m^3; fouling factor $= 0.0002$ m^2 K/W.

Neglect the thermal resistance of the pipe wall. $\qquad$ (98.8°C)

17.29 A condenser contains four tube passes with tubes 3 m long, 25 mm internal diameter, each pass containing 100 tubes. Cooling water enters the tubes at 20°C at the rate of 80 kg/s when the shell side vapour is at 50°C. Before cleaning, the fouling factor on the water side is 0·0005 m^2 K/W; the outside of the tubes may be taken to be clean.

Neglecting the thermal resistance of the fluid film on the outside of the tubes and the thermal resistance of the tube wall, calculate, using the data below:

(i)   the effectiveness of the heat exchanger;
(ii) the condensation rate;
(iii) the fouling factor required on the water side if the effectiveness is to be increased to 0·7 for the same mass flow rate of water.

For heat transfer in the tubes: $Nu = 0.023Re^{0.8}Pr^{1/3}$.

Data: latent heat of vapourization for shell side fluid = 300 kJ/kg; mean properties of water for the temperature range considered: density = 1000 kg/m^3; specific heat = 4·19 kJ/kg K; thermal conductivity = 0·6 W/m K; viscosity = 0.9×10^{-3} kg/m^3.

$$(0.337; \ 11.3 \ \text{kg/s}; \ 0.000 \ 049 \ m^2 \ \text{K/W})$$

17.30 In a closed cycle gas turbine plant air from the compressor enters one side of a compact heat exchanger at 150°C at a mass flow rate of 10 kg/s. The air leaving the turbine enters the heat exchanger at 504°C and flows in counter-flow to the air. The heat exchanger has a flow area of 0·144 m^2 and an effective heat transfer area of 115·2 m^2 per unit length in the direction of flow on both the hot and cold sides of the heat exchanger. Calculate the required length of the heat exchanger to obtain an effectiveness of 0·7.

Assume that the heat exchanger surfaces are clean and neglect the thermal resistance of the separating plates. For flow of air in the heat exchanger passages assume, $Nu = 0.023Re^{0.8}Pr^{0.3}$ based on an equivalent diameter given by 4 × (flow area)/(heated surface area per unit length); take the properties at the mean temperature between the cold air inlet and the hot air inlet.

$$(1.257 \ \text{m})$$

17.31 A wall 0·6 m high by 3 m wide is maintained at 47°C in an atmosphere at 15°C. Calculate the heat lost by natural convection neglecting end effects. For natural convection from a vertical flat surface take

$$Nu = 0{\cdot}509(Pr)^{1/2}(Pr + 0{\cdot}952)^{-1/4}(Gr)^{1/4}$$

(where all properties are at the surface temperature).

Take $\beta = 1/T$, where T K is the absolute air temperature. Neglect radiation. $\qquad$ (222·5 W)

17.32 Recalculate problem 7.18 using the approximations,

$$h = 1{\cdot}42\left(\frac{\theta}{l}\right)^{1/4} \quad \text{for} \quad 10^4 < Gr < 10^9$$

or $\qquad h = 1{\cdot}25(\theta)^{1/3} \quad \text{for} \quad 10^9 < Gr < 10^{12}$

(where h is in W/m² K, θ is in K, and l is in m). $\qquad$ (221 W)

17.33 A pipe containing dry saturated steam at 177°C is 150 mm bore and has a 50 mm thickness of 85% magnesia covering. The steam velocity is 6 m/s, and the heat transfer coefficient may be found from,

$$Nu = 0{\cdot}023(Re)^{0{\cdot}8} \times (Pr)^{0{\cdot}4}$$

(where all properties are at mean bulk temperature).

The atmospheric temperature is 17°C and the heat transfer coefficient for natural convection from a horizontal cylinder is given approximately by

$$h = 1{\cdot}42\left(\frac{\theta}{d}\right)^{1/4}$$

(where h is in W/m² K, θ is in K, and d is in m).

The pipe is 7 mm thick and the thermal conductivity of the pipe metal is 50 W/m K. Taking the thermal conductivity of 85% magnesia covering as 0·06 W/m K, calculate the temperature of the outside surface of the lagging, and the heat lost per m length. Use arithmetic mean areas for the pipe wall and lagging, and use a trial and error method to evaluate the temperature difference between the lagging surface and the air. Neglect radiation.

$$(44{\cdot}9°C; 105{\cdot}1 \text{ W})$$

17.34 Circular cross-section studs of radius 10 mm, length 100 mm, thermal conductivity 25 W/m K are attached to a flat surface with their axes perpendicular to the surface on a square pitch of 30 mm. The primary surface is at 300°C.

A fluid at 50°C is forced across the surface such that the mean heat transfer coefficient is 100 W/m² K. Calculate the heat loss per m² of primary surface. Assume that the heat transfer coefficient is the same for the primary surface and for the rod surfaces.

(77·66 kW)

17.35 A flat surface at a temperature of 300°C has rectangular section cooling fins perpendicular to the surface projecting into a fluid at 20°C. There are 12·5 fins per 100 m and the fins have a thickness of 3 mm and a length of 30 mm.

The thermal conductivity of the fin material is 26 W/m K and the heat transfer coefficient for all surfaces may be taken as 40 W/m² K.

Calculate:

(i)   the fin efficiency;
(ii) the heat loss from one square metre of the flat surface;
(iii) the temperature at the tip of each fin.

(77·5%; 72·1 kW/m²; 206·9°C)

17.36 A cylindrical electrode of radius, r, length, L, is immersed in a liquid which remains at a constant temperature when the current density in the electrode is J. The heat transfer coefficient, h, from the outside surface of the electrode may be assumed to be constant over the entire surface.

Assuming steady state conditions derive the differential equation,

$$\frac{d^2\theta}{dx^2} - \frac{2h}{rk}\theta + \frac{J^2 s}{k} = 0$$

(where θ = temperature difference between the electrode and the fluid at any distance x from the end of the electrode; k = thermal conductivity of the electrode material; s = electrical resistivity of the electrode material).

Hence show that for the case when the heat loss through the lead and support at each end is a fraction y, of the total electrical input, then:

$$\theta = \frac{J^2 s}{m^2 k}\left\{1 - \frac{mLy \cosh m(x - L/2)}{\sinh mL/2}\right\}$$

(where $m = (2h/rk)^{1/2}$).

17.37 A cylindrical storage tank, 1 m diameter by 1·2 m long, has an outside surface temperature of 60°C, and an emissivity of 0·9. Calculate the heat loss by radiation when the tank is in a large room, the walls of which are at 15°C. Calculate also the reduction in the heat loss by radiation if the tank is painted with aluminium paint of emissivity 0·4. Assume that the tank is a grey body.

(1474 W; 819 W)

17.38 A copper pipe at 260°C is in a large room at 15°C. Calculate the heat loss per m² of pipe surface area by radiation, taking the emissivity of copper as 0·61 at 260°C, and as 0·56 at 15°C. Assume that the absorptivity of a surface depends only on the temperature of the source of radiation.   (2571·5 W/m²)

17.39 Calculate the heat transferred per m² of surface area by radiation between two brick walls a short distance apart, when the temperatures of the surfaces are 30°C and 15°C. The emissivity of brick may be taken as 0·93, and the surfaces may be assumed to be grey.   (76·2 W/m²)

17.40 A thermos-flask consists of an inner cylindrical vessel of 60 mm outside diameter and an outer cylindrical vessel of 65 mm inside diameter. Both surfaces are of polished silver, emissivity 0·02. Calculate the heat loss per mm length of the flask when it contains boiling water and the temperature of the outside surface is 17°C. Neglect the thermal resistance of the metal walls of the flask. (N.B. Polished surfaces reflect specularly and hence in this case the surfaces act as large parallel planes.)   (0·00133 W)

17.41 A gas turbine can-type combustion chamber of 0·3 m diameter reaches a temperature of 500°C when undergoing a test in large surroundings at 15°C. The emissivity of the steel surface is 0·79. Calculate the percentage reduction in the radiant heat loss by enclosing the combustion chamber with a cylindrical screen of 0·6 m diameter, the inside and outside surfaces of which are painted with aluminium paint of emissivity 0·4.   (61·2%)

17.42 In a muffle furnace the floor, 4·5 m by 4·5 m, is constructed of refractory material (emissivity = 0·7). Two rows of oxidized steel tubes are placed 3 m above and parallel to the floor, but for the purpose of analysis these can be replaced by a 4·5 m by 4·5 m plane having an effective emissivity of 0·9. The average temperatures for the floor and tubes are 900°C and 270°C respectively. Taking the geometric factor for radiation from floor to tubes as

0·32, calculate:

(a) the net heat transfer to the tubes;
(b) the mean temperature of the refractory walls of the furnace, assuming that these are well-insulated.

(1009 kW; 688°C)

17.43 A circular plate of radius 0·1 m is at a temperature of 500°C in a large room, the walls of which are at 10°C. The air in the room is at a mean temperature of 15°C. A small, spherical thermocouple junction is placed at a distance of 0·1 m from the centre of the plate. Show that the temperature recorded by the thermocouple is approximately 100°C. The heat transfer coefficient from the thermocouple to the air is 25·6 W/m² K, and the plate surface may be assumed to be black for thermal radiation. Neglect conduction through the thermocouple leads. The geometric factor is given in Table 17·3.

17.44 An electric heater 25 mm diameter and 0·3 m long is used to heat a room. Calculate the electrical input to the heater if the bulk of the air in the room is at 20°C, if the walls are at 15°C, and if the surface of the heater is at 540°C. For convective heat transfer from the heater, assume that,

$$Nu = 0·4(Gr)^{1/4}$$

(where all properties are at mean film temperature and $\beta = 1/T$, where T K is the bulk temperature of the air).

Take the emissivity of the heater surface as 0·55 and assume that the surroundings are black. (481 W)

17.45 Calculate the radiation heat transfer coefficient for the flat plate in problem 17.24, assuming that the surroundings are large and are at the air temperature, and compare this with the heat transfer coefficient for convection. Take the emissivity of the plate surface as 0·6. (28·75 W/m² K; $h_r/h = 1·05$)

17.46 Calculate the radiation heat transfer coefficient for the vertical wall in problem 17.31, assuming that the wall radiates into black surroundings at 15°C, and the emissivity of the wall surface is 0·93. Compare this value with the heat transfer coefficient for convection.
(5·96 W/m² K; $h_r/h = 1·542$)

17.47 A hot-water heater 150 mm wide by 1·2 m long by 1 m high is at a surface temperature of 50°C in surroundings at 20°C. The

walls of the room are at 13°C. The surface area of the heater is 7 m²
and the heat transfer coefficient for convection is given by,

$$h = 1 \cdot 25(\theta)^{1/3}$$

(where h is in W/m² K, and θ is in K).

Calculate the heat transferred from the heater. Take the emissivity
of the heater as 0·95 and assume that it is completely surrounded by
black surroundings. (1·505 kW)

REFERENCES

17.1 ECKERT, E. R., and DRAKE, R. M., *Analysis of Heat and Mass Transfer*
(McGraw-Hill, 1972).

17.2 KERN, D. Q., *Process Heat Transfer* (McGraw-Hill, 1950).

17.3 KAYS, W. M. and LONDON, A. L., *Compact Heat Exchangers* (National
Press, California, 1984).

17.4 WALKER, G., *Industrial Heat Exchangers* (McGraw-Hill, 1982).

17.5 KNUDSEN, J. G., and KATZ, D. L., *Fluid Dynamics and Heat Transfer*
(McGraw-Hill, 1979).

17.6 BAYLEY, F. J., OWEN J. M., and TURNER, A. B., *Heat Transfer* (Nelson,
1972).

17.7 ROGERS, G. F. C., and MAYHEW, Y. R., *Thermodynamic and Transport
Properties of Fluids, SI Units* (Basil Blackwell, 1980).

17.8 WELTY, J. R., *Engineering Heat Transfer* (John Wiley, 1978).

17.9 WONG, H. Y., *Heat Transfer for Engineers* (Longman, 1977).

18. Reciprocating Internal-Combustion Engines

Theoretical power cycles are considered in Chapter 6, and the p-v diagrams analysed are similar to those obtained from actual reciprocating engines. There are, however, fundamental mechanical and thermodynamic differences between the cycles, which make comparison less valuable than might be expected.

In the theoretical cycles there is no chemical change in the working fluid, which is assumed to be air, and the heat exchanges in the cycle are made externally to the working fluid. In the practical cycle the heat supply is obtained from the combustion of a fuel in air and thus the air charge is consumed during combustion and the combustion products must be exhausted from the cylinder before a fresh charge of air can be induced for the next cycle. The practical cycle consists of the exhaust and induction processes together with the compression and expansion processes as in the theoretical cycle. Further differences between the ideal and the actual cycles are discussed in Section 18.8.

The reciprocating engine mechanism consists of a piston which moves in a cylinder and forms a movable gas-tight plug, a connecting rod and a crankshaft (see fig. 18.1). If the engine has more than one cylinder then the cylinders, pistons, etc., are identical, and all the connecting rods are fastened to a common crankshaft. The angular positions of the crank-pins are such that the cylinders contribute their power strokes in a selected and regular sequence. By means of this arrangement the reciprocating motion of the piston is converted to rotary motion at the crankshaft.

There are many types and arrangements of engines, and some classification is necessary to describe a particular engine adequately. The methods of classification are:

(a) By the fuel used and the way in which the combustion is initiated. Petrol engines and gas engines have spark ignition (S.I.). Diesel engines or oil engines have compression ignition (C.I.). In the S.I. engine the air and the fuel are mixed before compression.

In the C.I. engine the air only is compressed, and the fuel is injected into the air which is then at a sufficiently high temperature to initiate combustion.

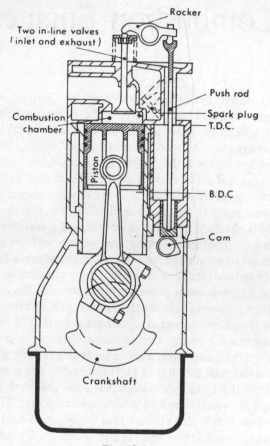

Fig. 18.1

(b) By the way in which the cycle of processes is arranged. This is defined by the number of complete strokes of the piston required for one complete cycle. The stroke of the piston is the distance it moves from the position most extreme from the crankshaft to that nearest it. This takes place over half a revolution of the crankshaft. In petrol engine practice the extreme positions of the piston are referred to as *top dead centre* (TDC), and *bottom dead centre* (BDC) (see fig. 18.1).

In oil-engine practice they are referred to as *outer dead centre* and *inner dead centre*, respectively. An engine which requires four strokes of the piston (i.e. two revolutions of the crankshaft) to complete its cycle, is called a *four-stroke cycle engine*. An engine which requires only two strokes of the piston (i.e. one crankshaft revolution) is called a *two-stroke cycle engine*.

In all reciprocating internal-combustion engines the gases are induced into and exhausted from the cylinder through ports, the opening and closing of which are related to the piston position. In a two-stroke engine the ports can be opened or closed by the piston itself, but in the four-stroke engine a separate shaft, called the cam-shaft, is required; this is driven from the crankshaft through a 2 to 1 speed reduction. The cams on this shaft operate valves, called poppet valves, either directly or by means of push rods. Modern high-speed petrol engines have two camshafts, one operating the exhaust valves, and the other operating the inlet valves. The timing of the valves and the point of ignition are fundamental to the engine performance and the specified timing is a result of compromise between the many factors involved, and is determined empirically. The beginning and end of each process does not coincide with the TDC and BDC positions, although nominally each process may be associated with a piston stroke. The timing of the valves can be indicated on a *p-V* diagram but is more conveniently represented by means of a timing diagram (see fig. 18.3), in terms of crankshaft angle.

18.1 Four-stroke cycle

Fig. 18.2 shows a typical *p-V* diagram for a S.I. petrol engine. The individual strokes are:

1–2 *Induction stroke*

The air plus fuel charge is induced into the cylinder as the piston moves from TDC to BDC. Due to the movement of the piston the pressure in the cylinder is reduced to a value below the atmospheric pressure, and air flows through the induction system because of this pressure difference. On its way to the cylinder the air passes through the carburettor in which the metered amount of petrol is added to the air. Nominally the inlet valve closes at point 2, but in fact this does not occur until the piston has moved part of the way along the return stroke.

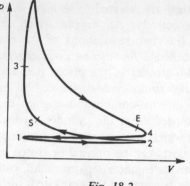

Fig. 18.2

2–3 *Compression stroke*

With both valves closed the charge is compressed by the piston. At the TDC position the charge occupies the volume above the piston, which is called the clearance volume, and consists mainly of the volume of the combustion chamber. The spark is timed to occur at a point such as S, which is before TDC. There is a time delay between S and the actual commencement of combustion. The combustion process occurs mainly at almost constant volume, and there is a large increase in pressure and temperature of the charge during this process.

3–4 *Working stroke*

The hot, high-pressure gas expands, pushing the piston down the cylinder. It would appear that this expansion should proceed to completion at 4, but in order to assist in exhausting the gaseous products the exhaust valve opens at some point E which is before BDC. At this point the pressure is about 3·5 bar or higher and about 60% of the gas is exhausted between E and 4 as the pressure in the cylinder falls to nearly atmospheric pressure.

4–1 *Exhaust stroke*

The returning piston clears the swept volume of exhaust gas, and the pressure during this stroke is slightly higher than atmospheric pressure. In a normally aspirated engine as described, the clearance volume cannot be exhausted, and at the commencement of the next cycle this volume is full of exhaust gas at about atmospheric pressure. The mixture which is compressed thus consists of the fresh air plus fuel mixture, diluted by a quantity of exhaust gas from the previous cycle.

It should be remembered that the maximum volume of fresh charge which can be induced is equal to the swept volume, V_s, but the actual mass induced in practice is less than the maximum possible, for reasons which will be considered later.

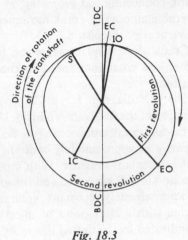

A typical timing diagram for a four-stroke petrol engine is shown in fig. 18.3, and the angular positions in terms of crank angle position are quoted in relation to the TDC and BDC positions of the piston. The points on the diagram are as follows:

Fig. 18.3

IO   Inlet valve opens.
     The actual position is between 10° before TDC and 15° after TDC.

IC   Inlet valve closes.
     This occurs 20° to 40° after BDC to take advantage of the momentum of the rapidly moving gas.

 S   Spark occurs.
     This is 20° to 40° before TDC when the ignition is fully advanced, and is at TDC when the ignition is fully retarded.

EO   Exhaust valve opens.
     The average value of this position is about 50° before BDC, but it is greater than this in racing-car engines.

EC   Exhaust valve closes.
     This occurs 0° to 10° after TDC.

There may be an overlap between IO and EC such that both valves are open at the same instant.

It should be appreciated that the values of angular position quoted are average ones, and considerable differences occur between different engines. The points shown are the normal opening and closing positions for each valve. The time required to open and close the valves means that each valve will be fully open for a crank angle movement much less than that indicated by the timing diagram. Another point to be borne in mind is shown by a consideration of crank-connecting rod geometry; at the dead centre positions a considerable amount of crank movement produces small corresponding movement of the piston.

Crank angle displacements can be translated into time values, if the engine rotational speed is known and can be assumed to be constant.

The p-V diagram for a C.I. engine with mechanical, or solid, injection of the fuel is shown in fig. 18.4. The ideal cycle for the engine is the dual combustion cycle (see Section 6.8) in which the heat input is partly at approximately constant volume, and partly at approximately constant pressure. With modern high-speed engines the fuel injection is well advanced and the greater proportion of fuel is burnt at approximately constant volume. A diagram from the same engine with a later point of injection would be flatter at the top, as indicated by the dotted line in fig. 18.4. The shape of the diagram is also influenced by the design of the combustion chamber.

With engines having the fuel injected by means of an air blast, the p-V diagram is similar to that shown in fig. 18.5. In this cycle the air which enters the cylinder with the fuel on injection, helps to maintain a constant pressure over the early part of the return stroke and this increases the area of the diagram. This method of injection is now practically obsolete due to the difficulties and cost of supplying the high-pressure air required.

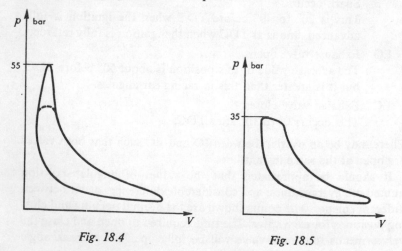

Fig. 18.4   Fig. 18.5

Fig. 18.6 shows a typical timing diagram for a four-stroke oil engine which has average values for the valve positions as follows:

IO   Up to 30° before TDC.
IC   Up to 50° after BDC.
EO   About 45° before BDC.
EC   About 30° after TDC.
Injection About 15° before TDC.

18.2 Two-stroke cycle

Fig. 18.7 represents the cylinder of a two-stroke petrol engine with crankcase compression. As the piston ascends on the compression stroke the next charge is drawn into the crankcase C through the

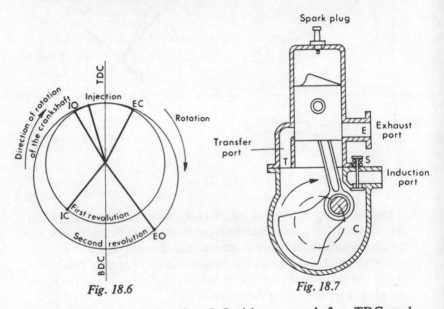

Fig. 18.6   Fig. 18.7

spring-loaded automatic valve, S. Ignition occurs before TDC, and at TDC the working stroke begins. As the piston descends through about 80% of the working stroke, the exhaust port, E, is uncovered by the piston and exhaust begins. The transfer port, T, is uncovered later in the stroke due to the shape of the piston or the position of the port in relation to the port E, and the charge in the crankcase C, which has been compressed by the descending piston, enters the cylinder through the port T.

The piston can be shaped to deflect the fresh gas across the cylinder to assist the scavenging of the cylinder; this is called *cross-flow scavenge*. As the piston rises, the transfer port, T, is closed slightly before the exhaust port, E, and after E is closed compression of the charge in the cylinder begins. The *p-V* diagram and the timing diagram for a two-stroke petrol engine are shown in fig. 18.8a and 18.8b.

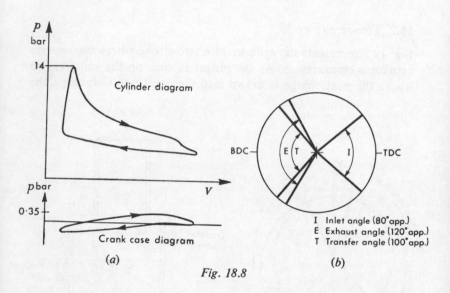

Fig. 18.8

Instead of the spring-loaded valve, S, a design with a third port may be used. This is an induction port controlled by the piston, and through which the mixture is drawn into the crankcase.

The above description of the two-stroke cycle applies also to C.I. engines with the exception that air only is compressed, and the sparking plug is replaced by a fuel injector.

Crankcase compression has been described but the scavenging and charging of the cylinder may be achieved by other means. A separately phased pump cylinder with its piston driven from the crankshaft may be used. A positive displacement compressor or blower driven from the engine is a third way of charging the cylinder.

The deflector piston, which is unbalanced and can cause 'rattle', may be dispensed with and a flat piston used. The scavenging is then obtained by using two transfer ports which divert the incoming air up the cylinder. This is called 'reverse flow', or 'inverted flow', and the system is called *loop scavenge*.

In engines which have simple inlet ports and poppet or sleeve valve controlled exhaust ports, the inlet and exhaust ports are placed at opposite ends of the cylinder and the fresh charge sweeps along the cylinder towards the exhaust port. This is called *uni-flow scavenge* and is applied with great mechanical simplicity in opposed piston engines.

For several reasons the two-stroke cycle has more application in the C.I. field than in the S.I. field, especially for stationary constant speed engines. In such engines a number of ingenious arrangements have been patented in an attempt to dispense with the scavenge blower. Some designs have used the Kadenacy effect, which employs the high vacuum created by suddenly releasing the exhaust gas through large area, sharp-edged ports. With constant-speed engines it is possible to 'tune' the exhaust system such that the high-pressure exhaust gas leaving one cylinder can be used to 'pack' another cylinder which is on the early part of its induction stroke.

18.3 Other types of engines

In the early days of development of the four-stroke engine one of the difficulties was the noisy poppet valve mechanism. As an alternative the *sleeve valve* became popular. A sliding sleeve is fitted in between the piston and the cylinder, the movement of the sleeve being controlled by an overhung crank-pin driven from a shaft at half crankshaft speed. The movement of this valve controls the inlet and exhaust ports in the cylinder. For an authoritative discussion of this type of engine, reference 18.1 should be consulted.

Engines have been developed which are called *multi-fuel engines*; such engines will run on any petroleum fuel from diesel fuel to premium petrol. The main application of such engines is for military purposes and it is unlikely that they will have extensive commercial application.

The *dual fuel engine* is of considerable industrial interest. Some diesel engines, naturally aspirated or turbo-charged (see Section 18.12), can be used as dual fuel engines. They are started on diesel oil and then run on an available gaseous fuel such as methane, natural gas, sewage gas, coal gas, etc. The combustion process requires a pilot injection of oil which amounts to 7 to 10% of the full power supply when running as a diesel engine. The change over from diesel fuel to gas can be done automatically or manually. To supply the pilot injection of oil a second set of pumps is required which delivers fuel to the standard injectors.

A considerable amount of effort has been put into the development of the *free piston engine*. In this type of engine the crankshaft and connecting rod are dispensed with, and two opposed but connected pistons operating in the same cylinder are used. The free piston engine is described more fully in Section 18.14. One of the main applications of the engine is that instead of having a separate air compressor driven by an I.C. engine the compressor and the engine can be combined with the result that the intermediate rotating shafts are eliminated and a more compact unit is obtained. This is especially important for portable air compressors which are used extensively. Another use for the free piston engine is as a 'gasifier' from which the gaseous products of combustion are exhausted at a suitable pressure and allowed to expand through a gas turbine. No power output is taken from the free piston engine, the power output of the unit being that obtained from the turbine. The potential field for the smaller size free piston engine is regarded as being with road, rail, and tracked vehicles, earth moving equipment, high speed marine craft, cargo ships, and in generating stations.

Some classification of I.C. engines has been given but this is not exhaustive. The applications for such engines are wide, both in the type of duty to be performed and in the power required. Modern developments promise the application of I.C. engines not only as individual units but as part of a complete plant to suit some specialized purpose.

18.4 Criteria of performance

An engine is selected to suit a particular application, the main consideration being its power/speed characteristics. Important additional factors are initial capital cost, and running cost. In order that different types of engines or different engines of the same type, may be compared, certain performance criteria must be defined. These are obtained by measurement of the quantities concerned during bench tests, and calculation is by standard procedures. The results are plotted graphically in the form of performance curves.

(a) *Indicated power (i.p.)*

This is defined as the rate of work done by the gas on the piston as evaluated from an indicator diagram obtained from the engine. An indicator diagram has the form shown in fig. 18.9. For low-speed engines (less than about 400 rev/min), a mechanical indicator is used (see reference 18.12), and a special design, the Maihak type 'S', has been developed to be

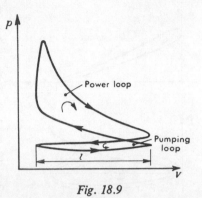

Fig. 18.9

used on engines running at speeds up to 2400 rev/min. Fig. 18.9 shows both the power and the pumping loops, but these would not be obtained usefully on the same indicator card. It is more usual to obtain the two loops separately. The power loop is obtained using a heavy spring fitted to the indicator and the pumping loop is obtained with a light spring fitted. The diagrams are sometimes referred to as *heavy spring* and *light spring* diagrams. On the heavy spring diagram the pumping loop is hardly distinguishable from the atmospheric line, which is always marked on the diagram as a datum.

The mean effective pressure has been defined in Section 6.9 and may be applied here.

Net work done per cycle

$\propto$ (area of power loop − area of pumping loop)

Therefore indicated mean effective pressure, p_i bar, is given by

$$p_i = \frac{\text{net area of diagram in mm}^2}{\text{length of diagram in mm}} \times \text{spring constant} \qquad 18.1$$

(the spring constant is given in bar per mm of vertical movement of the indicator stylus).

Considering one engine cylinder,

$$\text{Work done per cycle} = p_i \times A \times L$$

(where A = area of piston; L = length of stroke).

$$\text{Work done per min} = \text{Work done per cycle} \times \text{cycles per min}$$

or $\qquad\qquad$ i.p. $= p_i A L \times (\text{cycles/min})$

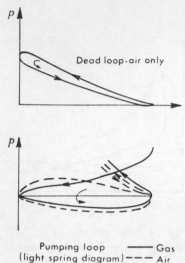

Fig. 18.10

The number of cycles per minute depends on the type of engine; for four-stroke engines the number of cycles per minute is N/2, and for two strokes the number of cycles is N, where N is the engine speed. Some gas engines are governed on the *hit-and-miss* principle, which means that when the speed tends to rise above the governed value further induction strokes are omitted, giving a dead cycle, until the required speed is restored. In this case the actual number of working cycles must be counted over a measured period of time and the average, f cycles/min taken. The dead cycle and pumping loops for such an engine are shown in fig. 18.10. The formula for i.p. then becomes:

Four-stroke engines,

$$\text{i.p.} = \frac{p_i A L N n}{2} \qquad 18.2$$

Two-stroke engines,

$$\text{i.p.} = p_i A L N n \qquad 18.3$$

Hit-and-miss gas engines,

$$\text{i.p.} = p_i A L f n \qquad 18.4$$

(where n is the number of cylinders).

For engine speeds greater than about 2400 rev/min the mechanical indicator is not suitable, and high-speed indicators have been developed which do not have the detrimental inertia effects of the mechanical types. The two types in use are the Farnboro indicator (for speeds up to 4000 rev/min), which is described in reference 18.12, and the electronic engine indicator, which uses a cathode-ray

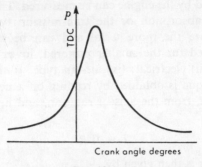

Fig. 18.11

oscilloscope, (CRO), to display the pressure trace obtained (see reference 18.12 also). A p-V diagram is not obtained directly be either of these types, but the pressure variation is obtained against a crank-angle base as shown in fig. 18.11.

The Farnboro diagram is obtained as an average of several hundred cycles on a special recording paper and a mean single diagram can be drawn through the plotted points. The output of the electronic indicator is obtained as a trace on the CRO for each successive cycle and a permanent record is obtained by photographing the trace.

Modern methods of instrumentation, data acquisition and analysis, based on recent advances in electronics, digital methods and the application of microprocessors, have become applicable to engine testing, development and control. These methods may be considered to have substantially replaced those described previously in this section which are still in use but are also useful in illustrating the principles involved. It is however certain that engine test and control systems will become increasingly 'computerized' and it is regretted that space requirements do not allow a comprehensive treatment of this very important aspect of engineering in this book. Readers are advised to obtain copies of the excellent publications by engine and equipment suppliers which will give a state-of-the-art description of the latest equipment and techniques.

(b) Brake power (b.p.)
This is the measured output of the engine. The engine is connected to a brake or dynamometer which can be loaded in such a way that

the torque exerted by the engine can be measured. The dynamometer may be of the absorption or the transmission type. Absorption dynamometers are the more usual and can be classified as: (i) friction type, used for the smaller powered, lower speed engines; (ii) hydraulic; (iii) electrical; (iv) air-fan type. With types (i), (ii), and (iii), the torque is obtained by reading off a net load, W, at a known radius, R, from the axis of rotation, and hence the torque, T, is given by

$$T = WR \qquad\qquad 18.5$$

The brake power is then given by

$$\text{b.p.} = 2\pi NT \qquad\qquad 18.6$$

In the transmission type of dynamometer the torque, T, transmitted by the driving shaft is measured directly, and the b.p. is obtained by substitution in equation 18.6. With air fans the torque is obtained from a calibration curve for the fan.

(c) *Friction power (f.p.) and mechanical efficiency, η_M*

The difference between the i.p. and the b.p. is the friction power (f.p.), and is that power required to overcome the frictional resistance of the engine parts,

i.e. $\qquad\qquad\qquad \text{f.p.} = \text{i.p.} - \text{b.p.} \qquad\qquad 18.7$

The mechanical efficiency of the engine is defined as

$$\text{Mechanical efficiency, } \eta_M = \frac{\text{b.p.}}{\text{i.p.}} \qquad\qquad 18.8$$

η_M usually lies between 80% and 90%.

Example 18.1

A single-cylinder four-stroke gas engine has a bore of 178 mm and a stroke of 330 mm, and is governed on the hit-and-miss principle. When running at 400 rev/min at full load, indicator cards are taken which give a working loop mean effective pressure of 6·2 bar, and a pumping loop mean effective pressure of 0·35 bar. Diagrams from the dead cycle give a mean effective pressure of 0·62 bar. The engine was run light at the same speed (i.e. with no load), and a mechanical counter recorded 47 firing strokes per minute. Calculate the full load b.p. and the mechanical efficiency of the engine.

The f.p. is taken as being constant at a given speed, and is independent of load. It is calculated in this problem from the no load test.

Net indicated mean effective pressure per working cycle is given by

$$\text{Net i.m.e.p.} = (6 \cdot 2 - 0 \cdot 35) = 5 \cdot 85 \text{ bar}$$

Also,   Working cycles per minute = 47

and   Dead cycles per minute $= \left(\dfrac{400}{2} - 47\right) = 153$

Therefore, since there is no b.p. output,

$$\text{f.p.} = (\text{net i.p.}) - (\text{pumping power of dead cycles})$$

Now,

$$\text{Area of piston} = \frac{\pi}{4}\left(\frac{178}{10^3}\right)^2 \text{ m}^2 \quad \text{and} \quad \text{stroke} = \frac{330}{10^3} \text{ m}$$

$$\therefore \text{ f.p.} = \left(\frac{5 \cdot 85 \times 10^5 \times \pi \times 178^2 \times 330 \times 47}{4 \times 10^6 \times 10^3 \times 10^3 \times 60}\right)$$

$$- \left(\frac{0 \cdot 62 \times 10^5 \times \pi \times 178^2 \times 330 \times 153}{4 \times 10^6 \times 10^3 \times 10^3 \times 60}\right) = 2 \cdot 46 \text{ kW}$$

At full load the engine fires regularly every two revs, and there are therefore 200 firing strokes per minute.

$$\therefore \text{ i.p.} = \frac{5 \cdot 85 \times 10^5 \times \pi \times 178^2 \times 330 \times 200}{4 \times 10^6 \times 10^3 \times 10^3 \times 60} = 16 \text{ kW}$$

Hence, $\text{b.p.} = (\text{i.p.} - \text{f.p.}) = 16 - 2 \cdot 46 = 13 \cdot 54 \text{ kW}$

$$\text{Mechanical efficiency} = \frac{\text{b.p.}}{\text{i.p.}} = \frac{13 \cdot 54}{16} = 0 \cdot 847 \quad \text{or} \quad 84 \cdot 7\%$$

The f.p. is very nearly constant at a given engine speed; and if the load is decreased giving lower values of b.p., then the variation in η_M with b.p. is as shown in fig. 18.12. At zero b.p. at the same speed the engine is developing just sufficient power to overcome the frictional resistance.

η_M depends on the i.p. and b.p., and is therefore found by

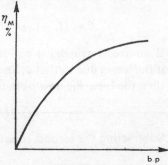

Fig. 18.12

evaluating these experimentally. A considerable amount of literature has been published dealing with mechanical efficiency and the analysis of engine power losses. A useful report on this subject is that of reference 18.2, and the conclusion is that no single method is satisfactory in every respect for the evaluation of mechanical efficiency. The four main methods are those given by Ricardo in reference 18.1, as follows:

(i) Measurement of the i.p. and the b.p. by the means already described in paragraphs (a) and (b) above.

(ii) Measurement of the b.p. at a given speed followed by 'motoring' of the engine with the fuel supply cut off. This method can only be used on engines with an electrical dynamometer, the dynamometer being used as a motor instead of as a generator in order to motor the engine at the firing speed. The torque can be measured under firing and under motoring conditions and the mechanical efficiency evaluated. The fact that an electrical dynamometer can be used to find the mechanical efficiency in this way is one of the main advantages of this type of dynamometer.

(iii) The Morse test: this is only applicable to multi-cylinder engines. The engine is run at the required speed and the torque is measured. One cylinder is cut out, by shorting the plug if an S.I. engine is under test, or by disconnecting an injector if a C.I. engine is under test. The speed falls because of the loss of power with one cylinder cut out, but is restored by reducing the load. The torque is measured again when the speed has reached its original value. If the values of i.p. of the cylinders are denoted by I_1, I_2, I_3, and I_4 (considering a four-cylinder engine), and the power losses in each cylinder are denoted by, L_1, L_2, L_3, and L_4, then the value of b.p., B, at the test speed with all cylinders firing is given by

$$B = (I_1 - L_1) + (I_2 - L_2) + (I_3 - L_3) + (I_4 - L_4)$$

If number 1 cylinder is cut out, then the contribution I_1 is lost; and if the losses due to that cylinder remain the same as when it is firing, then the b.p., B_1, now obtained at the same speed is

$$B_1 = (0 - L_1) + (I_2 - L_2) + (I_3 - L_3) + (I_4 - L_4)$$

Subtracting the second equation from the first gives

$$B - B_1 = I_1 \qquad\qquad 18.9$$

By cutting out each cylinder in turn the values I_2, I_3, and I_4 can be obtained from equations similar to 18.9.

Then, for the engine,

$$I = I_1 + I_2 + I_3 + I_4 \qquad 18.10$$

(iv) 'Willan's line': this method is applicable to C.I. engines only. At a constant engine speed the load is reduced in increments and the corresponding b.p. and gross fuel consumption readings are taken. A graph is then drawn of fuel consumption against b.p., as in fig. 18.13. The graph drawn is called the 'Willan's line' (analogous to

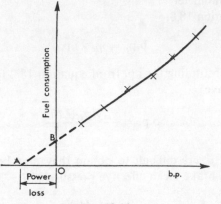

Fig. 18.13

Willan's line for a steam turbine), and is extrapolated back to cut the b.p. axis at the point A. The reading OA is taken as the power loss of the engine at that speed. The fuel consumption at zero b.p. is given by OB; and if the relationship between fuel consumption and b.p. is assumed to be linear, then a fuel consumption OB is equivalent to a power loss of OA.

The deficiencies of these methods are discussed in references 18.1 and 18.2, but a brief consideration will be given here. Method (i) has the difficulty that no indicator is yet available which will enable the i.p. to be measured accurately at high engine speeds. For method (ii) the friction and pumping losses are not the same during motoring and firing. Ricardo has stated that the motoring test may over-estimate the losses by as much as 20%. Method (iii), the Morse test, can be criticized for the same reason, since the cylinder which is cut out is really being motored. The mixture distribution between the cylinders is upset when one of them is cut out. If the cylinders have a common inlet manifold and a common exhaust manifold, then the pattern of the pressure variations in them will be disturbed and the breathing of the engine will thus be different. The Willan's line of method (iv) is not usually straight and the validity of the extrapolation is in doubt.

(d) *Brake mean effective pressure (b.m.e.p.), thermal efficiency and fuel consumption*

The b.p. of an engine can be obtained accurately and conveniently using a dynamometer.

From equation 18.8,

$$\text{b.p.} = \eta_M \times \text{i.p.}$$

Therefore, substituting for i.p. from equation 18.2 for a four-stroke engine, we have

$$\text{b.p.} = \frac{\eta_M \times p_i A L N n}{2}$$

Since η_M and p_i are difficult to obtain they may be combined and replaced by a brake mean effective pressure, p_b,

i.e.
$$\text{b.p.} = \frac{p_b A L N n}{2} \qquad 18.11$$

(where $p_b = \eta_M \times p_i$).

The b.m.e.p. may be thought of as that mean effective pressure acting on the pistons which would give the measured b.p. if the engine were frictionless. The b.m.e.p. is a useful criterion for comparing engine performance. Taking the two equations for b.p., equations 18.6 and 18.11, and putting them together we have

$$\frac{p_b A L N n}{2} = 2\pi N T$$

$$\therefore p_b = K \times T$$

(where K is a constant).

Therefore b.m.e.p. is directly proportional to the engine torque and is independent of the engine speed.

The power output of the engine is obtained from the chemical energy of the fuel supplied. The overall efficiency of the engine is given by the *brake thermal efficiency*, η_{BT},

i.e.
$$\eta_{BT} = \frac{\text{brake work}}{\text{energy supplied}}$$

$$\therefore \ \eta_{BT} = \frac{\text{b.p.}}{\dot{m}_f \times Q_{net,\,v}} \qquad 18.12$$

(where $\dot{m}_f$ is the mass of fuel consumed per unit time, and $Q_{net,\,v}$ is the lower calorific value of the fuel).

The *specific fuel consumption* (s.f.c.) is the mass of fuel consumed per kW developed per hour, and is a criterion of economical power production,

i.e.
$$\text{s.f.c.} = \frac{\dot{m}_f}{\text{b.p.}} \ \text{kg/kW h} \qquad 18.13$$

The *indicated thermal efficiency*, η_{IT}, is defined in a similar way to η_{BT},

i.e.
$$\eta_{IT} = \frac{\text{i.p.}}{\dot{m}_f \times Q_{net,\,v}} \qquad 18.14$$

Dividing equation 18.12 by equation 18.14 gives

$$\frac{\eta_{BT}}{\eta_{IT}} = \frac{\text{b.p.}}{\text{i.p.}} = \eta_M$$

$$\therefore \ \eta_{BT} = \eta_M \times \eta_{IT} \qquad 18.15$$

Example 18.2

A four-cylinder petrol engine has a bore of 57 mm and a stroke of 90 mm. Its rated speed is 2800 rev/min and it is tested at this speed against a brake which has a torque arm of 0·356 m. The net brake load is 155 N and the fuel consumption is 6·74 l/h. The specific gravity of the petrol used is 0·735 and it has a lower calorific value, $Q_{net,\,v}$, of 44 200 kJ/kg. A Morse test is carried out and the cylinders are cut out in the order 1, 2, 3, 4, with corresponding brake loads of 111, 106·5, 104·2, and 111 N, respectively. Calculate for this speed, the engine torque, the b.m.e.p., the brake thermal efficiency, the specific fuel consumption, the mechanical efficiency and the i.m.e.p.

Using equation 18.5,

$$\text{Torque} = T = WR = 155 \times 0·356 = 55·2 \ \text{N m}$$

Using equation 18.6,

$$\text{b.p.} = 2\pi NT = \frac{2\pi \times 2800 \times 55 \cdot 2}{60 \times 10^3} \text{ kN m/s} = 16 \cdot 2 \text{ kW}$$

From equation 18.11,

$$\text{b.m.e.p.} = \frac{\text{b.p.} \times 2}{ALNn} = \frac{16 \cdot 2 \times 2 \times 4 \times 60 \times 10^3}{\pi \times 0 \cdot 057^2 \times 0 \cdot 09 \times 2800 \times 4 \times 10^5}$$

$$= 7 \cdot 55 \text{ bar}$$

(where $A = \frac{\pi}{4} \times 0 \cdot 057^2 \text{ m}^2$, and $L = 0 \cdot 09$ m).

Using equation 18.12,

$$\eta_{\text{BT}} = \frac{\text{b.p.}}{\dot{m}_f \times Q_{\text{net, } v}} = \frac{16 \cdot 2}{0 \cdot 001 \, 377 \times 44 \, 200} = 0 \cdot 266 \text{ or } 26 \cdot 6\%$$

(where $\dot{m}_f = \frac{6 \cdot 74}{3600} \times 1 \times 0 \cdot 735 = 0 \cdot 001 \, 377$ kg/s).

Using equation 18.13,

$$\text{s.f.c.} = \frac{\dot{m}_f}{\text{b.p.}} = \frac{0 \cdot 001 \, 377 \times 3600}{16 \cdot 2} = 0 \cdot 306 \text{ kg/kW h}$$

Using equation 18.9 for each cylinder in turn, and substituting brake loads instead of the values of b.p. since the speed is constant, we have

$$I_1 = B - B_1 = 155 - 111 = 44 \text{ N}$$
$$I_2 = B - B_2 = 155 - 106 \cdot 5 = 48 \cdot 5 \text{ N}$$
$$I_3 = B - B_3 = 155 - 104 \cdot 2 = 50 \cdot 8 \text{ N}$$
$$I_4 = B - B_4 = 155 - 111 = 44 \text{ N}$$

Hence for the engine, the indicated load, I, is given by

$$I = I_1 + I_2 + I_3 + I_4 = 44 + 48 \cdot 5 + 50 \cdot 8 + 44 = 187 \cdot 3 \text{ N}$$

Therefore from equation 18.8,

$$\eta_{\text{M}} = \frac{\text{b.p.}}{\text{i.p.}} = \frac{155}{187 \cdot 3} = 0 \cdot 828 \quad \text{or} \quad 82 \cdot 8\%$$

From the definition of b.m.e.p. given by equation 18.11, we have

$$\text{b.m.e.p.} = \eta_{\text{M}} \times \text{i.m.e.p.}$$

i.e. $$\text{i.m.e.p.} = \frac{7 \cdot 55}{0 \cdot 828} = 9 \cdot 12 \text{ bar}$$

(e) *Volumetric efficiency*, η_V

The power output of an I.C. engine depends directly upon the amount of charge which can be induced into the cylinder. This is referred to as the *breathing capacity* of the engine and is expressed quantitatively by the *volumetric efficiency*, which is defined as for reciprocating compressors by equations 9.15 and 9.16. For I.C. engines, the volumetric efficiency is the ratio of the volume of air induced, measured at the free air conditions to the swept volume of the cylinder,

i.e. $$\eta_V = \frac{V}{V_s} \qquad\qquad 18.16$$

The air volume may be referred to *N.T.P.* to give a standard comparison, and the pressure and temperature used should be stated for each evaluation.

The power output of an engine depends on its capacity to breathe, and if a particular engine had a constant thermal efficiency then its output would be in proportion to the amount of air induced. The volumetric efficiency with normal aspiration is seldom above 80%, and to improve on this figure, *supercharging* is used. Air is forced into the cylinder by a blower or fan which is driven by the engine. More will be said about supercharging in Section 18.12.

The volumetric efficiency of an engine is affected by many variables such as compression ratio, valve timing, induction and port design, mixture strength, latent heat of evaporation of the fuel, heating of the induced charge, cylinder temperature, and the atmospheric conditions.

Example 18.3

In the example 18.2 an analysis of the dry exhaust showed no oxygen and negligible carbon monoxide. The engine was tested in an atmosphere at 1·013 bar and 15°C. Estimate the volumetric efficiency of the engine.

The condition of the exhaust implies a stoichiometric air/fuel ratio which for petrols can be taken to be 14·5/1.

From example 18.2,

$$\dot{m}_f = 0.001\ 377 \text{ kg/s}$$

$$\therefore \text{ Air mass flow} = 14.5 \times 0.001\ 377 = 0.019\ 95 \text{ kg/s}$$

$$\therefore \text{ Volume drawn in, } \dot{V} = \frac{0.019\ 95 \times 0.287 \times 10^3 \times 288}{10^5 \times 1.013}$$

$$= 0.0163 \text{ m}^3/\text{s}$$

Now,

$$\text{Swept volume of engine} = ALn \text{ m}^3/\text{cycle} = \frac{ALnN}{2} \text{ m}^3/\text{min}$$

i.e. $$V_s = \frac{\pi \times 0.057^2 \times 0.09 \times 4 \times 2800}{4 \times 2 \times 60} = 0.0214 \text{ m}^3/\text{s}$$

Then using equation 18.16,

$$\eta_v = \frac{\dot{V}}{V_s} = \frac{0.0163}{0.0214} = 0.76 \quad \text{or} \quad 76\%$$

18.5 Engine output and efficiency

The power output of an engine depends on the conditions under which it is tested. In order to make reported performances acceptable, and comparable, standard procedures are established which define the quantities to be measured, the methods of measurement to be used, and the procedure to be adopted for reporting. The standards which apply in Britain are defined in British Standard (BS) 5514 Parts 1–6, 'Reciprocating internal combustion engines', 1982, which is equivalent to the corresponding standards ISO 3046 of the International Organization for Standardisation (ISO). In the United States the standards are those of the Society of Automotive Engineers (SAE) and are described in the SAE Handbook, 1984, Volume 3 or its replacement. In West Germany the standards are those of the Deutsche Industrie Norm (DIN) and in Britain also the power and torque outputs of engines are often quoted to the DIN procedures. Standards are withdrawn or updated from time to time and the different standards, between which equivalent tests now exist, may become fewer as a single international standard emerges.

The power output changes with the atmospheric conditions of pressure, temperature and humidity and if the test conditions are not those of the relevant standard the readings of power and fuel consumption taken must be corrected to the defined conditions and procedures of the standard. The power output also depends on the form in which the engine is tested with regard to the auxiliaries with which it is fitted, i.e. fully or partly equipped. Diesel engine power, torque and fuel consumption values or complete characteristics are readily available; for petrol engines maximum values of power (kW)

Table 18.1 Comparison of the Induction Conditions for Different Standards

Reference test conditions	Test procedure					
	BS 5514	DIN 6270	SAE J1349	BS AU141a	BS 649*	DIN 70020
Barometric pressure mbar	1000	981	1000	1013	1000	1013
Vapour pressure mbar	21	25	13.33	NIL	20	NIL
Air temperature K	300	293	298	293	302.4	293

* Obsolete but included for reference purposes.

and torque (N m) are quoted but fuel consumption figures are not so readily come by. For private vehicle use the overall vehicle performance figures are quoted as power and torque to the specified standard, usually DIN, and the fuel consumption figures to official tests under the Passenger Car Fuel Consumption Order 1983 in litres/ 100 km at the specified test speeds of 90 and 120 km/h (56 and 75 mph).

For example, for one particular engine the power (in kW) and torque (in N m) outputs quoted to different test standards were respectively 69 and 359 to BS 5514; 72 and 377 to DIN 70020; 79 and 369 to BS AU 141a and 72·6 and 366 to BS 649 showing a considerable variation. This range of variation of the main performance parameters indicates the necessity for a rationalization of engine test standards.

The following table, 18.2, details the power and torque performances of engines (mainly petrol) to DIN ratings and the fuel consumptions of vehicles in which they are fitted. Some basic engines appear in different forms with different carburation or fuel injection systems, have alternative forms of ignition and may even be supercharged by means of a turbocharger (see Section 18.12). Similarly, the same engine may be available in different forms of vehicle, e.g. saloon, estate or cabriolet which have different aerodynamic characteristics with corresponding vehicle performance figures. The comparison of engine power outputs by the criterion power/volume should be done with the understanding that for a given total engine capacity the small capacity multi-cylinder engine will give a higher value than an engine with fewer, larger cylinders because the volume is swept more frequently for equal piston speeds.

The last three engines listed are different versions of the same engine and are the power ratings for intermittent running. The continuous running characteristics for the same engines are shown in fig. 18.52, (page 870).

The specific fuel consumptions at maximum power vary between 200 and 250 g/kW h with somewhat lower values (by 5 to 10 g/kW h at the maximum torque condition).

The details quoted in Table 18.3 are for the smaller size diesel engines for industrial and commercial vehicle use. Engines in the higher power range, 1000 to 3000 kW, are in service in fast patrol boats, for electrical power generation on warships, on offshore oil

Table 18.2 Some Private Vehicle Engine and Road Performances (1985 Figures)

Engine size, l (cylinders)	Type and (CR)	Maximum output		kW/l	Max. bmep bar	Vehicle fuel consumption l/100 m (mpg)		
		Power kW (speed) rev/min	Torque (N m) (speed) rev/min			56 mph (90 km/h)	75 mph (120 km/h)	Simulated urban drive
0·96 (4)	Petrol (8·5)	33 (5750)	68 (3700)	34·5	8·9	5·2 (54·3)	7·1 (39·8)	7·0 (40·4)
1·6 (4)	Petrol (8·5)	58 (5800)	125 (3300)	36·3	9·45	5·6 (50)	7·6 (37)	8·9 (32)
1·6 (4)	Petrol FI (8·5)	77 (6000)	138 (4800)	48·1	9·65	6·2 (45·6)	7·9 (37·5)	10·3 (27·7)
2·3 (V6)	Petrol (9·0)	84 (5300)	176 (3000)	36·5	9·6	6·8 (41·5)	8·7 (32·5)	12·6 (22·4)
2·3 (4)	Diesel (22·2)	49 (4200)	139 (2000)	21·2	7·6	5·05 (56)	6·7 (42·2)	8·35 (33·8)
2·8 (V6)	Petrol FI (9·2)	110 (5700)	216 (4000)	39·3	9·7	7·75 (36·5)	10·2 (27·7)	14·9 (19)
2·8 (V6)	Petrol TC (9·2)	151 (5000)	353 (3500)	54	15·8	—	—	—

FI – fuel injected; TC – turbocharged; CR – compression ratio; l – litres

Table 18.3   Some Diesel Engine Performances to BS 5514/DIN 6270 (1985 Figures)

Capacity (l)	Cylinders	CR	Breathing	Max. power kW at speed (rev/min)	Max. torque N m at speed (rev/min)	Max. bmep atm.	Power kW/l
2·9	3	17·3	Normal	29 at 2000	158 at 1600	7	9·9
3·3	3	16·3	Normal	43 at 2200	211 at 1400	8·3	13
4·4	4	16·3	Normal	57·5 at 2100	284 at 1600	8·6	13
4·4	4	15·6	Turbocharged	68·7 at 2100	359 at 1600	11	15·6
6·6	6	16·3	Normal	90 at 2300	436 at 1400	7·8	13·6
6·6	6	15·6	Turbocharged	103 at 2200	500 at 1600	11	15·6
6·6	6	15·6	Turbo + intercooling	130 at 2200	665 at 1600	13	19·7

CR – compression ratio; l – litres

rigs, in rail traction including high speed trains and in submarine applications.

Diesel engine competition with the petrol engine for the private car market is an interesting study exercise. The principles of the diesel engine were well known at the start of the century and prior to the 1939 war a few diesel engined cars had been built as conversions from petrol engines. The advantages of the diesel of better fuel consumption and longer engine life were well known particularly for use in taxis which are used for short journeys and have high annual mileages. For such applications the savings in fuel cost favours the diesel engine but its relative progress has been slow. In the mid 1970s only about 300 diesel cars were in use in the U.K., but the number rose to about 6000 in 1980 and reached 14 500 in 1982. The performance of an early (1974) diesel engined vehicle compared with that of a smaller petrol engine e.g. a 1·8 l diesel compared with a 1·3 l petrol engine but gave similar journey times and overall fuel costs of about half those of the petrol engine. The noise level of the diesel engine was not acceptable and the diesel was regarded as being sluggish in comparison with the petrol engine. Modern diesel engine design has narrowed the gap between the two with engines becoming lighter, quieter and more responsive. The better fuel consumption of the diesel makes it an attractive power unit for the taxi and vehicle fleet owner, provided the price of diesel fuel remains sufficiently lower than that of petrol. Another important factor is the frequency and cost of service work required by the diesel and petrol equivalents. Diesel engines required more frequent servicing but the intervals between servicing are approaching equality.

18.6 Performance characteristics

The testing of I.C. engines consists of running them at different loads and speeds and taking sufficient measurements for the performance criteria to be calculated. As well as the measurements required for the criteria of Section 18.4 the air flow is required to give the air/fuel ratio and the combustion products can be analysed using the Orsat apparatus, or one of the more elaborate methods (see Section 15.8).

An energy balance is sometimes presented for an engine, and the heat taken by the cooling water is obtained by measuring the rate of flow of the water and its temperature rise. The outlet temperature of the cooling water is usually limited to about 80°C to prevent the formation of steam pockets. To estimate the energy of the exhaust gas, an exhaust calorimeter can be fitted; this is simply a heat exchanger in which the exhaust gas is cooled by circulating water, the rate of flow and temperature rise of which are measured. In order to avoid condensation of the steam in the gas, the gas is not usually cooled below about 50°C.

The items usually included in an energy balance and expressed as percentages of the energy supplied by the fuel (i.e. $\dot{m}_f \times Q_{net, v}$), are as follows:

(a) b.p.; (b) the heat to cooling water; (c) the energy of the exhaust referred to inlet conditions, or as obtained by an exhaust calorimeter; (d) unaccounted losses obtained by difference, and which include radiation and convection losses, etc.

The energy balance usually presented is not an accurate account of the energy distribution, but it is a useful one. The b.p. is conveniently and accurately measured and the percentage of the input energy to the b.p. is the most important item in the balance. The heat transferred to the cooling water can be used as an indication of how much heat could be usefully obtained if this water is used for other heating purposes in a combined plant.

The energy to exhaust is best obtained by means of the exhaust calorimeter as described. Ideally the exhaust gas should be cooled to the temperature of the inlet air, and the heat taken by the cooling water in the calorimeter per minute would give item (c) of the balance. The temperature at which the exhaust gas enters the calorimeter is most likely not that at which it passes through the exhaust valve, and some of the energy to the exhaust will have been taken by the cooling water or lost to the atmosphere. To obtain item (c) by calculation is more speculative since the gas is chemically different from the inlet air and the mass flow has increased due to the addition of the fuel. The error involved in choosing the datum is likely to be less than that produced by using an inaccurate value for the exhaust temperature, which is not easy to obtain with accuracy. It is sufficient to write

$$\text{Energy to exhaust} = (\dot{m}_a + \dot{m}_f)h_e - \dot{m}_a h_a$$

(where $\dot{m}_a$ and $\dot{m}_f$ are the air and fuel mass flows, h_e is the enthalpy of the exhaust gas (dry exhaust + steam), reckoned from 0°C, and h_a

is the enthalpy of the air at inlet reckoned from 0°C). A suitable value for c_p for the dry exhaust gas must be calculated or assumed.

For a diesel engine at full load typical values would be: to b.p. 35%; to cooling water 20%; to exhaust 35%; to radiation, etc., 10%. The heat to the jacket cooling water and exhaust can be utilized in industries which have heating loads such as space heating and hot water systems, and which require either steam or hot water for process work. The heat to the jacket water is recoverable and about 18% of the total energy supplied can be recovered from the exhaust gas.

The most elementary power test is that which gives the power/ speed and the torque/speed characteristics, as shown in fig. 18.14. The test is carried out at constant throttle setting in the petrol engine, and at constant fuel pump setting in the C.I. engine.

In fig. 18.14 are shown the engine power characteristics for both i.p. and b.p. As the speed increases from the lower values the two curves are similar, the difference between the i.p. and the b.p. at any speed being the f.p., which increases with speed. Both curves show maximum values, but they occur at different speeds. The i.p. falls after the maximum because of a reduction in volumetric efficiency with increased speed. This is influenced by gas temperatures, valve timing, valve mechanism dynamics, and the pressure pulsation patterns in the induction and exhaust manifolds. This fall in volumetric efficiency affects also the b.p. curve, but this is further decreased by

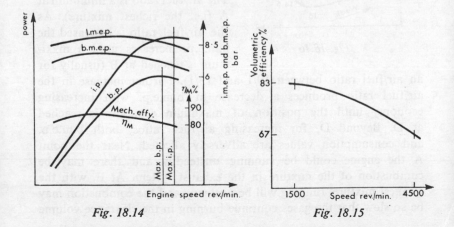

Fig. 18.14   Fig. 18.15

an increase in the f.p. This latter effect is predominant since the b.p. reaches its maximum at a lower speed than the i.p.

The variation of volumetric efficiency with speed is indicated in fig. 18.15, and that of mechanical efficiency with speed in fig. 18.14. The various methods used for the determination of mechanical efficiency are discussed in Section 18.4.

(i) *S.I. engines*

These engines are *quantity governed* by the opening or closing of a throttle valve which regulates the mass flow of charge to the cylinders. Some gas engines are throttled by alteration of the lift of the admission valve, and this can be controlled from the engine governor. The governed speed can be adjusted to select any value in its range.

The petrol engine will operate on air/fuel ratios in the range 10/1 to 22/1, but not necessarily satisfactorily at the extremes. There is some variation between engines. An important test is to run the engine with the air/fuel ratio as the only variable. This is carried out at constant speed, constant throttle opening, and constant ignition setting. The specific fuel consumption is plotted to a base of b.m.e.p. and a 'hook curve' or consumption 'loop' is obtained. For a single-cylinder engine at full throttle the curve is sharply defined as in fig. 18.16. The air/fuel ratio is a minimum at A (i.e. the richest mixture). As the air/fuel ratio is increased the b.m.e.p. increases until a maximum is reached at B (usually for an air/fuel ratio between 10/1 and 13/1). Further increase in the air/fuel ratio produces a decrease in b.m.e.p. with increasing economy until the position of maximum economy is reached at D. Beyond D, for increasing air/fuel ratios, both b.m.e.p. and consumption values are adversely affected. Near the point A the engine could be running unsteadily and there may be combustion of the mixture in the exhaust system. At E, with the weakest mixture, running will be unsteady and the combustion may be so slow that the gases continue burning in the clearance volume

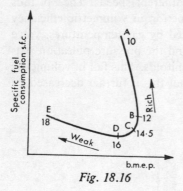

Fig. 18.16

until the next induction stroke begins; this causes *popping back* through the carburettor. Point C is the point of chemically correct or stoichiometric air/fuel ratio, and is about 14·5/1. The mixture strengths range between those at B and D, which are for maximum power and maximum economy respectively. The indicator diagrams corresponding to mixtures B, C, and D are shown in fig. 18.17.

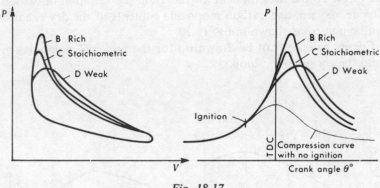

Fig. 18.17

For multi-cylinder engines the consumption loops are less distinct, but are generally similar in shape to that for the single-cylinder engine. This is also true for tests made at part throttle openings. A series of readings obtained at different throttle positions at constant speed is shown in fig. 18.18.

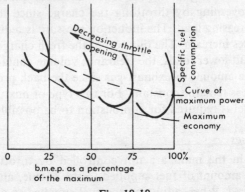

Fig. 18.18

In the above tests the ignition has been assumed to be constant, but other tests can be included to show the effect of ignition timing on the consumption loop. Alternatively the ignition setting can be adjusted at each mixture strength to give maximum power at the speed of the test; by this means the rate of pressure rise on combustion can be kept approximately constant.

B.m.e.p. and s.f.c. may be plotted against air/fuel ratio as shown in fig. 18.19. To the same base of air/fuel ratio the variation of carbon dioxide, oxygen, and carbon monoxide contents of the dry exhaust can be plotted as shown in fig. 18.20.

Energy balances can be drawn up for the principal points taken from the consumption loop.

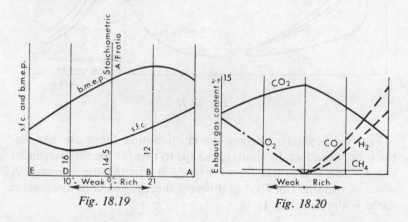

Fig. 18.19   Fig. 18.20

Testing the S.I. engine at part load shows the deficiency of the method of governing by throttling the charge since the efficiency falls with decreasing load. The induction pressure is reduced and the pumping losses increase. The dilution of the fresh charge by exhaust gas increases at lower loads, the clearance volume containing practically the same amount of exhaust gas since the back pressure in the exhaust process remains constant. For this type of mixture a greater amount of fuel is required for combustion to be possible.

(ii) *C.I. engines*

C.I. engines in the main are not controlled by throttling but by adjusting the amount of fuel supplied to the engine, and hence are *quality governed*. When adjusting the fuel supplied to a C.I. engine the limiting condition is given by the *smoke limit,* which is the appearance of black smoke in the exhaust. Engines should not be operated with mixtures rich enough to produce smoke, although such a mixture may give a greater power output. The efficiency under these conditions is low, and the engine soon becomes dirty. The smoke limit occurs at air/fuel ratios of about 16/1. The engine is tested at different speeds to the smoke limit, which can be observed visually or measured by a smoke meter. The values of torque, b.p., fuel consumption, and specific fuel consumption are then plotted against engine speed in rev/min.

A consumption loop for a C.I. engine has the form shown in fig. 18.21. This shows a minimum s.f.c. and therefore a maximum brake thermal efficiency, at part load (i.e. less than the maximum b.m.e.p.). The curve is reasonably flat over a wide range of values of b.m.e.p., which shows the virtue of the C.I. engine compared with the S.I. engine for part load operation, a condition often prevailing in road vehicle engines. The reduction in the thermal efficiency at part load is less for the C.I. engine than for the S.I. engine.

Mechanical efficiencies and minimum values of s.f.c. are plotted against speed to provide further characteristics.

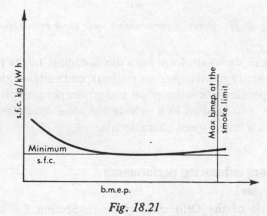

Fig. 18.21

For a given engine the individual characteristics described in this section are usually sufficient for normal use but if comparisons are to be made with other engines, perhaps of different types, then the more comprehensive presentation by means of a performance map is an advantage. Fig. 18.22 shows the form of such a map for a petrol engine

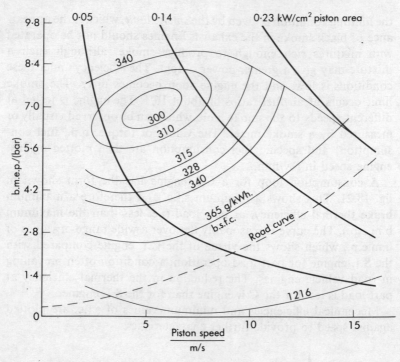

Fig. 18.22   Form of performance map for a petrol engine

and fig. 18.23 shows the form for a diesel engine. In the maps bmep (bars) is plotted against engine (or piston) speed and includes curves of constant specific fuel consumption and power per unit piston area. If the engine is to be used in a vehicle the road requirement can be included as a bmep/speed characteristic.

18.7 Factors influencing performance

(i) *S.I. engines*

The analysis of the Otto cycle given in Section 6.6 showed the dependence of thermal efficiency on compression ratio. A graph of air standard thermal efficiency against compression ratio is shown in fig. 18.24. This graph indicates the form engine development should take, and over the early years increases in compression ratio were made. However, between 1960 and 1985 the compression ratios have not increased greatly and are in the range 9–10 for production

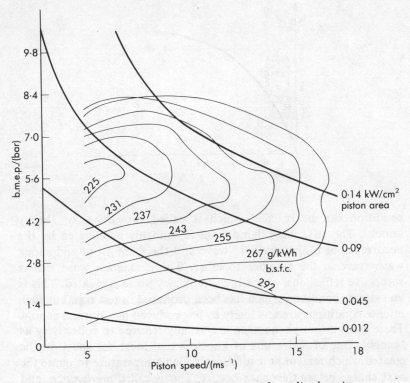

Fig. 18.23   Form of performance map for a diesel engine

vehicles in the U.K. The ability to use higher ratios has depended on the provision of better quality fuels and of improved designs of combustion chamber. The main features of the combustion chamber are the distances to be travelled by the flame after initiation of combustion, and the gas flow pattern established.

It is evident that if a petrol/air mixture is compressed sufficiently it will ignite spontaneously. This suggests one limit to compression ratio if controlled combustion is to be obtained from spark ignition. However, before this limit is reached for the whole charge, spontaneous ignition can occur in the unburnt charge after combustion has commenced normally. The unburnt gas, compressed by the advancing flame front is raised in temperature and may reach the point of *self-ignition*. This produces an uncontrolled combustion and its occurrence may be heard as a knocking sound. A critical

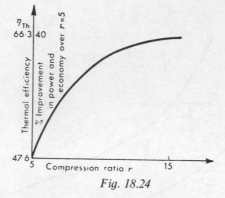

Fig. 18.24

condition can be reached which is called *detonation*, or 'heavy knock'. The advancing flame front is suddenly accelerated by the occurrence of a high-pressure wave and the flame front and shock wave traverse the cylinder together. The *detonation wave* suffers successive reflections, and a high-frequency noise is created. This is an extreme condition which has been produced in test rigs, but such intense conditions are less likely to be produced in a normal engine. These combustion phenomena are usually referred to collectively as '*knock*'. One of the results of knock is that local hot spots can be created which remain at a sufficiently high temperature to ignite the next charge before the spark occurs. This is called *pre-ignition*, and can help to promote further knocking. The result is a noisy, overheated, and inefficient engine, and perhaps eventual mechanical failure. The chemical behaviour during this type of combustion is still not fully understood, although a considerable amount of empirical data is available. The pressure/crank angle diagram for normal combustion is shown in fig. 18.25 with maximum pressure occurring at 10° to 12° after TDC, and a rate of pressure rise of 1·38 bar per degree of crank angle, with a compression ratio of 8/1. The spark occurs at the point S on the normal compression curve, but there is a *delay period* between the occurrence of the spark and a noticeable departure of the pressure curve from that of normal compression. This is a time delay which is independent of engine speed so that as the engine speed is increased the point S must occur earlier in the cycle to obtain the best position of the peak pressure. This *ignition advance* can be accomplished manually, but in practice is controlled automatically by a mechanism in the distributor which is sensitive

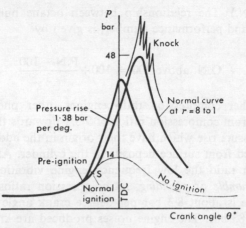

Fig. 18.25

to engine speed; an additional control is sometimes obtained at small throttle openings by a pressure connection from the distributor to the induction manifold.

The compression ratio which can be utilized depends on the fuel to be used and a scale has been developed against which the knock tendency of a fuel can be rated. The rating is given as an *octane number*. The fuel under test is compared with a mixture of iso-octane (high rating) and normal heptane (low rating), by volume. The octane number of the fuel is the percentage of octane in the reference mixture which knocks under the same conditions as the fuel. The number obtained depends on the conditions of the test and the two main methods in use (the *research* and the *motor* methods; see reference 18.3) give different ratings for the same fuel. The motor test is carried out at the higher temperature and gives the lower rating. The difference between the two is taken as a measure of the temperature sensitivity of the fuel. High octane fuels (up to 100) can be produced by refining techniques, but it is done more cheaply, and more frequently, by the use of anti-knock *additives*, such as *tetraethyl lead*. (An addition of 1·1 cm³ of tetraethyl lead to one litre of 80 octane petrol increases the octane number to 90.) Fuels have been developed which have a higher anti-knock rating than iso-octane and this has led to an extension of the octane scale. Aviation conditions of operation lead to another scale which gives a better indication of the detonation characteristics; this is the *performance number* (P.N.). The relationship between octane number (O.N.), above 100, and performance number, is given by

$$O.N. \text{ above } 100 = 100 + \frac{P.N. - 100}{3}$$

With higher compression ratio engines other phenomena are observed. From compression ratios of 9·5/1 upwards there are high rates of pressure rise which have their origin in the additional flame fronts started from surface deposits in the cylinder. At about 9·5/1 compression ratio the low frequency engine vibrations produced are called *rumble* or *pounding*. At compression ratios of 12/1 the pressure rise is about 8·3 bar per degree crank angle with a peak pressure of 83 bar. The engine noises produced are known as *thud* or *pressure rap*; surface ignition is not present and fuel characteristics have little influence. This field of development is one which brings many new problems which are more likely to be solved by the chemist than the engineer.

(ii) *C.I. engines*

The effect of compression ratio in the C.I. engine is somewhat simpler than in the S.I. engine. For combustion to occur at the temperature produced by the compression of the air a compression ratio of 12/1 is required. The efficiency of the cycle increases with higher values of compression ratio and the limit is a mechanical one imposed by the high pressures developed in the cylinder, a factor which adversely affects the power/weight ratio. The normal range of compression ratios is 13/1 to 17/1, but may be anything up to 25/1.

The combustible mixture in the S.I. engine is formed before compression, but with the C.I. engine this mixture has to be formed after compression and after injection begins. This leads to delay periods in the C.I. engine which are greater than those in the S.I. engine (see fig. 18.26). The fuel droplets injected have to evaporate and mix with oxygen to give a combustible mixture. The delay period forms the first phase of the combustion process, and is dependent on the nature of the fuel. The second phase consists of the spread of flame from the initial nucleus to the main body of the charge. There is a rapid increase in pressure during this phase and the rate of pressure rise depends to some extent on the availability of oxygen to the

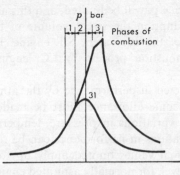

Fig. 18.26

fuel spray, which in turn depends on the turbulence in the cylinder. The main factor, however, is that of the delay period. A long delay period means more combustible mixture has had time to form, and so more charge will be involved in the initial combustion. As the speed increases the rate of pressure rise in this phase also increases. This is because the delay period is a function of time if surrounding conditions remain constant, and at the higher engine speeds more mixture will be formed in the delay period. The initial rapid combustion can give rise to rough running and a characteristic noise called *diesel knock*. During the third phase of combustion the fuel burns as it is injected into the cylinder, and this phase gives more controlled combustion than that of phase two. One of the main factors in a controlled combustion is the swirl which is induced by the design of the combustion chamber. It has been stated that the delay period depends on the nature of the fuel, and a fuel with a short delay period, or high ignitability, is required. The *ignitability* of a fuel oil is indicated by its *cetane number*, and the procedure for obtaining it is similar to that for obtaining the octane number of petrols. Reference mixtures of cetane ($C_{16}.H_{34}$) (high ignitability), and α-methyl-naphthalene ($C_{11}.H_{10}$) (low ignitability), are used. The mixture is made by volume and the ignitability of the test fuel is quoted as the percentage of cetane in the reference mixture which has the same ignitability. For higher speed engines the cetane number required is about 50, for medium speed engines about 40, and for slow speed engines about 30. For details of the standard tests see reference 18.3.

The air/fuel ratios used in C.I. engines lie between 20/1 and 25/1. As these mixtures are much weaker than the stoichiometric proportion then the i.m.e.p. will be limited, and this also means that for a given fuel consumption the swept volume of the engine will be greater than that of the equivalent S.I. engine. For more detailed work on the combustion process in I.C. engines see references 18.1 and 18.4.

Engines are affected in performance by the atmosphere in which they operate and some allowance must be made on performance figures quoted for variations in pressure, temperature, and relative humidity. The variations in performance can be represented graphically, but the normal values quoted apply up to 30°C and 150 m altitude from sea level, for normally aspirated engines. The reduction in output per 300 m of altitude above 150 m is about 3%, and for every·5 K above 30°C the reduction is also about 3%. The reduction in output for changes in relative humidity can be up to 6% depending on the conditions.

18.8 Comparison of real cycles with the air standard cycle

Some of the differences between practical and ideal cycles were discussed as an introduction to this chapter. Other differences are evident from a comparison of the indicator diagram, such as the rounding off of the corners of the diagram due to valve throttling and the fact that the combustion process is not truly at constant volume or at constant pressure. The working fluid is not air throughout; it starts off the combustion process as a mixture of air and fuel vapour, and this mixture changes as combustion proceeds. One of the properties of a perfect gas is that its specific heats remain constant, but in the I.C. engine there is a considerable change in these values due to the high temperatures reached. At the higher temperatures dissociation takes place (see Section 15.8), and the work output is reduced. The maximum pressure and temperature reached in the practical cycle are much less than those obtained by calculation based on the theoretical cycle, assuming the same heat input. Heat losses from a heat engine must be avoided but with a real engine some form of cooling is essential so that temperatures will not be reached such that there will be failure of the engine materials. The compression and expansion processes are thus not adiabatic but are of the form $pv^n = $ constant; the assumption of internal reversibility is nevertheless a good approximation. A comparison of the

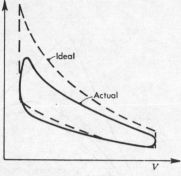

Fig. 18.27

real cycle and the ideal cycle is shown in fig. 18.27 for a petrol engine; the pumping loop of the practical cycle has been omitted.

A most important criterion is the temperature (and hence pressure) which can be attained by the 'constant volume' combustion phase. Fig. 18.28 shows the temperatures reached on the combustion of mixtures of fuel and air of different strengths between 50% weak and, for a compression ratio of 5 to 1, 50% rich. The maximum temperature should theoretically be produced by a stoichiometric mixture strength but in fact it occurs at about 20% rich. The shape of the temperature curve is due to the dissociation of CO_2 and H_2O into CO, H_2 and O_2 which is slight at a temperature of 2000°C but increases rapidly above that. In engine combustion the dissociation of CO_2 affects engine behaviour more than that of H_2O which is relatively slight. Combustion theory, as described in Chapter 15, allows the dissociated proportions of CO, H_2 and O_2 to be calculated and represented in the combustion equations. Hence the energy released on combustion can be calculated. If the formation of nitrogen oxide which occurs at the higher temperatures is taken into account the temperatures attained are even lower, due to the absorption of energy to form NO as shown in fig. 18.28. At the higher compression ratios of modern engines higher temperatures and pressures are attained and dissociation occurs, not only to a greater extent for NO, but also to some common radicals mainly OH, H and O, e.g. at 3000°C a stoichiometric mixture of octane and air at equilibrium will contain OH (1·4%); H (0·3%), O (0·3%) and NO (0·3%).

Mixture strength, particularly for a spark-ignited engine can vary considerably due to the way in which the mixture is created and

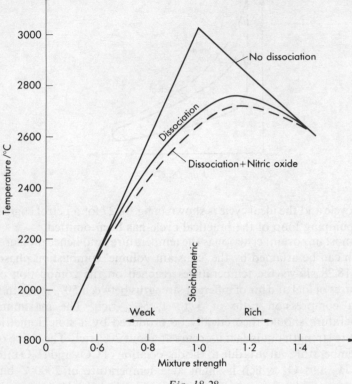

Fig. 18.28

distributed. It is not a controlled process as it is in the diesel engine. One of the results of this is a considerable variation in the cycle to cycle and cylinder to cylinder performance of petrol engines.

An important parameter which is open to selection is that of the spark timing as described in Sections 18.6 and 18.7. Fig. 18.29 shows the form of the variation of the maximum pressure reached in the cylinder with the advance of the ignition before the top dead-centre (TDC) position.

The factors dealt with above indicate the complexity of the processes of the internal combustion engine that give the engine a character which has fascinated many distinguished engineers over lifetimes of service and continues to do so. Calculation becomes more complex as the information necessary is increased manifold, the relationship between data is involved and the time taken for

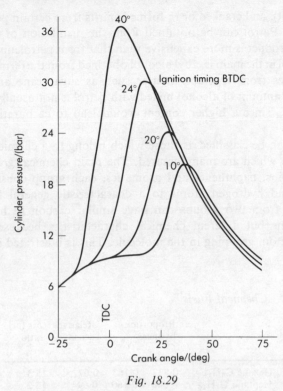

Fig. 18.29

computation increases rapidly. To this end charts have been prepared from established information to describe the behaviour of different fuels at different mixture ratios including the effects of dissociation and different proportions of recycling exhaust gas etc. (see reference 18.11). In serious work this information will be included as data stored in a computer ready for use in the analysis or computation of different engine cycles and running conditions.

18.9 Properties of fuels for I.C. engines

Some of the requirements of the fuels used in I.C. engines have been indicated by the earlier sections of this chapter, and these will be added to in this section.

Fuels suitable for petrol engines are obtained from several sources. The majority are obtained from petroleum; these consist of straight petrol distilled from crude petroleum, natural gasoline (which is a light spirit), and cracked or re-formed spirits from certain petroleum fractions. Petrol can be obtained from the distillation of oil shale but the product is more expensive than that from petroleum.

Alcohol in the main is ethyl alcohol obtained from the fermentation of residues from vegetable matter such as sugar cane and sugar beet. The amount of alcohol mixed with petrol is not usually greater than 20%, since a higher content would lead to carburation difficulties.

Fuels can be classified as those which belong to a chemical series and those which are manufactured. The main chemical groups are the paraffins, naphthenes and aromatics, each group consisting of carbon and hydrogen atoms to a characteristic general formula. Members from two groups can have similar carbon to hydrogen proportions but different chemical characteristics because of the different atom grouping in the molecule. This is illustrated in Table 18.4.

Table 18.4   Chemical Fuels

Fuel group and general formula	Fuel	Proportions by mass		Relative density	Air/fuel ratio	LCV kJ/kg
		C	H			
Paraffins C_nH_{2n+2}	Hexane C_6H_{14}	0·837	0·163	0·67	15·3	45 124
	Heptane C_7H_{16}	0·840	0·160	0·69	15·2	44 660
Naphthenes C_nH_{2n}	Cyclo-hexane C_6H_{12}	0·857	0·143	0·79	14·7	43 030
	Cyclo-heptane C_7H_{14}	0·857	0·143	0·78	14·7	43 960
	Cyclo-octane C_8H_{16}	0·857	0·143	0·74	14·7	43 960
Aromatics C_nH_{2n-6}	Benzene C_6H_6	0·923	0·077	0·88	13·2	40 700
	Toluene C_7H_8	0·913	0·087	0·87	13·4	41 054
	Xylene C_8H_{10}	0·905	0·095	0·95	13·6	41 868

The common manufactured fuels are as follows:

Petrol: a refined distillate of petroleum; the boiling point is not usually greater than 200°C.

Kerosene: a refined distillate of petroleum which boils between 150°C and 300°C.

Benzole: a refined distillate of coal tar which consists mainly of benzene and toluene.

Gas oil: an unrefined distillate after kerosene fractions have been removed.

Diesel fuel: mixtures of gas oils from various crude oils, or blends with fuel oils.

Blends of alcohol: petrol with about 15% alcohol and 15% benzole.

Kerosene is commonly called 'paraffin', but as this is the name of a chemical series as mentioned earlier, and as the commercial paraffin contains members of other series, the name kerosene is to be preferred. Petrol contains members from several different series and special additives. The actual content of the straight petrols depends on the geographical origin of the petroleum.

The latent heat of evaporation of fuels is an important quantity in both C.I. and S.I. engines. With S.I. engines this factor together with the volatility of the fuel determines the density of the charge in the cylinder; the density is inversely proportional to the absolute temperature in the cylinder at the closing of the inlet valve. This temperature depends on the latent heat of the fuel and the amount of heating which takes place during induction. Most liquid fuels have similar latent heats, the exception being alcohol which has a high latent heat, a higher density of charge, and a marked increase in power compared with other fuels. In C.I. engines the latent heat has no bearing on volumetric efficiency due to the fact that the fuel is injected after compression; this is detrimental when the calorific value and delay period are considered since the latent heat necessary for evaporation must come from the compressed air.

Fuel *volatility* is of importance in S.I. engines from the point of view of ease of starting and the heating required to provide an even distribution of mixture to different cylinders.

In C.I. engines the volatility of the fuel influences the time taken for a combustible envelope to form round the fuel droplets, and hence influences the delay period.

The calorific value of a fuel is important in determining the amount of fuel required, but does not indicate how much power can be obtained from it. Engines using gaseous fuels are affected more by variation in calorific value than petrol and diesel engines. The power output depends on the heating value of the mixture of fuel and air and the stoichiometric mixtures of all hydrocarbon fuels give the same heating value per m^3 at the same reference, temperature, and pressure, within close limits.

Tetraethyl lead has been mentioned as an anti-knock additive in petrols. Other additives are oxidation inhibitors, corrosion inhibitors, combustion control compounds and anti-icing additives to counteract freezing in the carburettor.

The properties of fuels are determined by standard tests the procedure for which may be obtained from reference 18.3.

18.10 Fuel systems

The purpose of an engine fuel system is to provide the cylinder with a mixture of air and fuel in the correct proportions for the engine requirements at any particular instant. There are basically two methods available, one is called *carburation* and is used for petrol engines and the other is a type of *fuel injection* which is a characteristic method for diesel engines. There are many different designs of each and only the basic principles will be dealt with in this book.

The petrol engine for automotive purposes has been developed on the basis of the carburettor although petrol injection has been used for a long time for aircraft and special engines, such as military vehicles and racing cars, for which cost is not a main consideration. The carburettor is a simple, cheap device which has served its purpose for many years but the trend to higher powered, multi cylinder engines has shown the single carburettor system to be inadequate. As a consequence multi-choke carburettors and twin or triple carburettor layouts have been used to meet increasingly sophisticated engine requirements but an active interest has developed in the potential of several proposed designs of petrol injection systems.

The fundamentally different methods of charge ignition by spark and by compression in the petrol and diesel engines respectively have dictated different means of fuel supply for the two engines. However, the features of the oil injection system from the point of view of control, accurate fuel metering and good fuel consumption characteristics in comparison with the carburated petrol engine with its poor fuel consumption, particularly at part load, have created the belief that ultimately petrol injection would be the preferred method if a sufficiently simple system could be produced at a cost which would make it competitive with the carburettor. In recent years the added requirement for engines to meet exhaust gas emission regulations (see

Section 18.13) has increased the demand for accurate fuel metering and engine control. These factors have increased the interest in petrol injection although the earlier regulations were more satisfactorily met by carburated engines than by those with the first petrol injection systems. In general the fundamental problem is to measure, or compute accurately, the mass flow rate of air into the engine at any instant and to mix the correct amount of petrol into it in such a way that the air and fuel mixture produced is right for the engine running condition. If the air flow is steady and at constant temperature and pressure the problem is fairly straightforward but in an engine air flow rates, pressures and temperatures change and the engine is in dynamic operation with phases of acceleration, deceleration and over-run (when the throttle is closed and the vehicle wheels are turning the engine). Thus a fuel supply system, to be successful, must be able to cope with a wide range of running conditions and demands.

(a) Carburation

The term carburation covers the whole process of supplying continuously to a petrol engine a mixture of vapourized fuel and air which is suitable to each engine condition of load, speed, and temperature. In Section 18.5 it was seen that the air/fuel ratio varies between maximum economy and maximum power conditions, and the stoichiometric air/fuel ratio is not adequate to all demands. The function of the carburettor is to measure out the correct proportions of liquid fuel and air for the particular engine condition. The liquid fuel must be 'atomized' at the carburettor (i.e. broken up into a fine spray to assist in the evaporation of the fuel, so that the mixture entering the cylinders is homogeneous). The necessary latent heat of vaporization must be supplied to the mixture and this is not the function of the carburettor but of the whole induction tract. The metering process is carried out at the carburettor but the actual mixture ratio, its condition and distribution between cylinders, depends also on the design of the complete induction system and the temperatures therein.

The perfect carburettor would supply the air/fuel ratio required at all speeds and throttle openings no matter what the climatic conditions or the rate at which the demand was changing. A consideration of all the factors which influence the final mixture burned in the cylinder (i.e. engine condition required, mechanical characteristics of the engine, the physical differences between the constituents of the mixture, the rapid fluctuation in demand, and the temperature and humidity variation) shows the difficulty in obtaining the ideal carburettor. The carburettor is a highly developed component which is produced in a number of different designs each having its own refinements, with the object of supplying to the engine the air/fuel mixture it needs. It is not the object of this short treatment to discuss the different designs in detail. For further work on this subject the reader is referred to reference 18.5. The elementary metering process will be considered, since this is of fundamental importance to all types.

In fig. 18.30 a simplified carburation system is shown. The petrol pump, either electrically driven or mechanically driven by the crankshaft, pumps petrol from the tank to the float chamber of the carburettor. The function of the float chamber is to maintain a constant

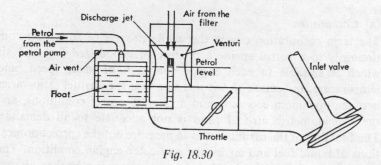

Fig. 18.30

level of petrol in the chamber by shutting off the supply from the pump when this level is about to be exceeded. The float chamber is vented to atmosphere through a small hole in the cover, hence the pressure on the surface of the petrol is constant and equal to that of the atmosphere. The air is induced by the depression created by the piston moving downwards in the engine cylinder, and after passing through a filter, enters the carburettor at about atmospheric pressure. The petrol engine is quantity governed which means that when less power is required at a particular speed the amount of charge delivered to the cylinders is reduced. This is achieved by means of a throttle valve of the butterfly type which is situated in the air inlet. The air on induction enters the venturi or choke tube. This is a tube of decreasing cross-section which reaches a minimum at the throat or choke of the venturi, which is shaped to give the minimum resistance to the air flow. The petrol discharge jet is situated at the throat and is subject to the air pressure there. The pressure at the throat is below atmospheric since the air velocity has been increased from that at inlet to the carburettor to a maximum at the throat. Thus the two petrol surfaces, that in the float chamber and that at the discharge jet, are subject to different pressures. This pressure difference acting on the petrol column causes the petrol to flow into the air stream, and the rate of flow is controlled or metered by the size of the smallest section in the petrol passage. This is provided by the main jet and the size of this jet is chosen to give the required engine performance, and is an empirical selection. The pressure at the throat at the fully open throttle condition lies usually between 38 and 50 mm Hg below atmospheric, and seldom exceeds 76 mm Hg below atmospheric.

In the carburettor described the choke or throat has a constant area and the pressure changes with throttle opening and engine speed. It is referred to as the *fixed choke* type. Another type, of which the S.U. carburettor is the main example, is the *constant vacuum* type. In this type the pressure at the throat and thus the air velocity, remains constant, but the area of the throat is varied. Similarly the area of the petrol orifice or jet must also vary, and this is achieved by means of a tapered needle, attached to a piston. The needle moves in the orifice thus forming a discharge annulus for the petrol (see fig. 18.31).

With the elementary form of carburettor described, the mixture would become richer as the air flow increased, and this must be 'corrected'. In the modern carburettor this is done by fitting the main metering jet about 1 in below the petrol level, and it is called a submerged jet (see fig. 18.32). The jet is situated at the bottom of a well, the sides of which have small holes which are in communication with the atmosphere. The second object of this arrangement is to achieve atomization which is not obtained with the elementary system described previously. Air is drawn through the holes in the well, the petrol is emulsified, and the pressure difference across the petrol column is not as great as in the elementary carburettor. On starting, the petrol in the well is at the level of that in the float chamber. On opening the throttle this petrol, being subject to the low throat pressure, is drawn into the air. This continues with decreasing

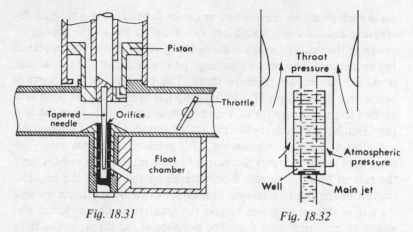

Fig. 18.31   Fig. 18.32

mixture richness as the holes in the central tube are progressively uncovered. Normal flow then takes place from the main jet.

A petrol engine operates in service mainly at part throttle and a carburettor which gives a correct full throttle delivery would not be able to meet part throttle requirements. As the throttle is closed the point of lowest pressure in the induction system moves from the venturi, at full throttle, to the engine side of the throttle when the engine is idling with the throttle practically closed. The position of minimum pressure changes between these limits. At the idling condition the pressure in the induction tract is 400 to 460 mm Hg below atmospheric. The main jet will not supply petrol when the engine is idling because there is no longer sufficient depression at the throat. Another petrol supply is provided which delivers to the engine side of the throttle. This is indicated diagrammatically in fig. 18.33.

A rich mixture is required for starting and idling, but as the throttle is opened the demand is for a weaker mixture. There is a merging of the idling and main jet deliveries, but in order to avoid flat spots and be able to deliver the required mixture throughout the range of throttle positions, it is usual to provide a third 'progression' system. On acceleration the throttle is opened suddenly and it is necessary to provide a rich mixture. If the part throttle characteristics of the carburettor are satisfactory the fuel for acceleration is best provided, in the interests of economy, by a separate accelerating device. The means of doing this vary and the types may be classified as using an *accelerating pump*, or an *accelerating well* (see reference 18.5).

Calculation of the air/fuel ratio for a simple carburettor

The most elementary arrangement as shown in fig. 18.34 will be considered. The air from the atmosphere is induced through the

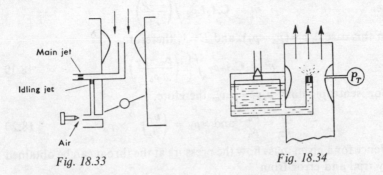

Fig. 18.33　　　　　　　　　　　*Fig. 18.34*

carburettor by the pressure difference across it created when the piston moves on its induction stroke. The velocity of the air increases as it passes through the venturi and reaches a maximum at the choke. The pressure also changes and is minimum at the choke if the choke is the minimum area in the induction tract. This will be so unless the throttle is closed sufficiently to present a smaller area to the flow than that of the choke. If the flow process between the atmosphere and the choke is isentropic, then the velocity of flow of air through the choke will be given by equation 10.30,

i.e.
$$C_2 = \frac{\dot{m}v_2}{A_2} = E\sqrt{\left[2T_1 c_p\left\{1 - \left(\frac{p_2}{p_1}\right)^{(\gamma-1)/\gamma}\right\}\right]}$$

The mass flow of air is given by equation 10.32,

i.e.
$$\dot{m} = \mathbf{C}A_2 E\rho_1\sqrt{\left[2c_p T_1\left\{\left(\frac{p_2}{p_1}\right)^{2/\gamma} - \left(\frac{p_2}{p_1}\right)^{(\gamma+1)/\gamma}\right\}\right]}$$

In this application $p_2 = p_T$, $p_1 = p_a$, and $\rho_1 = \rho_a$; the approach velocity is negligible also, therefore $E = 1$;

$$\therefore \quad C_T = \sqrt{\left[2T_a c_p\left\{1 - \left(\frac{p_T}{p_a}\right)^{(\gamma-1)/\gamma}\right\}\right]} \qquad \text{18.17}$$

and,

$$\dot{m} = \mathbf{C}A_T\rho_a\sqrt{\left[2c_p T_a\left\{\left(\frac{p_T}{p_a}\right)^{2/\gamma} - \left(\frac{p_T}{p_a}\right)^{(\gamma+1)/\gamma}\right\}\right]} \qquad \text{18.18}$$

(where the subscript a refers to the free air, and the subscript T refers to the throat).

If the pressure drop $(p_a - p_T)$ is sufficiently small, then the approximate result of equation 10.32 can be used,

i.e.
$$\dot{m} = \mathbf{C}A_2 E\rho_2 \sqrt{\left(\frac{2\,\Delta p}{\rho_1}\right)}$$

In this case $\Delta p = (p_a - p_T)$, and $E = 1$, therefore,

$$\dot{m} = \mathbf{C}A_T\rho_T \sqrt{\left\{2\left(\frac{p_a - p_T}{\rho_a}\right)\right\}} \qquad 18.19$$

For isentropic flow, $p_T v_T^\gamma = p_a v_a^\gamma$, therefore,

$$\frac{p_T}{\rho_T^\gamma} = \frac{p_a}{\rho_a^\gamma} \quad \text{and} \quad \rho_T = \left(\frac{p_T}{p_a}\right)^{1/\gamma} \times \rho_a \qquad 18.20$$

Hence for a given mass flow the pressure at the throat can be obtained by trial and error from

$$\dot{m} = \mathbf{C}A_T\rho_a\left(\frac{p_T}{p_a}\right)^{1/\gamma} \sqrt{\left\{2\left(\frac{p_a - p_T}{\rho_a}\right)\right\}} \qquad 18.21$$

i.e.
$$\dot{m} = \mathbf{C}A_T\sqrt{(2\rho_a)} \sqrt{\left\{\left(\frac{p_T}{p_a}\right)^{2/\gamma} \times (p_a - p_T)\right\}} \qquad 18.22$$

The quantity $\sqrt{\left\{\left(\frac{p_T}{p_a}\right)^{2/\gamma} \times (p_a - p_T)\right\}}$ can be evaluated for a fixed value of p_a and selected values of p_T, giving a table from which the value of p_T which satisfies equation 18.22 can be obtained.

The pressure difference acting on the petrol in the simple carburettor is the same as that for the air. The mass flow of petrol, $\dot{m}_f$, is given by equation 10.32 as,

$$\dot{m}_f = \mathbf{C}'A_j\rho_f \sqrt{\left\{2\left(\frac{p_a - p_T}{\rho_f}\right)\right\}}$$

(where subscript f refers to the fuel, and subscript j to the jet).

The coefficient, $\mathbf{C}'$, for the petrol jet must not be confused with the coefficient, $\mathbf{C}$, for the choke tube, in equations 18.18, 18.19, 18.21, and 18.22.

As the fluid is incompressible, then $\rho_1 = \rho_2 = \rho_f$.

$$\therefore \quad \dot{m}_f = \mathbf{C}'A_j\sqrt{\{2\rho_f(p_a - p_T)\}} \qquad 18.23$$

Thus the air/fuel ratio, $\dot{m}/\dot{m}_f$, is given, using the approximate expression for $\dot{m}$, from equation 18.23, as,

$$\text{Air/fuel ratio} = \frac{\dot{m}}{\dot{m}_f} = \frac{C A_T \rho_T \sqrt{\{2(p_a - p_T)\}}}{C' A_1 \sqrt{\{2\rho_f (p_a - p_T)\rho_a\}}}$$

i.e. $$\text{Air/fuel ratio} = \frac{C A_T \rho_T}{C' A_1 \sqrt{(\rho_f \rho_a)}} \qquad 18.24$$

As p_T is reduced so ρ_T is reduced, as shown by equation 18.20,

$\rho_T = \left(\dfrac{p_T}{p_a}\right)^{1/\gamma} \rho_a$, and so the air/fuel ratio decreases. This means

that for the simple carburettor, as the mass flow of air increases the mixture becomes richer in petrol.

If the approximate method is valid, as it is for small values of the pressure drop, $(p_a - p_T)$, then $\rho_a = \rho_T$, and the air flow can be considered incompressible. The effect of compressibility is not great in the usual order of the quantities concerned. For a throat air velocity of 76 m/s and a corresponding pressure difference of 25 mm of Hg, the air/fuel ratio as calculated neglecting compressibility would require a correction of 1·5% to increase the richness.

When the approximation is not valid, then equations 18.18 and 18.23 can be combined to give the true air/fuel ratio.

The coefficient C is practically constant over the pressure ratios used, and the usual value quoted is 0·85. This value is hardly affected by the amount of fuel flowing.

Since in an actual carburettor the petrol surface is below the top of the jet to prevent spilling when stationary, the expression for $\dot{m}_f$ must be modified to take this into account. If the height between the petrol surface and the top of the jet is Z, then the pressure difference becomes,

$$(p_a - p_T - Zg\rho_f)$$

It is usual to indicate the size of the carburettor by quoting the diameter of the choke tube in millimetres, and the jet size in hundredths of a millimetre (i.e. jet no. 50 has a diameter of 50/100 mm). The size is not an indication of the discharge characteristics and the jet should be calibrated against standard reference jets. The jets thus calibrated are stamped with a number which is the flow in millilitres per minute as determined by comparison with a standard jet under a head of 500 mm of pure benzol.

Example 18.4

A proposed petrol engine of 1·71 l capacity is to develop maximum power at 5400 rev/min. The volumetric efficiency at this speed is assumed to be 70% and the air/fuel ratio is 13/1. Two carburettors are to be fitted and it is expected that at peak power the air speed at the choke will be 105 m/s. The coefficient of discharge for the venturi is assumed to be 0·85 and that of the main petrol jet is 0·66. An allowance should be made for the emulsion tube, the diameter of which can be taken as 1/2·5 of the choke diameter. The petrol surface is 6·5 mm below the choke at this engine condition. Calculate the sizes of a suitable choke and main jet. The specific gravity of petrol is 0·75. Atmospheric pressure and temperature are 1·013 bar and 15°C respectively,

$$\text{Swept volume} = 1·71 \text{ l} = 0·00 \, 171 \text{ m}^3$$

By equation 18.16,

$$\text{Volume of air induced} = \eta_V \times V_s = \frac{0·7 \times 0·00 \, 171 \times 5400}{2 \times 60}$$

$$= 0·0538 \text{ m}^3/\text{s}$$

(Assuming that the engine is working on the four-stroke cycle.)

i.e. each carburettor delivers an air flow of $0·0538/2 = 0·0269 \text{ m}^3/\text{s}$.

$$\therefore \dot{m} = \frac{1·013 \times 10^5 \times 0·0269}{0·287 \times 10^3 \times 288} = 0·0329 \text{ kg/s}$$

For compressible flow, using equation 18.17,

$$C_T = \sqrt{\left[2T_a c_p \left\{1 - \left(\frac{p_T}{p_a}\right)^{(\gamma-1)/\gamma}\right\}\right]}$$

i.e. $$105 = \sqrt{\left[2 \times 288 \times 10^3 \times 1·005 \left\{1 - \left(\frac{p_T}{p_a}\right)^{0·286}\right\}\right]}$$

$$105 = 761·6 \times \sqrt{\left\{1 - \left(\frac{p_T}{p_a}\right)^{0·286}\right\}}$$

$$\therefore \frac{p_T}{p_a} = 0·9354$$

$$\therefore p_T = 0·9354 \times 1·013 = 0·9475 \text{ bar}$$

Volume flow of air at choke, $\dot{V}_T = 0.0269 \times \left(\dfrac{p_a}{p_T}\right)^{1/\gamma} = 0.0281 \text{ m}^3/\text{s}$

Therefore,

$$A_T = \frac{\dot{V}_T}{C_T \times 0.85} = \frac{0.0281}{105 \times 0.85} = 0.000\,315 \text{ m}^2$$

(where coefficient of discharge is 0.85),

i.e.   Nominal choke area = 315 mm²

Then, $\dfrac{\pi}{4}(D^2 - d^2) = 315$

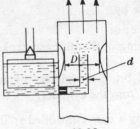

Now $d = D/2.5$ (see fig. 18.35).

$$\therefore \frac{\pi}{4}\left(D^2 - \frac{D^2}{2.5^2}\right) = 315$$

$$\therefore D = \sqrt{\left(\frac{4 \times 315}{\pi \times 0.84}\right)} = 21.83 \text{ mm}$$

Using equation 18.23,

Fig. 18.35 $\dot{m}_f = C'A_j\sqrt{\{2\rho_f(p_a - p_T - Zg\rho_f)\}}$

and $\dot{m}_f = \dfrac{\dot{m}}{13} = \dfrac{0.0329}{13} = 0.002\,53 \text{ kg/s}$

For the petrol, the pressure difference across the main jet is given by

$$(p_a - p_T - Zg\rho_f) = 1.013 - 0.9475 - \frac{6.5 \times 0.75 \times 10^3 \times 9.806}{10^3 \times 10^5}$$

$$= 1.013 - 0.9475 - 0.0005$$

$$= 0.065 \text{ bar}$$

$$\therefore 0.002\,53 = 0.66A_j\sqrt{(2 \times 0.75 \times 10^3 \times 0.065 \times 10^5)}$$

i.e. $A_j = \dfrac{0.002\,53 \times 10^3}{0.66 \times 3.122} \text{ mm}^2 = 1.225 \text{ mm}^2$

$$\therefore D_j = \sqrt{\left(\frac{4}{\pi} \times 1.225\right)} = 1.25 \text{ mm}$$

$\therefore$ Jet number = 125   (nearest size to be selected)

(b) *Fuel injection: diesel engines*

The function of a fuel injection system is to meter the fuel accurately and uniformly to the engine cylinders under all operating conditions from idling to full load. The timing of the injection should be accurate enough to give the required combustion characteristics. The fuel from the tank is filtered before passing to the pump, and the metered fuel is then passed to the injector which is fitted in the engine cylinder. The 'jerk pump' system is the one which is used almost universally over the whole range of oil engines, and will be the one described here. The jerk pump is a piece of precision equipment, and consists of a barrel with plunger, the close fit required to prevent leakage at the high pressures reached is obtained by lapping. The plunger is driven from the camshaft and the control of the amount of fuel to be delivered can be made by

(a) using a plunger with a variable stroke; or

(b) measuring the quantity at the beginning of the plunger stroke and spilling back the excess; or

(c) using a constant stroke plunger and bringing delivery to an end by suddenly spilling off the fuel from the cylinder.

The spill of the fuel may be controlled by a cam-controlled spill-valve (see reference 18.6), or a port control. Pumps with port control are produced by a number of manufacturers and their action will be described.

A simplified injection system is shown in fig. 18.36. The spring-loaded injector needle is set to lift at a predetermined pressure in the delivery line which contains oil at a high residual pressure. There are complex pressure and velocity variations set up in the system, the changes in the pressure being propagated in the oil at the speed of sound in the oil. The effect of such variations will not be considered here (see references 18.7 and 18.8 for a full treatment of the subject). The plunger is provided with a helical groove, the upper edge of which controls the uncovering of the spill port. The timing of the spill is thus decided by the shape of the helix, and this is most important. The part of the helix presented to the port is varied by rotating the plunger in the barrel, and a means must be provided for this to be done automatically or manually while the engine is running. An extreme position of rotation of the plunger in the barrel gives the position at which the pump will not deliver fuel to the engine, and this is the 'stop' position.

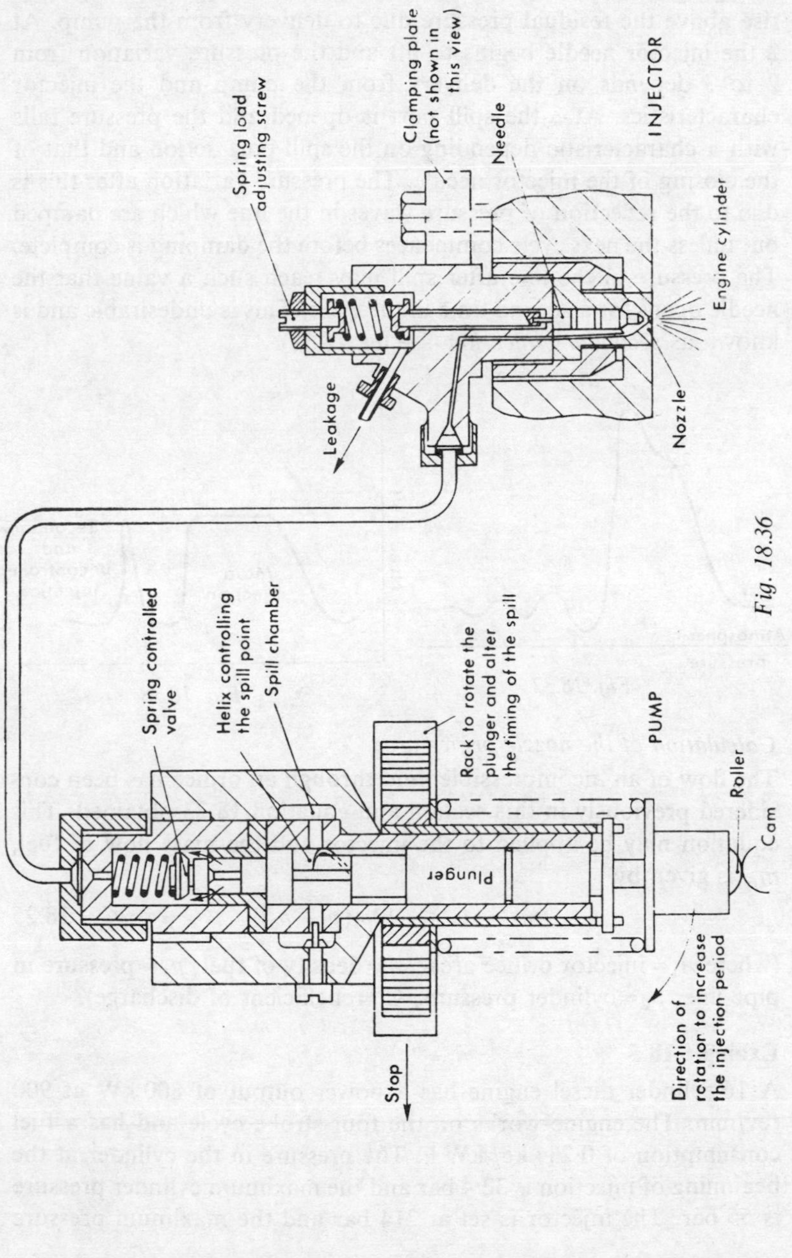
Spring load adjusting screw
Clamping plate (not shown in this view)
Needle
INJECTOR
Engine cylinder
Nozzle
Leakage
Spring controlled valve
Helix controlling the spill point
Spill chamber
Rack to rotate the plunger and alter the timing of the spill
PUMP
Roller
Cam
Plunger
Stop
Direction of rotation to increase the injection period

Fig. 18.36

Fig. 18.37 indicates the way in which the pressure in the fuel line changes during the injection cycle. At point 1 the pressure begins to rise above the residual pressure due to delivery from the pump. At 2 the injector needle begins to lift and the pressure variation from 2 to 3 depends on the delivery from the pump and the injector characteristics. At 3 the spill port is opened and the pressure falls with a characteristic depending on the spill port action and that of the closing of the injector needle. The pressure variation after this is due to the reflection of pressure waves in the line which are damped out unless the next cycle commences before the damping is complete. The pressure in the line after spill may reach such a value that the needle opens for a second time in the cycle. This is undesirable and is known as *secondary injection* (see fig. 18.38).

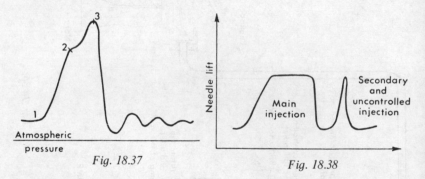

Fig. 18.37

Fig. 18.38

Calculation of the nozzle orifice area

The flow of an incompressible fluid through an orifice has been considered previously in this section and equation 18.23 obtained. This equation may be applied to the injector and the mass flow of fuel, $\dot{m}_f$, is given by

$$\dot{m}_f = \mathbf{C}' A_1 \sqrt{\{2\rho_f(p_1 - p_2)\}} \qquad 18.25$$

(where A_1 = injector orifice area; ρ_f = density of fuel; p_1 = pressure in pipe line; p_2 = cylinder pressure; $\mathbf{C}'$ = coefficient of discharge).

Example 18.5

A 16-cylinder diesel engine has a power output of 800 kW at 900 rev/min. The engine works on the four-stroke cycle and has a fuel consumption of 0·234 kg/ kW h. The pressure in the cylinder at the beginning of injection is 32·4 bar and the maximum cylinder pressure is 55 bar. The injector is set at 214 bar and the maximum pressure at the injector is expected to be about 600 bar. The coefficient of discharge for the injector, obtained from steady flow tests, is 0·6. The specific gravity of the fuel is 0·85. Calculate the orifice area required per injector if the injection takes place over 10° crank angle.

$$\text{b.p. per cylinder} = \frac{800}{16} = 50 \text{ kW}$$

$$\therefore \text{ fuel consumption per cylinder} = 50 \times 0.234 = 11.7 \text{ kg/h}$$
$$= 0.195 \text{ kg/min}^-$$

i.e. $$\text{Fuel to be injected per cycle} = \frac{0.195}{900/2} = 0.000\ 434 \text{ kg}$$

$$\text{Time for injection} = \frac{10}{360} \times \frac{60}{900} = 0.001\ 85 \text{ s}$$

$$\therefore \dot{m}_f = \frac{0.000\ 434}{0.001\ 85} = 0.234 \text{ kg/s}$$

An approximation must be taken for the effective pressure difference, $(p_1 - p_2)$, and this will be taken as the average pressure difference over the injection period.

$$\text{Pressure difference at beginning} = 214 - 32.4 = 181.6 \text{ bar}$$

$$\text{Pressure difference at end} = 600 - 55 = 545 \text{ bar}$$

i.e.

$$\text{Average pressure difference} = \frac{181.6 + 545}{2} = 363.3 \text{ bar}$$

Using equation 18.25,

$$\dot{m}_f = C' A_J \sqrt{\{2\rho_f(p_1 - p_2)\}}$$

where

$$(p_1 - p_2) = 363.3 \times 10^5 \text{ N/m}^2 \quad \text{and} \quad \rho_f = 0.85 \times 10^3 \text{ kg/m}^3$$

$$\therefore\ 0.234 \text{ kg/s} = A_J \times 0.6 \times \sqrt{(2 \times 0.85 \times 10^3 \times 363.3 \times 10^5)}$$

$$\therefore\ A_J = 0.157 \times 10^{-5} \text{ m}^2 = 1.57 \text{ mm}^2$$

i.e. $$\text{Area of injector orifice} = 1.57 \text{ mm}^2$$

(c) *Fuel injection: petrol engines*

At first sight it may appear that the application of diesel engine experience with injection systems to the petrol engine would produce a satisfactory mechanical petrol injection system. However, the different requirements of the two types of engine and the increased knowledge and use of electrical and electronically operated valves and transducers have made it evident that a petrol injection system would be wholly or mainly an electronic system. The development of electronic devices continues with increased component reliability, greater design sophistication, more miniaturization and reduced cost giving flexible systems which are simple to locate, robust in operation and easy to replace.

There are certain basic parameters to be decided upon for each system such as:

Manifold or in-cylinder injection Direct injection into the cylinder is attractive from the point of view of efficient fuel distribution but the injector used would be subject to high pressure and temperature conditions which it would not experience in the manifold. Injection into the manifold allows more time for the fuel and air to mix, giving better combustion, uses simpler injectors and requires easier access to the engine particularly if it was not originally designed for injection.

Continuous or pulsed injection The injector can be used to spray fuel continuously into the manifold and this requires a fuel flow rate varying by 50 to 1 between idle and maximum speed running, and accurate mixture preparation requires good control on air flow measuring and fuel metering. With pulsed injection the fuel is injected near each cylinder inlet valve in each cycle by an injector which is solenoid operated for a measured period of time. The range of pulse durations required is about 5 to 1 and 'time' is a simple quantity, conveniently and accurately measured electronically, by which the demands of several engine parameters can be computed and expressed as a single controlling value. The very small pulse durations required at idling speed, of the order of 1 ms, set the criterion of quality of design for the injector.

It would appear that the ability to time the injection pulse anywhere in the induction stroke would be a powerful facility but experience shows that the effect of such timing on power output and hydrocarbon emissions is not great. This allows some relaxation on the injector design as it enables injectors to be grouped to receive the same pulse, over a longer duration, and so injection occurs into different cylinders at different points in the cycle or by 'non-timed injection' procedure.

The layout of a typical injection system is illustrated in fig. 18.39 and includes provision for the various engine parameters which it is

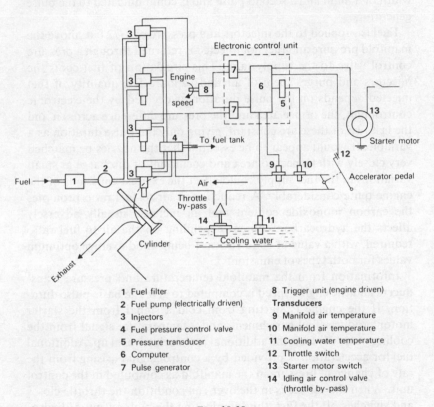

1 Fuel filter	8 Trigger unit (engine driven)
2 Fuel pump (electrically driven)	**Transducers**
3 Injectors	9 Manifold air temperature
4 Fuel pressure control valve	10 Manifold air temperature
5 Pressure transducer	11 Cooling water temperature
6 Computer	12 Throttle switch
7 Pulse generator	13 Starter motor switch
	14 Idling air control valve (throttle by-pass)

Fig. 18.39

believed it is necessary to take into account. Alternative, and possibly simpler, designs could employ fewer parameters and so eliminate the need for some of the transducers, etc, described in this illustration. For instance, in this case the steady air flow requirement is computed from measurements of manifold pressure, inlet air temperature and engine speed and a single reading of air mass flow rate would simplify considerably the requirements of the system. The system shown takes into account in addition the parameters of cooling water temperature, starter motor action, a measurement of the rate of change of manifold pressure by electronic differentiation of the pressure signal and a throttle-actuated switch which cuts off fuel supply when the throttle is closed. Information from each of the transducers or switches involved is 'computed' electronically to give a single pulse width or a main and a second pulse and is communicated to the pulse generators.

Fuel is supplied to the injectors at a pressure of 1·72 bar above the manifold pressure and fuel not injected is returned through a pressure control valve to the supply tank. This circulation of fuel cools the injectors and purges them of air and vapour. The quantity of fuel injected depends on the pulse duration as decided by the electronic control unit, the orifice area and the pressure difference across it and the last two of these are constant, giving only the pulse duration as a variable. It would appear to be essential that the nozzles be matched very closely with respect to area and coefficient of discharge as small variations in the fuel supply could affect the exhaust emissions of the engine quite considerably. A reduction in air-to-fuel ratio promotes the carbon monoxide content and an increase usually adversely affects the hydrocarbon content depending on the air-to-fuel ratio required, with a value of 16 to 16·5 to 1 being good average optimum values for both types of emission.

Information from the manifold temperature and pressure transducers and the engine speed is computed to give the basic pulse duration. If the engine is starting from cold a signal from the starter motor produces fuel enrichment, and a temperature signal from the cooling water controls the additional fuel during warm up. Additional fuel for acceleration is provided by a control signal arising from the rate of change of pressure in the manifold as computed in the control unit. When the vehicle is in the over-run condition the throttle closes and switches off the fuel, thus eliminating the high output of hydrocarbons in the exhaust gas obtained with a carburated engine during this phase of operation. A throttle by-pass valve allows an increase in the air supply during idling when the throttle is in the closed position and the opening of this is controlled by cooling water temperature.

One of the developments which can utilize petrol injection is that of *charge stratification*. The objective of this is to initiate combustion in a region where the mixture is rich and to produce a flame that can then travel easily through the rest of the charge which is, on the whole, weak. One such system employs two injectors, one supplying the main combustion chamber and the other a small directly connected auxiliary chamber in which the spark plug is fitted. In the main chamber the air to fuel ratio is 2·18 to 1, in the auxiliary it is 0·66 to 1 and overall it is 1·64 to 1 relative to the stoichiomatic ratio.

If the contents of this section are considered in relation to that of section 18.13 on engine emissions a reasonable picture of the petrol engine and its stage of development should be obtained.

18.11 Measurement of air/fuel ratio and volumetric efficiency

The practical determination of the air/fuel ratio consists of measuring the rate at which air and fuel are consumed by the engine.

Several means are available for the measurement of fuel consumption and one simple arrangement is shown in fig. 18.40. The measur-

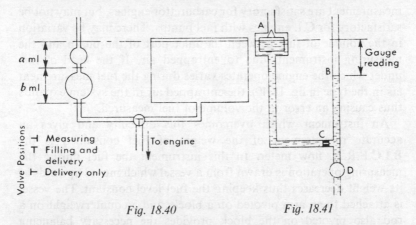

Fig. 18.40   Fig. 18.41

ing vessel consists of two reservoirs of known capacity in series, the capacities being measured between marks on the connecting capillary tubes. The fuel level thus falls quickly past the marks on the small bore capillary tube. Fuel from the tank or measuring vessel passes through a three-way valve to the engine. The three positions of the valve are shown in fig. 18.40. Provision must be made to allow the fuel to fill the measuring vessel up to the level in the tank. The time is taken for the consumption of a known volume of fuel, and thus the rate of fuel consumption can be determined. This arrangement has a number of possible variations which are suitable for particular purposes.

There are types of instruments available called flow meters, which give the rate of flow directly and the calibration can be made in litres per hour or kg/h as required. The Amal flow meter is used widely in industry and is illustrated diagrammatically in fig. 18.41. A float-controlled constant-level tank A is connected to a vertical glass tube B. The connecting tube carries an orifice at C and a tap D is fitted between the instrument and the engine. When no fuel is being delivered to the engine, the fuel is at the same level in A and B. As fuel is supplied to the engine the level in B falls until the fuel taken by the engine is equal to that passing through the orifice, and the level in B reaches an equilibrium position. The gauge reading is proportional to the fuel consumption and the calibration is an empirical one. Each instrument has two tubes such as B, each with its own range of flow.

Such instruments as those described which depend on volumetric measurement are satisfactory for carburettor engines, but may not be satisfactory for C.I. engines with fuel pumps. There may be variation in the volume of fuel between the inlet port of the pump and the measuring instrument, due to entrapped air. If the head of fuel under which the engine operates varies during the fuel measurement (as in the type in fig. 18.40), the entrapped air in the system expands thus causing an error in the volume of fuel measured.

An instrument which overcomes this difficulty and gives an accurate measurement of the weight of fuel consumed is the B.I.C.E.R.A. flow meter. In this instrument the fuel during the measuring operation is drawn from a vessel which moves upwards as its weight decreases thus keeping the fuel level constant. The vessel is attached to an arm pivoted on a block, and a counterweight on a rod also pivoted on the block provides the necessary balancing torque. The weight of fuel consumed is measured by observing the upward movement of the vessel; this is done by an optical system which magnifies the movement of the vessel approximately 30 times.

The satisfactory measurement of air consumption is a more difficult task. It is essentially the measurement of the rate of flow of a compressible fluid, complicated by the fact that the flow is pulsating due to the cyclic nature of the engine. This prevents the use of an orifice in the induction pipe since a steady and reliable reading would not be obtained. The usual method of damping out the pulsations is to fit an air box of suitable volume to the engine with an orifice in the side of the box remote from the engine. The pressure difference across the orifice is steady if the system is correctly designed, and is measured by means of a suitable manometer (see fig. 18.42).

The criteria for the selection of the size of airbox have been established and are described in reference 18.9. It is a conclusion of this reference that the dimensions of a suitable airbox meter can be determined by the use of a non-dimensional factor U, defined by

$$U = \frac{1}{40 \cdot 94 \times 10^5} \left(\frac{\eta_V CVN^2 n^2}{Td^4 p^2} \right)$$

(where η_V = volumetric efficiency; V = engine swept volume, m³; C = airbox volume, m³; N = engine speed, rev/min; n = number of cylinders; T = air temperature, K; d = orifice diameter, m; p = strokes of the piston per induction stroke).

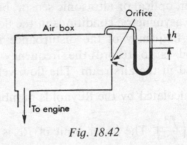

Fig. 18.42

For reasonable accuracy, U should not be less than 2·5, and the depression across the orifice should not exceed 100 to 150 mm of water. The calculation of air flow from orifice and manometer data has been described in Section 10.11.

Another design of air meter is available for the measurement of air flow; this is the Alcock viscous-flow air meter, and it is not subject to the errors of the simpler types of flow meters. With the airbox the flow is proportional to the square root of the pressure difference across the orifice. With the Alcock meter the air flows through a form of honeycomb so that the flow is viscous. The resistance of the element is directly proportional to the air velocity and is measured by means of an inclined manometer. Felt pads are fitted in the manometer connections to damp out fluctuations. The meter is shown diagrammatically in fig. 18.43. The accuracy is improved by fitting a damping vessel between the meter and the engine to reduce the effect of pulsations.

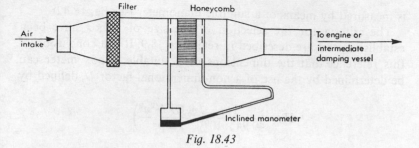

Fig. 18.43

An entirely different principle on which the velocity of an airflow can be measured is that of vortex shedding. As air flows over a bluff body, vortices or local regions of low pressure are created in the flow behind the body as shown in fig. 18.44. A sensor of some kind, e.g. a pressure transducer, optical or ultrasonic sensor, built into the bluff body, registers the asymmetric conditions in the flow during vortex shedding and the frequency of the disturbance is measured. An alternative method is to record the frequency by a hot wire anemometer placed in the airstream. The flow velocity C and the frequency f are calculated by the Reynolds number $\left(\dfrac{Cd}{v}\right)$ and the Strouhal number $\left(\dfrac{fd}{C}\right)$. The lower limit of Re is 1000 for vortex shedding to occur after which the Strouhal number is constant for a wide range of Reynolds number. Hence the volumetric flow rate is

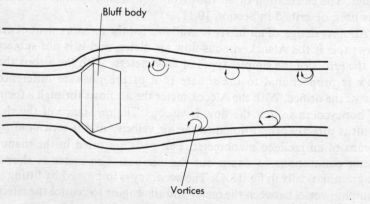

Fig. 18.44

given by $Q = K \times f$ where K is the calibration coefficient for the instrument which can be refined in value by more elaborate calibration techniques.

The need to be able to measure accurately the instantaneous mass flow rate of air into an engine, or possibly into each engine cylinder, is an important requirement of modern transducer and instrument design and engine development work. One method of development is to use the pressure drop across an orifice or throttle plate, accurately measure and record the pressure difference, the air pressure and temperature and then compute the air mass flow rate using modern microcomputing techniques. Another development is to seek new types of transducers or sensors which will respond, preferably in a linear manner, to the mass flow rate of air.

Such rapid response devices are particularly desirable when engine intakes suffer reversals in the direction of the air flow and one principle on which a suitable unit may be devised is that of *corona* discharge, see fig. 18.45. A thin wire maintained at a potential of 10 kV is fitted across the direction of the air flow and the discharge between the wire and the electrode collectors is disturbed by the air flow such that the discharge current to the electrodes is proportional to the mass flow rate of the air.

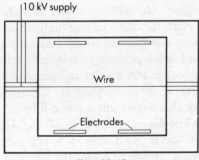

Fig. 18.45

18.12 Supercharging

Fig. 18.14 shows how the power output of an engine is affected by the reduction in volumetric efficiency at increased engine speed as shown in fig. 18.15. The purpose of supercharging is to raise the volumetric efficiency above that obtained with normal aspiration. This may be essential for engines for aircraft which have to operate at high altitude and are required to maintain the sea-level engine power output. In this case normal performance is being maintained in adverse conditions but the main attraction of supercharging engines is to obtain a high power output from a small engine hence giving a good power-to-weight ratio with a corresponding saving in space which is important in some applications such as road and rail vehicles, submarines, etc. In this case the engine is designed as a supercharged engine to withstand the higher loads and temperatures reached in supercharging compared with normal aspiration.

Greater benefits are to be expected from supercharging the diesel engine than from the petrol engine because of the different methods of charging the cylinders and the quite different combustion character-istics of the two types of engine. The diesel induces air only and the fuel is injected under pressure into the cylinder with self ignition of the fuel in the air; the petrol engine induces a mixture of air and fuel which is spark ignited and burns as described in Sections 18.6 and 18.7 giving fundamental combustion problems which do not occur with the diesel engine. To avoid charge detonation or 'knocking' in the petrol engine, giving uncontrolled combustion, the compression ratio may have to be reduced, an action which adversely affects the thermal efficiency of the engine. Alternatively a fuel of higher octane rating may be necessary. With the diesel engine higher boost pressures can give more satisfactory combustion conditions with a wider range of usable fuels, reduced delay periods, controlled pressure rise and an engine which is smoother and quieter in operation.

The main features of supercharging are illustrated in the $p - V$ diagrams for the idealized constant volume four-stroke cycle in fig. 18.46 and the plant line diagrams in fig. 18.47. Fig. 18.46(a) shows the normally aspirated cycle with line 1–5 representing both the induction and exhaust strokes at about the ambient air pressure p_a. The early applications of supercharging were for piston engined aircraft in which the 'blower' was driven mechanically from the engine as shown in fig. 18.47(a). The power output of the engine was increased by the higher flow of air, and hence the fuel consumed, but part of this increase in power was required to drive the blower. The effects on the $p-V$ diagram, as shown in fig. 18.46(b), are to increase the pressures (and temperatures) reached during the cycle and to give

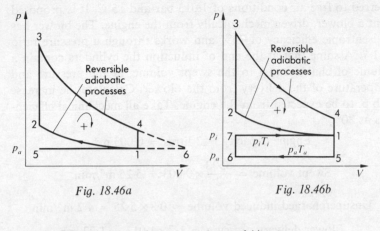

Fig. 18.46a   *Fig. 18.46b*

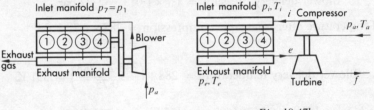

Fig. 18.47a   *Fig. 18.47b*

a positive pumping loop, 15671, to add to the main working loop 12341.

The power required to drive a blower mechanically connected to the engine must be subtracted from the engine output to obtain the net b.p. of the supercharged engine.

Then,

$$\text{i.m.e.p.} = \left(\frac{\text{area } 12341 + \text{area } 15671}{\text{length of diagram}} \right) \times \text{spring no.} \qquad 18.26$$

and $\text{b.p.} = (\eta_M \times \text{i.p.}) - (\text{power to drive blower}) \qquad 18.27$

(for mechanically driven blowers only).

Example 18.6

The average i.p. developed in a C.I. engine is 12·9 kW per m³ of free air induced per minute. The engine is a three-litre four-stroke engine running at 3500 rev/min, and has a volumetric efficiency of 80%, referred to free air conditions of 1·013 bar and 15°C. It is proposed to fit a blower, driven mechanically from the engine. The blower has an isentropic efficiency of 75% and works through a pressure ratio of 1·7. Assume that at the end of induction the cylinders contain a volume of charge equal to the swept volume, at the pressure and temperature of the delivery from the blower. Calculate the increase in b.p. to be expected from the engine. Take all mechanical efficiencies as 80%.

$$\text{Engine capacity} = 3 \text{ litres} = 0·003 \text{ m}^3$$

$$\text{Swept volume} = \frac{3500}{2} \times 0·003 = 5·25 \text{ m}^3/\text{min}$$

$$\text{Unsupercharged induced volume} = 0·8 \times 5·25 = 4·2 \text{ m}^3/\text{min}$$

$$\text{Blower delivery pressure} = 1·7 \times 1·013 = 1·72 \text{ bar}$$

$$\text{Temperature after isentropic compression} = 288 \times 1·7^{(1·4-1)/1·4}$$
$$= 335·2 \text{ K}$$

$$\therefore \text{ Blower delivery temperature} = 288 + \left(\frac{335·2 - 288}{0·75}\right) = 351 \text{ K}$$

The blower delivers 5·25 m³/min at 1·72 bar and 351 K.

$$\text{Equivalent volume at } 1·013 \text{ bar and } 15°C = \frac{5·25 \times 1·72 \times 288}{1·013 \times 351}$$
$$= 7·32 \text{ m}^3/\text{min}$$

$$\therefore \text{ Increase in induced volume} = 7·32 - 4·2 = 3·12 \text{ m}^3/\text{min}$$

$$\therefore \text{ Increase in i.p. from air induced} = 12·9 \times 3·12 = 40·2 \text{ kW}$$

Increase in i.p. due to the increased induction pressure

$$= \frac{(1·72 - 1·013) \times 10^5 \times 5·25}{10^3 \times 60} = 6·2 \text{ kW}$$

i.e.   Total increase in i.p. $= 40·2 + 6·2 = 46·4 \text{ kW}$

$$\therefore \text{ Increase in engine b.p.} = 0·8 \times 46·4 = 37·1 \text{ kW}$$

From this must be deducted the power required to drive the blower.

$$\text{Mass of air delivered by blower} = \frac{1·72 \times 10^5 \times 5·25}{60 \times 287 \times 351} = 0·149 \text{ kg/s}$$

Work input to blower $= \dot{m}c_p(351 - 288) = 0{\cdot}149 \times 1{\cdot}005 \times 63$

$$\therefore \text{ Power required } = \frac{0{\cdot}149 \times 1{\cdot}005 \times 63}{0{\cdot}8} = 11{\cdot}8 \text{ kW}$$

$$\therefore \text{ Net increase in b.p.} = 37{\cdot}1 - 11{\cdot}8 = 25{\cdot}3 \text{ kW}$$

Fig. 18.46(a) shows the start of the exhaust process at 4 at a pressure substantially greater than the ambient, p_a. This means that over 60% of the cylinder charge is suddenly exhausted by a free expansion which constitutes a considerable loss of the energy released on combustion; of about 30–40% between diesel and petrol engines. The attraction of 'turbocharging' is evident as the energy lost in this way is used to drive a turbine wheel integral with a compressor wheel which delivers compressed air or charge to the cylinder. The additional work available from the gas is indicated, after continuing the reversible adiabatic expansion line 3–4 down to the pressure p_a at 6, by area 4614. The physical arrangement is shown in fig. 18.47(b) and there is no mechanical connection to the engine. The turbocharger combination is a free running unit with approximately equal mass flow rates over the turbine and compressor wheels reaching an equilibrium speed in the range 20 000 to 80 000 rev/min.

The simplest form of the supercharged cycle shows constant pressures created in the inlet manifold, p_i, and in the exhaust manifold, p_e, and it is essential that $p_i > p_e$, see fig. 18.48(a). This pressure difference, $p_i - p_e$, can be utilized to scavenge residual gas from the combustion chamber if there is some overlap between the exhaust and inlet valve operation and particularly so for the diesel engine. This is called '*constant pressure*' supercharging and requires a large enough exhaust manifold to create a constant pressure supply to the turbine from a highly pulsating delivery from the engine cylinders. The $T - s$ diagram for the turbocharger is shown in fig. 18.48(b) and, using the methods of Chapter 12, the energy balance for the unit is obtained as follows:

The compressor work input,

$$W_c = \dot{m}_a c_{pa} T_a \left[\left(\frac{p_i}{p_a} \right)^{(\gamma_a - 1)/\gamma_a} - 1 \right] \bigg/ \eta_c$$

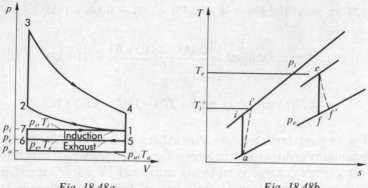

Fig. 18.48a

Fig. 18.48b

The turbine work output,

$$W_T = \dot{m}_e c_{pe} T_e \left[1 - \left(\frac{p_a}{p_e} \right)^{(\gamma_g - 1)/\gamma_g} \right] \times \eta_T$$

where η_c and η_T are the isentropic efficiencies of the compressor and turbine respectively.

For a balance of mass flow rates

$$\dot{m}_e = \dot{m}_a + \dot{m}_f, \qquad \frac{\dot{m}_e}{\dot{m}_a} = 1 + \frac{\dot{m}_f}{\dot{m}_a} = 1 + F/A$$

where $\dfrac{F}{A}$ = fuel to air ratio = $\dfrac{\dot{m}_f}{\dot{m}_a}$.

Also $W_c = W_T \times \eta_M$ where η_M = mechanical efficiency of the drive

$$\therefore \left[\left(\frac{p_i}{p_a} \right)^{(\gamma_a - 1)/\gamma_a} - 1 \right] = \left[1 - \left(\frac{p_a}{p_e} \right)^{(\gamma_g - 1)/\gamma_g} \right] \frac{c_{pe}}{c_{pa}} \left(\frac{T_e}{T_a} \right) (1 + F/A) \times \eta_o$$

where $\eta_o = \eta_M \times \eta_T \times \eta_C$ = the overall efficiency of the turbocharger.

This expression shows how the manifold pressure p_i depends mainly upon η_o and T_e as the effect of the F/A ratio is small and a set of characteristics can be drawn of p_i/p_a against η_o for different values of T_e. A set is obtained for each value of p_e and the minimum requirement to sustain the unit is $p_e = p_i$ as shown in fig. 18.49.

For example for $p_i/p_a = 2$, and $T_3 = 773$ K ($500°$C) $\eta_o = 42\%$. η_o is higher for lower values of p_e and higher values of p_i and T_3. The study of turbocharging should continue into the design of the turbocharger

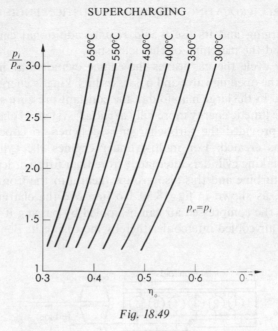

Fig. 18.49

unit to meet its service requirements but this is outside of the scope of this book and specialist references such as 18.10 should be consulted.

The usual arrangement for a turbocharger is a single stage centrifugal compressor driven by a single stage axial flow turbine, for the medium and large size engines for industrial, rail and marine applications, and by a radial flow turbine for the smaller engines used in automotive applications, transport vehicles and cars. It is somewhat against earlier expectations that most car manufacturers now include supercharged petrol engined cars in their product range with apparent overlap with their normally aspirated engines of different capacities. This is in spite of the fact that cars spend a great deal of their time at part throttle and that additional control is necessary to restrict the boost pressure and prevent the onset of knock by retarding the ignition.

The above description has been confined to the constant pressure charging of four stroke cycle engines. The blowing of the two stroke engine is attractive as the cycle does not include a separate exhaust stroke and a means of improving the scavenging process would improve the breathing and hence the power output. The two stroke is particularly sensitive to exhaust back pressure which is increased by turbocharging and its use would require additional care in port timing and the matching of characteristics.

In any cycle the gas leaves the engine cylinders at high speed through the opening valve and possesses high kinetic energy which is dissipated in the large manifold of the constant pressure system. To utilize the kinetic energy more fully a *'pulsed'* system of charging can be used provided the turbocharger is designed to cope with the conditions created. For multi-cylinder engines the cylinders are grouped taking cylinders alternately from their firing order for entry into the turbine and this also groups them into the front and rear cylinders, as shown in fig. 18.50. To improve the charging of the cylinders the compressed air can be cooled by passing it through a water or air cooled intercooler thereby increasing its density.

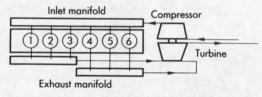

Fig. 18.50

The turbine and compressors are complex devices and further study will show that the complexities of design, matching of the units to each other and to the engine, etc. are such that the performance may not be satisfactory over the whole speed range. Compound engines have been proposed, and some have been built which are turbocharged and have a mechanical connection between the engine crankshaft and the turbocharger as shown in fig. 18.51(a). A fixed drive improves the low speed performance but restrains the turbine at high speeds. A variable speed drive is called for but this is an expensive addition. An extreme arrangement for a compound engine is shown in fig. 18.51(b) in which the engine drives the compressor only and the exhaust gas drives an independent turbine from which the power output of the combination is taken.

Fig. 18.52 shows the continuous running performance characteristics of a diesel engine in three modes; normally aspirated (a), turbocharged (b) and turbocharged with intercooling (c). The engines are of $6 \cdot 6$ l capacity (bore = stroke = 112 mm) with a compression

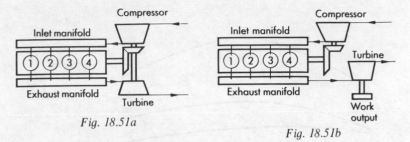

Fig. 18.51a

Fig. 18.51b

ratio of 16·3 for (a) and 15·6 for (b) and (c). The maximum cylinder pressures were 90, 121 and 138 bar respectively for (a), (b) and (c).

18.13   Engine emissions and legal requirements

An atmospheric phenomenon called 'smog' was experienced on an increasing scale in the State of California, U.S.A., around the city of Los Angeles until it became a severe social problem with political and subsequently commercial and engineering consequences which have had a fundamental effect upon the design and operation of the internal combustion engine.

Smog is a type of light fog which is unpleasant and a cause of irritation to the eyes and nasal passages, and although not affecting visibility greatly does affect vegetation and has caused serious economic losses in horticulture and agriculture. Smog is created by the action of sunlight on hydrocarbons in the atmosphere and the main cause of hydrocarbons (HC) is the exhaust gases of motor vehicles, the rapid increase in volume of town traffic causing the increase in smog. It was against a background of increasing public nuisance and distress that, in 1960, the State of California legislated against the motor vehicle and stated the procedures against which vehicles would be tested and the levels of pollutants which would have to be met by vehicles to be sold in that State. The procedures were extensive and covered the types of emissions to be controlled, hydrocarbons (HC) and carbon monoxide (CO), the sources of emissions (exhaust, crankcase, carburettor and fuel tank), the test procedures, the instrumentation and the test equipment to be used.

The initial aim of the Californian Motor Vehicle Pollution Control Board was to reduce the total exhaust emissions to pre 1940 levels by 1980. This was to be done by legislation against new vehicles, as no

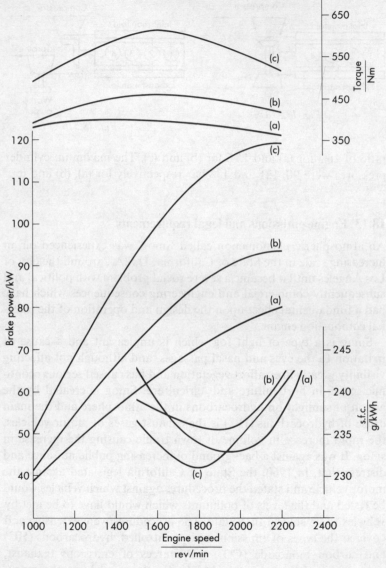

Fig. 18.52

requirement was made for existing vehicles to be modified, and included the effects of a continuous increase in the number of vehicles on the road. Although mainly American manufacturers of vehicles were affected so were the manufacturers of imported vehicles and these were penalized by having to meet prescribed rates of emission output, and not total amounts, as their engines were generally of small capacity.

The legislation was introduced before the process of smog creation was understood and it was believed that carbon monoxide and unburned hydrocarbons were the cause. Mixture strengths were weakened to reduce the CO and HC output but the smog levels were unaffected and it was later established that the oxides of nitrogen (No_x) and the unburned hydrocarbons combined photochemically in the strong sunlight to create the smog. The Californian legislation was followed by that of the U.S.A. federal government through the Environmental Protection Agency (EPA) although smog was not the primary concern in other parts of the U.S.A. where the atmospheric conditions for its creation did not occur. The Californian and Federal authorities each have emission standards.

The introduction of emission controls for motor vehicles has spread in the wake of the legislation introduced in the U.S.A. to other countries including Japan, the E.E.C., Canada, Sweden, the U.K., Australia and Finland. In Europe the most significant effort has been to reduce the use of leaded fuels as a means of raising the knock rating of petrol. The reduction of the lead content in fuels is a response to the demand to reduce the quantity of airborne lead particles in the atmosphere, 90% of which are attributed to petrol consumption. Inhaled lead is a smaller contributor to the consumed lead in the human body than food or water but it is not negligible. The lead content of fuel can be reduced if either a lower knock rating or an increase in the cost of fuel processing is accepted. Premium petrol has a research octane number of 98·5 with a lead concentration of 0·4 g/l. To reduce this to 0·15 g/l would require an increase in crude oil consumption of 0·5% by mass at the refinery a much more expensive operation than the proportion suggests but which can be avoided by accepting the octane rating of 95·5 as a consequence of the reduction in the lead content to 0·15 g/l. This in turn could lead to a reduction in useable compression ratio, and hence thermal efficiency of the engine, and an increase in fuel consumption. In Japan and North America lead has been removed from petrol because it poisoned the catalyst of the chemical converters fitted to engines to reduce the exhaust emissions and not for environmental reasons related to the lead itself.

The legislation of each country requires progressive reductions to be made in the HC, CO and NO_x content of the exhaust gas from vehicles and this anticipates that the technology will be available to achieve the levels defined by the time stated. The requirement in the U.S.A. in 1970 was to obtain a 90% reduction in emissions from the 1970 levels within a stated time scale. The aims of the different countries are not directly comparable as there are differences in the sampling and measuring techniques over test cycles which are selected to represent the driving patterns of the country concerned. The diesel engine is low on HC and CO emissions but is higher on NO_x and particulate emission, the latter being ten times that of a comparable petrol engine using lead for fuel and the technical objective is to reduce the levels without losing on fuel economy.

Over the years during which emission control has applied the fuel consumption of the smaller engined vehicles has improved, but the larger engines gave better emission figures at the expense of fuel economy as the engine had to be 'detuned' by adjusting mixture strength and ignition timing to meet the emission requirements. The optimum settings were restored when catalytic reactors were fitted to the engines to reduce the emission levels chemically by units fitted external to the engine combustion systems.

Some approaches to emission control such as increasing the air/fuel ratio and recirculated part of the exhaust gas through the cylinders reduces the CO and HC output, controls the NO_x and does not increase the fuel consumption but is comparatively cheap in the additional facilities it requires on the engine. The modification to basic engine parameters such as those affecting mixture preparation and spark ignition involve small costs, but others which require thermal reactors or catalytic converters to be added as additional units can involve large cost increases. Advanced aspects of engine design involving stratified charge, with torch ignition or open chamber systems, are expensive to produce but have considerable effects on emission levels, increasing the possibilities if taken together with exhaust gas recirculation and a catalytic converter.

A vehicle without emission control contributes to the pollution from several sources with unburned HC coming from the fuel tank vent and the carburettor bowl vent to the atmosphere; from piston blow-by into the crankcase and leaks from fuel lines as well as from the exhaust gas itself which puts, in addition, CO, CO_2, NO_x and particulates into the atmosphere. Crankcase emissions are dealt with by venting all crankcase fumes directly into the engine intake system to be burned in the engine cylinders.

The exhaust gas emissions are affected by many of the engine variables which also control the completeness of the combustion process, an appreciation of which is essential to an understanding of emission control. The way in which engines are used in the vehicles has a complex effect on emissions and so it was necessary to devise representative test methods to reproduce on the road running as closely as possible on the test rig. As mentioned earlier, different countries have different test procedures and the following comments can only be regarded as being general to the problem.

A chassis dynamometer is used to load a vehicle to reproduce the road condition with regard to wind and inertia loads, the latter being experienced during acceleration and deceleration. This is not easy to achieve as the tyre/roller contact is quite different from that between tyre and road and the engine operating parameters may not be the same as those experienced on the road for apparently corresponding conditions.

The drive cycle includes periods of acceleration, deceleration, steady state cruising and idling and the exhaust gases are continuously analysed for CO, CO_2, NO and HC (as a hexane equivalent). The analyses are recorded and may be interpreted by a weighting procedure to give an estimate of the total mass emitted over the test cycle. Acceleration periods have high weight factors and idle periods have low. More recent techniques described as 'constant volume sampling' include the collection of quantities of gas in bags either during or over all of the test cycle for subsequent analysis to a mass basis.

As the test procedures have developed and experience has been gained the test techniques have become established in relation to specified instrumentation. Exhaust analysers are of the following type and use:

Analyser	Gases
1. Non dispersive infra red (NDIR)	$CO\%$, $CO_2\%$
2. Flame ionization detector (FID)	HC ppm
3. Chemiluminescent	NO_x ppm
4. Paramagnetic	$O_2\%$

ppm—parts per million

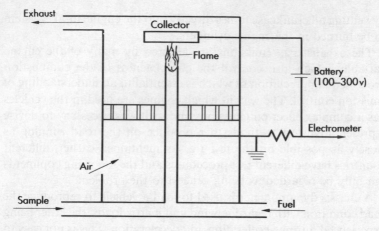

Fig. 18.53

The NDIR and Paramagnetic analysers have been described in the general treatment of the subject in Section 15.7. The NDIR was used originally for HC detection against hexane as the reference gas and for NO_x against nitric oxide. The FID is a more accurate HC detector giving the total concentration and the operating principle is illustrated in fig. 18.53. The sample analysed is mixed with a special burner fuel which may be hydrogen, hydrogen + helium or hydrogen + nitrogen. A polarizing voltage exists between the burner and the collector which causes a migration of ions in the flame and so a current to flow in the collector circuit. The current is proportional to the rate of ion formation which depends on the HC concentration in the gases and is detected by a suitable electrometer and displayed as an analogue output. The FID gives a rapid, accurate and continuous reading of total HC concentration for levels as low as 1 part in 10^9.

The chemiluminescent analyser shown in fig. 18.54 is preferred to the NDIR for the measurement of NO_x. The nitrous oxides are converted to NO before analysis by passing the gas sample over a heated catalyst such as stainless steel, graphite or molybdenum at about $600°C$. The sample is now mixed with ozone in the ozonizer where the reaction gives NO_2 and O_2 but some of the NO_2 contains an excess of energy within its atoms. This NO_2 then changes to the ground state with the emission of light (chemiluminescence) which is amplified then measured by a photomultiplier tube and the signal is

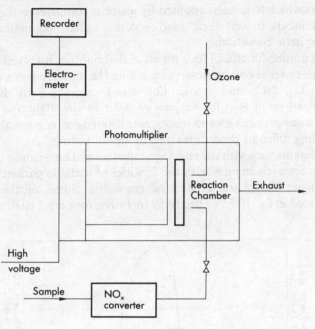

Fig. 18.54

displayed as an analogue reading of the NO_x content of the original sample.

A zirconia cell is used in research work to measure the oxygen concentration in a gas sample and is an absolute method as its output signal is directly related to the oxygen concentration of a gas sample. Paramagnetic analysers are usually used for the measurement of O_2 in the exhaust gas analysers employed in laboratories and engine test houses concerned with engine exhaust emissions.

The methods of sampling the gas and the calibration of the instruments (as all of them except the zirconia cell work on a comparison with known reference gases) are important techniques which have to be reliable. Emission control is a complex matter which was brought into existence by legislation and has had to develop as a science since. The different countries, e.g. Japan and Sweden, have emission control programmes to meet their own requirements albeit related to the American experience and procedures as the pioneers in this particular subject. The E.E.C. countries in Europe have their set of tests which have been adopted by member countries to their own requirements to suit their road systems, vehicle population and atmospheric conditions.

The number of official tests for all of the countries involved is high and the cover is comprehensive including the measurements of HC, CO, CO_2, NO_x and smoke (for diesel engines) with different classifications of tests for vehicles of different size, light and heavy duty passenger vehicles and trucks, petrol and diesel engines, the tests including different procedures and modes of testing.

Emissions vary with the engine parameters and tests can be carried out on research engines with the facilities of variable parameters to investigate the characteristics of emissions. Some of these are illustrated in fig. 18.55 to fig. 18.59 inclusive, for petrol engines, and

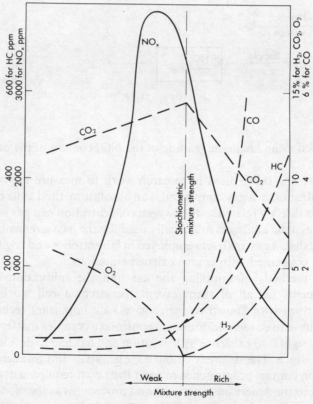

Fig. 18.55

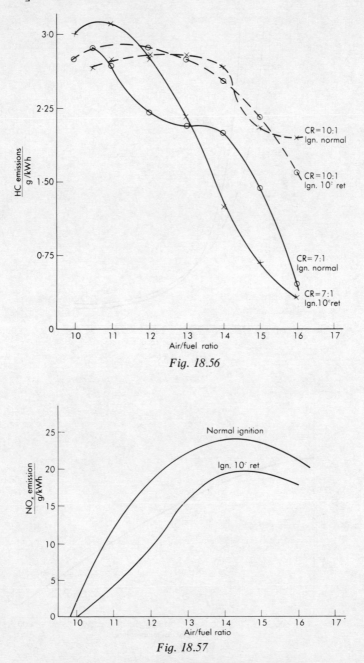

Fig. 18.56

Fig. 18.57

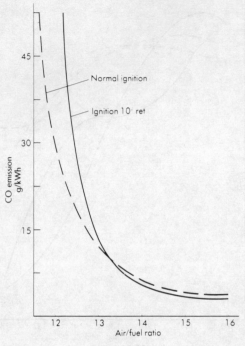

Fig. 18.58

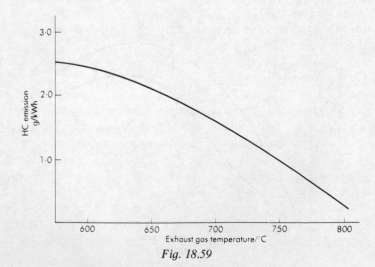

Fig. 18.59

serve to show the complexity of the processes involved. It is probably true to say that the legal emission requirements have added a new dimension to the development of the internal combustion engine and have probably led to more fundamental research into the process of internal combustion since the onset of legislation than was done before.

The pollution problem is closely linked to that of fuel economy with the increase in cost of fuel and oil and the need to conserve natural resources. It has therefore become essential to rationalize the experience of generations of engine builders and integrate modern, analytical, computational and experimental methods into the science of internal combustion engineering. Section 18.15 describes some of the modern developments in I.C. engines which are a result of advanced understanding of the basic processes and modern techniques. Further progress can be made by studying research papers on individual topics, (see references 18.13, 18.14 and 18.15).

18.14 Alternative forms of I.C. engines

In recent years there have been several attempts to produce power units which would be superior to the I.C. engine in its conventional forms. Some attempts have been merely to improve the breathing of the reciprocating engine by alternative designs of the valve mechanism. The more ambitious projects have had the object of basic improvements, and have included engines which have a fundamentally different geometry. The most promising engine with this object is the N.S.U. Wankel rotary engine.

This rotary, positive displacement engine which was invented by Felix Wankel and first introduced in the late 1950s by N.S.U. Motorenwerke of Neckarsulm, West Germany, is the most successful of many proposed rotary engine designs. For many years inventors have worked to produce an engine which would fulfil all the promises of the rotary engine concept, with improved power-to-weight ratio, and compete successfully with the reciprocating engine. A simplified representation of the Wankel engine is shown in fig. 18.60 (page 880) and the engine will be only briefly described here.

The rotor, which has a profile defined by three circular arcs PQ, QS and SP, is attached through a roller bearing to an eccentric which is an integral part of the engine main shaft. The bearing (and rotor) centre is at an eccentricity e to the mainshaft centre. The rotor

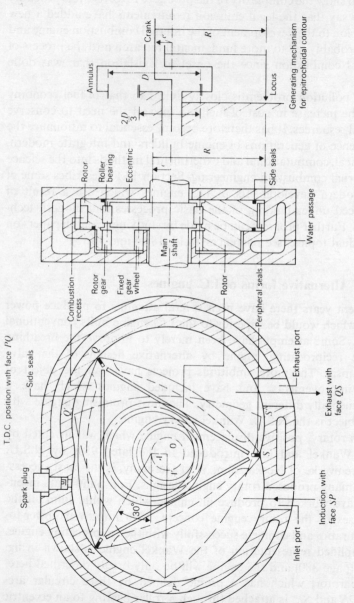

Fig. 18.60

radius R to the apex points P, Q and S is the generator radius for the enclosing 'cylinder' or housing. The profile of the cylinder is of epitrochoidal form and is followed by the rotor by means of the epicyclic gear formed by the gear wheel which is fixed to the casing and the internal gear of the rotor. The pitch circle diameter of the fixed gear wheel is two thirds that of the rotor gear wheel. With this mechanism the rotor turns at one third of the speed of the mainshaft (or eccentric). As each of the three faces of the rotor is concerned in turn with one power cycle, for each revolution of the rotor the main-shaft receives one working 'stroke' per revolution. Thus the single rotor Wankel compares with a single cylinder two stroke cycle engine.

The rotor is shown in fig. 18.60 in two positions, one in dotted outline and the other in full line; for the former position the eccen-tric OA' ($= e$) and the generator radius $A'P'$ ($= R$) are in line to show OP' equal to half the length of the major axis ($= R + e$). The rotor turns clockwise and to give the position P the mainshaft (or eccentric) has turned through 90° from OA' and the rotor through 30° (i.e. 90/3). By continuing this construction for other mainshaft positions the epitrochoidal profile can be generated and is repeated for apex S which is displaced 120° with regard to P and similarly for Q. For the full line position OS forms half of the minor axis length ($= R - e$). In the diagram the pitch circle of the bearing coincides with that of the rotor gear for simplicity of drawing.

The three parts of the cylinder are sealed by peripheral seals at P, Q and S and side seals which are held in the flanks of the rotor on both sides and these constitute a sealing system corresponding to that of the piston rings and valves in the reciprocating engine. The peri-pheral seals take gas loads only and do not influence the movement of the rotor.

In the full line position shown PQ is at top dead centre and defines the minimum or clearance volume of the engine. This volume can be varied by forming a recess in the rotor. The maximum cylinder vol-ume will be shown when S takes up a position just on completion of its movement across the inlet port. The difference between the two volumes gives the swept volume of the engine per rotor face. As PQ moves towards the position occupied by QS the working stroke is performed and exhaust begins when the cylinder space is uncovered by the seal as it passes over the exhaust port. Continued movement of the rotor opens the space to the inlet and the induction of the fresh charge and so there is a considerable overlap between exhaust and inlet phases and this can be considered to be a basic limitation of the design. When the apex seal passes over the inlet port, compression of the charge begins up to the *TDC* position after which the cycle is repeated. To the right of fig. 18.60 a 'skeleton' figure shows the essentials of the 'generating' mechanism.

The basic design is attractive and offers advantages in comparison with a reciprocating four-stroke engine with regard to compactness for a given power output, fewer working parts, lower mass, lack of vibrations and elimination of the poppet valve. There have been many other advantages claimed for the engine but there have been many development problems which are not yet solved to the extent that the automotive market shows a significant conversion to the rotary engine.

Some of the criticisms are basic and include the belief that the combustion space is the wrong shape (long and narrow) for good combustion and the limitations on porting are a disadvantage for good breathing particularly at low speeds. These are fundamental to the epitrochoidal geometry which also limits the attainable compression ratio thereby making a diesel version impossible without increased complication such as two cylinders working in series. Another outcome of the geometry is the re-entrant shape or 'cusp' on the minor axis which the peripheral seal finds difficult to follow at high speed because of the reversal of acceleration experienced. The effectiveness of the sealing is held to be in doubt in spite of the considerable development work which has taken place on these. It is also believed that the rotor bearing is subject to adverse conditions, being close to the hot rotor surface without adequate cooling. Solutions to these problems have been applied such as dual, phased ignition and side porting as well as peripheral ports and the use of different seal designs and materials. The assessment and development of the Wankel engine has been undertaken by companies in many countries of the world and for many applications other than for motor vehicle purposes, for much smaller engines and much larger power, lower speed units.

The S.I. petrol engine is predominant in the light power and high speed field, and for automobile purposes, but is receiving some competition from the high speed diesel engine. They have geometries which are basically identical and both are subject to continuous development; it is not likely that they will be replaced to any great extent in their particular field by any other unit for some time. Other power units may appear for particular purposes and may be modifications to the conventional lay-outs, or they may be combined units.

The *free piston* engine is a unit of interest and is referred to in Section 18.3 in an introductory manner. Two versions are referred to: the free piston air compressor and the free piston gasifier. The free piston air compressor cycle will be described and the gasifier action can be deduced from it. The free piston engine is usually constructed as an opposed-piston, two-stroke diesel cycle with a conventional fuel injection system, but it could be a gas engine or have spark ignition. In fig. 18.61 the engine/compressor unit is

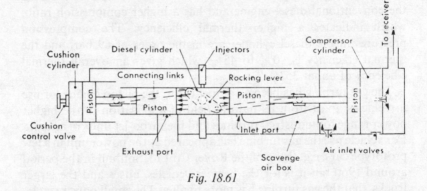

Fig. 18.61

shown on its power stroke. The two diesel pistons are connected mechanically and externally by lightly loaded links through a centrally situated rocking lever. Alternatively the connection may be by a twin rack and pinion arrangement. On the power stroke the air in the compressor cylinder is compressed and then delivered to the receiver. At the same time scavenge air enters the compressor cylinder on the underside of the piston, and the air in the cushion cylinder is compressed. Towards the end of this stroke the exhaust port is uncovered and exhaust begins. Slightly later in the stroke air enters from the scavenge air box to complete the exhaust process and to charge the cylinder. The piston then returns on the compression stroke in the diesel cylinder due to the pressure of the air in the cushion cylinder assisted by the air pressure in the compressor cylinder. The air from the underside of the compressor piston is forced into the scavenge air box during this stroke and is ready for the next scavenge process. At the inner dead centre, ignition occurs and the cycle is then repeated.

When used as a gasifier the two pistons have identical cushion cylinder arrangements and no power is taken from the engine. The high pressure exhaust gas, which is at a moderate temperature (< 480°C), is taken from the engine to an expansion turbine from which the whole of the mechanical power output is taken. This combined unit gives an overall thermal efficiency which is higher than that obtainable with a complete turbine unit fitted with a heat exchanger. The output of the gasifier is given as a *gas power* (g.p.), which is calculated on the assumption of isentropic expansion of the gas leaving the cylinder, and is the potential power.

The free piston engine can utilize a wider range of fuel oils than the conventional diesel engine and has a higher compression ratio, which indicates a higher thermal efficiency. The compression pressures in the diesel cylinder are in the order of 69 bar, and the thermal efficiency is 40% to 45%, which gives an overall thermal efficiency of about 35%.

The gas turbine has been a competitor to the piston engine for use in vehicles apart from its natural field of application in the higher power range for industrial turbines and the turbo-jet units of aircraft. For vehicle use the gas turbine has appeared as the power unit in a few prototype cars, e.g. those of the Rover Co Ltd, Solihull in the period around 1960 but it is for the heavier vehicles, buses and the larger trucks, that the gas turbine has more to offer. The small power single-shaft turbines could not give competitive fuel consumptions to the petrol engine without a highly effective heat exchanger and even then, although the full load figures were good, the part load fuel consumption was poor. The advantages offered by the gas turbine are small volume and mass for a given power output giving a good power to weight ratio; a light and relatively simple and robust construction; a potential for burning a wide range of fuels giving a running cost advantage if cheaper fuels are available; and low emission output. Power units can be scaled down to give the required power output but the inherent source of losses in the engine, e.g. essential running clearances cannot be scaled down and so some losses are proportionately increased.

In the search for an alternative power unit to the petrol and diesel engines interest in the Stirling engine, see Section 6.10, has had some revival. It has external continuous combustion, high thermal efficiency, quiet operation, good torque at low speed and low emissions but requires a highly effective heat exchanger, as for the gas turbine, which must be cheap in construction and reliable in operation at high temperatures (> 1000°C) over the life of the engine.

18.15 Modern developments in I.C. engines

Although competitive engines may have advantages over the reciprocating petrol and diesel engines there is still no serious replacement for the quantity produced power units required for cars and vehicles and this is likely to remain so long as the existing fuels are available. The petrol and diesel engines have been subject to continuous development to improve their performance particularly with respect to fuel consumption and specific power output. The constant volume cycle analysis in Section 6.6 shows the fundamental dependence of thermal efficiency on compression ratio. This study can be extended to deal with air/fuel cycles by the selection of appropriate values of γ for the mixture. As the fuel content is increased the value of γ falls and a plot of the thermal efficiency against compression ratio for mixture strengths 0·8, 1·0 and 1·2 are drawn in fig. 18.62, showing the increase in efficiency obtained with the weaker mixtures. The development

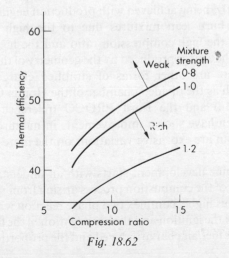

Fig. 18.62

problem is to incorporate these principles into practical engine design and prototype engines, radically designed and shown to give improved performances, have appeared only very slowly as production engines. The simple reciprocating piston-in-cylinder arrangements offers scope for development mainly in the combustion chamber, the induction and exhaust system, the preparation and movement of the combustible charge and in the timing and quality of the ignition. The use of multiple valves, e.g. two inlet and two exhaust or two inlet and one exhaust, improves the charging and exhausting of the cylinder if the added complexity and cost are acceptable; the former arrangement requires two overhead camshafts and the latter can be done with one. In addition to the quantity of gas induced, the movement of the charge is important with the gas swirling as the piston descends on induction and through compression until the gas is forced by an increasing amount of 'squish' into the combustion chamber as it is squeezed between the rising piston and the flat face of the exposed cylinder head. The purpose of combustion chamber design is to obtain the necessary characteristics and the various shapes are described as bathtub, hemispherical, pentroof, etc. The Fireball design of M. G. MAY (reference 18.13) is an advanced form of combustion chamber in which the main part of the volume is concentrated around the exhaust valve and charge movement is encouraged by the shape of path from the inlet valve side to the main part of the chamber round the exhaust valve. The design has shown good improvements in the average fuel consumption figures, with prospects of 20% being achieved with production engines, because of its ability to burn lean mixtures due to the high temperatures obtained with the high compression ratio and the highly turbulent movement of the charge produced by the geometry of the combustion chamber. There are other forms of stratified charge/combustion chambers, such as the divided chamber of the Honda CVCC engine (reference 18.14) and the Ford PROCO (reference 18.15) open chamber, which have shown improvements in mixture control and performance but are expensive variations on the more conventional engines.

Modern engine development starts with an understanding of the characteristics of the combustion process arising from the theoretical modelling of this highly complex two phase reaction which describes the initiation of the ignition and the propagation of the flame through a mixture of the fuel/air charge induced and the proportion of residual exhaust gas in the cylinder. This must describe the nature of the flame front, the rate of burn, the charge turbulence and combustion kinetics. The equations established are complex and require a great deal of computing time and even on modern main frame machines times up to 1·5 hours are reported. The predictions by such programmes have to be compared with engine results and then the models are further refined to improve the correlation between theory and practice.

Such complex theoretical investigations have only become possible because of the availability of adequate computing facilities but the impact of the computer and microprocessor have not ended there. The computer is a powerful support to all analytical and experimental work and can be involved directly in the experimental work for data collection, analysis and control of the experiment. Once the characteristics are known, from theory or experiment, the combustion process must be controlled and increasingly the microprocessor is in evidence as the brain of the engine management system. In the early 1980s engines appeared in production vehicles with programmed ignition and electronic fuel control and the inclusion of knock sensors to warn the ignition control unit of the onset of 'knock' which then retards the ignition of individual cylinders to avoid detonation. Similarly, during warm up the output of an electrically heated hot spot in the inlet manifold is automatically reduced as the engine warms up and is switched off at the normal running temperature. Fuel supply is controlled by micro-chip logic as an automatic choke, low speed control and overrun cut-off. Apart from the control of the basic engine parameters computer electronic control also leads to simpler ignition devices of more robust construction than the traditional contact breaker type requiring less maintenance and giving longer life.

The experimental investigation of engine performance has made increasing demands for refined techniques, improved transducers and instrumentation for the measurement of gas flow, turbulence, pressure, temperature and charge composition throughout the process. It is necessary to be able to take the measurements in production engines under test bed conditions and also, although it is more difficult, in the vehicle in which the engine is used if that is its application. A modern technique to allow such measurements to be made on the test bed is that which employs lasers such that no intrusion is made into the combustion process, a fault of the earlier techniques which by the presence of transducers or other measuring devices changed the combustion system.

One of the disadvantages of the spark ignition engine is the cycle-to-cycle and cylinder-to-cylinder variation in performance which is readily illustrated by taking a continuous sample of pressure/crank angle diagrams. Investigation has shown that a significant factor in the cycle to cycle variation is the size of the eddies created by the induction system. The smaller the eddy size variation about a critical mean size the more consistent the combustion process but the grids used to control the generation of eddies, although improving the efficiency of combustion, cause restrictions to gas flow and hence reduce the full throttle volumetric efficiency and power output. Greater knowledge of the induction and combustion processes supported by reliable experimental measurements should lead to further improvements in engine power output and fuel efficiency. For further reading see references 18.16–18.18.

18.16 Alternative fuels for vehicle power units

Alternative fuels for spark ignition and compression ignition engines are available and their use varies from being commercially available in some countries to being more experimental in nature or of restricted use as in racing cars. Alternative fuels are required to be as available as the normal petrol and diesel fuels at similar or less cost, and so the overall economic feasibility from wide natural availability, to preparation and networking through established petrol stations has to be acceptable in addition to the technical equality with established fuels. Alternative fuels may also require changes in engine design and construction to handle their different characteristics.

An alternative fuel for S.I. engines is liquefied petroleum gas (LPG), a mixture of propane and butane, which is available in association with natural gas and some crude oils. It is a by-product of petrol production and has previously been burned to waste in the production of natural gas. The fuel is lead free and its technology is known; required engine modifications are few but the storage of the fuel creates difficult problems. Gasohol is 90% (by volume) unleaded petrol and 10% ethanol. Other alternative fuels are compressed or liquefied methane and other alcohols and ethers, either neat or blended into petrols, are used as octane improving agents. Ethanol, which is 92·5 to 93·8% (by mass) of alcohol, is used for cars manufactured in Brazil. Methanol (methyl alcohol) is used in European fuels and is a racing engine fuel. These fuels, if they form more than 15% of the petrol require engine changes including those to the lubricating oils to be used. This subject is discussed in detail in reference 18.18.

Natural gas, which is mainly methane and has high reserve quantities, is attractive as a fuel for S.I. engines giving few combustion problems, is lead free, has a high octane rating and engine conversion is simple. The gas is stored as liquified (LNG) fuel or compressed (CNG) gaseous fuel and the problems with both arise from the storage requirements and it is unlikely that the use of such fuels will become great.

PROBLEMS

18.1 A quality governed four-stroke, single-cylinder gas engine has a bore of 146 mm and a stroke of 280 mm. At 475 rev/min and full load the net load on the friction brake is 433 N, and the torque arm is 0·45 m. The indicator diagram gives a net area of 578 mm² and a length of 70 mm with a spring rating of 0·815 bar per mm. Calculate the i.p., b.p., and mechanical efficiency.

(12·5 kW; 9·69 kW; 77·5%)

18.2 A two-cylinder, four-stroke gas engine has a bore of 380 mm and a stroke of 585 mm. At 240 rev/min the torque developed is 5·16 kN m. Calculate: (a) the b.p.; (b) the mean piston speed in m/s; (c) the b.m.e.p.   (129·8 kW; 4·68 m/s; 4·89 bar)

18.3 The engine of problem 18.2 is supplied with a mixture of coal gas and air in the proportion of 1 to 7 by volume. The estimated volumetric efficiency is 85% and the $Q_{net, p}$ of the coal gas is 16 800 kJ/m³. Calculate the brake thermal efficiency of the engine.

(27·4%)

18.4 A four-cylinder racing engine of capacity 2·495 litres has a bore of 94 mm and a compression ratio of 12/1. When tested against a dynamometer with a torque arm of 0·461 m a maximum load of 622 N was obtained at 5000 rev/min, and at the peak speed of 6750 rev/

min the load was 547 N. The minimum fuel consumption was 17·2 ml/s at a speed of 5000 rev/min, the specific gravity of the fuel being 0·735, and $Q_{net, v} = 44\,200$ kJ/kg. Calculate the maximum b.m.e.p. the maximum b.p., the minimum specific fuel consumption and the maximum brake thermal efficiency at maximum torque, and compare this latter answer with the air standard efficiency.

(14·5 bar; 178 kW; 0·303 kg/kW h; 26·9%; 63%)

18.5 A three-cylinder, direct-injection, water-cooled, two-stroke oil engine with two horizontally opposed pistons per cylinder has a bore of 82·6 mm and each piston has a stroke of 102 mm. The engine was tested against a brake with a torque arm of 0·381 m. The following results were taken on a variable speed test:

Speed rev/min	Brake load N	Fuel kg/min
1000	60·78	0·146
1100	61·46	0·157
1200	62·14	0·172
1300	62·14	0·185
1400	62·14	0·201
1500	62·14	0·216
1600	61·6	0·229
1700	60·9	0·241
1800	59·65	0·252

Plot curves of torque, b.p., and specific fuel consumption against speed. Convert the torque curve to a b.m.e.p. curve by calculation of the appropriate scale factor.

18.6 A four-cylinder, four-stroke diesel engine has a bore of 212 mm and a stroke of 292 mm. At full load at 720 rev/min the b.m.e.p. is 5·93 bar and the specific fuel consumption is 0·226 kg/kW h. The air/fuel ratio as determined by exhaust gas analysis is 25/1. Calculate the brake thermal efficiency and the volumetric efficiency of the engine. Atmospheric conditions are 1·01 bar and 15°C, and $Q_{net, v}$, for the fuel may be taken as 44 200 kJ/kg.   (36%; 76·5%)

18.7 The engine of problem 18.6 is to be used as a dual fuel engine. It is to burn methane (calorific value 33 480 kJ/m³ at 1·013 bar and 15°C), and has a pilot injection of oil of 10% of the input when running as a diesel engine. The air/fuel ratio for the oil is 25/1 as before, and for the methane 8·5/1. If the volumetric efficiency and the power output remain the same, what is the brake thermal efficiency of the engine when running on the dual fuel? (23%)

18.8 A four-cylinder petrol engine with a bore of 63 mm and a stroke of 76 mm was tested at full throttle at 3000 rev/min over a range of mixture strengths. The following readings were taken during the test:

Brake load N	162	165·5	169	170	169	162	159
Fuel consumption ml/s	2·08	2·04	2·17	2·5	2·84	3·395	3·56

The relative density of the fuel is 0·724. Calculate the corresponding values of b.m.e.p. and specific fuel consumption in kg/kW h. Plot a consumption loop and obtain from it the corresponding values for maximum power and maximum economy. The b.p. is given by $WN/26\,830$, where W is the brake load in N and N is the engine speed in rev/min.

(8·03 bar; 0·33 kg/kW h; 7·9 bar; 0·284 kg/kW h)

18.9 For the test outlined in problem 18.8 an air box was fitted which had an orifice of 41·65 mm diameter with a discharge coefficient of 0·6. The corresponding manometer readings in mm of water were:

Manometer mm water	33·5	33·5	33·5	33·8	33·8	34·25	34·8

Take the density of the air at inlet as 1·215 kg/m³. Plot to a base of air/fuel ratio graphs of b.p. and b.m.e.p. What are the air/fuel ratios at maximum power and maximum economy? (12·8/1; 15·7/1)

18.10 A four-cylinder petrol engine has an output of 52 kW at 2000 rev/min. A Morse test is carried out and the brake torque readings are 177, 170, 168, and 174 N m, respectively. For normal running at this speed the specific fuel consumption is 0·364 kg/kW h. The $Q_{net, v}$ of the fuel is 44 200 kJ/kg. Calculate the mechanical and brake thermal efficiencies of the engine. (82%; 22·4%)

18.11 A V-8 four-stroke petrol engine is required to give 186·5 kW at 4400 rev/min. The brake thermal efficiency can be assumed to be 32% at the compression ratio of 9/1. The air/fuel ratio is 12/1 and the volumetric efficiency at this speed is 69%. If the stroke to bore ratio is 0·8, determine the engine displacement required and the dimensions of the bore and stroke. The $Q_{net, v}$ of the fuel is 44 200 kJ/kg, and the free air conditions are 1·013 bar and 15°C.

(5·12 litres; 100·6 mm; 80·5 mm)

18.12 A four-cylinder, four-stroke diesel engine develops 83·5 kW at 1800 rev/min with a specific fuel consumption of 0·231 kg/kW h, and air/fuel ratio of 23/1. The analysis of the fuel is 87% carbon and 13% hydrogen, and the $Q_{net, v}$, is 43 500 kJ/kg. The jacket cooling water flows at 0·246 kg/s and its temperature rise is 50 K. The exhaust temperature is 316°C. Draw up an energy balance for the engine. Take $R = 0·302$ kJ/kg K and $c_p = 1·09$ kJ/kg K for the dry exhaust gas, and $c_p = 1·86$ kJ/kg K for superheated steam. The temperature in the test house is 17·8°C, and the exhaust gas pressure is 1·013 bar.   (b.h.p. 35·8%; cooling water 22·1%; exhaust 25·4%; radiation and unaccounted 16·7%)

18.13 An eight-cylinder, four-stroke diesel engine of 229 mm bore, 304 mm stroke, and compression ratio 14/1, has an output of 375 kW at 750 rev/min. The volumetric efficiency is 78% and the mechanical efficiency is 90%, and the air-fuel ratio is 25/1. If the i.m.e.p. for the pumping loop is 0·345 bar, calculate the i.m.e.p. for the working loop. The engine is now fitted with an exhaust driven turbo-blower which delivers air to the cylinders at 1·43 bar. The compression ratio is reduced to 13/1 and the measured volumetric efficiency is 102%. It can be assumed that the exhaust pressure remains constant and equal to 1·013 bar, and that the i.m.e.p. of the main loop is directly proportional to the mass of air induced. Calculate the b.p. which can be expected if the speed and mechanical efficiency remain the same. Compare the specific fuel consumption for the two cases, if the air/fuel ratio for the supercharged engine is 26·8/1. The free air conditions are 1·013 bar and 15°C.

(7·0 bar; 515 kW; 0·23 kg/kW h; 0·204 kg/kW h)

18.14 (a) A four-stroke petrol engine with a swept volume of 5·7 litres has a volumetric efficiency of 75% when running at 3000 rev/min. The engine is fitted with a carburettor which has a choke diameter of 38 mm. Assuming the conditions of a simple carburettor and neglecting the effects of compressibility, calculate the pressure and the air velocity at the choke. Take the coefficient of discharge at the throat as 0·85, and take the atmospheric conditions as 1·013 bar and 15°C.

(b) If the air/fuel ratio is 14/1 and the relative density of the fuel is 0·8, calculate the petrol jet size assuming a coefficient of discharge of 0·7. (0·9378 bar; 94·3 m/s; 222)

18.15 A six-cylinder, four-stroke compression ignition engine of 75 mm bore and 100 mm stroke has a brake power output of 110 kW at 3750 rev/min. The volumetric efficiency at this operating condition referred to ambient conditions of 1·013 bar and 20°C is 80%.

The engine is now fitted with a mechanically driven supercharger which has an isentropic efficiency of 0·7 and a pressure ratio of 1·6. The supercharged version has a volumetric efficiency of 100% referred to the supercharger delivery pressure and temperature. If it is assumed that the indicated power developed per m³/min of induced air at the ambient conditions is the same for normal aspiration and supercharging calculate the net increase in brake power to be expected from the supercharged engine. Take the mechanical efficiency of the engine as 80% in both cases and the mechanical efficiency of the drive from engine to supercharger as 95%.

(64·1 kW)

REFERENCES

18.1 RICARDO, H. R., *The High Speed Internal Combustion Engine* (Blackie, 1960).

18.2 *The Mechanical Efficiency of I.C. Engines*, MIRA Report No. 1958/5.

18.3 *Institute of Petroleum Standards for Petroleum and its Products*, Part 1, *Methods for Analysis and Testing;* Part 2, *Methods for Rating Fuels; Engine Tests* (Wiley, 1984).

18.4 PYE, D. R., *The Internal Combustion Engine* (O.U.P., 1953).

18.5 FISHER, C. H., *Carburation*, vols 1 and 2 (Chapman and Hall, 1953).

18.6 HADDAD, S., and WATSON, N., *Principles and performance in diesel engineering* (Ellis Horwood, 1984).

18.7 GIFFEN, E., and ROWE, A. W., *Pressure Calculations for Oil-engine Fuel-injection Systems*, Proc. Inst. Mech. Eng., 1939, vol. 141.

18.8 WASSENAAR, H., *Injection Phenomena in High-speed Diesel Engines*, Proc. Inst. Mech. Eng., Automobile Division, 1954–55.

18.9 KASTNER, L. J., *Investigation of the Air-box Method of Measuring* the Air Consumption of I.C. Engines, Proc. Inst. Mech. Eng., 1947, vol. 157.

18.10 WATSON, N., and JANOTA, M. S., *Turbocharging the I.C. engine* (Macmillan, 1984).

18.11 TAYLOR, C. F., *The Internal Combustion Engine in Theory and Practice*, vol. 1, *Thermodynamics, fluid flow, performance*, vol. 2, *Combustion, fuels, materials, design* (M.I.T. Press, 1977).

18.12 GREENE, A. B., and LUCAS, G. G., *The Testing of Internal Combustion Engines* (The English Universities Press, 1969).

18.13 MAY, M. C., *The potential of the high compression homogeneous charge spark ignited four stroke combustion engine*. Paper C204/79. I. Mech. E. Conference. Passenger car power plant of the future. October 1979.

18.14 DATE, T., YAGI, S., ISHIZUYA, A., and FUJI, I., *Research and development of the Honda CUCC engine*. SAE paper 740605, August 1974.

18.15 SIMKO, A., CHOMA, M. A., and REPCO, L. L., *Exhaust emission control by the Ford Programmed Combustion Process*. PROCO SAE paper 780699.

18.16 ANNAND, W. J. D., and ROE, G. E., *Gas flow in the I.C. engine* (Foulis, 1974).

18.17 BENSON, R. S., and WHITEHOUSE, N. D., *Internal combustion engines*. Vols. 1 and 2 (Pergamon Press, 1979).

18.18 I. Mech. E. Conference publications: *Diesel engines for passenger cars and light duty vehicles*, 1982; *Combustion in engineering*, vol. 1, 1983; *Turbocharging and turbochargers*, 1982.

Nomenclature

A	area
a	velocity of sound; acceleration; non-flow availability function
BDC	bottom dead centre
BS	British Standard
Bi	Biot number
b	steady-flow availability function
b.m.e.p.	brake mean effective pressure
b.p.	brake power
C	velocity; thermal capacity
C	temperature on the Celsius (or Centigrade) scale
C	coefficient of discharge
CHP	combined heat and power
C.I.	compression ignition
C_p	molar heat at constant pressure
C_v	molar heat at constant volume
c	specific heat
c_p	specific heat at constant pressure
c_v	specific heat at constant volume
D, d	bore; diameter
E	velocity of approach factor; emissive power; total energy
e	base of natural logarithms
F	force; geometric factor
F.I.	fuel injection
Fo	Fourier number
f	friction factor; interchange factor; flow velocity; frequency
f.p.	friction power
G	irradiation
Gr	Grashof number
g	gravitational acceleration
H	enthalpy
HC	hydrocarbons
ΔH_0	enthalpy of combustion
h	specific enthalpy; heat transfer coefficient
h_{f_g}	latent heat
I.C.	internal combustion

i	intensity of radiation; meshpoint
i.m.e.p.	indicated mean effective pressure
i.p.	indicated power
J	current density; radiosity
j	mesh point
K	equilibrium constant
K	temperature on the Kelvin scale (i.e. Celsius absolute)
k	thermal conductivity; isentropic index for steam; blade velocity coefficient
L	stroke; fundamental dimension of length
LMTD	log mean temperature difference
l	length; characteristic linear dimension
M	molar mass; fundamental dimension of mass
Ma	Mach number
m	mass
$\dot{m}$	rate of mass flow
N	rotational speed
NDIR	non dispersive infra red
NTU	number of transfer units
Nu	Nusselt number
n	polytropic index; number of kilomoles; number of cylinders; nozzle arc length
O.N.	octane number
P	power; perimeter
P.N.	performance number
Pr	Prandtl number
p	absolute pressure; blade pitch
p_m	mean effective pressure
p_i	indicated mean effective pressure
p_b	brake mean effective pressure
Q	heat; rate of heat transfer; fundamental dimension of heat
q	rate of heat transfer per unit area
q_g	internal heat generation rate per unit volume
R	specific gas constant; thermal resistance; radius; ratio of thermal capacities
R.F.	reheat factor
R_0	molar gas constant
Re	Reynolds number
r	radius; steam engine expansion ratio
r_p	pressure ratio
r_v	compression ratio
S	entropy; steam consumption
S.I.	spark ignition

St	Stanton number
s	specific entropy
s.f.c.	specific fuel consumption
T	absolute temperature; torque; fundamental dimension of time
TDC	top dead centre
t	temperature; fundamental dimension of temperature; blade thickness
U	internal energy; overall heat transfer coefficient
ΔU_0	internal energy of combustion
u	specific energy
V	volume
$\dot{V}$	rate of volume flow
v	specific volume
W	work; rate of work transfer; brake load; weight
w	specific weight; velocity of whirl
X	temperature on any arbitrary scale
x	dryness fraction; nozzle pressure ratio; length
Z	height above a datum level
z	number of stages

Greek Symbols

α	angle of absolute velocity; nozzle angle; absorptivity; thermal diffusivity
β	blade angle; coefficient of cubical expansion
γ	ratio of specific heats, c_p/c_v
δ	expansibility correction; film thickness
ε	emissivity; effectiveness
η	efficiency
θ	temperature difference
λ	wave-length; eigenvalue
μ	dynamic viscosity
ν	kinematic viscosity
ρ	density; reflectivity
σ	Stefan-Boltzmann constant
τ	transmissivity; shear stress; time
ϕ	relative humidity; angle
ω	specific humidity; solid angle
ψ	percentage saturation

898 APPENDIX

Suffixes

AS	air standard
a	absolute velocity; air; atmospheric
a	aircraft velocity
a_i	absolute velocity at inlet
a_e	absolute velocity at exit
B	black body
BT	brake thermal
b	blade velocity; black body
C	compressor; cold
c	critical value; condensate; clearance
d	diagram; dew point
e	exit; exhaust
F	fluid
F	fin
f	saturated liquid; fuel; film
fg	change of phase at constant pressure
g	saturated vapour; gases
gr	gross
H	high-pressure stage; hot
hp	heat pump
I	intercooler
IT	indicated thermal
i	inlet; a constituent in a mixture; intermediate; inside surface; mesh point; indicated
i	injector
j	mesh point
j	jet
L	low-pressure stage
M	mechanical
m	mean
N	normal
net	net
o	overall; outside; zero or reference condition
P	product of combustion
p	constant pressure
R	reactant
r	relative velocity; radiation
r	refrigerator
r_i	relative velocity at inlet
r_c	relative velocity at exit
s	vapour; swept volume; stage; steam

T	throat
T	turbine
t	total heat or stagnation conditions
V	volumetric
v	constant volume
w	water; wall
λ	monochromatic value at wave-length, λ
ϕ	radiation at angle ϕ

Index